T0339923

ITERATIVE SOLUTION METHODS

ITERATIVE SOLUTION METHODS

OWE AXELSSON

Faculty of Mathematics and Informatics
University of Nijmegen, The Netherlands

CAMBRIDGE
UNIVERSITY PRESS

CAMBRIDGE UNIVERSITY PRESS
Cambridge, New York, Melbourne, Madrid, Cape Town, Singapore,
São Paulo, Delhi, Dubai, Tokyo, Mexico City

Cambridge University Press
The Edinburgh Building, Cambridge CB2 8RU, UK

Published in the United States of America by
Cambridge University Press, New York

www.cambridge.org
Information on this title: www.cambridge.org/9780521555692

© Cambridge University Press 1996

This publication is in copyright. Subject to statutory exception
and to the provisions of relevant collective licensing agreements,
no reproduction of any part may take place without the written
permission of Cambridge University Press.

First published 1994
Reprinted 1996
First paperback edition 1996

A catalogue record for this publication is available from the British Library

Congress Cataloging-in-Publication Data is available.

ISBN 978-0-521-44524-5 Hardback
ISBN 978-0-521-55569-2 Paperback

Cambridge University Press has no responsibility for the persistence or
accuracy of URLs for external or third-party internet websites referred to in
this publication, and does not guarantee that any content on such websites is,
or will remain, accurate or appropriate. Information regarding prices, travel
timetables, and other factual information given in this work is correct at
the time of first printing but Cambridge University Press does not guarantee
the accuracy of such information thereafter.

Contents

Preface

Algorithms for the solution of linear systems of algebraic equations arise in one way or another in almost every scientific problem. This happens because such systems are of such a fundamental nature. For example, nonlinear problems are typically reduced to a sequence of linear problems, and differential equations are discretized to a finite dimensional system of equations.

The present book deals primarily with the numerical solution of linear systems. The solution algorithms considered are mainly iterative methods. Some results related to the estimate of eigenvalues (of importance for estimating the rate of convergence of iterative solution methods, for instance), are also presented. Both the algorithms and their theory are discussed. Many phenomena that can occur in the numerical solution of the above problems require a good understanding of the theoretical background of the methods. This background is also necessary for the further development of algorithms. It is assumed that the reader has a basic knowledge of linear algebra such as properties of sets of linearly independent vectors, elementary matrix algebra, and basic properties of determinants.

The first six or seven chapters and Appendix A can be (and have been) used as a textbook for an introductory course in numerical linear algebra, but this material demands students who are not afraid of theory. The theory is presented so that it can be followed even in selfstudy. Chapters 2, 3, and 4 contain much theoretical background for numerical linear algebra, the more difficult parts of which have been indicated by an asterisk. Some readers may wish to postpone study of some of this material until it is used in later chapters. To further help the reader, the definitions in each chapter are collected at the beginning of the chapter and appear in addition in the text at relevant places. Each of the above chapters and Appendix A contain a great number of exercises that either further illustrate the theory and the algorithms or, in many cases, give a fully programmed, step-by-step presentation of methods not treated in the text itself. These latter exercises can be given as homework assignments to students who want, for example, to take a supplementary course for extra credits. Topics discussed in these exercises include generalized inverses and singular values, the Aasens factorization algorithm, modified Gram-Schmidt orthogonalization, QR factorization methods, generalized eigenvalue problems, constrained optimization problems, and logarithmic norms.

The second half of the book presents recent results in the iterative solution of linear systems, mainly using preconditioned conjugate gradient methods. The latter have become well-established techniques since their early but rare use in the sixties and the beginning of the seventies. These chapters give research- or application-oriented students a thorough background that enables them not only to use these algorithms but also to derive and analyse algorithms for new types of problems and make them able, for instance, to read and apply new algorithms presented in numerical linear algebra journals.

This book has undergone many rewritings and has been developed over a long period of the author's teaching of numerical linear algebra. It reflects material that he has found most important or interesting, but this does not mean that topics not covered in the book cannot find use in practice. For instance, clearly many algorithms for the solution of eigenvalue problems and for linear least squares problems not discussed in the book are frequently used in practice. A thorough treatment of them would, however, require a second volume. Readers interested in these topics can consult the reference lists following each chapter.

Also in the interest of keeping the present volume at a reasonable size, certain solution methods that have attracted much recent interest, such as multilevel methods and domain decomposition methods, have not been presented.

Finally, any comments readers have on the contents of this book would be highly appreciated by the author.

Acknowledgments

During the writing of this book, the author benefited from the kind assistance of many colleagues, friends, and students. First of all I want to mention Allan Barker of the Technical University, Lyngby, Denmark, who made many valuable suggestions regarding presentation of the material and offered assistance in correct use of the English language. Radim Blaheta of the Mining Institute at Ostrava, Czech Republic; Ludwig Elsner of the University of Bielefeld, Germany; Igor Kaporin and Michail Makarov, both of the University of Moscow, Russia; Ivo Marek of the Charles University at Prague, Czech Republic; and Wim Lenferink, Hao Lu, Maya Neytcheva, and Ben Polman of the University of Nijmegen read parts of the manuscript and supplied the author with many helpful comments. In addition, Ben Polman and Maya Neytcheva helped with incorporating the figures into the text.

Of my students, Robert Hardin and Michael Bekker identified many opportunities for improvement of the manuscript.

It is a pleasure to express my gratitude to the secretarial staffs of the Department of Mathematics at the University of Nijmegen, whose expertise was indispensable in the writing of this book. In particular, Willy van de Sluis and Tarcise Bökkerink devoted countless hours to typing and retyping large parts of the manuscript.

Finally, I am indebted to the staff of Cambridge University Press and their associates for their fine work and cooperation throughout the publication process.

Nijmegen   *Owe Axelsson*
January 1993

1

Direct Solution Methods

The need to solve large linear systems of algebraic equations arises in almost any mathematical model, as illustrated in the instances below. In particular, we go into some detail regarding electrical networks. The most common method used to solve such linear systems is based on factorization of the matrix in triangular factors, which is discussed and shown to be equivalent to the Gaussian elimination method (learned in almost any elementary course in algebra).

We present this method in a form that can be the basis for a computer algorithm. When one wants to solve very large systems of equations, it is important to know how computational complexity grows with the size of the problem. This topic is also addressed, including the case where the matrix has a special structure in the form of a bandmatrix. The solution of tridiagonal systems is considered in particular detail. In addition, some basic dimension theory for matrices, such as the relation between rank and the dimension of the nullspace, is discussed and derived using the factored form of the matrix.

The following definitions or notations are introduced in this chapter:

Definition 1.1

   (a) The *range* of the mapping $\widetilde{A} : \mathbb{R}^n \to \mathbb{R}^m$ is

$$\mathcal{R}(\widetilde{A}) = \{\tilde{\mathbf{y}} = \widetilde{A}\tilde{\mathbf{x}}; \ \tilde{\mathbf{x}} \in \mathbb{R}^n\}.$$

   (b) The *nullspace* of $\widetilde{A}$ is $\mathcal{N}(\widetilde{A}) = \{\tilde{\mathbf{x}} \in \mathbb{R}^n; \ \widetilde{A}\tilde{\mathbf{x}} = \mathbf{0}\}$.

Definition 1.2

   (a) If $\mathbf{b} \in \mathcal{R}(A)$, then $A\mathbf{x} = \mathbf{b}$ is said to be a *consistent* system of linear equations. Otherwise, (if $\mathbf{b} \notin \mathcal{R}(A)$), it is said to be *inconsistent*.

   (b) If A is square of order n and $\dim \mathcal{N}(A) = 0$, then A is said to be *nonsingular* or *regular*.

1.1 Introduction: Networks and Structures

A network consists of a set of nodes and a set of edges connecting certain pairs of nodes. Each node is connected to at least one other node. In a physical network, nodes are connected by some device, such as resistors in an electrical network, pipes in a gas pipeline network, and bars, beams, or similar devices in a frame structure.

A source, such as an electromotive voltage, gaswell, or outer pressure, is present to drive the currents, gasflow, or stresses (strains), respectively, through the network. A linear system of algebraic equations, usually with the same number of equations as unknowns, arises. The unknowns may be the potential—i.e., voltage or pressure at the nodes—or the rate of exchange of the potential along the edges (current, strains).

Example 1.1 (Electric network) To be specific, consider the case of an electric network consisting of a set V of nodes (vertices) and a set $L = \{(i, j)\}$ of edges, where (i, j) denotes the edge connecting nodes i and j. In general, not all nodes are connected. At a subset $V_0 \subset V$, the voltage is prescribed. The remaining set of nodes, $V \setminus V_0$, are called "free." Given resistances $r_{i,j}$ at the edges $(i, j) \in L$, we want to find the resulting voltage v_i at the free nodes.

A remarkable phenomenon in nature is the principle of minimal energy loss. Applied in the present context, it means that the distribution of electrical currents in the network will be such that total heat loss is minimized. As the heat loss along edge (i, j) is $(v_i - v_j)^2/r_{i,j}$, this means that

$$\sum_{(i,j)\in L} (v_i - v_j)^2/r_{i,j},$$

where v_i takes the given values for all $i \in V_0$, is minimized. This is a real valued function $f(v_1, v_2, \ldots, v_N)$ of the variables v_i at the free nodes, assuming N such nodes. Taking the partial derivatives with respect to these variables, we get the stationary equations

$$(1.1) \quad \frac{\partial f}{\partial v_i} = 2 \sum_{j,(i,j)\in L} \frac{1}{r_{i,j}}(v_i - v_j) = 0, \quad i \in V \setminus V_0 = \{1, 2, \ldots, N\}.$$

This is, in fact, Kirchoff's law of electrical currents, implying that the sum of all currents entering and leaving a (free) node is zero. It gives us a system of N linear equations in the N unknowns $\{v_i\}_{i=1}^N$. Later in this book it shall be proved that the corresponding matrix is nonsingular. (It is a so-called irreducibly diagonally dominant matrix with diagonal entries d_i, where

$d_i = \sum_{j,(i,j)\in L} r_{i,j}^{-1}$, and with off-diagonal entries $-r_{i,j}^{-1}$, $i \neq j$, $i \in V \setminus V_0$, with strong inequality, $d_i > \sum_{k=1}^{N} r_{i,k}^{-1}$, for any free node i connected to a constrained node, $j_0 \in V_0$.)

Hence this system has a unique solution. Let the network consist of n nodes and m edges and let $1, 2, \ldots, n$ be a numbering (ordering) of the nodes and $1, 2, \ldots, m$ a numbering of the edges. In (1.1) we sum over the nodes. An alternative and interesting way to express (1.1) is by summing over the edges (cf. Strang [1986]). Let B be the set of edges (branches) and let $B_i^{(+)}$ and $B_i^{(-)}$ be the subsets of B of branches entering node i with $k > i$ and $k < i$, respectively, where k is the other node number of the branch.

Let $\tilde{x}_i$ be the current in branch i, directed from the lower node index to the higher. (Hence $\tilde{x}_i$ may be negative.) Similarly let $\tilde{v}_i$ be the potential difference at branch i, that is, $\tilde{v}_i = v_{j_i} - v_{k_i}$, where $k_i, j_i, k_i > j_i$ are the nodes of branch i. Hence

$$\tilde{\mathbf{v}} = E\mathbf{v},$$

where E is a matrix of order $m \times n$, which has zero entries everywhere except two entries per row, one entry $+1$ and the other -1.

Ohm's law states that

$$R\tilde{\mathbf{x}} = \tilde{\mathbf{v}}, \quad \text{or} \quad \tilde{\mathbf{x}} = R^{-1}\tilde{\mathbf{v}},$$

where R is a diagonal matrix with entries r_{k_i,j_i}, $i = 1, 2, \ldots, m$. (1.1) can also be written

$$\sum_{j \in B_i^{(-)}} \tilde{x}_j - \sum_{j \in B_i^{(+)}} \tilde{x}_j = 0, \quad i = 1, 2, \ldots, n,$$

which is readily seen to be equivalent with

$$E^T \tilde{\mathbf{x}} = \mathbf{0},$$

where E^T is the transpose of E.

The above relations can be described in a diagram as shown below:

$$
\begin{array}{ccc}
\boxed{\text{potential } \mathbf{v} \text{ at nodes}} & \xrightarrow{E^T R^{-1} E \mathbf{v}} & \boxed{0 = E^T \tilde{\mathbf{x}} = E^T R^{-1} \tilde{\mathbf{v}} = E^T R^{-1} E \mathbf{v}} \\[2mm]
E \downarrow & & \uparrow E^T \\[2mm]
\boxed{\text{potential difference } \tilde{\mathbf{v}} \text{ at branches}} & \xrightarrow[\text{Ohm's law}]{R^{-1}} & \boxed{\tilde{\mathbf{x}} = R^{-1}\tilde{\mathbf{v}}}
\end{array}
$$

By direct computation it can be shown that the matrix $A = E^T R^{-1} E$, which has order $n \times n$, has positive diagonal and nonpositive off-diagonal entries. Note also that $Ae = 0$, where $e = (1, 1, \ldots, 1)^T$ because $Ee = 0$. In addition, $A = A^T$, so A is symmetric. On the other hand, given any symmetric matrix A of order $n \times n$ with positive diagonal and nonpositive offdiagonal entries, where $Ae = 0$, we can associate a network with A where the branches between two nodes i, j correspond to the nonzero entries of A. It can be shown that by letting R^{-1} hold all the values of the nonzero offdiagonal entries of A and letting E have entries 0 and one entry $+1$ and one -1 in positions corresponding to the nonzero elements of A, we can make $A = E^T R^{-1} E$. In fact, the so-called matrix graph of A and the network are topologically identical. (For a definition of matrix graphs, see a discussion later in this chapter and in Chapter 4.)

Remark Similar systems are found in networks of gas pipelines and frame structures, for example. In general, $r_{i,j}$ depends on the current (or the corresponding variable) through the edge (i, j), and (1.1) is then a nonlinear system of equations. When this dependence is neglected, however, (1.1) becomes a linear system.

Example 1.2 (Tomography) Consider a plate of *inhomogeneous* materials which is part of a three-dimensional body. It is not possible to observe the plate from above, but only from the boundary. This would be the case if we want to observe a plane section through a human body, for instance. Assume for simplicity that the plane is square and subdivided into n^2 small squares, say 9 (i.e., $n = 3$). (See Figure 1.1.) By sending X-rays with known intensities I_0 through the plate and measuring the intensity of the outgoing X-ray, we want to determine the damping factors x_i in the different squares.

This technique, called *tomography*, has been used since about 1973 in medicine (in cancer research, for instance). It is effective because each tissue (material within each square) has its own damping factor. If an X-ray is sent through the three top squares in Figure 1.1, we get

$$I_1 = I_0 e^{-x_1} e^{-x_2} e^{-x_3} = I_0 e^{-(x_1 + x_2 + x_3)},$$

assuming that the damping factor is exponential.

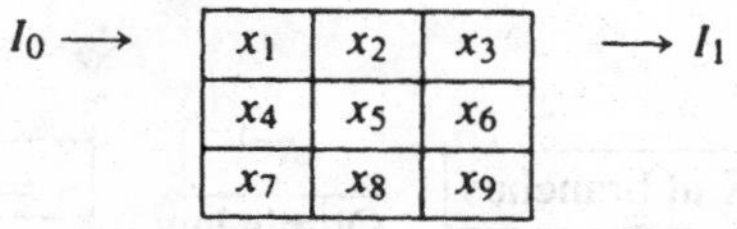

$$I_0 \longrightarrow \boxed{\begin{array}{|c|c|c|} \hline x_1 & x_2 & x_3 \\ \hline x_4 & x_5 & x_6 \\ \hline x_7 & x_8 & x_9 \\ \hline \end{array}} \longrightarrow I_1$$

Figure 1.1.   X-rays through a plate

Let $I_1 = I_0 e^{-b}$. Since I_1 is measured, we can determine b, and we have the linear relation $x_1 + x_2 + x_3 = b$. If we send three horizontal and three vertical X-rays we get the following linear algebraic system:

$$
\begin{array}{rrrrrrrrrl}
x_1 & + & x_2 & + & x_3 & & & & & = b_1 \\
& & & & & x_4 + x_5 + x_6 & & & & = b_2 \\
& & & & & & x_7 + x_8 + x_9 & & & = b_3 \\
x_1 & & & + & x_4 & & + x_7 & & & = b_4 \\
& & x_2 & & + x_5 & & & + x_8 & & = b_5 \\
& & & & x_3 & + x_6 & & & + x_9 & = b_6.
\end{array}
$$

Hence we have six equations but nine unknowns—i.e., the system is under-determined. Accordingly, it has no unique solution. We may send five more X-rays, now through the diagonals. The system then becomes overdetermined. In practice one sends X-rays along directions incremented by a small angle. A unique solution may be determined, for instance, by a least squares approximation method.

Example 1.3 (Diffusion) Consider a tube with a liquid of concentration x_i, $i = 0, 1, \ldots, n + 1$, in $n + 2$ different cells. Initially the cells are assumed to be closed by impervious walls and to have concentration a_i, $i = 0, 1, \ldots, n + 1$. At time $t = 0$ the walls become permeable and the concentration begins to diffuse between the cells. Assume that the left and right endcells have a fixed concentration $a_0 = 0$ and $a_{n+1} = 1$, respectively (see Figure 1.2).

To find the concentration in the other cells, we use the fact that the rate of diffusion $dx_i(t)/dt$ for a certain time is proportional to the difference of concentrations at time t, i.e. (for $n = 3$)

$$
\frac{dx_1(t)}{dt} = c[(x_2 - x_1) - (x_1 - x_0)]
$$

$$
\frac{dx_2(t)}{dt} = c[(x_3 - x_2) - (x_2 - x_1)]
$$

$$
\frac{dx_3(t)}{dt} = c[(x_4 - x_3) - (x_3 - x_2)],
$$

x_0	x_1	x_2	x_3	x_4

Figure 1.2. Diffusion in a tube ($n = 3$)

where c is a positive constant. This is a linear system of ordinary differential equations of first order. In matrix form we get

$$\frac{d\mathbf{x}(t)}{dt} = c[A\mathbf{x}(t) + \mathbf{f}], \quad t > 0, \quad x_i(0) = a_i, \quad i = 1, 2, \ldots, n,$$

where

$$A = \begin{bmatrix} -2 & 1 & 0 \\ 1 & -2 & 1 \\ 0 & 1 & -2 \end{bmatrix}, \quad \mathbf{x}(t) = \begin{bmatrix} x_1(t) \\ x_2(t) \\ x_3(t) \end{bmatrix} \quad \text{and} \quad \mathbf{f} = \begin{bmatrix} a_0 \\ 0 \\ a_4 \end{bmatrix}.$$

This is an initial value problem. It can readily be seen that the solution $\mathbf{x}(t)$ has a *steady state* solution $\mathbf{x}(\infty) = \mathbf{b}$, satisfying $A\mathbf{b} + \mathbf{f} = \mathbf{0}$ (because $d\mathbf{x}/dt = \mathbf{0}$ then). From this we easily find $b_i = i/(n + 1)$, $i = 1, 2, 3$. Note in passing that the steady state, or state of equilibrium, is independent of the initial values a_1, a_2, and a_3.

Discrete diffusion systems are examples of more general compartment models, which are discussed next.

Example 1.4 (Compartment models) Compartment models are devices for describing the circulation of various elements in nature, in the human body, or in other systems. In a compartment system we have n compartments $X_1, X_2, \ldots, X_n$, which contain quantities (concentrations) $x_1, x_2, \ldots, x_n$ of some "matter." For natural reasons we must have $x_i \geq 0$ for all i. Then there is a "rule," a differential equation, which governs the circulation of matter among the compartments, as well as leakage from one or more compartments and also a possible external supply of matter to the system.

Let us consider three examples. First, diffusion and circulation of carbon dioxide in nature can be discussed within the framework of compartment models. A diffusive process can be described by compartment models; for a one-dimensional example, see Figure 1.2.

Second, in pharmacology the distribution of a drug in the human body is often discussed in terms of a compartment system. The various organs of the human body and the circulatory system are considered compartments. In medical compartment models, one often assumes that the transport between two compartments is governed by the so-called Fick's law, which leads to linear differential systems.

Third, there is an analogy between continuous-time Markov chains with a finite number of states and closed compartment systems (i.e., systems with no loss of mass). For a reference on compartmental analysis, see Jacquez (1972).

Let us consider linear compartment models with no external supply of matter that are governed by systems of the form

$$\frac{d\mathbf{x}}{dt} = A(t)\mathbf{x}, \quad t > 0, \quad \mathbf{x}(0) = \mathbf{x}_0 \geq \mathbf{0}.$$

It is implied by the model that matter can flow (directly) from compartment X_k into X_i, if and only if $a_{ik} > 0$. Hence we have $a_{ik} \geq 0$, $i \neq k$. Let $\mathbf{e} = (1, 1, \ldots, 1)^T$. Then the total mass in the system is $(\mathbf{x}, \mathbf{e}) = \sum_{i=1}^{n} x_i$. Since $d/dt(\mathbf{x}, \mathbf{e}) = (A(t)\mathbf{x}, \mathbf{e}) = (\mathbf{x}, A^T(t)\mathbf{e})$, it is readily seen that if we wish the total mass always to be constant, we must assume $A^T(t)\mathbf{e} = \mathbf{0}$; i.e., the column sums of $A(t)$ must be zero for each t. Note that in this case there is no final steady state, but the total matter (mass) remains constant. Such systems are called *conservative*.

Similarly, if we wish the total mass to be nonincreasing, we must assume that $A^T(t)\mathbf{e} \leq \mathbf{0}$; i.e., the column sums of $A(t)$ must be nonpositive. Since $a_{ik}(t) \geq 0$, $i \neq k$, this implies that $a_{kk}(t) \leq 0$, $t \geq 0$. If there is at least one positive $a_{i,k}(t)$ in every column of $A(t)$, then $a_{kk}(t) < 0$, $k = 1, 2, \ldots, n$. Clearly, the positive orthant is invariant for the system—i.e., the solution stays in this orthant, for $\mathbf{x}(t) \geq \mathbf{0}$, for all $t > 0$.

Note: For other applications of matrix analysis in structural mechanics, electricity, and mathematical programming, see Kron (1959), Asplund (1966), Guillemin (1949), and Dantzig (1963).

1.2 Gaussian Elimination and Matrix Factorization

We shall now present a common solution method for systems of linear algebraic equations. A linear algebraic system of m equations with n unknowns can be written

$$a_{11}x_1 + a_{12}x_2 + \cdots + a_{1n}x_n = b_1$$
$$a_{21}x_1 + a_{22}x_2 + \cdots + a_{2n}x_n = b_2$$

(1.2)
$$\vdots$$

$$a_{m1}x_1 + a_{m2}x_2 + \cdots + a_{mn}x_n = b_m$$

or, in compact matrix form,

$$A\mathbf{x} = \mathbf{b},$$

where $A = [a_{ij}]$ is a matrix of order $m \times n$ and $\mathbf{x} = [x_i]$ and $\mathbf{b} = [b_i]$ are n-dimensional and m-dimensional column vectors, respectively. We assume that $m \geq 2$.

With a matrix A we associate two main problems: (a) the solution of $A\mathbf{x} = \mathbf{b}$, and (b) the determination of one or more eigenvalues of A in the case $m = n$, i.e., the calculation of numbers λ such that $A\mathbf{x} = \lambda\mathbf{x}$ for some nonzero vector $\mathbf{x}$.

In practice it is essential to use efficient solution methods, because the order of the matrix can be very large. In particular, fast methods are needed for large sparse systems of equations, but such methods will be presented in later chapters of this book. Consider now the Gaussian elimination method for (1.2). We may assume that $a_{11} \neq 0$, because if $a_{11} = 0$, we can make a permutation of rows—i.e., a renumbering of the equations—to find such a nonzero coefficient. The entry a_{11} is called the first *pivot entry*. Denote the coefficients in the matrix by $a_{ij}^{(1)} = a_{ij}$ and the right-hand side by $b_i^{(1)} = b_i$; that is, we want to solve

$$A^{(1)}\mathbf{x} = \mathbf{b}^{(1)}.$$

Elimination of the First Column: Eliminate the first variable from the remaining equations by multiplying the first row (called the *pivot row*) by $-a_{i1}^{(1)}/a_{11}^{(1)}$ and adding it to the ith row, $i = 2, \ldots, m$. We get

$$a_{11}^{(1)}x_1 + a_{12}^{(1)}x_2 + \cdots + a_{1n}^{(1)}x_n = b_1^{(1)} \text{ (first pivot row)}$$

$$a_{22}^{(2)}x_2 + \cdots + a_{2n}^{(2)}x_n = b_2^{(2)}$$

$$\vdots$$

$$a_{m2}^{(2)}x_2 + \cdots + a_{mn}^{(2)}x_n = b_m^{(2)},$$

where

(1.3a) $$a_{ij}^{(2)} = a_{ij} - a_{i1}a_{11}^{-1}a_{1j}, \quad 2 \leq i \leq m, \quad 2 \leq j \leq n$$

and

$$b_i^{(2)} = b_i - a_{i1}a_{11}^{-1}b_1.$$

Note that for $j = 1$ we have $a_{i1}^{(2)} = 0$, $2 \leq i \leq m$. We may now assume that $a_{22}^{(2)} \neq 0$, since if $a_{22}^{(2)} = 0$ we can make a row permutation. Or, if necessary, if $a_{i2}^{(2)} = 0$, $i = 2, 3, \ldots, m$, we can make a permutation of columns (i.e., we can renumber the unknowns) to make $a_{22}^{(2)} \neq 0$ or, in general, both a row and a column permutation. $a_{22}^{(2)}$ is called the second pivot entry. Note that

(1.3b) $$a_{22}^{(2)} = a_{22} - a_{11}^{-1} a_{21} a_{12} = a_{11}^{-1} \det \begin{bmatrix} a_{11} & a_{12} \\ a_{21} & a_{22} \end{bmatrix}.$$

Elimination of the Second Column: Repeat the above elimination, this time in the second column below the diagonal, of $A^{(2)} \mathbf{x} = \mathbf{b}^{(2)}$, where

$$A^{(2)} = \begin{bmatrix} a_{11}^{(1)} & a_{12}^{(1)} & a_{13}^{(1)} & \cdots & a_{1n}^{(1)} \\ 0 & a_{22}^{(2)} & a_{23}^{(2)} & \cdots & a_{2n}^{(2)} \\ \vdots & & & & \\ 0 & a_{m2}^{(2)} & a_{m3}^{(2)} & \cdots & a_{mn}^{(2)} \end{bmatrix} \text{(second pivot row)}, \quad \mathbf{b}^{(2)} = \begin{bmatrix} b_1^{(1)} \\ b_2^{(2)} \\ \vdots \\ b_m^{(2)} \end{bmatrix}.$$

$(m \times n)$

Elimination of the Remaining Columns: Continue to repeat this elimination (with row and column permutations, when required) until

$$A^{(t)} \mathbf{x} = \mathbf{b}^{(t)},$$

where (in case $r < m$)

(1.4) $$A^{(t)} = \begin{bmatrix} a_{11}^{(1)} & a_{12}^{(1)} & a_{13}^{(1)} & & \cdots & \cdots & a_{1n}^{(1)} \\ 0 & a_{22}^{(2)} & a_{23}^{(2)} & & \cdots & \cdots & a_{2n}^{(2)} \\ & & a_{33}^{(3)} & & \cdots & \cdots & a_{3n}^{(3)} \\ & & & \ddots & & & \\ & & & & a_{rr}^{(r)} & & a_{rn}^{(r)} \\ & & & & 0 & \cdots & 0 \\ & & & & \vdots & & \\ 0 & & & & 0 & \cdots & 0 \end{bmatrix}$$

$(m \times n)$

Here $a_{ss}^{(s)} \neq 0$, $1 \leq s \leq r \leq \min(m, n)$ and $t = r$ or $t = r + 1$. Example 1.5 below illustrates various cases where t takes values r or $r + 1$.

At each stage of the elimination we have the relation

$$a_{ij}^{(s+1)} = a_{ij}^{(s)} - a_{is}^{(s)} a_{ss}^{(s)-1} a_{sj}^{(s)}, \quad s + 1 \leq i \leq m, \quad s + 1 \leq j \leq n$$

to compute the new entries. These entries are, in fact, so-called Schur complements (see Chapter 3) of the matrices

$$\begin{bmatrix} a_{ss}^{(s)} & a_{sj}^{(s)} \\ a_{is}^{(s)} & a_{ij}^{(s)} \end{bmatrix}.$$

Similarly,

$$b_i^{(s+1)} = b_i^{(s)} - a_{is}^{(s)} a_{ss}^{(s)^{-1}} b_s^{(s)}, \quad s+1 \le i \le m.$$

We shall now present an alternative form of Gaussian elimination.

1.2.1 The Equivalence Between Elimination and Triangular Matrix Factorization

Let

$$L_1 = [l_{ij}] = \begin{bmatrix} 1 & & & 0 \\ -a_{21}/a_{11} & 1 & & \\ \vdots & & & \\ -a_{m1}/a_{11} & 0 & \cdots & 1 \end{bmatrix} m \times m \quad ,$$

that is,

$$l_{i1} = -a_{i1}^{(1)}/a_{11}^{(1)}, \quad 2 \le i \le m.$$

Lemma 1.1 $L_1 A^{(1)} = A^{(2)}$ and $\mathbf{b}^{(2)} = L_1 \mathbf{b}^{(1)}$.

Proof

$(L_1 A^{(1)})_{ij} = \sum_{k=1}^m l_{ik} a_{kj}^{(1)} = l_{i1} a_{1j}^{(1)} + 1 \cdot a_{ij}^{(1)} = -a_{i1}^{(1)} a_{11}^{(1)^{-1}} a_{1j}^{(1)} + a_{ij}^{(1)} = a_{ij}^{(2)}.$
Similarly for the vector $\mathbf{b}$. $\diamond$

In the same way, if

$$L_2 = \begin{bmatrix} 1 & & & & 0 \\ 0 & 1 & & & \\ 0 & (-a_{32}^{(2)}/a_{22}^{(2)}) & 1 & & \\ \vdots & & & & \\ 0 & (-a_{m2}^{(2)}/a_{22}^{(2)}) & 0 & \cdots & 1 \end{bmatrix} m \times m \quad ,$$

then $L_2 A^{(2)} = A^{(3)}$ (that is, $L_2 L_1 A^{(1)} = A^{(3)}$ etc.) and, finally,

$$L_{t-1} \ldots L_2 L_1 A^{(1)} = A^{(r)} = U,$$

where U is upper triangular. Hence

(1.5a) $$A = LU,$$

where

(1.5b) $$L = L_1^{-1} L_2^{-1} \dots L_{t-1}^{-1}.$$

Example 1.5

(a) If $A^{(1)} = \begin{bmatrix} 1 & 1 \\ 0 & 1 \\ 0 & 0 \end{bmatrix}$, then $A^{(2)} = A^{(1)}$ and $r = 2, t = r$.

(b) If $A^{(1)} = \begin{bmatrix} 1 & 1 & 1 \\ 0 & 1 & 1 \end{bmatrix}$, then $A^{(2)} = A^{(1)}$ and $r = 2, t = r$.

In (a) and (b) there is actually no need for any elimination, so $L_1 = I$, the identity matrix.

(c) If $A^{(1)} = \begin{bmatrix} 2 & 1 & 1 \\ 2 & 2 & 3 \end{bmatrix}$, then

$$L_1 = \begin{bmatrix} 1 & 0 \\ -1 & 1 \end{bmatrix}, \quad A^{(2)} = \begin{bmatrix} 2 & 1 & 1 \\ 0 & 1 & 2 \end{bmatrix} \text{ and } r = 2, \ t = r.$$

(d) If $A^{(1)} = \begin{bmatrix} 2 & 2 \\ 1 & 2 \\ 1 & 3 \end{bmatrix}$, then

$$L_1 = \begin{bmatrix} 1 & 0 & 0 \\ -\frac{1}{2} & 1 & 0 \\ -\frac{1}{2} & 0 & 1 \end{bmatrix}, \quad A^{(2)} = \begin{bmatrix} 2 & 2 \\ 0 & 1 \\ 0 & 2 \end{bmatrix}, \quad L_2 = \begin{bmatrix} 1 & 0 & 0 \\ 0 & 1 & 0 \\ 0 & -2 & 1 \end{bmatrix},$$

$$A^{(3)} = \begin{bmatrix} 2 & 2 \\ 0 & 1 \\ 0 & 0 \end{bmatrix}, \text{ and } r = 2, \ t = 3.$$

Lemma 1.2 *L, as defined by (1.5), is lower triangular.*

Proof An easy calculation (using $L_1^{-1} L_1 = I$) reveals that

$$L_1^{-1} = \begin{bmatrix} 1 & & & 0 \\ a_{21}^{(1)}/a_{11}^{(1)} & 1 & & \\ \vdots & & \ddots & \\ a_{m1}^{(1)}/a_{11}^{(1)} & 0 & \dots & 1 \end{bmatrix}_{m \times m}$$

and, similarly, for L_i^{-1}, $i = 2, \dots, t - 1$. Further,

$$L_1^{-1} L_2^{-1} = \begin{bmatrix} 1 & & & & & 0 \\ a_{21}^{(1)}/a_{11}^{(1)} & 1 & & & & \\ \vdots & \vdots & \ddots & & & \\ a_{m1}^{(1)}/a_{11}^{(1)} & a_{m2}^{(2)}/a_{22}^{(2)} & 0 & \cdots & & 1 \end{bmatrix}_{m \times m}$$

and by induction,

$$L = \begin{bmatrix} 1 & & & & & & 0 \\ a_{21}^{(1)}/a_{11}^{(1)} & 1 & & & & & \\ \vdots & \vdots & \ddots & & & & \\ a_{m1}^{(1)}/a_{11}^{(1)} & a_{m2}^{(2)}/a_{22}^{(2)} & \cdots & a_{m,t-1}^{(t-1)}/a_{t-1,t-1}^{(t-1)} & 0 & \cdots & 1 \end{bmatrix}_{m \times m} .$$

(1.6)

Corollary 1.3

(a) *Gaussian elimination (with pivoting) is equivalent to the multiplication of A (correspondingly row- and column-permuted) by the matrix L^{-1}, where L is defined in (1.6).*

(b) *A can be written as a product LU of two triangular factors L, U in (1.5a), where L is the lower triangular matrix in (1.6) and $U = A^{(r)}$ is the upper triangular matrix defined in (1.4), and where the entries are defined by*

$$a_{ij}^{(s+1)} = a_{ij}^{(s)} - a_{is}^{(s)} a_{ss}^{(s)-1} a_{sj}^{(s)}.$$

We now want to determine r in (1.4)—i.e., the largest s for which $a_{ss}^{(s)} \neq 0$. Let

$$\widehat{A}_k = \begin{bmatrix} a_{11} & a_{12} & \cdots & a_{1k} \\ a_{21} & a_{22} & \cdots & a_{2k} \\ \vdots & & & \\ a_{k1} & a_{k2} & \cdots & a_{kk} \end{bmatrix}$$

be the kth main submatrix of the matrix A. From the proof of the relation $LU = A$, we realize that

$$\widehat{L}_k \widehat{U}_k = \widehat{A}_k,$$

where

$$\widehat{U}_k = \begin{bmatrix} a_{11}^{(1)} & a_{12}^{(1)} & \cdots & a_{1k}^{(1)} \\ & a_{22}^{(2)} & \cdots & a_{2k}^{(2)} \\ & & \ddots & \\ 0 & & & a_{kk}^{(k)} \end{bmatrix},$$

that is, an upper triangular matrix of order k and $\widehat{L}_k$ is a lower triangular matrix, $1 \le k \le r$.

Hence, using the well-known theorem for the determinant of a product of matrices, we find

$$\det \widehat{A}_k = \det \widehat{L}_k \det \widehat{U}_k = 1 \cdot a_{11}^{(1)} a_{22}^{(2)} \ldots a_{kk}^{(k)},$$

so

$$a_{11}^{(1)} = \det \widehat{A}_1$$

and

$$a_{kk}^{(k)} = \det \widehat{A}_k / \det \widehat{A}_{k-1}, \quad k = 2, 3, \ldots, r.$$

Hence we have proved the following theorem.

Theorem 1.4 *The pivot entries $a_{kk}^{(k)}$, $k = 1, 2, \ldots, r$ are nonzero if and only if the main submatrices $\widehat{A}_k$, $k = 1, 2, \ldots, r$ are nonsingular.*

If we have made any permutations of rows and columns during the elimination process, then by A we here understand the corresponding matrix

$$P_1 A P_2,$$

where P_1, P_2 are permutation matrices of order $m \times m$ and $n \times n$, respectively. (A permutation matrix has exactly one nonzero entry (unity) in each row and column. Permutation matrices are convenient for describing reorderings of a linear system. If P_1 and P_2 are permutation matrices, then $Ax = b \Longleftrightarrow (P_1 A P_2)(P_2^T x) = P_1 b$. Permutation matrices will be further discussed in Chapter 4.

In fact, it can be seen that pivoting during the factorization leads to the same result as factoring without pivoting the correspondingly permuted matrix.

Example 1.6 Let

$$P_1 = \begin{bmatrix} 0 & 1 & 0 & \cdots & 0 \\ 1 & 0 & 0 & \cdots & \\ 0 & 0 & 1 & \cdots & \\ \vdots & & & & \\ 0 & 0 & 0 & \cdots & 1 \end{bmatrix}.$$

Then $P_1 A$ equals A except that the first and second rows of A are permuted, and with

$$P_2 = \begin{bmatrix} 0 & 0 & 1 & \cdots & 0 \\ 0 & 1 & 0 & \cdots & 0 \\ 1 & 0 & 0 & \cdots & 0 \\ \vdots & & & & \\ 0 & 0 & 0 & \cdots & 1 \end{bmatrix},$$

$A P_2$ equals A except that the first and third columns of A have been permuted.

Example 1.7 If we factor the matrix

$$A = \begin{bmatrix} 1 & -1 & 0 \\ -1 & 2 & -1 \\ 0 & -1 & 1 \end{bmatrix}$$

as a product of two triangular factors L, U, then the value of r in (1.4) is found to be $r = 2$.

1.3 Range and Nullspace

1.3.1 Linear Mappings

A real matrix of order $m \times n$ is just a representation of a linear mapping $\widetilde{A}$ on $\mathbb{R}^n \to \mathbb{R}^m$ in a given coordinate system, as will be clarified in Chapter 2. Given $\tilde{x} \in \mathbb{R}^n$, we have $\tilde{y} = \widetilde{A}\tilde{x}$, where $\tilde{y} \in \mathbb{R}^m$. If $x = \sum x_i \mathbf{e}_i^{(n)}$ is the representation of $\tilde{x}$ in a Euclidean coordinate system, where $\mathbf{e}_i^{(n)}$ is the ith Euclidean coordinate vector in $\mathbb{R}^n$, i.e., the jth component

$$(\mathbf{e}_i^{(n)})_j = \delta_{ij} = \begin{cases} 1, & j = i \\ 0, & j \neq i \end{cases}$$

and if $\mathbf{y} = \sum_i y_i \mathbf{e}_i^{(m)}$ is the representation of $\tilde{\mathbf{y}}$, where $\mathbf{e}_i^{(m)}$ is the ith Euclidean coordinate vector in $\mathbb{R}^m$, then with $\mathbf{x} = \mathbf{e}_j^{(n)}$ we let

$$a_{ij} = (\mathbf{y})_i = (\tilde{A}\mathbf{x})_i = (\tilde{A}\mathbf{e}_j^{(n)})_i, \quad 1 \le i \le m, \quad 1 \le j \le n.$$

$A = [a_{ij}]$ is the customary representation of the mapping $\tilde{A}$. In another co-ordinate system $\tilde{A}$ would have another representation. Consider now the case $m = n$ (of a square matrix A). Vectors $\tilde{\mathbf{x}}$ that are mapped in their own direction times a scalar—i.e., for which there exists some number λ (real or complex) such that $\tilde{A}\tilde{\mathbf{x}} = \lambda\tilde{\mathbf{x}}$, $\tilde{\mathbf{x}} \ne \mathbf{0}$—are called *eigenvectors* of $\tilde{A}$. λ is called an *eigenvalue* of $\tilde{A}$. The set of all eigenvalues of $\tilde{A}$ is called the *spectrum* of $\tilde{A}$, denoted by $S(\tilde{A})$.

Since λ depends on the mapping and not on its representation, the spectrum is independent of the coordinate system used for the representation of $\tilde{A}$ and we have $S(A) = S(\tilde{A})$ for any representation A of $\tilde{A}$. Properties of eigenvalues and eigenvectors will be discussed further in Chapter 2.

Let us now consider the range and nullspace of a linear mapping.

Definition 1.1

(a) The *range* of the mapping $\tilde{A} : \mathbb{R}^n \to \mathbb{R}^m$ is

$$\mathcal{R}(\tilde{A}) = \{\tilde{\mathbf{y}} = \tilde{A}\tilde{\mathbf{x}};\ \tilde{\mathbf{x}} \in \mathbb{R}^n\}.$$

(b) The *nullspace* of $\tilde{A}$ is

$$\mathcal{N}(\tilde{A}) = \{\tilde{\mathbf{x}} \in \mathbb{R}^n;\ \tilde{A}\tilde{\mathbf{x}} = \mathbf{0}\}.$$

Theorem 1.5 *Let the integer r be defined by (1.4). Then*

(a) $\dim \mathcal{R}(\tilde{A}) = r$, *also called the rank of* $\tilde{A}$,
(b) $\dim \mathcal{N}(\tilde{A}) = n - r$, *i.e., rank*$(\tilde{A}) + \dim \mathcal{N}(\tilde{A}) = n$.

Proof

(a) For the proof we can use any matrix representation of $\tilde{A}$. From $U = A^{(t)}$, defined in (1.4), and (1.5a) we find $U\mathbf{x} = \sum_{k=1}^r \alpha_k \mathbf{u}_k$ for some scalars α_k, $1 \le k \le r$, where $\mathbf{u}_k = [a_{1k}^{(1)}, a_{2k}^{(2)}, \ldots, a_{kk}^{(k)}, 0, \ldots, 0]^T$ is the kth column of U, $1 \leq k \leq r$. We note that $\mathbf{u}_1, \mathbf{u}_2, \ldots, \mathbf{u}_r$ are linearly independent, but that the remaining columns of U depend linearly on the first r columns. Since L is nonsingular, $L\mathbf{u}_1, \ldots, L\mathbf{u}_r$ is also a linearly independent set of vectors. Hence,

$$LU\mathbf{x} = \sum_{k=1}^{r} \alpha_k L\mathbf{u}_k,$$

so dim $\mathcal{R}(\widetilde{A}) \leq r$. But for any $\mathbf{b} \in \text{SPAN}\{L\mathbf{u}_1, L\mathbf{u}_2, \ldots, L\mathbf{u}_r\}$ there exists a vector $\mathbf{x} \in \mathbb{R}^n$ such that $U\mathbf{x} = L^{-1}\mathbf{b}$ or, equivalently, $A\mathbf{x} = LU\mathbf{x} = \mathbf{b}$. Hence dim $\mathcal{R}(\widetilde{A}) = r$.

(b) In the same way, we get

$$A\mathbf{x} = \mathbf{0} \iff U\mathbf{x} = \mathbf{0} \iff \alpha_1 = \alpha_2 = \ldots = \alpha_r = 0,$$

where r is defined above. These r constraints leave $n - r$ free variables, corresponding to the columns of U that do not contain pivots. Hence dim $\mathcal{N}(\widetilde{A}) = n - r$.   ◇

Theorems 1.4 and 1.5 show that the rank of a matrix A is equal to the maximal order of the nonsingular main submatrices $\widehat{A}_k$ for any permutation PAQ of A, where P and Q are permutation matrices.

Example 1.8a For

$$A = \begin{bmatrix} 1 & -1 & & & & 0 \\ -1 & 2 & -1 & & & \\ & & \ddots & \ddots & & \\ & & & -1 & 2 & -1 \\ 0 & & & & -1 & 1 \end{bmatrix}_{n \times n}$$

we find

$$L = \begin{bmatrix} 1 & & & & 0 \\ -1 & 1 & & & \\ & \ddots & \ddots & & \\ & & -1 & 1 & 0 \\ 0 & & & -1 & 1 \end{bmatrix}_{n \times n}, U = \begin{bmatrix} 1 & -1 & & & 0 \\ & 1 & -1 & & \\ & & \ddots & \ddots & \\ & & & 1 & -1 \\ 0 & & & & 0 \end{bmatrix}_{n \times n}.$$

Hence $r = \dim \mathcal{R}(\widetilde{A}) = n - 1$ and $\dim \mathcal{N}(\widetilde{A}) = 1$.

Definition 1.2

(a) If $\mathbf{b} \in \mathcal{R}(A)$, then $A\mathbf{x} = \mathbf{b}$ is said to be a *consistent* system of linear equations. Otherwise [if $\mathbf{b} \notin \mathcal{R}(A)$], it is said to be *inconsistent*.

(b) If A is square of order n and $\dim \mathcal{N}(A) = 0$, then A is said to be *nonsingular*.

Theorem 1.6

(a) *If A has order $m \times n$ and $r = \mathrm{rank}(A) < n$, and if $A\mathbf{x} = \mathbf{b}$ is consistent, then there are infinitely many solutions.*

(b) *If $A\mathbf{x} = \mathbf{b}$ is inconsistent, there is no solution.*

(c) *If $r = n$ and if $A\mathbf{x} = \mathbf{b}$ is consistent (in particular if $r = n = m$), then there is exactly one solution.*

(d) *There is a solution to every $\mathbf{b} \in \mathbb{R}^m$ if and only if $r = m$.*

Proof

(a) Since $A\mathbf{x} = \mathbf{b}$ is consistent, there is at least one solution $\hat{\mathbf{x}}$ (a so-called particular solution). Since $\dim \mathcal{N}(A) = n - r > 0$, the homogeneous system of equations $A\mathbf{x} = \mathbf{0}$ has infinitely many solutions. Finally, note that $\hat{\mathbf{x}} + \mathbf{x}_0$, for any $\mathbf{x}_0 \in \mathcal{N}(A)$, is also a solution of $A\mathbf{x} = \mathbf{b}$. (It can also be seen that any solution is of the form $\mathbf{x} = \hat{\mathbf{x}} + \mathbf{x}_0$.)

(b) Follows by definition.

(c) Since $\dim \mathcal{N}(A) = n - r = 0$, $A\mathbf{x} = \mathbf{0}$ has only the trivial solution $\mathbf{x} = \mathbf{0}$. Hence, if $\mathbf{x}^{(1)}$, $\mathbf{x}^{(2)}$ are two solutions such that $\mathbf{x}^{(1)} \neq \mathbf{x}^{(2)}$, then $A(\mathbf{x}^{(1)} - \mathbf{x}^{(2)}) = \mathbf{0}$, so $\mathbf{x}^{(1)} - \mathbf{x}^{(2)} = \mathbf{0}$, which is a contradiction.

(d) Left as an exercise.   ◇

Example 1.8b Let A be given as in Example 1.8a. Then $\mathcal{N}(A)$ has dimension 1 and consists of multiples of the vector $\mathbf{e} = [1, 1, \ldots, 1]^T$. Any right-hand vector $\mathbf{b}$ is consistent if and only if $\mathbf{e}^T \mathbf{b} = 0$, i.e., $\sum_{i=1}^{n} b_i = 0$.

The proof of this follows from the following lemma.

Lemma 1.7

(a) *For a matrix A (in general, nonsymmetric) of order $m \times n$, $A\mathbf{x} = \mathbf{b}$ is consistent if and only if $\mathbf{v}^T \mathbf{b} = 0$ for all vectors $\mathbf{v} \in \mathcal{N}(A^T)$.*

(b) *$\dim \mathcal{N}(A^T) = m - r$.*

The proof is left as an exercise.   ◇

Remark It can readily be seen that the statement in Lemma 1.7(a) is equivalent to $\mathcal{R}(A) = \mathcal{N}(A^T)^\perp$, i.e., the range of A is equal to the orthogonal complement of $\mathcal{N}(A^T)$. The *orthogonal complement* of a vector subspace $V \in \mathbb{R}^m$ denoted by $V^\perp$ is defined by $V^\perp = \{\mathbf{u} \in \mathbb{R}^m; \mathbf{u}^T \mathbf{v} = 0 \text{ for all } \mathbf{v} \in V\}$. Note that $V^\perp$ is itself a subspace of $\mathbb{R}^m$, and dim $V^\perp = m - \text{dim } V$. Hence dim $\mathcal{R}(A) = m - \text{dim } \mathcal{N}(A^T) = r$, which is in agreement with Theorem 1.5(a).

Using the factorization (1.5a) we can readily solve a consistent system $A\mathbf{x} = \mathbf{b}$. We then use forward and back substitutions:

1. *Forward substitution:* Solve $L\mathbf{y} = \mathbf{b}$, by computing in order $y_1, y_2, \ldots,$ y_m. Here $y_{r+1} = \cdots = y_m = 0$, if $r < m$.
2. *Back substitution:* Solve $U\mathbf{x} = \mathbf{y}$, by computing in order $x_r, x_{r-1}, \ldots,$ x_1, thereby letting $x_{r+1}, \ldots, x_n$ be free parameters when $r < n$.

1.4 Practical Considerations

We now discuss some aspects of the triangular factorization method that are of importance for practical implementation of the method.

1.4.1 Numerical Stability

Although all pivot entries are nonzero, in a numerical calculation we also have to be aware of pivot entries that are almost zero; otherwise, round-off (due to finite precision arithmetic) can cause unacceptable errors.

Example 1.9 Consider

$$\varepsilon x_1 + x_2 = 1, \quad 0 < \varepsilon \ll 1$$

$$x_1 + x_2 = 2.$$

By elimination or by using the factorization

$$\begin{bmatrix} \varepsilon & 1 \\ 1 & 1 \end{bmatrix} = \begin{bmatrix} 1 & 0 \\ 1/\varepsilon & 1 \end{bmatrix} \begin{bmatrix} \varepsilon & 1 \\ 0 & 1 - 1/\varepsilon \end{bmatrix},$$

we get $(1 - \frac{1}{\varepsilon})x_2 = 2 - \frac{1}{\varepsilon}$ and, hence, $x_2 = (1 - 2\varepsilon)/(1 - \varepsilon)$. In a machine representation of this number, $x_2 = 1$ if ε is sufficiently small. Hence $\varepsilon x_1 = 0$, $x_1 = 0$, whereas the correct solution is $x_1 = 1/(1 - \varepsilon)$.

This example shows that rounding errors can have disastrous effects when they arise as a result of division by pivots that are too small. A remedy for small pivot entries, of course, is to perform permutations. We can use *partial pivoting* (row interchanges) or *complete pivoting*. In the first case we choose as pivot any element $a_{ik}^{(k)}$, $k \le i \le n$ in column k satisfying

$$\left| a_{ik}^{(k)} \right| = \max_{k \le p \le n} \left| a_{pk}^{(k)} \right|,$$

and in the second case we choose as pivot any element $a_{ij}^{(k)}$, $k \le i, j \le n$ satisfying

$$\left| a_{ij}^{(k)} \right| = \max_{k \le p, q \le n} \left| a_{pq}^{(k)} \right|.$$

In the first case we perform row interchanges ($i \leftrightarrow p$); in the second case, we need also to perform column interchanges.

However, permutations often destroy any useful (sparsity) structure A may have. Fortunately, the matrices one comes across in practice frequently have a property such that there is no need for permutations for the purpose of stability. For instance, for diagonally dominant matrices or, more generally, for block H-matrices, pivoting is not necessary. This topic will be further discussed in Chapter 7. For rounding error analysis in the solution of linear systems, see Appendix A.

1.4.2 Fill-in

In many important problems the order of matrix A is very large, but A has a sparse structure, meaning that most entries a_{ij} of A are zero. However, during the factorization of A, entries in $A^{(k)}$, $k \ge 1$ that are zero can become replaced by nonzero at later stages. Such entries are called *fill-in*. For instance, if $a_{ij}^{(k)} = 0$ but $a_{ik}^{(k)} a_{kk}^{(k)-1} a_{kj}^{(k)}$ is nonzero—i.e., if both $a_{ik}^{(k)}$ and $a_{kj}^{(k)}$ are nonzero—then

$$a_{ij}^{(k+1)} = -a_{ik}^{(k)} a_{kk}^{(k)-1} a_{kj}^{(k)}$$

is an example of a fill-in entry. There is a tendency toward increasing fill-in during the stages, because old fill-in entries are likely to cause new fill-in entries at later stages.

A handy tool in the discussion of these questions when A is an $n \times n$ matrix with a symmetric sparsity structure (i.e., $a_{ij} \ne 0$ if and only if $a_{ji} \ne 0$) is the *graph* of A (Parter, 1961). To construct the graph of A, we draw n nodes, labeling them $1, \ldots, n$, and draw a line from node i to node j for all index pairs (i, j) such that $a_{ij} \neq 0$. Clearly, it suffices to do this for pairs with $i > j$.

Let $G^{(k)}$ be the matrix graph for $A^{(k)}$. At stage k, upon elimination of the variable x_k from the subsequent equations, the new graph $G^{(k+1)}$ of the remaining system is obtained from the previous graph $G^{(k)}$ by (a) eliminating node k and (b) pair-wise connecting by lines all points which in $G^{(k)}$ were directly connected to node k. Here any entry that becomes zero by cancellation is considered as nonzero. A fill-in entry therefore occurs at stage k in position (i, j) if there is no edge between nodes i and j but there are edges between nodes k and i and between nodes k and j.

To avoid a large number of fill-in entries, one can try to permute the matrix to an ordering of graph nodes for which less fill-in will occur. This topic, however, is outside the scope of this presentation. (For details, see George and Liu, 1981, for instance.) Alternatively, we can use iterative solution methods that do not change the matrix; this will be the topic of later chapters.

1.4.3 Computational Complexity

Consider now the number of operations needed to factor a square matrix A and to solve $A\mathbf{x} = \mathbf{b}$. Assuming that all pivot entries are nonzero, and, for simplicity, counting a division as one multiplication and an addition, we find that the number of multiplications and the number of additions is each as follows:

$$(1.7) \qquad \sum_{k=n}^{1} (k-1)k = \frac{1}{3}n(n^2 - 1) \sim \frac{1}{3}n^3, \quad n \to \infty.$$

Here the factor $k - 1$ is the number of remaining rows to be eliminated and k is the number of entries in the pivot row. (The actual number of divisions required to compute the reciprocals of the pivot entries is $n - 1$.) In this estimate we have considered A to be a full matrix— i.e., we have not taken advantage of possible zero entries.

In the following we shall use the operation count "flop," meaning a (floating point) multiplication followed by an addition (or subtraction) and including two fetches of data. This definition is natural in numerical linear algebra because additions and multiplications usually go together in matrix and vector computations. Hence the factorization of a matrix A needs about $\frac{1}{3}n^3$ flops. If the matrix A is *symmetric*—i.e., $a_{ij} = a_{ji}$, $i, j = 1, \ldots, n$—and can be factorized with nonzero pivot entries d_i, $i = 1, \ldots, n$, then it suffices to compute L

Table 1.1. *CPU-time for LU factorization*

n	CPU time (sec.)
100	0.33
1000	330
10000	33.10^4 sec $\simeq$ 3.8 days!

and the diagonal part D of U, since $A = LU = LDL^T$, where L^T is the transpose of L. Then the asymptotic operation count for the factorization of A is $\sim \frac{1}{6}n^3$.

Naturally it is of interest to know how large general (full and nonsymmetric) systems of equations can be solved using the method of elimination on a common serial computer. Assuming that the central processing (CPU) time is 1 μs (= 10^{-6} sec.) per flop, which was typical for the fastest mainframe computers in the seventies, the total time needed for the factorization is given in Table 1.1. More modern computers are not much faster, unless one is able to implement the algorithms to utilize vector and parallel computers.

Besides floating point operations, integer and logical operations can occur. Matrix computations usually involve a large amount of integer arithmetic for computing memory addresses, and this can take a significant part of the total computer time. In addition, memory traffic can be a bottleneck. In particular, for modern computers it is important to pay as much attention to the flow of data as to the amount of arithmetic. This topic, however, is outside the scope of this presentation.

At any rate, for such administrative ("red-tape") operations we perhaps have to add about the same amount of time. Hence, even on fast computers, only a few thousand equations represent a practical upper bound on the size of the problem. In addition, we have to store $n^2 = 10^6$ numbers for A if $n = 1000$, whereas the primary memory capacity—at least on smaller and average-size computers—is much less. Hence we have to use secondary storage, which means that memory traffic and the read-write (I/O) operations may increase the computer time even further.

In a computer implementation, we note that we may store the new entries [cf. below (1.4)]

$$(1.8) \quad a_{ij}^{(k+1)} = a_{ij}^{(k)} - (a_{kk}^{(k)})^{-1} a_{ik}^{(k)} a_{kj}^{(k)}, \quad k+1 \leq i \leq m, \quad k+1 \leq j \leq n$$

in the space originally occupied by the old entries. The off-diagonal entries of

L can be stored in the corresponding positions of A. Hence no extra space is required.

In many applications A is a *band-matrix*—i.e., a matrix for which $a_{ij} = 0$ if $|i - j| > q$ for some positive integer q (which is called the *semibandwidth*). It is left as an exercise to prove that the asymptotic operation count then is $\sim \frac{1}{2}(q + 1)^2 n$ for the factorizations of a symmetric matrix.

1.4.4 Forward and Back Substitutions

To solve $A\mathbf{x} = \mathbf{b}$, we make use of the factorization (1.5a). Then, we first solve

(1.9) $L\mathbf{y} = \mathbf{b}$ *(forward substitution)*

and subsequently solve

(1.10) $U\mathbf{x} = \mathbf{y}$ *(back substitution)*.

The latter is the same as $A^{(t)}\mathbf{x} = \mathbf{b}^{(t)}$. Each of these steps needs $\sum_{k=1}^{n} k = \frac{1}{2}n(n + 1) \sim \frac{1}{2}n^2$ flops and, for a bandmatrix, about $(q + 1)n$ flops. *The total computational complexity for a single linear (unsymmetric) system thus is* $\sim \frac{1}{3}n^3 + n^2$ flops and, for a bandmatrix, $\sim (q + 1)^2 n + 2qn$ flops, $n \to \infty$. For a symmetric system, where $A = LDL^T$ and L has unit diagonal entries we solve in order $L\mathbf{y} = \mathbf{b}$, $\tilde{\mathbf{y}} = D^{-1}\mathbf{y}$, $L^T\mathbf{x} = \tilde{\mathbf{y}}$; the computational complexity is $\sim 1/6n^3 + n^2$ flops and, for a bandmatrix, $\sim 1/2(q + 1)^2 n + 2qn$ flops, $n \to \infty$. Incidentally, to save computer time, let the factorization take the form $LD^{-1}L^T$; then there is no need to perform any divisions in the forward and backward solution method. As an example of this, see the subsection "A Division Free Form."

If, for stability reasons, we need to use pivoting—i.e., row and/or column interchanges—then the permuted matrix $P_1 A P_2$, where P_1, P_2 are corresponding permutation matrices (for a definition, see above Example 5), is no longer symmetric. A possible way around this is to perform *symmetric pivoting*, so that we seek a permutation P such that the LDL^T factorization of $\tilde{A} = PAP^T$ is numerically stable. However, the diagonal of $\tilde{A}$ is then just a permutation of A's diagonal. Hence, if all of A's diagonal entries are small, then numerical instability will occur for any choice of P.

There is a remedy for this problem. Aasen (1971) has shown that any symmetric (even indefinite) matrix can be factored in a numerically stable way as LTL^T, where T is tridiagonal. Aasen shows that there exists a permutation matrix P such that $PAP^T = LTL^T$, where L is unit-triangular and $|l_{ij}| \leq 1$ (see Exercise 31). The computational complexity of this algorithm is

$n^3/6$. Once obtained, the factorization can be used to solve a linear system as follows:

$$Ly = Pb, \quad T\tilde{y} = y, \quad L^T\tilde{x} = \tilde{y}, \quad x = P^T\tilde{x}.$$

Here $O(n^2)$ flops are required. The solution of tridiagonal systems will be discussed later in this chapter. There is a similar approach by Bunch and Kaufman (1977) where a permutation is found so that $PAP^T = LDL^T$, where L is unit-lower-triangular and D is a matrix on block diagonal form consisting of 1-by-1 and 2-by-2 matrix blocks.

In some applications—for example, in structural engineering and when solving certain time-dependent partial differential equations—we are faced with the problem of solving a set of (p) linear equations with the same matrix. Clearly, the factorization of A is then performed only once and the total work per right-hand side is about

(1.11) $$\frac{1}{3p}n^3 + n^2 \text{ flops}$$

and, for a bandmatrix, is

(1.12) $$\frac{1}{p}(q + 1)^2 n + 2qn \text{ flops}.$$

For a symmetric matrix, the first terms in (1.11, 1.12) should be divided by 2. The elimination may also be performed in a more compact way without storing the intermediate quantities $a_{ij}^{(k-1)}$. The corresponding formulas (referred to as Crout's or Dolittle's methods) can be derived by explicitly identifying the entries in $A = LU$ in the following order [we may augment A with an $(n + 1)$st column initially equal to the right-hand side vector $\mathbf{b}$]. For $i = 1, 2, \ldots, n$:

$$u_{ij} = a_{ij} - \sum_{k=1}^{i-1} l_{ik}u_{kj}, \quad j = i, \ldots, n$$

$$l_{ji} = [a_{ji} - \sum_{k=1}^{i-1} l_{jk}u_{ki}]/u_{ii}, \quad j = i + 1, \ldots, n.$$

1.4.5 Matrix Inversion

For the inversion of a nonsingular matrix, we may solve the linear equations $Ax_i = e_i$, $1 \le i \le n$, where e_i is the ith column of the identity matrix. Then, because $AA^{-1} = I$, $\mathbf{x}_i$ is the ith column of A^{-1}. Hence we may compute A^{-1} by the solution of n linear equations with the same matrix A. It follows that the total number of operations now becomes about $1/3n^3 + n \cdot n^2 = 4/3n^3$. If we modify the algorithm to make use of the large number of zeros among $\mathbf{e}_i$, it turns out that we may reduce the number of operations to about n^3 flops. The reason for this is that the right-hand side $\mathbf{e}_i$ is changed only when the elimination has reached the ith row of A, so we work, in fact, only with a lower triangular right-hand-side matrix. Hence, the first $(i - 1)$ components of the vector $\mathbf{y}^{(i)} = L^{-1}\mathbf{e}^{(i)}$ are zero; thus, the forward substitution requires only $\sim \sum_{k=1}^{n} \frac{1}{2}k^2 \sim \frac{1}{6}n^3$ flops and the total operation count therefore becomes (asymptotically) $n^3/3 + n^3/6 + n(n^2/2) = n^3$.

Incidentally, note that this is the same operation count as for the multiplication of two matrices! Similarly, for a symmetric matrix, the operation count for the computation of the inverse becomes $\frac{1}{6}n^3 + \frac{2}{6}n^3 = \frac{1}{2}n^3$ (i.e., even less than for the multiplication of two matrices). Here we have used the symmetry of A^{-1}; that is, it suffices to compute the lower triangular part of A^{-1}. At the backsubstitution we compute $\mathbf{x}_j^{(i)} = U^{-1}\mathbf{y}_j^{(i)}$, $j = i, i + 1, \ldots, n$.

When solving a system with many (p) right-hand sides, one might expect to gain by first computing the inverse and then making p matrix-vector multiplications. Each such multiplication costs n^2 flops—i.e., the same as a forward and a back substitution. As we have seen, however, the cost of an LU factorization is only about a third of the cost of the calculation of the inverse. For a parallel computer, it can be advisable to use the inverse.

Example 1.10 Factor the matrix C in Example 2.6. Then compute its inverse by the solution of $C\mathbf{x}_i = \mathbf{e}_i$, $i = 1, 2, 3$, to show that it takes the form given in that example. Note that here we need permutations.

1.5 Solution of Tridiagonal Systems of Equations

In practical problems, one frequently comes across tridiagonal systems of equations,

$$(1.13) \qquad \begin{bmatrix} a_{1,1} & a_{1,2} & & 0 \\ a_{2,1} & a_{2,2} & a_{2,3} & \\ \vdots & & & \\ 0 & & a_{n,n-1} & a_{n,n} \end{bmatrix} \begin{bmatrix} u_1 \\ u_2 \\ \vdots \\ u_n \end{bmatrix} = \begin{bmatrix} b_1 \\ b_2 \\ \vdots \\ b_n \end{bmatrix}.$$

This frequent occurrence motivates a special discussion of them. As we have seen in Section 1.3, such systems can be solved easily by elimination, using the bandmatrix form. For the special (tridiagonal) matrix in (1.13), we shall see that the factorization can best be performed as

$$A = LD^{-1}R = \begin{bmatrix} d_1 & & & 0 \\ a_{2,1} & d_2 & & \\ & \ddots & \ddots & \\ 0 & & a_{n,n-1} & d_n \end{bmatrix} \begin{bmatrix} d_1^{-1} & & & 0 \\ & d_2^{-1} & & \\ & & \ddots & \\ 0 & & & d_n^{-1} \end{bmatrix} \begin{bmatrix} d_1 & a_{1,2} & & 0 \\ & d_2 & a_{2,3} & \\ & & \ddots & \ddots \\ 0 & & & d_n \end{bmatrix}$$

(1.14)

or

$$A = LU = \begin{bmatrix} d_1 & & & 0 \\ a_{2,1} & d_2 & & \\ & \ddots & \ddots & \\ 0 & & a_{n,n-1} & d_n \end{bmatrix} \begin{bmatrix} 1 & (d_1^{-1}a_{1,2}) & & 0 \\ & 1 & (d_2^{-1}a_{2,3}) & \\ & & \ddots & \\ 0 & & & 1 \end{bmatrix}.$$

We find

$$LU = \begin{bmatrix} d_1 & a_{1,2} & & & 0 \\ a_{2,1} & (d_2 + a_{2,1}d_1^{-1}a_{1,2}) & a_{2,3} & & \vdots \\ & \cdots & & & \\ 0 & \cdots & & a_{n,n-1} & (d_n + a_{n,n-1}d_{n-1}^{-1}a_{n-1,n}) \end{bmatrix}.$$

From $A = LU$ it then follows that we have only to determine the diagonal entries d_i. They can be determined by the recursion,

(1.15) $d_i = a_{i,i} - a_{i,i-1}d_{i-1}^{-1}a_{i-1,i}$, $i = 2, 3, \ldots, n$, where $d_1 = a_{1,1}$.

It follows from Theorem 1.4 that $d_i \neq 0$ if and only if the main submatrices of A are nonsingular.

For the forward and back substitutions, we get

(1.16a) $L\mathbf{y} = \mathbf{b}$ or $y_1 = b_1/d_1$, $y_i = (b_i - a_{i,i-1}y_{i-1})/d_i$, $i = 2, \ldots, n$,

(1.16b) $U\mathbf{x} = \mathbf{y}$ or $x_n = y_n$, $x_i = y_i - a_{i,i+1}x_{i+1}/d_i$, $i = n-1, \ldots, 1$.

1.5.1 A Division-Free Form

In order to avoid the divisions that occur in (1.16 a,b) for each forward and back substitution, we may use the alternative form of factorization,

$$A = LDR = \begin{bmatrix} d_1^{-1} & & & 0 \\ a_{2,1} & d_2^{-1} & & \\ & & \ddots & \\ 0 & & a_{n,n-1} & d_n^{-1} \end{bmatrix} \begin{bmatrix} d_1 & & & 0 \\ & d_2 & & \\ & & \ddots & \\ 0 & & & d_n \end{bmatrix} \begin{bmatrix} d_1^{-1} & a_{1,2} & & 0 \\ & d_2^{-1} & a_{2,3} & \\ & & \ddots & \\ 0 & & & d_n^{-1} \end{bmatrix}$$

(1.17)

or

$$A = LU = \begin{bmatrix} d_1^{-1} & & & \\ a_{2,1} & d_2^{-1} & & \\ & & \ddots & \\ & & a_{n,n-1} & d_n^{-1} \end{bmatrix} \begin{bmatrix} 1 & (d_1 a_{1,2}) & & 0 \\ 0 & 1 & (d_2 a_{2,3}) & \\ & & \ddots & \\ 0 & & & 1 \end{bmatrix}.$$

We find now that these entries d_i are the reciprocals of the previous values and are determined by the recursion

(1.18) $d_i = (a_{i,i} - a_{i,i-1} d_{i-1} a_{i-1,i})^{-1}, \quad i = 1, 2, \ldots, n,$

where $d_0 = 0$.

For the forward and back substitutions we get now

(1.19a) $L\mathbf{y} = \mathbf{b}$ or $y_1 = d_1 b_1, \quad y_i = d_i(b_i - a_{i,i-1} y_{i-1}), \quad i = 2, \ldots, n,$

and

(1.19b) $U\mathbf{x} = \mathbf{y}$ or $x_n = y_n, \quad x_i = y_i - d_i a_{i,i+1} x_{i+1}, \quad i = n - 1, \ldots, 1.$

This division free algorithm is of importance when solving linear systems on presently available vector-computers, for instance, where a division operation takes significantly longer than a multiplication. The number of multiplications and divisions in the recursion (1.15) is $2n$, whereas it is $3n$ in (1.18). There are $2n$ multiplications and $2n$ divisions in the forward and back substitutions in (1.16 a,b), compared with $4n$ multiplications in (1.19 a,b).

If we compute and store the entries $d_i^{-1} a_{i,i+1}$ and $d_i a_{i,i+1}$, respectively, in U, the operation counts for the forward and back substitutions are diminished by n. The total number of multiplications for the solution of a single linear system, using the division free variant, is $6n$. In addition, some subtractions occur. The total work thus is proportional to n, the number of unknowns, and is hence close to optimal (minimal) on a sequential computer.

1.5.2 Block Tridiagonal Matrices

Frequently matrices of the form (1.13), where each entry a_{ij} is a block matrix, arise naturally in practice. The diagonal blocks are assumed to be square. The factorization methods (1.14) and (1.17) are still applicable, but these recursions also require the computation of the inverse of d_{i-1} and of $(a_{i,i} - a_{i,i-1}d_{i-1}a_{i-1,i})$, respectively.

Assume for simplicity of exposition that all matrix blocks are $m \times m$. To compute the recursion (1.18) we need, at each step, two matrix-matrix multiplications, which cost m^3 flops (assuming that the matrices $a_{i,i-1}$ and $a_{i-1,i}$ are full) and one matrix inversion, which costs also m^3 flops. Hence, if the block order of the matrix is n (and hence the total order is $N = mn$), we need a total of $3nm^3$ flops for the recursion.

The forward and back solution steps now cost four matrix-vector multiplications—i.e., $4nm^2$. Note that there is no need to solve linear systems (with matrices of order m), during the forward and back substitution steps, in the division free (inverse free) variant.

It can readily be seen that any bandmatrix can be partitioned in block tridiagonal form. If the semibandwidth is $m + 1$, then the complexity of the above method for a block-tridiagonal matrix turns out to be $2nm^3$ flops for the recursion (because $a_{i,i-1}$ and $a_{i-1,i}$ are triangular matrices in this case). The total computational complexity becomes $2nm^3$ for the factorization and $3nm^2$ for the solution. This compares somewhat less favorably with the operation count for a bandmatrix method, where the computational complexity was found to be $m(m + 1)^2 n$ and $2m(m + 1)n$ flops.

However, the solution method for the division free form for the block tridiagonal matrix can be more efficiently implemented on parallel and vector computers.

Frequently the matrices $a_{i,i-1}$ and $a_{i-1,i}$ are very sparse. The above block matrix factorization method thus needs storage essentially only for the matrices d_i, which is, in total, only about half the amount of storage required for the pointwise factorization method.

1.5.3 The Recursive Odd-Even Reordering Method

We shall now describe an algorithm that can take significantly less computational effort. Consider a central difference approximation for a two-point boundary value problem. Let us reorder the points in an odd-even way—i.e., take first the odd-numbered points and then the even-numbered ones. Then, if n is an odd number, the new ordering is as given in Figure 1.3. It can readily be seen that the corresponding permuted matrix, of order n, takes the form

$$\widetilde{A}_0 = \begin{bmatrix} x & & & & & x & & \\ & x & & & & x & x & \\ & & x & & & & x & x \\ & & & x & & & & x & x \\ & & & & x & & & & x \\ x & x & & & & x & & \\ x & x & & & & & x & \\ & x & x & & & & x & \\ & & x & x & & & & x \end{bmatrix},$$

where we have indicated nonzero entries by x. Here we eliminate the first group of unknowns (the originally odd ones), resulting in a new tridiagonal matrix of the form

$$A_1 = \begin{bmatrix} x & x & & \\ x & x & x & \\ & x & x & x \\ & & x & x \end{bmatrix}.$$

The order of the new matrix is $(n-1)/2$, which we assume to be an even number. Here we reorder the unknowns in the same odd-even way as before to reach a permuted system $\widetilde{A}_1$ of the same form as $\widetilde{A}_0$, but of order $(n-1)/2$. Then, by elimination of the first block of corresponding unknowns, we get a new tridiagonal system of order $(n-1)/4$. The odd-even renumbering and successive elimination may be repeated recursively on the remaining systems, which order decrease, until we eventually reach a system with only one unknown. For later purpose, we call this unknown the (first) *separator*.

This is solved, and the computation of the remaining unknowns can easily be performed by nesting up the equations (back substitution)—i.e., recursively substituting the computed values of the second group of unknowns for the computation of the first group in each of the systems.

Alternatively, when the first separator unknown has been computed, we may start from the beginning and compute the solution of the two uncoupled problems that arise when we use the computed value of the separator to decouple

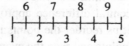

Figure 1.3.   Odd-even reordering

the original linear system in two systems. This computation can be repeated, and at each stage we then solve a sequence of smaller and smaller-size problems. (If $n = 2^r$, then the size of the subproblems decreases at about $n/2^s$, $s = 1, 2, \ldots, r$.) *The advantage with this latter approach is that we do not need to store any of the intermediate matrices.*

We leave it to the reader to establish that the number of operations for the odd-even reordering method is the same as when we solve the tridiagonal band matrix that arises when we use a natural ordering. However, if we have a system with *constant coefficients along the subdiagonals*, then the new entries in the reduced matrices are also constant *along the subdiagonals. Hence we need to compute the entries in only one row.* This means that the overall complexity for the elimination work for the successive matrices is (approximately) $(\log_2 n)/n$ of the work for the natural ordering.

Remark A similar reduction of computational complexity would occur even for a general nonconstant coefficient case if we have a computer with $n/2$ parallel processors at our disposal. Note, however, that the elimination work associated with the right-hand side in general does not change with the reordering; in total it is $(n - 1)$ flops. The complexity for the factorization part of the recursive odd-even reordering method, however, is only $O(\log n)$. This method is illustrated by the following example.

Example 1.11 Consider the central difference method for $-u'' = f$, $0 < x < 1$, where $f(x) = x$ and $u(0) = u(1) = 0$. [The solution is $u(x) = 1/6x(1 - x^2)$.] If $n = 7$ (i.e., $h = 1/8$), we get $A_0 u_h = F_0$, where $(F_h)_i = ih^3$, $i = 1, 2, \ldots, n$. After reordering we have $\widetilde{A}_0 \tilde{u}_h = \widetilde{F}_0$, where

$$
\widetilde{A}_0 = \begin{bmatrix} 2 & & & & -1 & & \\ & 2 & & & -1 & -1 & \\ & & 2 & & & -1 & -1 \\ & & & 2 & & & -1 \\ -1 & -1 & & & 2 & & \\ & -1 & -1 & & & 2 & \\ & & -1 & -1 & & & 2 \end{bmatrix}, \quad \widetilde{F}_0 = h^3 \begin{bmatrix} 1 \\ 3 \\ 5 \\ 7 \\ 2 \\ 4 \\ 6 \end{bmatrix}.
$$

After eliminations, the multiplication of the left- and right-hand sides by 2, and reordering, we get

$$
\widetilde{A}_1 = \begin{bmatrix} 2 & & -1 \\ & 2 & -1 \\ -1 & -1 & 2 \end{bmatrix}, \quad \widetilde{F}_1 = h^3 \begin{bmatrix} 8 \\ 24 \\ 16 \end{bmatrix}.
$$

After elimination, only one unknown (originally u_4) remains. We get $\tilde{u}_6 = u(1/2) = 32h^3 = 1/16$.

We now use u_4 as a separator and solve the two split problems separately for $0 < x < 1/2$ and for $1/2 < x < 1$, respectively, each with three unknowns. We leave it to the interested reader to perform the details. It is clear that this method is also a fast way to solve constant coefficient problems by hand computation or on (small) personal computers.

Exercises

1. Compute the solution ($x_i = i$, $i = 1, 2, 3$) of

    $$2x_1 + 4x_2 = 10$$

    $$3x_1 + 7x_2 + x_3 = 20$$

    $$-x_2 + 2x_3 = 4,$$

    using first (1.15) and then the division free form (1.18) for the LU factorization of the matrix.

2. Find a solution to the linear equation system in Example 1.2, where $b_i = i$, $i = 1, 2, \ldots, 6$.

3. (a) Find the LDL^T factorization of

    $$A = \begin{bmatrix} 1 & -1 & 0 \\ -1 & 2 & -1 \\ 0 & -1 & 2 \end{bmatrix}.$$

    (b) Find the operation counts for computation of the solution of a symmetric $n \times n$ tridiagonal matrix, using an LDL^T factorization.

4. Assuming a condition similar to that in Theorem 1.4, show how a matrix can be factored where the first factor is upper triangular and the second lower triangular.

5. Determine the rank and nullspace of the matrix

    $$A = \begin{bmatrix} 2 & 1 & 4 & 7 \\ 0 & 1 & 2 & 1 \\ 2 & 2 & 6 & 8 \\ 4 & 4 & 14 & 16 \end{bmatrix}.$$

6. Prove that if $B = C^{-1}AC$ and C is nonsingular, then for each eigensolution λ, $\mathbf{x}$ of A, there is an eigensolution μ, $\mathbf{y}$ of B, where $\mathbf{y} = C^{-1}\mathbf{x}$ and

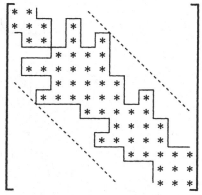

Figure 1.4.   The band and envelope of a symmetric matrix

$\mu = \lambda$. Hence, similarly equivalent matrices (see Chapter 2 for a definition of this concept) have the same spectrum.

7. Let $m(i)$ denote the column number of the first nonzero entry in the ith row of a symmetric band matrix A. The entries a_{ij} and a_{ji}, where $j = m(i), m(i) + 1, \ldots, i$, form the following pattern:

$$
\begin{array}{cccc}
 & & & \times \\
 & & & \times \\
 & & & \times \\
\times & \times & \times & \times
\end{array}
$$

The positions of these entries for $i = 1, 2, \ldots, n$ define the *envelope* of K. As an example, the envelope of the matrix in Figure 1.4 is bounded by solid lines. Compute the computational complexity for the LU factorization of A and for solving linear systems with A, using the factored form. Then show that the fill-in is confined to the envelope.

8. The *growth factor* in Gaussian elimination is defined as

$$
\rho = \max_{i,j,k} |a_{ij}^{(k)}| / \max_{i,j} |a_{i,j}|,
$$

where $a_{i,j}^{(k)}$ denotes the i, j element before the kth step of elimination. It can be shown that a large growth factor indicates the possibility of large relative errors due to round-off during the elimination.

(a) Consider a matrix of order n of the form shown below and its LU factorization (here shown for $n = 5$):

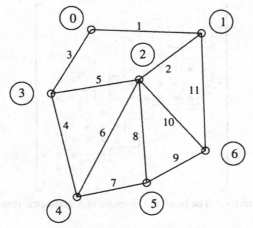

Figure 1.5.   An electric network

$$
\begin{bmatrix}
1 & & & & 1 \\
-1 & 1 & & & 1 \\
-1 & -1 & 1 & & 1 \\
-1 & -1 & -1 & 1 & 1 \\
-1 & -1 & -1 & -1 & 1
\end{bmatrix}
=
\begin{bmatrix}
1 & & & & \\
-1 & 1 & & & \\
-1 & -1 & 1 & & \\
-1 & -1 & -1 & 1 & \\
-1 & -1 & -1 & -1 & 1
\end{bmatrix}
\begin{bmatrix}
1 & & & & 1 \\
 & 1 & & & 2 \\
 & & 1 & & 4 \\
 & & & 1 & 8 \\
 & & & & 16
\end{bmatrix}
$$

Show that $\rho = 2^{n-1}$.

(b) A matrix is called diagonally dominant if $|a_{ii}| \geq \sum_{j \neq i} |a_{ij}|$, $i = 1, 2, \ldots, n$ and a Z-matrix if $a_{ij} \leq 0$, $i \neq j$. (For further related definitions, see Chapter 6.) Show that the growth factor in Gaussian elimination of a diagonally dominant Z-matrix with positive diagonal entries is unity.

Remark: It has been conjectured (Wilkinson, 1965) that when using complete pivoting, the maximum growth factor in Gaussian elimination on a real matrix of order n is equal to n. Numerical tests (Edelman, 1991) show that the conjecture is false—i.e., there exist matrices for which the growth factor is bigger than n, even if complete pivoting is used.

9. Consider an electric network with resistances and node orderings as shown in Figure 1.5. Write the matrices E^T, R^{-1}, E and compute their product. Explain why the resulting matrix could have been written up directly. (The $E^T R^{-1} E$ form is useful when R is nonlinear—i.e., depends on the solution—and when we solve $E^T R^{-1} E$ by iteration, as we shall see.)

10. *The star-delta transformation:* Consider first two simple networks,

where two resistors r_1 and r_2 are connected in *series* and in *parallel*, respectively:

Show that the total resistance between a and b becomes $r = r_1 + r_2$ and $r = \frac{1}{\frac{1}{r_1}+\frac{1}{r_2}}$, respectively. (Note, incidentally, that in the latter case $\frac{1}{r} = \frac{1}{r_1} + \frac{1}{r_2}$, so the so-called *conductances*—i.e., the inverse of the resistances—are additive.)

An efficient method in practical calculations with networks is the so-called *star-delta transformation*. If a vertex d is joined to just three vertices—say a, b, and c—by edges with resistances A, B, and C, then d is the center of a *star*. Since there is no sink or source at d, we can eliminate d to replace it with the *delta-network* as shown below:

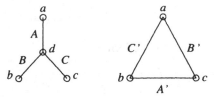

Using Gaussian elimination of the variable v_d, show that the resulting resistances of the delta network are

$$A' = S/A, \quad B' = S/B, \quad C' = S/C,$$

where

$$S = AB + AC + BC.$$

Hence $A' = B + C + (BC/A)$ etc. On the other hand, if we add the node d, going from the delta to the star network, show that the new conductances satisfy the same type of relation, for instance:

$$\alpha = \beta' + \gamma' + \beta'\gamma'/\alpha',$$

where $\alpha, \beta, \ldots$ etc., denote conductances.

11. As an application of the star-delta transformation, let us calculate the total resistance of a tetrahedron across an edge, in which the resistances are as shown below. Verify the resistances of the intermediate networks.

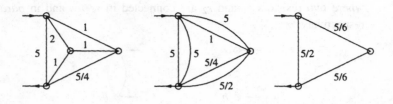

$$(s = 2 + 2 + 1 = 5)$$

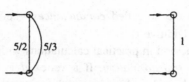

12. Show that Kirchoff's laws for an electric network are

$$R\tilde{x} - Ev = 0 \quad \text{(voltage law)}$$

$$E^T\tilde{x} \quad = 0 \quad \text{(current law)},$$

which can be written in the form

$$\begin{bmatrix} R & -E \\ -E^T & 0 \end{bmatrix} \begin{bmatrix} \tilde{x} \\ v \end{bmatrix} = \begin{bmatrix} 0 \\ 0 \end{bmatrix}.$$

Show that these equations also arise from the constrained minimization problem

$$\min \frac{1}{2}\tilde{x}^T R\tilde{x} \quad \text{subject to } E^T\tilde{x} = 0,$$

using the Lagrange multiplier method, and that the voltage v then plays the role of a Lagrange multiplier.

13. *Matrix computations:*
     (a) Verify the associative law for matrix multiplication, which states that

$$(AB)C = A(BC),$$

where A, B, and C have orders $m \times n$, $n \times p$ and $p \times q$, respectively.
     (b) Verify by an example that the commutative law for matrix multiplications does not hold in general. That is, find square matrices A, B

such that

$$AB \neq BA.$$

(c) Verify by an example that $AB = 0$ does not necessarily imply that either $A = 0$ or $B = 0$.

(d) The transpose of the $m \times n$ matrix $A = [a_{ij}]$ $i = 1, \ldots, m$, $j = 1, \ldots, n$, is the $n \times m$ matrix (denoted A^T), obtained by interchanging the rows and columns of A: $(A^T)_{ij} = a_{ji}$, $i = 1, \ldots, n$, $j = 1, \ldots, m$. Verify that:

    (i) $(A + B)^T = A^T + B^T$,

    (ii) $(A^T)^T = A$,

    (iii) $(AB)^T = B^T A^T$. That is, the transpose of the product of two matrices is the product of the transposes in the reverse order.

14. Let A, B be nonsingular matrices of the same order. Show that

    (a) $(A^{-1})^{-1} = A$,

    (b) $(AB)^{-1} = B^{-1}A^{-1}$,

    (c) $(A^{-1})^T = (A^T)^{-1}$.

Remark: Frequently one uses the notation $A^{-T} = (A^{-1})^T$. *Hint for (b):* Consider the corresponding mappings and their inverses—i.e., it is not necessary to compute the matrix products for the proof.

15. Suppose that A, B, and $A + B$ are all nonsingular matrices of order n. Show that $A^{-1} + B^{-1}$ is also nonsingular and that

$$(A^{-1} + B^{-1})^{-1} = A(A + B)^{-1}B = B(A + B)^{-1}A.$$

16. Let A, B be square full matrices of order n, where A is nonsingular. To compute $Q = A^{-1}B$, we can use either of the following methods:

(i.1) Compute A^{-1}.

(i.2) Form $Q = A^{-1}B$.

    or

(ii.1) Factor A as LU.

(ii.2) Solve n systems to find Q in $AQ = B$.

Show that the first method requires asymptotically $2n^3$ flops, while the second requires only $\frac{4}{3}n^3$ flops.

17. (a) Show that if $A = [a_{.1}, a_{.2}, \ldots, a_{.n}]$, where $a_{.i}$, $i = 1, \ldots, n$ are the column vectors of A and $\mathbf{x} = [x_1, x_2, \ldots, x_n]^T$, then

$$A\mathbf{x} = x_1 a_{.1} + x_2 a_{.2} + \ldots + x_n a_{.n}.$$

(b) Show that if A is an $m \times n$ matrix and $B = [\mathbf{b}_{.1}, \mathbf{b}_{.2}, \ldots, \mathbf{b}_{.p}]$ is an

$n \times p$ matrix, then

$$AB = [A\mathbf{b}_{.1}, A\mathbf{b}_{.2}, \ldots, A\mathbf{b}_{.p}].$$

(c)   Show that if

$$A^{-1} = [\mathbf{x}^{(1)}, \mathbf{x}^{(2)}, \ldots, \mathbf{x}^{(n)}],$$

then the $\mathbf{x}^{(i)}$ are the solutions of the equations

$$A\mathbf{x}^{(i)} = \mathbf{e}^{(i)}, \quad i = 1, 2, \ldots, n,$$

where $\mathbf{e}^{(i)}$ is the ith unit coordinate vector.

18.   Show that if A is invertible, then there exists precisely one matrix B such that $AB = BA = I$.

19.   (a)   Let A be symmetric and positive definite (see Chapter 3 for a definition of positive definiteness). Let L be defined as in Lemma 1.2 and let $\mathbf{l}_r$ be the rth column of L. Show that the LDL^T factorization of A can be decomposed as

$$A = \mathbf{l}_1 d_1 \mathbf{l}_1^T + \mathbf{l}_2 d_2 \mathbf{l}_2^T + \cdots + \mathbf{l}_n d_n \mathbf{l}_n^T,$$

where $d_i > 0$.

     (b)   Use $A^{-1} = L^{-T} D^{-1} L^{-1}$ to show a similar decomposition of A^{-1}. How can this be used when solving linear systems $A\mathbf{x} = \mathbf{b}$ on a parallel computer? Note, however, that the number of nonzero components of the vectors $\mathbf{l}_r$ is $n - r + 1$ at most, so the balance of workload between processors can be a problem.

20.   (a)   Show that $A = \begin{bmatrix} a & b \\ c & d \end{bmatrix}$ is invertible if and only if $ad - bc \neq 0$, in which case

$$A^{-1} = \frac{1}{ad - bc} \begin{bmatrix} d & -b \\ -c & a \end{bmatrix}.$$

     (b)   Let A and B be partitioned in 2×2 submatrices or "blocks":

$$A = \begin{bmatrix} A_{11} & A_{12} \\ A_{21} & A_{22} \end{bmatrix} \qquad B = \begin{bmatrix} B_{11} & B_{12} \\ B_{21} & B_{22} \end{bmatrix}.$$

Verify that AB can be calculated block-wise,

$$AB = \begin{bmatrix} A_{11}B_{11} + A_{12}B_{21} & A_{11}B_{12} + A_{12}B_{22} \\ A_{21}B_{11} + A_{22}B_{21} & A_{21}B_{12} + A_{22}B_{22} \end{bmatrix},$$

assuming that the submatrices involved in the products have compatible orders.

21. Let B be a skewsymmetric matrix—i.e., $B^T = -B$. If

$$A = (I + B)(I - B)^{-1},$$

show that

$$A^T A = A A^T = I.$$

(It will be seen in Chapter 2 that $I - B$ has no zero eigenvalue.)

Hint: Show first that $I + B$ and $I - B$ commute.

22. *Generalized inverses, full-column rank case:* The column rank of a matrix A is the dimension of the vector space spanned by the column vectors of A. The row rank is defined similarly. A matrix has full-column (row) rank if the column (row) vectors are linearly independent. Clearly $m \geq n$ ($m \leq n$) if a matrix has full-column (row) rank. Let A be an $m \times n$ real matrix, $m \geq n$ of rank n, and let $s = \mathbf{r}^T \mathbf{r}$, where $\mathbf{r} = \mathbf{b} - A\mathbf{x}$ and $\mathbf{x} \in \mathbb{R}^n$, $\mathbf{b} \in \mathbb{R}^m$.

(a) Show that

$$s = \mathbf{b}^T \mathbf{b} - 2\mathbf{b}^T A\mathbf{x} + \mathbf{x}^T A^T A\mathbf{x}$$

and that $\partial s / \partial x_j = 0$, $j = 1, \ldots, n$ gives

$$A^T A\mathbf{x} = A^T \mathbf{b}.$$

(This equation is called the *normal equation* associated with the system $A\mathbf{x} = \mathbf{b}$.)

(b) Show that $A^T A$ is nonsingular and that

$$\mathbf{x} = (A^T A)^{-1} A^T \mathbf{b}$$

minimizes s, the sum of squares of residuals.

(c) Let $A^\dagger = (A^T A)^{-1} A^T$. Show that:

(i) $A^\dagger A = I_n$, the n-dimensional unit matrix.

(ii) $A^\dagger A A^\dagger = A^\dagger$, $A A^\dagger A = A$.

(iii) If $m = n$, then $A^\dagger = A^{-1}$.

Note: $A^\dagger$ is called the (left-hand side) generalized inverse of A.

23. Let A be an $m \times n$ real matrix and let $\mathbf{x}_0$ be a solution of $A\mathbf{x} = \mathbf{b}$, and let $\mathbf{u}$ be a solution of $A\mathbf{x} = \mathbf{0}$.

(a) Show that $\mathbf{x}_0 + \mathbf{u}$ is also a solution of $A\mathbf{x} = \mathbf{b}$.

(b)   Show that $Au = 0$ has only the trivial solution if and only if A has full-column rank.

(c) $Ax = b$ has infinitely many solutions if rank $(A) < n$.

The next exercise shows how, among all solutions of $Ax = b$, the one with the smallest norm can be computed. To show this, consider first a vector variable $x \in \mathbb{R}^n$ and let $f = f(x) : \mathbb{R}^n \to \mathbb{R}$. Let

$$\underline{\nabla} f = \left[\frac{\partial f}{\partial x_i} \right]_{i=1}^n .$$

Show that:

(d)   If $f = x^T y$, where y is an n-dimensional vector, then $\underline{\nabla} f = y$.

(e)   If A is an $n \times n$ symmetric matrix and $f = x^T A x$, then $\nabla f = 2Ax$.

24.   *Generalized inverses, full row rank case:*

(a)   Let A be an $m \times n$ real matrix, $m \le n$, of rank m. Show that the solution of the equation $Ax = b$ that minimizes $x^T x$ is given by

$$x = A^T (AA^T)^{-1} b.$$

Hints:

(i)   Consider the method of Lagrange multipliers—i.e., the stationary relations for

$$f(x, \underline{\lambda}) = \tfrac{1}{2} x^T x + \underline{\lambda}^T (b - Ax),$$

where $\underline{\lambda} \in \mathbb{R}^m$ is the vector of Lagrange multipliers. (The factor $1/2$ is introduced for convenience.) Show that the relations

$$\frac{\partial f}{\partial x_j} = 0, \quad j = 1, \ldots, n, \qquad \frac{\partial f}{\partial \underline{\lambda}_i} = 0, \quad i = 1, \ldots, m$$

yield

$$x = A^T \underline{\lambda}$$

and

$$Ax = b,$$

respectively.

(ii)   Show that AA^T is nonsingular and that

$$\underline{\lambda} = (AA^T)^{-1} b, \text{ and } x = A^T (AA^T)^{-1} b.$$

(b) Let $A^\dagger = A^T(AA^T)^{-1}$. Show that $AA^\dagger = I_m$, the m-dimensional unit matrix, and that

$$AA^\dagger A = A, \quad A^\dagger AA^\dagger = A^\dagger.$$

(c) Show that if $m = n$, then $A^\dagger = A^{-1}$.

Note: $A^\dagger$ above is called the (right-hand side) generalized inverse of A.

25. (Noble, 1969) Show that if $A = BC$, where A, B, C have orders $m \times n$, $m \times k$, and $k \times n$ and all three matrices are of rank k, then the vector $\mathbf{x}$, which minimizes $\mathbf{r}^T\mathbf{r}$, where $\mathbf{r} = \mathbf{b} - A\mathbf{x}$, *and* $\mathbf{x}^T\mathbf{x}$, is given by $\mathbf{x} = A^\dagger\mathbf{b}$, where $A^\dagger$, the generalized inverse, satisfies

$$A^\dagger = C^\dagger B^\dagger = C^T(CC^T)^{-1}(B^T B)^{-1}B^T.$$

26. (a) Show that $A^\dagger$ (called the *Moore-Penrose pseudo inverse*) in the previous exercise satisfies:
   (i) $AA^\dagger = B(B^T B)^{-1}B^T$.
   (ii) $A^\dagger A = C^T(CC^T)^{-1}C$.
   (iii) $A^\dagger AA^\dagger = A^\dagger$.
   (iv) $AA^\dagger A = A$.
   (v) $AA^\dagger$ and $A^\dagger A$ are symmetric.

   (b) Show that if $A\mathbf{x} = \mathbf{b}$ is a consistent system, then any solution of $A\mathbf{x} = \mathbf{b}$, where A has order $m \times n$, is given by

$$\mathbf{x} = A^\dagger\mathbf{b} + (I - P)\mathbf{z},$$

   where $\mathbf{z}$ is an arbitrary vector in $\mathbb{C}^n$ and

$$P = A^\dagger A.$$

   Note: For much further information regarding least squares problems, see Björck, 1988.

27. Find the generalized inverse of

$$A = \begin{bmatrix} -1 & 0 & 1 & 2 \\ -1 & 1 & 0 & -1 \\ 0 & -1 & 1 & 3 \\ 0 & 1 & -1 & -3 \\ 1 & -1 & 0 & 1 \\ 1 & 0 & -1 & -2 \end{bmatrix}.$$

(a) Perform the factorization

$$
A = LU = \begin{bmatrix} 1 & & & & & \\ 1 & 1 & & & \text{\O} & \\ 0 & -1 & 1 & & & \\ 0 & 1 & 0 & 1 & & \\ -1 & -1 & 0 & 0 & 1 & \\ -1 & 0 & 0 & 0 & 0 & 1 \end{bmatrix} \begin{bmatrix} -1 & 0 & 1 & 2 \\ 0 & 1 & -1 & -3 \\ 0 & 0 & 0 & 0 \\ 0 & 0 & 0 & 0 \\ 0 & 0 & 0 & 0 \\ 0 & 0 & 0 & 0 \end{bmatrix}
$$

and then deduce from Theorem 1.5 that the rank of A is 2.

(b) Show that

$$
A = BC = \begin{bmatrix} 1 & 0 \\ 1 & 1 \\ 0 & -1 \\ 0 & 1 \\ -1 & -1 \\ -1 & 0 \end{bmatrix} \begin{bmatrix} -1 & 0 & 1 & 2 \\ 0 & 1 & -1 & -3 \end{bmatrix}.
$$

(c) Compute

$$
C^\dagger = C^T(CC^T)^{-1}, \quad B^\dagger = (B^T B)^{-1} B^T
$$

to show that

$$
A^\dagger = C^\dagger B^\dagger = \frac{1}{102} \begin{bmatrix} -15 & -18 & 3 & -3 & 18 & 15 \\ 8 & 13 & -5 & 5 & -13 & -8 \\ 7 & 5 & 2 & -2 & -5 & -7 \\ 6 & -3 & 9 & -9 & 3 & -6 \end{bmatrix}.
$$

28. *The inf-sup criterion:* Let B be an $m \times n$ matrix, where $m \le n$, and let

$$
\sigma = \inf_{\substack{y \in \mathbb{R}^m \\ y \ne 0}} \sup_{\substack{x \in \mathbb{R}^n \\ x \ne 0}} \frac{y^T Bx}{\|y\| \, \|x\|}.
$$

Show that:

(a) if rank $(B) = m$, then $\sigma = \frac{1}{\|B^\dagger\|} > 0$.

(b) if rank $(B) < m$, then $\sigma = 0$.

Hint: If rank $(B) = m$, let $x = B^\dagger y$, where $B^\dagger$ is the generalized inverse of Exercise 24. Show that $Bx = y$ and

$$
\frac{y^T Bx}{\|y\| \, \|x\|} = \frac{\|y\|}{\|B^\dagger y\|} \ge \frac{1}{\|B^\dagger\|}.
$$

That is, $\sigma \geq 1/\|B^\dagger\|$, and the inequality is sharp. If rank $(B) \leq m -$ 1, then let $\mathbf{y} \in R(B)^\perp = \mathcal{N}(B^T)$, where $R(B)^\perp$ denotes the orthogonal complement of range (B). Note that $\mathcal{N}(B^T)$ is not void. Then, for any $\mathbf{x}$, we have $\mathbf{y}^T B \mathbf{x} = 0$ and

$$\sup_{\mathbf{x}} \frac{\mathbf{y}^T B \mathbf{x}}{\|\mathbf{y}\| \, \|\mathbf{x}\|} = 0 \quad \text{for all } \mathbf{y} \in \mathcal{N}(B^T).$$

29. It follows from Exercise 22 that in the full-column rank case, the solution of the minimization problem

$$\min_{\mathbf{x} \in \mathbb{R}^n} \|A\mathbf{x} - \mathbf{b}\|_2, \quad A \in \mathbb{R}^{m,n}, \quad \mathbf{b} \in \mathbb{R}^m,$$

is the solution of the normal equations

$$A^T A \mathbf{x} = A^T \mathbf{b}.$$

In Exercise 24, it was shown that in the full-row rank case the solution to the dual problem

$$\min_{\mathbf{x} \in \mathbb{R}^n} \|\mathbf{x}\|_2, \quad \text{subject to } A\mathbf{x} = \mathbf{b}$$

satisfies

$$AA^T \underline{\lambda} = \mathbf{b}, \quad \mathbf{x} = A^T \underline{\lambda}.$$

Show that the first problem can be written as the augmented system

$$\begin{bmatrix} -I & A \\ A^T & 0 \end{bmatrix} \begin{bmatrix} \mathbf{r} \\ \mathbf{x} \end{bmatrix} = \begin{bmatrix} \mathbf{b} \\ 0 \end{bmatrix},$$

where $\mathbf{r} = A\mathbf{x} - \mathbf{b}$, and that the second can be written

$$\begin{bmatrix} 0 & A \\ A^T & -I \end{bmatrix} \begin{bmatrix} \underline{\lambda} \\ \mathbf{x} \end{bmatrix} = \begin{bmatrix} \mathbf{b} \\ 0 \end{bmatrix}.$$

In both cases, the matrix is square and symmetric, but indefinite if $A \neq 0$.

30. *Projection matrices:*
     (a) A projection matrix P is a matrix for which $P^T = P$ and $P^2 = P$. Show that if A (of order $m \times n$) is a matrix with full-column rank $(= n)$, then $P = A(A^T A)^{-1} A^T$ is a projection matrix.
     (b) Show that $\mathbf{x} = (A^T A)^{-1} A^T \mathbf{b}$ is a solution of $A\mathbf{x} = P\mathbf{b}$ and that $P\mathbf{b}$ is the projection of $\mathbf{b} \in \mathbb{R}^m$ $(m \geq n)$ onto the range of A.

(c)   Show that $A\mathbf{x}$ is orthogonal to $\mathbf{r} = \mathbf{b} - A\mathbf{x}$ and that

$$\|A\mathbf{x}\|^2 + \|\mathbf{r}\|^2 = \|\mathbf{b}\|^2,$$

or, equivalently, that

$$\|\mathbf{b} - A\mathbf{x}\|^2 = \|\mathbf{b}\|^2 - \|P\mathbf{b}\|^2.$$

[Hence $(\|\mathbf{b}\|^2 - \|P\mathbf{b}\|^2)^{\frac{1}{2}}$ is the residual error in the least square method.]

(d)   Show that the property that $A\mathbf{x}$ is orthogonal to $\mathbf{r}$, where $\mathbf{x}$ is the solution of the normal equations, holds even if A does not have full-column rank.

31.   *Aasen's algorithm:* Let A be symmetric of order n and partitioned as

$$A^{(1)} = A = \begin{bmatrix} a_{11} & a_{12} & \mathbf{u}^T \\ a_{21} & & \\ \mathbf{u} & & A_{22} \end{bmatrix},$$

and assume that $a_{12}\ (= a_{21}) \neq 0$. Here A_{22} has order $n - 1$. Let

$$L_1 = \begin{bmatrix} 1 & & \\ 0 & 1 & \\ \mathbf{0} & \mathbf{l}_1 & I \end{bmatrix}$$

where $\mathbf{l}_1 = -\frac{1}{a_{21}}\mathbf{u}$, and let $\mathbf{v}^{(1)} = \begin{bmatrix} a_{21} \\ \mathbf{u} \end{bmatrix}$.

(a)   Consider the congruence transformation $L_1 A^{(1)} L_1^T$ (for a definition of congruence transformations, see Chapter 3) and show that

$$A^{(2)} = L_1 A^{(1)} L_1^T = \begin{bmatrix} a_{11} & a_{12} & \mathbf{0}^T \\ a_{21} & & \\ \mathbf{0} & & \widetilde{A}^{(2)} \end{bmatrix},$$

where $\widetilde{A}^{(2)}$ has order $n - 1$ and

$$\widetilde{A}^{(2)} = \begin{bmatrix} 1 & \mathbf{0} \\ \mathbf{l}_1 & I \end{bmatrix} A_{22} \begin{bmatrix} 1 & \mathbf{l}_1^T \\ \mathbf{0} & I \end{bmatrix}.$$

(b)   Then, partition $\widetilde{A}^{(2)}$ similarly, compute $\widetilde{L}_2$ of order $n - 1$, and extend it to L_2 of order n. Continuing, show that after at most $n - 2$ steps, a tridiagonal matrix $A^{(r)}, r \leq n - 1$ has been found such that

$$A^{(r)} = L_{r-1} \ldots L_1 A^{(1)} L_1^T \ldots L_{r-1}^T.$$

(c) Show that at the sth stage after $s - 1$ steps, $1 < s < r$, we have

$$A^{(s)} = \begin{bmatrix} T^{(s-1)} & & \mathbf{0}^T \\ & & \mathbf{v}^{(s)T} \\ \mathbf{0} & \mathbf{v}^{(s)} & A_{22}^{(s)} \end{bmatrix},$$

where $\mathbf{v}^{(s)} = \left(a_{s+1,s}^{(s)}, \dots, a_{ns}^{(s)} \right)^T$ and $T^{(s-1)}$ is tridiagonal of order s. (Here $\mathbf{0}$ indicates an $(n - s) \times (s - 1)$ matrix.)

(d) *A stabilized version of Aasens algorithm:* For $s = 1, 2, \dots$, let $\hat{a}_{s+1,s}^{(s)} = \max_{s+1 \le i \le n} |a_{is}^{(s)}|$, and permute

$$\begin{bmatrix} 0 & \mathbf{v}^{(s)T} \\ \mathbf{v}^{(s)} & A_{22}^{(s)} \end{bmatrix}$$

symmetrically to

$$\begin{bmatrix} 0 & \hat{\mathbf{v}}^{(s)T} \\ \hat{\mathbf{v}}^{(s)} & \widehat{A}_{22}^{(s)} \end{bmatrix},$$

where $\hat{\mathbf{v}}_{s+1}^{(s)} = \hat{a}_{s+1,s}^{(s)}$. Then perform the factorization of

$$\begin{bmatrix} T^{(s-1)} & & \mathbf{0}^T \\ & & \hat{\mathbf{v}}^{(s)T} \\ \mathbf{0} & \hat{\mathbf{v}}^{(s)} & \widehat{A}_{22}^{(s)} \end{bmatrix}$$

and repeat. Then show that all entries $l_{ij}^{(s)}$ of L_s, $s = 1, 2, \dots, r - 1$ satisfy $|l_{ij}^{(s)}| \le 1$.

(e) Show that at the sth stage, there are $n - (s + 1)$ divisions and $n - (s + 1) + (n - (s + 1))(n - (s + 1) + 1)$ additions (or subtractions) and multiplications. Then, show that there is a total of $1/2(n - 2)(n - 1)$ divisions and $(n - 2)(n - 1)(1/3n + 1/2)$ additions and multiplications for the $n - 2$ steps. This is about the same as for the usual Gaussian elimination without the exploitation of symmetry.

(f) Try to improve the algorithm so as to reduce the operation count to $1/6n^3 + 3/2n^2 - 20/3n + 6$ multiplications, $1/6n^3 + n^2 - 25/6n + 3$ additions and $1/2n^2 - 3/2n + 1$ divisions. (The reader may wish to consult Aasen [1971].)

32. Factor $A = \begin{bmatrix} 0 & 1 & 1 \\ 1 & 2 & 1 \\ 1 & 1 & 2 \end{bmatrix}$ in the form $LDL^T = \begin{bmatrix} I & 0 \\ \mathbf{v}^t & 1 \end{bmatrix} \begin{bmatrix} D_1 & 0 \\ 0 & d_2 \end{bmatrix}$

$\begin{bmatrix} I & \mathbf{v} \\ \mathbf{0} & 1 \end{bmatrix}$, where D is a block diagonal matrix consisting of a 2-by-2 block and a 1-by-1 block.

33. Let A_n be tridiagonal of order n, $A_n = \text{tridiag}(a_{i,i-1}, a_{ii}, a_{i,i-1})$ and let $A = LD^{-1}R$ be its factorization in the form (1.14). Let A_i, $1 \le i \le n$ denote the principal submatrices of A_n and let $\sigma_i = \det(A_i)$.

(a) Show that the above factorization exists if and only if $\sigma_i \ne 0$, $1 \le i \le n - 1$, and that $d_i = \sigma_i/\sigma_{i-1}$, where d_i are the entries in D.

(b) Using expansion by minors, show that

$$\sigma_i = a_{ii}\sigma_{i-1} - a_{i,i-1}\sigma_{i-2}a_{i-1,i},$$

where $i = 2, \ldots, n$, $\sigma_1 = a_{11}$, $\sigma_0 = 1$.

(c) Show that the above recursion relation holds even if some $\sigma_i = 0$.

(d) If $\det(A_n) \ne 0$, show that A_i and A_{i-1} cannot both be singular.

References

J. O. Aasen (1971). On the reduction of a symmetric matrix to tridiagonal form. *BIT* **11**, 233–242.

S. O. Asplund (1966). *Structural Mechanics, Classical and Matrix Methods*. Englewood Cliffs, N.J.: Prentice Hall.

A. Ben-Israel, and T.N.E. Greville (1974). *Generalised Inverses: Theory and Applications*. New York: Wiley.

Å. Björck (1988), *Least Squares Methods: Handbook of Numerical Analysis; Vol. 1, Solution of Equations in* $\mathbb{R}^N$, Amsterdam: Elsevier, North Holland.

J. R. Bunch and L. Kaufman (1977). Some stable methods for calculating inertia and solving symmetric linear systems. *Math. Comp.* **31**, 162–179.

G. B. Dantzig (1963). *Linear Programming and Extensions*. Princeton, N.J.: Princeton Univ. Press.

A. Edelman (1991). Editor's note. *SIAM J. Matrix Anal. Appl.* **12**.

A. George and J. W. Liu (1981). *Computer Solution of Large Sparse Positive Definite Systems*. Englewood Cliffs, N.J.: Prentice Hall.

G. H. Golub and C. F. van Loan (1989). *Matrix Computations*, 2nd ed. Baltimore, Maryland: Johns Hopkins Univ. Press.

E. A. Guillemin (1949). *The Mathematics of Circuit Analysis*. New York: Wiley.

P. Henrici (1962). *Discrete Variable Methods in Ordinary Differential Equations*. New York: Wiley.

J. A. Jacquez (1972). *Compartmental Analysis in Biology and Medicine*. Amsterdam: Elsevier.

H. B. Keller (1968). *Numerical Methods for Two-Point Boundary Value Problems*. London: Blaisdell.

G. Kron (1959). *Tensors for Circuits*. New York: Dover.

B. Noble (1969). *Applied Linear Algebra*. Englewood Cliffs, N.J.: Prentice Hall.

S. V. Parter (1961). The use of linear graphs in Gauss elimination. *SIAM Rev.* **3**, 119–130.

G. Strang (1976). *Linear Algebra*. New York: John Wiley.

G. Strang (1986). *Introduction to Applied Mathematics*. Cambridge, Mass.: Wellesley Cambridge Press.

J. Wilkinson (1965). *The Algebraic Eigenvalue Problem*. Oxford: Clarendon.

2
Theory of Matrix Eigenvalues

Let us consider first a $n \times n$ matrix A as defining a linear mapping in $\mathbb{C}^n$ or $\mathbb{R}^n$, w.r.t. a fixed coordinate system. A number $\lambda \in \mathbb{C}$, for which $Ax = \lambda x$, where $x \neq 0$, is said to be an eigenvalue of A and x is said to be an eigenvector corresponding to λ; hence x is a vector, which is mapped by A onto its own direction. We show that there is at least one such vector for every square matrix. First, some fundamental concepts and properties in the theory of eigenvalues are presented. We prove that the eigenvalues of A are the zeros of $\varphi(\lambda) = \det(A - \lambda I)$, a polynomial in λ called the characteristic polynomial of A. We prove that $\varphi(A) = 0$, and we consider the polynomial $m(\lambda)$, the polynomial of minimal degree for which $m(A) = 0$.

Selfadjoint and unitary matrices play an important role in applications, and we derive properties of the eigensolutions of such matrices. If the matrix B of order n defines the same mapping as the matrix A, but with respect to another basis in $\mathbb{C}^n$ or $\mathbb{R}^n$, we can write B as $B = C^{-1}AC$, where C is a nonsingular matrix. We prove that B and A have the same eigenvalues (i.e., the eigenvalues are independent of the particular basis) and consider matrices A for which there exists a matrix C such that B is a triangular or even a diagonal matrix. We discuss normal and H-normal matrices and show, among other things, that H-normal matrices are precisely the class of matrices which are diagonalizable by a similarity transformation.

The following definitions are introduced in this chapter:

Definition 2.1 If $Ax = \lambda x$, where $x \neq 0$ and λ is a complex (possibly real) scalar, then x is an *eigenvector*, λ an *eigenvalue*, and the pair x, λ an *eigensolution* of A. The set of all eigenvalues of A is called the *spectrum* of A, denoted $S(A)$.

Definition 2.2 The polynomial

$$\varphi(\lambda) = \det(A - \lambda I)$$

46

is called the *characteristic polynomial* of A, and the equation $\det(A - \lambda I) = 0$ is the *characteristic equation* of A.

Definition 2.3 The sum of the eigenvalues of a matrix A is called the *trace* of A and denoted $\operatorname{tr}(A)$.

Definition 2.4 A polynomial $g(\lambda)$, not identically zero is said to be an *annihilating polynomial* of A if $g(A) = 0$.

Definition 2.5 Let $m(\lambda)$ be an annihilating polynomial of minimal degree. Then $m(\lambda)$ is called a *minimal polynomial* of A.

Definition 2.6 If $A = A^T$, A is *symmetric*; if $A = \overline{A}$, then A is *real*. If $A = A^* = \overline{A}^T$, A is *Hermitian* or *selfadjoint*.

Definition 2.7 The *innerproduct* $(\mathbf{u}, \mathbf{v})$ of two vectors $\mathbf{u}$ and $\mathbf{v}$ is

$$(\mathbf{u}, \mathbf{v}) = \mathbf{u}^* \mathbf{v} = \overline{\mathbf{u}}^T \mathbf{v} = \sum \overline{u}_j v_j.$$

If $(\mathbf{u}, \mathbf{v}) = 0$, then $\mathbf{u}$ and $\mathbf{v}$ are said to be *orthogonal*.

Definition 2.8 A is said to be orthogonal if A is real and $A^T A = I$, and *unitary* if $A^* A = I$.

Definition 2.9 Let C be an arbitrary, nonsingular matrix of the same order as A. Then the matrices A and $C^{-1}AC$ are said to be *similar*.

Definition 2.10 Let U be a unitary matrix of the same order as A. Then the matrices A and U^*AU are said to be *unitarily similar*.

Definition 2.11 A matrix is said to be *normal* if it commutes with its Hermitian transpose—i.e., if $AA^* = A^*A$.

Definition 2.12 Let H be a Hermitian matrix with positive eigenvalues. Then $A' = H^{-1}A^*H$ is said to be the *H-adjoint of* A and A is said to be *H-normal* if A commutes with A' for some H.

Definition 2.13 Let A be H-normal for a certain Hermitian positive definite matrix H. The polynomial $\hat{p}(A)$ of smallest degree $n(A, H)$, such that $A' = \hat{p}(A)$, is called the *H-normal polynomial* of A, and $n(A, H)$ is called the *H-normal degree*.

2.1 The Minimal Polynomial

As was seen in Chapter 1, a linear mapping $\mathbb{R}^n \to \mathbb{R}^m$ of the n-dimensional with respect to given basis vectors in $\mathbb{R}^n$ and $\mathbb{R}^m$, usually the unit coordinate vectors, is represented by a rectangular array (matrix) of numbers:

$$A = \begin{bmatrix} a_{11} & a_{12} & \cdots & a_{1n} \\ a_{21} & a_{22} & \cdots & a_{2n} \\ \vdots & \vdots & & \vdots \\ a_{m1} & a_{m2} & \cdots & a_{mn} \end{bmatrix}.$$

The representation is changed, however, when the basis is changed. In this chapter we consider only square matrices—i.e., $m = n$, and n is then called the *order* of the matrix. Similarly, $\mathbf{x} = (x_1, \ldots, x_n)^T$ denotes $\mathbf{x} = \sum_{i=1}^{n} x_i \mathbf{v}_i$, where $\{\mathbf{v}_i\}_{i=1}^{n}$ are the basis vectors.

In many scientific problems in physics, mathematics, engineering, and economics, one is interested in vectors $\mathbf{x}$, which are mapped by A onto a direction equal to some scalar λ times the vector—i.e., onto a vector $\mathbf{y} = \lambda\mathbf{x}$. Such vectors are solutions of the homogeneous system

$$A\mathbf{x} = \lambda\mathbf{x}.$$

We give the following definitions.

Definition 2.1 If $A\mathbf{x} = \lambda\mathbf{x}$, where $\mathbf{x} \neq \mathbf{0}$ and λ is a complex (possibly real) scalar, then $\mathbf{x}$ is said to be an *eigenvector*, λ an *eigenvalue*, and the pair $\mathbf{x}$, λ an *eigensolution* of A. The set of all eigenvalues of A is called the *spectrum* of A, denoted $S(A)$.

Example 2.1 $\mathbf{x}_1 = \begin{bmatrix} 1 \\ -2 \end{bmatrix}$ and $\lambda_1 = 1$ is an eigensolution of the mapping $A = \begin{bmatrix} 5 & 2 \\ 2 & 2 \end{bmatrix}$, because $A\mathbf{x}_1 = \begin{bmatrix} 1 \\ -2 \end{bmatrix} = \lambda_1\mathbf{x}_1$ and $\mathbf{x}_1 \neq \mathbf{0}$. Here we have used the representation of A in the basis $\begin{bmatrix} 1 \\ 0 \end{bmatrix}, \begin{bmatrix} 0 \\ 1 \end{bmatrix}$ for $\mathbb{R}^2$.

Example 2.2 Let's assume that the eigenvectors $\mathbf{v}_i$ of A corresponding to the eigenvalues λ_i are a basis of $\mathbb{R}^n$. Then A maps the vector $\mathbf{x} = \sum_{i=1}^{n} c_i \mathbf{v}_i$ onto the vector

$$A\mathbf{x} = \sum c_i A\mathbf{v}_i = \sum c_i \lambda_i \mathbf{v}_i.$$

Hence, with respect to the basis $\mathbf{v}_1, \mathbf{v}_2, \ldots, \mathbf{v}_n$, this mapping has the simple representation in the form of a diagonal matrix,

$$A = \begin{bmatrix} \lambda_1 & & & 0 \\ & \lambda_2 & & \\ & & \ddots & \\ 0 & & & \lambda_n \end{bmatrix}.$$

In order to compute the eigenvalues we may (in practice only for small values of n) use the equation $\det(A - \lambda I) = 0$, where I is the identity matrix—i.e., $I = [\delta_{ij}]$ and $\delta_{ij} = 1$, $i = j$, $\delta_{ij} = 0$, $i \neq j$.

Theorem 2.1 λ *is an eigenvalue of the matrix A if and only if* $\det(A - \lambda I) = 0$.

Proof We have $Ax = \lambda x$, $x \neq 0 \iff Ax - \lambda x = 0$, that is

$$(A - \lambda I)x = 0, \quad x \neq 0.$$

This is a linear homogeneous system of equations with a nontrivial solution. Hence $A - \lambda I$ is singular—i.e., $\det(A - \lambda I) = 0$. $\diamond$

From the definition of a determinant we find that

$$\det(A - \lambda I) = \det \left(\begin{bmatrix} a_{11} - \lambda & a_{12} & \cdots & a_{1n} \\ a_{21} & a_{22} - \lambda & & a_{2n} \\ \vdots & & \ddots & \\ a_{n1} & a_{n2} & & a_{nn} - \lambda \end{bmatrix} \right)$$

$$= \alpha_n - \alpha_{n-1}\lambda + \cdots + (-1)^{n-1}\alpha_1\lambda^{n-1} + (-1)^n\lambda^n$$

for certain scalars $\alpha_n, \alpha_{n-1}, \ldots, \alpha_1$ where, for instance,

$$\alpha_n = \det(A), \quad \alpha_1 = \sum_{i=1}^{n} a_{ii}.$$

The first relation is proved by letting $\lambda = 0$ and the second using a well-known expansion theorem for determinants.

Definition 2.2 The polynomial

$$\varphi(\lambda) = \det(A - \lambda I)$$

is called the *characteristic polynomial* of A, and the equation $\det(A - \lambda I) = 0$ the *characteristic equation* of A.

Theorem 2.1 shows that the zeros of the characteristic polynomial are the eigenvalues of A. We realize that if a matrix has only real entries, then its characteristic polynomial has only real coefficients, and therefore the eigenvalues of a real matrix are either real or complex conjugate. Further, the fundamental theorem of algebra and Theorem 2.1 show that a matrix of order n has exactly n eigenvalues (some of which may be multiple).

Definition 2.3 The sum of the eigenvalues of a matrix A is called the *trace* of A, denoted tr(A)—i.e., $\text{tr}(A) = \sum_{i=1}^{n} \lambda_i$.

Theorem 2.2 *The eigenvalues* $\lambda_i, i = 1, 2, \ldots, n$ *of A satisfy*

a) $$\prod_{i=1}^{n} \lambda_i = \det(A)$$

b) $$\sum_{i=1}^{n} \lambda_i = \text{tr}(A) = \sum_{i=1}^{n} a_{ii}.$$

Proof This proof follows from the above relations for α_n and α_1 and a well-known relation between the zeros and coefficients of a polynomial (Newton's theorem). ◇

The identities in Theorem 2.2 show that $\det(A)$ and $\sum a_{ii}$ are independent of the basis because the eigenvalues are uniquely determined by the corresponding mapping.

Theorem 2.3 *Let p be an arbitrary polynomial. Then*

$$A\mathbf{v} = \lambda\mathbf{v} \implies p(A)\mathbf{v} = p(\lambda)\mathbf{v}.$$

That is, if λ is an eigenvalue of A corresponding to an eigenvector $\mathbf{v}$*, then $p(\lambda)$ is an eigenvalue of $p(A)$ corresponding to the same eigenvector.*

Proof $A\mathbf{v} = \lambda\mathbf{v}$ implies $A^2\mathbf{v} = A(\lambda\mathbf{v}) = \lambda A\mathbf{v} = \lambda^2\mathbf{v}$ and, similarly, $A^i\mathbf{v} = \lambda^i\mathbf{v}, i = 1, 2, \ldots$. Hence, if $p(A) = \sum_{i=0}^{k} \alpha_i A^i$, then

$$p(A)\mathbf{v} = \sum_{i=0}^{k} \alpha_i A^i\mathbf{v} = \sum_{i=0}^{k} \alpha_i \lambda^i\mathbf{v} = p(\lambda)\mathbf{v}. \qquad ◇$$

Before we present the following theorem, let us recall the concept of the cofactor (or adjugate) matrix A^c. The entries of A^c, which we denote A_{ij}, are equal to the determinant of the matrix that remains from A when row i and column j are deleted, multiplied by $(-1)^{i+j}$. Using Cramer's rule,

$$A(A^c)^T = [\sum_{k=1}^{n} a_{ik} A_{jk}^c] = [\delta_{ij} \det(A)] = \det(A)I,$$

where δ_{ij} are the Kronecker numbers, $\delta_{ij} = 0$, $i \neq j$, $\delta_{ii} = 1$, and I is the identity matrix.

Remark 2.4 If A is nonsingular—i.e., if $\det(A) \neq 0$—we may use Cramer's rule to compute the inverse matrix A^{-1}:

$$A^{-1} = \frac{1}{\det(A)} (A^c)^T.$$

(For an alternative proof of this formula, see exercises 19 and 20.)

If the entries of the cofactor matrix are computed independently, the operation count for computing the cofactor matrix and $\det(A)$ is too high. Hence the straightforward application of the method above is useless from a practical point of view, even for quite small matrices. However, there exists a more efficient method where expressions occurring in several entries are computed only once (for details, see Baur, Strassen, 1983).

Theorem 2.5 (Cayley-Hamilton) *If A has the characteristic polynomial $\varphi(\lambda)$, then $\varphi(A) = 0$.*

Proof If the eigenvectors $v_1, v_2, \ldots, v_n$ of A are a basis of $\mathbb{R}^n$, the proof is simple. In this case, we have $x \in \mathbb{R}^n$ for an arbitrary vector, and the relation $x = \sum_{i=1}^n c_i v_i$ for some scalars c_i, so by Theorem 2.3,

$$\varphi(A)x = \varphi(A) \sum c_i v_i = \sum c_i \varphi(A) v_i = \sum c_i \varphi(\lambda_i) v_i.$$

Since by Theorem 2.1, $\varphi(\lambda_i) = 0$, therefore have

$$\varphi(A)x = 0 \quad \forall x \in \mathbb{R}^n, \text{ that is }, \varphi(A) = 0.$$

In the general case, by Cramer's rule we have

$$(A - \lambda I)(A - \lambda I)^{c^T} = \det(A - \lambda I)I = \varphi(\lambda)I,$$

where

$$(A - \lambda I)^{c^T} = [c_{ji}(\lambda)],$$

$$c_{ij}(\lambda) = (-1)^{i+j} \det \left(\underset{(i)}{\begin{bmatrix} (a_{11} - \lambda) & a_{12} & & \overset{(j)}{a_{1n}} \\ a_{21} & (a_{22} - \lambda) & & \\ & & \ddots & \\ a_{n1} & & & (a_{nn} - \lambda) \end{bmatrix}} \right).$$

Hence, all cofactor entries $c_{ij}(\lambda)$ are polynomials of degree $\leq (n-1)$, so $(A - \lambda I)^{c^T}$ is a matrix polynomial of degree $\leq (n-1)$ in λ, that is,

$$(A - \lambda I)^{c^T} = B_{n-1} - \lambda B_{n-2} + \cdots + (-\lambda)^{n-1} B_0,$$

where $B_0, B_1, \ldots, B_{n-1}$ are matrices that do not depend on λ.
Hence, we have

$$(A - \lambda I)(B_{n-1} + \cdots + (-\lambda)^{n-1} B_0) = \varphi(\lambda) I = [(-\lambda)^n + \sum_{i=0}^{n-1} \alpha_{n-i}(-\lambda)^i] I.$$

If we identify the coefficients of λ^i, we get

$$B_0 = I$$

$$A B_0 + B_1 = \alpha_1 I$$

$$\vdots$$

$$A B_{n-2} + B_{n-1} = \alpha_{n-1} I$$

$$A B_{n-1} = \alpha_n I$$

If we multiply the above from the left by $(-A)^n$, $(-A)^{n-1}, \ldots, -A$, and I, respectively, and sum them up, we get

$$0 = (-A)^n + \alpha_1(-A)^{n-1} + \cdots + \alpha_n I = \varphi(A). \qquad \diamond$$

Remark 2.6 By use of Newton's theorem (see Excercise 21) one can show that

$$\alpha_k = \frac{1}{k} \operatorname{tr}(A B_{k-1})$$
$$\phantom{\alpha_k = \frac{1}{k} \operatorname{tr}(A B_{k-1})} \qquad k = 1, 2, \ldots, n$$
$$B_k = -A B_{k-1} + \alpha_k I$$

where $B_0 = I$ and $B_n = 0$. These formulas are called *Faddeev's* or *Leverrier's formulas* (see Faddeev, 1963).

Since, in particular, $A B_{n-1} = \alpha_n I$, we have if $\alpha_n \neq 0$,

$$A^{-1} = \frac{1}{\alpha_n} B_{n-1},$$

which provides a method to compute the inverse of a matrix. For full matrices, it is readily seen that the operation count is $(n - 1)n^3$. (For further interesting comments regarding the derivation of the above method, see Barnett, 1989.)

Definition 2.4 A polynomial $p(\lambda)$, not identically zero, is said to be an *annihilating polynomial* of A if $p(A) = 0$.

It follows from Cayley-Hamilton's theorem that for every matrix A there exists an annihilating polynomial. Clearly such a polynomial is not unique, because if ψ has the property $\psi(A) = 0$, so has every polynomial divisible by ψ.

Definition 2.5 Let $m(\lambda)$ be an annihilating polynomial of minimal degree. Then $m(\lambda)$ is called a *minimal polynomial* of A.

Theorem 2.7

(a) *The characteristic polynomial $\varphi(\lambda)$ of A is divisible by the minimal polynomial $m(\lambda)$ of A.*

(b) *λ is eigenvalue of $A \Longleftrightarrow m(\lambda) = 0$.*

Proof Let $\varphi(\lambda) = c(\lambda)m(\lambda) + r(\lambda)$, where $r(\lambda)$ is the remainder in the division of $\varphi(\lambda)$ by $m(\lambda)$. Hence, degree $r(\lambda) <$ degree $m(\lambda)$. From

$$\varphi(A) = c(A)m(A) + r(A)$$

we get $r(A) = 0$, since both $\varphi(A)$ and $m(A)$ are zero matrices. But since degree $r(\lambda) <$ degree $m(\lambda)$, then $r(\lambda) \equiv 0$, else $m(\lambda)$ is not a minimal polynomial. This proves (a). To prove (b) let $\mathbf{x}, \lambda$ be an eigensolution of A. Then, by Theorem 2.3,

$$A\mathbf{x} = \lambda\mathbf{x}, \quad \mathbf{x} \neq \mathbf{0}, \quad \text{so} \quad \mathbf{0} = m(A)\mathbf{x} = m(\lambda)\mathbf{x},$$

that is, $m(\lambda) = 0$.

Note finally that every zero of $m(\lambda)$ is also a zero of $\varphi(\lambda)$ and is, therefore, an eigenvalue. $\diamond$

Similarly, it follows that any annihilating polynomial is divisible by the minimal polynomial.

Example 2.3 If $A^2 = I$, but $A \neq \pm I$, then $\lambda^2 - 1$ is the minimal polynomial of A. Hence, the only possible eigenvalues of such a matrix A are ± 1 (whatever the order of A). An example of such a matrix is

$$A = \begin{bmatrix} 0 & 1 & 0 \\ 1 & 0 & 0 \\ 0 & 0 & 1 \end{bmatrix}.$$

Example 2.4 For a diagonal matrix $A = \text{diag}(d_i)$ of order n, where $d_1, d_2, \ldots,$ $d_r, r \leq n$ are distinct numbers, but $d_r = d_{r+1} = \cdots = d_n$, the polynomial $m(\lambda) = \prod_{i=1}^{r} (\lambda - d_i)$, is minimal.

Example 2.5 If $N = \begin{bmatrix} 0 & 1 & 0 \\ 0 & 0 & 1 \\ 0 & 0 & 0 \end{bmatrix}$, then $N^2 = \begin{bmatrix} 0 & 0 & 1 \\ 0 & 0 & 0 \\ 0 & 0 & 0 \end{bmatrix}$, and $N^3 = 0$.

Since $\alpha_0 I + \alpha_1 N + \alpha_2 N^2 \neq 0$ for all scalars $\alpha_0, \alpha_1, \alpha_2$, at least one of which is nonzero, we have $m(\lambda) = \lambda^3$. Hence the only eigenvalue of N is 0 (multiplicity 3). There is only one eigenvector $\mathbf{x}^T = [a_1, 0, 0]$, $a_1 \neq 0$, a scalar.

2.2 Selfadjoint and Unitary Matrices

Notation If $A = [a_{ij}]$, we let $\overline{A} = [\overline{a}_{ij}]$, the complex conjugate of A, $A^* = \overline{A}^T = [\overline{a}_{ji}]$, the Hermitian transpose, or adjoint, of A.

Definition 2.6 If $A = A^T$, A is said to be *symmetric*; if $A = \overline{A}$, then A is real. If $A = A^*$, A is *Hermitian* or *selfadjoint*.

Theorem 2.8

(a) *A and A^T have the same spectrum.*

(b) *λ is an eigenvalue of A if and only if $\overline{\lambda}$ is an eigenvalue of $\overline{A}$ (and of A^*).*

Proof

(a) With reference to the well-known property that the determinant of the transpose of a matrix equals the determinant of the matrix, we find $\det(A^T - \lambda I) = \det((A - \lambda I)^T) = \det(A - \lambda I)$, which shows the first part.

(b) Since $Ax = \lambda x \Longleftrightarrow \overline{Ax} = \overline{\lambda}\overline{x}$, we find that $\overline{\lambda}$ is an eigenvalue of $\overline{A}$ if λ is an eigenvalue of A. Since $A^* = \overline{A}^T$, it follows from (a) that $\overline{\lambda}$ is an eigenvalue of A^*. ◇

Definition 2.7 The *innerproduct*, (u, v), of two vectors u and v is

$$(u, v) = u^* v = \overline{u}^T v = \sum \overline{u}_j v_j.$$

If $(u, v) = 0$, u and v are said to be *orthogonal*. We observe that

$$(u, u) = \sum |u_j|^2 \geq 0$$

$$(u, u) = 0 \Longleftrightarrow u = 0$$

$$(u, v_1 + v_2) = (u, v_1) + (u, v_2)$$

$$(u, \lambda v) = \lambda(u, v) = (\overline{\lambda}u, v)$$

$$(u, v) = \overline{(v, u)}.$$

The following relation will later be seen to be very useful.

Theorem 2.9 $(u, Av) = (A^* u, v)$.

Proof $(u, Av) = \sum_{i=1}^{n} \overline{u}_i (\sum_{j=1}^{n} a_{ij} v_j) = \sum_{j=1}^{n} (\overline{\sum_{i=1}^{n} \overline{a}_{ij} u_i}) v_j = (A^* u, v)$. ◇

Theorem 2.10 *If x and y are eigenvectors of A and A^*, respectively, corresponding to different eigenvalues λ and μ of A, then x and y are orthogonal.*

Proof Note first that $A^* y = \overline{\mu} y$. Then,

$$\lambda(y, x) = (y, \lambda x) = (y, Ax) = (A^* y, x) = (\overline{\mu}y, x) = \mu(y, x),$$

that is

$$(\lambda - \mu)(y, x) = 0 \quad \text{i.e.,} \quad (y, x) = 0. \qquad ◇$$

Theorem 2.11 *If A is selfadjoint—in particular if A is real and symmetric—then: (a) the eigenvectors corresponding to different eigenvalues are orthogonal and (b) the spectrum $S(A)$ is real.*

Proof

(a) If $A = A^*$ and if $\mathbf{x}_1, \mathbf{x}_2$ are eigenvectors of A corresponding to different eigenvalues λ_1, λ_2, then these are also eigenvalues of A^*. Hence by Theorem 2.10, $\mathbf{x}_1, \mathbf{x}_2$ are orthogonal.

(b) If $A\mathbf{x} = \lambda\mathbf{x}, \mathbf{x} \neq \mathbf{0}$, then

$$\lambda(\mathbf{x}, \mathbf{x}) = (\mathbf{x}, A\mathbf{x}) = (A^*\mathbf{x}, \mathbf{x}) = (A\mathbf{x}, \mathbf{x}) = (\lambda\mathbf{x}, \mathbf{x}) = \bar{\lambda}(\mathbf{x}, \mathbf{x}).$$

That is, $\lambda = \bar{\lambda}$, because $(\mathbf{x}, \mathbf{x}) \neq 0$ if $\mathbf{x} \neq \mathbf{0}$.   ◇

Definition 2.8 A is said to be *orthogonal* if $\overline{A} = A$ and $A^T A = I$. A is said to be *unitary* if $A^* A = I$.

We note that $A^{-1} = A^T$ and $A^{-1} = A^*$ if A is orthogonal and unitary, respectively. Note also that if $A^* A = I$ then, since A is square, $AA^* = I$. If A is real, then A is orthogonal $\Longleftrightarrow A$ is unitary.

Theorem 2.12 *If A is unitary, then $\lambda \in S(A) \Rightarrow |\lambda| = 1$.*

Proof If $A\mathbf{x} = \lambda\mathbf{x}, \mathbf{x} \neq \mathbf{0}$, then

$$(A\mathbf{x}, A\mathbf{x}) = (A^*A\mathbf{x}, \mathbf{x}) = (\mathbf{x}, \mathbf{x}).$$

But $(A\mathbf{x}, A\mathbf{x}) = (\lambda\mathbf{x}, \lambda\mathbf{x}) = \lambda\bar{\lambda}(\mathbf{x}, \mathbf{x}) = |\lambda|^2(\mathbf{x}, \mathbf{x})$, that is, $|\lambda| = 1$.   ◇

2.3 Matrix Equivalence (Similarity Transformations)

So far we have let every $(n \times n)$ matrix $A = [a_{ij}]$ represent a linear mapping in $\mathbb{C}^n$ or $\mathbb{R}^n$ w.r.t. a fixed (orthonormal) coordinate system with basis $(\mathbf{e}_1, \mathbf{e}_2, \ldots, \mathbf{e}_n)$ where $(\mathbf{e}_i)_j = \delta_{ij}$. If $\mathbf{x} = \sum_{i=1}^n x_i\mathbf{e}_i$ is mapped onto $\mathbf{y} = \sum_{i=1}^n y_i\mathbf{e}_i$ by A, then

$$A\mathbf{x} = \mathbf{y} \quad \text{or} \quad \sum_j a_{ij}x_j = y_i, \quad i = 1, 2, \ldots, n.$$

Let a new coordinate system in $\mathbb{C}^n$ be defined by the basis vectors

$$\mathbf{v}_k = \sum_{i=1}^n c_{ik}\mathbf{e}_i = \begin{bmatrix} c_{1k} \\ \vdots \\ c_{nk} \end{bmatrix}, \quad k = 1, 2, \ldots, n, \quad \det([c_{ik}]) \neq 0.$$

Then the mapping is represented in the new coordinate system by a matrix B, and we have

$$\mathbf{x} = \sum_k \xi_k \mathbf{v}_k \iff \mathbf{x} = \sum_k \sum_i \xi_k c_{ik} \mathbf{e}_i = \sum_i \left(\sum_k c_{ik} \xi_k \right) \mathbf{e}_i$$

so

$$x_i = \sum_k c_{ik} \xi_k \text{ or } \mathbf{x} = C\underline{\xi}, \quad \underline{\xi} = C^{-1}\mathbf{x}.$$

Here

$$C = [c_{ik}] = [\mathbf{v}_1, \mathbf{v}_2, \dots, \mathbf{v}_n].$$

Since

$$A\mathbf{x} = \mathbf{y} \iff AC\underline{\xi} = C\underline{\eta} \iff C^{-1}AC\underline{\xi} = \underline{\eta},$$

where $\mathbf{y} = \sum_k \eta_k \mathbf{v}_k$, we realize that the mapping in the coordinate system $\mathbf{v}_1, \mathbf{v}_2, \dots, \mathbf{v}_n$ is represented by $B = C^{-1}AC$, $C = [\mathbf{v}_1, \dots, \mathbf{v}_n]$, $\det(C) \neq 0$. We call $C^{-1}AC$ a *similarity transformation* of A and C a *similarity transformation matrix*.

Example 2.6 A linear mapping is represented by the matrix

$$A = \begin{bmatrix} 0 & 1 & 1 \\ 1 & 0 & -1 \\ -1 & -1 & 0 \end{bmatrix}$$

w.r.t. the basis $\mathbf{e}_1^T = [1, 0, 0]$, $\mathbf{e}_2^T = [0, 1, 0]$, $\mathbf{e}_3^T = [0, 0, 1]$. Which matrix B will then represent the same mapping w.r.t. the basis $\mathbf{v}_1^T = [0, 1, -1]$, $\mathbf{v}_2^T = [1, -1, 1]$, and $\mathbf{v}_3^T = [-1, 1, 0]$? Here we have $C = \begin{bmatrix} 0 & 1 & -1 \\ 1 & -1 & 1 \\ -1 & 1 & 0 \end{bmatrix}$ and

$$B = C^{-1}AC = \begin{bmatrix} 1 & 1 & 0 \\ 1 & 1 & 1 \\ 0 & 1 & 1 \end{bmatrix} \begin{bmatrix} 0 & 1 & 1 \\ 1 & 0 & -1 \\ -1 & -1 & 0 \end{bmatrix} \begin{bmatrix} 0 & 1 & -1 \\ 1 & -1 & 1 \\ -1 & 1 & 0 \end{bmatrix}$$

$$= \begin{bmatrix} 1 & 0 & 0 \\ 0 & 0 & 0 \\ 0 & 0 & -1 \end{bmatrix}.$$

It can be checked that $\mathbf{v}_i$ are eigenvectors of A.

Definition 2.9 Let C be an arbitrary, nonsingular matrix of the same order as A. Then the matrices A and $C^{-1}AC$ are said to be *similar*.

It is readily seen that this is an equivalence relation (see Exercise 2.9). We shall now show that irrespective of which basis vectors we have chosen, the resulting matrix has the same spectrum. Hence the spectrum is determined by the mapping.

Theorem 2.13 *Similarly equivalent matrices have the same minimal polynomial, the same characteristic polynomial, and, hence, the same spectrum.*

Proof

(a) If $B = C^{-1}AC$, then $B^k = (C^{-1}AC)(C^{-1}AC)\ldots(C^{-1}AC) = C^{-1}A^kC$. Hence

$$m_A(B) = C^{-1}m_A(A)C = 0,$$

where $m_A(\lambda)$ is the minimal polynomial of A. Similarly, $m_B(A) = Cm_B(B)C^{-1} = 0$. Hence, as indicated by Theorem 2.7, $m_A(x)$ is divisible by $m_B(x)$ and vice versa, so $m_A(\lambda) = m_B(\lambda)$.

(b) $\quad \varphi_B(\lambda) = \det(B - \lambda I) = \det(C^{-1}AC - \lambda I)$

$$= \det[C^{-1}(A - \lambda I)C] = \det(A - \lambda I) = \varphi_A(\lambda). \quad \diamond$$

Now let both $e_1, e_2, \ldots, e_n$ and $v_1, v_2, \ldots, v_n$ be orthonormal bases in $\mathbb{C}^n$, and let

$$U = [v_1, v_2, \ldots, v_n].$$

Then,

$$U^*U = [(v_i, v_k)] = [\delta_{ik}] = I,$$

that is, U is a unitary matrix. From the above it follows that if A represents a linear mapping in the orthonormal basis $e_1, e_2, \ldots, e_n$, then $B = U^{-1}AU = U^*AU$ represents the same mapping in the orthonormal basis $v_1, v_2, \ldots, v_n$, where U is a unitary matrix.

Definition 2.10 If U is a unitary matrix of the same order as A, then the matrices A and U^*AU are said to be *unitarily similar*.

We define orthogonally equivalent matrices in a similar way, representing the same mapping in $\mathbb{R}^n$ in two orthonormal bases.

Theorem 2.14

(a) *A is selfadjoint $\Longleftrightarrow U^{-1}AU$ is selfadjoint, where U is an arbitrary unitary matrix.*

(b) *A is real symmetric $\Longleftrightarrow O^{-1}AO$ is real symmetric, where O is an arbitrary orthogonal matrix.*

Proof

(a) Since $U^* = U^{-1}$ and $U^{**} = U$, we have

$$(U^{-1}AU)^* = (U^*AU)^* = U^*A^*U^{**} = U^{-1}A^*U.$$

Hence

$$A = A^* \Rightarrow (U^{-1}AU)^* = U^{-1}AU$$

and

$$(U^{-1}AU)^* = U^{-1}AU \implies U^{-1}A^*U = U^{-1}AU \implies A^* = A$$

(where we have multiplied by U and U^{-1} from the left- and right-hand sides, respectively).

(b) is proved in an analogous way. ◇

Theorem 2.15 (Schur's lemma) *Any square matrix is unitarily similar to a triangular matrix.*

Proof The theorem states that for any square matrix A there exists a unitary matrix U, such that U^*AU is triangular. Since the diagonal entries of a triangular matrix are the eigenvalues of that matrix, Theorem 2.13 shows that

$$U^{-1}AU = U^*AU = \begin{bmatrix} \lambda_1 & b_{12} & \cdots & b_{1n} \\ 0 & \lambda_2 & \cdots & b_{2n} \\ & & \ddots & \\ 0 & \cdots & & \lambda_n \end{bmatrix},$$

where λ_i are the eigenvalues of A.

The theorem will be proved by induction. Clearly, it is true for $n = 1$. Assuming that it is true for order $n - 1$, we now prove that it is true for order n. Let $\mathbf{u}_1, \lambda_1$ be an eigensolution of A, where $\mathbf{u}_1$ is normalized—that is, $(\mathbf{u}_1, \mathbf{u}_1) = 1$. Choose an orthonormal basis $\mathbf{u}_1, \mathbf{u}_2, \ldots, \mathbf{u}_n$ in $\mathbb{C}^n$ (for instance, by Gram-Schmidt's orthonormalizing process; see Exercise 23). Then $U = [\mathbf{u}_1, \mathbf{u}_2, \ldots, \mathbf{u}_n]$ is unitary and we have

$$U^*AU = U^*[A\mathbf{u}_1, A\mathbf{u}_2, \ldots, A\mathbf{u}_n]$$

$$= \begin{bmatrix} \mathbf{u}_1^* \\ \mathbf{u}_2^* \\ \vdots \\ \mathbf{u}_n^* \end{bmatrix} [\lambda_1\mathbf{u}_1, A\mathbf{u}_2, \ldots, A\mathbf{u}_n] = \begin{bmatrix} \lambda_1 & c_{12} & \cdots & c_{1n} \\ 0 & & & \\ \vdots & & A_1 & \\ 0 & & & \end{bmatrix},$$

where $c_{1j} = (\mathbf{u}_1, A\mathbf{u}_j)$, $j = 2, \ldots, n$, and where A_1 is of order $n - 1$.

Let now U_1 be a unitary matrix of order $n - 1$ such that $U_1^*A_1U_1$ is triangular, and let

$$U_2 = \begin{bmatrix} 1 & 0 & \cdots & 0 \\ 0 & & & \\ \vdots & & U_1 & \\ 0 & & & \end{bmatrix}.$$

Then U_2 and, hence, UU_2 are unitary, and we get

$$(UU_2)^*A(UU_2) = U_2^*(U^*AU)U_2 = \begin{bmatrix} \lambda_1 & b_{12} & \cdots & b_{1n} \\ 0 & & & \\ \vdots & & U_1^*A_1U_1 & \\ 0 & & & \end{bmatrix},$$

where $[\lambda_1, b_{12}, \ldots, b_{1n}] = [\lambda_1, c_{12}, \ldots, c_{1n}]U_2$.   ◇

Corollary 2.16

   (a) *Every selfadjoint matrix is unitarily equivalent to a real diagonal matrix.*
   (b) *Every real symmetric matrix is orthogonally equivalent to a real diagonal matrix.*

Proof

   (a) By Schur's lemma, there is a unitary matrix U, such that

$$U^*AU = \begin{bmatrix} \lambda_1 & b_{12} & \cdots & b_{1n} \\ 0 & \lambda_2 & & \\ \vdots & & \ddots & \vdots \\ 0 & & & \lambda_n \end{bmatrix}.$$

Hence, since A is selfadjoint,

$$\begin{bmatrix} \bar{\lambda}_1 & 0 & \cdots & 0 \\ \bar{b}_{12} & \bar{\lambda}_2 & & \\ \vdots & & \ddots & \vdots \\ \bar{b}_{1n} & & & \bar{\lambda}_n \end{bmatrix} = (U^*AU)^* = U^*A^*U$$

$$= U^*AU = \begin{bmatrix} \lambda_1 & b_{12} & \cdots & b_{1n} \\ 0 & \lambda_2 & & \\ \vdots & & \ddots & \vdots \\ 0 & & & \lambda_n \end{bmatrix},$$

so $\lambda_i = \bar{\lambda}_i$ and $b_{ij} = 0$, $i \neq j$, that is,

$$U^*AU = \text{diag}(\lambda_i).$$

(b) is proved in an analogous way. ◇

This corollary is of fundamental importance in many applications. It shows that all eigenvalues of a selfadjoint matrix are real. In addition, Theorem 2.11 shows that eigenvectors corresponding to different eigenvalues are orthogonal. (This is also seen from the proof of Schur's lemma, because the eigenvectors are the columns of U and $U^*U = I$.)

Example 2.6 shows that there exist nonsymmetric matrices that are similar to a diagonal matrix. We would like to know when a matrix is similar to a diagonal matrix.

Theorem 2.17 *A matrix A of order n is similar to a diagonal matrix D if and only if the eigenvectors of A form a basis for $\mathbb{C}^n$ (i.e., the eigenvector space of A is complete).*

Proof Let $v_i, \lambda_i, i = 1, 2, \ldots, n$ be eigensolutions of A, and let $T = [v_1, v_2, \ldots, v_n]$. Then

$$AT = [Av_1, Av_2, \ldots, Av_n] = [\lambda_1 v_1, \lambda_2 v_2, \ldots, \lambda_n v_n] = TD,$$

where $D = \text{diag}(\lambda_i)$. Since $v_1, v_2, \ldots, v_n$ are linearly independent, T is nonsingular and, hence, $T^{-1}AT = D$, which proves the sufficiency part of the theorem.

If, on the other hand, $T^{-1}AT = D$, where $T = [t_1, t_2, \ldots, t_n]$ and $D = \text{diag}(d_i)$, then $AT = TD$—that is, $At_i = d_i t_i$. But this means that $t_i, d_i, i = 1, 2, \ldots, n$ are eigensolutions of A. Since T is nonsingular, $\det(T) \neq 0$, so $t_1, t_2, \ldots, t_n$ are linearly independent. ◇

Theorem 2.18 *If A is selfadjoint (or real symmetric), then A is unitarily (or orthogonally) equivalent to a diagonal matrix* $D : A = UDU^*$, $D = \text{diag}(\lambda_i)$, $U = [\mathbf{v}_1, \mathbf{v}_2, \ldots, \mathbf{v}_n]$, *where* $\mathbf{v}_i, \lambda_i, i = 1, 2, \ldots, n$ *are eigensolutions of A, and* $\mathbf{v}_i$ *are normalized.*

Proof This follows from Corollary 2.16 and the proof of Theorem 2.17. ◇

Corollary 2.19 *If A is Hermitian and has positive eigenvalues (such a matrix is called Hermitian positive definite, see Chapter 3), then a matrix B exists where* $B^2 = A$ *and B is positive definite. We write* $B = A^{1/2}$ *and call* $A^{1/2}$ *the square root of A. (For the definition of positive definite matrices, see Chapter 3.)*

Proof Since A is selfadjoint there exists a unitary matrix U such that $A = UDU^*$, where $D = \text{diag}(\lambda_i)$ and λ_i are the eigenvalues of A. Let $B = UD^{1/2}U^*$, where $D^{1/2} = \text{diag}(\lambda_i^{1/2})$. Clearly B is positive definite. Further, $B^2 = UD^{1/2}U^*UD^{1/2}U = UDU^* = A$. ◇

2.4* Normal and *H*-Normal Matrices

Definition 2.11 A matrix A is said to be *normal* if it commutes with its Hermitian transpose—i.e., if $A^*A = AA^*$.

Lemma 2.20 *A is normal if and only if A is unitarily similar to a diagonal matrix.*

Proof We show first that if A is normal, then A has a complete set of eigenvectors that can be made orthonormal. Theorem 2.14 shows that A is then unitarily similar to a diagonal matrix. Schur's lemma shows that there exists a unitary matrix U, such that

$$U^*AU = T,$$

where T is upper triangular. Hence $U^*A^*U = T^*$, and

$$U^*AUU^*A^*U = TT^*,$$

or

$$U^*AA^*U = TT^*.$$

Similarly,

$$U^*A^*AU = T^*T.$$

Now A is normal, so $AA^* = A^*A$ and, hence, $T^*T = TT^*$. But since T is upper triangular,

$$(T^*T)_{ij} = \sum_{k=1}^{\min(i,j)} \bar{t}_{ki} t_{kj}$$

and

$$(TT^*)_{ij} = \sum_{k=\max(i,j)}^{n} t_{ik} \bar{t}_{jk}.$$

Hence, $(T^*T)_{ij} = (TT^*)_{ij}$, so, in particular, $(T^*T)_{ii} = (TT^*)_{ii}$, which shows that

$$\sum_{k=1}^{i} \bar{t}_{ki} t_{ki} = \sum_{k=i}^{n} t_{ik} \bar{t}_{ik}$$

or

$$\sum_{k=1}^{i} |t_{ki}|^2 = \sum_{k=i}^{n} |t_{ik}|^2.$$

This shows by induction for $i = 1, 2, \ldots, n$ that $t_{ik} = 0$, $k \neq i$, so T is diagonal. Theorem 2.14 shows now that A has a complete set of eigenvectors and, since U is unitary, A is then unitarily similar to a diagonal matrix. On the other hand, if $U^*AU = T$, where U is unitary and T is diagonal, then $A = UTU^*$ and $AA^* = UTT^*U^* = UT^*TU^* = A^*A$, so A is normal. $\diamond$

Clearly every symmetric (Hermitian) matrix is normal, but there exist non-symmetric (non-Hermitian) matrices that are normal. Lemma 2.20 shows that the set of normal matrices is a subset of matrices that are similar to a diagonal matrix. The above two statements are illustrated in the next example.

Example 2.7

(a) A computation shows that $A = \begin{bmatrix} 1 & 2 \\ -2 & 1 \end{bmatrix}$, which is nonsymmetric, is normal.

(b) Any unitary matrix U is normal, because $U^*U = I = UU^*$, but in general U is not Hermitian.
(c) The matrix in Example 2.6, which, as has been shown there, is similar to a diagonal matrix, is not normal because a computation shows that $AA^T \neq A^TA$.

We shall now extend the class of normal matrices to the class of H-normal matrices.

Definition 2.12 Let H be a Hermitian matrix with positive eigenvalues. Then $A' = H^{-1}A^*H$ is said to be the *H-adjoint of* A, and A is said to be *H-normal* if A commutes with A' for some H.

Note that the I-adjoint is the standard adjoint matrix and that if A is I-normal, then A is normal. We shall prove the interesting result that H-normal matrices comprise the class of matrices that are diagonalizable by a similarity transformation. Hence, the matrix in Example 2.6 is H-normal for some H, where $H \neq I$.

Theorem 2.21 *Let H be a Hermitian positive definite matrix. The following properties are equivalent:*

(a) *A is H-normal.*
(b) *$A' = H^{-1}A^*H$ commutes with A for some H.*
(c) *A is diagonalizable by a similarity transformation (or, equivalently, the eigenvector space of A is complete).*
(d) *There exists a polynomial p and an H such that $A' = p(A)$.*

Proof (a) and (b) are equivalent by Definition 2.12. Assume that (d) holds, that is, $H^{-1}A^*H = p(A)$, for some polynomial p. Then,

$$H^{-1}A^*HA = p(A)A = Ap(A) = AH^{-1}A^*H,$$

so

$$A'A = AA',$$

where $A' = H^{-1}A^*H$, which implies (b). If (b) holds, i.e., $H^{-1}A^*H$ commutes with A, that is

$$H^{-1}A^*HA = AH^{-1}A^*H,$$

then

$$H^{-1/2}A^*H^{1/2}H^{1/2}AH^{-1/2} = H^{1/2}AH^{-1/2}H^{-1/2}A^*H^{1/2},$$

where we have used the fact that $H^{1/2}$ exists (see Corollary 2.19), because H is Hermitian positive definite. Hence,

$$C^*C = CC^*,$$

where $C = H^{1/2}AH^{-1/2}$. Hence, C is normal—i.e., by Lemma 2.20, C has a complete eigenvector space. Hence, $A = H^{-1/2}CH^{1/2}$ has also a complete eigenvector space, and Theorem 2.17 implies that A is similarly equivalent to a diagonal matrix, which is property (c).

Finally, if (c) holds, there exists a nonsingular matrix S such that $S^{-1}AS = D$, where D is diagonal. Then let $H = (SS^*)^{-1}$. We know that D is normal, because $D^*D = DD^*$, D being diagonal. Let $D = \text{diag}(\lambda_1, \ldots, \lambda_n)$ and let $p(\lambda)$ be the interpolation polynomial which takes the values $\bar{\lambda}_i$ at the points $\lambda_i, i = 1, \ldots, n$ (possible multiple values are taken only once). Then $p(D) = D^*$ that is,

$$p(S^{-1}AS) = S^*A^*S^{*-1}$$

or

$$p(A) = SS^*A^*(SS^*)^{-1}$$

or

$$p(A) = H^{-1}A^*H = A',$$

which is (d). ◇

For the special case ($H = I$) of normal matrices, we get the following classical result.

Corollary 2.22 (Gantmacher, 1959) *The following properties are equivalent:*

   (a) A commutes with its Hermitian transpose, A^*.

   (b) A is diagonalizable by a unitary matrix U (or, equivalently, the eigenvector space of A is, or can be made, orthonormal), i.e.,

$$U^*AU = D, \text{ where } U^*U = I$$

   and D is diagonal.

   (c) There exists a polynomial p, such that $A^* = p(A)$.

Proof This follows from Theorem 2.21, with $S = U$ and $H = S^*S = I$. ◇

Let $<\mathbf{u}, \mathbf{v}> = \mathbf{u}^* H \mathbf{v}$ be the inner product defined by the Hermitian positive definite matrix H. Note that $<A'\mathbf{u}, \mathbf{v}> = <\mathbf{u}, A\mathbf{v}>$, that is, A' is the adjoint to A with respect to this inner product.

By Theorem 2.21 there exists a polynomial p such that $A' = p(A)$, if A is H-normal.

Definition 2.13 Let A be H-normal for a certain Hermitian positive definite matrix H. The polynomial $\hat{p}$ of smallest degree, $n(A, H)$, such that $A' = \hat{p}(A)$, is called the H-*normal polynomial* of A, and $n(A, H)$ is called the H-*normal degree*.

The following theorem gives a lower bound on the H-normal degree of A.

Theorem 2.23 (cf. Faber and Manteuffel, 1984.) *Let A be H-normal with degree $n(A, H) > 1$. Then $n(A, H) \geq \deg(m(A))^{1/2}$, where $\deg(m(A))$ is the degree of the minimal polynomial of A.*

Proof Let $p(\cdot)$ be the H-normal polynomial to A, that is,

$$p(A) = H^{-1} A^* H.$$

Then

$$A^* = H p(A) H^{-1}$$

or

$$A = H^{-1} \overline{p}(A^*) H = \overline{p}(H^{-1} A^* H).$$

Note: $\overline{p}(\cdot)$ denotes the polynomial with the complex conjugate coefficients of $p(\cdot)$. Hence, $A = \overline{p}(p(A))$ or $q(A) \equiv \overline{p}(p(A)) - A = 0$, that is, $q(A)$ is an annihilating polynomial. Since the degree of p is larger than one, the degree of $m(A)$, the minimal polynomial, satisfies $\deg(m(A)) \leq n^2(A, H)$. ◇

Exercises

1. Prove that
   (a) $\det(A) = 0 \iff \lambda = 0$ is an eigenvalue of A.
   (b) $\det(cA) = c^n \det(A)$, where A has an order n and c is a constant.

   (c) $\begin{cases} A\mathbf{v}_1 = \lambda \mathbf{v}_1 \\ A\mathbf{v}_2 = \lambda \mathbf{v}_2 \end{cases} \implies \begin{cases} A(\alpha \mathbf{v}_i) = \lambda(\alpha \mathbf{v}_i), \quad i = 1, 2 \\ A(\mathbf{v}_1 + \mathbf{v}_2) = \lambda(\mathbf{v}_1 + \mathbf{v}_2). \end{cases}$

(d) The eigenvectors in $\mathbb{R}^n$ corresponding to an eigenvalue λ, together with the 0-vector, form a linear subspace of $\mathbb{R}^n$.

2. (a) Show by computing the minimal polynomial that the matrix

$$A = \begin{bmatrix} 1 & 2 & 2 \\ 2 & 1 & 2 \\ 2 & 2 & 1 \end{bmatrix}$$

has two distinct eigenvalues. Find these by solving the minimal polynomial equation.

(b) Let $A = \begin{bmatrix} d & b & c \\ b & d & a \\ c & a & d \end{bmatrix}$. Show that $\det(A - \lambda I)$ is a polynomial of degree three in the variable λ. Compute the coefficients of this polynomial.

3. Let $\rho(A) = \max_i |\lambda_i|$.
   (a) Show that $\rho(A^2) = [\rho(A)]^2$.
   (b) Let A be a square matrix of order n with eigenvalues $\{\lambda_i\}_{i=1}^n$. Prove by using Schur's lemma that the eigenvalues of A^k are $\{\lambda_i^k\}_{i=1}^n$.

4. Show that the characteristic polynomial of A^2 is divisible by $m(\lambda)m(-\lambda)$ where $m(\lambda)$ is the minimal polynomial of A.

5. (a) Prove that if A^{-1} or B^{-1} exists and A and B have the same order, then AB and BA have the same spectrum.
   (b) Let A be a square matrix. Show that A^*A and AA^* have identical spectra.

6. *Defective matrices:* Let A be a square matrix of order n. If A does not possess n linearly independent eigenvectors, we say that A is *defective*. Show that:
   (a) No matrix with n different eigenvalues can be defective.
   (b) No normal matrix can be defective.
   (c) $A = \begin{bmatrix} 1 & 1 \\ 0 & 1 \end{bmatrix}$ is defective.
   (d) There exists a matrix (which?) with an eigenvalue of multiplicity n that is not defective.
   (e) There exists a matrix with an eigenvalue of multiplicity n with an eigenvector space of dimension 1.

7. Prove that for square matrices A and B of order n, we have:
   (a) $\operatorname{tr}((AB)^k) = \operatorname{tr}((BA)^k)$.
   (b) $\det(AB - \lambda I) = \det(BA - \lambda I)$, $\lambda \in \mathbb{C}$.

8. Compute an orthogonal matrix O such that $O^{-1}AO$ is triangular, where

$$A = \begin{bmatrix} 2 & 2 & 1 \\ 1 & 3 & 1 \\ 1 & 2 & 2 \end{bmatrix}.$$

Hint: Show first that $\lambda = 1$ is an eigenvalue.

9. If A and B are similarly equivalent, we write $A \sim B$. Show that $\sim$ is an equivalence relation, that is:
    (a) $A \sim A$ (reflexivity).
    (b) $A \sim B \Rightarrow B \sim A$ (symmetry).
    (c) $A \sim B$ and $B \sim C \Longrightarrow A \sim C$ (transitivity).

10. Let A be antihermitian, that is, $A^* = -A$. Prove that the eigenvalues of A are zero or purely imaginary.
    Hint: Use Theorem 2.8.

11. (a)   Prove that a square matrix A has a complete set of orthonormal eigenvectors if and only if $AA^* = A^*A$, that is, A commutes with A^* (that is, A is normal).

    (b)   Show that a normal matrix A can be decomposed into

$$A = UDU^* = \sum \lambda_i \mathbf{v}_i \mathbf{v}_i^*,$$

    where $D = \operatorname{diag}(\lambda_i)$, $U = [\mathbf{v}_1, \ldots, \mathbf{v}_n]$ is unitary. (This is known as the *spectral decomposition theorem*.)
    Hint: Use the proof of Lemma 2.20.

12. Show that $A = [a_{i,j}]$ of order $n \times n$, where

$$a_{i,j} = \left(\frac{2}{n+1}\right)^{\frac{1}{2}} \sin \frac{ij\pi}{n+1}, \quad i, j = 1, \ldots, n$$

    is orthogonal.

13. Show that (anti) Hermitian, real (anti) symmetric, and unitary matrices are normal.

14. (a)   Prove that $AB = BA$ if A and B have the same complete set of eigenvectors.

    (b)   Show that with $A = \begin{bmatrix} 1 & 0 \\ 0 & 1 \end{bmatrix}$, $B = \begin{bmatrix} 1 & 1 \\ 0 & 1 \end{bmatrix}$, then $AB = BA$, but the eigenvector space for B is not complete.

    Remark: Part (b) shows that the two statements in part (a) are not equivalent, because $AB = BA$ does not imply that A and B have the same (complete) eigenvector space. However, it will be seen in Lemma 7.18 that if A, B are Hermitian of the same order n, then $AB = BA$ if and only if A and B have a common (complete) set of eigenvectors.

15. (a) Show that the nonzero parts of the spectrum of AB and BA are identical for any $m \times n$ matrix A and $n \times m$ matrix B.

    Hint: Consider the matrices

    $$\begin{bmatrix} I & 0 \\ -B & \mu I \end{bmatrix} X \quad \text{and} \quad \begin{bmatrix} \mu I & -A \\ 0 & I \end{bmatrix} X,$$

    where $X = \begin{bmatrix} \mu I & A \\ B & \mu I \end{bmatrix}$. By taking the determinants, show that

    $$\mu^n \det X = \mu^m \det(\mu^2 I - BA)$$

    and

    $$\mu^m \det X = \mu^n \det(\mu^2 I - AB).$$

    (b) Show that

    $$\det(I_m + AB) = \det(I_n + BA).$$

16. Let A be real symmetric, with normalized eigensolutions $(\lambda_i, \mathbf{v}_i)_{i=1}^n$. Show that the solution of the linear system $A\mathbf{x} = \mathbf{b}$ is $\mathbf{x} = \sum_{i=1}^n \lambda_i^{-1} \beta_i \mathbf{v}_i$, where $\beta_i = (\mathbf{b}, \mathbf{v}_i)$ (called the *Fourier coefficients* of $\mathbf{b}$).

17. Let $A \in \mathbb{R}^{n \times n}$ (a real matrix) and assume that

    $$A(\mathbf{x} + i\mathbf{y}) = (\lambda + i\mu)(\mathbf{x} + i\mathbf{y}),$$

    where $\lambda, \mu \in \mathbb{R}$, $\mathbf{x}, \mathbf{y} \in \mathbb{R}^n$, that is, $(\mathbf{x} + i\mathbf{y}, \lambda + i\mu)$ is an eigensolution to A. Show that

    $$\lambda_{\min}[\tfrac{1}{2}(A + A^T)] \le \lambda \le \lambda_{\max}[\tfrac{1}{2}(A + A^T)],$$

    that is, the real parts of the spectrum of a real matrix are bounded by the extreme eigenvalues of the symmetric part of the matrix.

    Hint: First show that

    $$A\mathbf{x} = \lambda\mathbf{x} - \mu\mathbf{y}, \quad A\mathbf{y} = \mu\mathbf{x} + \lambda\mathbf{y},$$

    then take the inner products with $\mathbf{x}$ and $\mathbf{y}$, respectively, repeat this for the transposed equations, and sum up.

18. A real square matrix is said to be *idempotent* if $A = A^2$. Let A be symmetric and idempotent and B be symmetric and positive semi-definite. Assume further that $A + B$ is idempotent. Then show that $AB = BA = 0$.

Hint: Show first that $(\mathbf{x}, \mathbf{x}) \geq (A\mathbf{x}, \mathbf{x})$ by considering $((I - A)\mathbf{x}, (I - A)\mathbf{x})$. Show next that the only eigenvalues of A are 0 and 1. (Consider the minimal polynomial of A.) Show that $((I - A)\mathbf{x}, \mathbf{x}) - (B\mathbf{x}, B\mathbf{x}) \geq (BA\mathbf{x}, \mathbf{x}) + (AB\mathbf{x}, \mathbf{x})$ by use of $(\mathbf{x}, \mathbf{x}) \geq ((A + B)\mathbf{x}, (A + B)\mathbf{x})$. (Note that $A + B$ also is idempotent.) Choose now $\mathbf{x}$ such that $A\mathbf{x} = 1 \cdot \mathbf{x}$, $\mathbf{x} \neq 0$. Show that $B\mathbf{x} = \mathbf{0}$ by first showing that $(BA\mathbf{x}, \mathbf{x}) \geq 0$ and $(AB\mathbf{x}, \mathbf{x}) \geq 0$. (It follows also that if $A\mathbf{x} = 0 \cdot \mathbf{x}$, $\mathbf{x} \neq \mathbf{0}$, then $(B\mathbf{x}, B\mathbf{x}) \leq (\mathbf{x}, \mathbf{x})$.) Let $[(1, \mathbf{v}_i)]$ and $[(0, \mathbf{w}_j)]$ be the eigensolutions of A; then, for an arbitrary vector $\mathbf{x}$, $\mathbf{x} = \sum_i \alpha_i \mathbf{v}_i + \sum_j \beta_j \mathbf{w}_j$, $AB\mathbf{x} = \sum \beta_j AB\mathbf{w}_j = \mathbf{0}$, and $BA\mathbf{x} = \sum \alpha_i B\mathbf{v}_i = \mathbf{0}$.

19. *Cramer's rule:* Let $A(j \to \mathbf{x})$ denote the matrix obtained by replacing the jth column of A of order n, with $\mathbf{x} \in \mathbb{C}^n$. Derive Cramer's rule, which can be stated: If A is nonsingular, the jth component of the solution of $A\mathbf{x} = \mathbf{b}$, $\mathbf{x}, \mathbf{b} \in \mathbb{C}^n$ satisfies $x_j = \det[A(j \to \mathbf{b})]/\det(A)$, $j \in \mathbb{N} = \{1, 2, \ldots, n\}$.

    Hint: For any $j \in \mathbb{N}$, $A\mathbf{x} = \mathbf{b}$ is equivalent to $AI(j \to \mathbf{x}) = A(j \to \mathbf{b})$. Now take determinants and note that $\det(I(j \to \mathbf{x})) = x_j$. Credit for this neat proof is due S. M. Robinson (1970), *Math. Mag.* **43**, 94–95.

20. Let A be nonsingular. Show that

$$A^{-1} = \frac{1}{\det A}(A^c)^T.$$

    Hint: Use the previous exercise with, in order, $\mathbf{b} = \mathbf{e}_i$ (the unit coordinate vectors), $i = 1, 2, \ldots, n$.

21. (a) *Newton's formula:* Consider the nth degree polynomial

$$P_n(x) = \prod_{j=1}^{n}(x - x_j) = \sum_{k=0}^{n} \sigma_k x^{n-k}, \quad \sigma_0 \equiv 1.$$

    Define $S_k = x_1^k + x_2^k + \ldots + x_n^k$, $k = 1, 2, \ldots, n$. Then Newton's formula gives the following relation between the S_k and the σ_k:

$$k\sigma_k + S_k + \sum_{i=1}^{k-1} S_i \sigma_{k-i} = 0.$$

    Prove this relation.

    Hint: For $x \neq x_i$, $i = 1, 2, \ldots, n$, we have

$$P_n'(x) = \frac{d}{dx} \prod_{j=1}^{n}(x - x_j) = \sum_{j=1}^{n} \frac{P_n(x)}{x - x_j} = \sum_{j=1}^{n} \frac{P_n(x) - P_n(x_j)}{x - x_j}.$$

Further,

$$\frac{P_n(x) - P_n(x_j)}{x - x_j} = \sum_{k=0}^{n} \sigma_k \left(\frac{x^{n-k} - x_j^{n-k}}{x - x_j} \right) = \sum_{k=0}^{n-1} \sigma_k \left(\frac{x^{n-k} - x_j^{n-k}}{x - x_j} \right)$$

$$= \sum_{k=0}^{n-1} \sigma_k (x^{n-k-1} + x_j x^{n-k-2} + \ldots + x_j^{n-k-2} x + x_j^{n-k-1})$$

$$= \sum_{p=0}^{n-1} (\sigma_p + \sigma_{p-1} x_j + \ldots + \sigma_1 x_j^{p-1} + x_j^p) x^{n-p-1}.$$

Hence,

$$P_n'(x) = n x^{n-1} + \sum_{p=1}^{n-1} (n \sigma_p + S_1 \sigma_{p-1} + \ldots + S_{p-1} \sigma_1 + S_p) x^{n-p-1}.$$

(b) Consider the characteristic polynomial of A:

$$\phi(\lambda) = (-\lambda)^n + \sum_{i=0}^{n-1} a_{n-i} (-\lambda)^i.$$

Show that $a_k = \frac{1}{k} \operatorname{tr}(AB_{k-1})$ where $B_k = -AB_{k-1} + a_k I$, $k = 1, 2, \ldots, n-1$, $B_0 = I$.

Hint: Expand AB_{k-1} as a polynomial in A.

22. Let H be Hermitian positive definite and let the inner product be defined by

$$\langle \mathbf{u}, \mathbf{v} \rangle = \mathbf{u}^* H \mathbf{v}.$$

Show that:

(a) $\langle \mathbf{u}, \mathbf{v} \rangle = \overline{\langle \mathbf{v}, \mathbf{u} \rangle}$.

(b) $\langle \mathbf{u}, \mathbf{v} + \mathbf{w} \rangle = \langle \mathbf{u}, \mathbf{v} \rangle + \langle \mathbf{u}, \mathbf{w} \rangle$.

(c) $\langle c\mathbf{u}, \mathbf{v} \rangle = \bar{c} \langle \mathbf{u}, \mathbf{v} \rangle$.

(d) $\langle \mathbf{u}, \mathbf{u} \rangle \geq 0$, $\langle \mathbf{u}, \mathbf{u} \rangle = 0$ if and only if $\mathbf{u} = \mathbf{0}$.

23. (a) *Gram-Schmidt Orthogonalization* (Schmidt, 1907): We want to replace the linearly independent vectors $\mathbf{v}^{(1)}, \mathbf{v}^{(2)}, \ldots, \mathbf{v}^{(n)}$ one by one with mutually orthogonal vectors $\mathbf{u}^{(1)}, \mathbf{u}^{(2)}, \ldots \mathbf{u}^{(n)}$, that span the same subspace $\mathbb{C}^n$. Then let $\mathbf{u}^{(1)} = \mathbf{v}^{(1)}$, and for $k = 1, 2, \ldots, n-1$ take

$$\mathbf{u}^{(k+1)} = \mathbf{v}^{(k+1)} - \sum_{i=1}^{k} \frac{(\mathbf{u}^{(i)}, \mathbf{v}^{(k+1)})}{(\mathbf{u}^{(i)}, \mathbf{u}^{(i)})} \mathbf{u}^{(i)}.$$

Show that $(\mathbf{u}^{(k+1)}, \mathbf{u}^{(j)}) = 0$, $j = 1, 2, \ldots, k$. Show that the algorithm requires mn^2 flops.

(b) *QR factorization of A:* Let $A = [\mathbf{a}^{(1)}, \mathbf{a}^{(2)}, \ldots, \mathbf{a}^{(n)}]$ be an $m \times n$ matrix, let $\mathbf{q}^{(1)} = \mathbf{a}^{(1)}/\|\mathbf{a}^{(1)}\|$, and, for $k = 1, \ldots, n - 1$, let

$$\tilde{\mathbf{q}}^{(k+1)} = \mathbf{a}^{(k+1)} - \sum_{j=1}^{k}(\mathbf{q}^{(j)}, \mathbf{a}^{(k+1)})\mathbf{q}^{(j)},$$

$$\mathbf{q}^{(k+1)} = \tilde{\mathbf{q}}^{(k+1)}/\|\tilde{\mathbf{q}}^{(k+1)}\|.$$

Show that for a certain matrix R,

$$A = [\mathbf{q}^{(1)}, \ldots, \mathbf{q}^{(n)}]\begin{bmatrix} r_{11} & \cdots & r_{1n} \\ & \ddots & \\ 0 & & r_{nn} \end{bmatrix} = QR,$$

where Q is orthogonal $(Q^T Q = I)$ and R is upper triangular. Show that the factorization is uniquely determined if $r_{kk} > 0$, $k = 1, 2, \ldots, n$. Show also that this method breaks down at stage k if and only if $\mathbf{a}^{(k)}$ is linearly dependent on the previous vectors $\mathbf{a}^{(1)}, \ldots, \mathbf{a}^{(k-1)}$, in which case $r_{kk} = 0$.

(c) *Modified Gram-Schmidt:* Even if the vectors $\mathbf{a}^{(1)}, \mathbf{a}^{(2)}, \ldots, \mathbf{a}^{(k)}$ are linearly independent, they may be "nearly" dependent, showing up in some small values of $\|\tilde{q}^{(k+1)}\|$ and r_{kk}. This causes numerical instability, and the computed matrix Q can be far from orthogonal.

The following modified algorithm was shown by Björck (1969) to be numerically stable. At the beginning of the kth stage, it computes $(\mathbf{q}^{(1)}, \mathbf{a}^{(k)})\mathbf{q}^{(1)}$ and subtracts that off immediately, leaving

$$\mathbf{a}^{(k,1)} = \mathbf{a}^{(k)} - (\mathbf{q}^{(1)}, \mathbf{a}^{(k)})\mathbf{q}^{(1)}.$$

Now $\mathbf{a}^{(k,1)}$ (instead of the original $\mathbf{a}^{(k)}$) is projected onto $\mathbf{q}^{(2)}$, and that projection is subtracted:

$$\mathbf{a}^{(k,2)} = \mathbf{a}^{(k,1)} - (\mathbf{q}^{(2)}, \mathbf{a}^{(k,1)})\mathbf{q}^{(2)}.$$

Since

$$\mathbf{a}^{(k,2)} = [\mathbf{a}^{(k)} - (\mathbf{q}^{(1)}, \mathbf{a}^{(k)})\mathbf{q}^{(1)}] - \{\mathbf{q}^{(2)}, [\mathbf{a}^{(k)} - (\mathbf{q}^{(1)}, \mathbf{a}^{(k)})\mathbf{q}^{(1)}]\}\mathbf{q}^{(2)}$$

$$= \mathbf{a}^{(k)} - (\mathbf{q}^{(1)}, \mathbf{a}^{(k)})\mathbf{q}^{(1)} - (\mathbf{q}^{(2)}, \mathbf{a}^{(k)})\mathbf{q}^{(2)},$$

because $\mathbf{q}^{(1)}$ and $\mathbf{q}^{(2)}$ are orthogonal, $\mathbf{a}^{(k,2)}$ equals the corresponding vector in the ordinary Gram-Schmidt method. These repeated one-dimensional projections are continued and turn out to give less numerical cancellation; they are therefore more numerically stable. Show that the following code for the algorithm overwrites, a column at a time, the m by n matrix A by the matrix Q with orthonormal columns:

$$\text{for } k := 1, 2, \ldots, n$$

$$r_{kk} := \left(\sum_{l=1}^{m} a_{lk}^2 \right)^{\frac{1}{2}} ;$$

$$\text{for } i := 1, 2, \ldots, m$$

$$a_{ik} := a_{ik}/r_{kk};$$

$$\text{if } (k = n) \text{ go to end;}$$

$$\text{for } j = 1, 2, \ldots, k$$

$$r_{kj} := \sum_{l=1}^{m} a_{lk} a_{lj}$$

$$\text{for } i = 1, 2, \ldots, m$$

$$a_{ik+1} := a_{ik+1} - a_{ij} r_{kj};$$

$$\text{end}$$

The last line removes from the next $(k + 1)$ column of A the component in the direction of the next unit vector $\mathbf{q}^{(k+1)}$. The entries of $\mathbf{q}^{(k)}$ are the numbers a_{ik} computed in the fourth line. On the sixth line, the inner products $(\mathbf{q}^{(j)}, \mathbf{a}^{(k+1)})$ are computed and used as multipliers r_{kj} of $\mathbf{q}^{(j)}$, to be subtracted from $\mathbf{a}^{(k+1)}$, $j = 1, 2, \ldots, k$.

Historical remarks: The different versions of the Gram-Schmidt method have an interesting history (see Farebrother, 1988). A row-oriented version of modified Gram-Schmidt (MGS) was first described by Laplace (1812) and is equivalent to a method given by Bauer (1965). It is a special case of a more general algorithm attributed to Cauchy (1837). For further comments, see Björck (1989). It has recently been shown (see Björck and Paige, 1991) that the QR factorization of an $m \times n$ matrix A via MGS is numerically equivalent to that arising from Householder transformations applied to the matrix A augmented by an n by n zero matrix.

    (d)   Show that the modified Gram-Schmidt algorithm requires asymptotically mn^2 multiplications. (All vectors have length m.)

    (e)   Solve $Ax = b$, assuming rank$(A) = n$, using the QR factorization—i.e., show that $\mathbf{x} = R^{-1}Q^T\mathbf{b}$.

    (f)   Show that, to form $A^T A$ using symmetry, we need asymptotically $\frac{1}{2}mn^2$ operations. Then solve the system $A^T Ax = A^T\mathbf{b}$. Show that this method, the normal equation method, requires less work than the QR factorization method. It turns out, however, that forming $A^T A$ can cause large (relative) numerical errors, so the modified Gram-Schmidt is more stable.

24.   Let $\mathbf{v}, \mathbf{w} \in \mathbb{C}^n$ and let A be a nonsingular matrix of order n.

    (a)   Show that the matrix $\mathbf{wv}^T$ has one eigensolution $\mathbf{v}^T\mathbf{w}, \mathbf{w}$ and that the remaining eigensolutions are $0, \mathbf{u}$ for any vector in the set of $n - 1$ linearly independent vectors that are orthogonal to $\mathbf{v}$.

    (b)   Prove the identity

$$\frac{\det(A + \mathbf{uv}^T)}{\det A} = 1 + \mathbf{v}^T A^{-1}\mathbf{u}.$$

    Hint: $\det(A + \mathbf{uv}^T) = \det A \cdot \det(I + (A^{-1}\mathbf{u})\mathbf{v}^T)$. Then use Theorem 2.2.a, or Exercise 15b.

25.   Let A be nonsingular and $\mathbf{u}$ and $\mathbf{v}$ be two vectors such that $A + \mathbf{uv}^T$ is nonsingular. Show that

$$(A + \mathbf{uv}^T)^{-1} = A^{-1} - \frac{A^{-1}\mathbf{uv}^T A^{-1}}{1 + \mathbf{v}^T A^{-1}\mathbf{u}}.$$

26.   Show by direct computation the following extension of the previous exercise (frequently referred to as the *Sherman-Morrison formula*):

$$(A + UBV)^{-1} = A^{-1} - A^{-1}U(I_m + BVA^{-1}U)^{-1}BVA^{-1},$$

where A is an $n \times n$ nonsingular matrix, B an $m \times m$ matrix, U an $n \times m$ matrix, and V an $m \times n$ matrix. I_m is the identity matrix of order m. It is assumed that $I_m + BVA^{-1}U$ is nonsingular.

Hint: The computations are simplified if one shows first that the problem can be reduced to

$$(I_n + \widetilde{U}\widetilde{V})^{-1} = I_n - \widetilde{U}(I_m + \widetilde{V}\widetilde{U})^{-1}\widetilde{V},$$

where $\widetilde{U} = A^{-1}U, \widetilde{V} = BV$.

Remark: The Sherman-Morrison formula is useful when several, usually very large, subsystems are coupled with a small number of interconnections. Then A is a large block diagonal system, representing the subsystems, and U, B, V provide the adjustments reflecting the interconnections.

27. *The QR factorization method using Householder transformations:* There is an alternative to the LU factorization method, namely a method that computes an orthogonal matrix Q and an upper triangular matrix R to make $QR = A$, where A is the square matrix to be factorized. As we saw in Exercise 23, it can be based on Gram-Schmidt orthogonalization. It can also be computed using so-called *Householder transformation matrices,* which have the form:

$$H = H(\mathbf{v}) = I - \frac{2}{\|\mathbf{v}\|^2} \mathbf{v}\mathbf{v}^*,$$

where $\mathbf{v}$ is a nonzero vector in $\mathbb{C}^n$.

(a) Show that $H^* = H$ and that H is unitary—i.e., $H^* H = I$.

(b) Show that we do not need to compute $H(\mathbf{v})$ explicitly, because we need only to compute the action of $H(\mathbf{v})$:

$$[H(\mathbf{v})A]_{\cdot i} = \mathbf{a}_{\cdot i} - \frac{2}{\|\mathbf{v}\|^2}(\mathbf{v}^*\mathbf{a}_{\cdot i})\mathbf{v}.$$

(c) If $\mathbf{x}$ and $\mathbf{y}$ are any pair of distinct vectors such that $\|\mathbf{x}\| = \|\mathbf{y}\|$, then show that if $\mathbf{v} = \pm(\mathbf{x} - \mathbf{y})$, then

$$\mathbf{y} = H(\mathbf{v})\mathbf{x}.$$

This shows that $H(\mathbf{v})$ transforms $\mathbf{x}$ into $\mathbf{y}$.

(d) Let $\mathbf{a}_{\cdot 1}$ be the first column of a matrix A and let $\mathbf{e}_1$ be the first unit vector. Show that with $\mathbf{x} = \mathbf{a}_{\cdot 1}$, $\mathbf{y} = \xi_1 \mathbf{e}_1$, $\xi_1 = \|\mathbf{a}_{\cdot 1}\|$, $\mathbf{v} = \mathbf{x} - \mathbf{y}$, HA takes the form

$$HA = \begin{bmatrix} \xi_1 & * & \cdots & * \\ 0 & * & & * \\ \cdots & & & \\ 0 & * & & * \end{bmatrix},$$

where A has order n.

(e) Show how, by repeated use of Householder transformations, one can compute

$$R = H^{(n-1)}H^{(n-2)}\ldots H^{(1)}A,$$

where $H^{(i)}$ eliminates the entries of the ith column below the diagonal of $H^{(i-1)} \cdots H^{(1)} A$, so that R is upper triangular. Show also that $QR = A$, where

$$Q = H^{(1)} H^{(2)} \ldots H^{(n-1)}.$$

(f)   Note that we don't need to form Q explicitly. Also, the eliminated parts of A can be used to store the vectors defining the $H^{(i)}$ matrices (we need just n more storage locations). Show that, when R is computed, a linear system $Az = b$ can be solved by solving $Rz = \tilde{b}$, where $\tilde{b} = H^{(n-1)} \ldots H^{(1)} b$.

(g)   Show that the asymptotic computational cost of the QR factorization method is

$$\sim 2n^2 + 2(n-1)^2 + \ldots + 2 \sim \frac{2}{3}n^3 \text{ flops.}$$

In addition, we need to compute n square roots. The cost to compute the solution z is $O(n^2)$. Since A is an orthogonal transformation, it follows that the condition numbers of the intermediate matrices (say $A^{(k)}$) are not altered—i.e., $\text{cond}_2(A) = \text{cond}_2(A^{(k)})$, $k = 1, 2, \ldots, n-1$. Hence, the method is numerically stable. For further comments, see Wilkinson (1965).

28. (a)   Find the least squares solution to

$$\begin{bmatrix} 1 & 0 \\ 1 & 1 \\ 0 & 1 \\ -1 & -1 \\ -1 & 0 \end{bmatrix} \begin{bmatrix} x_1 \\ x_2 \end{bmatrix} = \begin{bmatrix} 1 \\ 1 \\ 1 \\ 1 \\ 1 \end{bmatrix} = b.$$

Write A as QR and compare

$$x = (A^T A)^{-1} A^T b \text{ with } x = R^{-1} Q^T b.$$

(b)   The projection matrix $P = A(A^T A)^{-1} A^T$ takes a much simpler form when $A = QR$. Find this! (Note that $Q^T Q = I$, but in general $QQ^T \neq I$.)

29.   Show that the inverse of $B = I - uv^T$ has the form $B^{-1} = I - cuv^T$ (cf. Exercise 25). Under what conditions on u and v is B singular?

30.   If v has components $(1/3, 2/3, 2/3)$, find:

(a)   The Householder matrix $H = I - 2vv^T$.

(b)   The eigenvalues of H.

(c) H^2.

(d) H^{-1}.

31.   *Parallel solution of matrix polynomial equations:* Let $p(\cdot)$ be a polynomial of degree m with distinct zeros r_i, $i = 1, 2, \ldots, m$, with leading coefficient equal to unity. Let A be a square matrix with eigenvalues not equal to any of the roots r_i. Show that the equation

$$p(A)\mathbf{x} = \mathbf{b}$$

has the solution

$$\mathbf{x} = \sum_{i=1}^{m} \frac{1}{p'(r_i)} \mathbf{x}_i,$$

where

$$\mathbf{x}_i = (A - r_i I)^{-1} \mathbf{b}$$

and

$$p'(r_i) = \prod_{\substack{j=1 \\ j \neq i}}^{m} (r_i - r_j).$$

Hence, the solution $\mathbf{x}$ can be computed as a sum involving m different vectors that are computed by solving m independent systems—i.e., that can be computed concurrently.

Hint: Show first, using the Lagrangian interpolation polynomial for the function $f(t) \equiv 1$, that

$$1 = \sum_{i=1}^{m} \frac{\prod_{j \neq i} (t - r_j)}{\prod_{j \neq i} (r_i - r_j)} = \sum_{i=1}^{m} \frac{p(t)}{p'(r_i)(t - r_i)}.$$

Hence,

$$\frac{1}{p(t)} = \sum_{i=1}^{m} \frac{1}{p'(r_i)} (t - r_i)^{-1}.$$

32.   *Quadratic and generalized eigenvalue problems:* Let A, B, and C be real symmetric; let A be positive definite and C positive semidefinite; and consider the quadratic eigenvalue problem,

$$A\mathbf{x} + \lambda B\mathbf{x} + \lambda^2 C\mathbf{x} = 0.$$

(a) Show that to any real eigenvalue of the quadratic eigenvalue problem there exists a real eigenvector.
Hint: Show that $A(\mathbf{x} + \bar{\mathbf{x}}) + \lambda B(\mathbf{x} + \bar{\mathbf{x}}) + \lambda^2 C(\mathbf{x} + \bar{\mathbf{x}}) = \mathbf{0}$.

(b) Let $(\lambda_1, \mathbf{x}_1), (\lambda_2, \mathbf{x}_2)$ be two eigensolution pairs with $\lambda_1 \neq \bar{\lambda}_2$.
Show that

$$\mathbf{x}_2^*(B + (\lambda_1 + \bar{\lambda}_2)C)\mathbf{x}_1 = \mathbf{0}$$

and

$$\mathbf{x}_2^* A \mathbf{x}_1 = \lambda_1 \bar{\lambda}_2 \mathbf{x}_2^* C \mathbf{x}_1.$$

In particular, if $C = 0$ (in which case the eigenvalue problem is
usually referred to as the *generalized eigenvalue problem*), then

$$\mathbf{x}_2^* B \mathbf{x}_1 = 0 \quad \text{and} \quad \mathbf{x}_2^* A \mathbf{x}_1 = 0.$$

Hint: Take the inner product of $\mathbf{x}_2^*$ with $A\mathbf{x}_1 + \lambda_1 B \mathbf{x}_1 + \lambda_1^2 C \mathbf{x}_1 =$
$\mathbf{0}$ and of $\mathbf{x}_2^* A + \bar{\lambda}_2 \mathbf{x}_2^* B + \bar{\lambda}_2^2 \mathbf{x}_2^* C = \mathbf{0}$ with $\mathbf{x}_1^*$, and subtract. This
shows the first part. Then take $\bar{\lambda}_2$ times the first inner product and
λ_1 times the second, and subtract. This shows the second part.

(c) Show that if A, B, and C are positive definite, then for any eigenvalue λ, $Re(\lambda) < 0$. If A is only symmetric (but not necessarily
positive definite), show that $Re(\lambda) < 0$ for any complex λ.

(d) If A is only symmetric but $C = 0$ and B is symmetric and positive
definite, then show that λ is real.

Remark: The above extends the results in Theorem 2.11.

33. Assume that $\lambda_0, \mathbf{x}_0$ is an eigensolution pair of the quadratic eigenvalue
problem in the previous exercise, where $\lambda_0 \neq 0$. Let

$$\widetilde{A} = A + \frac{\tau}{\lambda_0^2} A \mathbf{x}_0 \mathbf{x}_0^* A$$

$$\widetilde{B} = B - \frac{\tau}{\lambda_0}(C \mathbf{x}_0 \mathbf{x}_0^* A + A \mathbf{x}_0 \mathbf{x}_0^* C)$$

$$\widetilde{C} = C + \tau C \mathbf{x}_0 \mathbf{x}_0^* C.$$

Show that the quadratic eigenvalue problem

$$\widetilde{A}\mathbf{x} + \lambda \widetilde{B}\mathbf{x} + \lambda^2 \widetilde{C}\mathbf{x} = \mathbf{0}$$

has the same eigenvalues as

$$A\mathbf{x} + \lambda B\mathbf{x} + \lambda^2 C\mathbf{x} = \mathbf{0},$$

except that λ_0 is replaced by

$$\tilde{\lambda}_0 = [\lambda_0 + \frac{\tau}{\lambda_0}\mathbf{x}_0^* A\mathbf{x}_0]/[1 + \tau\mathbf{x}_0^* C\mathbf{x}_0].$$

Hint: Show first that

$$\tilde{A} + \lambda\tilde{B} + \lambda^2\tilde{C} = A + \lambda B + \lambda^2 C$$
$$+ \tau(\frac{1}{\lambda_0}A\mathbf{x}_0 - \lambda C\mathbf{x}_0)(\frac{1}{\lambda_0}\mathbf{x}_0^* A - \lambda\mathbf{x}_0^* C),$$

$$\frac{1}{\lambda_0}A\mathbf{x}_0 = -(B + \lambda_0 C)\mathbf{x}_0$$

and

$$(A + \lambda B + \lambda^2 C)\mathbf{x}_0 = (\lambda - \lambda_0)[B + (\lambda + \lambda_0)C]\mathbf{x}_0$$
$$= (\lambda - \lambda_0)[\lambda C - \frac{1}{\lambda_0}A]\mathbf{x}_0.$$

Show next that

$$\det(\tilde{A} + \lambda\tilde{B} + \lambda^2\tilde{C})$$
$$= \det(A + \lambda B + \lambda^2 C)\det[I_n - \frac{\tau}{\lambda - \lambda_0}\mathbf{x}_0(\frac{1}{\lambda_0}\mathbf{x}_0^* A - \lambda\mathbf{x}_0^* C)].$$

Then use the identity (see Exercise 15)

$$\det(I_n + UV) = \det(I_m + VU),$$

where U is $n \times m$ and V is $m \times n$, to show that

$$\det [I_n - \frac{\tau}{\lambda - \lambda_0} \mathbf{x}_0(\frac{1}{\lambda_0}\mathbf{x}_0^* A - \lambda \mathbf{x}_0^* C)]$$

$$= 1 + \frac{\tau}{\lambda - \lambda_0}(\lambda \mathbf{x}_0^* C \mathbf{x}_0 - \frac{1}{\lambda_0}\mathbf{x}_0^* A \mathbf{x}_0)$$

$$= \frac{1}{\lambda - \lambda_0}[\lambda + \lambda \tau \mathbf{x}_0^* C \mathbf{x}_0 - \lambda_0 - \frac{\tau}{\lambda_0}\mathbf{x}_0^* A \mathbf{x}_0]$$

$$= \frac{1}{\lambda - \lambda_0}(1 + \tau \mathbf{x}_0^* C \mathbf{x}_0)(\lambda - \tilde{\lambda}_0).$$

Remark: This method can be used to transform the smallest eigenvalue to a large eigenvalue for a similar eigenvalue problem. If an algorithm to compute the largest eigenvalue is available, then one can use such transformations to compute the smallest eigenvalues, one by one, in order of their size.

34. *Circulant matrices:*

$$C = \text{circ}(c_1, c_2, \ldots, c_n) = \begin{bmatrix} c_1 & c_2 & \cdots & c_n \\ c_n & c_1 & \cdots & c_{n-1} \\ \vdots & \vdots & \ddots & \vdots \\ c_2 & c_3 & \cdots & c_1 \end{bmatrix},$$

where c_i are complex numbers, is called a *circulant matrix*.

(a) Show that $CP = PC$, where P is the $n \times n$ permutation matrix

$$P = \text{circ}(0, 1, 0, \ldots, 0).$$

(b) Show that

$$C = \text{circ}(c_1, c_2, \ldots, c_n) = \sum_{k=1}^{n} c_k P^{k-1} \equiv p_C(P).$$

Hint: Show first that

$$P^2 = \text{circ}(0, 0, 1, 0, \ldots, 0), \quad P^3 = \text{circ}(0, 0, 0, 1, 0, \ldots 0),$$

etc., until $P^n = P^0 = I$.

(c) Let E be defined by the entries

$$E_{jk} = \frac{1}{\sqrt{n}} \exp\left(\frac{2\pi i(j-1)(k-1)}{n}\right), \quad \text{where } i^2 = -1.$$

Show that E is unitary—i.e., $EE^* = I_n$.
Hint:

$$(EE^*)_{j,k} = \frac{1}{n} \sum_{s=1}^{n} \exp\left(\frac{2\pi i(j-1)(s-1)}{n}\right) \exp\left(\frac{2\pi i(1-k)(s-1)}{n}\right)$$

$$= \frac{1}{n} \sum_{s=1}^{n} \exp\left(\frac{2\pi i(j-k)(s-1)}{n}\right) = \begin{cases} 1, & \text{if } j = k \\ 0, & \text{if } j \neq k \end{cases}$$

which can be shown by computing the sum of the nth roots of the unit number.

(d) Show that

$$E\Lambda E^* = P,$$

where $\Lambda = \text{diag}(1, w, w^2, \ldots, w^{n-1})$, $w = \exp(\frac{2\pi i}{n})$.
Hint:

$$(E^*CE)_{j,k} = \frac{1}{n} \sum_{s=1}^{n} w^{(j-1)(s-1)} w^{s-1} w^{(1-k)(s-1)}$$

$$= \frac{1}{n} \sum_{s=1}^{n} w^{(j-k+1)(s-1)} = \begin{cases} 1, & \text{if } j = k-1 \\ 0, & \text{otherwise,} \end{cases}$$

which is the form of the matrix P.

(e) Show next that the unitary transformation E^*CE of C is diagonal, and that

$$E^*CE = \text{diag}(\lambda_1, \ldots, \lambda_n),$$

where $\lambda_j = p_C(w^{j-1})$.

(f) Show that the columns of

$$E = \frac{1}{n} \begin{bmatrix} 1 & 1 & 1 & \cdots & 1 \\ 1 & w & w^2 & & w^{n-1} \\ \vdots & \vdots & \vdots & & \vdots \\ 1 & w^{n-1} & w^{2(n-1)} & & w^{(n-1)(n-1)} \end{bmatrix}$$

are the eigenvectors of C.

(g) Show that the inverse (if it exists) and the transpose of a circulant matrix are also circulant, and that circulant matrices commute.

35. Let $A = [a_{ij}]$ be a symmetric matrix and let

$$\mathbf{d} = \begin{bmatrix} a_{11} \\ a_{22} \\ \vdots \\ a_{nn} \end{bmatrix}, \quad P = \begin{bmatrix} |v_{11}|^2 & \cdots & |v_{1n}|^2 \\ & \cdots & \\ |v_{n1}|^2 & \cdots & |v_{nn}|^2 \end{bmatrix}, \quad \underline{\lambda} = \begin{bmatrix} \lambda_1 \\ \vdots \\ \lambda_n \end{bmatrix},$$

where $[v_{i,1}, \ldots, v_{i,n}]$ is the ith row of the unitary matrix V of the eigenvectors of A. Show that

   (a) $\sum_{i=1}^{n} |v_{ij}|^2 = \sum_{j=1}^{n} |v_{ij}|^2 = 1$ (such a matrix P is called a *doubly stochastic matrix*).

   (b) $\mathbf{d} = P\underline{\lambda}$.

36. Let λ_i, $i = 1, 2, \ldots, n$ be the eigenvalues of A and assume that $\lambda_i > 0$. Show that

   (a) $\frac{1}{n} \operatorname{tr}(A) \geq \det(A)^{\frac{1}{n}}$ or $\frac{1}{n} \sum_{i=1}^{n} \lambda_i \geq (\prod_{i=1}^{n} \lambda_i)^{\frac{1}{n}}$.

   (b) $(\prod_{i=1}^{n} \lambda_i)^{\frac{1}{n}} \geq \frac{n}{\sum_{i=1}^{n} \lambda_i^{-1}}$.

Hint: For (a) use Theorem 2.2 and the inequality between the arithmetic and geometric means (see Exercise 32, Appendix A). For (b) apply part (a) for the matrix A^{-1}, that is, $\frac{1}{n} \operatorname{tr}(A^{-1}) \geq \det(A^{-1})^{\frac{1}{n}}$.

References

S. Barnett (1989). Leverrier's algorithm: a new proof and extensions. *SIAM J. Matrix Anal. Appl.* **10**, 551–556.

F. L. Bauer (1965). Eliminations with weighted row combinations for solving linear equations and least squares problems. *Numer. Math.* **7**, 338–352.

W. Baur and V. Strassen (1983). The complexity of partial derivatives. *Theoretical Computer Science* **22**, 317–330.

Å. Björck (1967). Solving linear least squares problems by Gram-Schmidt orthogonalization. *BIT* **7**, 1–21.

Å. Björck (1989). Least squares methods. In *Handbook of Numerical Analysis*, vol. I, ed. P. Ciarlet and P. Lions. Amsterdam: Elsevier/North-Holland.

Å. Björck and C. C. Paige (1991). Loss and recapture of orthogonality in the modified Gram-Schmidt algorithm. *SIAM J. Matrix Anal.*, to appear.

A. Cauchy (1837). Mémoire sur l'interpolation. *J. de Mathématiques Pures et Appliquées* **2**, 193–205.

V. Faber and T. Manteuffel (1984). Necessary and sufficient conditions for the existence of a conjugate gradient method. *SIAM J. Numer. Anal.* **21**, 352–362.

D. K. Faddeev and V. N. Faddeeva (1963). *Computational Methods of Linear Algebra.* San Francisco and London: Freeman.

V. N. Faddeeva (1959). *Computational Methods of Linear Algebra*. New York: Dover.

R. W. Farebrother (1988). *Linear Least Squares Computations*, New York and Basel: Marcel Dekker.

J. N. Franklin (1968). *Matrix Theory*. Englewood Cliffs, N.J.: Prentice Hall.

F. R. Gantmacher (1959). *The Theory of Matrices*, vols. I & II. New York: Chelsea.

P. Lancaster and M. Tismenetsky (1985). *The Theory of Matrices*. Orlando, Fla.: Academic.

P. S. Laplace (1812). Théorie Analytiques des Probalités. Paris: Mme Courcier.

P. R. Halmos (1968). *Finite-dimensional Vector Spaces*. Princeton, N.J.: Van Nostrand.

E. Schmidt (1907). Zur Theorie der lineairen und nichtlineairen Integralgleichungen: I. Teil, Entwicklung willkürlicher Funktionen nach Systemen vorgeschriebener. *Mathematischer Annalen* **63**, 433–476.

J. Wilkinson (1965). *The Algebraic Eigenvalue Problem*. Oxford: Clarendon.

3

Positive Definite Matrices, Schur Complements, and Generalized Eigenvalue Problems

In this chapter, we define the concept of positive definite matrices and present some properties of real symmetric positive definite matrices. Next, some particularly important properties of Schur complement matrices are discussed, condition numbers for positive definite matrices are analyzed, and some estimates of eigenvalues of generalized eigenvalue problems based on the Courant-Fischer theorem are derived. We conclude with a discussion of congruence transformations and quasisymmetric matrices.

The following definitions are introduced in this chapter:

Definition 3.1
   (a) A matrix A is said to be *positive definite* (*positive semidefinite*) in $\mathbb{C}^n$ if its quadratic form is real and $(A\mathbf{x}, \mathbf{x}) > 0$ $[(A\mathbf{x}, \mathbf{x}) \geq 0]$ for all $\mathbf{x} \neq \mathbf{0}, \mathbf{x} \in \mathbb{C}^n$.
   (b) A real matrix A is said to be *positive definite* (*positive semidefinite*) in $\mathbb{R}^n$ if $(A\mathbf{x}, \mathbf{x}) > 0$ $[(A\mathbf{x}, \mathbf{x}) \geq 0]$, for all $\mathbf{x} \neq \mathbf{0}, \mathbf{x} \in \mathbb{R}^n$.
Definition 3.2 A matrix A is said to be *positive stable* if all its eigenvalues have positive real parts.
Definition 3.3
   (a) $B = 1/2(A + A^*)$ is called the *Hermitian* part of A.
   (b) $C = 1/2(A - A^*)$ is called the *anti-Hermitian* part of A.
      For a real matrix A,
   (c) $B = 1/2(A + A^T)$ is called the *symmetric* part of A.
   (d) $C = 1/2(A - A^T)$ is called the *antisymmetric* (also called the *skew-symmetric*) part of A.
Definition 3.4 Let $A = \begin{bmatrix} A_{11} & A_{12} \\ A_{21} & A_{22} \end{bmatrix}$. If A_{11} is nonsingular, we define

$$S \equiv A/A_{11} \equiv A_{22} - A_{21}A_{11}^{-1}A_{12},$$

84

which is called the *Schur complement* of A with respect to A_{11}.

Definition 3.5 Let A be symmetric, positive definite, and let $\lambda_{min}(A)$ and $\lambda_{max}(A)$ denote the extreme eigenvalues of A. Then by the *condition number* of A we mean

$$\text{cond}(A) = \lambda_{max}(A)/\lambda_{min}(A).$$

Definition 3.6 Let A be a square matrix and let C be nonsingular and of the same order as A. Then $C^T A C$ is called a *congruence transformation* of A.

Definition 3.7 A matrix A is said to be *quasisymmetric* if it is similar to a symmetric matrix.

Notation If A is symmetric and positive definite, we frequently write A is s.p.d.

3.1 Positive Definite Matrices

Definition 3.1

(a) A matrix A is said to be *positive definite (positive semidefinite) in* $\mathbb{C}^n$ if its quadratic form is real and $(A\mathbf{x}, \mathbf{x}) > 0 \;\; [(A\mathbf{x}, \mathbf{x}) \geq 0] \;\; \forall \mathbf{x} \neq \mathbf{0}$, $\mathbf{x} \in \mathbb{C}^n$.

(b) A real matrix A is said to be *positive definite (positive semidefinite) in* $\mathbb{R}^n$ if $(A\mathbf{x}, \mathbf{x}) > 0 \; [(A\mathbf{x}, \mathbf{x}) \geq 0] \; \forall \mathbf{x} \neq \mathbf{0}, \mathbf{x} \in \mathbb{R}^n$.

Lemma 3.1

(a) If $(A\mathbf{x}, \mathbf{x})$ is real for all $\mathbf{x} \in \mathbb{C}^n$, then A is Hermitian.

(b) A real matrix is positive definite if and only if its symmetric part $1/2(A + A^T)$ is positive definite.

Proof If $(A\mathbf{x}, \mathbf{x})$ is real, then

$$(A\mathbf{x}, \mathbf{x}) = \overline{(A\mathbf{x}, \mathbf{x})},$$

that is,

$$\sum_i \sum_j \overline{a_{ij} x_j} x_i = \sum_i \sum_j \overline{\overline{a_{ij} x_j} x_i} = \sum_i \sum_j a_{ij} x_j \bar{x}_i = \sum_j \sum_i a_{ji} x_i \bar{x}_j,$$

so

$$\sum_i \sum_j \overline{a}_{ij} x_i \overline{x}_j = \sum_i \sum_j a_{ji} \overline{x}_j x_i,$$

which implies

$$\sum_i \sum_j (\overline{a}_{ij} - a_{ji}) \overline{x}_j x_i = 0, \qquad \forall \mathbf{x} \in \mathbb{C}^n.$$

By choosing proper vectors, it can readily be seen that if for a matrix B, $(B\mathbf{x}, \mathbf{x}) = 0$, $\forall \mathbf{x} \in \mathbb{C}^n$, then $b_{ij} = 0$, $\forall i, j$, $1 \le i, j \le n$. Hence,

$$\overline{a}_{ij} = a_{ji}, \text{ that is } A = A^*.$$

Thus, a positive definite (or positive semidefinite) matrix on $\mathbb{C}^n$ is selfadjoint, which proves part (a). Now let A be real. By decomposing A as

$$A = \tfrac{1}{2}(A + A^T) + \tfrac{1}{2}(A - A^T)$$

and noting that, for any real vector $\mathbf{x}$,

$$(A\mathbf{x}, \mathbf{x}) = (\mathbf{x}, A^T \mathbf{x}) = (A^T \mathbf{x}, \mathbf{x}),$$

that is,

$$([A - A^T]\mathbf{x}, \mathbf{x}) = 0,$$

we find

$$(A\mathbf{x}, \mathbf{x}) = (\tfrac{1}{2}[A + A^T]\mathbf{x}, \mathbf{x}), \quad \text{for all } \mathbf{x} \in \mathbb{R}^n,$$

which proves part (b).   ◇

Theorem 3.2 *A is positive definite (positive semidefinite) on $\mathbb{C}^n$ if and only if A is Hermitian and its eigenvalues are positive (nonnegative). In particular, a positive definite matrix is regular (nonsingular).*

Proof If A is positive definite, Lemma 3.1 shows that A is Hermitian. Hence, by Theorem 2.16, there is a unitary matrix U that diagonalizes A—that is, $U^*AU = \text{diag}(\lambda_i)$ and λ_i are real. Let $\mathbf{x} = U\mathbf{y}$. Then

$$(A\mathbf{x}, \mathbf{x}) = (AU\mathbf{y}, U\mathbf{y}) = (U^*AU\mathbf{y}, \mathbf{y}) = \sum_{i=1}^n \overline{\lambda}_i |y_i|^2 = \sum_{i=1}^n \lambda_i |y_i|^2.$$

If $\mathbf{y} = [\delta_{1j}, \ldots, \delta_{nj}]^T$, then $(A\mathbf{x}, \mathbf{x}) = \lambda_j$ with $\mathbf{x} = U\mathbf{y}$, so the eigenvalues are positive since $(A\mathbf{x}, \mathbf{x}) > 0$. Since to every $\mathbf{y}$ there is exactly one $\mathbf{x}$ and vice versa, the converse of the theorem follows readily. $\diamond$

Matrices with eigenvalues with positive real parts are important in many applications—for instance, for the stability of solutions of systems of ordinary differential equations,

$$\frac{d\mathbf{x}(t)}{dt} + A\mathbf{x}(t) = \mathbf{0}, \quad t > 0, \ \mathbf{x}(0) = \mathbf{x}_0.$$

Here $\mathbf{x}(t)$ denotes a vectorial function and the solution is readily found to be $\mathbf{x}(t) = \exp(-tA)\mathbf{x}_0$, where $\exp(B)$ for any square matrix B is defined by

$$\exp(B) = \sum_{k=0}^{\infty} \frac{1}{k!} B^k.$$

Note further that $d/dt \exp(-tA) = -A \exp(-tA)$.

Definition 3.2 A matrix A is said to be *positive stable* if all its eigenvalues have positive real parts.

If A is positive stable, then one can show, using the Jordan canonical form, for instance, that $\|\mathbf{x}(t)\|$ is bounded for all bounded values of t, and $\| \exp(-tA)\| \to 0$, $t \to +\infty$, where $\mathbf{x}(t) = \exp(-tA)\mathbf{x}_0$.
    Note that any matrix A can be decomposed as

$$A = \frac{1}{2}(A + A^*) + \frac{1}{2}(A - A^*)$$

or as

$$A = \frac{1}{2}(A + A^T) + \frac{1}{2}(A - A^T).$$

In practice, the first decomposition is used for complex matrices and the second for real matrices.

Definition 3.3

(a) $B = 1/2(A + A^*)$ is called the *Hermitian* part of A,
(b) $C = 1/2(A - A^*)$ is called the *anti-Hermitian* part of A, and if A is real,

(c) $B = 1/2(A + A^T)$ is called the *symmetric part* of A,

(d) $C = 1/2(A - A^T)$ is called the *anti-symmetric* (also called the *skew-symmetric*) *part* of A.

Note that $B^* = B$, $C^* = -C$, and if A is real, $B^T = B$, $C^T = -C$.

Lemma 3.3 *If the Hermitian part of A is positive definite, then A is positive stable.*

Proof Let $\mathbf{x}$, λ be an eigensolution of A, that is, let $A\mathbf{x} = \lambda\mathbf{x}$, $\mathbf{x} \neq \mathbf{0}$. Then we also have $\mathbf{x}^*A^* = \bar{\lambda}\mathbf{x}^*$. Multiplying these equations from the left and right by $\mathbf{x}^*$ and $\mathbf{x}$, respectively, and summing up, we find

$$\mathrm{Re}(\lambda) = \tfrac{1}{2}(\lambda + \bar{\lambda}) = \mathbf{x}^*\tfrac{1}{2}(A + A^*)\mathbf{x}/\mathbf{x}^*\mathbf{x},$$

so $\mathrm{Re}(\lambda) > 0$. Since this is true for each eigenvalue λ, it follows that A is positive stable.   ◇

The proof shows that $\mathbf{x}^*(A + A^*)\mathbf{x}$ need be positive only for the eigenvectors of A for A being positive stable to hold. Hence, in particular, if λ is a multiple eigenvalue of A and the dimension of the space spanned by the eigenvectors corresponding to λ is smaller than the multiplicity of λ (A is then defective), then $A + A^*$ need not be positive definite on the whole space.

Example 3.1 Let $A = \begin{bmatrix} 2 & -1 \\ 1 & 0 \end{bmatrix}$, which has a multiple eigenvalue but is defective. As $1/2(A + A^*) = \begin{bmatrix} 2 & 0 \\ 0 & 0 \end{bmatrix}$, it suffices to note that $1/2\mathbf{v}_1^T(A + A^*)\mathbf{v}_1 = 1$, where $\mathbf{v}_1 = (1, 1)^T$ is an eigenvector to A; we see that A must be positive stable.

The converse statement to Lemma 3.3 does not hold—i.e., there exist matrices that are positive stable but for which the Hermitian part is not positive definite, as the following example shows.

Example 3.2 $A = \begin{bmatrix} 2 & -3 \\ -\frac{1}{2} & 1 \end{bmatrix}$ has eigenvalues that are positive but has a Hermitian (symmetric) part, $\frac{1}{2}(A + A^T) = \begin{bmatrix} 2 & -7/4 \\ -7/4 & 1 \end{bmatrix}$, that is indefinite.

There is, however, an extension of the inner product, $\langle \mathbf{x}, \mathbf{y} \rangle \equiv \mathbf{x}^*K\mathbf{y}$, defined by a s.p.d. matrix K, such that the positive stable matrix class is identical to the class of matrices that have a positive definite Hermitian part with respect to this inner product.

Theorem 3.4 (Lyapunov, 1897) *A is positive stable if and only if there exists a Hermitian matrix K positive definite in* $\mathbb{C}^n$ *such that* $W \equiv A^*K + KA$ *is positive definite in* $\mathbb{C}^n$.

Proof Let A be positive stable, let W be a given Hermitian positive definite matrix, and let $K = \int_0^\infty \exp(-tA^*)W \exp(-tA)dt$. Then, as is readily seen, K is Hermitian positive definite and

$$A^*K + KA = -\int_0^\infty \frac{d}{dt}[\exp(-tA^*)W\exp(-tA)]dt$$

$$= \exp[(-tA^*)W\exp(-tA)]|_\infty^0 = W,$$

because $\lim_{t\to\infty} \exp(-tA) = 0$, i.e., K must be a solution of the so-called Lyapunov equation,

$$A^*K + KA = W,$$

where W is Hermitian positive definite. Conversely, let $\mathbf{x}, \lambda$ be an eigensolution—i.e., let $A\mathbf{x} = \lambda\mathbf{x}$. Then $\mathbf{x}^*A^* = \bar{\lambda}\mathbf{x}^*$, and if $W = A^*K + KA$, where K and W are s.p.d., we obtain

$$\mathbf{x}^*W\mathbf{x} = (\bar{\lambda} + \lambda)\mathbf{x}^*K\mathbf{x},$$

so $\text{Re}(\lambda) > 0$. Since λ is an arbitrary eigenvalue, $\text{Re}(\lambda) > 0$ for any eigenvalue.
$\diamond$

A useful concept in practice is positive definiteness in $\mathbb{R}^n$.

Let A be real and note that

$$(A\mathbf{x}, \mathbf{x}) = (\mathbf{x}, A\mathbf{x}) = (\mathbf{x}, 1/2[A + A^T]\mathbf{x}) \quad \text{if } \mathbf{x} \in \mathbb{R}^n.$$

Recall that A is called positive definite if $(A\mathbf{x}, \mathbf{x}) > 0 \ \forall \mathbf{x} \neq \mathbf{0}, \ \mathbf{x} \in \mathbb{R}^n$, or equivalently, its symmetric part is positive definite. A comparison theorem for real positive definite matrices follows.

Lemma 3.5

(a) *Let* A, B *be real, symmetric, and positive definite of order n, and let* $(\mathbf{x}, A\mathbf{x}) \geq (\mathbf{x}, B\mathbf{x}) \ \forall \mathbf{x} \in \mathbb{R}^n$. *Then*

$$(\mathbf{x}, B^{-1}\mathbf{x}) \geq (\mathbf{x}, A^{-1}\mathbf{x}) \qquad \forall \mathbf{x} \in \mathbb{R}^n.$$

In particular, if $(\mathbf{x}, A\mathbf{x}) \geq (\mathbf{x}, \mathbf{x})$ $\forall \mathbf{x} \in \mathbb{R}^n$, then

$$(\mathbf{x}, A^{-1}\mathbf{x}) \leq (\mathbf{x}, \mathbf{x}) \qquad \forall \mathbf{x} \in \mathbb{R}^n.$$

(b) Let A be real and have order n and let $A^S = 1/2(A + A^T)$ denote the symmetric part of A. Then, for any real positive definite matrix A,

$$0 \leq (\mathbf{x}, (A^{-1})^S\mathbf{x}) \leq (\mathbf{x}, (A^S)^{-1}\mathbf{x}) \qquad \forall \mathbf{x} \in \mathbb{R}^n.$$

Proof

(a) Note first that since A is s.p.d., its square root $A^{\frac{1}{2}}$ exists and

$$(\mathbf{x}, A\mathbf{x}) \geq (\mathbf{x}, \mathbf{x}) \implies (\mathbf{y}, \mathbf{y}) \geq (\mathbf{y}, A^{-1}\mathbf{y}), \forall \mathbf{y} = A^{\frac{1}{2}}\mathbf{x}, \mathbf{x} \in \mathbb{R}^n.$$

Hence, using the notation $A \underset{\text{p.d.}}{\geq} I$, if $A - I$ is positive semidefinite, we see that $I \underset{\text{p.d.}}{\geq} A^{-1}$. Similarly, $A \underset{\text{p.d.}}{\geq} B \implies I \underset{\text{p.d.}}{\geq} A^{-\frac{1}{2}}BA^{-\frac{1}{2}}$, that is, $A^{\frac{1}{2}}B^{-1}A^{\frac{1}{2}} \underset{\text{p.d.}}{\geq} I$, which shows that $B^{-1} \underset{\text{p.d.}}{\geq} A^{-1}$.

(b) Let $A^Q = 1/2(A^T - A)$ and note that $(A^Q)^T = -A^Q$, that is, A^Q is skew symmetric. We have

$$\left(\mathbf{x}, \left[(A^{-1})^S\right]^{-1}\mathbf{x}\right) = \left(\mathbf{x}, \left[\tfrac{1}{2}(A^{-1} + A^{-T})\right]^{-1}\mathbf{x}\right)$$

$$= \left(\mathbf{x}, A^T\left[\tfrac{1}{2}(A + A^T)\right]^{-1}A\mathbf{x}\right)$$

$$= (\mathbf{x}, (A^S + A^Q)(A^S)^{-1}(A^S - A^Q)\mathbf{x})$$

$$= (\mathbf{x}, A^S\mathbf{x}) - (\mathbf{x}, A^Q(A^S)^{-1}A^Q\mathbf{x})$$

$$= (\mathbf{x}, A^S\mathbf{x}) + (A^Q\mathbf{x}, (A^S)^{-1}A^Q\mathbf{x}) \geq (\mathbf{x}, A^S\mathbf{x}) \qquad \forall \mathbf{x} \in \mathbb{R}^n,$$

because $(A^S\mathbf{x}, \mathbf{x}) > 0$, $\forall \mathbf{x} \neq 0$ shows that the eigenvalues of A^S are positive—that is, the eigenvalues of $(A^S)^{-1}$ are positive, so $((A^S)^{-1}\mathbf{x}, \mathbf{x}) > 0$ $\forall \mathbf{x} \neq 0$. Hence,

$$A^S \underset{\text{p.d.}}{\leq} [(A^{-1})^S]^{-1}$$

and, by (a), $(A^{-1})^S \underset{\text{p.d.}}{\leq} (A^S)^{-1}$. $\diamond$

For symmetric matrices, we can give more detailed information about the eigenvalues than that in Theorem 3.2. Note, first, that if A is an arbitrary matrix, then A^*A is selfadjoint. Furthermore, since $(A^*Ax, x) = (Ax, Ax) = ||Ax||_2^2 \geq 0 \ \forall x$, A^*A is positive semidefinite. Hence, the eigenvalues of A^*A are nonnegative. If, in addition, A is nonsingular, then A^*A is positive definite, because then $||Ax|| = 0$ implies $Ax = 0$ or $x = 0$, since A^{-1} exists. Hence, there can be no eigenvalue equal to zero.

Theorem 3.6 *Let A be a real symmetric matrix of order n. Then A is positive definite if and only if:*

> *(a) $(Ax, x) > 0 \ \forall x \neq 0$, $x \in \mathbb{R}^n$.*
> *(b) The eigenvalues of all principal submatrices are positive.*
> *(c) The first principal submatrices $\widehat{A}_k, k = 1, 2, \ldots$ have positive determinants (called principal minors).*
> *(d) All the pivots u_{ii} in the LU factorization of A satisfy $u_{ii} > 0$.*
> *(e) There exists a nonsingular matrix C such that $A = CC^T$.*

Proof Condition (a) defines a positive definite matrix. We prove (a) $\Rightarrow$ (b) $\Rightarrow$ (c) $\Rightarrow$ (d) $\Rightarrow$ (e) $\Rightarrow$ (a). Let $\widehat{A}_k$ be a principal submatrix of A of order k, and let $x^T = [x_k^T, 0]$, $x_k \in \mathbb{R}^k$. Then

$$(Ax, x) = (\widehat{A}_k x_k, x_k), \quad k = 1, 2, \ldots, n.$$

Hence, if A is positive definite, so is $\widehat{A}_k$. By Theorem 3.2, all eigenvalues $\lambda_i^{(k)}$ of $\widehat{A}_k$ are positive; this proves (a) $\Rightarrow$ (b). Since $\det(\widehat{A}_k) = \prod_{i=1}^k \lambda_i^{(k)}$, (b) $\Rightarrow$ (c). Because all principal minors are positive, Theorem 1.1 shows that there is a triangular factorization $\widehat{A}_k = \widehat{L}_k \widehat{U}_k$, and $\det(\widehat{A}_k) > 0$, $\det(\widehat{L}_k) = 1 \Rightarrow \det(\widehat{U}_k) > 0$. But the diagonal entries of $\widehat{U}_k$ are the pivots d_i. Hence, by induction, $d_1 > 0 \Rightarrow d_2 > 0$ etc., so $A = LDL^T$, where D is diagonal and positive definite. Thus (c) $\Rightarrow$ (d) and by letting $C = LD^{\frac{1}{2}}$, (d) $\Rightarrow$ (e). Finally if $A = CC^T$, where C is nonsingular, then

$$(Ax, x) = ||C^Tx||_2^2 > 0 \quad \forall x \neq 0. \qquad \diamond$$

This theorem holds also in the complex case for the Hermitian matrices $A = A^*$. For semidefinite matrices the relations are as shown in the following theorem.

Theorem 3.7 *Let A be a real symmetric matrix of order n. Then A is positive semidefinite if and only if:*

(a) $(Ax, x) \geq 0 \ \forall x \in \mathbb{R}^n$.

(b) $A + \varepsilon I$ is positive definite for all $\varepsilon > 0$.

(c) The eigenvalues of all principal submatrices (i.e., not only the first principal submatrices) are nonnegative.

(d) The principal submatrices have nonnegative determinants.

(e) When A is properly permuted, $\tilde{A} = PAP^T$, there exists a triangular decomposition $\tilde{A} = LDL^T$, where D is a diagonal matrix with nonnegative entries and L is lower triangular with unit diagonal entries.

(f) There exists a square matrix C such that $\tilde{A} = CC^T$.

Proof The proof follows the proof of the previous theorem and is left as an exercise.   ◇

For a list of many more properties that are equivalent to positive (semi) definiteness, see Fiedler and Pták (1966).

We note from Theorem 3.7 that if $\mathrm{rank}(A) = r$, then the quadratic form of A is a sum of r squares,

$$(Ax, x) = (\Lambda Qx, Qx) = \sum_{i=1}^{n} ||\lambda_i^{\frac{1}{2}} q_i x||_2^2,$$

where exactly r of the λ_i are $\neq 0$, where Q is unitary, $QAQ^* = \Lambda = \mathrm{diag}(\lambda_i)$, and q_i is the ith column of Q. Note also that in this theorem it does not suffice in (c) to consider only the first principal submatrices $\widehat{A}_k$, as was the case in the previous theorem. The simple example $A = \begin{bmatrix} 0 & 0 \\ 0 & -1 \end{bmatrix}$ shows this, where $\det(\widehat{A}_1) = \det(\widehat{A}_2) = 0$, but where A is negative semidefinite. Hence, we have to assume that (c) holds, or (which is equivalent) that the principal submatrices $\widehat{A}_k$ of $\tilde{A} = PAP^T$, where P is any permutation matrix, have nonnegative eigenvalues.

3.2 Schur Complements

Consider a matrix partitioned into two-by-two block form

(3.1)
$$A = \begin{bmatrix} A_{11} & A_{12} \\ A_{21} & A_{22} \end{bmatrix},$$

where A_{ii}, $i = 1, 2$ are square matrices.

Definition 3.4 If A_{11} is nonsingular, we define

$$S \equiv A/A_{11} \equiv A_{22} - A_{21} A_{11}^{-1} A_{12},$$

which is called the *Schur complement* (of A with respect to A_{11}).

The term Schur complement seems to have first appeared in Haynsworth (1968). A survey paper by Quellette (1981) contains many references, and the book by Fiedler (1986) gives some recent developments in Schur complements.

Note that S is the matrix that results when we eliminate the block A_{21} using block Gaussian elimination with A_{11} as pivot block. In fact, the block matrix triangular factorization of A is readily found to be

$$(3.2) \qquad A = \begin{bmatrix} I & 0 \\ A_{21} A_{11}^{-1} & I \end{bmatrix} \begin{bmatrix} A_{11} & 0 \\ 0 & S \end{bmatrix} \begin{bmatrix} I & A_{11}^{-1} A_{12} \\ 0 & I \end{bmatrix}$$

From this we find the following forms of the inverse of A (Banachiewicz, 1937):

$$A^{-1} = \begin{bmatrix} I & -A_{11}^{-1} A_{12} \\ 0 & I \end{bmatrix} \begin{bmatrix} A_{11}^{-1} & 0 \\ 0 & S^{-1} \end{bmatrix} \begin{bmatrix} I & 0 \\ -A_{21} A_{11}^{-1} & I \end{bmatrix}$$

or

$$(3.3) \qquad A^{-1} = \begin{bmatrix} A_{11}^{-1} + A_{11}^{-1} A_{12} S^{-1} A_{21} A_{11}^{-1} & -A_{11}^{-1} A_{12} S^{-1} \\ -S^{-1} A_{21} A_{11}^{-1} & S^{-1} \end{bmatrix}.$$

If both A_{11}^{-1} and A_{22}^{-1} exist, an alternative expression for A^{-1} is

$$(3.4) \qquad A^{-1} = \begin{bmatrix} S_1^{-1} & -A_{11}^{-1} A_{12} S_2^{-1} \\ -S_2^{-1} A_{21} A_{11}^{-1} & S_2^{-1} \end{bmatrix},$$

where

$$S_i = A/A_{jj}, \quad i \neq j, \ i, j = 1, 2.$$

This is shown by permuting A or using the relation

$$(A_{11} - A_{12} A_{22}^{-1} A_{21})(I + A_{11}^{-1} A_{12} S_2^{-1} A_{21}) A_{11}^{-1} = I,$$

which an elementary computation reveals. The following relation establishes a major characterization of Schur complements for symmetric or Hermitian positive definite matrices.

Theorem 3.8 *Let A be Hermitian positive definite, A_{11} be regular, and $\mathbf{x} = \begin{bmatrix} \mathbf{x}_1 \\ \mathbf{x}_2 \end{bmatrix}$ be a partitioning of $\mathbf{x}$ consistent with the partitioning of A. Then*

    (a) A_{11} and A_{22} are Hermitian and positive definite.
    (b) $\mathbf{x}^* A\mathbf{x} \geq \mathbf{x}_2^* S\mathbf{x}_2$.
    (c) For any $\mathbf{x}_2$, $\min_{\mathbf{x}_1} \mathbf{x}^* A\mathbf{x} = \mathbf{x}_2^* S\mathbf{x}_2$.
    (d) $\min_{\mathbf{x}, \mathbf{x}_2 \neq 0} \mathbf{x}^* A\mathbf{x} = \min_{\mathbf{x}_2 \neq 0} \mathbf{x}_2^* S\mathbf{x}_2$.

Proof To prove (a) note that $\mathbf{x}^* A\mathbf{x} = \mathbf{x}_1^* A_{11}\mathbf{x}_1$ if $\mathbf{x}_2 = 0$ and if $\mathbf{x} = (\mathbf{x}_1, \mathbf{x}_2)$ is partitioned consistently with the partitioning of A. Hence, it follows that $\mathbf{x}_1^* A_{11}\mathbf{x}_1 > 0 \ \forall \mathbf{x}_1 \neq \mathbf{0}$, so A_{11} is positive definite. Similarly A_{22} is positive definite. Next, (3.2) and $A_{12}^* = A_{21}$ show that

$$\mathbf{x}^* A\mathbf{x} = (\mathbf{x}_1 + A_{11}^{-1} A_{12}\mathbf{x}_2)^* A_{11}(\mathbf{x}_1 + A_{11}^{-1} A_{12}\mathbf{x}_2) + \mathbf{x}_2^* S\mathbf{x}_2 \geq \mathbf{x}_2^* S\mathbf{x}_2,$$

which shows (b). (An alternative proof using matrix algebra: Show that $A = \begin{bmatrix} 0 & 0 \\ 0 & S \end{bmatrix} + Q^* Q$, where $Q = [A_{11}^{\frac{1}{2}}, A_{11}^{-\frac{1}{2}} A_{12}]$.)

Parts (c) and (d) follow from the above by choosing $\mathbf{x}_1 = -A_{11}^{-1} A_{12}\mathbf{x}_2$.   ◇

Corollary 3.8′ *Let A be Hermitian and partitioned in the form (3.1). Then A is positive definite if and only if A_{ii} and $S_i = A/A_{ii}$, $i = 1$, or $i = 2$ are positive definite.*

Proof Theorem 3.5(a,d) shows the "only if" part. Conversely, if A_{ii} and S_i are positive definite for $i = 1$, then the proof of Theorem 3.5 shows that 3.5(d) holds—i.e., A is positive definite. The corresponding result holds for $i = 2$. ◇

Theorem 3.8(d) shows in particular that if A is Hermitian and positive definite, then S is Hermitian and positive definite. In fact we even have a corresponding statement for positive definite, but not necessarily symmetric matrices.

Theorem 3.9 *Let A be real and positive definite and partitioned in a two-by-two block matrix form. Then*

    (a) The inverse of A is positive definite.
    (b) A_{11} and A_{22} are positive definite.
    (c) The Schur complements $S_i = A/A_{jj}$, $i \neq j$, $i, j = 1, 2$ are positive definite.

Proof To prove part (a) we use

$$x^T A^{-1} x = y^T A^T y,$$

where $y = A^{-1}x$. Hence, if

$$y^T A^T y > 0 \quad \forall y \neq 0, \quad x = Ay,$$

then $x^T A^{-1} x > 0 \ \forall x \neq 0$. Since $y^T Ay > 0 \ \forall y \neq 0, y \in \mathbb{R}^n$ implies $y^T A^T y > 0 \ \forall y \neq 0$, the above shows that if A is positive definite, then A^{-1} is also positive definite.

Next note that A_{ii} and S_i are real matrices. Further, since

$$x^T Ax = x_1^T A_{11} x_1 \forall x = (x_1, 0), x \in \mathbb{R}^n$$

it follows that A_{11} is positive definite. Similarly, A_{22} is positive definite, which proves part (b). Next it follows by (b) that the Schur complement matrices S_i exist and, since A is positive definite, its inverse in the form (3.4) of A^{-1} exists. Hence part (b) shows that the diagonal blocks S_i^{-1} are positive definite, so part (a) shows that S_i are also positive definite. $\diamond$

Finally, note that this theorem extends the corresponding well-known theorem for symmetric or Hermitian matrices (see Corollary 3.8'). It follows from Theorem 3.9(c) that when we use Gaussian elimination on a positive definite matrix, the pivot (block matrix) entries are always positive (positive definite). This implies a certain 'stability' of the elimination process.

3.3 Condition Numbers

In subsequent chapters, we shall frequently be using the following concept.

Definition 3.5 Let A be s.p.d. and let

$$\lambda_{\min}(A) \quad \text{and} \quad \lambda_{\max}(A)$$

denote the extreme eigenvalues of A. Then, by the *condition number* of A we mean

$$\text{cond}(A) = \lambda_{\max}(A)/\lambda_{\min}(A).$$

Remark 3.10 Note that Theorem 3.2 shows that the eigenvalues of an s.p.d. matrix are positive. A general definition of condition numbers of matrices based on matrix norms can be found in Appendix A. The above condition number is actually the spectral condition number, corresponding to the Euclidean norm. When no ambiguity can arise, we call this simply the condition number. The following relations will be of further interest.

Lemma 3.11 (Kantorovič inequality, 1948) *For any s.p.d. matrix B, the following inequalities are valid:*

$$(3.5) \qquad 1 \le \frac{\mathbf{x}^T B \mathbf{x} \mathbf{x}^T B^{-1} \mathbf{x}}{(\mathbf{x}^T \mathbf{x})^2} \le \frac{[1 + \text{cond}(B)]^2}{4 \,\text{cond}(B)} \quad \forall \mathbf{x} \in \mathbb{R}^n.$$

Moreover, the upper bound is sharp—i.e.,

$$(3.6) \qquad \sup_{\mathbf{x} \ne 0} \frac{\mathbf{x}^T B \mathbf{x} \mathbf{x}^T B^{-1} \mathbf{x}}{(\mathbf{x}^T \mathbf{x})^2} = \frac{[1 + \text{cond}(B)]^2}{4 \,\text{cond}(B)}.$$

Proof Let B have order n, let $\lambda_1 \le \lambda_2 \le \ldots \le \lambda_n$ be the eigenvalues of B, and let $\mathbf{v}_1, \mathbf{v}_2, \ldots, \mathbf{v}_n$ be the corresponding orthonormal eigenvectors. Consider first the following expression:

$$\phi(\lambda) = \lambda^2 - (\lambda_1 + \lambda_n)\lambda + \lambda_1 \lambda_n, \quad \lambda_1 \le \lambda \le \lambda_n.$$

Note that λ_1, λ_n are roots of $\phi(\lambda)$ and that $\phi(\lambda) \le 0$, $\lambda_1 \le \lambda \le \lambda_n$. Hence, Theorem 3.2 shows that the matrix $\phi(B)$ and, hence, also the matrix

$$B - (\lambda_1 + \lambda_n)I + \lambda_1 \lambda_n B^{-1}$$

are negative semi-definite—i.e.,

$$(3.7) \qquad \mathbf{x}^T[B - (\lambda_1 + \lambda_n)I + \lambda_1 \lambda_n B^{-1}]\mathbf{x} \le 0 \qquad \forall \mathbf{x}.$$

Consider now the real function

$$f(\lambda) = (\mathbf{x}^T B \mathbf{x})\lambda^2 - (\lambda_1 + \lambda_n)\mathbf{x}^T \mathbf{x}\lambda + \lambda_1 \lambda_n \mathbf{x}^T B^{-1} \mathbf{x},$$

where $\mathbf{x} \ne 0$ is arbitrary. We have

$$f(0) = \lambda_1 \lambda_n \mathbf{x}^T B^{-1} \mathbf{x},$$

so $f(0) > 0$, since B^{-1} is positive definite. Further, by (3.7), $f(1) \le 0$. Hence, $f(\lambda) = 0$ has a real root, which shows that the discriminant is nonnegative:

$$(\lambda_1 + \lambda_n)^2 (\mathbf{x}^T \mathbf{x})^2 - 4\lambda_1 \lambda_n \mathbf{x}^T B \mathbf{x} \mathbf{x}^T B^{-1} \mathbf{x} \ge 0.$$

Hence,

$$\frac{\mathbf{x}^T B \mathbf{x} \mathbf{x}^T B^{-1} \mathbf{x}}{(\mathbf{x}^T \mathbf{x})^2} \le \frac{(1 + \lambda_n/\lambda_1)^2}{4\lambda_n/\lambda_1},$$

which is the upper bound part of (3.5).

To show (3.6), let $\mathbf{x} = \mathbf{v}_1 + \mathbf{v}_n$. Then, using the orthogonality of the normalized eigenvectors, we find

$$\mathbf{x}^T B \mathbf{x} = \lambda_1 + \lambda_n, \quad \mathbf{x}^T B^{-1} \mathbf{x} = \lambda_1^{-1} + \lambda_n^{-1}, \quad (\mathbf{x}^T \mathbf{x})^2 = 4,$$

and (3.6) follows by an elementary computation.

And, finally, the lower bound part of (3.5) can be shown by writing $\mathbf{x} = \sum_{i=1}^{n} \eta_i \mathbf{v}_i$, $\eta_i = \mathbf{x}^T \mathbf{v}_i$, $\xi_i = \eta_i^2$ and noting that $\mathbf{x}^T B \mathbf{x} = \sum_1^n \xi_i \lambda_i$, $\mathbf{x}^T B^{-1} \mathbf{x} = \sum_1^n \xi_i \lambda_i^{-1}$. Then, by the Cauchy inequality (noting that $\xi_i \ge 0$ and $\lambda_i > 0$), it follows readily that

$$\sum_1^n \xi_i \lambda_i \sum_1^n \xi_i \lambda_i^{-1} \Big/ \sum_1^n \xi_i^2 \ge 1. \qquad \diamond$$

The following important relation between the condition number of A in (3.1) and that of its Schur complement S is valid.

Lemma 3.12 *Let A be s.p.d. and partitioned into the form (3.1), where A_{11} is regular, and let S be the Schur complement, $A_{22} - A_{21} A_{11}^{-1} A_{12}$. Then*

(a) $\lambda_{\min}(A) \le \lambda_{\min}(A_{11}) \le \lambda_{\max}(A_{11}) \le \lambda_{\max}(A)$ *and, hence,* $\text{cond}(A_{11}) \le \text{cond}(A)$.

(b) $\lambda_{\min}(A) \le \lambda_{\min}(S) \le \lambda_{\max}(S) \le \lambda_{\max}(A)$ *and, hence,* $\text{cond}(S) \le \text{cond}(A)$.

Proof Note first that for a s.p.d. matrix A there is an orthogonal matrix V, such that $V^T A V = \Lambda = \text{diag}(\lambda_1, \ldots, \lambda_n)$, where λ_i are the eigenvalues of A. Hence, with $\mathbf{y} = V^T \mathbf{x}$,

$$\mathbf{x}^T A \mathbf{x} = \mathbf{y}^T \Lambda \mathbf{y} = \sum_{i=1}^{n} \lambda_i y_i^2.$$

This shows

$$\lambda_{\min}(A) = \min_{\mathbf{x} \neq 0} \frac{\mathbf{x}^T A \mathbf{x}}{\mathbf{x}^T \mathbf{x}}, \quad \lambda_{\max}(A) = \max_{\mathbf{x} \neq 0} \frac{\mathbf{x}^T A \mathbf{x}}{\mathbf{x}^T \mathbf{x}}.$$

We have

$$\frac{x^T A x}{x^T x} = \frac{x_1^T A_{11} x_1}{x_1^T x_1}, \quad \text{for any } x = (x_1, 0).$$

Hence, in particular,

$$(3.8) \qquad \lambda_{min}(A) \le \frac{\hat{x}_1^T A_{11} \hat{x}_1}{\hat{x}_1^T \hat{x}_1} = \lambda_{min}(A_{11}),$$

where $\hat{x}_1$ is the eigenvector to A_{11} for $\lambda_{min}(A_{11})$. Similarly,

$$\lambda_{max}(A) \ge \lambda_{max}(A_{11}),$$

which together with (3.8) shows part (a). Similarly, (see the proof of Theorem 3.8) we have

$$\lambda_{min}(A) \le \frac{x^T A x}{x^T x} = \frac{x_2^T S x_2}{x_1^T x_1 + x_2^T x_2} \le \frac{x_2^T S x_2}{x_2^T x_2}$$

for any $x = (x_1, x_2)$, where $x_1 = -A_{11}^{-1} A_{12} x_2$ and x_2 is arbitrary. Hence,

$$\lambda_{min}(A) \le \lambda_{min}(S).$$

Similarly, noting that $A^{-1} = \begin{bmatrix} * & * \\ * & S^{-1} \end{bmatrix}$, we have by (3.8) $\lambda_{min}(A^{-1}) \le \lambda_{min}(S^{-1})$, that is,

$$\lambda_{max}(A) \ge \lambda_{max}(S),$$

which shows part (b).   ◇

Using a proper permutation of A, it can be seen that $\text{cond}(A_{11}) \le \text{cond}(A)$ holds when A_{11} is replaced by any principal submatrix.

As is shown in Appendix A, errors such as rounding errors can be amplified by the condition number of the matrix. It is therefore important to know that, as Lemma 3.12(b) shows, the condition number of the Schur complement of a s.p.d. matrix is never greater than the condition number of the given matrix.

3.4* Estimates of Eigenvalues of Generalized Eigenvalue Problems

We show next some estimates of eigenvalues of the generalized eigenvalue problem

(3.9) $$\lambda C\mathbf{x} = A\mathbf{x},$$

where C is s.p.d. and A is symmetric and positive semidefinite. These will be useful, for instance, when estimating eigenvalues and condition numbers of so-called preconditioned matrices in Chapter 10. Basic for these estimates is the following lemma.

Lemma 3.13 (Courant-Fischer) *Let A be Hermitian of order n and let $\mathbf{p}_1, \mathbf{p}_2, \ldots, \mathbf{p}_n$ be n orthonormal vectors. Further, let $\lambda_1 \le \lambda_2 \le \ldots \le \lambda_n$ be the eigenvalues of A and $\mathbf{v}_1, \mathbf{v}_2, \ldots, \mathbf{v}_n$ the corresponding orthonormal eigenvectors. Then:*

(a) $$\min_{\substack{\mathbf{x}^*\mathbf{x}=1 \\ \mathbf{x}\perp\mathbf{p}_1,\ldots,\mathbf{p}_{s-1}}} \mathbf{x}^* A\mathbf{x} \le \lambda_s$$
(b) $$\max_{\substack{\mathbf{x}^*\mathbf{x}=1 \\ \mathbf{x}\perp\mathbf{p}_n,\mathbf{p}_{n-1},\ldots,\mathbf{p}_{s+1}}} \mathbf{x}^* A\mathbf{x} \ge \lambda_s$$

and

(c) $$\max_{\substack{\mathbf{p}_i,\ 1\le i\le s-1 \\ \mathbf{p}_i^*\mathbf{p}_j=\delta_{ij}}} \min_{\substack{\mathbf{x}^*\mathbf{x}=1 \\ \mathbf{x}\perp\mathbf{p}_1,\ldots,\mathbf{p}_{s-1}}} \mathbf{x}^* A\mathbf{x} = \lambda_s$$
(d) $$\min_{\substack{\mathbf{p}_i,\ s+1\le i\le n \\ \mathbf{p}_i^*\mathbf{p}_j=\delta_{ij}}} \max_{\substack{\mathbf{x}^*\mathbf{x}=1 \\ \mathbf{x}\perp\mathbf{p}_n,\ldots,\mathbf{p}_{s+1}}} \mathbf{x}^* A\mathbf{x} = \lambda_s$$

and the extreme values are attained when $\mathbf{p}_i = \mathbf{v}_i$, $i = 1, \ldots, s - 1$, and $\mathbf{p}_i = \mathbf{v}_i$, $i = n, \ldots, s + 1$, respectively.

Proof Let $S = [\mathbf{v}_1, \mathbf{v}_2, \ldots, \mathbf{v}_n]$. Then $S^*S = I$, $SS^* = I$,

$$S^* A S = \Lambda = \operatorname{diag}(\lambda_1, \ldots, \lambda_n),$$

and

$$\mathbf{x}^* A\mathbf{x} = \mathbf{y}^* \Lambda \mathbf{y} = \sum_{i=1}^{n} \lambda_i |y_i|^2,$$

where $\mathbf{y} = S^*\mathbf{x}$, or $\mathbf{x} = S\mathbf{y}$. Note that $\mathbf{y}^*\mathbf{y} = \mathbf{x}^* SS^*\mathbf{x} = \mathbf{x}^*\mathbf{x} = 1$. Also $\mathbf{x} \perp \mathbf{p}_1, \ldots, \mathbf{p}_{s-1} \iff \mathbf{y} \perp \mathbf{q}_1 \ldots \mathbf{q}_{s-1}$ where $\mathbf{q}_i = S^*\mathbf{p}_i$ and $\mathbf{q}_i^*\mathbf{q}_j = \delta_{ij}$.
We then have

$$\mathbf{x}^* A\mathbf{x} = \lambda_s - [(\lambda_s - \lambda_1)|y_1|^2 + \ldots + (\lambda_s - \lambda_{s-1})|y_{s-1}|^2]$$

$$+ [(\lambda_{s+1} - \lambda_s)|y_{s+1}|^2 + \ldots + (\lambda_n - \lambda_s)|y_n|^2].$$

Since the second term is nonpositive, the third term nonnegative, and since $\Sigma|y_i|^2 = 1$, we see that $\min \mathbf{x}^* A\mathbf{x}$ is taken for some vector $\mathbf{x} = S\mathbf{y}$, where

$y_{s+1} = \ldots = y_n = 0$. Restricting $\mathbf{y}$ in this way leaves still s free parameters, which suffices to satisfy the constraints $\mathbf{x} \perp \mathbf{q}_1, \ldots, \mathbf{q}_{s-1}$. This shows

$$\min_{\substack{\mathbf{x} \perp \mathbf{p}_1, \ldots, \mathbf{p}_{s-1} \\ \mathbf{x}^*\mathbf{x}=1}} \mathbf{x}^* A \mathbf{x} \leq \lambda_s,$$

which is (a). If $\mathbf{x} = \mathbf{v}_s$ and $\mathbf{p}_i = \mathbf{v}_i$, $i = 1, \ldots, s - 1$, then $\mathbf{x} \perp \mathbf{p}_1, \ldots, \mathbf{p}_{s-1}$ and $\mathbf{x}^* A \mathbf{x} = \lambda_s$. This and part (a) show that

$$\max_{\mathbf{p}_i, 1 \leq i \leq s-1} \quad \min_{\substack{\mathbf{x} \perp \mathbf{p}_i, 1 \leq i \leq s-1 \\ \mathbf{x}^*\mathbf{x}=1}} \mathbf{x}^* A \mathbf{x} = \lambda_s,$$

which is (c). Parts (b) and (d) are proved analogously. ◇

Corollary 3.14 *Let A and B be symmetric matrices, and let $\lambda_i(A)$, $\lambda_i(A + B)$ denote the ith eigenvalue, where we order the eigenvalues in an increasing order. Then*

(a) $\lambda_i(A) + \lambda_{\min}(B) \leq \lambda_i(A + B) \leq \lambda_i(A) + \lambda_{\max}(B)$.
(b) If $\lambda_{\max}(B)$ is nonnegative and A is positive definite, then

$$\lambda_i(AB) \leq \lambda_i(A)\lambda_{\max}(B).$$

(c) If $\lambda_{\min}(B)$ is nonnegative and A is positive definite, then

$$\lambda_i(AB) \geq \lambda_i(A)\lambda_{\min}(B).$$

Proof Let $\mathbf{v}_1, \ldots, \mathbf{v}_n$ denote the eigenvectors of A. The Courant-Fischer lemma shows that for any $\mathbf{x}$, $\mathbf{x}^T\mathbf{x} = 1$,

$$\lambda_i(A + B) \geq \min_{\mathbf{x} \perp \mathbf{v}_1, \ldots, \mathbf{v}_{i-1}} \mathbf{x}^T(A + B)\mathbf{x}$$

$$\geq \min_{\mathbf{x} \perp \mathbf{v}_1, \ldots, \mathbf{v}_{i-1}} \mathbf{x}^T A \mathbf{x} + \min_{\mathbf{x}^T\mathbf{x}=1} \mathbf{x}^T B \mathbf{x}$$

$$= \lambda_i(A) + \lambda_{\min}(B),$$

which shows the lower bound of (a). Similarly,

$$\lambda_i(AB) = \lambda_i(A^{\frac{1}{2}}BA^{\frac{1}{2}})$$

$$\leq \max_{\mathbf{x} \perp \mathbf{v}_n,\ldots,\mathbf{v}_{i+1}} \left\{ \frac{\mathbf{x}^T A^{\frac{1}{2}} B A^{\frac{1}{2}} \mathbf{x}}{\mathbf{x}^T A \mathbf{x}} \cdot \frac{\mathbf{x}^T A \mathbf{x}}{\mathbf{x}^T \mathbf{x}} \right\}$$

$$\leq \max_{\mathbf{x} \neq 0} \frac{(A^{\frac{1}{2}}\mathbf{x})^T B (A^{\frac{1}{2}}\mathbf{x})}{(A^{\frac{1}{2}}\mathbf{x})^T (A^{\frac{1}{2}}\mathbf{x})} \max_{\mathbf{x} \perp \mathbf{v}_n,\ldots,\mathbf{v}_{i+1}} \frac{\mathbf{x}^T A \mathbf{x}}{\mathbf{x}^T \mathbf{x}},$$

$$= \lambda_i(A)\lambda_{\max}(B),$$

which shows (b). The proofs of (c) and the upper bound of (a) follow the same lines.   ◇

Consider now the generalized eigenvalue problem (3.9). Note that since C is s.p.d., the eigenvalues of (3.9) are identical to the eigenvalues $\lambda_i(C^{-1}A)$. We order the eigenvalues in increasing order.

Theorem 3.15 *Let* $R = C - A$, *where* R, C, A *are matrices of order n, and assume that*

$$\lambda_{\min}(R) \leq 0 \leq \lambda_{\max}(R).$$

Then

(3.10)
$$1 - \lambda_{n-i+1}(C^{-1})\lambda_{\max}(R) \leq \lambda_i(C^{-1}A)$$
$$\leq 1 + \lambda_i(C^{-1})\lambda_{\max}(-R), i = 1, 2, \ldots, n$$

and, in particular,

$$\lambda_{\max}(C^{-1}A) \leq 1 + \lambda_{\max}(-R)/\lambda_{\min}(C).$$

Proof Since $A = C - R$, we have

$$C^{-1}A = I - C^{-1}R = I + C^{-1}(-R).$$

Hence, Corollary 3.14(b) shows that

$$\lambda_i(C^{-1}A) = 1 + \lambda_i(C^{-1}(-R)) \leq 1 + \lambda_i(C^{-1})\lambda_{\max}(-R),$$

where we have used the relation

$$\lambda_{\max}(-R) = -\lambda_{\min}(R),$$

this quantity being nonnegative by assumption. Similarly,

$$\lambda_i(C^{-1}A) = 1 - \lambda_{n-i+1}(C^{-1}R) \geq 1 - \lambda_{n-i+1}(C^{-1})\lambda_{\max}(R). \qquad \diamond$$

Consider next the matrix $\lambda C - A$, as a function of λ. This can be written in the form

$$(3.11) \qquad \lambda C - A = \left(\frac{\lambda}{\mu} - 1\right) A + \frac{\lambda}{\mu} R_\mu,$$

where μ is a positive number, $R_\mu = \mu C - A$, and where we let $0 < \lambda < \mu$. This decomposition of $\lambda C - A$ can be used to show the following alternative bounds on $\lambda_i(C^{-1}A)$.

Theorem 3.16 *Assume that μ is sufficiently large so that $\lambda_{\min}(\mu C - A) \geq 0$. Then*

$$(3.12) \qquad \frac{\mu\lambda_i(A)}{\lambda_i(A) + \lambda_{\max}(\mu C - A)} \leq \lambda_i(C^{-1}A) \leq \frac{\mu\lambda_i(A)}{\lambda_i(A) + \lambda_{\min}(\mu C - A)}.$$

In particular, assuming that $\lambda_{\min}(R) \geq 0$, for $\mu = 1$ we have

$$(3.13) \qquad \frac{\lambda_i(A)}{\lambda_i(A) + \lambda_{\max}(R)} \leq \lambda_i(C^{-1}A) \leq \frac{\lambda_i(A)}{\lambda_i(A) + \lambda_{\min}(R)}$$

and, as $\mu \to \infty$, we find

$$(3.14) \qquad \lambda_i(A)/\lambda_{\max}(C) \leq \lambda_i(C^{-1}A) \leq \lambda_i(A)/\lambda_{\min}(C).$$

Proof Corollary 3.14(a) and (3.11) show that for any λ, $0 < \lambda < \mu$,

$$\lambda_{n-i+1}(\lambda C - A) \leq \left(1 - \frac{\lambda}{\mu}\right) \lambda_{n-i+1}(-A) + \frac{\lambda}{\mu}\lambda_{\max}(R_\mu)$$

$$(3.15)$$

$$= -\left(1 - \frac{\lambda}{\mu}\right) \lambda_i(A) + \frac{\lambda}{\mu}\lambda_{\max}(R_\mu).$$

Note now that $\lambda_{n-i+1}(\lambda C - A) = 0$ if and only if $\lambda = \lambda_i(C^{-1}A)$. Also note that the right-hand part of (3.15) has a zero, $\underline{\lambda}_i$, in the interval $(0, \mu)$, where

$$\underline{\lambda}_i = \mu\lambda_i(A)[\lambda_i(A) + \lambda_{\max}(R_\mu)]^{-1}.$$

It is readily seen that this quantity is a lower bound of $\lambda_i(C^{-1}A)$. Similarly,

$$
\lambda_{n-i+1}(\lambda C - A) \geq \left(1 - \frac{\lambda}{\mu}\right)\lambda_{n-i+1}(-A) + \frac{\lambda}{\mu}\lambda_{\min}(R_\mu)
$$

(3.16)

$$
= -\left(1 - \frac{\lambda}{\mu}\right)\lambda_i(A) + \frac{\lambda}{\mu}\lambda_{\min}(R_\mu),
$$

and the right-hand part of (3.16) has a zero, $\overline{\lambda}_i$, where

$$
\overline{\lambda}_i = \mu\lambda_i(A)[\lambda_i(A) + \lambda_{\min}(R_\mu)]^{-1},
$$

and this is an upper bound of $\lambda_i(C^{-1}A)$. The remaining statements follow at once. $\diamond$

The next theorem is helpful when in a generalized eigenvalue problem (3.9) both matrices A and C are singular with a common vector in their nullspaces.

Theorem 3.17 *Let A and C, of order n, be singular, and assume that there is a nontrivial vector $\mathbf{v}$ in $\mathcal{N}(A) \cap \mathcal{N}(C)$. Assume that $v_i \neq 0$ for some i, $1 \leq i \leq n$. Then, for any eigenvector $\mathbf{x} \neq c\mathbf{v}$ (for any constant c) of the generalized eigenvalue problem,*

$$
\lambda C\mathbf{x} = A\mathbf{x},
$$

the corresponding eigenvalue λ is also an eigenvalue of the generalized eigenvalue problem

$$
\lambda\widehat{C}\hat{\mathbf{x}} = \widehat{A}\hat{\mathbf{x}},
$$

where $\widehat{C}$ and $\widehat{A}$ are obtained from C and A by deleting the ith rows and ith columns.

Proof Let $\mathbf{x} = \tilde{\mathbf{x}} + (x_i/v_i)\mathbf{v}$. Then $\tilde{x}_i = 0$ and $A\mathbf{x} = \widetilde{A}\tilde{\mathbf{x}}$, $C\mathbf{x} = \widetilde{C}\tilde{\mathbf{x}}$, where $\widetilde{A}$, $\widetilde{C}$ are obtained from A, C by zeroing out the ith columns. Neglecting the ith rows we then get

(3.17) $$\lambda\widehat{C}\hat{\mathbf{x}} = \widehat{A}\hat{\mathbf{x}},$$

where $\hat{\mathbf{x}} \neq \mathbf{0}$, because $\mathbf{x} \neq c\mathbf{v}$. Hence, λ is an eigenvalue of the generalized eigenvalue problem (3.17). $\diamond$

Note that $\widehat{A}, \widehat{C}$ have order $n - 1$. If $\widehat{A}, \widehat{C}$ are singular with a nonzero vector in $\mathcal{N}(\widehat{A}) \cap \mathcal{N}(\widehat{C})$, then the above reduction process can be repeated, and eventually we obtain matrices with $\mathcal{N}(\widehat{A}) \cap \mathcal{N}(\widehat{C}) = \{0\}$, containing just the zero vector.

Generalized eigenvalue problems of the above type occur for finite element matrices—for instance, where the matrices A, C correspond to a lower (say linear) and higher (say quadratic) polynomial approximation, respectively. For details, see Axelsson and Barker (1984).

Remark 3.18 Theorem 3.16 relates the eigenvalues of $C^{-1}A$ to those of C and to A. These relations will turn out to be useful when we analyze the condition numbers and eigenvalue distributions of preconditioned matrices in Chapter 10. For further information regarding generalized eigenvalue problems, see Chapter 9.

3.5 Congruence Transformations

We recall from Chapter 2 that a similarity transformation $S^{-1}AS$ preserves the eigenvalues of A. We now consider a transformation that preserves the sign of the eigenvalues.

Definition 3.6 Let A be a square matrix and let C be nonsingular and of the same order as A. Then $C^T AC$ is called a *congruence transformation* of A.

Note that if C is orthogonal—i.e., if $C^T C = I$, then $C^T AC$ is both a congruence transformation and a similarity transformation. We shall now see that a congruence transformation preserves the signs of the eigenvalues.

Theorem 3.19 *Assume that A is symmetric, and let C be nonsingular. Then $C^T AC$ has the same number of positive eigenvalues, the same number of negative eigenvalues, and the same number of zero eigenvalues as A.*

Proof We may assume that A is nonsingular, because otherwise we consider $A + \varepsilon I$ (which is nonsingular if ε is nonzero but $|\varepsilon|$ is small enough) and let $\varepsilon \to 0$. Recall the property that the eigenvalues depend continuously on the coefficients of the matrix (see Appendix A), and let

$$C(t) = tC + (1 - t)Q,$$

where Q is orthogonal, $Q^T Q = I$; that makes $C(t)$ nonsingular for $0 \leq t \leq 1$. (The existence of such a Q is established below). Then $C(0) = Q$ and

$C(1) = C$. The eigenvalues of $C(t)^T A C(t)$ depend continuously on t as t increases from 0 to 1. Since $C(t)$ is nonsingular, none of these eigenvalues can become zero (or change sign). Therefore, the number of positive eigenvalues and the number of negative ones is the same for $C(t)^T A C(t)$ for all t in the interval $0 \leq t \leq 1$. In particular, this is true for $C^T A C$ $(t = 1)$ and $Q^T A Q = Q^{-1} A Q$ $(t = 0)$. But the latter matrix has the same eigenvalues (by similarity equivalence) as A.

It remains to prove that it is possible to choose an orthogonal matrix such that $C(t) = tC + (1 - t)Q$ is nonsingular for $0 \leq t \leq 1$. To this end, let $Q = [\mathbf{q}_1, \mathbf{q}_2, \ldots, \mathbf{q}_n]$ be orthogonal to the columns of $C = [\mathbf{c}_1, \mathbf{c}_2, \ldots, \mathbf{c}_n]$ (i.e., $\mathbf{q}_i^T \mathbf{c}_j = 0$ for $i > j$), so $Q^T C$ is triangular. This is equivalent to the construction of the Gram-Schmidt process. By a proper choice of signs $\pm \mathbf{q}_i$, the diagonal of $Q^T C$ is positive. Hence, with this choice of Q, $C(t) = Q[t Q^T C + (1 - t)I]$ is nonsingular, since both factors are so. $\diamond$

We now apply this theorem to the signs of the pivots for a symmetric matrix.

Theorem 3.20 *For any symmetric matrix A properly permuted to allow for $r = \text{rank}(A)$ nonzero pivot entries, the signs of the pivots agree with the signs of the eigenvalues—that is, the eigenvalue matrix Λ and the pivot matrix D have the same number of positive entries, negative entries, and zero entries.*

Proof Since A is symmetric, there is an orthogonal matrix Q such that $A = Q \Lambda Q^T$. By the triangular factorization of A, (see Chapter 1) we have $A = L D L^T$, where $\text{diag}(L) = I$. Hence,

$$D = L^{-1} A (L^T)^{-1} = L^{-1} Q \Lambda Q^T (L^T)^{-1},$$

so $D = C^T \Lambda C$, where $C = (L^{-1} Q)^T$, and Theorem 3.19 shows that the signs of the eigenvalues of D and Λ are the same. $\diamond$

Note that Theorem 3.20 shows, in particular, that the pivot entries of a symmetric and nonsingular matrix are nonzero. It shows also that the LU factorization of a square matrix can be used when determining the number of positive, negative, and zero eigenvalues of a matrix. As has been shown in Chapter 1, in the case of zero eigenvalues we must use permutations of the matrix to get nonzero pivot entries as long as possible.

3.6 Quasisymmetric Matrices

Definition 3.7 A matrix A is said to be *quasisymmetric* if it is similar to a symmetric matrix.

We shall consider a class of nonsymmetric matrices A that are similarly equivalent to symmetric matrices by a diagonal transformation D—that is, $D^{-1}AD$ is symmetric. We call such matrices *diagonally quasisymmetric*. Since the diagonal of $D^{-1}AD$ is the same as that of A, in the following discussion we may assume with no loss of generality that there are only zeroes on the diagonal of A.

Example 3.3 Let

$$A = \begin{bmatrix} 0 & b_1 & & & \emptyset \\ c_1 & 0 & b_2 & & \\ & c_2 & 0 & \ddots & \\ & & \ddots & \ddots & b_{n-1} \\ \emptyset & & & c_{n-1} & 0 \end{bmatrix}$$

be a tridiagonal matrix where $b_i c_i > 0$, $i = 1, 2, \ldots, n-1$, that is, b_i and c_i are nonzero numbers with the same sign, and let $\tilde{A} = D^{-1}AD$. Then the entries $\tilde{a}_{ij}$ of $\tilde{A}$ satisfy $\tilde{a}_{ij} = 0$, $|i - j| \neq 1$, and

$$\tilde{a}_{i,i+1} = d_i^{-1} b_i d_{i+1}, \, \tilde{a}_{i+1,i} = d_{i+1}^{-1} c_i d_i,$$

so

$$\tilde{a}_{i,i+1} = \tilde{a}_{i+1,i} \Rightarrow d_{i+1}^2 = \frac{c_i}{b_i} d_i^2.$$

Letting $d_i > 0$, $d_1 = 1$, we get

$$d_{i+1} = \sqrt{\frac{c_i}{b_i}} \, d_i, \quad i = 1, 2, \ldots, n-1 \text{ and}$$

$$D^{-1}AD = \begin{bmatrix} 0 & v_1\sqrt{b_1 c_1} & & & \emptyset \\ v_1\sqrt{b_1 c_1} & 0 & v_2\sqrt{b_2 c_2} & & \\ & v_2\sqrt{b_2 c_2} & 0 & \ddots & \\ & & \ddots & \ddots & \\ \emptyset & & & v_{n-1}\sqrt{b_{n-1} c_{n-1}} & 0 \end{bmatrix},$$

where $v_i = \begin{cases} 1 & \text{if } b_i > 0 \\ -1 & \text{if } b_i < 0 \end{cases}$.

Assume now that A is diagonally quasisymmetric—i.e., there is a diagonal matrix D such that $D^{-1}AD$ is symmetric. Consider the block-tridiagonal matrix

$$\mathcal{A} = \begin{bmatrix} A & b_1 I & & & \emptyset \\ c_1 I & A & b_2 I & & \\ & c_2 I & A & \ddots & \\ & & \ddots & \ddots & b_{n-1} I \\ \emptyset & & & c_{n-1} I & A \end{bmatrix}.$$

Let $d_{i+1} = \sqrt{c_i/b_i}\, d_i$, and let $i = 1, 2, \ldots, n-1$, and let

$$\mathcal{D} = \begin{bmatrix} d_1 D & & & \emptyset \\ & d_2 D & & \\ & & \ddots & \\ \emptyset & & & d_n D \end{bmatrix}.$$

Then

$$\mathcal{D}^{-1}\mathcal{A}\mathcal{D} = \begin{bmatrix} \tilde{A} & v_1\sqrt{b_1 c_1}\, I & & & \emptyset \\ v_1\sqrt{b_1 c_1}\, I & \tilde{A} & v_2\sqrt{b_2 c_2}\, I & & \\ & v_2\sqrt{b_2 c_2}\, I & \tilde{A} & \ddots & \\ & & \ddots & \ddots & v_{n-1}\sqrt{b_{n-1} c_{n-1}}\, I \\ \emptyset & & & v_{n-1}\sqrt{b_{n-1} c_{n-1}}\, I & \tilde{A} \end{bmatrix}$$

where $\tilde{A} = D^{-1}AD$, which is symmetric. Hence $\mathcal{D}^{-1}\mathcal{A}\mathcal{D}$ is symmetric, so $\mathcal{A}$ is quasisymmetric.

Example 3.4 Consider the central difference approximation matrix A for the second-order elliptic problem $-u_{xx} - u_{yy} + a(x)u_x + b(y)u_y$ on the unit square $(0, 1)^2$, where the stepsize $h \leq 2/\max\{|a|, |b|\}$. Here a does not depend on y and b does not depend on x. Show that if one uses a natural ordering of the meshpoints along lines parallel to the x-axis, then A becomes quasisymmetric. Hence, the eigenvalues of A are real.

Exercises

1. Prove that the following matrix is positive definite:

$$A = \begin{bmatrix} 1 & 1+i & -1 \\ 1-i & 6 & -3+i \\ -1 & -3-i & 11 \end{bmatrix}.$$

2. (a) Let $A = \begin{bmatrix} 2 & -1 & 0 \\ -1 & 2 & -1 \\ 0 & -1 & 2 \end{bmatrix}$. Factor A in the form LDL^T and use this decomposition to prove that

$$\mathbf{x}^T A \mathbf{x} = 2(x_1 - \frac{1}{2}x_2)^2 + \frac{3}{2}(x_2 - \frac{2}{3}x_3)^2 + \frac{4}{3}x_3^2.$$

Use this to show that A is positive definite.

(b) Let A be a matrix with all entries equal to one. Show that A can be written as $A = \mathbf{e}\mathbf{e}^T$, where $\mathbf{e}^T = (1, 1, \ldots, 1)$, and that A is positive semidefinite. Find the eigensolutions of A.

3. (a) Let $A = \begin{bmatrix} 2 & -1 & 0 \\ -1 & 2 & -1 \\ 0 & -1 & 0 \end{bmatrix}$. Using the LDL^T factorization of A and Theorem 3.20, show that A has two positive and one negative eigenvalue.

(b) Using the LDL^T factorization in (a), find the smallest value of ε for which $A = \begin{bmatrix} 2 & -1 & 0 \\ -1 & 2 & -1 \\ 0 & -1 & \varepsilon \end{bmatrix}$ becomes positive semidefinite.

4. Let $A = \begin{bmatrix} 2 & -1 & & & \\ -1 & 2 & -1 & & 0 \\ & \ddots & \ddots & \ddots & \\ & & -1 & 2 & -1 \\ 0 & & & -1 & 0.8 & -1 \\ & & & & -1 & 1 \end{bmatrix}$ have order 10. Show that A

has 9 positive and one negative eigenvalues.

5. Let A be a real matrix.

(a) If A is positive definite and C is real and nonsingular, show that $B = C^T A C$ is positive definite.

(b) If A is symmetric and positive definite, show that A^k, where k is an arbitrary integer (positive or negative), is positive definite.

6. Let A and B be two positive definite Hermitian matrices. Show that

$$C = A^{-1} + B^{-1} - 4(A + B)^{-1}$$

is Hermitian and positive semidefinite.

Hint: Consider first the matrices

$$\widetilde{C} = A^{\frac{1}{2}} C A^{\frac{1}{2}}, \quad \widetilde{B} = A^{-\frac{1}{2}} B A^{-\frac{1}{2}},$$

and show that the problem can be reduced to

$$\widetilde{C} = I + \widetilde{B}^{-1} - 4(I + \widetilde{B})^{-1}.$$

Since $\widetilde{B}$ is Hermitian and positive definite, it suffices to consider an arbitrary eigensolution λ, $\mathbf{x}$ of $\widetilde{B}$, (i.e., $\widetilde{B}\mathbf{x} = \lambda\mathbf{x}$). Show that then

$$\widetilde{C}\mathbf{x} = \mu\mathbf{x},$$

where $\mu = 1 + \lambda^{-1} - 4(1 + \lambda)^{-1}$, and that μ is positive except when $\lambda = 1$ ($\mu = 0$).

7. Let A be symmetric and positive definite. Prove that the Rayleigh's quotient,

$$R(\mathbf{x}) = \frac{(A\mathbf{x}, \mathbf{x})}{(\mathbf{x}, \mathbf{x})}, \qquad \mathbf{x} \neq \mathbf{0},$$

is minimized by an eigenvector $\mathbf{v}_1$ corresponding to the smallest eigenvalue λ_1, and

$$\min_{\mathbf{x} \in R^n} R(\mathbf{x}) = \frac{(A\mathbf{v}_1, \mathbf{v}_1)}{(\mathbf{v}_1, \mathbf{v}_1)} = \lambda_1 \quad \text{(Rayleigh's principle)}.$$

8. Find the minimum, if there is one, of
   (a) $R_1(x, y, z) = 2x^2 - 2xy + 2y^2 - 2xz + z^2 - 3x + 4y - 2z$.
   (b) $R_2(x, y) = x^2 - 2xy + y^2$.
   Hint: Write R_1 and R_2 in the form $\mathbf{x}^T A \mathbf{x}$.

9. (a) Assume that A is reducible to a diagonal matrix. Show that

$$\det(\exp(A)) = e^{\text{tr}(A)}.$$

   (b) Show that

$$\exp(A + B) = \exp(A)\exp(B)$$

   if and only if $AB = BA$.
   Hint: Consider the power series of $\exp(tA), \exp(tB)$, and $\exp(t(A + B))$. For example,

$$\exp(tA) = I + tA + \frac{1}{2}t^2 A^2 + \cdots.$$

Assuming that

$$\exp(tA)\exp(tB) = \exp(t(A+B)),$$

form the matrix product

$$(I + tA + \frac{1}{2}t^2A^2 + \cdots)(I + tB + \frac{1}{2}t^2B^2 + \cdots)$$

and equate the coefficients of t^2 with that of the corresponding term of $\exp[t(A+B)]$ to obtain

$$\frac{1}{2}(A+B)^2 = AB + \frac{1}{2}A^2 + \frac{1}{2}B^2.$$

It follows that $AB = BA$. Conversely, if $AB = BA$, then B commutes with $\exp(tA)$ and

$$\frac{d}{dt}\left[\exp(tA)\exp(tB)\right] = (A+B)\exp(tA)\exp(tB).$$

Thus,

$$X(t) = \exp(tA)\exp(tB) \quad \text{solves } X'(t) = (A+B)X(t), \quad X(0) = I,$$

and this shows that

$$X(t) = \exp[t(A+B)].$$

10. Let $\mathbf{a}, \mathbf{b}$ be given nonzero vectors in $\mathbb{R}^n$.
    (a) Let L be the line $t\mathbf{a}$, $-\infty < t < \infty$ (i.e., $\mathbf{a}$ lies in L). Show that the shortest distance from $\mathbf{b}$ to the line L is

    $$\|\mathbf{b} - \frac{(\mathbf{a}, \mathbf{b})}{(\mathbf{a}, \mathbf{a})}\mathbf{a}\|^2 = \frac{(\mathbf{b}, \mathbf{b})(\mathbf{a}, \mathbf{a}) - (\mathbf{a}, \mathbf{b})^2}{(\mathbf{a}, \mathbf{a})}$$

    and use this result to derive the Schwarz inequality:

    $$|(\mathbf{a}, \mathbf{b})| \leq (\mathbf{a}, \mathbf{a})^{\frac{1}{2}}(\mathbf{b}, \mathbf{b})^{\frac{1}{2}}.$$

    (b) Show that $(\mathbf{a}, \mathbf{b}) = \cos\theta[(\mathbf{a}, \mathbf{a})(\mathbf{b}, \mathbf{b})]^{\frac{1}{2}}$, where θ is the angle between $\mathbf{a}$ and $\mathbf{b}$.
    (c) Let $\mathbf{a} \in \mathbb{R}^n$. Show that by making a proper choice of $\mathbf{b}$,

    $$(|a_1| + |a_2| + \cdots + |a_n|)^2 = n|\cos\theta|(|a_1|^2 + |a_2|^2 + \cdots + |a_n|^2)$$
    $$\leq n(\sum |a_i|^2).$$

(d)   Determine when equality holds in the above inequalities.

11.   *A matrix trace inequality* (cf. Yang, 1988): Show that if A and B are two Hermitian positive definite matrices, then

(a) $\text{tr}(AB) > 0$.

(b) $[\text{tr}(AB)]^{\frac{1}{2}} \leq \frac{1}{2}[\varepsilon\ \text{tr}(A) + \frac{1}{\varepsilon}\text{tr}(B)]$, $\varepsilon > 0$.

Hint: Let U be a unitary matrix such that

$$U^*AU = D = \text{diag}(d_1, d_2, \ldots, d_n).$$

Then $\text{tr}(A) = \text{tr}(D)$ and

$$\text{tr}(AB) = \text{tr}(U^*ABU) = \text{tr}(DC),$$

where $C = U^*BU$. Note that C is Hermitian positive definite; hence, its diagonal entries c_{ii} are real and positive and $\text{tr}(C) = \text{tr}(B)$. Note that the diagonal entries of DC are $d_i c_{ii}$, so (a) follows from

$$\text{tr}(DC) = \sum d_i c_{ii} > 0.$$

Next note that for any sequences $\{a_i\}$, $\{b_i\}$ of positive numbers,

$$\left(\sum a_i b_i\right)^{\frac{1}{2}} \leq \frac{1}{2}\sum(a_i + b_i)$$

or, more generally,

$$\left(\sum a_i b_i\right)^{\frac{1}{2}} \leq \frac{1}{2}\sum(\varepsilon a_i + \frac{1}{\varepsilon}b_i),$$

where $\varepsilon > 0$. Now $a_i = d_i$, $b_i = c_{ii}$ leads to (b).

12.   *Cholesky decomposition:* Let A be s.p.d. Then a decomposition of A, $A = LL^T$, where L is lower triangular, is called a Cholesky decomposition of A (see Benoit, 1924; Zurmühl, 1961; and Doolittle, 1878).

(a)   Let A be partitioned in 2×2 block form, and assume that A is s.p.d. Then the Cholesky decomposition of A exists,

$$\begin{bmatrix} A_{11} & A_{12} \\ A_{21} & A_{22} \end{bmatrix} = \begin{bmatrix} U_{11}^T & 0 \\ U_{12}^T & U_{22}^T \end{bmatrix} \cdot \begin{bmatrix} U_{11} & U_{12} \\ 0 & U_{22} \end{bmatrix},$$

where $U_{11}^T U_{11} = A_{11}$ and $U_{12} = U_{11}^{-T} A_{12}$. Show that $S = U_{22}^T U_{22}$, where $S = A_{22} - A_{21}A_{11}^{-1}A_{12}$ is the Schur complement of A w.r.t. A_{11}.

(b)   Show that

$$\begin{bmatrix} A_{11} & A_{12} \\ A_{21} & A_{22} \end{bmatrix} = \begin{bmatrix} U_{11}^T & 0 \\ U_{12}^T & I \end{bmatrix} \begin{bmatrix} U_{11} & U_{12} \\ 0 & S \end{bmatrix}.$$

(c)   Compute the Cholesky decomposition of the matrix in Exercise 3.

13.   (a)   Let A be a square matrix partitioned in the form

$$A = \begin{bmatrix} A_{11} & A_{12} \\ A_{21} & A_{22} \end{bmatrix}$$

and assume that A_{11} and $S = A_{22} - A_{21} A_{11}^{-1} A_{12}$ are nonsingular. Show that

$$\det(A) = \det(A_{11})\det(S).$$

Hint: Let $B = \begin{bmatrix} A_{11} & 0 \\ A_{21} & S \end{bmatrix}$ and show that $B^{-1}A = \begin{bmatrix} I & * \\ 0 & I \end{bmatrix}$. (Here $*$ indicates a matrix, irrelevant for the present discussion.)

(b)   Show that the matrix $D - \mathbf{u}\mathbf{u}^T$, where D is diagonal of order n and $\mathbf{u} \in \mathbb{R}^n$, is positive definite if and only if $d_i > 0$, $i = 1, \ldots, n$ and $\sum_{k=1}^{n} u_k^2/d_k < 1$.
Hint: Use Exercise 2.24.

14.   *Matrix factorizations:* This and previous chapters have introduced the matrix decompositions mentioned below. Identify the matrix classes to which they are applicable:

(a)   The Cholesky factorization.

(b)   The LDL^T factorization, where L is lower triangular, with unit diagonal, and
   (i) D is diagonal.
   (ii) D is symmetric and tridiagonal (cf. Aasen, 1971).

(c) LU (or $LD\tilde{U}$) factorization, where L (and $\tilde{U}$) have unit diagonals and U is upper triangular.

(d) QR factorization, where Q is orthogonal and R is upper triangular with nonnegative diagonal entries (see Exercise 2.27).

(e) $U^*AU = D$, where U is unitary and D is diagonal.

(f) $S^{-1}AS = D$, where S is nonsingular and D is diagonal.

(g) $U^*AU = R$, where U is unitary and R is upper triangular.

15.   *The singular value decomposition* (see, for example, Golub and Van Loan, 1989, and Björck, 1989):

(a) A generalization of Schur's Lemma: Using an induction proof similar to that of Schur's Lemma (Theorem 2.15), follow steps 1–4 below to show that to any $m \times n$ matrix there exist unitary matrices U (of order m) and V (of order n) such that $A = UDV^*$. Here D has the form

$$D = \begin{bmatrix} D_r & 0 \\ 0 & 0 \end{bmatrix},$$

where D_r is real and diagonal, $D_r = \mathrm{diag}(\sigma_1, \sigma_2, \ldots, \sigma_r)$, and $\sigma_1 \geq \sigma_2 \geq \ldots \geq \sigma_r > 0$. The diagonal entries σ_i of D_r are called the *singular values* of A.

(i) Let v_1 be the normalized eigenvector of A^*A corresponding to the eigenvalue $\|A\|_2^2 = \rho(A^*A)$ of A^*A and also let $u_1 = \sigma_1^{-1} A v_1$, where $\sigma_1 = \|A\|_2$. Show that $\|u_1\|_2 = 1$ and $u_1^* A v_1 = \sigma_1$.

(ii) Write the unitary matrices V, U in the form $V = [v_1, V_1]$ and $U = [u_1, U_1]$. Show that $U_1^* A v_1 = \sigma_1 U_1^* u_1 = 0$ and that $u_1^* A V_1 = \sigma_1 v_1^* V_1 = 0$.

(iii) Show that $A_1 \equiv U^* A V$ has the form

$$A_1 = \begin{bmatrix} \sigma_1 & 0 \\ 0 & B \end{bmatrix},$$

where $B = U_1^* A V_1$ and B has order $(m-1) \times (n-1)$.

(iv) Proceed by induction to show that $A = UDV^*$.

(b) Show that $A^*A = VD^*DV^*$, $AA^* = U^*DD^*U$, and that the square of the singular values of a matrix A are the nonzero eigenvalues of the (Hermitian and positive semidefinite) matrices A^*A and AA^*. Show further that if $U = [u_1, \ldots, u_m]$, $V = [v_1, \ldots, v_n]$, then u_i and v_i are the corresponding eigenvectors. (These are called left and right *singular vectors* of A.)

(c) If A is a 2-by-2 matrix, show that the singular value decomposition

$$A = U \begin{bmatrix} \sigma_1 & 0 \\ 0 & \sigma_2 \end{bmatrix} V^*$$

leads to the following geometric result: the set

$$E = \{z : z = Ax, \; x \in \mathbb{R}^2, \; \|x\|_2 = 1\},$$

which defines an ellipse with semi-axes of length σ_1 and σ_2.

(d) Show that if $A = [\mathbf{a}_1, \mathbf{a}_2]$, $\mathbf{a}_i \in \mathbb{R}^2$, $\|\mathbf{a}_i\|_2 = 1$, and $\mathbf{a}_1^T \mathbf{a}_2 = \cos \phi$ ($0 < \phi < \pi$), then

$$A^T A = \begin{bmatrix} 1 & \cos \phi \\ \cos \phi & 1 \end{bmatrix},$$

which has eigenvalues $2\cos^2(\frac{1}{2}\phi)$, $2\sin^2(\frac{1}{2}\phi)$, so the singular values are $\sigma_1 = \sqrt{2}\cos(\frac{1}{2}\phi)$, $\sigma_2 = \sqrt{2}\sin(\frac{1}{2}\phi)$. Show that if $\phi \leq \sqrt{\varepsilon}_{fl}$, where ε_{fl} is the floating point precision, then the computed values in $A^T A$ will be equal to 1. It follows that the computed eigenvalues of $A^T A$ will be 2 and 0 and that the information of the smallest eigenvalue σ_2 is lost. Hence, forming $A^T A$ cannot be recommended in general for the purpose of computing singular values of A. In addition, even if A is sparse, $A^T A$ will be much less sparse (even full in exceptional cases).

(e) Another relation between the singular values of A and the eigenvalues of an associated Hermitian matrix: Show that if $A = UDV^*$, then

$$B = \begin{bmatrix} 0 & A^* \\ A & 0 \end{bmatrix} = \frac{1}{\sqrt{2}} \begin{bmatrix} V & V \\ U & -U \end{bmatrix} \begin{bmatrix} D & 0 \\ 0 & -D \end{bmatrix} \frac{1}{\sqrt{2}} \begin{bmatrix} V & V \\ U & -U \end{bmatrix}^*.$$

Show that the eigenvalues of B are $\pm\sigma_1, \pm\sigma_2, \ldots, \pm\sigma_r$ and 0 (with multiplicity $m + n - 2r$).

(f) It can be seen that the singular values of A are unique. The singular vectors are unique only when σ_j^2 is a simple eigenvalue of A^*A. For multiple singular values, the corresponding singular vectors can be chosen as any orthonormal basis for the unique subspace that they span. Show that once the singular vectors $\mathbf{v}_j$, $1 \leq j \leq r$ have been chosen, the vectors $\mathbf{u}_j$, $1 \leq j \leq r$ are uniquely determined from the relation

$$A\mathbf{v}_j = \sigma_j \mathbf{u}_j, \quad j = 1, \ldots, r.$$

Similarly, given $\mathbf{u}_j$, $1 \leq j \leq r$, the vectors $\mathbf{v}_j$, $1 \leq j \leq r$ are uniquely determined from

$$A^*\mathbf{u}_j = \sigma_j \mathbf{v}_j, \quad j = 1, \ldots, r.$$

(g) Show that the *singular value decomposition* $A = UDV^*$ implies that $A = \sum_{i=1}^r \sigma_i \mathbf{u}_i \mathbf{v}_i^*$.

16. If M is positive definite and symmetric, show that for $\varepsilon > 0$,

(a) $B(\varepsilon M + B^* B)^{-1} = (\varepsilon I + B M^{-1} B^*)^{-1} B M^{-1}$,

(b) $B(\varepsilon M + B^* B)^{-1} B^* = I - \varepsilon(\varepsilon I + B M^{-1} B^*)^{-1}$,

(c) $0 \le B(\varepsilon M + B^* B)^{-1} B^* < I$, and

(d) $0 < (I + \frac{1}{\varepsilon} B M^{-1} B^*)^{-1} \le I$ where the inequalities are to be interpreted in a positive (semi) definite sense.

Hint: For the proof of (a) use the Sherman-Morrison formula (see Exercise 2.26) or, alternatively, use (3.3) by writing

$$\begin{bmatrix} A & F \\ C & E \end{bmatrix} = \begin{bmatrix} M & B^* \\ B & -\varepsilon I \end{bmatrix}^{-1},$$

where both F and C are evaluated independently and then equated. Part (b) follows readily from (a).

17. Let $B^\dagger$ denote a matrix (generalized inverse) such that $B B^\dagger B = B$. Show that $I - A(A^* A)^\dagger A^*$ is positive semi-definite.

Hint: Use the previous exercise with $M = I$ and let $\varepsilon \to 0$. It must be shown that

$$\lim_{\varepsilon \to 0} (\varepsilon I + A^* A)^{-1} = (A^* A)^\dagger,$$

i.e., that the limes matrix is a generalized inverse. Note that then

$$A^* A(\varepsilon I + A^* A)^{-1} A^* A = A^* A - \varepsilon A^* (\varepsilon I + A A^*)^{-1} A$$

$$= A^* A - \varepsilon(I - (I + \frac{1}{\varepsilon} A^* A)^{-1}) \to A^* A, \quad \varepsilon \to 0.$$

Remark: The above shows also that $(A^* A)^\dagger$ is symmetric.

18. *Constrained optimization problems:* In linearly equality constrained optimization problems, one encounters linear systems of the form

$$\begin{bmatrix} 0 & A \\ A^T & M \end{bmatrix} \begin{bmatrix} \mathbf{x} \\ \mathbf{y} \end{bmatrix} = \begin{bmatrix} \mathbf{a} \\ \mathbf{b} \end{bmatrix}.$$

Here A is $m \times n$ and M is symmetric and positive semidefinite. We assume that $M + A^T A$ is positive definite—i.e., that $\mathcal{N}(M) \cap \mathcal{N}(A) = \{0\}$. Consider the following solution methods:

(a) If M is positive definite and A has full-row rank, show that

$$A M^{-1} A^T \mathbf{x} = A M^{-1} \mathbf{b} - \mathbf{a}; \quad M \mathbf{y} = \mathbf{b} - A^T \mathbf{x}.$$

Here $A M^{-1} A^T$ is s.p.d. Treating A and M as full matrices, compute the operation count to form and solve these systems.

(b) Show that $\begin{bmatrix} \mathbf{x} \\ \mathbf{y} \end{bmatrix}$ is also a solution of

$$\begin{bmatrix} 0 & A \\ A^T & (M + \frac{1}{\varepsilon} A^T A) \end{bmatrix} \begin{bmatrix} \mathbf{x} \\ \mathbf{y} \end{bmatrix} = \begin{bmatrix} \mathbf{a} \\ \mathbf{b} + \frac{1}{\varepsilon} A^T \mathbf{a} \end{bmatrix}.$$

Here, by assumption, $M + \frac{1}{\varepsilon} A^T A$ is positive definite (even if M is singular). Solve this in the same way as in part (a) and compute the operation count.

Remark: $AM^{-1}A^T$ or $A(M + \frac{1}{\varepsilon} A^T A)^{-1} A^T$ need not be sparse, even if A and M are sparse matrices. It can even be a full matrix.

19. Consider the matrix $C = \begin{bmatrix} -\varepsilon I_m & A \\ A^T & M \end{bmatrix}$, where $\varepsilon > 0$, A is $m \times n$, and M is symmetric and positive semidefinite. Show that

(a) $C = \begin{bmatrix} I_m & 0 \\ -\frac{1}{\varepsilon} A^T & I \end{bmatrix} \begin{bmatrix} -\varepsilon I_m & 0 \\ 0 & (M + \frac{1}{\varepsilon} A^T A) \end{bmatrix} \begin{bmatrix} I_m & -\frac{1}{\varepsilon} A \\ 0 & I_n \end{bmatrix}$,

$C^{-1} = \begin{bmatrix} I_m & \frac{1}{\varepsilon} A \\ 0 & I_n \end{bmatrix} \begin{bmatrix} -\frac{1}{\varepsilon} I_m & 0 \\ 0 & (M + \frac{1}{\varepsilon} A^T A)^{-1} \end{bmatrix} \begin{bmatrix} I_m & 0 \\ \frac{1}{\varepsilon} A^T & I_n \end{bmatrix}$.

(b) $C^{-1} \begin{bmatrix} 0 & A \\ A^T & M \end{bmatrix} = \begin{bmatrix} A(\varepsilon M + A^T A)^{-1} A^T & 0 \\ \varepsilon(\varepsilon M + A^T A)^{-1} A^T & I_n \end{bmatrix}$.

(c) Consider the system

$$\begin{bmatrix} 0 & A \\ A^T & M \end{bmatrix} \begin{bmatrix} \mathbf{x} \\ \mathbf{y} \end{bmatrix} = \begin{bmatrix} \mathbf{a} \\ \mathbf{b} \end{bmatrix}.$$

Show that

$$\begin{bmatrix} A(\varepsilon M + A^T A)^{-1} A^T \mathbf{x} \\ \mathbf{y} + \varepsilon(\varepsilon M + A^T A)^{-1} A^T \mathbf{x} \end{bmatrix} = C^{-1} \begin{bmatrix} \mathbf{a} \\ \mathbf{b} \end{bmatrix},$$

where the solution vector $\mathbf{x}$ is computed first, and that the operation count of this method is identical to the method in Exercise 18(b).

20. (J. F. Maître) Let

$$B = \begin{bmatrix} 0 & A \\ A^T & M \end{bmatrix}, \quad B_\varepsilon = \begin{bmatrix} 0 & A \\ A^T & M_\varepsilon \end{bmatrix}, \quad \text{and} \quad C_\varepsilon = \begin{bmatrix} -\varepsilon I_m & A \\ A^T & M \end{bmatrix},$$

where $M_\varepsilon = M + \frac{1}{\varepsilon} A^T A$, $\varepsilon > 0$, and assume that $A M_\varepsilon^{-1} A^T$ is nonsingular. Show that

$$B^{-1} = B_\varepsilon^{-1} + \frac{1}{\varepsilon} \begin{bmatrix} I_m & 0 \\ 0 & 0 \end{bmatrix}.$$

while

$$B = C_\varepsilon + \varepsilon \begin{bmatrix} I_m & 0 \\ 0 & 0 \end{bmatrix}.$$

Hint: Exercise 18(b) shows that

$$B^{-1} \begin{bmatrix} \mathbf{a} \\ \mathbf{b} \end{bmatrix} = B_\varepsilon^{-1} \begin{bmatrix} \mathbf{a} \\ \mathbf{b} \end{bmatrix} + B_\varepsilon^{-1} \begin{bmatrix} 0 \\ \frac{1}{\varepsilon} A^T \mathbf{a} \end{bmatrix}.$$

But it is readily seen that the solution of

$$B_\varepsilon \begin{bmatrix} \mathbf{z} \\ \mathbf{w} \end{bmatrix} = \begin{bmatrix} 0 \\ \frac{1}{\varepsilon} A^T \mathbf{a} \end{bmatrix}$$

is

$$\mathbf{z} = \frac{1}{\varepsilon} \mathbf{a}, \quad \mathbf{w} = \mathbf{0}.$$

21. Let f_i, $1 \le i \le n$ be real functions that are continuous on the interval $[0, 1]$. Show that if these functions are linearly independent, then the matrix $A = [a_{ij}]$, where $a_{ij} = \int_0^1 f_i(x) f_j(x) dx$, is positive definite. (Such a matrix is called a *Gramian matrix*.)

22. Compute the Gramian matrix A_n for $f_i = x^{i-1}$, $i = 1, \ldots, n$. Compute the condition number of A_n for $n = 3$ and $n = 4$ using a computer algebra system, for instance. Try to estimate the rate with which $\mathrm{cond}(A_n)$ increases with n.

23. *Computations with complex matrices:* Given a matrix $C \in \mathbb{C}^{n \times n}$, let $C = A + iB$, where $A, B \in \mathbb{R}^{n \times n}$ and $i = \sqrt{-1}$.

    (a) By straightforward matrix computations show that

    $$C^{-1} = (A + BA^{-1}B)^{-1} - iA^{-1}B(A + BA^{-1}B)^{-1}.$$

    Hint: Show first that $(A + iB)(I - iA^{-1}B) = A + BA^{-1}B$.

    (b) Let $\mathbf{z} = \mathbf{x} + i\mathbf{y}$, where $\mathbf{x}, \mathbf{y} \in \mathbb{R}^n$. Show how $C^{-1}\mathbf{z}$ can be computed by solving two systems with the real matrix $A + BA^{-1}B$, two matrix vector multiplications with $A^{-1}B$, and two vector additions. One may assume that A^{-1} is first computed and then $A^{-1}B$ formed, or that $A^{-1}B$ is formed by solving n real linear systems.

    (c) Consider the following alternative method to compute $\mathbf{u} + i\mathbf{v} = C^{-1}\mathbf{z}$, $\mathbf{u}, \mathbf{v} \in \mathbb{R}^n$. Show that

    $$C(\mathbf{u} + i\mathbf{v}) = \mathbf{x} + i\mathbf{y}$$

can be rewritten

$$\begin{bmatrix} A & -B \\ B & A \end{bmatrix}\begin{bmatrix} \mathbf{u} \\ \mathbf{v} \end{bmatrix} = \begin{bmatrix} \mathbf{x} \\ \mathbf{y} \end{bmatrix},$$

and show that

(i) $(A + BA^{-1}B)\mathbf{u} = \mathbf{x} + BA^{-1}\mathbf{y}.$

(ii) $A\mathbf{v} = \mathbf{y} - B\mathbf{u}.$

Hence, show that using this method, we need solve only one system with $A + BA^{-1}B$ in addition to two other systems with A (this latter cost can be neglected if we have formed A^{-1}) plus some vector additions.

24. *Interlocking eigenvalues lemma* (see Loewner, 1934): Let A be symmetric of order n with eigenvalues $\lambda_1 \leq \lambda_2 \leq \ldots \leq \lambda_n$. Let $\mathbf{a}$ be any vector in $\mathbb{R}^n$ and denote the eigenvalues of the matrix $A + \mathbf{aa}^T$ by $\mu_i, i = 1, \ldots, n$, where $\mu_1 \leq \mu_2 \leq \ldots \leq \mu_n$. Show that $\lambda_1 \leq \mu_1 \leq \lambda_2 \leq \mu_2 \leq \ldots \leq \lambda_n \leq \mu_n$.

Hint: Write $\mathbf{a} = \sum \alpha_i \mathbf{v}_i$, where $A\mathbf{v}_i = \lambda_i \mathbf{v}_i$, $\mathbf{v}_i \neq \mathbf{0}$, and use the Courant-Fischer theorem.

25. Let

$$A = \begin{bmatrix} A_0 & \mathbf{v} \\ \mathbf{v}^T & a \end{bmatrix},$$

where A_0 is a Hermitian matrix of order $n - 1$ and $\mathbf{v} \in \mathbb{C}^{n-1}$. Show that the eigenvalues of A_0 separate the eigenvalues of A. Hence, the eigenvalues of the leading principal minor of order $n - 1$ of a Hermitian matrix interlock the eigenvalues of the $n \times n$ matrix.

26. Let $A \in \mathbb{C}^{m,n}$ have singular values

$$\sigma_1 \geq \sigma_2 \geq \ldots \geq \sigma_p \geq 0,\ p = \min(m, n),$$

and let S be a subspace of $\mathbb{C}^n$ of dimension $\dim(S)$. Show that

$$\sigma_i = \min_{\dim(S)=n-i+1} \max_{\substack{\mathbf{x} \in S \\ \mathbf{x} \neq 0}} \|A\mathbf{x}\|_2 / \|\mathbf{x}\|_2.$$

Hint: Use the relation between the singular values of A and the eigenvalues of A^*A, and the Courant-Fischer theorem.

27. (Mirsky, 1960) Let $M_k^{m,n}$ denote the set of matrices in $\mathbb{C}^{m,n}$ of rank k. Assume that $A \in M_r^{m,n}$ and let $B \in M_k^{m,n}$, $k < r$ be such that

$$\|A - B\|_2 \leq \|A - X\|_2 \text{ for all } X \in M_k^{m,n}.$$

Then, if $A = UDV^* = \sum_{i=1}^{r} \sigma_i \mathbf{u}_i \mathbf{v}_i^*$ is the singular value decomposition of A, we have

$$B = \sum_{i=1}^{k} \sigma_i \mathbf{u}_i \mathbf{v}_i^* \quad \text{and} \quad \|A - B\|_2 = \sigma_{k+1}.$$

Hint: Use the previous exercise to show first that $\sigma_{k+1} = \|A - B\|_2$ where B is as given.

28. (Marek and Žitný, 1986) Let $A^\dagger$ be the Moore-Penrose pseudoinverse operator (see Exercise 1.24). Show that:
    (a) $A^\dagger A$ and $A A^\dagger$ are orthogonal projections.
    (b) $(A^\dagger)^\dagger = A$.
    (c) $(AB)^\dagger = B^\dagger A^\dagger$.
    (d) $(A^*)^\dagger = (A^\dagger)^*$.
    (e) $A^\dagger = (A^*A)^\dagger A^* = A^*(AA^*)^\dagger$.
    (f) $(AA^*)^\dagger = (A^\dagger)^* A^\dagger$, $(A^*A)^\dagger = A^\dagger (A^\dagger)^*$.
    (g) $AA^\dagger = (AA^*)^\dagger AA^* = AA^*(AA^*)^\dagger$,
        $A^\dagger A = A^*A(A^*A)^\dagger = (A^*A)^\dagger A^*A$.

29. (Showalter, 1967) Let A be $m \times n$, where $m \geq n$, and let rank $(A) = n$. Let

$$A^\dagger = \int_0^\infty [\exp(-tA^*A)]A^* dt,$$

show that $A^\dagger A = I$ and that $A A^\dagger$ is a projection operator, and hence that $A^\dagger$ is a generalized inverse of A.

30. (Kantorovič inequality) Let A be s.p.d. and for any $\mathbf{x} \in \mathbb{R}^n$, $\|\mathbf{x}\|_2 = 1$, let $\xi_i = (\mathbf{x}^T \mathbf{v}_i)^2$, $\phi(\mathbf{x}) = \sum_{i=1}^{n} \xi_i \lambda_i$ and $\psi(\mathbf{x}) = \sum_{i=1}^{n} \xi_i \lambda_i^{-1}$, where $\mathbf{v}_1, \mathbf{v}_2, \ldots, \mathbf{v}_n$ are the orthonormal eigenvectors of A and $\lambda_1, \lambda_2, \ldots, \lambda_n$ are the corresponding eigenvalues. Show that:
    (a) $\sum_{i=1}^{n} \xi_i = 1$, $\phi(\mathbf{x}) = \mathbf{x}^T A\mathbf{x}$, $\psi(\mathbf{x}) = \mathbf{x}^T A^{-1}\mathbf{x}$.
    (b) $1 \leq \phi(\mathbf{x})\psi(\mathbf{x})$ (*Hint:* first consider the case $n = 2$).
    (c) $\phi(\mathbf{x}) + \lambda_1\lambda_n\psi(\mathbf{x}) \leq \lambda_1 + \lambda_n$.
    (d) $\phi(\mathbf{x}) = \lambda$ for some $\lambda \in [\lambda_1, \lambda_n]$.
    (e) $\phi(\mathbf{x})\psi(\mathbf{x}) \leq \phi(\mathbf{x})[\lambda_1 + \lambda_n - \phi(\mathbf{x})]/(\lambda_1\lambda_n) \leq (\lambda_1 + \lambda_n)^2/(4\lambda_1\lambda_n)$.
    This is a slightly different proof of the Kantorovič inequality than the one given in Lemma 3.11.

31. Let $A = -A^T$ (i.e., A is antisymmetric). Show that $U = \exp(A)$ is orthogonal and has determinant 1.

32. Let A be an arbitrary complex matrix of order n with eigenvalues $\lambda_1, \lambda_2, \ldots, \lambda_n$, where $\operatorname{Re} \lambda_1 \leq \operatorname{Re} \lambda_2 \leq \ldots \leq \operatorname{Re} \lambda_n$, and let A have a

complete eigenvector space. Find a necessary and sufficient condition for $\| \exp(tA) \|_2$ to be bounded for all $t \geq 0$.

33. (Kreiss, 1959). Let A be an arbitrary complex matrix of order n such that $\| \exp(tA) \|_2$ is uniformly bounded for all $t \geq 0$. Show that there exists a positive scalar K such that

$$\|(A - zI)^{-1}\|_2 \leq \frac{K}{\operatorname{Re} z}$$

for any complex number z with $\operatorname{Re} z > 0$.

34. Let $P_0(x) = 1$, $P(x) = \alpha_1 - x$, and

$$P_k(x) = (\alpha_k - x)P_{k-1}(x) - \beta_{k-1}P_{k-2}(x), \quad k = 2, 3, \ldots,$$

where β_j, $j = 1, 2, \ldots$ are positive numbers. Show that:

(a) $P_k(x) = \det(A_k)$, where

$$A_k = \begin{bmatrix} \alpha_1 - x & \beta_1 & & 0 \\ 1 & \alpha_2 - x & \beta_2 & \\ & \ddots & \ddots & \ddots \\ 0 & & 1 & \alpha_k - x \end{bmatrix}.$$

(b) All zeroes of $P_k(x)$ are real, using the diagonal similarity transformation $\widetilde{A} = DAD^{-1}$, with a proper diagonal matrix D, showing that A is diagonally quasisymmetric.

Remark: Orthogonal polynomials satisfy a recursion as given above. Part (b) therefore shows, in particular, that orthogonal polynomials have real zeroes.

References

J. O. Aasen (1971). On the reduction of a symmetric matrix to tridiagonal form. *BIT* **11**, 233–242.

O. Axelsson (1992). Bounds of eigenvalues of preconditioned matrices. *SIAM J. Matrix Anal.* **13**, 847–862.

O. Axelsson and V. A. Barker (1984). *Finite Element Solution of Boundary Value Problems, Theory and Computation*. Orlando, Fla.: Academic.

T. Banachiewicz (1937). Zur Berechnung der Determinatien, wie auch den Inversen und zur darauf basierten Auflösung der Systeme linairen Gleichungen. *Acta Astron. Ser.* **C3**, 41–67.

C. Benoit (1924). Note sur une méthode de résolution des équations normales . . . . *Procédés du Commandant Cholesky, Bulletin Géodésique* **2**, 67–77.

A. Berman and R. J. Plemmons (1979). *Nonnegative Matrices in the Mathematical Sciences*. New York: Academic.

Å. Björck (1989). Least squares methods. In *Handbook of Numerical Analysis*, Vol. I (ed. P. Ciarlet and P. Lions). Amsterdam: Elsevier/North Holland.

M. H. Doolittle (1878). U.S. Coast and Geodetic Report, 115–120.

M. Fiedler and V. Pták (1966). Some generalizations of positive definiteness and monotonicity. *Numer. Math.* **9**, 163–172.

M. Fiedler (1986). *Special Matrices and Their Applications in Numerical Mathematics*. Dordrecht, The Netherlands: Nijhoff.

G. Golub and C. F. van Loan (1989). *Matrix Computations*, 2nd ed. Baltimore, Md.: John Hopkins Univ. Press.

E. Haynsworth (1968). On the Schur complement. *Basel Math. Notes* **20**.

R. A. Horn and C. R. Johnson (1985). *Matrix Analysis*. Cambridge: Cambridge Univ. Press.

L. V. Kantorovič (1948). Funkcional'nyi analiz i prikladnaja matematika. *Uspehi Mat. Nauk* **3 (28)**, 9–185.

C. Loewner (1934). Über monotone Matrix Funktionen. *Math. Zeitschr.* **38**, 177–216.

I. Marek and K. Žitný (1986). *Matrix Analysis for Applied Sciences*, Vol. 2. Leipzig: Teubner-Texte zur Mathematik, Band 84.

L. Mirsky (1960). Symmetric gauge functions and unitarily invariant norms. *Quart. J. Math.* **11**, 50–59.

H. Moore (1920). On the reciprocal of the general algebraic matrix. *Bull. Amer. Math. Soc.* **26**, 394–395.

R. Penrose (1956). On best approximate solution of linear matrix equations. *Proc. Cambridge Phil. Soc.* **52**, 17–19.

D. V. Quelette (1981). Schur complements and statistics. *Lin. Alg. Appl.* **36**, 187–295.

D. Showalter (1967). Representation and computation of the pseudoinverse. *Proc. Amer. Math. Soc.* **18**, 584–586.

G. Strang (1976). *Linear Algebra and Its Applications*. Orlando, Fla.: Academic.

Y. Yang (1988). A matrix trace inequality. *J. Math. Anal. Applic.* **133**, 573–574.

R. Zurmühl (1961). Matrizen und Ihre Technische Anwendungen. Berlin: Springer Verlag.

4

Reducible and Irreducible Matrices and the Perron-Frobenius Theory for Nonnegative Matrices

Various topics of matrix theory, in particular, those related to nonnegative matrices (matrices with nonnegative entries) are considered in this chapter. We introduce the concepts of reducible and irreducible matrices and matrix graph theory (the concepts of directed and strongly connected graphs), and show the equivalence between irreducibility of a matrix and the connectivity of its directed graph. This enables us, among other things, to strengthen the Gershgorin theorem for estimating the location of eigenvalues of irreducible matrices. In order to determine if a symmetric matrix is positive definite, we need information regarding the signs of its eigenvalues. Also, in order to determine the rate of convergence of certain iterative methods to solve linear systems of algebraic equations, we need to know—as we shall see in later chapters—some information regarding the location of the eigenvalues of the iteration matrix.

The Perron-Frobenius theorem, showing that the spectral radius $\rho(A)$ is an eigenvalue corresponding to a positive eigenvector, if A is nonnegative and irreducible, is presented. It will be seen in some of the following chapters that the concept of numerical radius can give sharper estimates of the norm of the powers of matrices, for instance, than the spectral radius can. Some results relating the numerical radius with the norm of the matrix and with the spectral radius of the symmetric part of nonnegative matrices are presented. Finally, some estimates of the spectral radius of a nonnegative matrix are derived using geometric averages of pairs of symmetrically located entries.

The concepts discussed in this chapter have many important applications, such as in Leontief closed input-output models, which will be shown here, and for iterative solution methods discussed in later chapters.

The following definitions are introduced in this chapter:

Definition 4.1 Let $\pi := \{1, 2, \ldots, n\} \rightarrow \{1, 2, \ldots, n\}$ be a one-to-one mapping—i.e., π is a permutation. A matrix $P = [p_{ij}]_{i,j=1}^n$ is called a *permutation matrix*, if there exists a permutation π, such that

$$p_{ij} = \begin{cases} 1, & j = \pi(i) \\ 0, & j \neq \pi(i). \end{cases}$$

Definition 4.2 A matrix $A = [a_{ij}] \in \mathbb{R}^{n,n}$ is said to be *reducible* if there exists a nonempty subset $S \subset N = \{1, 2, \ldots, n\}$ with $S \neq N$, such that $a_{ij} = 0$ for all index pairs (i, j) where $i \in S$ and $j \in N \setminus S$.

Definition 4.3 A directed graph is said to be *strongly connected* if, to each ordered pair of disjoint points P_i, P_j, there exists a directed path in the graph, $\overrightarrow{P_{i_0}, P_{i_1}}, \overrightarrow{P_{i_1}, P_{i_2}}, \ldots, \overrightarrow{P_{i_{r-1}}, P_{i_r}}$ with $i_0 = i, i_r = j$.

Definition 4.4 A matrix A is said to be *strictly diagonally dominant* if $|a_{i,i}| > \sigma_i$, $i = 1, 2, \ldots, n$ where $\sigma_i = \sum_{j \neq i} |a_{i,j}|$, and *irreducibly diagonally dominant* if A is irreducible and

    (a) $|a_{i,i}| \geq \sigma_i, i = 1, 2, \ldots, n$.

    (b) $|a_{k,k}| > \sigma_k$ for at least one index k.

Definition 4.5 A real matrix B is said to be *nonnegative*, denoted $B \geq 0$, if its entries $b_{i,j} \geq 0$.

Definition 4.6 Let A be a matrix of order n. Then the *Rayleigh quotient* of A for a vector $\mathbf{x} \neq \mathbf{0}$ is $q(\mathbf{x}) = (\mathbf{x}^T A \mathbf{x})/(\mathbf{x}^T \mathbf{x})$.

Definition 4.7

    (a) The *field of values*, or *numerical range*, of a matrix A is the set

$$W(A) = \{(\mathbf{x}^* A \mathbf{x}; \ \mathbf{x} \in \mathbb{C}^n, \ \mathbf{x}^* \mathbf{x} = 1\}.$$

    (b) The *numerical radius* of A is

$$r(A) = \sup\{|\mathbf{x}^* A \mathbf{x}|; \ \mathbf{x} \in \mathbb{C}^n, \ \mathbf{x}^* \mathbf{x} = 1\}.$$

4.1 Reducible and Irreducible Matrices

Definition 4.1 Let $\pi : \{1, 2, \ldots, n\} \to \{1, 2, \ldots, n\}$ be a one-to-one mapping—i.e., π is a permutation. A matrix $P = [p_{i,j}]_{i,j=1}^n$ is called a *permutation matrix*, if there exists a permutation π such that

$$p_{i,j} = \begin{cases} 1, & j = \pi(i) \\ 0, & j \neq \pi(i). \end{cases}$$

Note that each row and each column of a permutation matrix has precisely one nonzero entry. A left-hand (right-hand) multiplication of a square matrix A with P will permute its rows (columns) according to the permutation π.

The following elementary fact is valid.

Lemma 4.1 *Any permutation matrix is orthogonal—i.e., $P^T = P^{-1}$.*

Proof We find

$$(PP^T)_{i,j} = \sum_{k=1}^{n} p_{i,k} p_{j,k} = p_{i,\pi(i)} p_{j,\pi(i)} = \delta_{ij}. \qquad \diamond$$

Definition 4.2 A matrix $A = [a_{i,j}] \in \mathbb{R}^{n,n}$ is said to be *reducible* if there exists a nonempty subset $S \subset N = \{1, 2, \ldots, n\}$ with $S \neq N$, such that $a_{i,j} = 0$ for all index pairs (i, j) where $i \in S$ and $j \in N \setminus S$. A matrix is said to be *irreducible* if it is not reducible.

Theorem 4.2 *A matrix A is reducible if and only if there exists a permutation matrix P such that*

$$(4.1) \qquad PAP^T = \begin{bmatrix} A_{1,1} & A_{1,2} \\ 0 & A_{2,2} \end{bmatrix},$$

where $A_{1,1}$ and $A_{2,2}$ are square matrices.

Proof If A is reducible, let $P = [e_{i_1}, \ldots, e_{i_r}, e_{i_{r+1}}, \ldots, e_{i_n}]$, where $e_{i_k} = (0, \ldots, 1, \ldots 0)^T$ is the i_kth coordinate vector and where $i_k \neq i_l, k \neq l$, where $S = \{i_{r+1}, \ldots, i_n\}$. Then $(PAP^T)_{s,t} = a_{kl} = 0, s = i_k \in S, t = i_l \in N \setminus S$, $s = r+1, \ldots, n, t = 1, \ldots, r$, that is, PAP^T has the form (4.1). Conversely, if $B = PAP^T$ has the form (4.1) for some permutation matrix P, we see immediately that $A = P^T BP$ is reducible. $\qquad \diamond$

Example 4.1 If

$$A = \begin{bmatrix} a_{1,1} & \cdots & a_{1,r} & \cdots & a_{1,n} \\ & \vdots & & & \\ 0 & \cdots & 0\, a_{r,r}\, 0 & \cdots & 0 \\ & \vdots & & & \\ a_{n,1} & \cdots & a_{n,r} & \cdots & a_{n,n} \end{bmatrix},$$

then A is reducible. In this case, we have $S = \{r\}$. In fact with the permutation matrix $P = I_{r,n}$, which permutes the rth and nth rows, we get

$$
PAP^T =
\begin{bmatrix}
a_{1,1} & \cdots & a_{1,n} & \vdots & a_{1,n-1} & \vdots & a_{1,r} \\
\vdots & & & & & & \vdots \\
a_{n,1} & \cdots & a_{n,n} & \cdots & a_{n,n-1} & \vdots & a_{n,r} \\
\vdots & \vdots & \vdots & \vdots & \vdots & \vdots & \vdots \\
0 & \cdots & 0 & \cdots & 0 & \vdots & a_{r,r}
\end{bmatrix}.
$$

Assume that A is reducible and permuted to the reduced form (4.1). Then $Ax = y$, that is,

$$
\begin{bmatrix} A_{1,1} & A_{1,2} \\ 0 & A_{2,2} \end{bmatrix}
\begin{bmatrix} \mathbf{x}^{(1)} \\ \mathbf{x}^{(2)} \end{bmatrix}
=
\begin{bmatrix} \mathbf{y}^{(1)} \\ \mathbf{y}^{(2)} \end{bmatrix},
$$

implies that

$$
A_{1,1}\mathbf{x}^{(1)} + A_{1,2}\mathbf{x}^{(2)} = \mathbf{y}^{(1)}
$$

$$
A_{2,2}\mathbf{x}^{(2)} = \mathbf{y}^{(2)}.
$$

Hence, if A_{22} is nonsingular, we may first solve $\mathbf{x}^{(2)}$ from a reduced linear system of equations and then substitute the solutions in the first set of equations to compute $\mathbf{x}^{(1)}$.

Frequently it is difficult to see directly from a given matrix whether it is reducible or not. As we shall now see, this task can sometimes be simplified if we use matrix graphs. To each square matrix of order n we associate n distinct points $P_1, P_2, \ldots, P_n$, which we may place in arbitrary positions in a plane. For every entry $a_{i,j} \neq 0$, we join P_i, P_j with a (curved) line directed from P_i towards P_j, which we denote $\overrightarrow{P_i P_j}$. The resulting figure is called the *directed graph of A*. Note that the directed graph is independent of the value of the nonzero entries. Hence, when discussing matrix graphs, it suffices to consider the so-called "Boolean matrix" associated with A, which has entries 1 or 0 only (an entry 1 in position i, j if and only if $a_{i,j} \neq 0$). Furthermore, it is readily seen that for any symmetric reordering $B = PAP^T$ of A, where P is a permutation matrix, the directed graphs of A and B differ only with respect to the labelling of points.

Definition 4.3 A directed graph is said to be *strongly connected* if to each ordered pair of disjoint points P_i, P_j there exists a directed path in the graph, $\overrightarrow{P_{i_0}P_{i_1}}, \overrightarrow{P_{i_1}P_{i_2}}, \ldots, \overrightarrow{P_{i_{r-1}}P_{i_r}}$ with $i_0 = i$, $i_r = j$.

Example 4.2 $A = \begin{bmatrix} 1 & 0 & 1 & 1 \\ 0 & 1 & 1 & 0 \\ 1 & 1 & 0 & 0 \\ 1 & 1 & 0 & 0 \end{bmatrix}$ has the directed graph

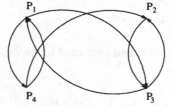

This graph is connected.

Example 4.3 $A = \begin{bmatrix} 1 & 1 & 1 \\ 1 & 1 & 1 \\ 0 & 0 & 0 \end{bmatrix}$ has the directed graph

There is no directed path from P_3 to P_1, for instance, so this graph is not connected. Let us now consider the following important theorem.

Theorem 4.3 (Varga, 1962) *A matrix is irreducible if and only if its directed graph is connected.*

Proof Let A be irreducible and let i_0 be an arbitrary index. Then there exists at least one index j for which $a_{i_0,j} \neq 0$, $j \neq i_0$, because otherwise A would be reducible (see Example 4.1). Let T be the set of indices j such that there exists a directed path from P_{i_0} to P_j, and let S be the remaining set, $S = N \setminus T$. Then T and S are disjoint and T is nonempty. Assume that S is nonempty and let $j \in T$, $k \in S$. Then $a_{jk} = 0$, because if $a_{j,k} \neq 0$, there would exist a directed path from P_{i_0} to P_k via P_j, $\overrightarrow{P_{i_0}P_{i_1}}, \overrightarrow{P_{i_1}P_{i_2}}, \ldots, \overrightarrow{P_{i_{r-1}}P_j}, \overrightarrow{P_jP_k}$, which means that $k \in T$. Now we note that $a_{j,k} = 0$ $\forall j \in T$, $k \in S$ and $S \cap T = \emptyset$, implies by definition that A is reducible. This contradiction implies that S must be empty and, hence, the matrix graph is strongly connected.

To prove the converse statement, assume that the graph is strongly connected. If A is reducible, then there exist nonempty disjoint sets S and T such that $a_{jk} = 0$, $j \in S$, $k \in T$. Because the graph is strongly connected, there exists a directed path $\overrightarrow{P_{i_0} P_{i_1}}, \ldots, \overrightarrow{P_{i_{r-1}} P_{i_r}}$, where $i_0 = j \in S$, $i_r = k \in T$. In the sequence $i_1, \ldots, i_{r-1}$ there is a final index i_s such that $i_{s-1} \in S$ and $i_s \in T$ (possibly $i_s = k$). But since $a_{i_{s-1}, i_s} \neq 0$, this contradicts the definition of S and T. Hence, A is irreducible. $\diamond$

Note that it follows from this theorem that the perhaps seemingly reducible matrix in Example 4.2 is, in fact, irreducible.

Example 4.4 $A = \begin{bmatrix} 2 & 1 & 0 & \ldots & 0 \\ 1 & 2 & 1 & \ldots & 0 \\ \ldots & & & & \\ 0 & \ldots & \ldots & 1 & 2 \end{bmatrix}$ has the directed graph

which is connected, so A is irreducible.

As an application of Theorem 4.3, we shall now show that for irreducible matrices, we can strengthen the classical Gershgorin theorem, which we present first. As we have seen, matrix graphs can be used to determine whether a matrix is irreducible or not. Matrix graphs can also be used to give an interpretation of fill-in for the Gaussian elimination method; see, for instance, the classical article by Parter (1961) and the comments on fill-in in Chapter 1.

4.2 Gershgorin Type Eigenvalue Estimates

The following famous and useful classical theorem to locate eigenvalues roughly, was first published in 1931, while the Perron-Frobenius theorem (see next section) was published already in 1907 (1912).

Theorem 4.4 (Gershgorin, 1931) *The spectrum $S(A)$ of the matrix $A = [a_{ij}]$ is enclosed in the union of the discs*

$$C_i = \left\{ z \in \mathbb{C}; |z - a_{ii}| \leq \sum_{j \neq i} |a_{ij}| \right\}, \quad 1 \leq i \leq n$$

and in the union of the discs

$$C_i' = \left\{ z \in \mathbb{C}; |z - a_{ii}| \leq \sum_{j \neq i} |a_{ji}| \right\}, \quad 1 \leq i \leq n.$$

That is, $S(A) \subset (\cup C_i) \cap (\cup C_i')$.

Proof Let $\mathbf{x}$, λ be an eigensolution—that is, $A\mathbf{x} = \lambda\mathbf{x}$, $\mathbf{x} \neq \mathbf{0}$, or

$$\sum_j a_{ij} x_j = \lambda x_i, \quad 1 \leq i \leq n,$$

where $\mathbf{x}^T = [x_1, x_2, \ldots, x_n]$. We let $\mathbf{x}$ be normalized such that $\|\mathbf{x}\|_\infty = 1$ and we assume that the ith component of $\mathbf{x}$ satisfies $|x_i| = 1$. We may let $x_i = 1$. Then

$$\sum_j a_{ij} x_j = \lambda x_i = \lambda$$

and

$$|\lambda - a_{ii}| = |\sum_{j \neq i} a_{ij} x_j| \leq \sum_{j \neq i} |a_{ij}|.$$

Since λ is an arbitrary eigenvalue, it follows that all of the eigenvalues lie in the union of the discs C_i, $1 \leq i \leq n$.

The second inequality follows from the first if we consider the matrix A^T instead and recall that $S(A) = S(A^T)$. $\diamond$

Example 4.5 Let

$$A = \begin{bmatrix} 1 & -2 & 0 \\ 1 & 2 & 0 \\ 0 & 0 & 2 \end{bmatrix}.$$

The eigenvalues of A are $\lambda_1 = 2$, $\lambda_{2,3} = (3 \pm i\sqrt{7})/2$. The set $(\cup C_i) \cap (\cup C_i')$ is bounded by the bold line in Figure 4.1. Here $\times$ indicates locations of the eigenvalues.

Example 4.6 If $A = \text{diag}(d_i)$, then the distance $|z - d_i| \leq 0$, so $\lambda_i = d_i$, $i = 1, 2, \ldots, n$. Hence the eigenvalues of a diagonal matrix are equal to the diagonal entries.

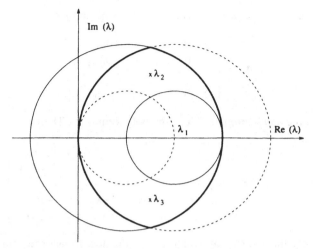

Figure 4.1.   Gershgorin's discs for Example 4.5.

Corollary 4.5 *If λ is an eigenvalue of A and $\mathbf{x}$ is a corresponding eigenvector, then*

$$\lambda \in (\bigcup_{i \in I} C_i) \cap (\bigcup_{j \in J} C_j'),$$

where

$$I = \{i; \, |x_i| = \|\mathbf{x}\|_\infty, \, A\mathbf{x} = \lambda\mathbf{x}\}$$

$$J = \{j; \, |y_j| = \|\mathbf{y}\|_\infty, \, A^T\mathbf{y} = \lambda\mathbf{y}\}.$$

Proof Left as an exercise.   ◇

Corollary 4.6 $\rho(A) \le \min\{\max_i \sum_{j=1}^n |a_{ij}|, \max_j \sum_{i=1}^n |a_{ij}|\}$.

Proof $|\lambda - a_{ii}| \le \sum_{j \ne i} |a_{ij}| \Longrightarrow |\lambda| \le |a_{ii}| + \sum_{j \ne i} |a_{ij}|$.   ◇

Example 4.7 Let $A = \begin{bmatrix} 1 & 2 \\ 4 & 3 \end{bmatrix}$.

Using Corollary 4.6 we have here $\rho(A) \le \min(5, 7) = 5$. Since

$$\det(A - \lambda I) = \lambda^2 - 4\lambda - 5 \text{ we see that } \lambda_1 = 5 \text{ and } \lambda_2 = -1.$$

Example 4.8 Let $P_n(x) = x^n + a_1 x^{n-1} + \cdots + a_n$ and let

$$A_n = \begin{bmatrix} 0 & 0 & \cdots & 0 & -a_n \\ 1 & 0 & \cdots & 0 & -a_{n-1} \\ & & & & \\ 0 & 0 & \cdots & 1 & -a_1 \end{bmatrix}.$$

A_n is called the *companion matrix* to the polynomial P_n. Then

$$\det(A_n - \lambda I) = \det \begin{bmatrix} -\lambda & 0 & \cdots & & -a_n \\ 1 & -\lambda & & & -a_{n-1} \\ \cdots & & & & \\ 0 & 0 & \cdots & -\lambda & -a_2 \\ 0 & 0 & \cdots & 1 & (-\lambda - a_1) \end{bmatrix}.$$

If we multiply the last row by λ, add this to the next to last row, multiply the result by λ, and add to the line above, etcetera, the determinant remains the same and we get

$$\det(A_n - \lambda I) = (-1)^n P_n(\lambda).$$

Hence, we may use Gershgorin's theorem for the companion matrix in order to locate (usually roughly) the zeros of a given polynomial. We note that if $P_n(z) = 0$, then z lies in the union of the discs

$$|z| \leq 1 + \max_{2 \leq i \leq n-1} |a_i|, \quad |z| \leq |a_n|, \quad |z + a_1| \leq 1$$

as well as in the union of the discs

$$|z| \leq 1, \quad |z + a_1| \leq \sum_{i=2}^{n} |a_i|.$$

Example 4.9 The zeros of the equation $z^3 + 2z^2 - 3z - 1 = 0$ satisfy $|z| \leq 1 + 3$ as well as $|z + 2| \leq 4$. Hence, the real part of z satisfies $-4 \leq \mathrm{Re}(z) \leq 2$.

Theorem 4.7 *Let A be an irreducible matrix of order n and let*

$$C_i = \{z \in \mathbb{C}; \ |z - a_{ii}| \leq \sigma_i\}, \quad \text{where } \sigma_i = \sum_{\substack{j=1 \\ j \neq i}}^{n} |a_{ij}|.$$

(a) *Then λ is an eigenvalue on the boundary ∂K of $K = \bigcup_{i=1}^{n} C_i$ only if $|\lambda - a_{ii}| = \sigma_i, i = 1, \ldots, n$, that is, only if all discs have the point λ as a common intersection.*

(b) *Let $\mathbf{x}$ be an eigenvector corresponding to an eigenvalue λ on the boundary ∂K of K. Then $|x_i| = \|\mathbf{x}\|_\infty, i = 1, 2, \ldots, n$.*

Proof Let λ be an eigenvalue on ∂K and let $\mathbf{x} = (x_1, x_2, \ldots, x_n)^T$ be the associated eigenvector, normalized such that $\|\mathbf{x}\|_\infty = 1$. If $1 = |x_r| \geq |x_i|, i = 1, \ldots, n$, then

$$|\lambda - a_{rr}| \leq \sum_{j \neq r} |a_{rj}| \, |x_j| \leq \sigma_r.$$

But $\lambda \in \partial K$ implies $|\lambda - a_{r,r}| = \sigma_r$, so we must have

$$\sum_{j \neq r} |a_{rj}| |x_j| = \sum_{j \neq r} |a_{rj}|.$$

Hence, $|x_j| = 1$ for all j such that $a_{r,j} \neq 0$. Since A is irreducible, it follows from Theorem 4.3 that there exists a directed path

$$\overrightarrow{P_r P_{r_1}}, \ \overrightarrow{P_{r_1} P_{r_2}}, \ldots, \ \overrightarrow{P_{r_m} P_j}$$

for each j, $j = 1, 2, \ldots, n$ and for any r. Hence

$$a_{r,r_1} \neq 0, \quad a_{r_1,r_2} \neq 0, \ldots, \quad a_{r_m,j} \neq 0 \text{ and}$$

$$|\lambda - a_{j,j}| = \sigma_j, \quad j = 1, 2, \ldots, n,$$

that is, λ is a point in the intersection of the discs C_j. This shows part (a). The proof above shows also that $|x_j| = 1$, $j = 1, 2, \ldots, n$, which implies part (b).

Example 4.10 Let $A = \begin{bmatrix} 1 & 1 & & & 0 \\ 1 & 2 & 1 & & \\ & \ddots & \ddots & \ddots & \\ & & 1 & 2 & 1 \\ 0 & & & 1 & 1 \end{bmatrix}$. Then $0 \in \partial K$ and $\lambda = 0$

is an eigenvalue [with eigenvector $\mathbf{v} = (1, -1, 1, \ldots, \pm 1)^T$]. Theorem 4.7 shows that then we must have $0 \in \cap_{i=1}^n C_i$, and indeed, this is readily seen to hold. In fact, there are only two different discs,

$$C_1 = \{z \in \mathbb{C}; \ |z - 1| \leq 1\} \ \text{and}$$

$$C_2 = \{z \in \mathbb{C}; \ |z - 2| \leq 2\}.$$

The following example shows that the assumption of irreducibility is essential for the statement in Theorem 4.7—i.e., if the matrix is reducible, then there can exist an eigenvalue $\lambda \in \partial K$ even though not all discs intersect in λ.

Example 4.11

(a) Let $A = \begin{bmatrix} 2 & -2 & 0 \\ -1 & 1 & 0 \\ 0 & 0 & 1 \end{bmatrix}$, which is reducible. Note that the Gershgorin

discs C_1, C_2 intersect in $\lambda = 0$, which is an eigenvalue, but $\lambda \notin C_3 = \{1\}$.

(b) Let $A = \begin{bmatrix} 2 & -2 & 0 \\ -1 & (1 + \varepsilon) & -\varepsilon \\ 0 & -\varepsilon & 1 \end{bmatrix}$, where $0 < \varepsilon < 1$, which is irreducible.

Here there is no eigenvalue λ for which $\lambda \in \partial K$ and also $\cap \partial C_i = \emptyset$.

(c) Let $\varepsilon = 1$ in (b). Then $\lambda = 0 \in \partial K$ and all three Gershgorin discs intersect in λ.

Note that the converse of Theorem 4.7 does not hold—i.e., all discs can have a common intersection but this does not have to be an eigenvalue. This is shown by the following example.

Example 4.12 Let $A = \begin{bmatrix} 1 & 1 & 0 \\ 1 & 2 & 1 \\ 1 & 2 & 3 \end{bmatrix}$. Here $0 \in \partial K$, but it is readily seen that

$\det(A) \neq 0$, so 0 is not an eigenvalue of A.

For irreducible matrices we can now strengthen Corollary 4.6.

Corollary 4.8 *Let $A = [a_{i,j}]$ be irreducible and let $v = \max_j(\sigma_j + |a_{j,j}|)$ where σ_j is defined in Theorem 4.7. If there exists at least one index k for which $|a_{kk}| + \sigma_k < v$, then $\rho(A) < v$.*

Proof We have $\sum_{j=1}^{n} |a_{i,j}| \le v$, $i = 1, \ldots, n$, but there is a k such that $\sum_{j=1}^{n} |a_{k,j}| < v$. Hence, the intersection of the circles $|\lambda - a_{i,i}| = \sigma_i = \sum_{j \ne i} |a_{i,j}|$ does not contain a point ξ with $|\xi| = v$. Because the discs $\{\lambda; |\lambda - a_{i,i}| \le \sigma_i\}$ are subsets of the set $\{\lambda; |\lambda| \le v\}$, it follows from Theorem 4.7 that there is no point on $|\lambda| = v$ such that λ is an eigenvalue of A. Hence, $\rho(A) = \max |\lambda| < v$. $\diamond$

Definition 4.4 The matrix A is said to be strictly *diagonally dominant* if $|a_{i,i}| > \sigma_i$, $i = 1, 2, \ldots, n$ and *irreducibly diagonally dominant* if A is irreducible and

(a) $|a_{i,i}| \ge \sigma_i$, $i = 1, 2, \ldots, n$
(b) $|a_{k,k}| > \sigma_k$, for at least one index k.

Example 4.13 The matrix in Example 4.4 is irreducibly diagonally dominant. The next theorem is frequently used in practical applications.

Theorem 4.9 *If $A = [a_{i,j}]$ is strictly diagonally dominant or irreducibly diagonally dominant, then A is nonsingular. If, in addition, its diagonal entries are positive, i.e., $a_{i,i} > 0$, then $Re(\lambda_i) > 0$ for all eigenvalues λ_i of A.*

Proof Let A be strictly diagonally dominant. Then $\lambda = 0 \notin \{\lambda; |\lambda - a_{i,i}| \le \sigma_i\}$, and by Gershgorin's theorem $\lambda = 0$ cannot be an eigenvalue of A; hence, $\det(A) \ne 0$. If $a_{i,i} > 0$, then the discs $\{\lambda; |\lambda - a_{i,i}| \le \sigma_i\}$ lie in the right half of the complex plane; hence, $Re(\lambda_i) > 0$ for every eigenvalue of A. If A is irreducibly diagonally dominant, then the theorem has a proof similar to that of Corollary 4.8. $\diamond$

Corollary 4.10 *If A is real symmetric with positive diagonal entries and strictly diagonally dominant or irreducibly diagonally dominant, then A is positive definite.*

Proof A real symmetric matrix has only real eigenvalues and is positive definite if and only if all eigenvalues are positive, so the statement follows from Theorem 4.9. $\diamond$

Example 4.14

(a) The matrix in Example 4.4 is positive definite.
(b) The five-point difference operator $(-\Delta_h^{(5)})$ on a uniform mesh on a rectangular region with at least one boundary meshpoint a Dirichlet point, is positive definite.

4.3 The Perron-Frobenius Theorem

Definition 4.5 A real matrix A is said to be *nonnegative*, denoted $A \geq 0$, if $a_{i,j} \geq 0$. For nonnegative matrices, a famous and important theorem by Perron and Frobenius is valid. (Perron seems to have considered only positive matrices.) Let $\rho(A) = \max |\lambda(A)|$, that is, $\rho(A)$, called *spectral radius*, denote the maximum absolute value of any eigenvalue of A.

Theorem 4.11 (Perron, 1907, and Frobenius, 1912) *Let $A \geq 0$ be a square irreducible matrix. Then:*

(a) $\rho(A)$ is an eigenvalue of A.
(b) A positive eigenvector $\mathbf{v} > \mathbf{0}$ corresponds to $\rho(A)$.
(c) $\rho(A)$ increases when any entry of A increases.
(d) $\rho(A)$ is a simple eigenvalue of A.

Proof We shall prove properties (a) and (b) only. To prove (a), we consider for simplicity only symmetric matrices. Let λ be an eigenvalue and $\mathbf{v}$ a real eigenvector of A. Then

$$|\lambda| = \frac{|\mathbf{v}^T A \mathbf{v}|}{\mathbf{v}^T \mathbf{v}} \leq \sum_{i,j=1}^{n} a_{ij} |v_i| \, |v_j| / \sum_{i=1}^{n} |v_i|^2 = \frac{\hat{\mathbf{v}}^T A \hat{\mathbf{v}}}{\hat{\mathbf{v}}^T \hat{\mathbf{v}}},$$

where $\hat{\mathbf{v}} = (|v_1|, |v_2|, \ldots, |v_n|)^T$. Hence if λ is an eigenvalue that satisfies $|\lambda| = \rho(A)$, then, as A is symmetric, Lemma 3.13 shows that

$$(4.2) \qquad |\lambda| = \sup_{\mathbf{x}} \frac{|\mathbf{x}^T A \mathbf{x}|}{\mathbf{x}^T \mathbf{x}} = \frac{|\mathbf{v}^T A \mathbf{v}|}{\mathbf{v}^T \mathbf{v}} \leq \frac{\hat{\mathbf{v}}^T A \hat{\mathbf{v}}}{\hat{\mathbf{v}}^T \hat{\mathbf{v}}} \leq |\lambda|.$$

This shows that we have equalities everywhere in (4.2) and, in particular, that

$$\frac{\hat{\mathbf{v}}^T A \hat{\mathbf{v}}}{\hat{\mathbf{v}}^T \hat{\mathbf{v}}} = |\lambda|.$$

This and Lemma 3.13(d) show that $\rho(A)$ is an eigenvalue of A (property (a)) and that $\hat{\mathbf{v}}$ is the corresponding eigenvector. Hence we have shown that the eigenvector to $\rho(A)$ has nonnegative entries, and it remains to show that it has only nonzero entries. ◇

The proof will be based on the following lemma.

Lemma 4.12 (Wielandt, 1950) *Let A be a square, nonnegative, and irreducible matrix of order $n > 1$. Then for any $\varepsilon > 0$, the matrix $B = (\varepsilon I + A)^{n-1}$ is positive.*

Proof We show that for any vector $\mathbf{x} \in \mathbb{R}^n$, $\mathbf{x} \geq \mathbf{0}$, $\mathbf{x} \neq \mathbf{0}$ we have $B\mathbf{x} > \mathbf{0}$. Given such a vector $\mathbf{x}$, let $\mathbf{x} = \begin{bmatrix} \mathbf{x}^{(1)} \\ \mathbf{0} \end{bmatrix}$ be a (possibly permuted) partitioning of $\mathbf{x}$ such that $x_j^{(1)} > 0$, $j = 1, \ldots, s$, $s < n$. Let $A = \begin{bmatrix} A_{11} & A_{12} \\ A_{21} & A_{22} \end{bmatrix}$ be the corresponding partitioning (block decomposition) of A. Then

$$(\varepsilon I + A)\mathbf{x} = \begin{bmatrix} \varepsilon \mathbf{x}^{(1)} + A_{11}\mathbf{x}^{(1)} \\ A_{21}\mathbf{x}^{(1)} \end{bmatrix} = \begin{bmatrix} \mathbf{y}^{(1)} \\ \mathbf{y}^{(2)} \end{bmatrix}.$$

Since $\mathbf{x}^{(1)} > \mathbf{0}$, we have $\mathbf{y}^{(1)} > \mathbf{0}$ and $\mathbf{y}^{(2)} = A_{21}\mathbf{x}^{(1)} \geq \mathbf{0}$. Furthermore, $\mathbf{y}^{(2)} \neq \mathbf{0}$, since otherwise $A_{21} = 0$ (which is excluded because A is irreducible). Hence, $\mathbf{y} = \begin{bmatrix} \mathbf{y}^{(1)} \\ \mathbf{y}^{(2)} \end{bmatrix}$ has at least $s + 1$ nonzero (i.e., positive) components. Similarly, by a proper permutation and partitioning it is seen that

$$(\varepsilon I + A)^2 \mathbf{x} = (\varepsilon I + A)\mathbf{y}$$

has at least $s + 2$ positive components. Finally, $(\varepsilon I + A)^{n-1}\mathbf{x}$ will have all n components positive. This shows that $B > 0$.

The proof of positivity of the vector $\hat{\mathbf{v}}$ now follows readily. If $n = 1$, then it follows from the definition of an eigenvector. Let $n \geq 2$. Then $A\hat{\mathbf{v}} = \rho(A)\hat{\mathbf{v}}$, $\rho(A) > 0$ and Lemma 4.12 show that

$$\mathbf{0} < (\varepsilon I + A)^{n-1}\hat{\mathbf{v}} = [\varepsilon + \rho(A)]^{n-1}\hat{\mathbf{v}}.$$

But $\varepsilon + \rho(A)$ is a positive number. Hence $\hat{\mathbf{v}} > \mathbf{0}$. ◇

If we don't know whether A is irreducible, then the following extension of Theorem 4.11 can be used.

Theorem 4.13 (Gantmacher, 1959 and Varga, 1962) *Let $A \geq 0$ be a square matrix. Then:*

> *(a) $\rho(A)$ is an eigenvalue of A. Further, $\rho(A) > 0$ unless A is reducible and in addition the reduced form (4.1) of A is strictly triangular, in which case $\rho(A) = 0$,*
> *(b) To $\rho(A)$ corresponds an eigenvector $\mathbf{v} \geq \mathbf{0}$,*
> *(c) $\rho(A)$ does not decrease when any entry of A increases.*

Proof See Varga, 1962, Theorem 2.7, p. 46.   ◇

Remark 4.14 If A is nonnegative, $\rho(A)$ is called the Perron-Frobenius root of A. Corollaries 4.6 and 4.8 can be used to estimate $\rho(A)$ for matrices A, which are not necessarily nonnegative. Some improved estimates of the Perron-Frobenius root for nonnegative matrices will be presented in Section 4.5.

4.4 Rayleigh Quotient and Numerical Range

Definition 4.6 Let A be a matrix of order n. Then the *Rayleigh quotient* of A for a vector $\mathbf{x} \neq \mathbf{0}$ is

$$q(\mathbf{x}) = (\mathbf{x}^* A\mathbf{x})/(\mathbf{x}^*\mathbf{x}).$$

Note that $q(\cdot)$ is a homogeneous function—i.e., $q(\alpha\mathbf{x}) = q(\mathbf{x})$ for any scalar $\alpha \in \mathbb{C}$, and that $q(\mathbf{x}) = q(\hat{\mathbf{x}}) = \hat{\mathbf{x}}^* A\hat{\mathbf{x}}$, where $\hat{\mathbf{x}} = ||\mathbf{x}||^{-1}\mathbf{x}$.

Definition 4.7

> (a) The *field of values*, or *numerical range*, of a matrix A is the set
> $$W(A) = \{\mathbf{x}^* A\mathbf{x}; \ \mathbf{x} \in \mathbb{C}^n, \mathbf{x}^*\mathbf{x} = 1\}.$$
>
> (b) The *numerical radius* of A is
> $$r(A) = \sup\{|\mathbf{x}^* A\mathbf{x}|; \ \mathbf{x} \in \mathbb{C}^n, \mathbf{x}^*\mathbf{x} = 1\}.$$

Note that $W(A)$ is a subset of $\mathbb{C}$, that it is the image of the unit sphere under the mapping that associates with each unit vector $\mathbf{x}$ the complex number $\mathbf{x}^* A\mathbf{x}$, and that $r(A)$ is a matrix norm:

> (a) $r(A) = 0$ if and only if $A = 0$.

(b) $r(\alpha A) = |\alpha| r(A)$, for any scalar $\alpha \in \mathbb{C}$.

(c) $r(A + B) \leq r(A) + r(B)$.

All the above properties are elementary. Note that this norm is *not multiplicative*—i.e., there exist matrices A, B for which $r(AB) > r(A)r(B)$. To show this consider the following example.

Example 4.15 Let

$$A = \begin{bmatrix} 0 & 1 \\ 0 & 0 \end{bmatrix}, \quad B = \begin{bmatrix} 0 & 0 \\ 1 & 0 \end{bmatrix}.$$

Here $|\mathbf{x}^* A \mathbf{x}| = |\bar{x}_2 x_1| \leq 1/2(|x_1|^2 + |x_2|^2)$ and $\sup_{\mathbf{x}^* \mathbf{x} = 1} |\mathbf{x}^* A \mathbf{x}| = 1/2$. Hence $r(A) = r(B) = 1/2$ but $r(AB) = 1$.

There are, however, some important and very useful properties of $W(A)$ and $r(A)$, as we now shall see.

Theorem 4.15

(a) $S(A) \subset W(A)$, *where $S(A)$ denotes the spectrum of A.*

(b) $\rho(A) \leq r(A)$, *and* $\|A\|_2 = \rho(A) = r(A)$ *for a normal matrix A.*

(c) $r(A) \leq \|A\| \leq 2\, r(A)$, *where $\|\cdot\|$ is the spectral norm.*

(d) $W(A) = W(UAU^*)$ *for any unitary matrix U, i.e., the numerical range is invariant under unitary similarity transformations.*

(e) *For any normal matrix A, we have* $W(A) = W(\operatorname{diag}(\lambda_i))$, $\lambda_i \in S(A)$.

Proof For any eigensolution $\mathbf{x}$, λ of A, i.e., for which $A\mathbf{x} = \lambda \mathbf{x}$, $\mathbf{x}^* \mathbf{x} = 1$, we have $\mathbf{x}^* A \mathbf{x} = \lambda \mathbf{x}^* \mathbf{x} = \lambda$, which shows (a). Also

$$\max_{\lambda \in S(A)} |\lambda| \leq \max_{\mathbf{x}^* \mathbf{x} = 1} |\mathbf{x}^* A \mathbf{x}|,$$

so $\rho(A) \leq r(A)$. Also note that

$$r(A) = \sup_{\|\mathbf{x}\| = 1} |\mathbf{x}^* A \mathbf{x}| \leq \sup_{\|\mathbf{x}\| = 1} (\|A\mathbf{x}\| \, \|\mathbf{x}\|) = \|A\|,$$

so

(4.3) $$r(A) \leq \|A\|.$$

If A is normal, Corollary 2.22 shows that there exists a unitary matrix U, such that $U^* A U = D = \operatorname{diag}(\lambda_1, \ldots, \lambda_n)$, $\lambda_i \in S(A)$. Then $A = U D U^*$

and $A^*A = UD^*DU^*$, which shows that $\|A\| = \rho(A^*A)^{1/2} = \rho(D^*D)^{1/2} = \rho(A)$, so by (4.3) and the inequality $\rho(A) \le r(A)$, we find $\rho(A) = r(A)$. This completes the proof of (b).

To complete the proof of (c) we write A as a sum of normal matrices,

$$A = \frac{1}{2}(A + A^*) + \frac{1}{2}(A - A^*)$$

and observe that $W(A) = W(A^*)$, so $r(A) = r(A^*)$. Since by (b), $\|A + A^*\| = r(A + A^*)$ and $\|A - A^*\| = r(A - A^*)$, we get

$$\|A\| \le \frac{1}{2}\|A + A^*\| + \frac{1}{2}\|A - A^*\| = \frac{1}{2}r(A + A^*) + \frac{1}{2}r(A - A^*)$$

$$\le r(A) + r(A^*) = 2r(A),$$

where the last inequality follows from property of $r(A)$ (the triangle inequality). To prove (d), note that

$$(UAU^*\mathbf{x}, \mathbf{x}) = (AU^*\mathbf{x}, U^*\mathbf{x}) = (A\mathbf{y}, \mathbf{y}), \quad \mathbf{y} = U^*\mathbf{x}.$$

Since U is unitary we have, $\|\mathbf{y}\|_2 = \|\mathbf{x}\|_2$. Finally, by choosing U properly (as in the proof of (b)), we get

$$W(A) = W(\text{diag}(\lambda_i)), \quad \lambda_i \in S(A). \qquad \diamond$$

Remark 4.16 A stronger statement than that in Theorem 4.15(a) is, in fact, valid: For the convex hull of $S(A)$, we have

$$\text{conv } S(A) \subset W(A).$$

For a proof and a generalization to more general fields, see Zenger (1968). For a normal matrix, one can prove that $W(A)$ coincides with the convex hull of $S(A)$ (see Exercise 4.3). This proof generalizes the statement in part (b) on normal matrices, because given the set of eigenvalues $\{\lambda_i\}$ in $\mathbb{C}$, we can simply construct the polygon through this set which contains $S(A)$ to obtain the field of values of the normal matrix A. Since $r(A) = \sup[|(A\mathbf{x}, \mathbf{x})|; \ (\mathbf{x}, \mathbf{x}) = 1]$ must be taken at a corner point of the convex hull, and since the corner points consist of eigenvalues λ_i, we have $\rho(A) = \max_i |\lambda_i| = r(A)$. It should be mentioned that there exist nonnormal matrices (of order $n \ge 3$) for which $\rho(A) = r(A)$ (see, for instance, Goldberg and Tadmor, 1982).

Part (c) of Theorem 4.15 shows that $r(A)$ and the spectral norm of A are closely related. As a further property of $r(A)$, we will show the interesting relation,

$$r(A^m) \le [r(A)]^m, \quad m = 0, 1, \ldots$$

We will then need the following lemmas.

Lemma 4.17 (Pearcy, 1966) *Let $w_k = \exp(i2k\pi/m)$ and $\mathbf{x}^k = \prod_{\substack{j=1 \\ j \ne k}}^{m} (I - w_j A)\mathbf{x}$, $k = 1, 2, \ldots, m$, where $\mathbf{x} \in \mathbb{C}^n$ is a normalized vector and A is a matrix (of order n), and let $f(z) = 1 - z^m$. Then:*

(a) *the following polynomial identities hold:*

$$f(z) = (-1)^{m-1} \prod_{k=1}^{m} (\overline{w}_k - z) = \prod_{k=1}^{m} (1 - zw_k),$$

$$\frac{1}{1 - z^m} = \frac{1}{m} \sum_{k=1}^{m} \frac{1}{1 - zw_k} \quad (z \ne \overline{w}_k)$$

and, hence,

$$1 = \frac{1}{m} \sum_{k=1}^{m} \prod_{\substack{j=1 \\ j \ne k}}^{m} (1 - zw_j).$$

(b)

$$(I - w_k A)\mathbf{x}^k = (I - A^m)\mathbf{x}$$

and

$$\frac{1}{m} \sum_{k=1}^{m} \mathbf{x}^k = \mathbf{x}.$$

(c)

$$(4.4) \qquad 1 - (A^m \mathbf{x}, \mathbf{x}) = \frac{1}{m} \sum_{k=1}^{m} [(\mathbf{x}^k, \mathbf{x}^k) - w_k(A\mathbf{x}^k, \mathbf{x}^k)]$$

Proof To prove (a), note that $\prod_{k=1}^{m} w_k = (-1)^{m-1}$ and $f'(\overline{w}_k) = -mw_k$. (b) follows by elementary computations. To prove (c) note that $(\mathbf{x}^k, \mathbf{x}^k) - w_k(A\mathbf{x}^k, \mathbf{x}^k) = ((I - w_k A)\mathbf{x}^k, \mathbf{x}^k)$ and use (b). $\diamond$

Lemma 4.18 *Let A be a square matrix. Then the following two properties are equivalent:*

(a) $r(A) \leq 1$ *implies* $r(A^m) \leq 1$, $m = 1, 2, \ldots$
(b) $r(A^m) \leq [r(A)]^m$, $m = 1, 2, \ldots$

Proof Clearly (b) implies (a). To prove that (a) implies (b), note that with $B = r(A)^{-1}A$ (and $A \neq 0$) we have $r(B) \leq 1$ and, hence, $r(B^m) \leq 1$, so $r(A)^{-m} r(A^m) \leq 1$. ◇

Theorem 4.19 *For any square matrix A, $r(A) \leq 1$ implies $r(A^m) \leq 1$, $m = 1, 2, \ldots$, or, equivalently, for any square matrix A, $r(A^m) \leq r(A)^m$.*

Proof Let A in (4.4) be replaced by $B = e^{i\theta} A$, $\theta \in \mathbb{R}$. This yields

$$1 - e^{im\theta}(A^m \mathbf{x}, \mathbf{x}) = \frac{1}{m} \sum_{k=1}^{m} [(\mathbf{x}^k, \mathbf{x}^k) - w_k e^{i\theta}(A\mathbf{x}^k, \mathbf{x}^k)].$$

Note that since by hypothesis $r(A) \leq 1$, the real part of the expression in brackets is nonnegative. Since θ is arbitrary, this implies $|(A^m \mathbf{x}, \mathbf{x})| \leq 1$ for all $\mathbf{x}$, $(\mathbf{x}, \mathbf{x}) = 1$, that is, $r(A^m) \leq 1$. The remainder of the theorem follows by Lemma 4.18. ◇

Next we relate the numerical radius and the spectral radius for nonnegative matrices.

Corollary 4.20 (Goldberg and Tadmor, 1982) *Let A be a nonnegative matrix of order n. Then*

$$r(A) = \rho(\frac{1}{2}[A + A^T]).$$

Proof Since $B = 1/2(A + A^T)$ is symmetric, we have

$$\rho(B) = \sup_{\|\mathbf{x}\|=1} |(B\mathbf{x}, \mathbf{x})| = \frac{1}{2} \sup_{\|\mathbf{x}\|=1} |[(A + A^T)\mathbf{x}, \mathbf{x}]|$$

$$\leq \frac{1}{2}[r(A) + r(A^T)] = r(A).$$

This is valid for any matrix. Next, we follow the proof of Theorem 4.11. Since

(4.5) $\qquad r(A) = \max\{|(A\mathbf{x}, \mathbf{x})|; \ \mathbf{x} \in \mathbb{C}^n, (\mathbf{x}, \mathbf{x}) = 1\},$

there is a unit vector $\mathbf{x}^0 = (x_1, \ldots, x_n)^T \in \mathbb{C}^n$ such that $r(A) = |(A\mathbf{x}^0, \mathbf{x}^0)|$. Let $\mathbf{y} = (|x_1|, \ldots, |x_n|)^T$. Note that $\mathbf{y}^T \mathbf{y} = 1$.

Since A is nonnegative and $\mathbf{y}$ is a unit vector, we have

$$r(A) = |(A\mathbf{x}^0, \mathbf{x}^0)| \leq (A\mathbf{y}, \mathbf{y}) \leq r(A),$$

so

$$r(A) = \max\{(A\mathbf{x}, \mathbf{x}); \ \mathbf{x} \in \mathbb{R}^n, (\mathbf{x}, \mathbf{x}) = 1\}.$$

Note that here there are no absolute value signs as in (4.5) and $\mathbf{x} \in \mathbb{R}^n$.

Similarly, since B is nonnegative, then

$$r(B) = \max\{(B\mathbf{x}, \mathbf{x}); \ \mathbf{x} \in \mathbb{R}^n, (\mathbf{x}, \mathbf{x}) = 1\}.$$

Since $(A\mathbf{x}, \mathbf{x}) = (\mathbf{x}, A^*\mathbf{x})$ we have $(B\mathbf{x}, \mathbf{x}) = (A\mathbf{x}, \mathbf{x})$ for all $\mathbf{x} \in \mathbb{R}^n$, so $r(A) = r(B)$ and Theorem 4.15(b) now completes the proof. $\diamond$

Remark 4.21 Lemma 4.18 shows that $r(A^k) \leq [r(A)]^k \leq 1, k = 1, 2, \ldots$, for any square matrix, if $r(A) \leq 1$. As a historic remark, Lax and Wendroff (1964) analyzed the stability of difference schemes and proved that $r(A) \leq 1$ implies $r(A^k) \leq \gamma(n)$, where $\gamma(n) \to \infty$ as $n \to \infty$. Berger (1965) and Pearcy (1966) proved that $\gamma(n)$ could be replaced by 1.

The matrix class for which $\rho(A) = r(A)$ has been characterized as follows: $\rho(A) = r(A)$ if and only if $r(A^k) = [r(A)]^k$, k being any integer exceeding the degree of the minimal polynomial of A. For references to this and similar results, see Goldberg (1979). This result has been generalized (see Li, 1987) to so-called p, k-spectral radius, p, k-numerical radius, etcetera, where, for example, the p, k-spectral radius is

$$\rho_{p,k}(A) = \max \left\{ \left(\sum_{j=1}^{k} |\lambda_{i_j}|^p \right)^{1/p}; \ 1 \leq i_1 < \ldots < i_k \leq n \right\}.$$

Note that $\rho_{p,1}(A) = \rho(A)$.

4.5* Some Estimates of the Perron-Frobenius Root of Nonnegative Matrices

Let A be a nonnegative matrix. It follows from Theorem 4.15(b) and Corollary 4.20 that

$$\rho(A) \le r(A) = \rho(\frac{1}{2}[A + A^T]).$$

As we have seen, this bound is sharp if A is a normal matrix. To get a lower bound of $\rho(A)$, note first that for a nonnegative normal matrix Theorem 4.15(b) and Corollary 4.20 show that

$$\rho(A) = r(A) = \max_{\mathbf{x} \ne 0}\{\mathbf{x}^T A\mathbf{x}/\mathbf{x}^T\mathbf{x}\},$$

so

$$\rho(A) \ge \mathbf{x}^T A\mathbf{x}/\mathbf{x}^T\mathbf{x} \quad \text{for any vector } \mathbf{x} \ne \mathbf{0}.$$

In particular see, for instance, Kolotilina (1989),

$$(4.6) \qquad \rho(A) = r(A) \ge \mathbf{e}^T A\mathbf{e}/\mathbf{e}^T\mathbf{e} = \frac{1}{n}\sum_{i=1}^{n}\sigma_i,$$

where $\mathbf{e} = (1, 1, \ldots, 1)^T$ and $\sigma_i = \sum_{j=1}^{n} a_{ij}$.

However, this bound is not, in general, valid for a nonnormal matrix. To see this, consider

$$A = \begin{bmatrix} 1 & 1 \\ \alpha & 1 \end{bmatrix}, \quad 0 < \alpha < 1.$$

Then $n = 2$ and $\frac{1}{2}\sum_{i=1}^{n}\sigma_i = \frac{1}{2}(3 + \alpha)$. Since $\rho(A) = 1 + \sqrt{\alpha}$, the lower bound would require

$$1 + \sqrt{\alpha} \ge \frac{1}{2}(3 + \alpha), \quad \text{i.e. } \alpha = 1,$$

in which case A would be symmetric (and hence normal).

The entries of $B = \frac{1}{2}(A + A^T)$ are arithmetic averages of the entries of A, that is, $b_{ij} = \frac{1}{2}(a_{ij} + a_{ji})$. We can call B an *arithmetic mean symmetrization* of A. Similarly, for a nonnegative matrix A, the matrix $G = G(A)$, where

$$g_{ij} = \{a_{ij}a_{ji}\}^{\frac{1}{2}},$$

has entries equal to the geometric averages of the entries of A, and we call G the *geometric mean symmetrization* of A.

As has been shown by Schwenk (1986), the correct generalization of (4.6) to nonnormal matrices is with the geometric symmetrization matrix G. Before we show this, we present an elementary lemma.

Lemma 4.22 *Let A be a nonnegative square matrix of order n. Then:*

(a) $G(A) = A$, *if A is symmetric.*

(b) $G(DAD^{-1}) = G(A)$, *for any nonsingular diagonal matrix D of order n.*

(c) $G(PAP^T) = PG(A)P^T$ *for any permutation matrix P of order n.*

Proof If A is symmetric, then $g_{ij} = \{a_{ij}a_{ji}\}^{\frac{1}{2}} = a_{ij}$, which shows (a). For the i, jth entry of DAD^{-1} we have

$$(DAD^{-1})_{i,j} = d_i a_{ij} d_j^{-1},$$

so

$$\{G(DAD^{-1})\}_{i,j} = \{(d_i a_{i,j} d_j^{-1})(d_j a_{ji} d_i^{-1})\}^{\frac{1}{2}} = \{a_{ij}a_{ji}\}^{\frac{1}{2}},$$

which shows (b). Finally, let P be defined by the permutation (mapping) $i \rightarrow \pi(i), i = 1, 2, \ldots, n$. Then

$$\{G(PAP^T)\}_{i,j} = \{(PAP^T)_{i,j}(PAP^T)_{j,i}\}^{\frac{1}{2}}$$

$$= \{a_{\pi(i),\pi(j)} a_{\pi(j),\pi(i)}\}^{\frac{1}{2}},$$

which is $\{PG(A)P^T\}_{i,j}$.   ◇

Theorem 4.23 *Let A be a nonnegative matrix and let $G(A) = [g_{ij}]$, where $g_{ij} = \{a_{ij}a_{ji}\}^{1/2}$. Let P be the permutation matrix that takes A into the reducible form (called the* normal form*)*

(4.7) $$PAP^T = \begin{bmatrix} A_{11} & A_{12} & \cdots & A_{1m} \\ 0 & A_{22} & \cdots & A_{2m} \\ \vdots & & & \\ 0 & 0 & & A_{mm} \end{bmatrix}, \quad 1 \le m \le n,$$

where the diagonal blocks A_{kk}, $k = 1, \ldots, m$ are irreducible. (In the case $m = 1$, A itself is irreducible.) Then:

(a) *The spectral radius of A is bounded by the spectral radii of the geometric mean and arithmetic mean symmetrizations of A, that is,*

(4.8) $$\rho(G(A)) \le \rho(A) \le \rho(\tfrac{1}{2}[A + A^T]).$$

    (b) The upper bound in (4.8) is sharp if A is normal.

    (c) The lower bound in (4.8) is sharp if and only if there exists a block
       matrix $A_{k_0k_0}$ for which $\rho(A_{k_0k_0}) = \max_k \rho(A_{kk})$ and which is diago-
       nally quasisymmetric.

Proof The upper bound follows from Corollary 4.20, and property (b) follows
from Theorem 4.15(b).

To prove the lower bound in (4.8), we follow the proof of Kolotilina (1989)
and assume first that A is irreducible. Since A is nonnegative, the Perron-
Frobenius theorem shows that there is a vector $\mathbf{v} = (v_1, v_2, \ldots, v_n)^T > \mathbf{0}$, for
which

$$(4.9) \qquad\qquad A\mathbf{v} = \rho(A)\mathbf{v}.$$

Let $\mathbf{y} = (v_1^{-1}, \ldots, v_n^{-1})^T$ and let $D = \mathrm{diag}(d_1, \ldots, d_n)$ be a nonzero nonnegative diagonal matrix. Multiplying (4.9) by $\mathbf{y}^T D^2$ we obtain

$$(4.10) \qquad \rho(A) = \mathbf{y}^T D^2 A\mathbf{v}/\mathbf{y}^T D^2 \mathbf{v} = \sum_{i,j=1}^{n} d_i^2 a_{ij} v_j v_i^{-1} \Big/ \sum_{i=1}^{n} d_i^2.$$

Since

$$\mathbf{y}^T D^2 A\mathbf{v} = \sum_i d_i^2 a_{ii} + \sum_{i<j} (d_i^2 a_{ij} v_j/v_i + d_j^2 a_{ji} v_i/v_j),$$

the arithmetic-geometric mean inequality, $\sqrt{ab} \le \frac{1}{2}(a+b)$, where $a \ge 0, b \ge 0$, leads to

$$\mathbf{y}^T D^2 A\mathbf{v} \ge \sum_i d_i^2 a_{ii} + 2\sum_{i<j} \sqrt{a_{ij}a_{ji}} d_i d_j$$

$$= \sum_{i,j} d_i \sqrt{a_{ij}a_{ji}} d_j$$

$$= \mathbf{d}^T G(A)\mathbf{d},$$

where $\mathbf{d} = (d_1, \ldots, d_n)^T$. Now (4.10) shows that

$$(4.11) \qquad\qquad \rho(A) \ge \mathbf{d}^T G(A)\mathbf{d}/\mathbf{d}^T \mathbf{d}.$$

Furthermore, since $G(A)$ is also nonnegative (but possibly reducible), Theorem 4.13 shows that there is a

$$\mathbf{z} = (z_1, z_2, \ldots, z_n)^T \ge \mathbf{0},$$

for which

(4.12) $$G(A)\mathbf{z} = \rho(G(A))\mathbf{z}.$$

Substituting $\mathbf{d} = \mathbf{z}$ in (4.11), we obtain

$$\rho(A) \geq \rho(G(A)),$$

which establishes the lower bound part of (4.8) for irreducible matrices.

Assume now that $\rho(A) = \rho(G(A))$. Then (4.12) shows that we have equality in (4.11) with $\mathbf{d} = \mathbf{z}$, and the last part of the proof of Theorem 4.11 shows that, in fact, $\mathbf{z} > \mathbf{0}$. The equality in (4.11) implies, further, the equality in the arithmetic-geometric mean inequality above (i.e., that $a = b$), so

$$z_i^2 a_{ij} v_j / v_i = z_j^2 a_{ji} v_i / v_j,$$

that is,

$$\sqrt{a_{ij} a_{ji}} = \frac{z_j}{v_j} a_{ji} \frac{v_i}{z_i},$$

or

$$G(A) = \Delta A \Delta^{-1},$$

where $\Delta = \text{diag}(z_1/v_1, \ldots, z_n/v_n)$. Hence A is diagonally similar to $G(A)$, which is symmetric. Conversely, assume that for a positive diagonal matrix Δ the matrix $B = \Delta A \Delta^{-1}$ is symmetric. Then $B = G(B)$, and Lemma 4.22(b) shows that $G(B) = G(A)$. Hence, the eigenvalues of $G(A)$ and A coincide, and, in particular, $\rho(A) = \rho(G(A))$. This completes the proof of (a), and the sharpness of the lower bound if A is irreducible and diagonally quasisymmetric.

In the general (reducible) case, we consider the form PAP^T in (4.7). Lemma 4.22(c) shows that

$$PG(A)P^T = G(PAP^T).$$

Therefore, applying the above result to the irreducible blocks of PAP^T, we find

$$\rho(G(A)) = \rho(G(PAP^T)) = \max_k \rho(G(A_{kk})) \leq \max_k \rho(A_{kk}) = \rho(A),$$

because, by assumption, there is a block A_{k_0, k_0}, for which

$$\max_k \rho(A_{kk}) = \rho(A_{k_0, k_0}),$$

and this block is irreducible and diagonally quasisymmetric. The proof above for the irreducible case shows also that this is a necessary condition, which completes the proof of (c).   ◇

Using (4.11) we find the following readily computable lower bound.

Corollary 4.24 *Let A be a nonnegative $n \times n$ matrix. Then*

$$\rho(A) \geq \frac{1}{n} e^T G(A) e,$$

where $e = (1, 1, \ldots, 1)^T$. This bound is sharp if $(G(A)e)_i = c$, a constant, for $i = 1, 2, \ldots, n$.

Remark The lower bound in (4.8) has been generalized in (Elsner et al., 1988) to certain Hadamard products of nonnegative matrices.

Finally, we present a classical application of the Perron-Frobenius theory.

4.6 A Leontief Closed Input-Output Model

Consider an economic system consisting of a finite number of "industries" $1, 2, \ldots, k$. Over some fixed period of time, each industry produces an "output" of some goods, which we assume is completely utilized in a predetermined manner by the k industries (a closed system). Suppose we are faced with the problem of allocating "prices" to these k outputs so that for each industry, total expenditures equals total income. Such a structure would represent an equilibrium state for the economy.

Let p_i be the price charged by the ith industry for its total output and e_{ij} the fraction of the total output of the jth industry purchased by the ith industry, $i, j = 1, 2, \ldots, k$. By definition, we have

(a) $p_i \geq 0$,

(b) $e_{ij} \geq 0$, and

(c) $\sum_{i=1}^{k} e_{ij} = 1$.

Let the price vector be $\mathbf{p} = (p_1, \ldots, p_k)^T$ and the exchange or input-output matrix $E = [e_{ij}]$. By (c), all column sums of E are unity.

The expenditure of the ith industry is $\sum_{j=1}^{k} e_{ij} p_j$. In order for this to equal its income, we must have $\sum_{j=1}^{k} e_{ij} p_j = p_i$, $i = 1, 2, \ldots, k$, that is,

$$E\mathbf{p} = \mathbf{p}$$

or

$$(I - E)\mathbf{p} = \mathbf{0}.$$

This is a homogeneous linear system for the price vector $\mathbf{p}$. Since

$$(I - E^T)\mathbf{e} = \mathbf{0}, \mathbf{e} = (1, 1, \ldots, 1)^T,$$

we see that $\lambda = 1$ is a root of the equation $\det(I - \lambda E^T) = 0$ that equals the spectral radius. Hence, $\det(I - E^T) = 0$ and $\det(I - E) = 0$, so $E\mathbf{p} = \mathbf{p}$ has a nontrivial solution $\mathbf{p} \neq \mathbf{0}$. But we would like to know if there is a positive solution, that is, if $\mathbf{p} > \mathbf{0}$. The affirmative answer to this question is found in the Perron-Frobenius theory: E is either irreducible or reducible to a block diagonal matrix, $E = \text{diag}(E_{i,i})$ where $E_{i,i}$ are irreducible. To prove the latter part, assume that E is reducible to the form $\begin{bmatrix} E_{1,1}^{(T)} & E_{1,2}^{(S)} \\ 0 & E_{2,2} \end{bmatrix}$, where $E_{1,2}^{(S)} \neq 0$. Then there would exist a group (T) of industries that receive some goods from the complementary group (S) but do not deliver any goods to this group. Hence, total income for S (as a unit) would be zero, but expenditure would be positive because $\mathbf{p}^{(2)} > \mathbf{0}$, where $\mathbf{p} = \begin{bmatrix} \mathbf{p}^{(1)} \\ \mathbf{p}^{(2)} \end{bmatrix}$, since E_{22} is irreducible. Hence, there is no equilibrium state for the economy, contrary to the assumption. Continuing in this manner, we see that either E is irreducible or E is reducible to a block diagonal matrix where the blocks $E_{i,i}$ are irreducible. This shows that $\mathbf{p} > \mathbf{0}$.

Exercises

1. Let A be the difference matrix for $(-h^2 \Delta_h^{(5)})$ on a unit square with Dirichlet boundary conditions. If $A = D - R$, $D = 4I$, then $B = D^{-1}R = I - 1/4A \geq 0$. Show by use of the Perron-Frobenius theorem that $\rho(B)$ is an eigenvalue of B and that the corresponding eigenvector is $(\sin \pi x_i \sin \pi y_j)$, $1 \leq i, j \leq h^{-1} - 1$, $x_i = hi$, $x_j = hj$, and, hence, is positive. (The value of $\rho(B)$ is $\cos \pi h$.)

2. Let A be a matrix with real diagonal entries. Show that for any eigenvalue λ of A,

$$|\text{Im}(\lambda)| \leq \max_i \sum_{j \neq i} |a_{ij}|.$$

3. Show that the field of values of a normal matrix $A \in \mathbb{C}^{n \times n}$ coincides with the convex hull of its spectrum.

Hint: Theorem 4.15(e) shows that $W(A) = W(D)$, $D = \text{diag}(\lambda_i)$, so it remains to show that the field of values for a diagonal matrix D coincides with the convex hull of its diagonal entries. But $W(D) = (D\mathbf{x}, \mathbf{x}) = \sum \lambda_i |x_i|^2 \, \forall \mathbf{x}$, where $\sum |x_i|^2 = 1$. This is, indeed, a convex linear combination of the form $\sum \theta_i \lambda_i$, where $\theta_i \geq 0$, $\sum \theta_i = 1$, that is, $W(D)$ forms the convex hull of the set of (complex) numbers λ_i.

4. Consider the nilpotent matrix $N = \begin{bmatrix} 0 & 1 & 0 & 0 \\ 0 & 0 & 1 & 0 \\ 0 & 0 & 0 & 1 \\ 0 & 0 & 0 & 0 \end{bmatrix}$. (For a nilpotent matrix N, there exists a positive integer s such that $N^s = 0$.) Show that the numerical radius satisfies $r(N) < 1$, $r(N^2) \leq 1/2$ and $r(N^3) = 1/2$, and, hence, $r(N)r(N^2) < r(N \cdot N^2) = r(N^3)$. This establishes that the multiplicative rule $r(AB) \leq r(A)r(B)$ can be violated for the numerical radius even if A and B commute or even when A and B are powers of the same matrix. (However, in Theorem 4.19 it is shown that $r(A^m) \leq [r(A)]^m$.)

5. Show that the constant λ, $\lambda \in \mathbb{R}$, that minimizes $\|A\mathbf{x} - \lambda\mathbf{x}\|$ for a general square matrix and any given $\mathbf{x}$, is

$$\lambda = \frac{1}{2}(\mathbf{x}, (A^* + A)\mathbf{x})/(\mathbf{x}, \mathbf{x}).$$

Note that this is the Rayleigh quotient if A is Hermitian.

6. Let $\mathbf{x}$ be normalized so that $(\mathbf{x}, \mathbf{x}) = 1$ and let $q = (A\mathbf{x}, \mathbf{x})$, the Rayleigh quotient, where A is Hermitian. Show that:
    (a) $q^2 = \|A\mathbf{x}\|^2 - \varepsilon^2$ where $\varepsilon = \|A\mathbf{x} - q\mathbf{x}\|$.
        Hint: $(A\mathbf{x} - q\mathbf{x}, \mathbf{x}) = 0$.
    (b) $f(\lambda) = \|A\mathbf{x} - \lambda\mathbf{x}\|^2$, $\lambda \in R$, takes its minimum value for $\lambda = q$.

7. *Some closure properties of matrix operations:* Let A and B be Hermitian positive definite (h.p.d.) matrices.
    (a) Show that $A + B$ is h.p.d.—i.e., the h.p.d. matrix class is closed under matrix addition.
    (b) Show that the matrix product has positive eigenvalues.
        Hint: Write $A^{-\frac{1}{2}}(AB)A^{\frac{1}{2}} = A^{\frac{1}{2}}BA^{\frac{1}{2}}$ and use the similarity equivalence and congruence transformation theorems.
    (c) Show that AB is Hermitian if and only if A and B commute.
    (d) AB is not positive definite in general, because its Hermitian part, $1/2(AB + BA)$ can have negative eigenvalues. Show this using the following matrices:

$$A = \begin{bmatrix} a & -1 \\ -1 & 1 \end{bmatrix}, B = \begin{bmatrix} 1 & -1 \\ -1 & b \end{bmatrix},$$

where $a > 1, b > 1$.

(e) Show that the Kronecker product $C = A \otimes B$ (i.e., $C_{ij} = a_{ij}B$) is h.p.d.

(f) (Schur, 1911) Show that the Hadamard product $C = A \odot B$ (i.e., $c_{ij} = a_{ij}b_{ij}$) is h.p.d.

Hint: Show that the Hadamard matrix is a principal submatrix of the Kronecker matrix.

(g) Show that any symmetric positive semidefinite matrix can be written in the form

$$A = v_1 v_1^* + \cdots + v_n v_n^*,$$

(h) where $v_i \in \mathbb{C}^n$ and $\{v_1, \ldots, v_n\}$ is an orthogonal set of nonzero vectors.

Hint: Write $A = U \Lambda U^*$ and let $v_i = \lambda_i^{\frac{1}{2}} u_{\cdot i}$, where $u_{\cdot i}$ is the ith column of U.

Remark 4.25 (c), (e), and (f) show that the h.p.d. matrix class is closed under the Kronecker and Hadamard multiplications but not under the ordinary matrix multiplication.

8. (a) Show that the product of two symmetric matrices can have complex eigenvalues. Consider, for example,

$$A = \begin{bmatrix} 0 & 1 \\ 1 & 0 \end{bmatrix}, \quad B = \begin{bmatrix} 2 & 0 \\ 0 & -1 \end{bmatrix}.$$

Compare this with Exercise 7(b).

(b) Show that if $\frac{1}{2}(A + A^T)$ is positive definite, then $Re\lambda(A) > 0$ for any eigenvalue of A.

Hint: If $Ax = \lambda x$, then $x^* A^* = \bar{\lambda} x^*$. Form $x^* A x$ and $x^* A^* x$ and sum these expressions.

9. Show that the matrix class consisting of square matrices (of order n), having eigenvalues with positive real parts, is not closed under addition.

Hint: Consider, for example, $A = \begin{bmatrix} 2 & -2 \\ -1 & (1+\varepsilon) \end{bmatrix}$ and $B = A^T$, where

$0 < \varepsilon < \frac{1}{8}$.

10. If A is irreducible and nonnegative and B is nonnegative, show that $A + B$ is irreducible.

11. Let $A = \begin{bmatrix} 0 & a & 0 \\ 0 & 0 & b \\ c & 0 & 0 \end{bmatrix}$. Is A^k irreducible for $k = 1, 2, 3$?

12. Let A, B denote arbitrary nonnegative square matrices of order n. Prove or disprove the following statements:
    (a) A is irreducible if and only if $I_n + A$ is irreducible.
    (b) If A is irreducible, then A^T is irreducible.
    (c) If A and B are irreducible, then AB is irreducible.
    (d) Let A be irreducible, then A^k is irreducible for $k \geq 2$.

13. Let A be nonnegative and irreducible. Using a proof similar to the proof of Theorem 4.7, show that if $\rho(A)\mathbf{v} = A\mathbf{v}$, $\mathbf{v} \neq \mathbf{0}$, where $\mathbf{v} \geq \mathbf{0}$ and $\rho(A)$ lies on the boundary of ∂K, where K is the union of the Gershgorin's discs for A, then all components of $\mathbf{v}$ are equal.

14. By $A \geq 0$ we mean $a_{i,j} \geq 0$ $\forall i, j$, and by $B \geq A$ we mean $B - A \geq 0$. It is known that if $A \geq B \geq 0$, where A and B are irreducible matrices, then $\rho(A) \geq \rho(B)$. (Theorem 4.11.) Prove that if $A \geq 0$ is irreducible of order n, then either
    (a) $\sum_{j=1}^{n} a_{i,j} = \rho(A), i = 1, 2, \ldots, n$
    or
    (b) $\min_{1 \leq i \leq n} \sum_{j=1}^{n} a_{i,j} < \rho(A) < \max_{1 \leq i \leq n} \sum_{j=1}^{n} a_{i,j}$.

15. Show that any square matrix A is nonnegative (positive) if and only if the following holds:

$$\mathbf{x} \geq \mathbf{0}, \ \mathbf{x} \neq \mathbf{0} \implies A\mathbf{x} \geq \mathbf{0}(A\mathbf{x} > \mathbf{0}).$$

    Hint: Assume first that the above implication holds, but that $a_{jk} < 0$ for some j, k, and consider $\mathbf{e}_j^T A \mathbf{e}_k$. Show that this leads to a contradiction. The converse is, of course, trivial. Show then that the case of positive matrix can be treated in the same manner.

16. Show by induction that every square matrix A of order n can be transformed to the form

$$A = P \begin{bmatrix} A_{11} & \cdots & A_{1p} \\ & \vdots & \\ 0 & \cdots & A_{pp} \end{bmatrix} P^T,$$

    where P is a permutation matrix and the diagonal blocks are irreducible. This form is called the *normal form* of a reducible matrix.
    Hint: If A is irreducible, there is nothing to prove. If A is reducible, then
$A = Q \begin{bmatrix} A_{11} & A_{12} \\ 0 & A_{22} \end{bmatrix} Q^T$ for some permutation matrix Q. Now A_{11} and

A_{22} can each be treated in the same way, decomposing them in such a reducible form, unless the matrix is already irreducible. Finally, show that a product of permutation matrices is again a permutation matrix.

17. (Ivo Marek) Show that a square, nonnegative matrix A is irreducible if and only if the following holds. If $\mathbf{u} \geq \mathbf{0}$ and $a\mathbf{u} \geq A\mathbf{u}$ for $a > 0$ sufficiently large, then $\mathbf{u} > \mathbf{0}$.

18. Let $A = (a_{i,j})$ be given, let D be an arbitrary diagonal matrix with positive diagonal entries, and let

$$C_i = \{z; \, |z - a_{i,i}| \leq \sum_{j \neq i} |d_i a_{i,j} d_j^{-1}|\}$$

$$C_i' = \{z; \, |z - a_{i,i}| \leq \sum_{j \neq i} |d_i a_{j,i} d_j^{-1}|\}.$$

Show that
(a) $S(A) \subset (\bigcup_{i=1}^n C_i) \cap (\bigcup_{i=1}^n C_i')$
(b) $\rho(A) \leq \min_{d_1,\ldots,d_n, d_i > 0} \max_i \sum_{j=1}^n |d_i a_{i,j} d_j^{-1}|$
Hint: Apply Gershgorin's theorem to the matrix DAD^{-1}.

19. Consider the matrix

$$A = \begin{bmatrix} 10 & 1 & 1 \\ 1 & 1 & 1 \\ 1 & 1 & 1 \end{bmatrix}.$$

(a) Show by use of an appropriate diagonal matrix $D = \text{diag}(1, \delta, \delta)$ and part (a) of the previous exercise that the largest eigenvalue of A is bounded by $10 + 2/\delta \simeq 10.23$, where $\delta = 4 + \sqrt{18}$.
(b) Use Theorem 4.7 to show that, in fact, $\rho(A) = 10 + 2/\delta$, where $\delta = 4 + \sqrt{18}$.

20. Let $A = [a_{ij}]$ be real and symmetric of order n and satisfy
(a) $a_{ii} > 0, a_{ij} \leq 0, i \neq j$,
(b) $\sum_i a_{i1} > 0$
(c) $\sum_i a_{ij} = 0, j = 2, \ldots, n$
Prove that the eigenvalues of A are nonnegative.
Hint: Use Gershgorin's theorem.

21. Assume that in addition to the assumptions in the previous exercise, $a_{ij} \neq 0$ for at least one $i = i(j) > j, j = 2, 3, \ldots, n$.
(a) Prove that the eigenvalues of A are positive by choosing an appropriate diagonal matrix D, such that $\tilde{A} = DAD$ satisfies

$$\sum_j \tilde{a}_{ij} > 0, i = 1, \ldots, n.$$

Then use Gershgorin's theorem and a property of congruence transformations.

(b) As an alternative proof of the positivity of the eigenvalues, use Theorem 4.9.

22. Prove that if the union of m of the circles C_i and $C = \cup C_i$ is disjoint from the remaining $n - m$ circles C_i, then there are exactly m eigenvalues in C.

    Hint: Let $D = \text{diag}(A)$ and let the nondiagonal entries of D continuously approach those of A. What then happens to the radius of the circles C_i?

23. The real matrix $P = [p_{ij}]_{i,j=1}^n$ is said to be a *stochastic* matrix if $P \geq 0$ and $\sum_{j=1}^n p_{ij} = 1$, $i = 1, \ldots, n$. Show that a nonnegative matrix P is stochastic if and only if it has the eigenvalue 1, with eigenvector given by $\mathbf{u} = [1, 1, \ldots, 1]^T$, and, furthermore, that the spectral radius of a stochastic matrix is 1.

    Remark: For further reading about stochastic matrices, see Lancaster and Tismenetsky (1985).

24. Let A be nonnegative and nonsingular. Show that A^{-1} is nonnegative if and only if there exists a permutation matrix P such that $A = PD$, where D is diagonal.

25. Show that if $-A$ is a Z-matrix (for a definition, see Chapter 6), then $\exp(A) \geq 0$.

    Hint: Consider first $\exp(t(\alpha I + A))$ where α is sufficiently large so that $\alpha I + A \geq 0$. Then use the power expansion

    $$\exp(\alpha I + A) = I + (\alpha I + A) + \frac{1}{2!}(\alpha I + A)^2 + \ldots$$

    to show that $\exp(\alpha I + A) \geq 0$. Note, finally, that

    $$\exp(A) = e^{-\alpha} \exp(\alpha I + A).$$

26. Given a norm $\| \cdot \|$, consider the matrix class

    $$S_1 = \{A \in \mathbb{C}^{n \times n}; \ \forall \mathbf{x} \in \mathbb{C}^n, \ A\mathbf{x} \neq \mathbf{x} \Longrightarrow \|A\mathbf{x}\| < \|\mathbf{x}\|\}.$$

    Equivalently, $A \in S_1$ means that $\|A\mathbf{x}\| \leq \|\mathbf{x}\|$ holds for all $\mathbf{x} \in \mathbb{C}^n$ and fails to be strict only when $A\mathbf{x} = \mathbf{x}$. If $A \in S_1$ and $B \in S_1$, show that

    (a) $AB \in S_1$

    Hint: Note first that $\|A\| \leq 1$, $\|B\| \leq 1$. If $\|AB\mathbf{y}\| = \|\mathbf{y}\|$, then

    $$\|\mathbf{y}\| = \|AB\mathbf{y}\| \leq \|B\mathbf{y}\| \leq \|\mathbf{y}\|,$$

so $\|By\| = \|y\|$; that is, $By = y$. But then $\|Ay\| = \|y\|$, so $y = Ay = ABy$; that is, $AB \in S_1$.

(b)　$\mathcal{N}(I - AB) = \mathcal{N}(I - A) \cap \mathcal{N}(I - B)$.

Hint: If $y \in \mathcal{N}(I - AB)$, then $ABy = y$ and the above inequality shows that $By = y = Ay$; that is, $y \in \mathcal{N}(I - A) \cap \mathcal{N}(I - B)$. The converse inclusion is trivial.

27. Let $A = \begin{bmatrix} 2 & 1 & 1 \\ 4 & 5 & 2 \\ 9 & 8 & 7 \end{bmatrix}$. Find a lower and an upper bound of $\rho(A)$, using the arithmetic mean and geometric mean symmetrizations of A.

28. *Logarithmic norms:* Let $\| \cdot \|$ be a natural norm and let

$$\mu(A) = \lim_{\varepsilon \to 0+} \frac{\|I + \varepsilon A\| - 1}{\varepsilon}.$$

It can be seen that this limit exists. Show that:

(a)　If $A = z = x + iy$, where x and y are real numbers, then $\mu(A) = x = Re(z)$.

(b)　$\mu(cA) = c\mu(A)$, if $c \geq 0$.

(c)　$-\|A\| \leq \mu(A) \leq \|A\|$.

(d)　$\mu(A + B) \leq \mu(A) + \mu(B)$.

(e)　$|\mu(A) - \mu(B)| \leq \|A - B\|$.

(f)　$\lim_{\varepsilon \to 0+} \frac{ln\| \exp(\varepsilon A)\|}{\varepsilon} = \lim_{\varepsilon \to 0+} \frac{ln\|I + \varepsilon A\|}{\varepsilon} = \mu(A)$.

Hint: $\| \exp(\varepsilon A)\| = 1 + \|I + \varepsilon A + O(\varepsilon^2)\| - 1$ and $ln(1 + \varepsilon) = \varepsilon + O(\varepsilon^2)$.

Remark: $\mu(A)$ has been called *logarithmic norm* (Dahlquist, 1959, Lozinskij, 1958). Note, however, that it is not a matrix norm, because $\mu(A) = 0$ does not imply that $A = 0$. In fact, $\mu(A)$ can even be negative. For further information regarding logarithmic norms, see the following exercises, as well as Ström (1975).

29. (a)　If $\| \cdot \| = \| \cdot \|_\infty$, show that

$$\mu(A) = \max_i \{Re(a_{ii}) + \sum_{j \neq i} |a_{ij}|\}.$$

　　Hint: $\|I + \varepsilon A\|_\infty - 1 = \max_i [|1 + \varepsilon a_{ii}| - 1 + \varepsilon \sum_{j \neq i} |a_{ij}|]$.

(b)　If $\| \cdot \| = \| \cdot \|_1$, show that

$$\mu(A) = \max_j \{Re(a_{ii}) + \sum_{i \neq j} |a_{ij}|\}.$$

(c)　If $\| \cdot \| = \| \cdot \|_2$, show that

$$\mu(A) = \max_i \lambda_i(Re(A)),$$

where $\lambda_i(B)$ denotes the ith eigenvalue of B and where $Re(A) = \frac{1}{2}(A + A^*)$.

Hint: To prove (b), show first that

$$\|I + \varepsilon A\|_2^2 = \rho(I + \varepsilon(A^* + A) + \varepsilon^2 A^* A) \leq$$

$$1 + \varepsilon \max \lambda_i(A^* + A) + \varepsilon^2 \rho(A^* A),$$

and use then the continuity of matrix eigenvalues with respect to perturbations.

(d)   Show that for any natural matrix norm,

$$\max_i \lambda_i(Re(A)) \leq \mu(A) \leq \|A\|.$$

Hint: $\rho(I + \varepsilon A) - 1 \leq \|I + \varepsilon A\| - 1 \leq \varepsilon \|A\|$, and for any eigenvalue λ of A we have

$$\lim_{\varepsilon \to 0+} \frac{|1 + \varepsilon \lambda| - 1}{\varepsilon} = Re(\lambda).$$

30.   (a)   Consider the system of ordinary differential equations,

$$\frac{d\mathbf{x}(t)}{dt} = A(t)\mathbf{x}(t) + \mathbf{b}(t), \ t > 0, \ \mathbf{x}(0) = \mathbf{x}_0.$$

Show

$$\frac{d\|\mathbf{x}(t)\|}{dt} \leq \mu(A(t))\|\mathbf{x}(t)\| + \|\mathbf{b}(t)\|,$$

that is,

$$\|\mathbf{x}(t)\| \leq e^{\int_0^t \mu(A(s))ds}\|\mathbf{x}_0\| + \int_0^t e^{\int_s^t \mu(A(\xi))d\xi} \mathbf{b}(s)ds.$$

(b)   If

$$\frac{d\mathbf{x}}{dt} = \begin{bmatrix} 0 & -1 \\ 1 & -1 \end{bmatrix} \mathbf{x}(t) + \begin{bmatrix} 0 \\ t \end{bmatrix},$$

show that for $\|\cdot\| = \|\cdot\|_2$,

$$\mu(A) = 0$$

and

$$\|\mathbf{x}(t)\|_2 \leq \frac{1}{2}t^2 + \|\mathbf{x}_0\|.$$

31. (a) Let $S(A) = \{\lambda_i\}$ be the spectrum of A and let $\mu(A)$ be the logarithmic norm, corresponding to $\| \cdot \|_2$. Show that

$$\mu(-A) \leq \lambda_i \leq \mu(A).$$

(b) Let $\alpha(A) = \max_i Re(\lambda_i)$, called the *spectral abscissa* of A. Show that

$$e^{\alpha(A)t} \leq \| \exp(tA) \| \leq e^{\mu(A)t}.$$

Hint: Let $\mathbf{v}$ be an arbitrary unit vector, $\|\mathbf{v}\| = 1$, and let $\phi(t) = \mathbf{v}^* \exp(tA^*) \exp(tA)\mathbf{v}$. Then show that

$$\phi'(t) = [\exp(tA)\mathbf{v}]^*(A^* + A) \exp(tA)\mathbf{v} \leq 2\mu(A)\phi(t),$$

and

$$\sup_{\|\mathbf{v}\|=1} \phi(t) = \| \exp(tA) \|^2 \leq e^{2\mu(A)t}.$$

Next, note that

$$\| \exp(tA) \| \geq \max_i |e^{\lambda_i t}| = \max_i e^{Re\lambda_i t} = e^{\alpha(A)t}.$$

(c) Show that

$$\sup_{t \geq 0} \| \exp(tA) \| = 1 \iff \mu(A) \leq 0.$$

Hint: Part (b) shows the sufficiency part. The converse follows by

$$\sup_{t \geq 0} \| \exp(tA) \| = 1 \Rightarrow \phi(t) \leq 1, \quad t \geq 0 \; \forall \mathbf{v} \in \mathbb{C}^n, \; \|\mathbf{v}\| = 1,$$

so $\phi'(0) \leq 0$ or $\mathbf{v}^*(A^* + A)\mathbf{v} \leq 0 \; \forall \mathbf{v}, \; \|\mathbf{v}\| = 1$.

Remark: If A is a normal matrix, then $\mu(A) = \alpha(A)$; show this. Hence, for a normal matrix, $\| \exp(tA) \| = e^{\mu(A)t}$. This holds more generally for any diagonalizable matrix (see Ström, 1975).

32. Consider the matrix

$$A = (n + 1)^2 \begin{bmatrix} -2 & 1 & & & 0 \\ 1 & -2 & 1 & & \\ & \ddots & \ddots & \ddots & \\ 0 & & & 1 & -2 \end{bmatrix},$$

of order n. Show that:

(a)   For $\| \cdot \|_\infty$, $\|A\| = 4(n+1)^2$, $\mu(A) = 0$.

(b)   For $\| \cdot \|_1$, $\|A\| = 4(n+1)^2$, $\mu(A) = 0$.

(c)   For $\| \cdot \|_2$, $\|A\| = 4(n+1)^2 - \pi^2 + O(n^{-2})$, $\mu(A) = -\pi^2 + O(n^2)$.

(d)   For the spectral abscissa (see the previous exercise), $\alpha(A) = -\pi^2 + O(n^{-2})$, $n \to \infty$.

33.   Let $A = \begin{bmatrix} -1 & 5 \\ 0 & -1 \end{bmatrix}$. Show that:

(a) $\exp(tA) = e^{-t} \begin{bmatrix} 1 & 5t \\ 0 & 1 \end{bmatrix}$.

(b) $\| \exp(tA) \| \geq 5te^{-t}$.

(c) $\| \exp(tA) \| \geq \frac{5}{e}$.

34.   (Moler and Van Loan, 1978) Let $A = \begin{bmatrix} -49 & 24 \\ -64 & 31 \end{bmatrix}$.

(a)   Show that $\{\lambda_i(A)\} = \{-1, -17\}$.

(b)   Estimate a lower bound of

$$a = \max_{0 \leq t \leq 1} \| \exp(tA) \|$$

to show that $a \gg 1$.

Hint: Compute $\mu(A) = \rho(Re(A))$.

35.   Let A be nonsingular and let $B = A \, \mathrm{diag}(\lambda_1, \ldots, \lambda_n) A^{-1}$. Show that

$$\mathbf{d} = A \odot (A^{-1})^T \underline{\lambda},$$

where $\mathbf{d}^T = [b_{11}, b_{22}, \ldots, b_{nn}]$.

Hint: Use Cramer's rule.

Remark: If B is normal, then we can take A unitary to give an alternative proof of Exercise 2.35.

References

C. A. Berger (1965). On the numerical range of powers of an operator. Abstract No. 625-152, *Notices Amer. Math. Soc.* **12**, 590.

G. Dahlquist (1959). Stability and error bounds in the numerical integration of ordinary differential equations. *Transactions of the Royal Institute of Technology*, No. 130, Stockholm.

L. Elsner, C. R. Johnson, and J. A. Dias de Silva (1988). The Peron root of a weighted geometric mean of nonnegative matrices. *Lin. and Multilin. Alg.* **24**, 1–13.

G. Frobenius (1912). Über Matrizen aus nicht negativen Elementen. S.-B. Preuss. Akad. Wiss. Berlin, pp. 456–477.

F. R. Gantmacher (1959). *The Theory of Matrices*, vol I, II. New York: Chelsea.

A. Gershgorin (1931). Über die Abgrenzung der Eigenwerte einer Matrix. *Izv. Akad. Nauk SSSR*, Ser. Fiz.-Mat. **6**, 749–754.

M. Goldberg (1979). On certain finite dimensional numerical ranges and numerical radii. *Lin. and Multilin. Alg.* **7**, 329–342.

M. Goldberg and E. Tadmor (1982). On the numerical radius and its applications. *Lin. Alg. and Its Applic.* **42**, 263–284.

R. Horn and C. R. Johnson (1985). *Matrix Analysis*. New York: Cambridge Univ. Press.

L. Yu. Kolotilina. Lower bounds for the Perron root of a nonnegative matrix. *Report 8911*, Department of Mathematics, University of Nijmegen, The Netherlands.

P. Lancaster and M. Tismenetsky (1985). *The Theory of Matrices*, with Applications, 2nd ed. Orlando, Fla.: Academic.

P. D. Lax and B. Wendroff (1964). Difference schemes for hyperbolic equations with high order of accuracy. *Comm. Pure Appl. Math.* **17**, 381–391.

Chi-Kwong Li (1987). A generalization of spectral radius, numerical radius, and spectral norm. *Lin. Alg. and Its Applic.* **90**, 105–118.

S. M. Lozinskij (1958). Error estimates for numerical integration of ordinary differential equations, Part I. *Izv. Vys. Ucebn. Zaved Mathematika* **6**, 52–90 (in Russian).

C. B. Moler and C. F. van Loan (1978). Nineteen dubious ways to compute the exponential of a matrix. *SIAM Rev.* **20**, 801–837.

S. Parter (1961). The use of linear graphs in Gauss elimination. *SIAM Rev.* **3**, 119–130.

C. Pearcy (1966). An elementary proof of the power inequality for the numerical radius. *Michigan Mathematical J.* **13**, 285–291.

O. Perron (1907). Zur Theorie der Matrizen. *Math. Ann.* **64**, 248–263.

Lord Rayleigh (J. W. Strutt) (1899). On the calculation of the frequency of vibration of a system in its gravest mode, with an example from hydrodynamics. *Philos. Mag.* **47**, 556–572.

A. J. Schwenk (1986). Tight bounds on the spectral radius of asymmetric nonnegative matrices. *Lin. Alg. and Its Applic.* **75**, 257–265.

I. Schur (1911). Bemerkungen zur Theorie der beschränkten Bilinearformen mit unendlich vielen Veränderlichen. *J. Reine Angew. Math.* **140**, 1–28.

T. Ström (1975). On logarithmic norms. *SIAM J. Numer. Anal.* **12**, 741–753.

R. Varga (1962). *Matrix Iterative Analysis*. Englewood Cliffs, N.J.: Prentice Hall.

H. Wielandt (1950). Unzerlegbare, nicht negative Matrizen. *Math. Zeitschrift* **52**, 642–648.

Chr. Zenger (1968). On convexity properties of the Bauer field of values of a matrix. *Numer. Math.* **12**, 96–105.

5

Basic Iterative Methods and Their Rates of Convergence

Let us now consider various iterative methods for the numerical computation of the solution of a linear system of equations. Iterative methods for solving linear systems (originally by Gauss in 1823, Liouville in 1837, and Jacobi in 1845) embody an approach quite different from that behind direct methods such as Gaussian elimination (see Chapter 1). In 1823 Gauss wrote, "Fast jeden Abend mache ich eine neue Auflage des Tableau, wo immer leicht nachzuhelfen ist. Bei der Einförmigkeit des Messungsgeschäfts gibt dies immer eine angenehme Unterhaltung; man sieht daran auch immer gleich, ob etwas Zweifelhaftes eingeschlichen ist, was noch wünschenswert bleibt usw. Ich empfehle Ihnen diesen Modus zur Nachahmung. Schwerlich werden Sie je wieder direct eliminieren, wenigstens nicht, wenn Sie mehr als zwei Unbekannte haben. Das indirecte Verfahren läßt sich halb im Schlafe ausführen oder man kann während desselben an andere Dingen denken." (Freely translated, "I recommend this modus operandi. You will hardly eliminate directly anymore, at least not when you have more than two unknowns. The indirect method can be pursued while half asleep or while thinking about other things.")

The simplest iterative method has the form $\mathbf{x}^{l+1} = \mathbf{x}^l - \tau(A\mathbf{x}^l - \mathbf{b})$, $l = 0, 1, \ldots$, where $\mathbf{x}^0$ is given and τ is a parameter. Under certain conditions, the sequence $\{\mathbf{x}^l\}$ converges to the solution of $A\mathbf{x} = \mathbf{b}$. An advantage of iterative methods is their simplicity: We need only perform matrix-vector multiplications and vector additions. Furthermore, the sparsity of the matrix A can be fully exploited, and computer storage is required only for the nonzero entries of A and the vector $\mathbf{x}$, plus one or two more vectors. The entries of A do not even have to be known explicitly. For example, the matrix can be given as a direct sum of sparse matrices, and the matrix-vector product is then computed as the sum of the matrix-vector products with these matrices. Unless the sparsity of A is confined to a band around the main diagonal or to an envelope structure, direct methods need reorderings (permutations) to avoid an excessive amount of fill-in. In some problems, such as those for difference matrices arising from three-space dimensional partial differential equations, it turns out to be virtually impossible to gain much in decreasing the amount of fill-in using reordering methods. The border line between dense and sparse matrices can be somewhat inprecise, but we should treat a matrix as sparse if we can save space and/or computer time, whichever is more important, by employing methods that utilize sparsity.

The disadvantage with iterative methods is that the rate of convergence may be slow or they may even diverge, and we need to find a stopping test. In this and in subsequent chapters, we shall present convergence results and practical implementational aspects of various methods for some important classes of matrices. We present first some fundamental results for the convergence of basic iterative methods and for stationary iterative methods where two parameters have been introduced.

Next, we show for a second-order method for certain classes of problems that the rate of convergence can be improved by an order of magnitude over the first order method if these parameters are chosen properly.

Finally we present the Chebyshev iterative methods, where the parameters vary during the iterations. In the first order (one step) method, the parameters are computed from a closed expression, but for reasons of numerical stability, they must be chosen in a proper order. In the second order (two step) method, the parameters are computed from certain recursions, and this version is unconditionally stable. Chebyshev iterative methods require estimates of extreme eigenvalues. In a later chapter, another iterative method, with the same asymptotic convergence rate in a certain norm, the conjugate gradient method, will be presented, where such eigenvalue estimates are not required.

The following definitions are introduced in this chapter:

Definition 5.1 For a vector $\mathbf{x}$, the *residual*, denoted $\mathbf{r} = \mathbf{r}(\mathbf{x})$ of $A\mathbf{x} = \mathbf{b}$, is $\mathbf{r} = A\mathbf{x} - \mathbf{b}$.

Definition 5.2 Let B be the *iteration matrix*, $B = C^{-1}R$, where $R = C - A$. Then, given a matrix norm $\| \cdot \|$, $\|B^m\|$ is called the *convergence factor* for m steps and $R_m = \|B^m\|^{1/m}$ is called the *average convergence factor* (per step for m steps) for this norm.

Definition 5.3 $r_m = -\log_{10} R_m$ is called the *average rate of convergence* and $r = -\log_{10} \rho(B) = -\log_{10} R_\infty$ is called the *asymptotic rate of convergence*.

Definition 5.4

(a) A first-order, or *one-step iterative,* method for the solution of $A\mathbf{x} =$

b is defined by $C\mathbf{d}^{l+1} = -\tau_l \mathbf{r}^l$, $\mathbf{x}^{l+1} = \mathbf{x}^l + \mathbf{d}^{l+1}$, $l = 0, 1, \ldots$ where $\mathbf{r}^l = A\mathbf{x}^l - \mathbf{b}$, $\{\tau_l\}$ is a sequence of parameters and $\mathbf{x}^0$ is given. If $\tau_l = \tau$, $l = 0, 1, 2, \ldots$, then the method is called *stationary*, otherwise *nonstationary*.

(b) A second-order, or *two-step iterative*, method is defined by

$$C\mathbf{s}^l = \mathbf{r}^l, \quad \mathbf{x}^{l+1} = \alpha_l \mathbf{x}^l + (1 - \alpha_l)\mathbf{x}^{l-1} - \beta_l \mathbf{s}^l, \quad l = 0, 1, \ldots,$$

where $\{\alpha_l\}$, $\{\beta_l\}$ are sequences of parameters with $\alpha_0 = 1$.

5.1 Basic Iterative Methods

Consider a linear system of equations,

$$(5.1) \qquad\qquad\qquad A\mathbf{x} = \mathbf{b}$$

or

$$(5.1') \qquad\qquad \sum_{j=1}^{n} a_{ij} x_j = b_i, \quad i = 1, 2, \ldots, n,$$

in n equations and n unknowns, where A is nonsingular. Its solution is $\hat{\mathbf{x}} = A^{-1}\mathbf{b}$.

We shall consider various iterative solution methods to find $\hat{\mathbf{x}}$. Iterative solution methods give a sequence of vectors which approximate $\hat{\mathbf{x}}$.

Definition 5.1 For a vector $\mathbf{x}$, the *residual*, denoted $\mathbf{r} = \mathbf{r}(\mathbf{x})$ of (5.1), is $\mathbf{r} = A\mathbf{x} - \mathbf{b}$. Note that the norm of the residual is a measure of how accurately a vector $\mathbf{x}$ satisfies (5.1).

Iterative methods are, in particular, well suited for large systems of equations (n large) that are sparse—i.e., for which the number of nonzero entries, $a_{ij} \neq 0$, of A is small compared with the total number (n^2) of entries. Such linear systems arise, for instance, when we use difference methods, in particular for elliptic partial differential equation problems in two and three-space dimensions. In that case, the number of nonzero entries of A is proportional to n, typically like $5n$ or $7n$.

A *basic iterative method* has the form

$$(5.2) \qquad C\mathbf{d}^{l+1} = -\mathbf{r}^l, \quad \mathbf{x}^{l+1} = \mathbf{x}^l + \mathbf{d}^{l+1}, \qquad l = 0, 1, 2, \ldots,$$

where $\mathbf{r}^l$ is the residual, $\mathbf{r}^l = A\mathbf{x}^l - \mathbf{b}$ and $\mathbf{d}^{l+1}$ is the correction at stage l. $\mathbf{x}^0$ is an initial approximation, sometimes arbitrarily chosen, sometimes taken to be $\mathbf{x}^0 = C^{-1}\mathbf{b}$. C is a nonsingular matrix, sometimes called the *preconditioning* matrix, that may be chosen in various ways for a faster rate of convergence (see Chapters 6-9).

$A = C - R$ is called a *splitting* of A, and R is called the *defect* matrix of the splitting. We see that (5.2) has the alternative form

$$(5.3) \qquad C\mathbf{x}^{l+1} = R\mathbf{x}^l + \mathbf{b}, \qquad l = 0, 1, 2, \ldots.$$

Note that if we let $C = A$, then $R = 0$, and we get the exact solution $\hat{\mathbf{x}}$ after one step of this method. Clearly, in a practical method we must choose C such that the cost of computing $\mathbf{d}^{l+1}$ in (5.2), or $\mathbf{x}^{l+1}$ in (5.3), is relatively small; this means that solutions of systems of linear equations of the form $C\mathbf{x} = \mathbf{y}$ must be performed with little cost, typically of the order of one or a few matrix vector multiplications with A. This is the case, for instance, when C is a diagonal matrix or a sparse triangular matrix with about the same sparsity as A or a product of sparse triangular (lower and upper, for instance) matrices.

Note also that if $\mathbf{x}^0 = \hat{\mathbf{x}}$, then $\mathbf{x}^l = \hat{\mathbf{x}}$, $l = 1, 2, \ldots$. Hence, the solution of (5.1) satisfies the recursion (5.3); (5.3) is then said to be *consistent* with (5.1).

Example 5.1 Consider $A\mathbf{x} = \mathbf{b}$, where $A = \begin{bmatrix} 4 & -1 & 0 \\ -1 & 4 & -1 \\ 0 & -2 & 4 \end{bmatrix}$ and $\mathbf{b} = [3, 2, 2]^T$. Let $C = \begin{bmatrix} 4 & 0 & 0 \\ 0 & 4 & 0 \\ 0 & 0 & 4 \end{bmatrix}$. Then $R = \begin{bmatrix} 0 & 1 & 0 \\ 1 & 0 & 1 \\ 0 & 2 & 0 \end{bmatrix}$.

(a) With $\mathbf{x}^0 = [1, 1, 1]^T$, we find $\mathbf{x}^l = [1, 1, 1]^T$, $l = 1, 2, \ldots$.
(b) With $\mathbf{x}^0 = [0, 0, 0]^T$, the first vectors in (5.3) are $\mathbf{x}^1 = \frac{1}{4}[3, 2, 2]^T$, $\mathbf{x}^2 = \frac{1}{16}[14, 13, 12]$, $\mathbf{x}^3 = \frac{1}{64}[61, 58, 58]^T$. It can be seen that $\mathbf{x}^l \to [1, 1, 1]^T$ as $l \to \infty$.

To study the convergence of the sequence $\{\mathbf{x}^l\}$, we need the following basic lemmas.

Lemma 5.1 *Let T be a nonsingular matrix and let $\|\mathbf{x}\|_T = \|T\mathbf{x}\|_\infty$ and $\|A\|_T = \sup_{\mathbf{x}\neq 0}(\|A\mathbf{x}\|_T / \|\mathbf{x}\|_T)$, the induced matrix norm. Then*

(a) $\|A\|_T = \|TAT^{-1}\|_\infty$.
(b) *For every $\varepsilon > 0$ and matrix A, there is a nonsingular matrix T such*

that

$$\|A\|_T \le \rho(A) + \varepsilon.$$

Proof Since T is nonsingular, it follows readily that $\|\mathbf{x}\|_T$ is a vector norm. We have

$$\|A\|_T = \sup_{\mathbf{x}\neq 0} \frac{\|A\mathbf{x}\|_T}{\|\mathbf{x}\|_T} = \sup_{\mathbf{x}\neq 0} \frac{\|TA\mathbf{x}\|_\infty}{\|T\mathbf{x}\|_\infty} = \sup_{\mathbf{y}\neq 0} \frac{\|TAT^{-1}\mathbf{y}\|_\infty}{\|\mathbf{y}\|_\infty} = \|TAT^{-1}\|_\infty,$$

which proves part (a). To prove part (b), observe first that there exists a unitary matrix U, such that

$$B = UAU^{-1} = \begin{bmatrix} b_{11} & b_{12} & \dots & b_{1n} \\ & b_{22} & & b_{2n} \\ & & \ddots & \vdots \\ \emptyset & & & b_{nn} \end{bmatrix},$$

where $b_{ii} = \lambda_i \in \mathcal{S}(A)$. Let

$$D = D(\delta) = \begin{bmatrix} \delta^{-1} & & \emptyset \\ & \ddots & \\ \emptyset & & \delta^{-n} \end{bmatrix},$$

where $0 < \delta$. Then

$$(DBD^{-1})_{ij} = \begin{cases} 0 & , i > j \\ b_{ii} & , i = j, \\ b_{ij}\delta^{j-i} & , i < j \end{cases}$$

so, $\|DBD^{-1}\|_\infty \le \max_i\{|b_{ii}| + n\max_{j>i}\{|b_{ij}\delta^{j-i}|\}\}$.

Hence, we can get

$$\|DBD^{-1}\|_\infty \le \max_i |b_{ii}| + \varepsilon$$

by choosing $\delta = \delta_0$ sufficiently small. Let $D_0 = D(\delta_0)$ and $T = D_0 U$. Then

$$\|A\|_T = \|TAT^{-1}\|_\infty = \|D_0 U A U^{-1} D_0^{-1}\|_\infty$$

$$= \|D_0 B D_0^{-1}\|_\infty \le \max_i |b_{ii}| + \varepsilon = \rho(A) + \varepsilon. \qquad \diamond$$

Note that Lemma 5.1 says that there exist matrix norms that are arbitrarily close to the spectral radius of a given matrix. These can, however, correspond to an unnatural scaling of the matrix.

Lemma 5.2 *For an arbitrary square matrix A,*

(a) $\lim_{k \to \infty} A^k = 0 \iff \rho(A) < 1.$
(b) If $\rho(A) < 1$ then $(I - A)^{-1} = I + A + A^2 + \ldots$ is convergent.

Proof If $\rho(A) < 1$, then choose $\varepsilon > 0$, such that $\rho(A) + \varepsilon < 1$. By Lemma 5.1, there exists a matrix norm $\| \cdot \|_T$, where T depends on A and ε, such that

$$\|A\|_T \le \rho(A) + \varepsilon < 1.$$

Hence, $\|A^k\|_T \le \|A\|_T^k \to 0$, that is, $A^k \to 0$, $k \to \infty$. If, on the other hand, $\lim_{k \to \infty} A^k = 0$, then letting $\{\lambda, \mathbf{x}\}$ be an eigensolution of A we have $\lambda^k \mathbf{x} = A^k \mathbf{x} \to \mathbf{0}$, $k \to \infty$, that is, since $\mathbf{x} \ne \mathbf{0}$, $\lim \lambda^k = 0$.

This is true for all eigenvalues; hence, $\rho(A) = \max |\lambda| < 1$, which proves part (a).

To prove part (b), note that

$$(I - A)(I + A + \cdots + A^k) = I - A^{k+1},$$

so $\rho(A) < 1$ and part (a) show that $A^{k+1} \to 0$, which implies (b).   ◇

Theorem 5.3 *The sequence of vectors $\mathbf{x}^l$ in (5.3) converges to the solution of $A\hat{\mathbf{x}} = \mathbf{b}$ for any $\mathbf{x}^0$ if and only if $\rho(C^{-1}R) < 1$.*

Proof Let $\mathbf{e}^l = \hat{\mathbf{x}} - \mathbf{x}^l$. It follows from $A = C - R$ that $C\hat{\mathbf{x}} = R\hat{\mathbf{x}} + \mathbf{b}$ and from (5.3) that $C\mathbf{e}^{l+1} = R\mathbf{e}^l$. Hence, $\mathbf{e}^{l+1} = B\mathbf{e}^l$, $l = 0, 1, 2, \ldots$, where $B = C^{-1}R$. By recursion, $\mathbf{e}^m = B^m \mathbf{e}^0$. If $\rho(B) < 1$, then by Lemma 5.2(a), $B^m \to 0$, $m \to \infty$, so $\mathbf{e}^m \to \mathbf{0}$, $m \to \infty$. If $\rho(B) \ge 1$, let λ_i be an eigenvalue of B for which $|\lambda_i| = \rho(B)$, and let $\mathbf{e}^0$ be a corresponding eigenvector. Then $\mathbf{e}^m = B^m \mathbf{e}^0 = \lambda_i^m \mathbf{e}^0$ so $\mathbf{e}^m \not\to \mathbf{0}$, which proves the theorem.   ◇

We need also to measure the rate of convergence. Let $B = C^{-1}R$. We have $\mathbf{e}^m = B^m \mathbf{e}^0$, where $\mathbf{e}^m = \hat{\mathbf{x}} - \mathbf{x}^m$.

Definition 5.2 Let B be the *iteration matrix*, $B = C^{-1}R$, where $R = C - A$. Then, given a matrix norm $\| \cdot \|$, $\|B^m\|$ is called the *convergence factor* for m steps and $R_m = \|B^m\|^{1/m}$ is called the *average convergence factor* (per step for m steps) for this norm.

Unfortunately, in practice it is difficult to compute these numbers, because B^m is frequently expensive to compute, for instance, and $B = C^{-1}R$ may not even be explicitly available. One may, however, prove the following lemma.

Lemma 5.4 *Let B be a square matrix and $\| \cdot \|$ a matrix norm. Then there exist positive constants, $c, C \in \mathbb{R}$, such that*

$$(5.4) \qquad cm^{s-1}\rho(B)^m \leq \|B^m\| \leq Cm^{s-1}\rho(B)^m, \qquad m = 1, 2, \ldots,$$

where s is the order of the largest Jordan block that belongs to an eigenvalue λ with $|\lambda| = \rho(B)$.

Proof For a proof and a definition of Jordan blocks, see Exercise 5. See also D. Young, 1971 (Theorem 7.1, p. 85). $\diamond$

Note that (5.4) implies

$$(5.5) \qquad \|B^m\|^{1/m} \to \rho(B), \qquad m \to \infty.$$

Hence, asymptotically, the average convergence factor approaches the spectral radius. For a symmetric matrix B we have, in fact, $R_m \equiv \|B^m\|_2^{1/m} = \rho(B)$.

Caution: For unsymmetric matrices, however, $\rho(B)$ may be a far too optimistic estimate of $\|B^m\|^{1/m}$ for practical values of m, because even if $\rho(B) < 1$, $\|B^m\|$ does not have to converge *monotonically* to zero, as the following example shows.

Example 5.2 *Nonmonotone convergence:* Consider the matrix (a Jordan block of order two), $B = \begin{bmatrix} a & 4 \\ 0 & a \end{bmatrix}$, where $0 < a < 1$. We have $\rho(B) = a$, so $\rho(B) < 1$ and Theorem 5.3 show that the iterative method 5.3 converges if $B = C^{-1}R$. By induction we find

$$B^m = \begin{bmatrix} a^m & 4ma^{m-1} \\ 0 & a^m \end{bmatrix}$$

and

$$B^{m^T} B^m = \begin{bmatrix} a^{2m} & 4ma^{2m-1} \\ 4ma^{2m-1} & (a^{2m} + 16m^2 a^{2m-2}) \end{bmatrix}.$$

The largest eigenvalue λ_m of $B^{m^T} B^m$ is readily found to satisfy

$$\lambda_m \simeq a^{2m} + 8m^2 a^{2m-2}(1 + \sqrt{1 + a^2/4m^2}).$$

Since $\|B^m\|_2 = \lambda_m^{\frac{1}{2}}$, one finds for $a = 1 - \delta$, $0 < \delta \ll 1$, that $\|B^m\|_2$ increases until $m \simeq \delta^{-1}$, when $\|B^m\|_2 \simeq \{8\delta^{-2}e^{-2}\}^{\frac{1}{2}}$, and, for a certain m_0, $\|B^m\|_2^{1/m} \geq 1$ for all $m \leq m_0$. As an example, for $a = 0.99$, we find $m_0 = 805$!

We leave it to the interested reader to determine how many iterations are needed to get an iteration error in, say, maximum norm $\| \cdot \|_\infty$, smaller than the initial error, when $e^0 = (1, 1)^T$ and $a = 0.99$.

5.1.1 Lower and Upper Bounds of the Average Convergence Factor

It follows from Theorem 4.15(c) that

$$r(B^m) \le \| B^m \| \le 2r(B^m),$$

where $r(\cdot)$ is the numerical radius and the norm $\| \cdot \|$ is the spectral norm. Hence,

$$(5.4') \qquad r(B^m)^{1/m} \le \| B^m \|^{1/m} \le 2^{1/m}(r(B^m))^{1/m},$$

and we see that $r(B^m)^{1/m}$ becomes an increasingly accurate estimate of $\| B^m \|^{1/m}$ as $m \to \infty$. In particular, Theorem 4.19 shows that if $r(B) < 1$, then $r(B^m) \le [r(B)]^m < 1$ and $\| B^m \|^{1/m} \le 2^{1/m} r(B)$. (Observe that, in general, $r(B)$ is not a lower bound of $\| B^m \|^{1/m}$.) Observe that B is nonnegative ($B \ge 0$) in Example 5.2. Therefore (see Corollary 4.20), if B has order n,

$$r(B) = \max(B\mathbf{x}, \mathbf{x}) = \max \frac{1}{2}((B + B^T)\mathbf{x}, \mathbf{x}), \quad \forall \mathbf{x} \in \mathbb{R}^n, \|\mathbf{x}\|_2 = 1,$$

so

$$r(B) = \rho(\frac{1}{2}(B + B^T))$$

and we have

$$\frac{1}{2}(B^m + B^{m^T}) = a^m \begin{bmatrix} 1 & 2m/a \\ 2m/a & 1 \end{bmatrix}$$

and

$$\left\{ \rho(\frac{1}{2}(B^m + B^{m^T})) \right\}^{1/m} = a(1 + \frac{2m}{a})^{1/m}.$$

Hence,

$$\| B^m \|^{1/m} \ge r(B^m)^{1/m} = a(1 + \frac{2m}{a})^{1/m},$$

and this lower bound shows that $\| B^m \|^{1/m} \ge 1$, at least for $m \le m_1$, where $m_1 \simeq 740$ in Example 5.2.

The above discussion shows that *the numerical radius* is a more reliable function than the spectral radius to use when analyzing iterative methods for nonsymmetric matrices. Unfortunately, it generally is difficult to compute the numerical radius. However, if B is nonnegative, we can compute $r(B) = 1/2 \max \lambda(B + B^T)$ (Corollary 4.20) by using, for instance, a power iteration method for $B + B^T$. Note also that (5.4) and (5.4') show that

$$(cm^{s-1})^{\frac{1}{m}} \rho(B) \le \| B^m \|^{\frac{1}{m}} \le 2^{\frac{1}{m}} r(B)$$

and, hence,

$$\rho(B) = \lim_{m \to \infty} \| B^m \|^{\frac{1}{m}} \le r(B).$$

Since the spectral radius tells us the asymptotic rate of convergence, further analysis is still of interest.

Definition 5.3 $r_m = -\log_{10} R_m$, $R_m = \| B^m \|^{1/m}$ is called the *average rate of convergence*, and $r = -\log_{10} \rho(B) = -\log_{10} R_\infty$ is called the *asymptotic rate of convergence*.

Note that r_m and r are estimates of the number of new correct decimals per iteration. In practice, it is more convenient to estimate the natural logarithm (ln) of $\rho(B)$ and R_m. We recall that $\ln 10 \simeq 2.30$.

5.1.2 Stopping Tests

If $\|B\|$ is known and $\|B\| < 1$, we can use the following theorem to get a test when the iteration error is small enough.

Theorem 5.5 *Let* $\|B\| < 1$, $B = C^{-1}R$, *and* x^m *be defined by (5.3). Then*

(5.6) $$\| \hat{x} - x^m \| \le \frac{\|B\|}{1 - \|B\|} \| x^m - x^{m-1} \|, \quad m = 1, 2, \ldots.$$

Proof From (5.3) follows $x^{m+1} - x^m = B(x^m - x^{m-1})$ and, by recursion,

(5.7) $$x^{m+k+1} - x^{m+k} = B^{k+1}(x^m - x^{m-1}), \quad k = 0, 1, \ldots.$$

Note now that $x^{m+p} - x^m = \sum_{k=0}^{p-1}(x^{m+k+1} - x^{m+k})$. Hence, by the triangle inequality and by (5.7),

$$\|\mathbf{x}^{m+p} - \mathbf{x}^m\| \le \sum_{k=0}^{p-1} \|B^{k+1}\| \|\mathbf{x}^m - \mathbf{x}^{m-1}\|$$

$$\le \frac{\|B\| - \|B\|^{p+1}}{1 - \|B\|} \|\mathbf{x}^m - \mathbf{x}^{m-1}\|.$$

Letting $p \to \infty$ and noting that by Theorem 5.3, $\mathbf{x}^{m+p} \to \hat{\mathbf{x}}$, (5.6) follows. $\diamond$

Example 5.3 Let $A = \begin{bmatrix} 2 & -1 \\ -3 & 4 \end{bmatrix}$, $\mathbf{b} = \begin{bmatrix} 1 \\ 1 \end{bmatrix}$, $C = \begin{bmatrix} 2 & 0 \\ 0 & 4 \end{bmatrix}$, and $\mathbf{x}^0 = \begin{bmatrix} 0 \\ 0 \end{bmatrix}$.
Then with $\mathbf{x}^{(l+1)} = B\mathbf{x}^{(l)} + C^{-1}\mathbf{b}, l = 0, 1, \dots$ where $B = C^{-1}R, R = C - A$,
that is,

$$B = \begin{bmatrix} 0 & \frac{1}{2} \\ \frac{3}{4} & 0 \end{bmatrix},$$

we get

$$x_1^{(l+1)} = \frac{1}{2}x_2^{(l)} + \frac{1}{2}, \quad x_1^{(0)} = 0$$

$$x_2^{(l+1)} = \frac{3}{4}x_1^{(l)} + \frac{1}{4}, \quad x_2^{(0)} = 0,$$

that is,

$$x_1^{(1)} = \frac{1}{2}, \quad x_1^{(2)} = \frac{5}{8}, \quad x_1^{(3)} = \frac{13}{16}, \quad x_1^{(4)} = \frac{55}{64},$$

$$x_2^{(1)} = \frac{1}{4}, \quad x_2^{(2)} = \frac{5}{8}, \quad x_2^{(3)} = \frac{23}{32}, \quad x_2^{(4)} = \frac{55}{64}.$$

With $\|B\| = \|B\|_2$, we find $\|B\| = 3/4$, and Theorem 5.5 gives the estimate

$$\|\hat{\mathbf{x}} - \mathbf{x}^{(4)}\| \le \frac{\frac{3}{4}}{1 - \frac{3}{4}} \|\mathbf{x}^{(4)} - \mathbf{x}^{(3)}\| = 3 \cdot \frac{3\sqrt{10}}{64} = \frac{9}{64}\sqrt{10},$$

while the exact error is

$$\|\hat{\mathbf{x}} - \mathbf{x}^{(4)}\| = \frac{9}{64}\sqrt{2}.$$

As a practical test to stop the iterations on the basis of the size of some available quantity, we may, for instance, use one of the following:

(a) $\|\mathbf{r}^m\| = \|A\mathbf{x}^m - \mathbf{b}\| \le \varepsilon$

or

(b) $\|\mathbf{x}^m - \mathbf{x}^{m-1}\| \le \varepsilon$.

Then, in the first case, we have the iteration error $\|\hat{\mathbf{x}} - \mathbf{x}^m\| = \|A^{-1}\mathbf{r}^m\| \le \|A^{-1}\| \, \varepsilon$, and in the second case, if $\|B\| < 1$, (by Theorem 5.5), $\|\hat{\mathbf{x}} - \mathbf{x}^m\| \le \frac{\|B\|}{1-\|B\|}\varepsilon$. If we want a relative error $\|\hat{\mathbf{x}} - \mathbf{x}^m\|/\|\hat{\mathbf{x}}\| \le \delta$, then we have to choose ε accordingly—i.e., $\varepsilon \le \|\hat{\mathbf{x}}\|\delta/\|A^{-1}\|$ and $\varepsilon \le \frac{1-\|B\|}{\|B\|}\delta\|\hat{\mathbf{x}}\|$, respectively.

We may, alternatively, choose a relative iteration error-stopping test. If we want $\|\hat{\mathbf{x}} - \mathbf{x}^m\|/\|\hat{\mathbf{x}} - \mathbf{x}^0\| \le \delta$, then we choose $\varepsilon \le \frac{1-\|B\|}{\|B\|}\delta\|\hat{\mathbf{x}} - \mathbf{x}^0\|$. In practice, $\|A^{-1}\|$ is usually a large number and $\|B\|$ is close to 1. Hence, for any given δ, we must take ε much smaller. Since $\hat{\mathbf{x}}$ is not known, in practice we replace $\hat{\mathbf{x}}$ in these estimates (after some iterations have been performed) by the current approximation. Naturally, we need to estimate $\|A^{-1}\|$ and $\|B\|$, respectively.

Example 5.4 Consider the matrix A, defined by the difference operator $(-h^2\Delta_h^{(5)})$ on a unit square with Dirichlet boundary conditions. Let $C = 4I$ in (5.2). Then it can be seen that:

(a) $\|A^{-1}\| = \lambda_1^{-1} \simeq \frac{1}{2}(\pi h)^{-2}$ $[\lambda_1 = 4(1 - \cos \pi h)]$ and that $B = I - \frac{1}{4}A$, $\|B\| = \rho(B) = \cos \pi h$, where $\|\cdot\| = \|\cdot\|_2$.

(b) The number of iterations m needed to get a relative iteration error $\|\hat{\mathbf{x}} - \mathbf{x}^m\|/\|\hat{\mathbf{x}} - \mathbf{x}^0\| \le \varepsilon_0$ satisfies $m \simeq \ln\frac{1}{\varepsilon_0}/\ln\frac{1}{\rho_0} = O(h^{-2})$, $h \to 0$, where $\rho_0 = \|B\|$. To see this result use $\mathbf{e}^m = B^m\mathbf{e}^0$.

For stopping criteria based on computing componentwise error estimates, similar to the estimates discussed in Exercise 39, Appendix A (see Arioli et al., 1991).

Note that if $\mathbf{x}^0 = \mathbf{0}$, we get a relative error $\|\hat{\mathbf{x}} - \mathbf{x}^m\|/\|\hat{\mathbf{x}}\| \le \varepsilon_0$, and we find that the number of iterations for a relative error ε_0 is proportional to the number of correct digits $(\ln\frac{1}{\varepsilon_0})$ and to $(\ln\frac{1}{\rho_0})^{-1}$. Since $\lambda_1 = 4(1 - \rho_0)$, $\lambda_n = 4(1 + \rho_0)$, we get $\frac{1}{\rho_0} = (1 + \frac{\lambda_1}{\lambda_n})/(1 - \frac{\lambda_1}{\lambda_n})$. Hence, $(\ln\frac{1}{\rho_0})^{-1}$ is close to twice the condition number λ_n/λ_1. In practice, this is by far too large a number of iterations. We shall now consider improvements of the basic iterative method.

5.2 Stationary Iterative Methods

We show that the iterative method (5.2) can be generalized and improved by introducing parameters. For matrices with real and positive eigenvalues, the new methods, of first and second order, will be shown to be convergent for any initial vector if only the parameters take sufficiently small values. We also show the important result that for a second order, or two-step, iterative method, one may choose the corresponding stationary parameters such that the rate of convergence is improved by an order of magnitude.

Definition 5.4

(a) A *first-order*, or one-step iterative, *method* for the solution of (5.1) is defined by

$$(5.8) \quad C\mathbf{d}^{l+1} = -\tau_l\mathbf{r}^l, \quad \mathbf{x}^{l+1} = \mathbf{x}^l + \mathbf{d}^{l+1}, \quad l = 0, 1, \ldots,$$

where $\mathbf{r}^l = A\mathbf{x}^l - \mathbf{b}$, $\{\tau_l\}$ is a sequence of parameters and $\mathbf{x}^0$ is given. If $\tau_l = \tau, l = 0, 1, 2, \ldots$, then the method is called *stationary*; otherwise *nonstationary*.

(b) A *second-order*, or two-step iterative, *method* is defined by

$$(5.9) \quad C\mathbf{s}^l = \mathbf{r}^l, \quad \mathbf{x}^{l+1} = \alpha_l\mathbf{x}^l + (1 - \alpha_l)\mathbf{x}^{l-1} - \beta_l\mathbf{s}^l, \quad l = 0, 1, \ldots,$$

where $\{\alpha_l\}$, $\{\beta_l\}$ are sequences of parameters with $\alpha_0 = 1$. Note here that, for consistency, we must require that $\mathbf{x}^l = \hat{\mathbf{x}}$ satisfies the recursion. This explains the choice of the coefficient $(1 - \alpha_l)$. With $C = I$ we get

$$(5.8') \qquad\qquad \mathbf{x}^{l+1} = \mathbf{x}^l - \tau_l\mathbf{r}^l, \quad l = 0, 1, \ldots$$

$$(5.9') \quad \mathbf{x}^{l+1} = \alpha_l\mathbf{x}^l + (1 - \alpha_l)\mathbf{x}^{l-1} - \beta_l\mathbf{r}^l, \quad l = 0, 1, \ldots,$$

respectively.

5.2.1 Convergence Analysis

Consider the errors $\mathbf{e}^l = \hat{\mathbf{x}} - \mathbf{x}^l$. Note that $\mathbf{r}^l = -A\mathbf{e}^l$. From (5.8') follows

$$(5.10) \qquad\qquad \mathbf{e}^{l+1} = (I - \tau_l A)\mathbf{e}^l$$

and, by recursion,

$$(5.11) \qquad \mathbf{e}^m = \prod_{l=0}^{m-1}(I - \tau_l A)\mathbf{e}^0 = P_m(A)\mathbf{e}^0, \quad \mathbf{r}^m = P_m(A)\mathbf{r}^0$$

for some polynomial P_m of degree m such that $P_m(0) = 1$. Similarly, for (5.9'),

$$\mathbf{e}^{l+1} = \alpha_l\mathbf{e}^l + (1 - \alpha_l)\mathbf{e}^{l-1} - \beta_l A\mathbf{e}^l$$

and, because $\alpha_0 = 1$, we have

$$\mathbf{e}^m = P_m(A)\mathbf{e}^0, \quad \mathbf{r}^m = P_m(A)\mathbf{r}^0$$

for some polynomial

$$P_m(z) = 1 - \gamma_{m,1}z - \ldots - \gamma_{m,m}z^m$$

of mth degree, where $\gamma_{m,l}$, $l = 1, 2, \ldots, m$ are certain numbers, determined by α_l, and β_l, $l = 1, 2, \ldots, m - 1$. In the same manner, it follows from (5.8) and (5.9),

$$(5.12) \quad \mathbf{e}^m = P_m(C^{-1}A)\mathbf{e}^0, \quad \mathbf{r}^m = -P_m(AC^{-1})A\mathbf{e}^0 = P_m(AC^{-1})\mathbf{r}^0.$$

Consider, first, the case where A is symmetric and positive definite, with eigenvalues $\lambda_1, \ldots, \lambda_n$, and let $C = I$. Then $P_m(A)$ is symmetric with eigenvalues $P_m(\lambda_i)$, $i = 1, \ldots, n$. It follows from (5.11),

$$\|\mathbf{e}^m\|_2 = \|P_m(A)\mathbf{e}^0\|_2 \le \|P_m(A)\|_2 \|\mathbf{e}^0\|_2$$

$$= \max_{1 \le j \le n} |P_m(\lambda_j)| \|\mathbf{e}^0\|_2.$$

It is readily seen that for a certain $\mathbf{e}^0$, we have equality. Hence, we have convergence for any initial vector if and only if $P_m(\lambda_j) \to 0$, $m \to \infty$ $\forall \lambda_j$. Therefore, we make the important observation that the best choice of the parameters $\{\tau_l\}$ (and $\{\alpha_l\}$, $\{\beta_l\}$) for the fastest convergence, in general, is a function of the distribution of eigenvalues.

Example 5.5 Consider (5.8') with $\tau_l = \tau$. It is readily seen that if $\lambda_1 < 0 < \lambda_2$, then the method diverges for any τ, except for certain initial vectors. Consider now the stationary methods with a fixed set of parameters.

Notation: Let A be s.p.d. The weighted vector-norm is

$$\|\mathbf{x}\|_{A^\nu} = \{\mathbf{x}^T A^{2\nu} \mathbf{x}\}^{\frac{1}{2}},$$

where ν is a real number. The relation (5.12) shows that

$$\|\mathbf{e}^m\|_{A^\nu} = \|\mathbf{r}^m\|_{A^{\nu-1}},$$

and

$$(5.13) \qquad\qquad \|\mathbf{e}^m\|_{A^\nu} = \|\tilde{\mathbf{e}}^m\|_{A^{\nu-\frac{1}{2}}},$$

where $\tilde{\mathbf{e}}^m = A^{\frac{1}{2}}\mathbf{e}^m$.

5.2.2 The First-Order Stationary Iterative Method

Theorem 5.6 *Assume that C and A are s.p.d. and consider the first-order method (5.8) with a fixed parameter, i.e., $\tau_l = \tau$, $l = 0, 1, \ldots$. Assume, further, that $C^{-1}A$ has eigenvalues λ_j, with extreme eigenvalues λ_1, λ_n, where $0 < \lambda_1 < \lambda_n$, and let $\mathbf{e}^m = \hat{\mathbf{x}} - \mathbf{x}^m$. Then:*

(a)
$$\mathbf{e}^m = P_m(C^{-1}A)\mathbf{e}^0 = (I - \tau C^{-1}A)^m \mathbf{e}^0,$$

$$\|\mathbf{e}^m\|_{A^{\frac{1}{2}}} / \|\mathbf{e}^0\|_{A^{\frac{1}{2}}} \leq \rho(I - \tau C^{-\frac{1}{2}}AC^{-\frac{1}{2}})^m$$

$$= \max_{1 \leq j \leq n} |1 - \tau\lambda_j|^m = \max\{|1 - \tau\lambda_1|, |1 - \tau\lambda_n|\}^m,$$

so the method converges if $0 < \tau < 2/\lambda_n$.

(b) *Further,*

(5.14)
$$\min_\tau \rho(I - \tau C^{-\frac{1}{2}}AC^{-\frac{1}{2}}) = \min_\tau \{|1 - \tau\lambda_1|, |1 - \tau\lambda_n|\}$$

$$= (1 - \lambda_1/\lambda_n)/(1 + \lambda_1/\lambda_n).$$

This minimum value is taken for $\tau = \tau_{\text{opt}} = 2/(\lambda_1 + \lambda_n)$.

Proof A similarity transformation (5.12) shows that

$$\tilde{\mathbf{e}}^m = P_m(A^{\frac{1}{2}}C^{-1}A^{\frac{1}{2}})\tilde{\mathbf{e}}^0.$$

(5.13) shows that for $\nu = \frac{1}{2}$, and letting $\| \cdot \|$ denote the spectral norm,

$$\|\mathbf{e}^m\|_{A^{\frac{1}{2}}} / \|\mathbf{e}^0\|_{A^{\frac{1}{2}}} = \|\tilde{\mathbf{e}}^m\| / \|\tilde{\mathbf{e}}^{(0)}\|$$

$$\leq \|P_m(A^{\frac{1}{2}}C^{-1}A^{\frac{1}{2}})\|$$

$$= \rho(P_m(A^{\frac{1}{2}}C^{-1}A^{\frac{1}{2}}))$$

$$= \max_j |1 - \tau\lambda_j|^m,$$

where the last part follows from (5.10). The remainder of the theorem follows readily from Figure (5.1). The minimum value of $\rho(I - \tau A^{\frac{1}{2}}C^{-1}A^{\frac{1}{2}})$ is taken when $1 - \tau\lambda_1 = \tau\lambda_n - 1$. $\diamond$

Remark 5.7 For the stationary one-step method, we have

(5.15)
$$\mathbf{x}^{l+1} = B\mathbf{x}^l + \tilde{C}^{-1}\mathbf{b},$$

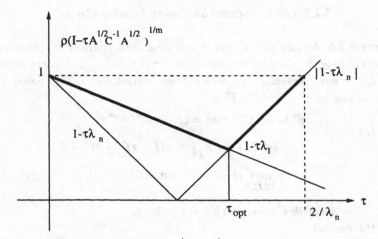

Figure 5.1. $\rho(I - \tau A^{\frac{1}{2}} C^{-1} A^{\frac{1}{2}})$ as a function of τ.

where B, the *iteration matrix*, is defined by

$$(5.16) \qquad B = I - \tilde{C}^{-1} A.$$

Here, $\tilde{C} = 1/\tau C$ contains the parameter τ; i.e., with no loss of generality, we may let $\tau = 1$.

5.2.3 *The Second-Order Stationary Iterative Method*

Consider now the second-order, or two-step stationary, iterative method (5.9), with $\alpha_l = \alpha$, $\beta_l = \beta$, $l = 1, 2, \ldots$, where $\mathbf{x}^1 = \mathbf{x}^0 - \beta_0 C^{-1} \mathbf{r}^0$. Then

$$(5.17) \qquad \begin{aligned} \mathbf{e}^{l+1} &= \alpha \mathbf{e}^l + (1 - \alpha) \mathbf{e}^{l-1} - \beta C^{-1} A \mathbf{e}^l, \quad l = 1, 2, \ldots \\ \mathbf{e}^1 &= \mathbf{e}^0 - \beta_0 C^{-1} A \mathbf{e}^0. \end{aligned}$$

We shall now find how α and β must be chosen for convergence and for optimal rate of convergence. We consider only the case where $C^{-1} A$ has real and positive eigenvalues, with extreme values λ_1, λ_n. Let

$$\mathbf{w}^l = \begin{bmatrix} \mathbf{e}^{l+1} \\ \mathbf{e}^l \end{bmatrix}.$$

It then follows from (5.17) that

$$\mathbf{w}^l = A \mathbf{w}^{l-1}, \quad l = 1, 2, \ldots,$$

where

$$A = \begin{bmatrix} (\alpha I - \beta C^{-1} A) & (1-\alpha)I \\ I & 0 \end{bmatrix}$$

and I is the identity matrix of order N. Hence, $\mathbf{w}^m = \mathcal{A}^m \mathbf{w}^0$, and it follows from Lemma 5.2 that the method converges for any $\mathbf{x}^0$ if and only if the spectral radius of $\mathcal{A}$, $\rho(\mathcal{A}) < 1$. Consider then the eigenvalue problem

$$A \begin{bmatrix} \mathbf{x} \\ \mathbf{y} \end{bmatrix} = \mu \begin{bmatrix} \mathbf{x} \\ \mathbf{y} \end{bmatrix}.$$

It follows readily that if $\lambda, \mathbf{x}$ is an eigensolution of $C^{-1} A$, that is, if $C^{-1} A \mathbf{x} = \lambda \mathbf{x}$, then

$$A \begin{bmatrix} \mathbf{x} \\ \mathbf{y} \end{bmatrix} = \begin{bmatrix} (\alpha - \beta \lambda)I & (1-\alpha)I \\ I & 0 \end{bmatrix} \begin{bmatrix} \mathbf{x} \\ \mathbf{y} \end{bmatrix} = \mu \begin{bmatrix} \mathbf{x} \\ \mathbf{y} \end{bmatrix}.$$

To every λ there exist, in general, two values of μ, $\mu_i = \mu_i(\lambda)$, $i = 1, 2$, which are eigenvalues of

$$\begin{bmatrix} \alpha - \beta \lambda & 1 - \alpha \\ 1 & 0 \end{bmatrix} \begin{bmatrix} \xi \\ \eta \end{bmatrix} = \mu \begin{bmatrix} \xi \\ \eta \end{bmatrix}, \qquad (\xi, \eta)^T \in \mathbb{C}^2,$$

and $(\mathbf{x}, \mathbf{y}) = (\xi \mathbf{x}, \eta \mathbf{x})$, $\xi = 1$. Hence,

$$\det \begin{bmatrix} (\alpha - \beta \lambda - \mu) & 1 - \alpha \\ 1 & (-\mu) \end{bmatrix} = 0$$

or

(5.18) $$\mu(\mu - \alpha + \beta \lambda) = 1 - \alpha,$$

that is,

$$\begin{cases} \mu_1 \mu_2 = \alpha - 1, \\ \mu_1 + \mu_2 = \alpha - \beta \lambda. \end{cases}$$

To study the roots of (5.18), we consider first the following lemma.

Lemma 5.8 *The second-degree equation $z^2 - \gamma z + \delta = 0$, where γ, δ are real, has roots z_1, z_2 with maximum moduli $z_0 = \max\{|z_1|, |z_2|\} < 1$ if and only if $|\delta| < 1$ and $|\gamma| < 1 + \delta$. If $\delta > 0$ is fixed but γ varies, then z_0 is smallest, $z_0 = \delta^{\frac{1}{2}}$, when $(\gamma/2)^2 \le \delta$, that is, when the roots are conjugate complex.*

Proof We have $z_1 z_2 = \delta$, so $z_0 = \max\{|z_1|, |z_2|\} \geq |\delta|^{\frac{1}{2}}$. If $(\gamma/2)^2 < \delta$ (i.e., in the first place, $\delta > 0$), then the roots are conjugate complex and $z_0 = \delta^{\frac{1}{2}}$. If $(\gamma/2)^2 \geq \delta$, then the roots are real.

Assume first that $z_0 < 1$. Then $|\delta| < 1$, and if $(\gamma/2)^2 < \delta$, then $|\gamma| < 2\delta^{\frac{1}{2}} \leq 1 + \delta$. If $(\gamma/2)^2 \geq \delta$, then

$$(5.19) \qquad z_0 = \frac{1}{2}|\gamma| + |(\frac{1}{2}\gamma)^2 - \delta|^{\frac{1}{2}},$$

which is an increasing function of $|\gamma|$. For $|\gamma| = 1 + \delta$, we get $z_0 = 1 (|z_1| = 1, |z_2| = \delta)$. Hence, $|\gamma| < 1 + \delta$ when $z_0 < 1$. If, conversely, $|\delta| < 1$ and $|\gamma| < 1 + \delta$, (5.19) shows that $z_0 < \max\{1, |\delta|^{\frac{1}{2}}\} = 1$.   ◇

Theorem 5.9 *Assume that the eigenvalues λ_i of $C^{-1}A$ are positive, with extreme values $\lambda_1, \lambda_n, \lambda_1 < \lambda_n$, and that the parameters α, β in the second-order method (5.9), with $\alpha_l = \alpha, \beta_l = \beta, l = 1, 2, \ldots \alpha_0 = 1$, are real-valued. Then:*

(a) *(5.9) converges if and only if $0 < \alpha < 2, \quad 0 < \beta < 2\alpha/\lambda_n$.*

(b) *The asymptotic convergence factor is smallest and equal to*

$$
\{\alpha_{\text{opt}} - 1\}^{\frac{1}{2}} = \left\{ \left[1 - \sqrt{1 - \rho_0^2} \right] \Big/ \left[1 + \sqrt{1 - \rho_0^2} \right] \right\}^{\frac{1}{2}}
$$
$$(5.20)$$
$$
= \left[1 - \sqrt{\frac{\lambda_1}{\lambda_n}} \right] \Big/ \left[1 + \sqrt{\frac{\lambda_1}{\lambda_n}} \right],
$$

for

$$(5.21) \quad \alpha = \alpha_{\text{opt}} = 2/[1 + (1 - \rho_0^2)^{\frac{1}{2}}], \quad \beta = \beta_{\text{opt}} = \frac{2\alpha}{\lambda_1 + \lambda_n},$$

where $\rho_0 = [1 - \lambda_1/\lambda_n]/[1 + \lambda_1/\lambda_n]$.

Proof By Theorem 5.3, the iterative method (5.17) converges if and only if the asymptotic convergence factor $\rho(\mathcal{A}) < 1$, that is, if and only if $\mu_0 = \max\{|\mu_1|, |\mu_2|\} < 1$ for all λ_i, $\lambda_1 \leq \lambda_i \leq \lambda_n$, where μ_i are the roots of (5.18), that is, of $\mu^2 - (\alpha - \beta\lambda)\mu + \alpha - 1 = 0$. Hence, by Lemma 5.8, we have convergence if and only if $-1 < \alpha - 1 < 1$ and $-\alpha < \alpha - \beta\lambda < \alpha$, that is, $0 < \alpha < 2$ and $0 < \beta < 2\alpha/\lambda_n$, which proves (a). If $0 < \alpha \leq \alpha_1, \alpha_1 > 1$ but small enough, the roots μ_1, μ_2 are real. For α fixed, it follows from (5.19) (with $\gamma = \alpha - \beta\lambda$, $\delta = \alpha - 1$), that the smallest value of μ_0 is taken when $\max_{\lambda_1 \le \lambda \le \lambda_n} |\alpha - \beta\lambda|$, is smallest—i.e., when (cf. Figure 5.1)

$$\alpha - \beta\lambda_1 = \beta\lambda_n - \alpha \quad \text{or} \quad \beta = 2\alpha/(\lambda_1 + \lambda_n).$$

For this value of β, we get $\alpha - \beta\lambda = \alpha\nu$, and

$$\mu_0 = \frac{1}{2}\alpha\nu + \left[1 - \alpha + (\frac{1}{2}\alpha\nu)^2\right]^{\frac{1}{2}},$$

where $\nu = 1 - 2\lambda/(\lambda_1 + \lambda_n)$. Note that for $\lambda_1 \le \lambda \le \lambda_n$, we have $-\rho_0 \le \nu \le \rho_0$, where $\rho_0 = (1 - \lambda_1/\lambda_n)/(1 + \lambda_1/\lambda_n)$, and

$$(5.22) \qquad f(\alpha) \equiv \max_{|\nu| \le \rho_0} \mu_0 = \frac{1}{2}\alpha\rho_0 + \left\{1 - \alpha + (\frac{1}{2}\alpha\rho_0)^2\right\}^{\frac{1}{2}}.$$

We find $f'(\alpha) \le 0$, $0 \le \alpha < \alpha_1$ and that $f(\alpha)$ is smallest for $\alpha = \alpha_1$, where α_1 is the smallest root of $1 - \alpha + (\frac{1}{2}\alpha\rho_0)^2 = 0$, that is,

$$\alpha_1 = \frac{2}{\rho_0^2}(1 - (1 - \rho_0^2)^{\frac{1}{2}}) = \alpha_{\text{opt}} = 2/[1 + (1 - \rho_0^2)^{\frac{1}{2}}].$$

It follows that for this value, $\alpha = \alpha_{\text{opt}}$, $\max_{|\nu| \le \rho_0} \mu_0$ is smallest, and that for $0 < \alpha \le \alpha_1$, $\max_{|\nu| \le \rho_0} |\mu_0| = \{\alpha_{\text{opt}} - 1\}^{\frac{1}{2}}$ and α_{opt} satisfies (5.20). If $\alpha > \alpha_1$, then $\mu_0 = \{\alpha - 1\}^{\frac{1}{2}}$; hence, $\min_\alpha \max_{|\nu| \le \rho_0} |\mu_0| = \{\alpha_{\text{opt}} - 1\}^{\frac{1}{2}}$ for all α. ◇

Remark 5.10 It follows from (5.22) that $f(\alpha) = \max_{|\nu| \le \rho_0} \mu_0$ varies for $0 \le \alpha \le 2$, as shown in Figure 5.2a.

It is readily seen that $f'(\alpha_1) = -\infty$. For $\alpha > \alpha_1$, we have $\mu_0 = (\alpha - 1)^{\frac{1}{2}}$ and the eigenvalues μ_1, μ_2 are distributed about a circle for $\lambda_1 \le \lambda \le \lambda_n$, as shown in Figure 5.2b. Note that it is better to overestimate α than to underestimate it, because $(\alpha - 1)^{\frac{1}{2}}$ increases much more slowly for increasing $\alpha > \alpha_{\text{opt}}$ than $f(\alpha)$ increases for $\alpha < \alpha_{\text{opt}}$.

To apply this theorem, we must estimate λ_1 and λ_n. In many problems it is easy to estimate λ_n from above by a matrix norm, for instance. Since usually $\lambda_1 \ll \lambda_n$, we can choose $\beta = 2\alpha/\lambda_n$. The parameter α, however, is more sensitive to the estimation of λ_1, and to estimate λ_1 accurately is usually more difficult. Various methods to estimate eigenvalues are discussed in Appendix A.

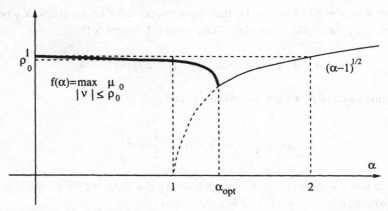

Figure 5.2a.   The asymptotic convergence factor $\rho(\mathcal{A})$ as a function of α.

It remains to choose β_0. This choice is arbitrary, but $\beta_0 = \tau_{\text{opt}}$ seems most reasonable. For $\beta_0 = 0$, on the other hand, we get $\mathbf{e}^1 = \mathbf{e}^0$, which simplifies the estimation of $\mathbf{w}^m$, because then $\mathbf{w}^{0^T} = (\mathbf{e}^{0^T}, \mathbf{e}^{0^T})$.

Let us now summarize the two types of stationary methods we have considered.

(a) The first-order method:

$$(5.23) \quad \mathbf{x}^{l+1} = \mathbf{x}^l - \tau C^{-1}\mathbf{r}^l, \quad \mathbf{r}^l = A\mathbf{x}^l - \mathbf{b}, \quad l = 0, 1, \dots,$$

where $\mathbf{x}^0$ is given and $\tau = \tau_{\text{opt}} = 2/(\lambda_1 + \lambda_n)$. The average convergence factor is

$$\rho_0 = [1 - \lambda_1/\lambda_n]/[1 + \lambda_1/\lambda_n].$$

(b) The second-order method:

$$
\begin{aligned}
(5.24) \quad & \mathbf{x}^{l+1} = \alpha\mathbf{x}^l + (1 - \alpha)\mathbf{x}^{l-1} - \beta C^{-1}\mathbf{r}^l, \\
& \mathbf{r}^l = A\mathbf{x}^l - \mathbf{b}, \quad l = 1, 2, \dots,
\end{aligned}
$$

where $\mathbf{x}^0$ is given,

$$\mathbf{x}^1 = \mathbf{x}^0 - \beta_0 C^{-1}\mathbf{r}^0;$$

$$\beta_0 = 2/(\lambda_1 + \lambda_n), \quad \beta = 2\alpha/(\lambda_1 + \lambda_n), \quad \alpha = 2/[1 + (1 - \rho_0^2)^{\frac{1}{2}}].$$

The average convergence factor is

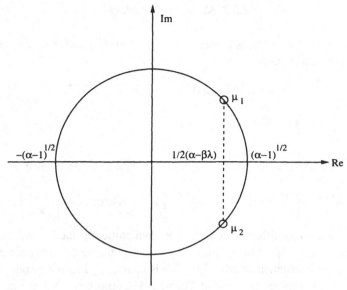

Figure 5.2b. Distribution of eigenvalues μ_i, $i = 1, 2$ for $\alpha \geq \alpha_{\text{opt}}$.

$$(\alpha - 1)^{\frac{1}{2}} = \left\{ \left[1 - (1 - \rho_0^2)^{\frac{1}{2}} \right] \bigg/ \left[1 + (1 - \rho_0^2)^{\frac{1}{2}} \right] \right\}^{\frac{1}{2}}$$

$$= \left[1 - \sqrt{\lambda_1/\lambda_n} \right] \bigg/ \left[1 + \sqrt{\lambda_1/\lambda_n} \right]$$

Note: Method (5.24) was first presented by Frankel in 1950.

5.2.4 Efficient Implementation

In the splitting $A = C - R$, sometimes R is much sparser than A, and it may then be more efficient in practice to use the alternative forms:

(5.23′) $\mathbf{x}^{l+1} = (1 - \tau)\mathbf{x}^l + \tau C^{-1}(R\mathbf{x}^l + \mathbf{b})$, $l = 0, 1, \ldots$.

(5.24′) $\mathbf{x}^{l+1} = (\alpha - \beta)\mathbf{x}^l + (1 - \alpha)\mathbf{x}^{l-1} + \beta C^{-1}(R\mathbf{x}^l + \mathbf{b})$, $l = 1, 2, \ldots$.

In both cases, $\mathbf{y}^l = C^{-1}(R\mathbf{x}^l + \mathbf{b})$ is computed by the solution of the linear system $C\mathbf{y}^l = R\mathbf{x}^l + \mathbf{b}$.

5.2.5 Numerical Stability

We now show that (5.24) is numerically stable. To this end, we consider the homogeneous recursion

$$\mathbf{x}^{l+1} = \alpha \mathbf{x}^l + (1-\alpha)\mathbf{x}^{l-1} - \beta F \mathbf{x}^l, \quad l = 1, 2, \ldots$$

$$\mathbf{x}^1 = \mathbf{x}^0 - \beta_0 F \mathbf{x}^0,$$

where $F = C^{-1}A$.

This recursion can be written as a system of first-order equations:

$$\mathbf{w}^{l+1} = \begin{bmatrix} (\alpha I - \beta F) & (1-\alpha)I \\ I & 0 \end{bmatrix} \mathbf{w}^l, \quad \text{where } \mathbf{w}^l = \begin{bmatrix} \mathbf{x}^{l+1} \\ \mathbf{x}^l \end{bmatrix}.$$

This homogeneous difference equation is asymptotically stable, and is therefore stable with respect to perturbations of the initial vector $\mathbf{x}^0$ if the eigenvalues of the coefficient matrix $\mathcal{A}$ (which is constant—i.e., independent of l) are < 1 in modulus. In the proof of Theorem 5.9, however, it has already been shown that $\rho(\mathcal{A}) < 1$.

Example 5.6 *Rate of Convergence:* Let $C = 4I$ and let A be the five-point difference matrix for a unit square with Dirichlet boundary conditions. Then $\lambda_1 = 1 - \cos \pi h$, $\lambda_n = 1 + \cos \pi h$, $\rho_0 = \cos \pi h \simeq 1 - (\pi h)^2/2$, $h \to 0$, and by (5.20),

$$(\alpha_{\text{opt}} - 1)^{\frac{1}{2}} = [(1 - \sin \pi h)/(1 + \sin \pi h)]^{\frac{1}{2}}$$

$$= \cos \pi h/(1 + \sin \pi h) \simeq 1 - \pi h, \quad h \to 0.$$

This is the smallest average convergence factor for the stationary two-step method (5.9) with $\alpha_l = \alpha$, $\beta_l = \beta$, $l = 1, 2, \ldots$ and with $\alpha \beta$ defined by (5.21). For this problem, we have $\alpha = \beta = 2/(1 + \sin \pi h)$.

Compare this result with the (average) convergence factor for the one-step stationary method (5.8), with $\tau_l = \tau = \tau_{\text{opt}}$. Then, by (5.14), this factor takes the value $\rho_0 = (1 - \lambda_1/\lambda_n)/(1 + \lambda_1/\lambda_n) \simeq 1 - (\pi h)^2/2$. Hence, the optimal two-step method converges with an order of magnitude faster rate than the optimal one-step stationary method! For $h = 1/64$, for instance, we get $(\alpha_{\text{opt}} - 1)^{1/2} \simeq 1 - \pi/64 \simeq 0.95$ and $\rho \simeq 0.9988$, respectively. Hence, for the latter method we get a very slow rate of convergence (about 2000 iterations for just one decimal digit!) For the second-order method, we need about 47 iterations for one decimal digit. The convergence rate can be further speeded up by an appropriate choice of the matrix C. (For details, see Chapters 7 and 8.)

5.3 The Chebyshev Iterative Method

In this section we shall consider variable parameter versions of iterative methods (5.23) and (5.24). It will be shown that the asymptotic rate of convergence of the new methods will be the same as for the second-order stationary method with optimal parameters, as given in Theorem 5.9(b), but it turns out that the Chebyshev iterative method is less sensitive to estimates of the extreme eigenvalues. Consider the first-order iterative method with $C = I$:

$$(5.25) \qquad \mathbf{x}^{l+1} = \mathbf{x}^l - \tau_l(A\mathbf{x}^l - \mathbf{b}), \quad l = 0, 1, \ldots.$$

Here A is s.p.d. and $\{\tau_l\}$ is a parameter set, the proper choice of which will give an accelerated convergence over the simplest choice, $\tau_l = \tau$, $l = 0, 1, 2, \ldots$. With $\tau = 2/(\lambda_n + \lambda_1)$, we have the smallest spectral radius $\rho(I - \tau A) = (1 - \lambda_1/\lambda_n)/(1 + \lambda_1/\lambda_n)$. The relative error in the Euclidean norm is then decreased to a number at most $\varepsilon > 0$, if $[(1 - \lambda_1/\lambda_n)/(1 + \lambda_1/\lambda_n)]^p \leq \varepsilon$, and this inequality is satisfied if

$$p \geq \frac{1}{2}\frac{\lambda_n}{\lambda_1} \ln\frac{1}{\varepsilon}.$$

The number of necessary iterations is thus, in general, directly proportional to the spectral condition number of A.

To get an accelerated convergence, we choose a suitable set $\{\tau_l\}$ in (5.25) for the purpose of minimizing the norm of the corresponding iteration matrix after p iterations. Then the errors satisfy

$$\mathbf{e}^p = Q_p(A)\mathbf{e}^0,$$

where

$$\mathbf{e}^p = \hat{\mathbf{x}} - \mathbf{x}^p,$$

$$(5.26) \qquad Q_p(\lambda) = \prod_{l=0}^{p-1}(1 - \tau_l\lambda),$$

and $Q_p(A)$ is the corresponding matrix polynomial. We observe that $Q_p(0) = 1$. We denote by π_p^1 and π_p^0 the set of polynomials of degree at most p, which take the values unity and zero, respectively, at the origin.

More generally, we would like to minimize the residual $\mathbf{r}^p = A\mathbf{x}^p - \mathbf{b}$ or the error $\mathbf{e}^p = -A^{-1}\mathbf{r}^p$ in the norm $\|\mathbf{e}^p\|_{A^\nu}$. Then we have

$$\|\mathbf{e}^p\|_{A^\nu} \leq \|Q_p(A)\|_{A^\nu}\|\mathbf{e}^0\|_{A^\nu}.$$

Here

$$\|Q_p(A)\|_{A^\nu} = \max_{\mathbf{x}\neq\mathbf{0}} \frac{\|Q_p(A)\mathbf{x}\|_{A^\nu}}{\|\mathbf{x}\|_{A^\nu}} = \max_{\mathbf{x}\neq\mathbf{0}} \frac{\|Q_p(A)A^\nu\mathbf{x}\|}{\|A^\nu\mathbf{x}\|}$$

$$= \max_{\mathbf{y}\neq\mathbf{0}} \frac{\|Q_p(A)\mathbf{y}\|}{\|\mathbf{y}\|} = \max |\mu_i|,$$

where $\{\mu_i\}$ are the eigenvalues of $Q_p(A)$. As we know, $\mu_i = Q_p(\lambda_i)$, so

$$\|Q_p(A)\|_{A^\nu} = \max_i |Q_p(\lambda_i)|,$$

which thus is what we want to minimize.

In general, the best polynomial approximation on a discrete set of points is not readily available. Hence, we simplify the approximation problem to the corresponding approximation problem for the continuous interval $[\lambda_1, \lambda_n]$. Then,

$$\|Q_p(A)\|_{A^\nu} = \max_i |Q_p(\lambda_i)| \leq \max_{\lambda_1 \leq \lambda \leq \lambda_n} |Q_p(\lambda)|.$$

This shows that the best approximation polynomial on the continuous interval provides us with an upper bound of $\|Q_p(A)\|_{A^\nu}$. It can be shown that there exist distributions of eigenvalues λ_i for which equality holds above.

It is well known (see Appendix B) that the least maximum is achieved by the Chebyshev polynomials, namely,

$$(5.27) \quad \min_{Q_p \in \pi_p^1} \max_{\lambda_1 \leq \lambda \leq \lambda_n} |Q_p(\lambda)| = \max_{\lambda_1 \leq \lambda \leq \lambda_n} \frac{|T_p[(\lambda_n + \lambda_1 - 2\lambda)/(\lambda_n - \lambda_1)]|}{T_p[(\lambda_n + \lambda_1)/(\lambda_n - \lambda_1)]}$$

$$= 1/T_p[(\lambda_n + \lambda_1)/(\lambda_n - \lambda_1)],$$

where T_p is the Chebyshev polynomial of first kind—i.e.,

$$(5.28) \quad T_p(z) = \frac{1}{2}\left[(z + \sqrt{z^2 - 1})^p + (z - \sqrt{z^2 - 1})^p\right]$$

or, equivalently,

$$T_p(z) = \begin{cases} \cos(p\cos^{-1}(z)), & \text{for } -1 \leq z \leq 1 \\ \cosh(p\cosh^{-1}(z)), & \text{for } z \geq 1 \\ (-1)^p \cosh(p\cosh^{-1}(-z)), & \text{for } z \leq -1. \end{cases}$$

(5.26) shows that the zeros of Q_p are $\lambda = 1/\tau_l$, $l = 0, 1, \ldots, p - 1$. Hence, the optimal parameters τ_l (for the approximation problem, simplified to an interval) are given by the inverses of the zeros of T_p, or,

$$(5.29) \qquad \frac{1}{\tau_l} = \frac{\lambda_n - \lambda_1}{2} \cos \theta_l + \frac{\lambda_n + \lambda_1}{2},$$

where

$$\theta_l = \frac{2 \cdot l + 1}{2p} \pi, \qquad l = 0, 1, \ldots, p - 1.$$

In practice, the smallest and largest eigenvalues are not known, so we need lower and upper bounds, $a(a > 0)$ and b, respectively, of these eigenvalues. Then we have to use the parameters

$$(5.29') \qquad \frac{1}{\tau_l} = \frac{b - a}{2} \cos \theta_l + \frac{b + a}{2}.$$

(5.27), (5.28) and an elementary computation show that

$$(5.30) \qquad \min_{Q_p \in \pi_p^1} \max_{\lambda_1 \le \lambda \le \lambda_n} |Q_p(\lambda)| = \frac{1}{T_p \left(\frac{\lambda_n + \lambda_1}{\lambda_n - \lambda_1} \right)} = 2 \frac{\sigma^p}{1 + \sigma^{2p}},$$

where $\sigma = (1 - \sqrt{\lambda_1/\lambda_n})/(1 + \sqrt{\lambda_1/\lambda_n})$. This implies that the asymptotic average reduction rate is $\lim_{p \to \infty} \{\min \max |Q_p(\lambda)|\}^{\frac{1}{p}} = \sigma$. Also, for any ε, $0 < \varepsilon < 1$, (5.30) shows that

$$(5.30') \qquad \|e^p\|_{A^\nu}/\|e^0\|_{A^\nu} \le \varepsilon$$

holds if

$$2 \frac{\sigma^p}{1 + \sigma^{2p}} \le \varepsilon.$$

An elementary computation shows that this result holds for any $p \ge p^*$, where

$$(5.31) \qquad p^* = \ln \left(\frac{1}{\varepsilon} + \sqrt{\frac{1}{\varepsilon^2} - 1} \right) / \ln \sigma^{-1}.$$

For $\varepsilon \ll 1$ and $\lambda_n/\lambda_1 \gg 1$, we find the following accurate upper bound:

$$(5.32) \qquad p^* \le \lceil \ln \frac{2}{\varepsilon} / \ln \sigma^{-1} \rceil \le \frac{1}{2} \left(\frac{\lambda_n}{\lambda_1} \right)^{\frac{1}{2}} \ln \frac{2}{\varepsilon}.$$

($\lceil a \rceil$ denotes the ceiling of a, that is, the smallest integer larger or equal to a.)

The Chebyshev iteration method can also be applied with a preconditioning matrix C if $C^{-1}A$ has positive eigenvalues. In particular, it is applicable if both C and A are s.p.d. Method (5.25) with τ_l defined by (5.29) is called the *first-order Chebyshev iterative method*.

The above shows the important result that the number of iterations of the Chebyshev iterative method increases at most as the square root of the condition number of $C^{-1}A$.

Remark 5.11 The above method defines the classic Chebyshev semi-iterative method (Golub and Varga, 1961, or Richardson, 1911 and 1925) method. Note that (5.29) shows that we have to choose the number of iteration steps, p, a priori; this can be done using (5.31) or (5.32). Unfortunately, there exist indices l, for which the matrices $I - \tau_l C^{-\frac{1}{2}} A C^{-\frac{1}{2}}$ have a spectral radius much larger than 1. This property causes the process to be numerically unstable (see Exercise 5.10), unless we use some particular ordering of the parameters so that large values of τ_l are balanced by small values. This topic is discussed by Lebedev and Finogenov (1971 and 1973) and will not be further discussed here. The disadvantages of having to choose p a priori, with possible numerical instability, can be eliminated in the following way.

5.3.1 The Second-Order Chebyshev Iterative Method

Consider the second-order method

(5.33)
$$x^{l+1} = \alpha_l x^l + (1 - \alpha_l)x^{l-1} - \beta_l r^l, \quad l = 1, 2, \ldots,$$
$$x^1 = x^0 - \tfrac{1}{2}\beta_0 r^0.$$

Let us show how to choose the parameter set $\{\alpha_l, \beta_l\}$ in order that this process, apart from rounding errors, gives the optimal result for every l, while the one-step Chebyshev process does this only for a predetermined p. With the notation introduced previously, we have

(5.34)
$$e^l = Q_l(A)e^0.$$

Since the recursion formula (5.33) is valid for all initial vectors, we get

$$Q_{l+1}(A) - \alpha_l Q_l(A) + \beta_l A Q_l(A) + (\alpha_l - 1)Q_{l-1}(A) = 0, \quad l = 1, 2, \ldots.$$

Comparing this with the recursion formula for the Chebyshev polynomials

$$T_0(z) = 1, \quad T_1(z) = z, \quad T_{l+1}(z) - 2zT_l(z) + T_{l-1}(z) = 0, \quad l = 1, 2, \ldots,$$

we find that if

$$\alpha_l = 2\tilde{b}T_l(\tilde{b})/T_{l+1}(\tilde{b}) = 1 + T_{l-1}(\tilde{b})/T_{l+1}(\tilde{b})$$

$$\beta_l = \frac{4}{b-a}T_l(\tilde{b})/T_{l+1}(\tilde{b}), \quad \tilde{b} = (b+a)/(b-a),$$

where $0 < a \le \lambda_1, b \ge \lambda_n$, then

(5.35)
$$Q_l(A) = T_l(Z)/T_l(\tilde{b}),$$
$$Z = \frac{1}{b-a}[(b+a)I - 2A],$$

which is the result we want. Elementary computations now give the recursion formulae

(5.36) $\qquad \alpha_l = \dfrac{a+b}{2}\beta_l, \quad \beta_l^{-1} = \dfrac{a+b}{2} - \left(\dfrac{b-a}{4}\right)^2 \beta_{l-1}, \quad l = 1, 2, \ldots,$

where $\beta_0 = 4/(a+b)$. Note that $\alpha_l > 1, l \ge 1$.

It is readily seen that the sequence $\{\beta_l\}$ decreases monotonically and converges to $\tilde{\beta} = 4/(\sqrt{b}+\sqrt{a})^2, l \to \infty$. This limit value can be seen to be equal to the optimal parameter β_{opt} in the stationary second-order method (see Theorem 5.9(b) and also Young, 1971). Incidentally, in a problem where $A = I - B$ and the eigenvalues of B belong to the interval $[-\rho, \rho], 1 > \rho = \rho(B)$ (the spectral radius of B), we have

$$a = 1 - \rho, \quad b = 1 + \rho,$$

and we find

(5.37) $\qquad\qquad \alpha = \beta = \tilde{\beta} = 2/\left(1 + \sqrt{1 - \rho^2}\right),$

which will be recognized as the parameter α_{opt} in the stationary method.

Finally, we will prove that the two-step version of the Chebyshev iterative method has the additional advantage over the one-step method of being numerically stable. Consider the homogeneous difference equation corresponding to the two-step formula (5.33), i.e.,

(5.38)
$$\mathbf{x}^{l+1} - \alpha_l\mathbf{x}^l + \beta_l A\mathbf{x}^l + (\alpha_l - 1)\mathbf{x}^{l-1} = 0, l = 1, 2, \ldots,$$
$$\mathbf{x}^1 = \mathbf{x}^0 - \frac{1}{2}\beta_0 A\mathbf{x}^0,$$

where $\mathbf{x}^0$ is an arbitrary vector. However, (5.34) and (5.35) show that

$$\mathbf{x}^l = T_l(\tilde{b})^{-1} T_l(Z) \mathbf{x}^0,$$

where $Z = 1/(b-a)[(b+a)I - 2A]$ and $\tilde{b} = (b+a)/(b-a)$. Hence $\tilde{b} > 1$, while the eigenvalues of Z belong to the interval $(-1, 1)$. This shows that $\|\mathbf{x}^l\|_2 < \|\mathbf{x}^0\|_2$, $l \geq 1$, that is, the recursion (5.38) is numerically stable for perturbations of the initial vector and also for rounding errors.

Method (5.33) with α_l, β_l defined by (5.36), is called the *second-order Chebyshev iterative method*.

Let us now summarize these interesting results for the second-order method (5.33).

Theorem 5.12 *Let C and A be s.p.d. and consider the second-order method,*

$$\mathbf{x}^{l+1} = \alpha_l \mathbf{x}^l + (1 - \alpha_l)\mathbf{x}^{l-1} - \beta_l \mathbf{r}^l, \quad \mathbf{r}^l = C^{-1}(A\mathbf{x}^l - \mathbf{b})$$

and $\mathbf{x}^1 = \mathbf{x}^0 - \beta_0 \mathbf{r}^0/2$, *where*

$$\alpha_l = \frac{a+b}{2}\beta_l, \quad \beta_l^{-1} = \frac{a+b}{2} - \left(\frac{b-a}{4}\right)^2 \beta_{l-1}, \quad l = 1, 2, \ldots$$

and $\beta_0 = 4/(a+b)$. *We have* $\lim_{l\to\infty} \beta_l = 4/(\sqrt{b} + \sqrt{a})^2$. *Let* $\mathbf{e}^l = \hat{\mathbf{x}} - \mathbf{x}^l$, *where* $A\hat{\mathbf{x}} = \mathbf{b}$. *Then*

$$\|\mathbf{e}^l\|_{A^\nu}/\|\mathbf{e}^0\|_{A^\nu} \leq 1/T_l\left(\frac{b+a}{b-a}\right) \leq 2\frac{\sigma^l}{1+\sigma^{2l}},$$

where $0 < a \leq \lambda_1$, $b \geq \lambda_n$, $\sigma = \left(1 - \sqrt{a/b}\right)/\left(1 + \sqrt{a/b}\right)$, *so the relative error converges monotonically to zero. We find*

$$\|\mathbf{e}^p\|_{A^\nu}/\|\mathbf{e}^0\|_{A^\nu} \leq \varepsilon,$$

if

$$p \geq \ln\left(\frac{1}{\varepsilon} + \sqrt{\frac{1}{\varepsilon^2} - 1}\right)/\ln\sigma^{-1},$$

so it suffices to perform

$$p^* = \lceil \frac{1}{2}\sqrt{\frac{b}{a}} \ln\frac{2}{\varepsilon}\rceil$$

iterations for a relative error ε. Finally, this method is numerically stable.

Proof As in previous derivations, we find

$$\|e^p\|_{A^\nu} \le \|Q_p(A^{\frac{1}{2}}C^{-1}A^{\frac{1}{2}})\|_{A^{\nu-\frac{1}{2}}} \|e^0\|_{A^\nu}$$

$$\le \max_{1 \le j \le N} |Q_p(\lambda_j)| \|e^0\|_{A^\nu},$$

and (5.30), (5.31), (5.32), and (5.36) establish the theorem. ◇

5.4* The Chebyshev Iterative Method for Matrices with Special Eigenvalue Distributions

Consider first the case where $C^{-1}A$ has complex eigenvalues, but with positive real parts and contained in an ellipse. As we shall see, in this case, the iteration parameters in the second-order method can be chosen the same as when the eigenvalues are real.

5.4.1 Complex Eigenvalues

Thus assume that the eigenvalues of A are contained in an ellipse in the right-half complex plane, symmetric with respect to the real axis,

$$S = \left\{ \zeta; \zeta = \left[\frac{b+a}{2} - \frac{b-a}{2}(\cos\theta + i\delta\sin\theta)/\sqrt{1-\delta^2} \right], \quad 0 \le \theta \le 2\pi \right\}.$$

Here $(a, 0)$ and $(b, 0)$ are the foci of the ellipse, δ is the eccentricity (that is, the ratio of the semiaxes), and $0 < a < b$. We assume that the ellipse does not contain the origin $(b + a)/2 > (b - a)/(2\sqrt{1 - \delta^2})$, that is,

$$\delta < 2\sqrt{\frac{a}{b}} \Big/ \left(1 + \frac{a}{b}\right).$$

This condition means that the ellipse must be narrow if $a \ll b$. Note that for $\delta = 0$, S is the interval $[a, b]$. The transformation

(5.39) $$P_1(\zeta) = z = (b + a - 2\zeta)/(b - a)$$

takes the given ellipse S into a new ellipse. This ellipse is

(5.39') $$E = \{z; z = (\cos\theta + i\delta\sin\theta)/\sqrt{(1 - \delta^2)}, \quad 0 \le \theta \le 2\pi\}.$$

The foci of this ellipse are $(-1, 0)$ and $(1, 0)$. Let $\rho_1 = \sqrt{[(1 + \delta)/(1 - \delta)]}$. Then an elementary computation shows that

$$E = \{z; \ z = \frac{1}{2}(\rho_1 + \rho_1^{-1})\cos\theta + i\frac{1}{2}(\rho_1 - \rho_1^{-1})\sin\theta$$

$$= \frac{1}{2}(\rho_1 e^{i\theta} + \rho_1^{-1}e^{-i\theta}), \ 0 \le \theta \le 2\pi\}.$$

Consider the Chebyshev polynomial

$$(5.40) \qquad T_k(z) = \frac{1}{2}\{[z + \sqrt{(z^2 - 1)}]^k + [z + \sqrt{(z^2 - 1)}]^{-k}\},$$

which is a polynomial of degree k in z, and note that $z + \sqrt{z^2 - 1} = \rho_1 e^{i\theta}$. Hence,

$$T_k(z) = \frac{1}{2}(\rho_1^k e^{ik\theta} + \rho_1^{-k}e^{-ik\theta})$$

and

$$\max_{z \in E} |T_k(z)| = \frac{1}{2}(\rho_1^k + \rho_1^{-k}),$$

where the maximum is taken for $\theta = 0$, for instance. Then

$$z = \frac{1}{2}(\rho_1 + \rho_1^{-1}) = 1/\sqrt{1 - \delta^2},$$

so

$$\max_{z \in E} |T_k(z)| = T_k(1/\sqrt{(1 - \delta^2)}).$$

The Chebyshev polynomials (5.40), where z is defined by (5.39), will be generated by the parameters

$$\tau_l^{-1} = \frac{b - a}{2}\cos(\theta_l) + \frac{b + a}{2}, \qquad \theta_l = \frac{2l - 1}{2p}\pi, \qquad l = 1, \ldots, p$$

when we use the first-order method (5.25), or by the parameters given in Theorem 5.9 when we use the second-order method. Hence, in the iterative method we use the same iteration parameters as for the real interval $[a, b]$, corresponding to $\delta = 0$. Recall that $(a, 0)$ and $(b, 0)$ are the foci of the ellipse. Considering the normalized polynomial $T_k[P_1(\zeta)]/T_k[P_1(0)]$, we find the average asymptotic convergence factor of the corresponding iterative method,

$$\rho \le \lim_{k\to\infty}\left\{\max_{\zeta\in S}\frac{|T_k(P_1(\zeta))|}{|T_k(P_1(0))|}\right\}^{1/k} = \lim_{k\to\infty}\left\{\frac{T_k[1/\sqrt{(1-\delta^2)}]}{T_k[(b+a)/(b-a)]}\right\}^{1/k}$$

(5.41)

$$= \rho_1/\lim_{k\to\infty}T_k(\alpha)^{\frac{1}{k}}$$

$$= \rho_1/[\alpha+\sqrt{\alpha^2-1}],$$

where

$$\alpha = \frac{(b+a)}{(b-a)}, \quad \rho_1 = \frac{1+\delta}{\sqrt{1-\delta^2}}.$$

Finally we get $\rho \le (1+\delta)/(\sqrt{1-\delta^2})(1-\sqrt{a/b})/(1+\sqrt{a/b})$. Note that under the stated condition on δ, we have $\rho < 1$. It is seen from (5.41) that the rate of convergence for complex eigenvalues in a narrow ellipse S, where $\delta << 2\sqrt{a/b}$, is about the same as for the real case, where $\delta = 0$.

The above analysis shows that for given values of a and b, the same Chebyshev iterative method converges for all ellipses with foci a and b and $0 \le \delta < 2\sqrt{a/b}/(1+a/b)$. As δ increases from its lower bound to its upper bound, the asymptotic convergence factor ρ increases from $(1-\sqrt{a/b})/(1+\sqrt{a/b})$ to the value 1. It is important to observe that the factor ρ gives only the asymptotic rate of convergence,

$$\{\|\mathbf{r}^k\|/\|\mathbf{r}^0\|\}^{1/k} \to \rho, \ k\to\infty.$$

Hence, as in the discussion in Section 5.1, it follows that even if $\rho < 1$, for nonsymmetric problems phases can occur during the iteration where the convergence is not monotone. If, however, the matrix is normal, the convergence is monotone.

Remark 5.13 As has been shown by Fischer and Freund (1990), the Chebyshev polynomials T_k are not always the best approximation polynomials on ellipses. For our purposes, however, the upper bounds provided by T_k are sufficiently accurate. If the eigenvalues of A have both negative and positive real parts, we can transform the system $A\mathbf{x} = \mathbf{b}$ into a system

(5.42) $$B\mathbf{x} = (A^*A + \alpha A)\mathbf{x} = \tilde{\mathbf{b}},$$

where $\tilde{\mathbf{b}} = A^*\mathbf{b} + \alpha\mathbf{b}$. If $|\alpha|$ is chosen sufficiently small, the eigenvalues of B will have positive real parts. It can be seen that if we let $\alpha = 0$, the condition number of B usually becomes much larger than for some $\alpha \neq 0$.

Let us now consider the Chebyshev iterative method for a symmetrix matrix A that is indefinite—i.e., has both negative and positive eigenvalues.

5.4.2 Indefinite Matrices

Assume that the eigenvalues of A are contained in two intervals, $S = [a, b] \cup [c, d]$, where $a < b < 0$ and $d > c > 0$. We shall also assume that the length of the intervals is equal—i.e., $d - c = b - a$. If this is not the case, we extend the shortest interval by decreasing a or increasing d sufficiently. In this case, the previously presented Chebyshev iterative method is not directly applicable because eigenvalues occur with different signs. As we shall see, however, the method is applicable on a transformed matrix.

Next consider (5.42) with—for convenience, in our case—an opposite sign,

$$B = A^2 - \alpha A,$$

where α is chosen so that the eigenvalues of B become positive and the condition number is smallest. This means that the parabola

$$P_2(\lambda) = (\lambda - \alpha)\lambda$$

must take positive values in S and that $P_2(A)$ must have the smallest condition number. An elementary derivation then shows that

$$\alpha = d + a,$$

or (which is the same)

$$\alpha = b + c.$$

The smallest value of $P_2(\lambda)$ in S then becomes $-bc$, the largest becomes $-ad$, and the condition number of $P_2(A)$ becomes

$$\text{cond}[P_2(A)] = \frac{d}{c} \cdot \left|\frac{a}{b}\right|.$$

If the intervals are symmetric with respect to the origin, then $\text{cond}[P_2(A)] = (d/c)^2$ and $\alpha = 0$, but if, say $|b| \gg c$, then

$$\text{cond}[P_2(A)] = \frac{d}{c} \cdot \frac{d - c + |b|}{|b|} \leq \frac{d}{c}\left(\frac{d}{|b|} + 1\right),$$

which is significantly smaller than $(d/c)^2$. The Chebyshev iterative method can now be applied to solve

$$(5.43) \qquad\qquad B\mathbf{x} = \tilde{\mathbf{b}},$$

where $B = A^2 - \alpha A$ and $\tilde{\mathbf{b}} = A\mathbf{b} - \alpha\mathbf{b}$. The eigenvalues of B are contained in the interval $[c|b|, d|a|]$.

5.4.3 Complex Eigenvalues in Two Ovals Oriented Along the Same Axis

Consider first the mapping of two real intervals $[a, b]$ and $[c, d]$ of equal length $b - a = d - c$ onto the interval $[-1, 1]$. Changing the previous notations somewhat, we let

$$\mu = \tfrac{1}{2}(d + a), \quad \alpha = \tfrac{1}{2}(b - a), \quad \gamma = \tfrac{1}{2}(c - a)$$

(see Figure 5.3). We assume that $0 \notin [a, b] \cup [c, d]$. The mapping is then

$$z = P_2(\zeta) = [(\mu - \zeta)^2 - \alpha^2 - \gamma^2]/(2\alpha\gamma).$$

Following Axelsson (1988), consider now the mapping of the ellipse E in (5.39$'$) by the inverse of this mapping,

$$(5.44) \qquad \zeta = P_2^{-1}(z) = \mu \pm \sqrt{(2\alpha\gamma z + \alpha^2 + \gamma^2)},$$

where the usual branch of the square root is chosen. We have

$$z = 1/2(\rho_1 e^{i\theta} + \rho_1^{-1} e^{-i\theta}) \text{ and } \rho_1 = \sqrt{[(1 + \delta)/(1 - \delta)]},$$

and we let

$$\sqrt{(\alpha^2 + \gamma^2 + 2\alpha\gamma z)} = e^{i\phi/2}\sqrt{r},$$

where r is positive. Hence,

$$r\cos\phi = u \equiv \alpha^2 + \gamma^2 + \alpha\gamma(\rho_1 + \rho_1^{-1})\cos\theta$$

$$r\sin\phi = v \equiv \alpha\gamma(\rho_1 - \rho_1^{-1})\sin\theta$$

or

$$\phi = \tan^{-1} v/u, \quad r = \sqrt{(u^2 + v^2)}.$$

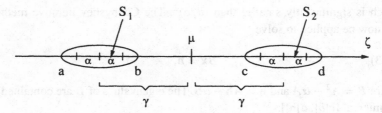

Figure 5.3.   Two ellipse-like ovals.

As θ runs through $0 \le \theta \le 2\pi$, then ζ in (5.44) runs through the boundary of two ellipse-like ovals (see Figure 5.3), the interior of which is the domain $S_1 \cup S_2$. We then have

$$\min_{P_{2k} \in \pi_{2k}^0} \max_{\zeta \in S_1 \cup S_2} |1 + P_{2k}(\zeta)| \le \max_{\zeta \in S_1 \cup S_2} |T_k[P_2(\zeta)]/T_k[P_2(0)]|$$

$$= \tfrac{1}{2}(\rho_1^k + \rho_1^{-k}) / \left| T_k \left(\frac{\mu^2 - \alpha^2 - \gamma^2}{2\alpha\gamma} \right) \right|$$

$$= \tfrac{1}{2}(\rho_1^k + \rho_1^{-k}) / \left| T_k \left(\frac{bc + ad}{bc - ad} \right) \right|$$

and, as $k \to \infty$, we obtain the asymptotic average reduction rate

$$\rho = \lim_{k \to \infty} \left[\tfrac{1}{2}(\rho_1^k + \rho_1^{-k}) / \left| T_k \left(\frac{bc + ad}{bc - ad} \right) \right| \right]^{1/k} = \frac{\rho_1 |1 - \sqrt{(ad/bc)}|}{1 + \sqrt{(ad/bc)}}.$$

The corresponding parameters in the Chebyshev iterative method can be determined as before for the ellipse E.

If the ovals are thin enough, so that $\delta \ll \sqrt{(a/b)}$, then the rate of convergence is essentially determined by the two real intervals $[a, b] \cup [c, d]$. Furthermore, if $c \gg \max[|a|, |b|]$, then the second cluster has little influence on the rate of convergence and the number of steps (iterations) is about twice that needed for the first cluster. Hence, the effective condition number is related to $4b/a$ rather than to d/a.

The case of two clusters appears, for instance, when solving constrained optimization problems by penalty methods. In certain cases—similar to that with Schur systems—solved by inner-outer iterations, small perturbations of the eigenvalues occur so that the eigenvalues, although originally real, have small imaginary parts.

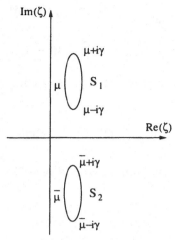

Figure 5.4.   Two ovals along an axis parallel to the imaginary axis.

5.4.4 Complex Eigenvalues in Two Ovals Along Axis Parallel to the Imaginary Axis

Let $P_2(\zeta) = [(\zeta - \operatorname{Re}\mu)^2 + \operatorname{Im}\mu)^2 + \gamma^2]/(2\gamma|\operatorname{Im}\mu|)$, where μ and γ are as defined in Figure 5.4. The inverse mapping $\zeta = P_2^{-1}(z)$ then maps E into two ovals. The best polynomial approximation is found to satisfy (see Axelsson, 1988),

$$\min_{P_{2k}\in\pi_{2k}^0} \max_{\zeta\in S_1\in S_2} |1 + P_{2k}(\zeta)| \le \max_{\zeta\in S_1\cup S_2} \left|\frac{T_k(P_2(\zeta))}{T_k(P_2(0))}\right| = \frac{\frac{1}{2}(\rho_1^k + \rho_1^{-k})}{T_k\left(\frac{|\mu|^2+\gamma^2}{2\gamma|\operatorname{Im}\mu|}\right)}$$

5.4.5 General Case

Following Lebedev (1969), we assume that, in general, there exists a polynomial $P_r(\zeta)$ of degree r, with real coefficients, which maps the eigenvalues of B contained in r different clusters of equal length into the ellipse E. Then

$$\min_{P_{kr}\in\pi_{kr}^0} \|[I + P_{kr}(B)]r^0\| \le \frac{\frac{1}{2}(\rho_1^k + \rho_1^{-k})}{T_k[P_r(0)]}\|r^0\|,$$

and the average convergence factor becomes

$$\rho = \lim_{k \to \infty} \{ \tfrac{1}{2}(\rho_1^k + \rho_1^{-k}) / T_k[P_r(0)] \}^{1/k}$$

$$= \rho_1 / \{ P_r(0) + \sqrt{[P_r(0)^2 - 1]} \}.$$

For a general set S of eigenvalues, we want to find the best approximation polynomial T_k, such that

$$\max_{z \in S} |T_k(z)| \leq \max_{z \in S} |P_k(z)|$$

for any polynomial normalized in the same way as $T_k(z)$. A common normalization is

$$P_k(z) = z^k - p_{k-1}(z),$$

i.e., $P_k(z)$ has a leading coefficient equal to 1, but for the analysis of iterative methods we need to use an interpolatory constraint, $P_k(0) = 1$. The best approximation polynomial for S, which is called the *Chebyshev approximation polynomial*, is known explicitly only for special sets S.

5.4.6 Orthogonal Matrices

If A is orthogonal, i.e., $A^T A = I$, then the eigenvalues of A are situated on the unit circle and no convergence acceleration using Chebyshev (or other polynomials) is possible. Naturally, in such a case we solve $Ax = b$ as

$$A^T A x = A^T b, \text{ that is, } x = A^T b.$$

Example 5.7 Consider the partial differential equation $\partial u / \partial t = \partial u / \partial x$, $t > 0$, $0 < x < 1$, $u(0) = 0$, where $u = u(t, x)$ and where $u(0, x)$ is given. Using central differences for $\partial u / \partial x$ and the so-called Crank-Nicholson method for the time-discretization leads to systems in the form

$$(I + \frac{\tau}{2} B) U(t + \tau) = (I - \frac{\tau}{2} B) U(t), \ t = 0, \tau, \ldots,$$

where $\tau > 0$ is the time step. Here B is skewsymmetric $B^T = -B$, and $A = (I + \frac{\tau}{2} B)^{-1} (I - \frac{\tau}{2} B)$ is orthogonal (see Exercise 1.21).

5.4.7 Historical Remark

Much effort has been devoted to finding the Chebyshev polynomial for the case when S is the union of two arbitrary disjoint intervals. Achieser (1932, p. 1190) represented $T_k(z)$ in terms of elliptic functions; Atlestam (1977) obtained a representation in terms of elliptic integrals; and, finally, Peherstorfer (1984) obtained a three-term recurrence formula. Fischer (1991) gives a unified presentation of this topic.

5.4.8 Estimates of Extremal Eigenvalues

For the efficient application of Chebyshev iterative methods, we need accurate estimates of the extreme eigenvalues. For an s.p.d. matrix, frequently the estimate $\max \lambda \leq \|A\|$—where $\|A\|$ is the maximum norm, for instance (see Appendix A)—is sufficiently accurate. An estimate of the smallest eigenvalue or of intermediate eigenvalues, in the case of indefinite matrices, is usually more troublesome. In some cases, such as for incomplete factorization preconditioning methods (see Chapters 7 and 9), such estimates are available and can be used for the Chebyshev iterative method. If A is s.p.d. and *monotone*, that is, $A^{-1} \geq 0$, then

$$\lambda_{\min}(A) = \frac{1}{\lambda_{\max}(A^{-1})} \geq \frac{1}{\|A^{-1}\|_\infty}.$$

But because $A^{-1} \geq 0$, we find

$$\|A^{-1}\|_\infty = \|A^{-1}\mathbf{e}\|_\infty, \quad \mathbf{e} = (1, 1, \ldots, 1)^T,$$

so

$$\lambda_{\min}(A) \geq \frac{1}{\|A^{-1}\mathbf{e}\|_\infty}.$$

The system $A\mathbf{y} = \mathbf{e}$ can be solved (approximately) to estimate its solution $\mathbf{y} = A^{-1}\mathbf{e}$. In general, however, it is difficult to estimate the eigenvalues, so one must use a parameter-free method, such as the conjugate gradient method (see Chapter 11). It will be seen that this method converges even faster in a certain norm than the Chebyshev iteration method.

Exercises

1. Consider $A\mathbf{x} = \mathbf{b}$, where

$$A = \begin{bmatrix} 2 & -1 & 0 \\ -\frac{1}{2} & 2 & -\frac{1}{2} \\ 0 & -1 & 2 \end{bmatrix} \text{ and } \mathbf{b} = \begin{bmatrix} 3 \\ 0 \\ 3 \end{bmatrix}.$$

(a) Compute the first four vectors in (5.3) with $C = 2I_3$ and $x^0 = [3, 0, 3]^T$.

(b) Show that the eigenvalues of $C^{-1}A$ are real. Using Gershgorin's disc theorem, estimate the smallest and largest eigenvalues of $C^{-1}A$ and use the Chebyshev first-order method (5.25) to compute the first four vectors, with $p = 4$ and parameters τ_l computed by (5.29′).

(c) Show that $\|B\|_\infty \le \frac{1}{2}$. Use Theorem 5.2 to estimate the error $\|\hat{\mathbf{x}} - \mathbf{x}^4\|_\infty$ in (a).

(d) Use (5.30) and (5.30′) to estimate the relative error of the Chebyshev iterative method, $\|\mathbf{e}^4\|_\infty / \|\mathbf{e}^0\|_\infty$, and use this to estimate $\|\mathbf{e}^4\|_\infty$.

2. Compute the spectral radius and the numerical radius for the iteration matrix in Example 5.3.

3. Let $B = \begin{bmatrix} 2 & -1 \\ 1 & 0 \end{bmatrix}$.

(a) Show that: $B^s = s \begin{bmatrix} 1 + \frac{1}{s} & -1 \\ 1 & \frac{1}{s} - 1 \end{bmatrix}$, $s = 1, 2, \dots$.

(b) Show that: $\rho(B^s) = 1$.

(c) Show that: $\frac{1}{s}\|B^s\| \to 2, s \to \infty$.

(d) Let $B = \begin{bmatrix} 2 & -0.99 \\ 0.99 & 0 \end{bmatrix}$. After how many iterations $\mathbf{e}^l = B\mathbf{e}^{l-1}$, $l = 1, 2, \dots$ is $\|\mathbf{e}^l\|_2 < \|\mathbf{e}^{(0)}\|_2$ if $\mathbf{e}^0 = (1, 1)^T$?

4. Let $A = C - R$, where C is nonsingular, be a splitting of A. Consider the more general splitting $C(\alpha) = (1 + \alpha)C$ and $R(\alpha) = C(\alpha) - A = R + \alpha C$. Let $\lambda_1 \le \lambda_2 \le \dots \le \lambda_n < 1$ be eigenvalues of $C^{-1}R$. Prove the statement: The iterative method $C(\alpha)\mathbf{x}^{l+1} = R(\alpha)\mathbf{x}^l + \mathbf{b}$, $l = 0, 1, \dots$ converges for any α, $\alpha > -(1 + \lambda_1)/2$, and $\min_\alpha \rho(C(\alpha)^{-1}R(\alpha)) = \rho(C(\alpha^*)^{-1}R(\alpha^*))$, where $\alpha^* = -(\lambda_1 + \lambda_n)/2$.

5. *An alternative proof of* $B^k \to 0 \iff \rho(B) < 1$: One can show that to every square matrix B, there exists a nonsingular matrix S that reduces the matrix to a block diagonal matrix called the *Jordan canonical form*:

$$J = S^{-1}BS = \begin{bmatrix} J_1 & & & 0 \\ & J_2 & & \\ & & \ddots & \\ 0 & & & J_r \end{bmatrix},$$

where each submatrix J_k, of order n_k, is of the form $J_k = \lambda_k$, if $n_k = 1$, and

$$J_k = \begin{bmatrix} \lambda_k & 1 & & & 0 \\ & \lambda_k & 1 & & \\ & & \ddots & \ddots & \\ & & & \lambda_k & 1 \\ 0 & & & & \lambda_k \end{bmatrix}, \text{ if } n_k \geq 2.$$

(For a proof, see, for example, Godeman, 1966 or Strang, 1980). Here $\{\lambda_k\}$ is the set of eigenvalues of B; two or more different Jordan blocks may correspond to the same eigenvalue.

(a) Show that B is diagonalizable if and only if $n_k = 1$, $k = 1, \ldots, r$.

(b) Assuming $n_k \geq 2$, compute explicitly the entries of the matrix B^l, using $B^l = SJ^lS^{-1}$.

(c) Show that $B^l \to 0 \iff \rho(B) < 1$.

(d) Show that the sequence of successive matrix powers $\{B^l\}_{l \geq 0}$ *converge* if and only if either $\rho(B) < 1$ or $\rho(B) \leq 1$. In the latter case the only eigenvalue of modulus 1 is equal to 1 (i.e., not to -1, for instance), *and* the dimension of the corresponding eigenspace (the geometric dimension) equals the (algebraic) multiplicity of this eigenvalue. (In other words, the Jordan blocks corresponding to $\lambda_k = 1$ have order 1.)

(e) Prove Lemma 5.3.

6. Let B be a square matrix satisfying $\|B\|_2 < 1$. Show the following:

(a) $I - B$ is nonsingular, and the eigenvalues of $I + 2(B - I)^{-1}$ have negative real parts.

(b) By way of example, show that there exists B, $\|B\|_2 < 1$ and a positive definite matrix H, such that $H - B^*HB$ is indefinite.

(c) There exists a positive definite Hermitian matrix H such that the matrix $H - B^*HB$ is positive definite (Stein's theorem).

(d) Given any matrix norm $\|.\|$, there exists an integer l such that $\|B^l\| < 1$.

(e) There exists a constant μ such that $\mu < 1$, and $|\operatorname{tr}(B^k)| \leq n\mu^k$ for every $k \geq 1$, where n is the order of B.

7. *A relation between the condition number of a matrix and the spectral radius of the associated Jacobi matrix:* Let A be an irreducibly diagonally dominant symmetric real matrix of order n with $a_{ii} > 0$, and let $\widetilde{A} = D^{-\frac{1}{2}} A D^{-\frac{1}{2}}$, where D is the diagonal of A. Clearly $\widetilde{A}$ is symmetric and has diagonal entries equal to 1. Show that:
    (a) The eigenvalues λ_i of $\widetilde{A}$ satisfy

    $$0 < a \equiv \lambda_1 \le \lambda_2 \le \ldots \le \lambda_n \equiv b < 2.$$

    (b) $a \le 1 \le b$. Let $B = I - \widetilde{A}$ be the Jacobi iteration matrix and let $\kappa = b/a$ denote the condition number of $\widetilde{A}$. Then show that:
    (c) $\rho_0 = \rho(B) = \max\{1 - a, \ b - 1\} < 1.$
    (d) $\kappa \le (1 + \rho_0)/(1 - \rho_0).$
    (e) $\kappa = (1 + \rho_0)/(1 - \rho_0)$ if $1 - a = b - 1.$
    Hints: For (b): The trace of $\widetilde{A}$ is n, hence, $\frac{1}{n} \sum_1^n \lambda_i = 1$. For (d): $1 - a \le \rho_0$ and $b - 1 \le \rho_0$.

8. Let $A = \begin{bmatrix} 3 & -1 & 0 & 0 \\ -1 & 3 & -1 & 0 \\ 0 & -1 & 3 & -1 \\ 0 & 0 & -1 & 2 \end{bmatrix}$.

    Solve $A\mathbf{x} = \mathbf{b}$, where $\mathbf{b} = [2, 1, 1, 1]^T$, using:
    (a) the nonstationary second-order Chebyshev iteration method.
    (b) the stationary second-order method.
    (c) Then show that the parameters β_l used in (a) converge monotonically, increasing to a certain limit value. Which?
    Hint: Show first that the spectrum of A is contained in the interval $[1, 5]$.

9. Consider the numerical solution of the system of ordinary differential equations $\frac{d\mathbf{x}(t)}{dt} + A\mathbf{x}(t) = \mathbf{b}$, $t > 0$, $\mathbf{x}(0) = \mathbf{x}_0$, a given vector, where A is s.p.d., using the Euler forward time-stepping method:

    $$\tilde{\mathbf{x}}(0) = \mathbf{x}_0; \quad \tilde{\mathbf{x}}(t + \tau_k) = (I - \tau_k A)\tilde{\mathbf{x}}(t) + \tau_k \mathbf{b}, \quad k = 0, 1, \ldots.$$

    Let $t_0 = 0$, $t_{k+1} = t_k + \tau_k$, $k = 0, 1, \ldots$, where $\tau_k > 0$ are time steps. Let $\hat{\mathbf{x}} = A^{-1}\mathbf{b}$ (the stationary solution of the differential equation). Show that $\hat{\mathbf{x}} - \tilde{\mathbf{x}}(t_{l+1}) = \prod_{k=0}^{l}(I - \tau_k A)(\hat{\mathbf{x}} - \mathbf{x}_0)$ and that the optimal time-steps to make $\|\hat{\mathbf{x}} - \hat{\mathbf{x}}(t_{l+1})\|$ as small as possible can be found from (5.24), using Chebyshev polynomials.

10. *The instability of the first-order Chebyshev iterative method:*
    (a) Show that the iteration parameters τ_l in (5.25) vary between $(\frac{\lambda_n + \lambda_1}{2} + \frac{\lambda_n - \lambda_1}{2} \cos \frac{\pi}{2p})^{-1}$ and $(\frac{\lambda_n + \lambda_1}{2} - \frac{\lambda_n - \lambda_1}{2} \cos \frac{\pi}{2p})^{-1}$.
    (b) Estimate the extreme eigenvalues of the iteration matrix $I - \tau_l A$

for any l, $0 \le l \le p - 1$. That is, compute $\mu_0 = \min_l(1 - \tau_l \lambda_n)$ and $\mu_1 = \max_l(1 - \tau_l \lambda_1)$ and determine the asymptotic behavior of μ_0, μ_1 as $p \to \infty$.

(c) Compute μ_0, μ_1 for $\lambda_n/\lambda_1 = 10^3$ and $p = 10$.

11. Show that:

(a) The matrix series

$$I + A + A^2 + \cdots$$

converges if and only if the matrix series

$$I + B + B^2 + \cdots$$

converges, where B is similar to A, that is $B = P^{-1}AP$, for some nonsingular matrix P.

(b) $I + A + A^2 + \cdots$, converges where

$$A = \begin{bmatrix} 0.1 & 0 & 1.0 \\ 0.2 & 0.3 & 0 \\ 0.2 & 0.1 & 0.4 \end{bmatrix}$$

by choosing P, defined in (a) as a suitable diagonal matrix, so that $\|B\|_\infty < 1$.

12. Let A be a matrix of order n, where all $a_{ik} = 0$, $i \le k$ (a so-called nilpotent matrix). Prove that

$$(I - A)^{-1} = I + A + A^2 + \cdots + A^{n-1}.$$

13. Let $B = I - A$, $A = 10^{-2} \begin{bmatrix} 1 & 1 & 2 \\ -1 & 2 & 0 \\ -2 & 1 & -1 \end{bmatrix}$.

Show that $\|B^{-1} - (I + A + A^2)\|_\infty < 0,7 \cdot 10^{-4}$.

14. *Newton's method:*

(a) Prove that if $\|I - AB_0\| = c < 1$ and if

$$B_k = B_{k-1} + B_{k-1}(I - AB_{k-1}), \quad k = 1, 2, \ldots,$$

then

$$\lim_{k \to \infty} B_k = A^{-1} \quad \text{and} \quad \|A^{-1} - B_k\| \le \|B_0\| \frac{c^{2^k}}{1 - c}.$$

Hint: Prove first that both A and B_0 are nonsingular and then consider $\|A^{-1} - B_k\| \le \|A^{-1}\| \, \|I - AB_k\|$. This is a Newton-type method to solve $X^{-1} - A = 0$.

(b)   Let A be a matrix that can be diagonalized by a similarity transformation, $A = T \ \text{diag}(d_1, \ldots, d_n) T^{-1}$, where $\text{Re}(d_i) \neq 0$. Show that the iterative method, $X_0 = A$; $X_{s+1} = \frac{1}{2}(X_s + X_s^{-1})$ converges to the so-called matrix sign function,

$$\text{sgn}(A) = T \ \text{diag}\,(\text{sgn}(d_1), \ldots, \text{sgn}(d_n)) \, T^{-1},$$

where

$$\text{sgn}(d_i) = \begin{cases} +1, & \text{when } \text{Re}(d_i) > 0 \\ -1, & \text{when } \text{Re}(d_i) < 0. \end{cases}$$

Show also that $X^2 = I$. The algorithm is Newton's method of computing the matrix solution of $X^2 = I$ with initial matrix A.

References

N. I. Achieser (1932). Über einige Funktionen, welche in zwei gegebenen Intervallen am wenigsten von Null abweichen I. Teil. *Bull. Acad. Sci. USSR S. VII* **9**, 1163–1202.

M. Arioli, I. S. Duff, and D. Rinz (1991). Stopping criteria for iterative solvers, *Report RAL-91-057*, Rutherford Appleton Laboratory. Didcot, United Kingdom: Chilton.

B. Atlestam (1977). Tschebycheff-polynomials for sets consisting of two disjoint intervals with application to convergence analysis. *Research Report 77.06 R*, Dept. Computer Sciences, Chalmers University of Technology and the University of Göteborg, Sweden.

O. Axelsson (1972). Lecture notes on iterative methods. Report 72.04, Department of Computer Sciences, Chalmers University of Technology, Göteborg, Sweden.

O. Axelsson (1973). Notes on the numerical solution of the biharmonic equation, *J. Inst. Math. Applic.* **11**, 213–226.

O. Axelsson (1988). A restarted version of a generalized preconditioned conjugate gradient method. *Comm. Appl. Num. Methods* **4**, 521–530.

B. Fischer (1991). Chebyshev polynomials for disjoint compact sets. Report of Institut für Angewandte Mathematik, Universität Hamburg, Germany.

B. Fischer and R. Freund (1990). Chebyshev polynomials are not always optimal. *J. Approx. Theory* **65**, 261–272.

S. P. Frankel (1950). Convergence rates of iterative treatments of partial differential equations. *Math. Tables Aids Comput.* **4**, 65–75.

C. F. Gauss (1823). Brief an Gerling vom 26 Dec. 1823, Werke Vol. 9, 278–281. A translation by G. E. Forsythe appears in *MTAC* **5** (1950), 255–258.

C. F. Gauss (1826). *Supplementum . . . , Werke*. Göttingen, **Vol. 4**, 55–93.

R. Godeman (1966). *Algebra*, London: Kershaw.

G. H. Golub and R. S. Varga (1961). Chebyshev semi-iterative methods, successive over relaxation iterative methods, and second-order Richardson Iterative Methods, Parts I and II. *Numer. Math.* **3**, 147–156, 157–168.

W. Hackbusch (1993). *Iterative Solution of Large Sparse Systems of Equations.* New York: Springer.

C. G. J. Jacobi (1845). Über eine neue Auflösungsart. *Astr. Nachr.* 22 **No. 523** (1845), 297–306. [Reprinted in his *Werke,* Vol. 3, p. 467].

V. I. Lebedev (1969). On iterative methods of solving operator equations with operator spectrum concentrated in several separate segments. *Zh. Vychisl. Matem. Fiz.* **9**, 1247–1252. [An English translation has appeared in R. S. Anderssen and G. H. Golub (1972), Richardson's non-stationary matrix iterative procedure. Rep. STAN-CS-72-304, Computer Science Dept., Stanford University.]

V. I. Lebedev and S. A. Finogenov (1971, 1973). On the order of the iteration parameters in the Chebyshev cyclic iteration method. *Zh. Vychisl. Mat. i Mat. Fiz.* **11** (1971), 425 and 13 (1973), 18.

J. Liouville (1837). Sur le développement des fonctions en series . . . , II. *J. Math. Pures Appl.* (1), **2**, 16–37.

T. A. Manteuffel (1977). The Tchebyshev iteration for nonsymmetric linear systems. *Numer. Math.,* **28**, 307–327.

F. Peherstorfer (1984). Extremal Polynome in der L^1- und L^2-Norm auf zwei disjunkten Intervallen. In *Approximation Theory and Functional Analysis,* ed. P. L. Butzer, R. L. Stens, and B. Sz. Nagy, ISNM **65**. Basel: Birkhäuser, 269–280.

L. F. Richardson (1911). The approximate arithmetical solution by finite differences of physical problems involving differential equations, with an application to the stresses in a masonry dam. *Trans. Roy. Soc.,* London A210 , 307–357.

L. F. Richardson (1925). How to solve differential equations approximately by arithmetic. *Math. Gazette* **12**, 415–421.

V. I. Smirnov and N. A. Lebedev (1968). Functions of a Complex Variable. Constructive Theory. London: Iliffe.

G. Strang (1980). *Linear Algebra and Its Applications,* 2nd ed. New York: Academic Press.

H. E. Wrigley (1963). Accelerating the Jacobi method for solving simultaneous equations by Chebychev extrapolation when the eigenvalues of the iteration matrix are complex. *Computer J.* **6**, 169–176.

D. M. Young (1954). On Richardson's method for solving linear systems with positive definite matrices. *J. Math. Phys.* **32**, 243–255.

D. M. Young (1971). Iterative solution of large linear systems. Orlando, Fla.: Academic.

6

M-Matrices, Convergent Splittings, and the SOR Method

A crucial task in the construction of efficient basic iterative methods is the choice of a splitting of the given matrix, $A = C - R$, into two matrices, C and R. At each iteration step of such a method, we must solve a linear system with C. This matrix is sometimes called a preconditioning matrix; it must be simple (in a certain respect) but still effective for the increase of the rate of convergence. The case where the corresponding preconditioning matrix is triangular or block-triangular and A is a so-called *M*-matrix is of particular importance because it appears frequently in practice. Let us first study some general properties of *M*-matrices and general families of convergent splittings. Several principles for comparing the rate of convergence of different splittings are also presented.

Even if a given matrix is not a *M*-matrix, we show next that any symmetric positive definite matrix can be reduced to a Stieltjes matrix by using a method of diagonal compensation of reduced positive entries. Splittings for the reduced matrix can be used as splittings for the original matrix.

By introducing a relaxation parameter (ω) as in the successive overrelaxation (SOR) method, we show that a proper choice of this parameter sometimes can speed up the rate of convergence by an order of magnitude. It turns out that the theory for this method can be based on some of the results already derived in the previous chapter. This classical theory is based on matrices that can be partitioned into block-tridiagonal form, possibly after a suitable permutation (they are said to have property A^π). We also present an alternative convergence theory for the SOR method for symmetric and positive definite matrices where there is no need to assume property A^π. The following definitions are introduced in this chapter:

Definition 6.1 A real matrix A is called *monotone* if $A\mathbf{x} \geq \mathbf{0}$ implies $\mathbf{x} \geq \mathbf{0}$.

Definition 6.2 A real square matrix $A = [a_{ij}]$ is called a (nonsingular) *M-matrix* if $a_{ij} \leq 0$ for $i \neq j$ and if it is monotone—i.e., if $A^{-1} \geq 0$. For any real or complex $n \times n$ matrix $A = [a_{ij}]$, we denote by $\mathcal{M}(A) = [m_{ij}(A)]$ the $n \times n$ real matrix, called the *comparison matrix*, defined by

$$m_{ij}(A) = \begin{cases} |a_{ij}|, & \text{if } i = j \\ -|a_{ij}|, & \text{if } i \neq j \end{cases}.$$

A real or complex matrix A is called a *H-matrix* if $\mathcal{M}(A)$ is a *M*-matrix.

Definition 6.3 A square matrix A is said to be *generalized diagonally dominant* if $|a_{ii}|x_i \geq \sum_{j \neq i} |a_{ij}|x_j$, $i = 1, 2, \ldots, n$ for some positive vector $\mathbf{x} = (x_1, x_2, \ldots, x_n)^T$ and to be *generalized strictly diagonally dominant* if this is valid with strict inequality. It is called *irreducibly generalized diagonally dominant* if:

(a) A is generalized diagonally dominant, with strict inequality for at least one row.

(b) A is irreducible.

Definition 6.4 A real matrix $A = [a_{i,j}]$ of order n with $a_{i,j} \leq 0$ for $i \neq j$ is said to be of *generalized positive type* if there exists a real vector $\mathbf{x} > \mathbf{0}$, such that

(a) $A\mathbf{x} \geq \mathbf{0}$ and $J(A) = \{i; (A\mathbf{x})_i > 0\} \neq \emptyset$, that is, the index set $J(A)$ is nonempty.

(b) If $i \notin J(A)$, there exists indices $i_1, i_2, \ldots, i_r \in \{1, 2, \ldots, n\}$, such that $a_{i_k, i_{k+1}} \neq 0$, $i_1 = i$, and $i_r \in J(A)$, or, equivalently, there exists a path in the directed graph of A from any index $i \notin J(A)$ to an index in $J(A)$.

It is said to be of *positive type* if (a) and (b) are valid with $\mathbf{x} = \mathbf{e}$.

Definition 6.5 $A = [a_{i,j}]$ is called a *Stieltjes matrix* (Stieltjes, 1887) if $a_{i,j} \leq 0$, for $i \neq j$ and A is symmetric and positive definite.

Definition 6.6 Let $C, R \in \mathbb{R}^{n,n}$. Then $A = C - R$ is called:

(a) a *regular splitting* if C is monotone and $R \geq 0$.

(b) a *weak regular splitting* if C is monotone and $C^{-1}R \geq 0$.

(c) a *nonnegative splitting* if C is nonsingular and $C^{-1}R \geq 0$.

(d) a *convergent splitting* if C is nonsingular and $\rho(C^{-1}R) < 1$.

Definition 6.7 The matrix A is said to have *property* (A^π) if A can be permuted by PAP^T into a form that can be partitioned into block-tridiagonal form, where the diagonal block-matrices D_i, $i = 1, \ldots, r$ are nonsingular.

6.1 *M*-Matrices

Let us consider convergent splittings—i.e., splittings for which the basic iterative method converges for any initial vector. We shall study, in particular, the class of *M*-matrices and show the equivalence with some other matrix properties that are more easily used in various applications. The *order relation* used in this chapter between real matrices and vectors of the same dimension is the usual componentwise order: with $A = [a_{i,j}]$ and $B = [b_{i,j}]$, then $A \leq B$ if $a_{i,j} \leq b_{i,j}$ for all i, j, while $A < B$ if $a_{i,j} < b_{i,j}$ for all i, j; A is said to be nonnegative (positive) if $A \geq 0$ $(A > 0)$.

Definition 6.1 A real matrix A is called *monotone* if $A\mathbf{x} \geq \mathbf{0}$ implies $\mathbf{x} \geq \mathbf{0}$.

Lemma 6.1 *A is monotone if and only if A is nonsingular with $A^{-1} \geq 0$.*

Proof Assume that A is monotone and let $A\mathbf{x} = \mathbf{0}$. Then $\mathbf{x} \geq \mathbf{0}$, because A is monotone. But $A\mathbf{x} = \mathbf{0} \Rightarrow A(-\mathbf{x}) = \mathbf{0}$; therefore (again, because A is monotone), $-\mathbf{x} \geq \mathbf{0}$ or $\mathbf{x} \leq \mathbf{0}$. Hence, $\mathbf{x} = \mathbf{0}$, that is, A is nonsingular. Let $\mathbf{e}_i$ denote the ith coordinate vector. Next, note that $A\,A^{-1}\mathbf{e}_i = \mathbf{e}_i \geq \mathbf{0}$, which shows that $A^{-1}\mathbf{e}_i \geq \mathbf{0}$, $i = 1, \ldots, n$, where n is the order of A. But $A^{-1}\mathbf{e}_i$ equals the ith column of A^{-1}, hence, $A^{-1} \geq 0$. The converse statement is trivial.   ◇

Definition 6.2 A real square matrix $A = [a_{i,j}]$ is called a (nonsingular) *M*-*matrix* if $a_{i,j} \leq 0$ for $i \neq j$ and if it is monotone—i.e., (by Lemma 6.1) if $A^{-1} \geq 0$. For any real or complex $n \times n$ matrix $A = [a_{i,j}]$, we denote by $\mathcal{M}(A) = [m_{i,j}(A)]$, the $n \times n$ real matrix, called the *comparison matrix*, defined by

$$m_{i,j}(A) = \begin{cases} |a_{i,j}|, & \text{if } i = j \\ -|a_{i,j}|, & \text{if } i \neq j \end{cases}.$$

A real or complex matrix A is called a *H-matrix* if $\mathcal{M}(A)$ is a *M*-matrix.

Example 6.1 If $A = \begin{bmatrix} 2 & -1 & 0 \\ -2 & 3 & 1 \\ -1 & 0 & 1 \end{bmatrix}$, then $\mathcal{M}(A) = \begin{bmatrix} 2 & -1 & 0 \\ -2 & 3 & -1 \\ -1 & 0 & 1 \end{bmatrix}$. As we shall see later, $\mathcal{M}(A)$ is a *M*-matrix. Hence, A is an *H*-matrix.

Definition 6.3 A square matrix A is said to be *generalized diagonally dominant* if

$$(6.1) \qquad |a_{i,i}|x_i \geq \sum_{j \neq i} |a_{i,j}|x_j, \quad i = 1, 2, \ldots, n$$

for some positive vector $\mathbf{x} = (x_1, x_2, \ldots, x_n)^T$ and *generalized strictly diagonally dominant* if (6.1) is valid with strict inequality. Recall from Definition 4.4 that it is called *(strictly) diagonally dominant* if (6.1) is valid for $\mathbf{x} = \mathbf{e} \equiv (1, 1, \ldots, 1)^T$ (strictly, if (6.1) is valid with strict inequality). It is called *irreducibly generalized diagonally dominant* if:

(a) A is generalized diagonally dominant, with strict inequality for at least one row.
(b) A is irreducible.

For later use, we note the important fact that if A is generalized diagonally dominant for a vector $\mathbf{x} > \mathbf{0}$, then AD is diagonally dominant where $D = \operatorname{diag}(x_1, x_2, \ldots, x_n)$.

Example 6.2

(a) $A = \begin{bmatrix} 5 & -1 & -1 \\ -1 & 2 & -1 \\ -1 & -1 & 1 \end{bmatrix}$ is generalized diagonally dominant for $\mathbf{x} = (1, 2, 3)^T$. Note that this matrix is singular.

(b) $A = \begin{bmatrix} 1 & -1 & 0 \\ -1 & 2 & -1 \\ 0 & -1 & 2 \end{bmatrix}$ is irreducibly diagonally dominant but also generalized strictly diagonally dominant, with $\mathbf{x} = (6, 5, 3)^T$, for instance.

Lemma 6.2 *Let $A = [a_{i,j}]$ be a $n \times n$ matrix, with $a_{i,j} \le 0$ for $i \ne j$ and $a_{i,i} > 0$. If A is strictly diagonally dominant, then A is a M-matrix.*

Proof Let $B = D^{-1}(D - A) = I - D^{-1}A$, where $D = \operatorname{diag}(A)$. Because A is strictly diagonally dominant, we must have $d_{i,i} = a_{i,i} > 0$. Hence, $b_{i,j} \ge 0$ and $\sum_{j=1}^{n} |b_{i,j}| < 1$. It follows from Corollary 4.6 that $\rho(B) < 1$. Hence, by Lemma 5.2(b),

$$A^{-1} = [D(I - B)]^{-1} = (I - B)^{-1}D^{-1} = (I + B + B^2 + \ldots)D^{-1} \ge 0,$$

which shows that A is a M-matrix. $\diamond$

Example 6.3 $A = \begin{bmatrix} 2 & -1 & 0 \\ -1 & 3 & -1 \\ 0 & -1 & 2 \end{bmatrix}$ is a M-matrix, because its off-diagonal entries are nonpositive and it is strictly diagonally dominant.

Note that if A is a M-matrix and $A\mathbf{x} = \mathbf{b}$ for some nonnegative vector $\mathbf{b}$, then $\mathbf{x} = A^{-1}\mathbf{b} \geq \mathbf{0}$. In particular, if A is a M-matrix and $A\mathbf{x} \geq \mathbf{0}$, then $\mathbf{x} \geq \mathbf{0}$. Similarly, if $A\mathbf{x} \leq \mathbf{0}$, then $A(-\mathbf{x}) \geq \mathbf{0}$, that is, $(-\mathbf{x}) \geq \mathbf{0}$, or $\mathbf{x} \leq \mathbf{0}$. More generally, if $A\mathbf{x} = \mathbf{b}$, $A\mathbf{y} = \mathbf{c}$, and $\mathbf{b} \leq \mathbf{c}$, then $\mathbf{x} \leq \mathbf{y}$.

Hence, M-matrices are associated with a so-called *discrete maximum principle*, which we state in a separate lemma because of its practical importance for solutions of, for example, certain difference equations.

Lemma 6.3 *Let* $A = [a_{i,j}]$ *be strictly diagonally dominant, where* $a_{i,j} \leq 0$, $i \neq j$ *and* $a_{i,i} > 0$. *If* $A\mathbf{x} \leq \mathbf{0}$ *for some* $\mathbf{x} \neq \mathbf{0}$, *then* $\max_i x_i \leq 0$ *(discrete maximum principle), that is,* $\mathbf{x} \leq \mathbf{0}$, *and* A *is a* M-*matrix.*

Proof We present here a direct proof, not making use of Lemma 6.2. Assume that for some i_0, $x_{i_0} = \max_i x_i \geq 0$. Then, as

$$(A\mathbf{x})_{i_0} = \sum_j a_{i_0,j} x_j = (a_{i_0,i_0} + \sum_{j \neq i_0} a_{i_0,j}) x_{i_0} + \sum_{j \neq i_0} a_{i_0,j}(x_j - x_{i_0}),$$

we find $0 \geq (A\mathbf{x})_{i_0} \geq 0$, because the first term above is nonnegative by the assumption of diagonal dominance and the second term is nonnegative because both of the factors, $a_{i_0,j}$, $j \neq i_0$ and $(x_j - x_{i_0})$, are nonpositive. Hence, we must have equality $(A\mathbf{x})_{i_0} = 0$, so $x_{i_0} = 0$ (because of the strict diagonal dominance). This shows the discrete maximum principle. Hence, $A\mathbf{x} \geq \mathbf{0}$ implies $\mathbf{x} \geq \mathbf{0}$, and, by Definition 6.2, A is a M-matrix. $\diamond$

Let us consider next the following important equivalence results.

Lemma 6.4

   (a) *A matrix* $A = [a_{i,j}]$, *with* $a_{i,j} \leq 0$, $i \neq j$, *is a* M-*matrix if and only if there exists a positive vector* $\mathbf{x}$, *such that* $A\mathbf{x}$ *is positive.*

   (b) *A matrix* $A = [a_{ij}]$ *is a* H-*matrix if and only if* A *is generalized strictly diagonally dominant.*

Proof If A is a M-matrix, then A is nonsingular and $A^{-1} \geq 0$. Let $\mathbf{x} = A^{-1}\mathbf{e}$, $\mathbf{e} = (1, 1, \ldots, 1)^T$. Then $\mathbf{x} \geq \mathbf{0}$. In fact, $\mathbf{x} > \mathbf{0}$, because if $x_i = 0$ for some i, then $\sum_{j=1}^n (A^{-1})_{i,j} = 0$, which would imply that all entries of the ith row were zero—that is, $\det(A^{-1}) = 0$, which contradicts the invertibility. Hence, $A\mathbf{x} = \mathbf{e} > \mathbf{0}$ for a vector $\mathbf{x} > \mathbf{0}$.

To prove the converse statement of part (a), assume that $A\mathbf{x} > \mathbf{0}$ for some $\mathbf{x} > \mathbf{0}$, let $D = \operatorname{diag}(x_1, x_2, \ldots, x_n)$, and let $B = AD$. Then $B\mathbf{e} = AD\mathbf{e} = A\mathbf{x} > \mathbf{0}$, which shows that B is strictly diagonally dominant. Furthermore, $b_{i,j} =$

$a_{i,j}x_j \leq 0$, $i \neq j$. Lemma 6.2 now shows that B is a M-matrix—that is, in particular, B is nonsingular and $B^{-1} \geq 0$. Hence, $A = BD^{-1}$ is nonsingular and $A^{-1} = DB^{-1} \geq 0$, so A is a M-matrix.

To prove part (b), assume first that A is a H-matrix. Then, by definition, $\mathcal{M}(A)$ is a M-matrix and part (a) shows that there exists a positive vector $\mathbf{x}$ such that $\mathcal{M}(A)\mathbf{x} > \mathbf{0}$, that is,

$$|a_{ii}|x_i > \sum |a_{ij}|x_j > 0, \quad i = 1, 2, \ldots, n,$$

which is (6.1) with strict inequality. Conversely, if (6.1) holds with strict inequality, then part (a) shows that $\mathcal{M}(A)$ is a M-matrix—that is, by definition, A is a H-matrix. $\diamond$

Definition 6.4 (Bramble and Hubbard, 1964) A real matrix $A = [a_{i,j}]$ of order n, with $a_{i,j} \leq 0$ for $i \neq j$, is said to be of *generalized positive type* if there exists a real vector $\mathbf{x} > \mathbf{0}$, $\mathbf{x} \in \mathbb{R}^n$, such that:

   (a) $A\mathbf{x} \geq \mathbf{0}$ and $J(A) = \{i; (A\mathbf{x})_i > 0\} \neq \emptyset$, that is, the index set $J(A)$ is nonempty.

   (b) If $i \notin J(A)$, there exist indices $i_1, i_2, \ldots, i_r \in \{1, 2, \ldots, n\}$, such that $a_{i_k, i_{k+1}} \neq 0$, $k = 1, 2, \ldots, r - 1$, $i_1 = i$, and $i_r \in J(A)$. Equivalently, there exists a path in the directed graph of A from any index $i \neq J(A)$ to an index in $J(A)$.

It is said to be of *positive type* if (a) and (b) are valid with $\mathbf{x} = \mathbf{e}$.

The matrix in Example 6.5, for instance, is of positive type. Here, $J(A) = \{n\}$.

We now extend the previous equivalence result [Lemma 6.4(a)] and show that many of the above properties are, in fact, equivalent to the statement that A is a M-matrix. This means that, in many instances, we may apply a simple check to see if a matrix is a M-matrix.

Theorem 6.5 (Beauwens, 1976, and Varga, 1976) *Let $A = [a_{i,j}]$ be a matrix of order n, with $a_{i,j} \leq 0$ for $i \neq j$. Then the following are equivalent:*

   (a) A is of generalized positive type.

   (b) There exists an $\mathbf{x} > \mathbf{0}$ and a permutation matrix P, such that $PA\mathbf{x} \geq \mathbf{0}$ and

$$\sum_{j=1}^{i} b_{i,j} x_j > 0, \quad i = 1, 2, \ldots, n, \quad \text{where} \quad b_{i,j} = (PA)_{i,j}.$$

(c) *A is generalized strictly diagonally dominant, and* $a_{ii} \geq 0$.

(d) *A is a M-matrix.*

Proof We shall prove that (a) $\Rightarrow$ (b) $\Rightarrow$ (c) $\Rightarrow$ (d) $\Rightarrow$ (c) $\Rightarrow$ (a). Assume, first, that A is of generalized positive type, and let P^0 be the permutation matrix that permutes the rows with positive sums, $\sum_{j=1}^{n} a_{i,j} x_j > 0$, to the last rows. Let $C = P^0 A P^{0^T}$ be the correspondingly symmetrically permuted matrix. If there exist rows with $\sum_{j=1}^{i} c_{ij} x_j = 0$ and $c_{ij} = 0$, $j = i + 1, \ldots, n$, permute these rows to the first rows. Let $\tilde{A} = PAP^T = [\tilde{a}_{i,j}]$ be the corresponding permutation of A. Since A is of generalized positive type, so is $\tilde{A}$, because $\tilde{A}\tilde{x} \geq 0$, where $\tilde{x} = Px > 0$, that is,

(6.2) $$\sum_{j=1}^{i} \tilde{a}_{i,j} \tilde{x}_j \geq \sum_{j=i+1}^{n} (-\tilde{a}_{i,j}) \tilde{x}_j \geq 0,$$

which is nonnegative because $\tilde{a}_{i,j} \leq 0$, $j \neq i$.

If there exist rows of $\tilde{A}$ with $\sum_{j=1}^{i} \tilde{a}_{i,j} \tilde{x}_j = 0$, then by (6.2), $\tilde{a}_{i,j} = 0$, $j = i + 1, \ldots, n$. Hence, $(Ax)_i = 0$ for these rows, and they would appear in a reducible block with no coupling to the last rows (with positive sums). But this contradicts part (b) of Definition 6.4. Hence, for all rows we must have $\sum_{j=1}^{i} \tilde{a}_{i,j} \tilde{x}_j > 0$, which proves (b) for $\tilde{A}$ and $\tilde{x}$. But $Bx = PAP^T Px = \tilde{A}\tilde{x}$ then shows that $\sum_{j=1}^{i} b_{i,j} x_j > 0$, which completes the proof of (b).

If (b) is valid, then, in particular, $v_1 := \sum_{j=1}^{n} a_{n,j} x_j > 0$. Clearly $a_{n,n} > 0$ and $x_n - v_1/a_{n,n} = \frac{1}{a_{n,n}} \sum_{j=1}^{n-1} (-a_{n,j}) x_j \geq 0$. Let $x_i^{(1)} = x_i$, $i = 1, 2, \ldots, n - 1$, $x_n^{(1)} = x_n - v_1/(2a_{n,n})$. Then $x_n^{(1)} > 0$ and $\sum_{j=1}^{n} a_{n,j} x_j^{(1)} = \sum_{j=1}^{n} a_{n,j} x_j - 1/2 v_1 = 1/2 v_1 > 0$. Further, if $v_2' = \sum_{j=1}^{n} a_{n-1,j} x_j > 0$, then

$$v_2 := \sum_{j=1}^{n} a_{n-1,j} x_j^{(1)} \geq v_2' > 0,$$

because $a_{n-1,n} \leq 0$ and $0 < x_n^{(1)} < x_n$. If, on the other hand, $v_2' = 0$, then $a_{n-1,n} \neq 0$, because, by assumption, $\sum_{j=1}^{n-1} a_{n-1,j} x_j > 0$. Hence,

$$v_2 = \sum_{j=1}^{n} a_{n-1,j} x_j^{(1)} = \sum_{j=1}^{n} a_{n-1,j} x_j + |a_{n-1,n}| \frac{v_1}{2a_{n,n}}$$

$$= v_2' + |a_{n-1,n}| \frac{v_1}{2a_{n,n}} > 0.$$

In either case, the row sum $v_2 > 0$. The other generalized row sums $\sum_j a_{i,j} x_j^{(1)}$ have not decreased as compared with the generalized row sums with $\mathbf{x}$. Now let $x_i^{(2)} = x_i^{(1)}$, $i \neq n - 1$, $x_{n-1}^{(2)} = x_{n-1}^{(1)} - v_2/(2a_{n-1,n-1})$. In a similar way as above, we find that $x_{n-1}^{(2)} > 0$ and $v_3 = \sum_{j=1}^{n} a_{n-2,j} x_j^{(2)} > 0$. The above modifications of the components of the vector are repeated and, eventually, we have $A\mathbf{x}^{(n)} > \mathbf{0}$ for some vector $\mathbf{x}^{(n)} > \mathbf{0}$. This proves (c). That (c) is equivalent to (d) is Lemma 6.4. Finally, note that (c) implies (a) [with $J(A) = \{1, \ldots, n\}$].

$\diamond$

Example 6.4

(a) The matrix $\mathcal{M}(A)$ in Example 6.1 is a M-matrix.

(b) $A = \begin{bmatrix} 6 & -1 & -1 \\ -1 & 2 & -1 \\ -1 & -1 & 1 \end{bmatrix}$ is a M-matrix.

To see this for the first example, observe that $\mathcal{M}(A)$ is of generalized positive type for the vector $\mathbf{x} = \mathbf{e}$. For the second example note, for instance, that it is generalized strictly diagonally dominant for $\mathbf{x} = (1, 2\frac{1}{4}, 3\frac{3}{8})^T$. Observe that the matrix in Example 6.2 is not a M-matrix, because it is singular. However, it is of a wider class that includes some singular matrices: A matrix A is called generalized M-matrix if $a_{ij} \leq 0$, $i \neq j$ and $A + \varepsilon I$ is a (nonsingular) M-matrix for all scalars $\varepsilon > 0$.

Example 6.5

(a) $A = \begin{bmatrix} 1 & -1 & & & \\ -1 & 2 & -1 & & \\ & \ddots & \ddots & \ddots & \\ & & 2 & -1 \\ & & & -1 & 2 \end{bmatrix}$ of order n is a M-matrix. It is of

positive type and also satisfies (b) in Theorem 6.5 with $\mathbf{x} = \mathbf{e}$ and $P = I$. We leave it to the interested reader to find a vector $\mathbf{x}$ for which it is generalized strictly diagonally dominant.

Hint: Try $\mathbf{x} = [1, (1 - \frac{1}{n})^2, \ldots, (\frac{1}{n})^2]^T$ and $\mathbf{x} = [1, 1 - (\frac{1}{n})^2, \ldots, 1 - (\frac{n-1}{n})^2]$.

(b) Any upper or lower triangular matrix A with $a_{i,j} \leq 0$, $i \neq j$, and $a_{ii} > 0$ is a M-matrix.

Remark 6.6 Beauwens (1976) showed that a matrix A is irreducibly diagonally dominant if and only if it is semistrictly diagonally dominant and irreducible. By *semistrict diagonal dominance* is meant that there exists a permutation matrix P such that $\widehat{A} = PAP^T$ is lower semistrictly diagonally dominant, i.e.,

$$|\hat{a}_{i,i}| \geq \sum_{j=1, j \neq i}^{n} |\hat{a}_{i,j}| \quad \text{and, moreover,}$$

$$|\hat{a}_{i,i}| > \sum_{j=1}^{i-1} |\hat{a}_{i,j}| \quad i = 1, 2, \ldots, n.$$

This follows from Theorem 6.5(a) and (b).

Now let us discuss a quite useful lemma.

Lemma 6.7 *Let B be a matrix of order n, let A be a $n \times n$ M-matrix, and let $C = B \odot A$, the so-called* Hadamard matrix product *defined by $c_{i,j} = b_{i,j} a_{i,j}$. If $0 \leq b_{i,j} \leq 1$, $i \neq j$, $b_{i,i} \geq 1$, $i = 1, 2, \ldots, n$, then C is also a M-matrix.*

Proof Since A is a M-matrix, there exists by Lemma 6.4 a positive vector $\mathbf{x}$, such that $A\mathbf{x} > \mathbf{0}$. Since $a_{i,i} > 0$ and $a_{i,j} \leq 0$, we get

$$(C\mathbf{x})_i = b_{i,i} a_{i,i} x_i + \sum_{j \neq i} b_{i,j} a_{i,j} x_j \geq a_{i,i} x_i + \sum_{j \neq i} a_{i,j} x_j > 0, \quad i = 1, 2, \ldots, n.$$

Hence, Lemma 6.4 shows that C is also a M-matrix.   ◇

Remark 6.8 An equivalent statement to Lemma 6.7 is: Let A be a M-matrix of order n, and let $\widehat{A} \geq A$. If $\hat{a}_{i,j} \leq 0$, $i \neq j$, then $\widehat{A} = [\hat{a}_{i,j}]$ is also a M-matrix.

6.1.1 Block Matrices

Here and in later chapters we shall deal with matrices partitioned in block form. Any partitioning of a n-dimensional vector $\mathbf{x} = (\mathbf{x}^{(1)}, \ldots, \mathbf{x}^{(m)})$ into block components $\mathbf{x}^{(k)}$ of dimensions n_k, $k = 1, \ldots, m$ (with $\sum_{k=1}^{m} n_k = n$)

is uniquely determined by a partitioning $\pi = \{\pi_k\}_{k=1}^m$ of the set of the first n integers, where π_k contains the integers $s_k + 1, \ldots, s_k + n_k$, $s_k = \sum_{j=1}^{k-1} n_j$.

The same partitioning π induces also a partitioning of any $n \times n$ matrix A into block matrix components $A_{i,j}$ of dimensions $n_i \times n_j$. A matrix that is block diagonal (triangular) relative to the π-partitioning will be referred to as *block diagonal (block triangular)*. Note that the matrices $A_{i,i}$ are square. The following results will be useful for later applications.

Theorem 6.9 *Let A be a M-matrix that is partitioned in block matrix form, $A = [A_{i,j}]$. Then:*

    (a) The matrices $A_{i,i}$ on the diagonal of A are M-matrices.
    (b) The block lower and upper triangular parts of A are M-matrices.

Proof To prove (a), let $B = [B_{i,j}]$ be partitioned consistently with the partitioning of A and let the entries $B_{i,i}$ be unity and the entries $B_{i,j}$, $i \neq j$, be zero. Then

$$B \odot A = \text{blockdiag}(A_{1,1}, \ldots, A_{m,m}),$$

and Lemma 6.5 shows that this is a M-matrix. In particular, each block is a M-matrix. Part (b) is proved similarly. $\diamond$

The following important result is also valid.

Theorem 6.10 *Let A be a M-matrix that is partitioned in two-by-two block form. Then the Schur complement*

$$S = A/A_{11} = A_{22} - A_{21}A_{11}^{-1}A_{12}$$

exists and is, itself, a M-matrix.

Proof Theorem 6.9 shows that A_{11} is a M-matrix, so A_{11} is nonsingular and A/A_{11} exists. In addition, $A_{11}^{-1} \geq 0$. Since A is a M-matrix, we have $A_{21} \leq 0$ and $A_{12} \leq 0$, so $A_{21}A_{11}^{-1}A_{12} \geq 0$. This shows that the off-diagonal entries of S are nonpositive. Next use the explicit form of the inverse (see 3.3), which shows that

$$A^{-1} = \begin{bmatrix} \bullet & \bullet \\ \bullet & S^{-1} \end{bmatrix}.$$

(The blocks not shown explicitly are irrelevant to the proof.) Hence, $A^{-1} \geq 0$ implies that $S^{-1} \geq 0$. $\diamond$

6.1.2 Positive Definiteness and Monotonicity

Next we present a relation between positive definiteness and monotonicity.

Theorem 6.11 *Let A be a real square matrix, with $a_{ij} \leq 0$, $i \neq j$. Then A is monotone and, hence, a M-matrix, if A is positive definite in $\mathbb{R}^n$.*

Proof If A is positive definite and of order n, then

(6.3) $$\mathbf{x}^T A \mathbf{x} > 0 \quad \text{for any nonzero } \mathbf{x} \in \mathbb{R}^n.$$

Let $\mathbf{x}$ be a vector such that $A\mathbf{x} \geq \mathbf{0}$, $\mathbf{x} \neq \mathbf{0}$. Following the proof of Stieltjes (1887), we shall show that then $\mathbf{x} \geq \mathbf{0}$. To this end, we split $\mathbf{x}$ into its positive and negative parts—i.e., we let $\mathbf{x} = \mathbf{x}^{(+)} - \mathbf{x}^{(-)}$, where $\mathbf{x}^{(+)} \geq \mathbf{0}$ and $\mathbf{x}^{(-)} \geq \mathbf{0}$. Note that $x_i^{(+)} x_i^{(-)} = 0$. (6.3) shows that $\mathbf{x}^{(-)^T} A \mathbf{x}^{(-)} \geq 0$, so

(6.4)
$$
\begin{aligned}
0 \geq & -\sum_i \sum_j a_{ij} x_i^{(-)} x_j^{(-)} \\
= & \sum_i \sum_j a_{ij} x_i^{(-)} (x_j^{(+)} - x_j^{(-)}) - \sum_i \sum_j a_{ij} x_i^{(-)} x_j^{(+)} \\
= & \sum_i x_i^{(-)} \sum_j a_{ij} x_j - \sum_i \sum_{j \neq i} a_{ij} x_i^{(-)} x_j^{(+)} \\
= & \sum_i x_i^{(-)} (A\mathbf{x})_i - \sum_i x_i^{(-)} \Big(\sum_{j \neq i} a_{ij} x_j^{(+)} \Big).
\end{aligned}
$$

However, $(A\mathbf{x})_i \geq 0$ by assumption, and $a_{ij} \leq 0$, $i \neq j$. Hence, (6.4) shows that

$$0 \geq -\mathbf{x}^{(-)^T} A \mathbf{x}^{(-)} \geq 0,$$

and it follows that $\mathbf{x}^{(-)^T} A \mathbf{x}^{(-)} = \mathbf{0}$ or $\mathbf{x}^{(-)} = \mathbf{0}$. Hence, $\mathbf{x} \geq \mathbf{0}$, that is, A is monotone.   ◇

The converse statement is not true, as can be seen by Example 3.2. It is also seen by considering $A = \begin{bmatrix} 1 & -3 \\ 0 & 1 \end{bmatrix}$, for example. Let $\mathcal{M}$ denote the class of M-matrices, Z the class of matrices A with $a_{ij} \leq 0$, $i \neq j$, and SP the class of positive definite matrices. Then Theorem 6.4 shows that $SP \cap Z \subset \mathcal{M}$, that is, a proper subset of $\mathcal{M}$. We shall now, however, show that $PR \cap Z \equiv \mathcal{M}$, where PR denotes the class of positive stable (also called positive real) matrices.

Theorem 6.12 *Let A be a real square Z matrix. Then A is a M-matrix if and only if A is positive stable (i.e., has eigenvalues with positive real parts).*

Proof If A is a M-matrix of order n, then Lemma 6.4 shows that there exists a $\mathbf{v} > \mathbf{0}$, such that $A\mathbf{v} > \mathbf{0}$, or equivalently, AV and also $V^{-1}AV$ are strictly diagonally dominant for $V = \text{diag}(v_1, v_2, \ldots, v_n)$, where $v_i > 0$. Then Gershgorin's disc theorem shows that the eigenvalues of $V^{-1}AV$ are contained in the right-hand half of the complex plane, so $\text{Re}(\lambda) > 0$ for any eigenvalue of $V^{-1}AV$ (that is, of A).

Conversely, assume $\text{Re}(\lambda) > 0$ for an eigenvalue of A. Since A is a Z-matrix, there exists a positive α such that $A = \alpha I_n - B$, where $B \geq 0$. By the Perron-Frobenius theorem, there exists an eigenvector $\mathbf{v} \geq \mathbf{0}$, such that $B\mathbf{v} = \rho(B)\mathbf{v}$. Then $A\mathbf{v} = (\alpha - \rho(B))\mathbf{v}$, so $\alpha - \rho(B)$ is an eigenvalue of A. Since the real parts of the eigenvalues are positive, we must have $\alpha > \rho(B)$. Consider $1/\alpha A = I_n - 1/\alpha B$. Here $\rho(1/\alpha B) < 1$ and $1/\alpha B \geq 0$, which shows that $(I - 1/\alpha B)^{-1} = I + 1/\alpha B + (1/\alpha B)^2 + \cdots \geq 0$, so $I_n - 1/\alpha B$ and, hence, A is monotone. $\diamond$

Remark 6.13 Ostrowski (1937) showed that for a Z-matrix A, the following properties are equivalent:

(a) A is a M-matrix.
(b) A is positive stable (Theorem 6.12).
(c) All principal minors of A are positive.

Fiedler and Ptak (1962) and Berman and Plemmons (1979) gave these and numerous additional equivalences.

Definition 6.5 $A = [a_{i,j}]$ is called a *Stieltjes matrix* (Stieltjes, 1887) if $a_{i,j} \leq 0$ for $i \neq j$ and A is symmetric and positive definite.

Corollary 6.14 *A symmetric Z-matrix A is a Stieltjes matrix if and only if it is a M-matrix.*

Proof This follows from Theorems 6.12 and 3.2. $\diamond$

Another interesting result for Stieltjes matrices is the following.

Theorem 6.15* *Let A be a Stieltjes matrix. Then, for any positive integer m there exists a Stieltjes matrix B such that $B^m = A$, that is, A has a root $A^{\frac{1}{m}}$ that is also a Stieltjes matrix.*

Proof Since A is symmetric and positive definite, the eigenvalues of A are real and positive. Let $V = [\lambda_{\max}(A)]^{-1}A$ and define $W = [w_{i,j}]$ by $W = I - V$. Note that

$$A = \lambda_{\max}(A)V = \lambda_{\max}(A)(I - W)$$

and

$$w_{i,j} = \begin{cases} 1 - a_{i,i}/\lambda_{\max}(A), & i = j \\ \\ - a_{i,j}/\lambda_{\max}(A), & i \neq j. \end{cases}$$

Clearly $a_{i,i} < \lambda_{\max}(A)$. [Note that (4.6) shows that $\lambda_{\max}(A) > \mathbf{e}_i^T A \mathbf{e}_i = a_{i,i}$ for any unit coordinate vector.] Since, in addition, $a_{i,j} \leq 0$, $i \neq j$, we have $w_{i,j} \geq 0$, that is, $W \geq \mathbf{0}$.

For any $t \in [0, 1]$, the series

$$(1 - t)^s = \sum_{k=0}^{\infty} (-1)^k \binom{s}{k} t^k,$$

where $s = 1/m$, is convergent. Hence, as

$$\|W\| = 1 - \lambda_{\min}(A)/\lambda_{\max}(A) < 1,$$

the corresponding matrix series is convergent, so we can define

$$V^{\frac{1}{m}} = (I - W)^{\frac{1}{m}} = \sum_{k=0}^{\infty} (-1)^k \binom{s}{k} W^k = I - R,$$

where $R = \sum_{k=1}^{\infty} \alpha_k W^k$ and where $\alpha_k = (-1)^{k+1}\binom{s}{k}$. It is readily seen that $\alpha_k \geq 0$ for any s, $0 < s \leq 1$. This, and $W \geq 0$, shows that $R \geq 0$, so $(V^{\frac{1}{m}})_{i,j} \leq 0$, $i \neq j$. Furthermore,

$$\|R\| = 1 - [1/\operatorname{cond}(A)]^{\frac{1}{m}} < 1,$$

so

$$V^{-\frac{1}{m}} = (I - R)^{-1} = \sum_{k=0}^{\infty} R^k \geq 0.$$

This shows that

$$A^{\frac{1}{m}} = \lambda_{\max}(A)^{\frac{1}{m}} V^{\frac{1}{m}}$$

is a M-matrix. Since V is symmetric, then W and R and, hence, also $V^{\frac{1}{m}}$ are symmetric. $\diamond$

6.2 Convergent Splittings

We return now to the question of convergence of the basic iterative method $C(\mathbf{x}^{l+1} - \mathbf{x}^l) = \mathbf{b} - A\mathbf{x}^l$, or

$$(6.5) \qquad C\mathbf{x}^{l+1} = R\mathbf{x}^l + \mathbf{b}, \quad l = 0, 1, \ldots,$$

where $A = C - R$ is a splitting of A.

Definition 6.6 Let $C, R \in \mathbb{R}^{n,n}$. Then $A = C - R$ is called:

    (a) a *regular splitting* (Varga, 1962), if C is monotone and $R \geq 0$.

    (b) a *weak regular splitting* (Ortega and Rheinboldt, 1970) if C is monotone and $C^{-1}R \geq 0$.

    (c) a *nonnegative splitting* (Beauwens, 1979, and Song, 1991) if C is nonsingular and $C^{-1}R \geq 0$.

    (d) a *convergent splitting* if C is nonsingular and $\rho(C^{-1}R) < 1$.

Clearly, a regular splitting is also a weak regular splitting, and a weak regular splitting is also a nonnegative splitting. The basic iterative method (6.5) is convergent if and only if $A = C - R$ is a convergent splitting (Theorem 5.3). The following general result was stated (without proof) by Beauwens (1979). This extends a classic theorem in Varga (1962).

Theorem 6.16 *Let $A = C - R$ be a nonnegative splitting of A, that is, such that C is nonsingular and $B = C^{-1}R$ is nonnegative. Then the following properties are equivalent:*

    (a) $\rho(B) < 1$, i.e., $A = C - R$ is a convergent splitting.

    (b) $I - B$ is monotone.

    (c) A is nonsingular and $G = A^{-1}R \geq 0$.

    (d) A is nonsingular and $\rho(B) = \rho(G)/[1 + \rho(G)]$, where $G = A^{-1}R$.

Proof If $\rho(B) < 1$, then $I - B$ is nonsingular and $(I - B)^{-1} = \sum_{k=0}^{\infty} B^k \geq 0$, which proves (b). If $(I - B)$ is monotone, then $A = C - R = C(I - B)$ is nonsingular and

$$(6.6) \qquad G = A^{-1}R = (I - B)^{-1}C^{-1}R = (I - B)^{-1}B \geq 0,$$

which proves (c).

Assume now that A is nonsingular and $G \geq 0$. Then $I - B = C^{-1}A$ is nonsingular. Since $B \geq 0$, there exists, by Varga's extension (1962) of the Perron-Frobenius theorem (see Theorem 4.13) an eigenvector $\mathbf{x} \geq \mathbf{0}$ (called the Perron vector) and an eigenvalue $\lambda = \rho(B)$ (the Perron root), such that $B\mathbf{x} = \rho(B)\mathbf{x}$. Hence, because $I - B$ is nonsingular, $\rho(B) \neq 1$. Further, (6.6) shows that $G\mathbf{x} = [1 - \rho(B)]^{-1}\rho(B)\mathbf{x}$. Thus, $G \geq 0$, $\mathbf{x} \geq \mathbf{0}$ implies $\rho(B) < 1$. This implies that $G = \sum_{k=1}^{\infty} B^k$ is convergent. $\mathbf{x}$ is a Perron vector for B, thus also for B^k and for G. This implies that $\rho(G) = \sum_{k=1}^{\infty} \rho(B)^k = \rho(B)/[1 - \rho(B)]$, that is, $\rho(B) = \rho(G)/[1 + \rho(G)]$, which proves (d).

Finally, (d) implies (a).   ◇

Corollary 6.17 *If $A = C - R$ is a weak regular splitting, then the splitting is convergent if and only if A is monotone.*

Proof If the splitting is convergent, i.e., if $\rho(B) < 1$, where $B = C^{-1}R$, then it follows by Theorem 6.16 that $I - B$ is monotone, so $A^{-1} = (I - B)^{-1}C^{-1} \geq 0$. If, on the other hand, A is monotone, then $(I - B)A^{-1} = C^{-1}$ shows that

$$(I + B + \cdots + B^k)C^{-1} = (I - B^{k+1})A^{-1}.$$

So, because $B \geq 0$ and $A^{-1} \geq 0$ for any positive vector $\mathbf{v}$, we have

$$(I + B + \cdots + B^k)C^{-1}\mathbf{v} \leq A^{-1}\mathbf{v}, \quad k = 0, 1, \ldots.$$

Since $C^{-1}\mathbf{v} > \mathbf{0}$, this implies that $I + B + \cdots + B^k$ is convergent and, in particular, $B^k \to 0$, $k \to \infty$, which, by Lemma 5.2(a), shows that $\rho(B) < 1$.   ◇

Remark 6.18 Note that in Theorem 6.16 we have not required that C or A be M-matrices; they do not even have to satisfy $a_{i,j} \leq 0$, $c_{i,j} \leq 0$, $i \neq j$. Note also the symmetry between the roles A and C play in Theorem 6.16 and Corollary 6.17, respectively:

(a) If C is nonsingular and $C^{-1}R$ is nonnegative, the splitting is convergent if and only if A is nonsingular and $A^{-1}R$ is nonnegative.

(b) If C is monotone and $C^{-1}R \geq 0$, the splitting is convergent if and only if A is monotone and $A^{-1}R \geq 0$.

Corollary 6.19 *If $A = C - R$ is a weak regular splitting (in particular, if it is a regular splitting) and if A is monotone, then the splitting is convergent—i.e., the iterative method $Cx^{l+1} = Rx^l + b$, $l = 0, 1, \ldots$ converges to the solution of $Ax = b$ for any initial vector x^0.*

Proof This proof follows by Corollary 6.17 and Theorem 5.3. ◇

Example 6.6

(a) The splitting

$$A = \begin{bmatrix} (1-\varepsilon) & -(1-\varepsilon) \\ -\varepsilon & \varepsilon \end{bmatrix}$$

$$= C - R = \begin{bmatrix} 1 & -1 \\ 0 & 2\varepsilon \end{bmatrix} - \begin{bmatrix} \varepsilon & -\varepsilon \\ \varepsilon & \varepsilon \end{bmatrix}, \quad 0 < \varepsilon \leq \frac{1}{2}$$

satisfies $C^{-1} \geq 0$, $C^{-1}R \geq 0$, but A is singular. Hence, A is not monotone.

(b) The splitting

$$A = \begin{bmatrix} (1-\varepsilon) & -(1-\varepsilon) \\ -\varepsilon & \varepsilon' \end{bmatrix}$$

$$= C - R = \begin{bmatrix} 1 & -1 \\ 0 & 2\varepsilon' \end{bmatrix} - \begin{bmatrix} \varepsilon & -\varepsilon \\ \varepsilon & \varepsilon' \end{bmatrix}, \quad 0 \leq \varepsilon < \varepsilon' \leq \frac{1}{2}.$$

is a weak regular splitting (but not a regular splitting): A is monotone and $\rho(C^{-1}R) < 1$.

(c) Show the same result as in part (b) for

$$A = \begin{bmatrix} 1 & -1 \\ -\frac{1}{2} & 1 \end{bmatrix}$$

$$= C - R = \begin{bmatrix} 1 & -(1+\varepsilon) \\ -\frac{1}{2} & 1-\varepsilon \end{bmatrix} - \begin{bmatrix} 0 & -\varepsilon \\ 0 & \varepsilon \end{bmatrix}, \quad 0 < \varepsilon < \frac{1}{3}.$$

6.3* Comparison Theorems

In order to compare the rate of convergence of different matrix splittings, we present some comparison results based mainly on a paper by Beauwens

(1979); see also Woźnicki (1973), Czordas and Varga (1984), and Marek and Szyld (1989). The following lemma will be useful.

Lemma 6.20 *Let $A = C - R$ be a convergent splitting of A. Then*

$$A^{-1} = C^{-1} + S,$$

where

$$S = C^{-1}RC^{-1} + (C^{-1}R)^2C^{-1} + \cdots$$

and

$$A^{-1}R = C^{-1}R + (C^{-1}R)^2 + \cdots.$$

Proof Note first that by the definition of a convergent splitting, C is nonsingular and $\rho(C^{-1}R) < 1$. Next,

$$A = C(I - C^{-1}R),$$

so

$$A^{-1} = (I - C^{-1}R)^{-1}C^{-1} = C^{-1} + C^{-1}RC^{-1} + (C^{-1}R)^2C^{-1} + \cdots. \quad \diamond$$

There are two classical comparison theorems: Let $A = C_i - R_i$, $i = 1, 2$ be two regular splittings. Then

$$\rho(C_1^{-1}R_1) \le \rho(C_2^{-1}R_2)$$

if either (Varga, 1962) $R_1 \le R_2$ or if (Woźnicki, 1973) $C_1^{-1} \ge C_2^{-1}$. These results can be seen to be special cases of Theorem 6.22 below. It is also seen from this theorem that the first condition ($R_1 \le R_2$) implies the second. At this point, we should note that the result of Woźnicki does not carry over to the case of weak regular splittings, as the following example shows:
Let $A = C_1 - R_1 = \begin{bmatrix} 1 & -(1+\varepsilon) \\ -\frac{1}{2} & (1-\varepsilon) \end{bmatrix} - \begin{bmatrix} 0 & -\varepsilon \\ 0 & \varepsilon \end{bmatrix}$ where $0 < \varepsilon < \frac{1}{3}$, and let $A = C_2 - R_2$, where $C_2 = A$ and $R_2 = 0$. Then

$$C_1^{-1} = \frac{2}{1-3\varepsilon}\begin{bmatrix} (1-\varepsilon) & (1+\varepsilon) \\ \frac{1}{2} & 1 \end{bmatrix}, \text{ while } C_2^{-1} = A^{-1} = 2\begin{bmatrix} 1 & 1 \\ \frac{1}{2} & 1 \end{bmatrix},$$

so $C_1^{-1} \ge C_2^{-1}$ but $\rho(C_1^{-1}R_1) > \rho(C_2^{-1}R_2)$, because $\rho(C_2^{-1}R_2) = 0$.

The next theorem, which is an extension of a theorem in Beauwens (1979) and shows that it suffices to compare the matrix splittings on a Perron vector, is based on similar ideas in Marek and Szyld (1989).

Theorem 6.21 *Let* $A = C_1 - R_1 = C_2 - R_2$ *be two convergent splittings where* $C_i^{-1} R_i$ *are nonnegative. Let* $G_i = A^{-1} R_i$, $i = 1, 2$. *Then*

(a) $G_i \geq 0$, $i = 1, 2$, *and*
(b) $(G_2 - G_1)G_1 \mathbf{x} \geq \mathbf{0}$, *where* $\mathbf{x} \geq \mathbf{0}$ *is the Perron vector of* G_1, *implies* $\rho(C_1^{-1} R_1) \leq \rho(C_2^{-1} R_2)$.

Proof By Lemma 6.20, $G_i = A^{-1} R_i \geq 0$, $i = 1, 2$, which proves (a). To prove (b), note that since $G_1 \mathbf{x} = \rho(G_1)\mathbf{x}$ holds, we have, by the assumption, $(G_2 - G_1)G_1 \mathbf{x} \geq \mathbf{0}$, so that

$$\rho(G_1)G_2 \mathbf{x} \geq \rho(G_1)^2 \mathbf{x}.$$

(i) If $\rho(G_1) = 0$ then $\rho(G_2) \geq \rho(G_1)$, since $\rho(G_2) \geq 0$.
(ii) If $\rho(G_1) > 0$, then $G_2 \mathbf{x} \geq \rho(G_1)\mathbf{x}$, so, in particular, with $G_2 = [g_{ij}]$,

$$(6.7) \qquad \sum_{j, x_j \neq 0} g_{ij} x_j \geq \rho(G_1) x_i, \quad x_i \neq 0.$$

Let $x_i = 0$, $i \in J$ and let $J' = \{i, 1 \leq i \leq n, i \notin J\}$. Let $\tilde{G} = \{g_{ij}\}$, $i, j \in J'$ and let $\tilde{D} = \mathrm{diag}(x_i)$, $i \in J'$. Then (6.7) shows that $\tilde{G}\tilde{D}\mathbf{e} \geq \rho(G_1)\tilde{D}\mathbf{e}$, $\mathbf{e} = (1, 1, \ldots, 1) \in \mathbb{R}^{n'}$, where n' is the dimension of J'. Hence, $\tilde{D}^{-1}\tilde{G}\tilde{D}\mathbf{e} \geq \rho(G_1)\mathbf{e}$, and Exercise 4.14 shows that $\rho(\tilde{G}) = \rho(\tilde{D}^{-1}\tilde{G}\tilde{D}) \geq \rho(G_1)$. Since $g_{ij} \geq 0$, Theorem 4.13 shows that $\rho(G_2) \geq \rho(G_1)$. Since $\rho/(1 + \rho)$ is a monotone function, Theorem 6.16(d) shows that

$$\rho(C_2^{-1} R_2) \geq \rho(C_1^{-1} R_1). \qquad \Diamond$$

The following theorem is an extension of theorems in Beauwens (1979), Woźnicki (1973), and Czordas and Varga (1984).

Theorem 6.22 *Let* $A = C_1 - R_1 = C_2 - R_2$ *be two convergent splittings of A. Then the following holds:*

(a) *If* $R_2 \geq R_1 \geq 0$ *and* C_i, $i = 1, 2$ *are monotone (i.e., the splittings are regular), then* $C_1^{-1} \geq C_2^{-1}$.

(b) *If $C_1^{-1} \geq C_2^{-1}$ and $R_1 x \geq 0$, then $(C_1^{-1} - C_2^{-1})R_1 x \geq 0$, where x is the Perron vector of $G_1 = A^{-1}R_1$.*

(c) *If $(C_1^{-1} - C_2^{-1})R_1 x \geq 0$, where x is the Perron vector of G_1, and if $A = C_i - R_i$, $i = 1, 2$ are weak regular splittings, then $\rho(C_1^{-1}R_1) \leq \rho(C_2^{-1}R_2)$.*

Proof

(a) Assume that $R_2 \geq R_1 \geq 0$. Then $R_i = C_i - A$, $i = 1, 2$ and $C_2 - C_1 = R_2 - R_1 \geq 0$. Hence,

$$C_1(C_1^{-1} - C_2^{-1})C_2 \geq 0$$

or, since, by assumption, $C_1^{-1} \geq 0$, $C_2^{-1} \geq 0$, premultiplying by C_1^{-1} and postmultiplying by C_2^{-1} yields

$$C_1^{-1} - C_2^{-1} \geq 0,$$

that is,

$$C_1^{-1} \geq C_2^{-1}.$$

(b) Follows by a straightforward multiplication.

(c) With $G_i = A^{-1}R_i$, we find

$$R_1(I + G_1) = R_1 + R_1 A^{-1}R_1 = (A + R_1)A^{-1}R_1 = C_1 G_1$$

and

$$I + G_2 = I + A^{-1}R_2 = A^{-1}(A + R_2) = A^{-1}C_2.$$

These relations show that

$$(I + G_2)(C_1^{-1} - C_2^{-1})R_1(I + G_1) = A^{-1}C_2(C_1^{-1} - C_2^{-1})C_1 G_1$$

$$(6.8) \qquad = A^{-1}(C_2 - C_1)G_1$$

$$= A^{-1}(R_2 - R_1)G_1 = (G_2 - G_1)G_1.$$

Lemma 6.20, $G_i \geq 0$ and (6.8) and the assumption show that

$$(G_2 - G_1)G_1 x = (1 + \rho(G_1))(I + G_2)(C_1^{-1} - C_2^{-1})R_1 x \geq 0,$$

where x is the Perron vector of G_1. By Theorem 6.21(b), this implies that

$$\rho(C_1^{-1}R_1) \leq \rho(C_2^{-1}R_2).$$ ◇

As can be seen from the proof of Theorem 6.21, one can replace the assumption (b) in the theorem by the weaker assumption that there exists $k > 0$, such that

$$(G_2^k - G_1^k)G_1 x \geq 0,$$

where $x \geq 0$ is the Perron vector of G_1.

Slightly different expressions, but without the weaker assumption involving the comparison for the Perron vector only, were considered by Czordas and Varga (1984).

Corollary 6.22′ *Let* $A = M_1 - N_1 = M_2 - N_2$ *be weak regular splittings. Then*

$$\rho(M_1^{-1}N_1) \leq \rho(M_2^{-1}N_2)$$

if any of the following holds:

(a) $N_2 \geq N_1 \geq 0$.
(b) $M_1^{-1} \geq M_2^{-1}$, $N_1 \geq 0$.

Proof This follows readily from Theorem 6.22(a) and (b), respectively.

◇

Several other comparison theorems can be found in Elsner (1989) and Song (1991).

6.3.1 Multisplitting Methods

To construct iterative methods for solving linear systems of algebraic equations on parallel computers, O'Leary and White (1985) proposed multisplittings of A. Let

$$A = C_i - R_i, \quad i = 1, 2, \ldots, k$$

be weak regular splittings, and let $E_i \geq 0$, $i = 1, 2, \ldots, k$ be diagonal matrices, such that $\sum_{i=1}^{k} E_i = I$. The iterative method

$$\mathbf{x}^{l+1} = \sum_{i=1}^{k} E_i C_i^{-1}(R_i \mathbf{x}^l + \mathbf{b}) = M \mathbf{x}^l + T \mathbf{b},$$

where

$$M = \sum_{i=1}^{k} E_i C_i^{-1} R_i, \quad T = \sum_{i=1}^{k} E_i C_i^{-1}$$

is called a *multisplitting method*. Note that

$$I - TA = I - \sum_{i=1}^{k}(E_i C_i^{-1} A) = I - \sum_{i=1}^{k} E_i C_i^{-1}(C_i - R_i)$$

$$= I - \sum_{i=1}^{k} E_i + \sum_{i=1}^{k} E_i C_i^{-1} R_i = \sum_{i=1}^{k} E_i C_i^{-1} R_i,$$

so

$$M = I - TA.$$

Note also that the k components $E_i C_i^{-1}(R_i \mathbf{x}^l + \mathbf{b})$, $i = 1, 2, \ldots, k$, of $\mathbf{x}^{l+1}$ can be computed in parallel. O'Leary and White (1985) show that for monotone matrices, the method converges for any $\mathbf{x}^0$ to the solution $\mathbf{x} = A^{-1}\mathbf{b}$ of $A\mathbf{x} = \mathbf{b}$. Following Elsner (1989), we present the next comparison theorem which gives a lower and an upper bound of the spectral radius $\rho(M)$ of M.

Theorem 6.23 *Let $A = C_i - R_i$, $i = 1, 2, \ldots, k$ be weak regular splittings, let $E_i \geq 0$ be diagonal matrices such that $\sum_{i=1}^{k} E_i = I$, and let*

$$M = \sum_{i=1}^{k} E_i C_i^{-1} R_i.$$

Further let $\underline{C}$ and $\overline{C}$ be lower and upper bounds of C_i, that is,

$$\underline{C} \leq C_i \leq \overline{C}, \quad i = 1, 2, \ldots, k.$$

Then, (a) if $A = \overline{C} - \overline{R}$ is a regular splitting, we have

$$\rho(M) \leq \rho(\overline{C}^{-1}\overline{R}),$$

and (b) if $A = \underline{C} - \underline{R}$ is a regular splitting, we have

$$\rho(M) \geq \rho(\underline{C}^{-1}\underline{R}).$$

Proof In the first case, apply Corollary 6.22'(c) to M_1, M_2, and N_2, where

$$M_1 = \left(\sum_{i=1}^{k} E_i C_i^{-1}\right)^{-1}, \quad M_2 = \overline{C}, \quad N_2 = \overline{R}.$$

As $C_i^{-1} \geq \overline{C}^{-1}$, we have

$$M_1^{-1} = \sum_{i=1}^{k} E_i C_i^{-1} \geq \sum_{i=1}^{k} E_i \overline{C}_i^{-1} = \overline{C}^{-1} = M_2^{-1},$$

so, using $\overline{R} \geq 0$, Corollary 6.22'(c) shows that

$$\rho(M_1^{-1}(M_1 - A)) \leq \rho(M_2^{-1}N_2) = \rho(\overline{C}^{-1}\overline{R}).$$

But

$$M_1^{-1}(M_1 - A) = I - M_1^{-1}A = I - TA = M,$$

which shows (a). In case (b) take

$$M_1 = \underline{C}, \quad N_1 = \underline{R} \quad \text{and} \quad M_2 = \left(\sum_{i=1}^{k} E_i C_i^{-1}\right)^{-1}$$

and apply Corollary 6.22'(b).   ◇

Note that it is not required in part (a) of the theorem that $A = \underline{C} - \underline{R}$. Similarly, in part (b) it is not required that $A = \overline{C} - \overline{R}$.

As in the case of Theorem 6.21 and 6.22, it is readily seen that this theorem can be extended to comparison for a special vector by letting $\underline{C}$, $\overline{C}$ be defined as matrices for which:

$(C_i^{-1} - \overline{C}^{-1})\overline{R}\mathbf{x} \geq \mathbf{0}$, where $\mathbf{x}$ is the Perron vector of $A^{-1}\overline{R}$, and

$(\underline{C}^{-1} - C_i^{-1})\underline{R}\mathbf{x} \geq \mathbf{0}$, where $\mathbf{x}$ is the Perron vector of $A^{-1}\underline{R}$.

For further information on multisplitting schemes, see White (1989) and Bai, and Wang (1993).

6.4* Diagonally Compensated Reduction of Positive Matrix Entries

In this section we consider symmetric positive definite matrices and analyze an approximation method that consists in reducing some or all positive offdiagonal entries, with subsequent modification (compensation) of the diagonal entries in rows (or columns) where a reduction has taken place. This compensation is done with respect to positive weights, defined by a positive vector. Such an approach permits us to reduce the problem of constructing a preconditioner for a s.p.d. matrix to the same problem for a M-matrix. (The latter problem is considered in the following chapter.) Frequently it turns out that the quality of this preconditioner—for instance, measured as the condition number—as a preconditioner to the original matrix is about as good as a preconditioner to the M-matrix on which its computation is based.

We now present results on convergent splittings and condition numbers for such preconditioners. The presentation is based on Axelsson and Kolotilina (1991). Given two matrices A, B and an s.p.d. preconditioner to B, the next lemma relates the eigenvalues of $M^{-1}A$ to those of $M^{-1}B$. For instance, if M is a preconditioner to B, then its quality as a preconditioner to A can be estimated from the following lemma.

Lemma 6.24 *Let A, B and M be symmetric and positive definite. Then*

$$\lambda_i(M^{-1}B)\lambda_{\max}(B^{-1}A) \geq \lambda_i(M^{-1}A) \geq \lambda_i(M^{-1}B)\lambda_{\min}(B^{-1}A),$$

where $\lambda_i(C)$ denotes the ith eigenvalue of C, and where the eigenvalues are numbered in a nondecreasing order.

Proof To prove the right-hand side inequality, we use matrix similarity and a Courant-Fischer type inequality for products of s.p.d. matrices to find

$$\lambda_i(M^{-1}A) = \lambda_i(M^{-1}B^{\frac{1}{2}}B^{-\frac{1}{2}}A) = \lambda_i[(B^{\frac{1}{2}}M^{-1}B^{\frac{1}{2}})(B^{-\frac{1}{2}}AB^{-\frac{1}{2}})]$$

$$\geq \lambda_i(M^{-1}B)\lambda_{\min}(B^{-1}A).$$

The left-hand side inequality is proved analogously. $\diamond$

Now let A be a s.p.d. matrix, let R be symmetric and nonnegative, and consider the matrix

$$B = A - R, \quad R \geq 0,$$

where the last inequality holds elementwise. Select next a positive vector $\mathbf{v}$ and define the matrix

$$M = D + B,$$

where D is a diagonal matrix defined by

$$D\mathbf{v} = R\mathbf{v}.$$

D is the compensation matrix for the reduced entry matrix R. Note that $D \geq 0$ and $M\mathbf{v} = A\mathbf{v}$. When M is a preconditioner to A, A becomes split as

$$A = M - (D - R).$$

Since $D \geq 0$, $R \geq 0$ and $(D - R)\mathbf{v} = \mathbf{0}$, $D - R$ is positive semidefinite. Hence, $M - A$ is positive semidefinite, and for any eigenvalue $\lambda_i(M^{-1}A)$ we have

$$\lambda_i(M^{-1}A) \leq 1.$$

Moreover, let

$$M = C - Q$$

be a symmetric splitting with a s.p.d. matrix C. Then, by Lemma 6.24,

$$\lambda_i(C^{-1}A) \leq \lambda_i(C^{-1}M)\lambda_{\max}(M^{-1}A) \leq \lambda_i(C^{-1}M),$$

i.e., we have proved the following theorem.

Theorem 6.25 (Convergent splittings) *Let A be a s.p.d. matrix and let $M = D + B$, $B = A - R$, $R \geq 0$, where all matrices are symmetric, D is diagonal, and $D\mathbf{v} = R\mathbf{v}$ for a positive vector $\mathbf{v}$. Let C be s.p.d. Then,*

$$0 < \lambda_i(C^{-1}A) \leq \lambda_i(C^{-1}M)$$

and, thus, the splitting

$$A = C - S,$$

(where $S = C - A$) is convergent if the splitting $M = C - Q$ is convergent. ◇

In the case of diagonally compensated reduction of all positive offdiagonal entries of A, the matrix B has no positive off-diagonal entries, so M is a Z-matrix. Since $M - A$ is positive semidefinite, M is, in fact, a Stieltjes matrix—i.e., a symmetric (positive definite) M-matrix. Thus, the next corollary holds.

Corollary 6.25' *Let* $A = B + R$, *where* $R \geq 0$ *contains all positive offdiagonal entries of* A *and* $diag(R) = 0$. *Then* $M = D + B$, *constructed as in Theorem 6.25, is a Stieltjes matrix.*   ◇

Furthermore, Theorem 6.25 shows that the problem of constructing a convergent splitting for an s.p.d. matrix can always be reduced to that for a Stieltjes matrix.

To estimate the condition number of $C^{-1}A$ compared with the condition number of $C^{-1}M$, we can use the following theorem.

Theorem 6.26 (Condition numbers) *Let* A, C, *and* M *be defined as in Theorem 6.25. Then*

(6.9)
$$\frac{1}{1 + \lambda_{\max}(A^{-1}(D - R))} \lambda_i(C^{-1}M) \leq \lambda_i(C^{-1}A) \leq \lambda_i(C^{-1}M)$$

and

$$\text{cond}(C^{-1}A) \leq \{1 + \lambda_{\max}[A^{-1}(D - R)]\} \text{cond}(C^{-1}M).$$

Proof The right-hand inequality (6.9) is shown in Theorem 6.25. For the left-hand inequality, use

$$\lambda_i(C^{-1}A) \geq \lambda_i(C^{-1}M)\lambda_{\min}(M^{-1}A)$$

and

$$\lambda_{\min}(M^{-1}A) = \inf_{\mathbf{x} \neq 0} \frac{\mathbf{x}^T A \mathbf{x}}{\mathbf{x}^T (A + D - R)\mathbf{x}} = \frac{1}{1 + \lambda_{\max}(A^{-1}(D - R))}. \quad ◇$$

6.4.1 Monotone Matrices

Our next result is, in principle, applicable only when $\|R\|$ is sufficiently small. It shows that if A is monotone and s.p.d, and if, in addition, either B is generalized diagonally dominant—i.e., diagonally dominant w.r.t. weights defined by the entries of $\mathbf{v}$—or if $\rho(A^{-1}R) < 1$, then the eigenvalues of $M^{-1}A$ can be bounded by the corresponding eigenvalues of $M^{-1}B$ times certain positive factors.

Note that a positive vector $B\mathbf{v}$ exists, in particular, if A is generalized diagonally dominant—i.e., $|a_{ii}|v_i \geq \sum_{j \neq i} |a_{ij}|v_j$, i.e. if A is a H-matrix.

Theorem 6.27 *Let A be a monotone s.p.d. matrix of order n, and let B and M be defined as in Theorem 6.25. Assume additionally that either*

> *(i) $B\mathbf{v} > \mathbf{0}$ for some positive vector $\mathbf{v}$ or*
> *(ii) $\rho(A^{-1}R) < 1$*

holds [it will be shown that (i) $\implies$ (ii)]. Then

> *(a)* $[1 - \rho(A^{-1}R)]^{-1}\lambda_i(M^{-1}B) \geq \lambda_i(M^{-1}A) \geq \frac{1}{1+\rho(A^{-1}R)}\lambda_i(M^{-1}B),$
>
> $i = 1, 2, \ldots, n$ *and* $\operatorname{cond}(M^{-1}A) \leq \frac{1+\rho(A^{-1}R)}{1-\rho(A^{-1}R)} \operatorname{cond}(M^{-1}B)$

and

> *(b)* $[1 + \rho(B^{-1}R)]\lambda_i(M^{-1}B) \geq \lambda_i(M^{-1}A) \geq \frac{1+\rho(B^{-1}R)}{1+2\rho(B^{-1}R)}\lambda_i(M^{-1}B),$
>
> $i = 1, 2, \ldots, n$ *and* $\operatorname{cond}(M^{-1}A) \leq [1 + 2\rho(B^{-1}R)]\operatorname{cond}(M^{-1}B).$

Proof The splitting $B = A - R$ shows that

$$(6.10) \qquad\qquad A^{-1}B = I - A^{-1}R,$$

where $A^{-1}R \geq 0$, because A is monotone and $R \geq 0$. If $B\mathbf{v} > \mathbf{0}$ for some vector $\mathbf{v} > \mathbf{0}$, then (6.10) shows that $\mathbf{v} - A^{-1}R\mathbf{v} = A^{-1}B\mathbf{v} > \mathbf{0}$, so $\mathbf{0} \leq A^{-1}R\mathbf{v} < \mathbf{v}$, that is,

$$\rho(A^{-1}R) \leq \|V^{-1}A^{-1}RV\|_\infty < 1,$$

where $V = \operatorname{diag}(v_1, v_2, \ldots, v_n)$. Hence *(i)* implies *(ii)*.

The fact that $\rho(A^{-1}R) < 1$, (6.10) shows that B is nonsingular and

$$B^{-1}A = (I - A^{-1}R)^{-1}.$$

that A is monotone and $A^{-1}R \geq 0$ shows that B is monotone, and

$$[1 - \rho(A^{-1}R)]^{-1} \geq \lambda_i(B^{-1}A) \geq [1 + \rho(A^{-1}R)]^{-1}.$$

Together with Lemma 6.24, this result shows part (a) of the theorem.

Finally, Theorem 6.16(d) shows that

$$\rho(A^{-1}R) = \rho(B^{-1}R)/[1 + \rho(B^{-1}R)],$$

so

$$[1 - \rho(A^{-1}R)]^{-1} = 1 + \rho(B^{-1}R)$$

and

$$[1 + \rho(A^{-1}R)]^{-1} = \frac{1 + \rho(B^{-1}R)}{1 + 2\rho(B^{-1}R)},$$

which shows part (b).   ◇

Remark 6.28 If A is irreducible, then, as can readily be seen, the assumption $B\mathbf{v} > \mathbf{0}$ can be replaced by $B\mathbf{v} \geq \mathbf{0}$.

Corollary 6.29 *If, under the assumptions of Theorem 6.27, all positive offdiagonal entries of A are reduced, then B is a Z-matrix:*

(a) $\lambda_{\min}(M^{-1}B) = 1/[1 + \rho(B^{-1}D)]$, *and*
(b) $\text{cond}(M^{-1}A) < 2[1 + \rho(B^{-1}D)]$.

Proof Note first that both B and M are M-matrices and the splitting $B = M - D$ is regular. Hence,

$$\lambda_{\min}(M^{-1}B) = 1 - \rho(M^{-1}D) = 1 - \rho(B^{-1}D)/[1 + \rho(B^{-1}D)]$$

$$= 1/[1 + \rho(B^{-1}D)],$$

which is part (a). Since $\lambda_{\max}(M^{-1}A) \leq 1$, then Theorem 6.27(b) shows that

$$1 \geq \lambda_i(M^{-1}A) > \tfrac{1}{2}\lambda_i(M^{-1}B),$$

so

$$\text{cond}(M^{-1}A) < 2[1 + \rho(B^{-1}D)],$$

which is part (b).   ◇

Remark 6.30 Because, under the conditions of Corollary 6.29, $B^{-1}D$ is nonnegative, its spectral radius—i.e., its Perron-Frobenius root—can be easily bounded using, for example, the two-sided bounds

$$\min_i(B^{-1}D\mathbf{e})_i \leq \rho(B^{-1}D) \leq \max_i(B^{-1}D\mathbf{e})_i,$$

where $\mathbf{e} = (1, 1, \ldots, 1)^T$.

We next consider two applications of the method of diagonal compensation.

Example 6.7 Let G be a nonnegative s.p.d. matrix of order n (such as the inverse of a monotone matrix) and let $[G]^{(p)}$ denote the bandmatrix part of G with symmetric semibandwidths p. We have

$$G = [G]^{(p)} + R, \quad R \geq 0.$$

Since the diagonal of R is zero, R is indefinite; it therefore can happen that $[G]^{(p)} = G - R$ also is indefinite. Hence, we can loose positive definiteness if we approximate a s.p.d. matrix with a band part of it. A remedy, however, is to use the method of diagonal compensation to modify $[G]^{(p)}$. Then let D be a diagonal matrix, such that

$$D\mathbf{v} = R\mathbf{v}$$

for some positive vector $\mathbf{v}$, and let

$$\widetilde{G} = [G]^{(p)} + D.$$

In order to show that $\widetilde{G}$ is s.p.d., let $V = \text{diag}(v_1, v_2, \ldots, v_n)$. Then $DV - RV$ and, also, $B = V(D - R)V$. The latter is symmetric and diagonally dominant, so its eigenvalues are nonnegative. But $V^{-1}BV^{-1} = D - R$ is a congruence transformation of B, so its eigenvalues are also nonnegative. Hence, $D - R$ is positive semidefinite.

Note now that

$$\widetilde{G} - G = [G]^{(p)} + D - G = D - R,$$

so $\widetilde{G} - G$ is positive semidefinite—that is, $\widetilde{G}$ is positive definite.

In some important applications, such as finite element applications, matrices arise that can be written as a direct sum of elementary matrices. We shall show that reduction and diagonal compensation can take place for these elementary matrices and that a spectral equivalence property holds between the original and the new matrix that is the direct sum of the reduced and compensated elementary matrices. The spectral equivalence property holds because the spectral bounds are valid independent of the number of elementary matrices.

Theorem 6.31 *Let $\{A_i\}_{i=1}^n$ be symmetric positive semidefinite (s.p.s.d.) matrices, let $A_i = M_i - N_i$, $A = \sum_{i=1}^n A_i$, $M = \sum_{i=1}^n M_i$, $N = \sum_{i=1}^n N_i$, where M_i are s.p.s.d. Assume, for some positive β_i and $\alpha_i \geq \beta_i$, that*

$$\beta_i M_i \leq N_i \leq \alpha_i M_i$$

where (in this theorem) the inequalities are in a positive semidefinite sense— i.e., if M_i, N_i have order n_i, then $\beta_i \mathbf{x}^T M_i \mathbf{x} \leq \mathbf{x}^T N_i \mathbf{x} \leq \alpha_i \mathbf{x}^T M_i \mathbf{x}$ for all $\mathbf{x} \in \mathbb{R}^{n_i}$. Then

$$(1 - \alpha)M \leq A \leq (1 - \beta)M,$$

where $\alpha = \max_i \alpha_i$, $\beta = \min_i \beta_i$.

Proof Since $N = \sum N_i$, we have

$$N \leq \sum \alpha_i M_i \leq \alpha \sum M_i = \alpha M$$

and, similarly, $N \geq \beta M$. Therefore,

$$\beta M \leq N \leq \alpha M$$

and, since $A = M - N$,

$$(1 - \alpha)M \leq A \leq (1 - \beta)M. \qquad \diamond$$

Theorem 6.31 shows that if all the splittings $A_i = M_i - N_i$ are convergent, then the splitting $A = M - N$ is also convergent, and $\rho(M^{-1}N) \leq \max_i \rho(M_i^{-1}N_i)$, assuming here that M_i are nonsingular. Hence, to construct a convergent splitting for the global matrix, it suffices to construct convergent splittings for the elementary matrices.

Example 6.8 Consider the bilinear form

$$a(u, v) = \int_0^1 u'v' dx,$$

which arises in the variational formulation of the boundary value problem $u'' = f$, $0 < x < 1$. The elementary stiffness matrix, with entries $a(\phi_i, \phi_j)$, resulting from piecewise quadratic basis functions on a unit interval, takes the form

$$A' = \begin{bmatrix} 7 & -8 & 1 \\ -8 & 16 & -8 \\ 1 & -8 & 7 \end{bmatrix}.$$

The diagonally compensated reduction of positive offdiagonal matrices yields the splitting

$$A' = M' - N' = 8\begin{bmatrix} 1 & -1 & 0 \\ -1 & 2 & -1 \\ 0 & -1 & 1 \end{bmatrix} - \begin{bmatrix} 1 & 0 & -1 \\ 0 & 0 & 0 \\ -1 & 0 & 1 \end{bmatrix},$$

or, after a symmetric permutation,

$$\tilde{A} = \tilde{M} - \tilde{N} = 8 \begin{bmatrix} 1 & 0 & -1 \\ 0 & 1 & -1 \\ -1 & -1 & 2 \end{bmatrix} - \begin{bmatrix} 1 & -1 & 0 \\ -1 & 1 & 0 \\ 0 & 0 & 0 \end{bmatrix}.$$

For any s.p.s.d. matrix B, partitioned in 2×2 block form,

$$B = \begin{bmatrix} B_{11} & B_{12} \\ B_{21} & B_{22} \end{bmatrix},$$

where $B_{i,i}$ are positive definite, it is readily seen that

$$B = \begin{bmatrix} S & 0 \\ 0 & 0 \end{bmatrix} + QQ^T,$$

where $S = B_{11} - B_{12}B_{22}^{-1}B_{21}$ and $Q = \begin{bmatrix} B_{12}B_{22}^{-\frac{1}{2}} \\ B_{22}^{\frac{1}{2}} \end{bmatrix}$. Hence, $B \geq \begin{bmatrix} S & 0 \\ 0 & 0 \end{bmatrix}$ in

a positive semidefinite sense. Applied to the matrix $\tilde{M}$, we find

$$\tilde{M} \geq 4 \begin{bmatrix} 1 & -1 & 0 \\ -1 & 1 & 0 \\ 0 & 0 & 0 \end{bmatrix},$$

so $\tilde{M} \geq 4\tilde{N}$ and, since $\tilde{N} \geq 0$, we find

$$0 \leq \tilde{N} \leq \tfrac{1}{4}\tilde{M}.$$

Hence, we can take $\alpha = 1/4$, $\beta = 0$ in Theorem 6.31. The global finite element matrix A is a sum of elementary matrices of the form $\tilde{A}$, corresponding to each subinterval (x_i, x_{i+1}), $x_i = ih$, $h = 1/(n+1)$, $i = 1, 2, \ldots, n$. For Dirichlet boundary conditions, it takes the form

$$A = \begin{bmatrix} 16 & -8 & & & \\ -8 & 14 & -8 & 1 & \\ & -8 & 16 & -8 & \\ & 1 & -8 & 14 & -8 & 1 \\ & & \cdot & \cdot & \cdot \end{bmatrix}.$$

The preconditioning matrix $M = \sum M_i$, resulting from the diagonally compensated complete reduction of offdiagonal positive entries of the element matrices, takes the form

$$M = \text{tridiag}(-8, 16, -8),$$

and Theorem 6.31 shows that the matrix $M^{-1}A$ has all its eigenvalues in the interval $[\frac{3}{4}, 1]$. (Incidentally, in this case the preconditioner M coincides, up to a factor 8, with the matrix resulting from approximation of the same differential equation by using piecewise linear basis functions.)

6.5 The SOR Method

We consider splittings based on the diagonal and triangular parts of A. By introducing a parameter (ω), we show that, in the latter case, under certain assumptions of A, the asymptotic convergence factor can be decreased by an order of magnitude, as was done for the second-order stationary method for proper values of the corresponding parameters in that method.

6.5.1 Regular Splittings

Let us assume that $a_{i,j} \le 0$, $i \ne j$ and that A is a M-matrix. Consider, first, the splitting $A = D_A - B_A$, where D_A is the diagonal part of A, and let $C = D_A$, $R = B_A$. Then $C^{-1} \ge 0$, $R \ge 0$, so $A = D_A - B_A$ is a regular splitting (Definition 6.6). More generally, this is also the case when A is partitioned into block matrix form:

$$A = \begin{bmatrix} D_1 & A_{1,2} & \cdots & A_{1,n} \\ A_{2,1} & D_2 & \cdots & A_{2,n} \\ \vdots & & & \\ A_{n,1} & A_{n,2} & \cdots & D_n \end{bmatrix}.$$

Here, D_i are square matrices. In fact, it follows from Theorem 6.9 that D_i are nonsingular and $D_i^{-1} \ge 0$.

Let $D_A = \text{diag}(D_1, D_2, \ldots, D_n)$, that is, let D_A be the block-diagonal part of A.

Then $D_A^{-1} \ge 0$ and the block matrix splitting, $A = D_A - B_A$, is also a regular splitting. From Corollary 6.17, it thus follows that the following iterative method, called the *(block) Jacobi method* (Gauss, 1823, and Jacobi, 1845) or *simultaneous iteration method*, is convergent:

$$D_A x^{l+1} = B_A x^l + \mathbf{b}, \quad l = 0, 1, \ldots,$$

and, likewise, its pointwise counterpart (where $A_{i,j} = a_{i,j}$ are scalars),

$$x_i^{(l+1)} = \frac{1}{a_{i,i}} [\sum_{j \neq i} (-a_{i,j}) x_j^{(l)} + b_i], \quad i = 1, 2, \ldots, n, \quad l = 0, 1, \ldots$$

$B = I - D_A^{-1} A = D_A^{-1} B_A$ is called the *Jacobi matrix*.

Note that even if A is not a M-matrix, we have for a diagonally dominant matrix A that $\rho(D_A^{-1} B_A) \leq \|D_A^{-1} B_A\|_\infty < 1$, so the Jacobi method converges.

Consider now the splitting $A = D - L - U$, where $D = D_A$ and L, U are the lower and upper block matrix triangular parts of A, respectively.

Then $C = D - L$ is a M-matrix, because $C^{-1} = [D(I - D^{-1}L)]^{-1} = [I + D^{-1}L + (D^{-1}L)^2 + \ldots]D^{-1} \geq 0$. Further, $U \geq 0$. Hence, $A = C - R, R = U$ is a regular splitting, and the following iterative method, called the *(block) Gauss-Seidel method* (Gauss, 1823, and Seidel, 1874) *or successive iteration method*, is convergent:

$$(D - L)\mathbf{x}^{l+1} = U\mathbf{x}^l + \mathbf{b}, \quad l = 0, 1, \ldots,$$

or

$$D_i \mathbf{x}_i^{l+1} = \sum_{j=1}^{i-1} (-A_{i,j}) \mathbf{x}_j^{l+1} + \sum_{j=i+1}^{n} (-A_{i,j}) \mathbf{x}_j^l + \mathbf{b}_i, \quad i = 1, 2, \ldots, n.$$

Note that in this method, the entries of $\mathbf{x}^{l+1}$ are computed in a fixed prescribed order (hence the name successive iteration). In the Jacobi method, on the other hand, the entries of $\mathbf{x}^{l+1}$ may be computed in any order, or for that matter, in parallel (hence the name simultaneous iteration method).

More generally, we have

(6.11) $\qquad (D - \omega L)\mathbf{x}^{l+1} = [(1 - \omega)D + \omega U]\mathbf{x}^l + \omega \mathbf{b}, \quad l = 0, 1, \ldots$

or

(6.12) $\qquad (\frac{1}{\omega}D - L)\mathbf{x}^{l+1} = [(\frac{1}{\omega} - 1)D + U]\mathbf{x}^l + \mathbf{b}, \quad l = 0, 1, \ldots,$

where $\omega \neq 0$ is a parameter called the *relaxation parameter*. The method is called the *successive relaxation method*. (Note that for $\omega = 1$, we get the Gauss-Seidel method.)

The pointwise counterpart of (6.11) is

(6.13) $\qquad x_i^{(l+1)} = (1 - \omega)x_i^{(l)} - \frac{\omega}{a_{i,i}} [\sum_{j=1}^{i-1} a_{i,j} x_j^{(l+1)} + \sum_{j=i+1}^{n} a_{i,j} x_j^{(l)} - b_i].$

Note that in the Jacobi method, we need to store two vectors, $\mathbf{x}^l$ and $\mathbf{x}^{l+1}$. In (6.13) we can store $x_i^{(l+1)}$ in the position of $x_i^{(l)}$, as soon as $x_i^{(l+1)}$ has been computed. Hence, in (6.11) it suffices to store one vector. Furthermore, as we shall see, (6.11) often converges faster than (6.9).

We can write (6.12) in the alternative form

$$(\frac{1}{\omega}D - L)(\mathbf{x}^{l+1} - \mathbf{x}^l) = -(A\mathbf{x}^l - \mathbf{b}), \quad l = 0, 1, \ldots.$$

This is, accordingly, a first-order method of the type in (5.2), with $C = 1/\omega D - L$. Note that (6.12) has the form (5.3), with $R = (1/\omega - 1)D + U$. The iteration matrix of (6.11) is

(6.14) $$L_\omega = (D - \omega L)^{-1}[(1 - \omega)D + \omega U].$$

We have $C^{-1} \geq 0$. If $\omega \leq 1$, then $R \geq 0$. Hence, for $\omega \leq 1$, $A = (1/\omega D - L) - [(1/\omega - 1)D + U]$ is a regular splitting, and by Corollary 6.17, (6.11) is convergent if A is monotone. As we shall see, however, the method frequently converges faster with $1 < \omega < 2$. This is called overrelaxation, and the method (6.11) is then called the *successive overrelaxation (SOR) method* (Frankel, 1950, and Young, 1950).

Let us now consider some results for the SOR-method that are valid in general—i.e., in particular, even if A is not a M-matrix.

Theorem 6.32 *For the iteration matrix L_ω, defined by (6.14), we have $\rho(L_\omega) \geq |\omega - 1|$. Hence, the relaxation method is divergent for $\omega \leq 0$ and for $\omega \geq 2$.*

Proof If λ_i are the eigenvalues of L_ω, we find

$$|\prod_{i=1}^n \lambda_i| = |\det[(1 - \omega)I + \omega D^{-1}U]| = |1 - \omega|^n.$$

Hence, for at least one eigenvalue, $|\lambda_i| \geq |1 - \omega|$.   ◇

The following theorem states an-easy-to check criterion for the convergence of the Gauss-Seidel method.

Theorem 6.33 *If A is diagonally dominant, $\sum_{j=2}^n |a_{1,j}| < |a_{1,1}|$ and, for at least one index, $j = j(i) < i$, we have $a_{i,j} \neq 0$, $i = 2, 3, \ldots, n$, then*

$$\rho(L_1) \leq \max_{1 \leq i \leq n} \nu_i < 1,$$

where

$$v_1 = \sum_{j=2}^{n} |a_{1,j}/a_{1,1}|, \; v_i = \sum_{j<i} |\frac{a_{i,j}}{a_{i,i}}|v_j + \sum_{j>i} |\frac{a_{i,j}}{a_{i,i}}|.$$

Proof We shall prove the stronger result, $\|L_1\|_\infty \le \max_{1\le i\le n} v_i < 1$. Let then **x** be an arbitrary vector with $\max_i |x_i| = 1$ and let $\mathbf{y} = L_1\mathbf{x}$. Then $(D - L)\mathbf{y} = U\mathbf{x}$ and $|y_1| \le \sum_{j\ge 2} |a_{1,j}/a_{1,1}| \, |x_j| \le v_1 < 1$.

Further, by induction,

$$|y_i| \le \sum_{j<i} |a_{i,j}/a_{i,i}||y_j| + \sum_{j>i} |a_{i,j}/a_{i,i}||x_j|$$

(6.15)
$$\le v_i \equiv \sum_{j<i} |a_{i,j}/a_{i,i}|v_j + \sum_{j>i} |a_{i,j}/a_{i,i}|$$

$$\le \sum_{j\ne i} |a_{i,j}/a_{i,i}| \le 1, \; i = 2, \ldots, n.$$

Hence, $\|L_1\|_\infty = \max |y_i| \le \max v_i < 1$, $i = 1, 2, \ldots, n$. ◇

For an extension of this theorem to nonlinear problems, see exercise 6. Similarly, it can be shown:

Theorem 6.33′ *If A is strictly diagonally dominant or irreducibly diagonally dominant, then $\rho(L_1) < 1$.*

It is left to the interested reader to verify that if $\max_{1\le i\le n} v_i < 1$, then the Gauss-Seidel method converges, even if A is not diagonally dominant (see Exercise 9). Consider the following example.

Example 6.9 The Gauss-Seidel method applied to the matrix

$$A = \begin{bmatrix} 6 & -2 & 0 \\ -1 & 2 & -1 \\ 0 & -6/5 & 1 \end{bmatrix}$$

is convergent, because one finds $v_1 = 1/3$, $v_2 = 2/3$, and $v_3 = 4/5$, that is, $\rho(L_1) < 4/5$. It is interesting to note that even the Jacobi method converges for this matrix. This observation will follow from Corollary 6.35 below.

6.5.2 Optimal Value of the Relaxation Parameter

Because the method (6.12) (for l large enough) converges faster if the asymptotic convergence factor $\rho(L_\omega)$ is smaller, we call the value $\omega = \omega_{\text{opt}}$, for which $\rho(L_\omega)$ is minimal, the *optimal relaxation parameter*. For the class of matrices we shall consider, we shall see that there exists a unique value ω_{opt}, that is, $\rho(L_{\omega_{\text{opt}}}) < \rho(L_\omega)$ $\forall \omega \neq \omega_{\text{opt}}$.

Definition 6.7 The matrix A is said to have *property* (A^π) if A can be permuted by PAP^T into a form that can be partitioned into block-tridiagonal form, that is,

$$(6.16) \qquad PAP^T = \begin{bmatrix} D_1 & U_1 & & & & \emptyset \\ L_1 & D_2 & U_2 & & & \\ & & \vdots & & & \\ & & & L_{r-2} & D_{r-1} & U_{r-1} \\ \emptyset & & & & L_{r-1} & D_r \end{bmatrix},$$

where the matrices D_i, $i = 1, \ldots, r$ are nonsingular.

For simplicity, in the sequel we assume that when A has property A^π, then A has already been permuted to the proper form.

Example 6.10 Standard difference approximation matrices for $u_{xx} + u_{yy}$ on rectangular domains have property (A^π). The block-tridiagonal form is taken when we use a row-wise, column-wise, or diagonal-wise ordering of the meshpoints. It can readily be seen that the diagonal-wise ordering for a rectangular domain, with M and N interior points along the sides of the rectangle, leads to a matrix with property (A^π) with diagonal matrices D_i, where for the standard five-point difference matrix

$$A = \begin{bmatrix} D_1 & L_1^T & & & & \emptyset \\ L_1 & D_2 & L_2^T & & & \\ & & \vdots & & & \\ & & & L_{r-2} & D_{r-1} & L_{r-1}^T \\ \emptyset & & & & L_{r-1} & D_r \end{bmatrix}, \qquad D_i = 4I_i,$$

and I_i is an identity matrix of order $\min(i, M)$, $i = 1, 2, \ldots, N$ $[h = 1/(N + 1)]$ and of order $(M + N - i)$, $i = N + 1, \ldots, M + N - 1 = r$. The matrices L_i have the form

$$(-L_i) = \begin{bmatrix} 1 & & & & \emptyset \\ 1 & 1 & & & \\ & \ddots & \ddots & & \\ & & 1 & 1 & \\ \emptyset & & & 1 & \end{bmatrix}, \quad i = 1, \ldots, M,$$

$$(-L_i) = \begin{bmatrix} 1 & & & \emptyset \\ 1 & 1 & & \\ & \ddots & \ddots & \\ \emptyset & & 1 & 1 \end{bmatrix}, \quad i = M+1, \ldots, N$$

and

$$(-L_i) = \begin{bmatrix} 1 & 1 & & \emptyset \\ & \ddots & \ddots & \\ \emptyset & & 1 & 1 \end{bmatrix}, \quad i = N+1, \ldots, M+N-1.$$

(We have assumed that $M \leq N$.) Hence, for this ordering the elliptic difference matrix has property (A^π) and, in addition, the matrices D_i are diagonal.

Since other orderings of the meshpoints correspond to some permutation of the above matrix, it follows that for any ordering there exists a permutation matrix P that takes the difference matrix A into the form (6.16) with diagonal matrices D_i. As will be seen later, the optimal relaxation parameter can be more easily estimated for matrices with property (A^π) that have diagonal matrices D_i.

Let $A = D - L - U$, where A has the form (6.16) and D, L, U are defined as before. We recall that the Jacobi and the relaxation iteration matrices are $B = D^{-1}(L + U)$ and $L_\omega = (D - \omega L)^{-1}[(1 - \omega)D + \omega U]$, respectively. The following lemma gives the relation between the eigenvalues of these matrices.

Lemma 6.34 (Young, 1950) *Assume that A has property (A^π) and let $\omega \neq 0$. Then:*

(a) *If $\mu \neq 0$ is any eigenvalue of B of multiplicity ν, then $-\mu$ is also an eigenvalue of B of multiplicity ν.*

(b) *If $\lambda \neq 0$ is an eigenvalue of L_ω and μ satisfies*

(6.17) $$\mu^2 = (\lambda + \omega - 1)^2/(\omega^2\lambda),$$

then μ is an eigenvalue of B.

(c) *If μ is an eigenvalue of B and λ satisfies*

$$\lambda + \omega - 1 = \omega\mu\lambda^{\frac{1}{2}},$$

then λ is an eigenvalue of L_ω.

Proof By Exercise 8, $\det(L + U + \mu D) = \pm \det(L + U - \mu D)$. Hence, $\det(D^{-1}L + D^{-1}U - \mu I)$ is some power of μ times a polynomial in μ^2, proving (a). Parts (b) and (c) follow from Exercise 8. $\diamond$

For a generalization of Lemma 6.34 to so-called *p*-cyclic matrices, see Varga (1962), Theorem 4.3, p. 126.

Corollary 6.35

(a) *If A has property (A^π), then $\rho(L_1) = \rho(B)^2$; further, if $\mu \neq 0$ is an eigenvalue of B, then also $(-\mu)$ is an eigenvalue of B.*
(b) *The Gauss-Seidel method converges if and only if the Jacobi method converges.*

Proof This proof follows from (6.17), because for $\omega = 1$, we have $\mu^2 = \lambda$. $\diamond$

We shall now derive a closed expression for ω_{opt} that is valid for matrices with property (A^π) and show that it depends only on the spectral radius of the Jacobi matrix B.

Theorem 6.36 *Assume that:*

(a) *A has property (A^π).*
(b) *The block Jacobi matrix $B = I - D_A^{-1}A$ has only real eigenvalues.*

Then the SOR method converges for any initial vector if and only if $\rho(B) < 1$ and $0 < \omega < 2$. Further, we have

$$\omega_{\text{opt}} = \frac{2}{1 + \sqrt{1 - \rho(B)^2}},$$

for which we find the asymptotic convergence factor

$$\min_\omega \rho(L_\omega) = \rho(L_{\omega_{\text{opt}}}) = \omega_{\text{opt}} - 1$$

$$= \left[1 - \sqrt{1 - \rho(B)^2}\right] \Big/ \left[1 + \sqrt{1 - \rho(B)^2}\right].$$

Proof Let $\lambda_1(\mu, \omega)$ and $\lambda_2(\mu, \omega)$ be eigenvalues of L_ω that by (6.17) correspond to eigenvalues $+\mu$ and $-\mu$, respectively, of B. Since the eigenvalues μ are real, it suffices to consider $\mu \geq 0$. It follows from (6.17) that

$$(6.18) \qquad \lambda^2 - 2\left(1 - \omega + \frac{1}{2}(\omega\mu)^2\right)\lambda + (\omega - 1)^2 = 0.$$

Hence, by Lemma 5.8, the SOR method converges if and only if $|\lambda| < 1$, that is, if and only if $2|1 - \omega + \frac{1}{2}(\omega\mu)^2| < 1 + (\omega - 1)^2$ and $(\omega - 1)^2 < 1$, that is, if and only if $\mu^2 < 1$ (i.e., $\rho(B) < 1$) and $0 < \omega < 2$.

It follows further by (6.18) that

$$\lambda_{1,2} = 1 - \omega + \frac{1}{2}(\omega\mu)^2 \pm \omega\mu\left\{1 - \omega + (\frac{1}{2}\omega\mu)^2\right\}^{\frac{1}{2}}$$

$$= \left[\frac{1}{2}\omega\mu \pm \left\{1 - \omega + (\frac{1}{2}\omega\mu)^2\right\}^{\frac{1}{2}}\right]^2.$$

As in the proof of Theorem 5.9, we get that

$$\min_\omega \max\{\lambda_1(\omega, \mu), \lambda_2(\omega, \mu)\}, \quad 0 < \mu \leq \mu_0 = \rho(B),$$

is taken when $1 - \omega + (\frac{1}{2}\omega\mu_0)^2 = 0$, that is, for $\omega = \omega_{\text{opt}}$. The convergence factor then is $\omega_{\text{opt}} - 1$. $\diamond$

Remark 6.37 Note that ω_{opt} equals α_{opt} in the second-order method. Note further that for $\omega = 1$ (the Gauss-Seidel method), we have (cf. Corollary 6.35) $\max\{\lambda_1(1, \mu), \lambda_2(1, \mu)\} = \rho_0^2$ under the stated assumptions. It follows, further, from Theorem 6.36 that under the stated properties of A and with $0 < \omega < 2$, the SOR method converges if and only if the Jacobi method converges.

Example 6.11 Consider a matrix A with property (A^π) for which the Jacobi matrix B has real eigenvalues μ, $-\rho_0 \leq \mu \leq \rho_0$ and $\rho_0 = \rho(B) = 1 - \delta$, $0 < \delta \ll 1$. Then the rate of convergence of the Jacobi method is $r = -\log_{10}\rho(B) = -\log_{10}(1 - \delta) \simeq \delta/\ln 10$. It follows from Corollary 6.35, Theorem 6.36, and Theorem 5.9 that the following convergence factors and rates of convergence are valid, as shown in Table 6.1.

Note that for $\delta \simeq 0.5 \cdot 10^{-2}$, for instance, the number of iterations for the Gauss-Seidel method is about 20 times more than for the optimal SOR method for the same iteration error. The result for the second-order method is valid for $C = D = \text{diag}(A)$ and $-\rho_0 \leq \mu \leq \rho_0$, where μ is an eigenvalue of B.

Table 6.1.   *Convergence Factors and Rates of Convergence
for Various Methods*

Method	Asymptotic Convergence Factor	Rate of Convergence
Jacobi	$\simeq 1 - \delta$	$\simeq \delta / \ln 10$
Gauss-Seidel	$\simeq 1 - 2\delta$	$\simeq 2\delta / \ln 10$
SOR $(\omega = \omega_{opt})$	$\simeq 1 - 2\sqrt{2\delta}$	$\simeq 2\sqrt{2\delta} / \ln 10$
Second-order method, $\alpha = \alpha_{opt}$	$\simeq 1 - \sqrt{2\delta}$	$\simeq \sqrt{2\delta} / \ln 10$

For other matrices C, this latter method may converge much faster (see Chapter 7).

Let us now consider the validity of the assumption in Theorem 6.36 that the eigenvalues of B are real and that $\rho(B) < 1$.

Theorem 6.38 *If A and D are symmetric and positive definite and A has property A^{π}, then the eigenvalues of $B = D^{-1}(L + L^T)$, where $A = D - L - L^T$, are real and $\rho(B) < 1$.*

Proof B is not symmetric in general, but it is spectrally equivalent to $\widetilde{B} = D^{\frac{1}{2}} B D^{-\frac{1}{2}} = D^{-\frac{1}{2}}(L + L^T)D^{-\frac{1}{2}}$, which is symmetric. (Here $D^{\frac{1}{2}}$ is the square root of the positive definite matrix D.) Hence, the eigenvalues of $\widetilde{B}$, and, accordingly, also those of B, are real.

Further, since A is positive definite, $\mathbf{x}^*(L + L^T)\mathbf{x} \leq \mathbf{x}^* D \mathbf{x}$ for all $\mathbf{x} \neq \mathbf{0}$, so the largest eigenvalue μ_0 of B is less than one. Lemma 6.34(a) shows that the nonzero eigenvalues occur in $\pm$ pairs—that is, $-\mu_0$ is also an eigenvalue of B and $\rho(B) = \mu_0 < 1$.   ◇

Remark 6.39 In many applications, the matrix A is quasisymmetric—i.e., there exists a matrix S such that $\widetilde{A} = SAS^{-1}$ is symmetric. If S is (block-)diagonal, then $\widetilde{A}$ is block-tridiagonal if A is so. Hence, B has real eigenvalues and $\widetilde{A}$ has property (A^{π}). Regarding quasisymmetric matrices, see Chapter 3.

6.5.3 Optimal Parameter for Point-Wise Relaxation Methods

If D_i, $i = 1, \ldots, r$ in (6.16) are diagonal matrices, then the point-wise relaxation method (6.13) is equivalent to the block-wise (6.11).

Example 6.12 If we use a diagonal-wise ordering of the node points in a square mesh for the standard five-point elliptic difference operator, we get a matrix with diagonal blocks D_i (see Example 6.10). Further, the Jacobi iteration matrix B has eigenvalues independent of the ordering of the meshpoints.

Lemma 6.40 *Let* $B = I - D^{-1}A$ *and* $\widetilde{B} = I - \widetilde{D}^{-1}\widetilde{A}$, *where* $\widetilde{A} = PAP^T$ *is a permutation of* A *and* D, $\widetilde{D}$ *are the diagonal (i.e., not the block-diagonal) parts of* A *and* $\widetilde{A}$ *respectively. Then the spectrum of* $\widetilde{B}$ *is identical to that of* B.

Proof It is readily seen that $\widetilde{D} = PDP^T$. For the corresponding Jacobi matrix, we have

$$\widetilde{B} = I - \widetilde{D}^{-1}\widetilde{A} = I - PD^{-1}AP^T = P(I - D^{-1}A)P^T = PBP^T.$$

Hence, $\widetilde{B}$ is a similarity transformation of B, so the spectrum is the same. $\diamond$

Lemma 6.40 shows, in particular, that $\rho(\widetilde{B}) = \rho(B)$. But for the elliptic difference matrix, we have $\rho(B) = \cos \pi h$, if h corresponds to the natural (rowwise) ordering. Hence, for the diagonal-wise, we have also $\rho(\widetilde{B}) = \cos \pi h$, and $\omega_{opt} = 2/[1 + \sin \pi h]$, and, in Table 6.1, we get the stated convergence rates with $\delta = 1 - \cos \pi h \simeq \frac{1}{2}(\pi h)^2$.

Remark 6.41 In Theorem 6.36 we assumed that the matrix A can be partitioned to a block tridiagonal form with nonsingular matrices D_i (A has property A^π). Young (1950) defines a matrix to be *consistently ordered* if $\alpha D^{-1}L + \alpha^{-1}D^{-1}U, \alpha \neq 0$ has eigenvalues that do not depend on α. It can be seen that a matrix with property (A^π) is consistently ordered (see Exercise 8). It can also be seen that a consistently ordered matrix with D a diagonal matrix has what Young (1950) calls *property A*—namely, it can be partitioned in the 2-by-2 block matrix form

$$\begin{bmatrix} \widehat{D}_1 & A_{12} \\ A_{21} & \widehat{D}_2 \end{bmatrix},$$

where $\widehat{D}_i$, $i = 1, 2$ are diagonal matrices.

It can be shown that for matrices with property A, we have the same results as in Theorem 6.36. Varga (1959, 1962) extends this further to p-cyclic matrices, $p \geq 2$, where a 2-cyclic matrix is identical to a matrix with property (A^π). In Young (1971) the optimal relaxation parameter is derived for the case where some eigenvalues of B are complex. For consistently ordered matrices, it is shown that $\rho(L_0) < 1$ if the eigenvalues $\mu = \alpha + i\beta$, α, β real, are contained in the interior of the ellipse $x^2 + (y/\delta)^2 = 1$, and ω satisfies $0 < \omega < 2/(1 + \delta)$. For a given ω, $0 < \omega < 2$, this holds if $\delta < (2 - \omega)/\omega$. It is readily seen that a matrix with property (A^π) with diagonal matrices D_i can be permuted to the form for a matrix with property A. Just take a permutation based on the odd-even reordering of the tridiagonal block structure. Then the matrix $\widehat{D}_1$ will consist of the odd-numbered blocks D_i, and $\widehat{D}_2$ will consist of the even-numbered blocks.

A typical example of matrices with property (A^π) with diagonal matrices D_i occurs for second-order elliptic difference equations for a rectangular mesh when we use the standard five-point difference approximation and a diagonal-wise ordering of mesh points. For this case, we get the canonical form for a matrix with property A if we use a so-called red-black (or white-black as on a chessboard) ordering of the meshpoints.

It can be shown that natural orderings (row-wise or column-wise) are consistent orderings. Here the diagonal matrix blocks do not have (point-wise) diagonal form. For such orderings, it can be shown that the spectral radius $\rho(B)$ of the Jacobi iteration matrix is somewhat smaller than for the property A^π ordering with diagonal matrices D_i. (See Exercise 11.)

6.5.4 SOR Theory for Symmetric Positive Definite Matrices

Let us now discuss some properties of the SOR method that are valid for symmetric positive definite matrices but that do not require property (A^π). For historical reference, we first state a result (see R. Varga, 1992, Theorem 3.6, p. 91) without proof.

Theorem 6.42 *If A is symmetric, D is positive definite, and $0 < \omega < 2$, then the relaxation method is convergent if and only if A is positive definite.*

Let A be a Stieltjes matrix. Kahan (1958) and Varga (1959) showed that even if A does not have property (A^π), the relaxation parameter $\omega = \omega_{\text{opt}} = 2/[1 + \{1 - \rho(B)^2\}^{\frac{1}{2}}]$ is still close to optimal, because they found

$$\omega_{\text{opt}} - 1 \le \rho(L_{\omega_{\text{opt}}}) \le \{\omega_{\text{opt}} - 1\}^{\frac{1}{2}}$$

and

$$\omega_{\text{opt}} - 1 \le \min_\omega \rho(L_\omega) < \{\omega_{\text{opt}} - 1\}^{\frac{1}{2}}.$$

There is equality, $\omega_{\text{opt}} - 1 = \rho(L_{\omega_{\text{opt}}})$ if and only if A is consistently ordered.

Note that $\omega_{\text{opt}} - 1$ therefore is the best possible asymptotic convergence factor [because $\rho(L_\omega) \ge \omega_{\text{opt}} - 1$ for any ω!], but that the asymptotic convergence factor is never worse than that for the second-order (Richardson's or

Frankel's) method. The rate of convergence of the latter is not worse than half of that for the optimal SOR-method. In fact, as we shall now see, for symmetric positive definite matrices (not necessarily Stieltjes matrices) we can derive a closed expression for the optimal parameter ω—or, rather, for an accurate approximation of it—as we did for matrices with property A^π. Let $A = D - L - L^*$, where A is Hermitian and positive definite. Consider the SOR iteration matrix for A,

$$(6.19) \qquad L_\omega = (\frac{1}{\omega}D - L)^{-1}[(\frac{1}{\omega} - 1)D + L^*]$$

and transform L_ω by a similarity transformation

$$(6.20) \qquad \widetilde{L}_\omega = D^{\frac{1}{2}}L_\omega D^{-\frac{1}{2}} = (\frac{1}{\omega}I - \widetilde{L})^{-1}((\frac{1}{\omega} - 1)I + \widetilde{L}^*),$$

where $\widetilde{L} = D^{-\frac{1}{2}}LD^{-\frac{1}{2}}$. Define $\widetilde{A} = D^{-\frac{1}{2}}AD^{-\frac{1}{2}}$. Then we consider the following theorem.

Theorem 6.43 *Let A be Hermitian and positive definite and let L_ω and $\widetilde{L}_\omega$ be defined by (6.19) and (6.20), respectively, and let $0 < \omega < 2$. Then*

$$(6.21) \qquad \rho(L_\omega)^2 = \rho(\widetilde{L}_\omega)^2 \le 1 - \frac{\frac{2}{\omega} - 1}{(\frac{1}{\omega} - \frac{1}{2})^2\delta^{-1} + \gamma + \frac{1}{\omega}},$$

where

$$\gamma = \sup_{x\neq 0}\left\{\left[\frac{|(x, \widetilde{L}x)|^2}{(x, x)} - \frac{1}{4}(x, x)\right]/(\widetilde{A}x, x)\right\},$$

and

$$\delta = \lambda_{\min}(\widetilde{A}) = \min_{x\neq 0}(\widetilde{A}x, x)/(x, x).$$

Further, if $|(x, \widetilde{L}x)| \le \frac{1}{2}(x, x)$, then

$$\omega^* = 2/[1 + \sqrt{2\delta}],$$

minimizes the upper bound (6.21) of $\rho(L_\omega)$ and we have

$$\rho(L_{\omega^*})^2 = (1 - \sqrt{\delta/2})/(1 + \sqrt{\delta/2}).$$

Proof

Let λ, $\mathbf{x}$ be an eigensolution of $\widetilde{L}_\omega$, i.e. $\widetilde{L}_\omega \mathbf{x} = \lambda \mathbf{x}$ and let $z = (\mathbf{x}, \widetilde{L}\mathbf{x})/(\mathbf{x}, \mathbf{x})$. Then

$$[(\frac{1}{\omega} - 1)I + \widetilde{L}^*]\mathbf{x} = \lambda(\frac{1}{\omega}I - \widetilde{L})\mathbf{x}$$

and

$$\lambda = (\frac{1}{\omega} - 1 + \bar{z})/(\frac{1}{\omega} - z).$$

Hence,

$$|\lambda|^2 = \left[(\frac{1}{\omega} - 1)^2 + 2(\frac{1}{\omega} - 1)Re(z) + |z|^2\right] \Big/ \left[(\frac{1}{\omega})^2 - \frac{2}{\omega}Re(z) + |z|^2\right]$$

$$= 1 - (\frac{2}{\omega} - 1)(1 - 2Re(z)) \Big/ \left[(\frac{1}{\omega} - \frac{1}{2})^2 + \frac{1}{\omega}(1 - 2Re(z)) + |z|^2 - \frac{1}{4}\right].$$

Note that

$$(\widetilde{A}\mathbf{x}, \mathbf{x})/(\mathbf{x}, \mathbf{x}) = 1 - (z + \bar{z}) = 1 - 2Re(z),$$

so

$$|\lambda|^2 = 1 - (\frac{2}{\omega} - 1) \Big/ \left\{ \left[(\frac{1}{\omega} - \frac{1}{2})^2 + \left|\frac{(\mathbf{x}, \widetilde{L}\mathbf{x})}{(\mathbf{x}, \mathbf{x})}\right|^2 - \frac{1}{4}\right] \frac{(\mathbf{x}, \mathbf{x})}{(\widetilde{A}\mathbf{x}, \mathbf{x})} + \frac{1}{\omega} \right\}.$$

Theorem 6.42 shows that $\rho(\widetilde{L}_\omega) < 1$ for $0 < \omega < 2$, so the second term is positive. Hence,

$$\rho(\widetilde{L}_\omega)^2 \leq 1 - (\frac{2}{\omega} - 1) \Big/ \left\{ \left[(\frac{1}{\omega} - \frac{1}{2})^2\delta^{-1} + \gamma\right] + \frac{1}{\omega} \right\},$$

where γ and δ are defined in the theorem. Further, if $|(\mathbf{x}, \widetilde{L}\mathbf{x})| \leq \frac{1}{2}(\mathbf{x}, \mathbf{x})$, then $\gamma \leq 0$ and

(6.22) $$\rho(\widetilde{L}_\omega)^2 \leq 1 - 2 \Big/ \left[(1 - \frac{\omega}{2})\frac{1}{\omega\delta} + (1 - \frac{\omega}{2})^{-1}\right].$$

It is readily seen that the upper bound in (6.22) is minimized for $\omega = \omega^* = 2/(1 + \sqrt{2\delta})$, and for this value of ω we get

$$\rho(\widetilde{L}_{\omega^*})^2 \leq (1 - \sqrt{\delta/2})/(1 + \sqrt{\delta/2}). \qquad \diamond$$

Note that $B = D^{-1}(L + L^*)$ and $\widetilde{B} = D^{1/2}BD^{-1/2} = \widetilde{L} + \widetilde{L}^*$. Hence,

$$\rho(B) = \rho(\widetilde{B}) = \max_{\mathbf{x}\neq 0} \mid \left(\mathbf{x}, [\widetilde{L} + \widetilde{L}^*]\mathbf{x}\right) \mid$$

$$= \max_{\mathbf{x}\neq 0} 2|\operatorname{Re} z| = 1 - \min_{\mathbf{x}\neq 0}(\widetilde{A}\mathbf{x}, \mathbf{x})/(\mathbf{x}, \mathbf{x}) = 1 - \delta.$$

If the conditions in Theorem 6.43 hold, then

$$\rho(L_{\omega^*}) \simeq 1 - \sqrt{\frac{\delta}{2}}, \quad \delta \to 0.$$

Comparing this with the estimate $\rho(L_{\omega_{\text{opt}}})$ in Table 6.1, we see that the convergence rate estimate based on (the upper bound of) $\rho(L_{\omega^*})$ is about 4 times as slow as for the estimate for matrices with property (A^π). If, however, we compare ω^* with ω_{opt}, which is valid for matrices with property (A^π), we find

$$\omega^* = \omega_{\text{opt}}\sqrt{\frac{1 + \rho(B)}{2}} = \omega_{\text{opt}}\sqrt{1 - \delta/2} \sim \omega_{\text{opt}}(1 - \delta/4), \delta \to 0,$$

which means that ω^* and ω_{opt} have the same asymptotic limit,

$$\lim_{\delta\to 0} \omega_{\text{opt}}(1 + \sqrt{2\delta}) = 2 = \omega^*(1 + \sqrt{2\delta}).$$

In practice it is advisable to choose $\omega = 2/\left(1 + \zeta\sqrt{\tilde{\delta}}\right)$, where $\tilde{\delta}$ has the same asymptotic behavior as δ (i.e., $\tilde{\delta} = O(h^2)$ for difference equations) and $\zeta > 0$ is a parameter. This will give the same $O(\sqrt{\delta})$ rate of convergence as for $\omega = \omega^*$. It is also readily seen that we can permit $\gamma \leq$ const, that is, $|(\mathbf{x}, L\mathbf{x})|^2 \leq 1/4(\mathbf{x}, \mathbf{x})^2 + \text{const}(\mathbf{x}, \widetilde{A}\mathbf{x})$, and still retain the asymptotic rate of convergence.

The following corollary gives further information regarding the constant γ.

Corollary 6.44 *Let A be real and s.p.d. Then*

(a) $\gamma \leq \frac{1}{4}\sup_{\mathbf{x}\in\mathbb{C}^n, (\mathbf{x},\mathbf{x})=1} \frac{|(\mathbf{x},(\widetilde{L}-\widetilde{L}^*)\mathbf{x})|^2}{(\mathbf{x},\widetilde{A}\mathbf{x})}.$

(b) *If, in addition, D is Stieltjes and L is nonnegative, then $\gamma \leq 0$ and the last part of Theorem 6.43 holds.*

Proof Let $\|\mathbf{x}\| = 1$. Note first that

$$z = (\mathbf{x}, \widetilde{L}\mathbf{x}) = \frac{1}{2}(\mathbf{x}, (\widetilde{L} + \widetilde{L}^*)\mathbf{x}) + \frac{1}{2}(\mathbf{x}, (\widetilde{L} - \widetilde{L}^*)\mathbf{x})$$

$$= \operatorname{Re} z + \operatorname{Im} z,$$

so

$$|z|^2 = \frac{1}{4}(\mathbf{x}, (\widetilde{L} + \widetilde{L}^*)\mathbf{x})^2 + \frac{1}{4}|(\mathbf{x}, (\widetilde{L} - \widetilde{L}^*)\mathbf{x})|^2.$$

But $\widetilde{A} = I - \widetilde{L} - \widetilde{L}^*$ is positive definite. Hence, $(\mathbf{x}, (\widetilde{L} + \widetilde{L}^*)\mathbf{x}) \le 1$ and

$$\gamma = \sup_{\substack{(\mathbf{x},\mathbf{x})=1}} (|z|^2 - \frac{1}{4})/(\widetilde{A}\mathbf{x}, \mathbf{x}) \le \frac{1}{4} \sup_{\substack{(\mathbf{x},\mathbf{x})=1}} |(\mathbf{x}, (\widetilde{L} - \widetilde{L}^*)\mathbf{x})|^2/(\widetilde{A}\mathbf{x}, \mathbf{x}),$$

which shows part (a). To show part (b), note that D^{-1} is nonnegative and take $D^{-\frac{1}{2}}$ nonnegative. Then $\widetilde{L} = D^{-\frac{1}{2}}LD^{-\frac{1}{2}}$ is nonnegative. Hence, Corollary 4.20 shows that

$$r(\widetilde{L}) = \max_{\substack{\mathbf{x} \in \mathbb{C}^n \\ (\mathbf{x},\mathbf{x})=1}} |(\mathbf{x}, \widetilde{L}\mathbf{x})| = \max_{\substack{\mathbf{x} \in \mathbb{R}^n \\ (\mathbf{x},\mathbf{x})=1}} (\mathbf{x}, \widetilde{L}\mathbf{x}) = \frac{1}{2} \max_{\substack{\mathbf{x} \in \mathbb{R}^n \\ (\mathbf{x},\mathbf{x})=1}} (\mathbf{x}, (\widetilde{L} + \widetilde{L}^*)\mathbf{x}),$$

so $|z| = |(\mathbf{x}, \widetilde{L}\mathbf{x})| \le r(\widetilde{L}) \le 1/2$.   ◇

The next corollary shows that the Jacobian iterative method is convergent (for any initial vector) if A has both property (A^π) and is Hermitian positive definite.

Corollary 6.45 *Let $A = D_A - L - U$ have property (A^π). Then $\rho(B) < 1$, where $B = D_A^{-1}(L + U)$.*

Proof Choose $\omega = 1$ in Theorem 6.43. This shows that $\rho(L_1) < 1$. Since A has property (A^π), Corollary 6.35 shows that $\rho(B) = \sqrt{\rho(L_1)}$.   ◇

Corollary 6.44(a) indicates that $\gamma \le$ const. does not generally hold because $(\mathbf{x}, (\widetilde{L} - \widetilde{L}^*)\mathbf{x})$ generally will be bounded below by some fixed number, even when $(\mathbf{x}, \widetilde{A}\mathbf{x})$ is close to zero. On the other hand, Corollary 6.44(b) explains why the SOR method works well for symmetric, positive definite matrices where L is nonnegative, even if they do not have property (A^π).

However, the above analysis has been based on the spectral radius of the iteration matrix. As we saw in Section 5.1, the spectral radius gives only information about the asymptotic rate of convergence. If one uses the spectral norm $\|L_\omega\|$ to estimate the rate of convergence, it will be seen in Chapter 7 that a stronger estimate, $\|L\|_2 \le 1/2$, is needed for the SSOR method, instead of $r(L) \le 1/2$.

Historic Remark 6.46 The relaxation method goes back to Southwell and his coworkers (see Southwell, 1940 and 1946). Young (1950) derived the optimal value of ω for consistently ordered matrices. The SOR theory for symmetric positive definite matrices was presented by Anderson in 1976 (see also Axelsson, 1977), following the similar theory for SSOR method (Axelsson, 1974). The SSOR method will be discussed in Chapter 9.

Exercises

1. Let $A = \begin{bmatrix} 4 & -1 & 0 \\ -1 & 4 & -2 \\ 0 & -1 & 4 \end{bmatrix}$ be block-partitioned in 2×2 and 1×1
    blocks, using the partitioning $\pi_1 = \{1, 2\}$, $\pi_2 = \{3\}$ of the set $\{1, 2, 3\}$.
    (a) Compute the spectral radius for the pointwise and block-wise Gauss-Seidel matrices, L_1 and L_1^B, respectively.
    (b) Compute three iterates for the linear equation $Ax = 0$, with initial vector $\mathbf{x}^0 = (1, 1, 1)^T$ for each of the methods.

2. Let $\xi_i = \sum_{\substack{j=1 \\ j \neq i}}^{n} |a_{i,j}/a_{i,i}| < 1$ for all i, and let $\{\mathbf{x}^l\}$ be the Jacobi iterates
    for $Ax = b$. Prove that

    $$\|\mathbf{x} - \mathbf{x}^l\|_\infty \leq \max_{1 \leq i \leq n} \frac{\xi_i}{1 - \xi_i} \|\mathbf{x}^l - \mathbf{x}^{l-1}\|_\infty.$$

3. For nondiagonally dominant matrices, it may happen that the Seidel method diverges even though the Jacobi method converges. Verify this
    for the case $A = I - B$, where $B = \frac{1}{2} \begin{bmatrix} 0 & -2 & 2 \\ -1 & 0 & -2 \\ -2 & -2 & 0 \end{bmatrix}$. Here $\rho(B) < $
    1, but $\rho(L_1) > 1$.

4. Consider $Ax = b$, where

    $$A = \begin{bmatrix} 2 & -1 & 0 \\ -1 & 2 & -1 \\ 0 & -1 & 2 \end{bmatrix}.$$

    Show that the Gauss-Seidel method converges and that $\rho(L_1) \leq 3/4$.

5. Consider $Ax = b$, where $A = -\Delta_h^{(5)}$, the five-point difference matrix on a unit square with Dirichlet boundary conditions. If the first row of A corresponds to a point with at least one neighboring point (S, W, N, or E) on the boundary, prove that the Gauss-Seidel method converges.

6. *Nonlinear Gauss-Seidel:* Consider a nonlinear system of n algebraic equations, $\mathbf{x} - f(\mathbf{x}) = \mathbf{0}$, $\mathbf{x} = (x_1, x_2, \ldots, x_n)^T$. Assume that

$$f \in [C^1(X)]^n, \mathbf{x} \subset \mathbb{R}^n$$

and that

$$\nu_1 = \max_{\mathbf{x} \in X} \left\{ \sum_{j=1}^n \left| \frac{\partial f_1(\mathbf{x})}{\partial x_j} \right| \right\} < 1,$$

$$\nu_i = \max_{\mathbf{x} \in X} \left\{ \sum_{j=1}^{i-1} \left| \frac{\partial f_i(\mathbf{x})}{\partial x_j} \right| \nu_j + \sum_{j=i}^n \left| \frac{\partial f_i(\mathbf{x})}{\partial x_j} \right| \right\} < 1, \quad i = 2, \ldots, n.$$

Now let $X = B(\mathbf{x}^0, 1)$, be a unit ball with center at $\mathbf{x}^0$. Prove that the Gauss-Seidel iterative method,

$$x_i^{l+1} = f_i(x_1^{(l+1)}, \ldots, x_{i-1}^{(l+1)}, x_i^{(l)}, \ldots, x_n^{(l)}),$$

$$i = 1, 2, \ldots, n, \quad l = 0, 1, \ldots$$

converges toward a solution $\mathbf{x} = f(\mathbf{x})$ (called fix-point). (Compare with Theorem 6.33.)

7. Let

$$A = \begin{bmatrix} 4 & -1 & -1 & 0 & 0 & 0 \\ -1 & 4 & -1 & 0 & 0 & 0 \\ 0 & 0 & 2 & -1 & 0 & 0 \\ 0 & 0 & -1 & 2 & -1 & -1 \\ 0 & 0 & 0 & 0 & 4 & -1 \\ 0 & 0 & 0 & 0 & -1 & 4 \end{bmatrix}$$

be partitioned in blocks with 2×2, 1×1, 1×1, and 2×2 diagonal blocks. Prove that the block-relaxation method converges for $0 < \omega < 2$. Determine the optimal value of ω and the corresponding rate of convergence.

8. Prove Lemma 6.34.

   Hint:

   (a)   Prove first by a suitable diagonal similarity transformation of $B(\alpha)$ that the eigenvalues of $B(\alpha) = \alpha L + \alpha^{-1} U$, $\alpha \neq 0$ are independent of α (and thus equal to the eigenvalues of $L + U$).

(b) Consider $\lambda\mathbf{x} = L_\omega\mathbf{x}$. Prove that

$$(\lambda^{\frac{1}{2}}D^{-1}L + \lambda^{-\frac{1}{2}}D^{-1}U)\mathbf{x} = \frac{\lambda + \omega - 1}{\omega\lambda^{\frac{1}{2}}}\mathbf{x}.$$

(c) Combine (a) and (b) to prove (6.17).

9. Let the coefficients a_{ij} of the matrix A, of order n, satisfy

$$v_1 = \sum_{j=2}^{n} |a_{ij}|/|a_{11}| < 1,$$

$$v_i = \sum_{j<i} \frac{|a_{ij}|}{|a_{ii}|} v_j + \sum_{j>i} \frac{|a_{ij}|}{|a_{ii}|} < 1, \quad i = 2, 3, \ldots, n.$$

Show that the Gauss-Seidel method converges.

10. Let $A = \begin{bmatrix} 5 & -1 & 0 \\ -1 & 2 & -1 \\ 0 & -\frac{3}{2} & 1 \end{bmatrix}$. Show that the Jacobi and Gauss-Seidel methods converge. (Note that this matrix is not diagonally dominant.)

11. Prove that for the five-point difference matrix A_h, for $-\Delta_h^{(5)}$ on a unit square with Dirichlet boundary data with N^2 interior points, we have :

(a) $\rho(B) = \cos \pi h \simeq 1 - \frac{1}{2}(\pi h)^2$, $h = 1/(N + 1)$ for the point-wise Jacobi iteration method.

(b) $\omega_{\text{opt}} = \frac{2}{1+\sin \pi h} \simeq 2(1 - \pi h)$ for point-wise block relaxation (consistently ordered) or for relaxation with diagonal-wise ordering.

(c) $\omega_{\text{opt}} = 2/[1 + \{1 + (\frac{\cos \pi h}{2-\cos \pi h})^2\}^{\frac{1}{2}}] \simeq 2(1 - \sqrt{2}\pi h)$, for block-wise SOR, corresponding to the natural orderings (cf. Varga, 1962, p. 204).

(d) Compute in cases (b) and (c) the rate of convergence.

(e) How many iterations are needed to compute a solution of $A_h\mathbf{u}_h = \mathbf{f}_h$ by relaxation with ω_{opt} with a relative iteration error $\varepsilon = 10^{-1}$?

12. *Comparison principle for monotone matrices:* Let A be monotone and consider the case where we know two vectors $\mathbf{v}$ and $\mathbf{w}$, such that $A\mathbf{v} \geq \mathbf{b} \geq A\mathbf{w}$. Then, if $\mathbf{u}$ is the solution of $A\mathbf{u} = \mathbf{b}$, prove that

$$A(\mathbf{v} - \mathbf{u}) \geq \mathbf{0}, \text{ that is, } \mathbf{v} \geq \mathbf{u}$$

and

$$A(\mathbf{u} - \mathbf{w}) \geq \mathbf{0}, \text{ that is, } \mathbf{u} \geq \mathbf{w}.$$

Hence $\mathbf{w}$ and $\mathbf{v}$ are lower and upper bounds, respectively, of $\mathbf{u}$: $\mathbf{w} \leq \mathbf{u} \leq \mathbf{v}$.

13. Let A be the elliptic difference matrix for $-\Delta_h^{(5)}$ on a unit square. Prove that the Perron eigenvector to A^{-1} is the first harmonic eigenvector of A, that is, the eigenvector with the largest wavelength.

14. Show that the rate of convergence of the second-order stationary iterative method is $-\log \rho_0 \approx \pi h$ for an elliptic difference problem with the 5-point difference matrix, $-\Delta_h^{(5)}$ on a unit square with Dirichlet boundary data, and $-\log \rho_0 \simeq \sqrt{3/2}\pi h$ for the 9-point difference matrix $-\Delta_h^{(9)}$ with coefficients $(-1, -4, -1, \ldots, 0, -4, 20, -4, 0, \ldots, 0, -1, -4, -1)$.

15. Let $A = [A_{i,j}]$ be a matrix partitioned in block matrix form, where $A_{ii} = D_i$, $i = 1, \ldots, n$ are square matrices. Let $C_i = C = \mathrm{diag}(D_1, \ldots, D_n)$ and $E_i = \mathrm{diag}(0, \ldots, I_i, \ldots, 0)$ where I_i is the identity matrix of the same order as D_i. Give conditions for convergence of the multisplitting method in Section 6.3 and apply Theorem 6.23 to obtain lower and upper bounds of its rate of convergence.

16. For $B \geq 0$, show the following:
     (a) If there exists $\mathbf{x} \geq \mathbf{0}$, $\mathbf{x} \neq \mathbf{0}$ and a scalar $\alpha > 0$, such that $\alpha \mathbf{x} \leq B\mathbf{x}$, then $\alpha \leq \rho(B)$.
     (b) If there exists $\mathbf{y} > \mathbf{0}$ and a scalar $\beta > 0$, such that $B\mathbf{y} \leq \beta \mathbf{y}$, then $\rho(B) \leq \beta$.
     (c) Let $B = \begin{bmatrix} 1 & 0 \\ 1 & 0 \end{bmatrix}$ and $\mathbf{y} = \begin{bmatrix} 0 \\ 1 \end{bmatrix}$. Show that $B\mathbf{y} \leq \beta \mathbf{y}$ for all $\beta > 0$.
     Hence (b) does not hold, in general, unless $\mathbf{y}$ is strictly positive.
     Hint: Since $\alpha > 0$, it is readily seen that (a) implies that B contains at least one positive entry in each row. Now diminish a positive entry in each row of B to get a matrix $\widetilde{B} \geq 0$, for which $\widetilde{B}\mathbf{x} = \alpha \mathbf{x}$. Why is $\alpha \leq \rho(\widetilde{B})$? Use continuity to show that $\rho(\widetilde{B}) \leq \rho(B)$. Part (b) is shown in a similar way, but here some entry in B is increased to attain $\widetilde{B}\mathbf{y} = \beta \mathbf{y}$.

17. Show that a Z-matrix is positive stable if and only if there exists a positive vector $\mathbf{x}$, such that $A\mathbf{x} > \mathbf{0}$.

18. *Closure properties of M-matrices:* Show that the ordinary matrix product, the Hadamard matrix product, and matrix addition are not closed operators for the class of M-matrices—i.e., show that there exist M-matrices A, B, such that:
     (a) $A \cdot B$ is not a M-matrix.
          Hint: Show that the sign pattern can fail if $n \geq 3$.
     (b) $A \odot B$ is not a M-matrix.

Note: Failure is due to loss of sign pattern.

(c) $A + B$ is not a M-matrix.

19. (a) Show that if A and B are M-matrices, such that $(A \cdot B)_{i,j} \leq 0$, $i \neq j$, then $A \cdot B$ is also a M-matrix.

(b) Show that if A and B are H-matrices, then $A \odot B$ is also a H-matrix.

Hint: Use Lemma 6.3 and the nonsingularity of B.

(c) Show that if A and B are M-matrices and $A + B$ is positive definite, then $A + B$ is a M-matrix.

(See also Exercise 4.7 for closure properties of the product of positive definite matrices $B(I + B^{-1}A)$.)

20. Let $A = [a_{ij}]$ be a real $n \times n$ matrix, with $a_{ij} \leq 0$ for $i \neq j$. Show that A is monotone if and only if it can be written in the form

$$A = LU,$$

where L and U are, respectively, lower and upper triangular $n \times n$ matrices with positive diagonal entries and nonpositive offdiagonal entries.

References

L. Anderson (1976), SSOR preconditioning of Toeplitz matrices. Ph.D. thesis, Department of Computer Sciences, Chalmers University of Technology, Göteborg, Sweden.

O. Axelsson (1974). On preconditioning and convergence acceleration in sparse matrix problems. CERN Technical Report 74-10, Data Handling Division, Geneva.

O. Axelsson (1977). Solution of linear systems of equations: iterative methods. In *Sparse Matrix Techniques*, ed. V. A. Barker. *Lecture Notes in Mathematics* #572. Berlin, Heidelberg, New York: Springer-Verlag.

O. Axelsson (1985). A survey of preconditioned iterative methods for linear systems of algebraic equations. *BIT* **25**, 166–187.

O. Axelsson and L.Yu. Kolotilina (1991). Diagonally compensated reduction and related preconditioning methods. Submitted.

Z. Bai and D. Wang (1993). Generalized matrix multisplitting relaxation methods and their convergence. Shanghai University of Science and Technology, to appear.

R. Beauwens (1976). Semistrict diagonal dominance. *SIAM J. Numer. Anal.* **13**, 109–112.

R. Beauwens (1979). Factorization iterative methods, M-operators and H-operators. *Numer. Math.* **31**, 335–357.

A. Berman and R. J. Plemmons (1979). *Nonnegative Matrices in the Mathematical Sciences*. Orlando, Fla.: Academic.

J. H. Bramble and B. E. Hubbard (1964). On a finite difference analogue of an elliptic boundary value problem which is neither diagonally dominant nor of nonnegative type. *J. Math. and Phys.* **43**, 117–132.

G. Csordas and R. Varga (1984). Comparisons of regular splittings of matrices. *Numer. Math.* **44**, 23–35.

L. Elsner (1989). Comparisons of weak regular splittings and multisplitting methods. *Numer. Math.* **56**, 283–289.

M. Fiedler and V. Ptak (1962). On matrices with non-positive off-diagonal elements and positive principal minors. *Czechoslovak Math. J.* **12**, 123–128.

M. Fiedler and V. Ptak, (1966). Some generalizations of positive definiteness and monotonicity. *Numer. Math.* **9**, 163–172.

S. P. Frankel (1950). Convergence rates of iterative treatments of partial differential equations. *Math. Tables Aids Comp.* **4**, 65–75.

C. F. Gauss (1823). Letter to Gerlin, 26 Dec. 1823, in *Werke*, **Vol. 9**, 278–281. A translation by G. E. Forsythe appears in *MTAC 5* (1950), 255–258.

C. F. Gauss (1826). *Supplementum . . . , Werke.* **Vol. 4**, 55–93, Göttingen.

C. G. J. Jacobi (1845). Über eine neue Auflösungsart. *Astr. Nachr.* **22**, No. 523 (1845), 297–306. [Reprinted in his *Werke*, Vol. 3, p. 467.]

W. Kahan (1958). Gauss-Seidel methods of solving large systems of linear equations. Ph.D. Thesis, Univ. of Toronto.

J. Liouville (1853). Sur le développement des fonctions en series . . . , II. *J. Math. Pures Appl. (1)*, **2**, 16–37.

I. Marek and D. B. Szyld (1989). Comparison theorem for weak splittings of bounded operators. Department of Computer Science, Duke University, CS-1989-9.

D. P. O'Leary and R. E. White (1985). Multisplittings of matrices and parallel solutions of linear systems. *SIAM J. Algebraic Discrete Methods* **6**, 630–640.

J. M. Ortega and W. C. Rheinboldt (1970). *Iterative Solution of Nonlinear Equations in Several Variables*. Orlando, Fla.: Academic.

A. M. Ostrowski (1937). Determinanten mit überwiegender Hauptdiagonale. *Comm. Math. Helv.* **10**, 69–96.

G. D. Poole and T. Boullion (1974). A survey of *M*-matrices. *SIAM Rev.* **16**, 419–427.

Ph. Seidel (1874). Münchener Akademische Abhandlungen, 2 Abh. 81–108.

Y. Song (1991). Comparisons of nonnegative splittings of matrices. *Lin. Alg. Appl.* **154–156**, pp. 433–455.

R. V. Southwell (1940). *Relaxation Methods in Engineering Science*. Oxford Univ. Press.

R. V. Southwell (1946). *Relaxation Methods in Theoretical Physics*. Oxford Univ. Press.

T. J. Stieltjes (1887). Sur les racines de l'équation $x_n = 0$. *Acta Math.* **9**, 385–400.

R. S. Varga (1959). Ordering of successive overrelaxation schemes. *Pacific J. Math.* **9**, 925–936.

R. S. Varga (1962). *Matrix Iterative Analysis*. Englewood Cliffs, N.J.: Prentice Hall.

R. S. Varga (1976). On recurring theorems on diagonal dominance. *Lin. Alg. Appl.* **13**, 1–9.

E. Wachspress (1966). *Iterative Solution of Elliptic Systems and Applications to the Neutron Diffusion Equations of Reactor Physics.* Englewood Cliffs, N.J.: Prentice-Hall.

R. E. White (1989). Multisplitting with different weighting schemes. *SIAM J. Matrix Anal. Appl.* **10**, 481–493.

J. Wilkinson (1965). *The Algebraic Eigenvalue Problem.* Oxford: Clarendon.

Z. Woźnicki (1973). Two-sweep iterative methods for solving large linear systems and their application to the numerical solution of multi-group multi-dimensional neutron diffusion equation. Doctoral dissertation, Institute of Nuclear Research. IBJ Report No. 1447 (CYFRONET) PM/A, Warzaw.

Z. Woźnicki (1979). AGA two-sweep iterative methods and their application in critical reactor calculations. *Nukleonika* **23**, 941–968.

David M. Young (1950). Iterative methods for solving partial difference equations of elliptic type. Doctoral thesis, Harvard University.

David M. Young (1971). *Iterative Solution of Large Systems.* Orlando, Fla.: Academic Press.

7

Incomplete Factorization Preconditioning Methods

In Chapter 5 we presented basic iterative methods, both of stationary and of nonstationary type. The parameters in the methods were chosen to accelerate their convergence. The ability to accelerate depends, however, on the eigenvalue distribution. As we have seen, for iteration matrices with real and positive eigenvalues, or with nearly real eigenvalues but positive real parts, the parameters can be chosen so that the rate of convergence is increased by an order of magnitude, such as in the Chebyshev iteration method. Certain cases of an indefinite or more general complex spectrum can also be handled.

The eigenvalue distribution depends on the matrix splitting method used. Some splitting methods, such as the SOR method, lead to a "dead-end"—i.e., an iteration matrix where no, or only minor, polynomial acceleration of convergence is possible, because the spectrum of the iteration matrix is typically located on a circle and, hence, powers of the iteration matrix are also located on a circle.

The purpose of this and the following chapters is to present some practically important splitting methods of A, that is, $A = C - R$ (where C is nonsingular), which can improve the eigenvalue distribution of the iteration matrix $C^{-1}A$ in such a way that the iterative method will converge much faster with the splitting than without it. For this reason, the matrix C has been called a *preconditioning matrix* (Evans, 1968, and Axelsson, 1972) and the associated iterative method is termed an acceleration method (Axelsson, 1974).

For practical reasons, the matrix C must have a form such that the solution of a linear system with coefficient matrix C requires little computational effort, typically of the order of the computational cost of one or a few matrix-vector products with A. In addition, the spectrum of $C^{-1}A$ must be considerably better (i.e., more clustered) than that of A.

Clearly, however, these are conflicting goals. If, for instance, we choose C equal to the identity matrix, there is no cost associated with C, but the

252

spectrum has not been improved at all. On the other hand, if we make the choice $C = A$, the spectrum consists of just one point, and any (consistent) iterative method will converge in one step; but now the cost required is equal to the cost of solving the system with A itself. In addition, the cost of computing C must be taken into account. The optimal choice of C to minimize the total computational effort, or the total computer time, can depend on factors such as the size of the problem, the sparsity pattern and spectrum of A, and even the computer architecture and the relative speed of various computer operations. It is therefore not possible to give a general answer about how to choose an optimal matrix C, but we can present various methods to construct C and give general guidelines for making a good choice.

Different classes of preconditioning methods exist, and these will be discussed in this and later chapters. In the present chapter, we consider the class of incomplete factorization methods, which is closely related to the full (Gauss) factorization method discussed in Chapter 1. As we saw there, any sparsity within the band, or more generally within the envelope, of a matrix will be lost during the factorization. The idea of incomplete factorization methods is to reject those "fill-in" entries which either occur in positions outside a chosen sparsity pattern or which are small relative to some diagonal entries. In this way, the factorization becomes approximate, or incomplete, and must therefore be coupled with an iterative solution method—i.e., it must be used as a preconditioning for such a method. (Incidentally, even a full factorization method will, in practice, be only approximate because of round-off errors which occur when we work with finite precision numbers. Commonly, one then uses an iterative improvement step with some simple iteration method.) When A is s.p.d., C can be made s.p.d. also, and, hence, the spectrum of $C^{-1}A$ remains positive. As we shall see, however, the methods can be applied also to more general classes of matrices.

Naturally, an important aspect of analyzing incomplete factorization methods is to prove their existence—i.e., to prove that the pivot entries are nonzero. This will be considered here for certain classes of matrices—namely M-matrices, positive definite matrices on block tridiagonal form, and block H-matrices. For each of these classes, we shall prove that the matrix that arises after an elimination step stays in the same class. (Incidentally, our analysis shows also the existence of the full (LU)-factorization method for these classes of matrices, thus complementing an existence result in Chapter 1.)

The present chapter first presents the incomplete factorization method of the LU or $LD^{-1}U$ form, where D is a diagonal matrix. We then consider the general case where A is partitioned in block matrix form, D is a blockdiagonal matrix, and L and U are lower and upper block triangular factors, respectively, which can have a general sparsity structure. The methods are constructed by recursively applying a two-by-two block partitioning of the current matrix where elimination is to take place and then computing an approximation of its Schur complement.

Since we can impose an (almost) arbitrary sparsity pattern on the matrices, the construction of the factors L, U of $C = LU$ (or of L, D, U of $C = LD^{-1}U$) and the solution of the systems with C that occur at every iteration step can be performed with relatively little computational effort as compared with the construction and solution of the corresponding factors of the full factorization method. In the latter, fill-in can grow rapidly from one elimination stage to the next, and it is essentially outside the control of the user. Clearly, some improvements can be achieved using reorderings (permutations) of the given matrix to reduce fill-in, but even the computation of a good permutation matrix needs computer time, which can offset the gains of having less fill-in. In our method, we can control the fill-in by specifying a priori chosen positions where it will be accepted, or by accepting fill-in entries only if their absolute value is sufficiently large compared with certain diagonal entries, for instance.

For incomplete factorization of block-matrices, an important aspect is approximation of the inverses of the pivot blocks that occur during the elimination. Methods to compute such approximations will be discussed in the next chapter. Another important aspect of incomplete factorization methods is their modification to satisfy certain (generalized) row sum or column sum criteria, or the relaxed versions of these; this topic will be discussed at relevant places of the present chapter.

The final section considers symmetrization of iterative methods and the rate of convergence of a particular form of symmetrization. This theory is applied to two important methods: the SSOR method (which is based on applying a forward, followed by a backward, SOR sweep) and the ADI method.

7.1 Point Incomplete Factorization

The incomplete factorization method has been described in earlier publications for square matrices but, as we shall see, the methods are equally applicable to the factorization of rectangular matrices of order $m \times n$, where $m \leq n$. This follows in the same way as for the full factorization method described in Chapter 1.

Consider, therefore, a matrix $A = [a_{ij}]$ of order $m \leq n$, which in practice will be sparse. Similar to the full Gaussian factorization method, an incomplete factorization method works recursively on a number of levels. The next pivot row is chosen on each level, and, in the standard implementation of the method, the entries of the column of the pivot element are eliminated by a proper choice of the entries of this column of the left triangular factor. In contrast to the full factorization method, however, we will not accept any nonzero entries in L or U in positions outside a sparsity structure S, which is a proper subset of the set $\{(i, j) ; 1 \leq i \leq m, 1 \leq j \leq n\}$. This sparsity structure can be chosen *a priori* (i.e., before the factorization takes place), a process we call factorization *by position*, or *dynamically* during the factorization, where, at the rth elimination stage, we typically neglect fill-in entries that are sufficiently small to satisfy

$$(7.1) \qquad |a_{ij}^{(r+1)}| \leq c|a_{ii}^{(r+1)} a_{jj}^{(r+1)}|^{\frac{1}{2}},$$

where $0 < c < 1$. (Note that $c = 0$ gives a full factorization and $c = 1$ gives a diagonal matrix if A is s.p.d.) This latter method has been called incomplete factorization *by value* (see Axelsson and Munksgaard, 1983). The method to be described is applicable also to block-matrices, where A has been partitioned into matrix blocks A_{ij}. If the order of each block is small, we can accept the cost of computing the exact inverse of the pivot blocks, and the algorithm below will then be directly applicable. (In fact, the full version of the algorithm, where all fill-in is accepted, describes both the exact point factorization and the exact block matrix factorization.) However, in a following section we shall present a more general algorithm for matrices partitioned into blocks, in which we also permit approximations of these matrix inverses. We therefore shall assume, first, that the entries a_{ij} are scalars.

Let $A^{(r)} = [a_{ij}^{(r)}]$ denote the matrix at the rth stage, where $A^{(1)} = A$ and $a_{r,r}^{(r)}$ is the current pivot entry. Naturally, we must assume that $a_{rr}^{(r)} \neq 0$. A standard method to achieve this is to permute rows or columns, or both, until a nonzero pivot entry (sufficiently large) has been found. Because this will change the given sparsity structure, however, we shall not use permutations. (If we use the method of incomplete factorization by value, such permutations can alternatively be used to obtain a smaller number of fill-in entries in the same way as for full factorization methods. This topic will not be discussed further here.)

As we have seen in Chapter 1, the basic step in a Gaussian elimination or factorization method is the computation

$$(7.2) \qquad a_{ij}^{(r+1)} := a_{ij}^{(r)} - a_{ir}^{(r)} a_{rr}^{(r)^{-1}} a_{rj}^{(r)}, \quad r+1 \leq i \leq m, r+1 \leq j \leq n.$$

The entries $a_{ij}^{(r+1)}$ of the matrix $A^{(r+1)}$ are defined as follows:

(a) By (7.2), for $r + 1 \leq i, j \leq n$

(b) $a_{ij}^{(r+1)} = a_{ij}^{(r)}$, for $1 \leq i \leq r, 1 \leq j \leq n$ and $1 \leq i \leq n, 1 \leq j \leq r - 1$

(c) $a_{ir}^{(r+1)} = 0, r + 1 \leq i \leq n$.

However, in the incomplete factorization method, nonzero entries $a_{ij}^{(r+1)}$ are accepted only if $(i, j) \in S^{(r)}$, where $S^{(r)} \subseteq S$ is an index set that at stage r defines the position where nonzero entries are accepted in positions where i and $j \geq r + 1$. In the method of incomplete factorization by value, $S^{(r)}$ will be defined implicitly by (7.1) during each factorization step. We shall here consider only symmetric sparsity patterns—i.e., if $(i, j) \in S, 1 \leq i \leq m$, then also $(j, i) \in S$. For simplicity, we will also assume that the sparsity pattern of A is equal to or a proper subset of S. In the following, even if the sparsity structure $S^{(r)}$ depends on the stage number r, we do not always indicate this, but simply write S in place of $S^{(r)}$. The incomplete factorization algorithm then takes the following form (described here in a pseudo-code):

(7.3)   **for** $r := 1$ **step** 1 **until** $m - 1$ **do**

      begin $d := 1/a(r, r)$;

           for $i := (r + 1)$ **step** 1 **until** m **do**

           begin if $(i, r) \in S$ **then**

                  begin $e := a(i, r) \times d$; $a(i, r) := e$;

                        for $j := (r + 1)$ **step** 1 **until** n **do**

                        if $(i, j) \in S \wedge (r, j) \in S$ **then**

                              $a(i, j) := a(i, j) - e \times a(r, j)$

                    end (j-loop)

                end

           end (i-loop)

      end (r-loop)

If $a_{rr}^{(r)} \neq 0, r = 1, 2, \ldots, m - 1$, we can continue the algorithm until the final stage, where $r = m - 1$, to form an approximate or incomplete matrix factorization LU, where L is a lower triangular matrix of order $m \times m$ consisting of entries $a_{ir}^{(r)}/a_{rr}^{(r)}, i = r, r + 1, \ldots, m, r = 1, 2, \ldots, m$, and where U is an upper triangular matrix, consisting of entries $a_{rj}^{(r)}, j = r, r + 1, \ldots, n, r = 1, 2, \ldots, m$. These matrices have the same form as the corresponding matrices in Chapter 1.

Example 7.1 Consider $A = A^{(1)} = \begin{bmatrix} 4 & -1 & -1 & 0 \\ -1 & 2\frac{1}{4} & 0 & -1 \\ -1 & 0 & 2\frac{1}{4} & -1 \\ 0 & -1 & -1 & 4 \end{bmatrix}$ and let $S =$

$\{(i, j); a_{ij} \neq 0\}$. Using the incomplete factorization method we find

$$A^{(2)} = \begin{bmatrix} 4 & -1 & -1 & 0 \\ 0 & 2 & 0 & -1 \\ 0 & 0 & 2 & -1 \\ 0 & -1 & -1 & 4 \end{bmatrix}, \quad L^{(1)} = \begin{bmatrix} 1 & & & \emptyset \\ -\frac{1}{4} & 1 & & \\ -\frac{1}{4} & 0 & 1 & \\ 0 & 0 & 0 & 1 \end{bmatrix},$$

$$A^{(3)} = \begin{bmatrix} 4 & -1 & -1 & 0 \\ 0 & 2 & 0 & -1 \\ 0 & 0 & 2 & -1 \\ 0 & 0 & -1 & \frac{1}{2} \end{bmatrix}, \quad L^{(2)} = \begin{bmatrix} 1 & & & \emptyset \\ -\frac{1}{4} & 1 & & \\ -\frac{1}{4} & 0 & 1 & \\ 0 & -\frac{1}{2} & 0 & 1 \end{bmatrix},$$

$$A^{(4)} = \begin{bmatrix} 4 & -1 & -1 & 0 \\ 0 & 2 & 0 & -1 \\ 0 & 0 & 2 & -1 \\ 0 & 0 & 0 & 3 \end{bmatrix}, \quad L^{(3)} = \begin{bmatrix} 1 & & & \emptyset \\ -\frac{1}{4} & 1 & & \\ -\frac{1}{4} & 0 & 1 & \\ 0 & -\frac{1}{2} & -\frac{1}{2} & 1 \end{bmatrix},$$

and $C = LU = L^{(3)}A^{(4)} = A + R$, where

$$R = \begin{bmatrix} 0 & 0 & 0 & 0 \\ 0 & 0 & \frac{1}{4} & 0 \\ 0 & \frac{1}{4} & 0 & 0 \\ 0 & 0 & 0 & 0 \end{bmatrix}.$$

Similarly, using the modification step (7.4) to be presented below, one finds

$$C = \widetilde{L}^{(3)}\widetilde{A}^{(4)} = A + \widetilde{R},$$

where

$$\tilde{A}^{(4)} = \begin{bmatrix} 4 & -1 & -1 & 0 \\ 0 & \frac{7}{4} & 0 & -1 \\ 0 & 0 & \frac{7}{4} & -1 \\ 0 & 0 & 0 & \frac{20}{7} \end{bmatrix}, \quad \tilde{L}^{(3)} = \begin{bmatrix} 1 & & & \emptyset \\ -\frac{1}{4} & 1 & & \\ -\frac{1}{4} & 0 & 1 & \\ 0 & -\frac{4}{7} & -\frac{4}{7} & 1 \end{bmatrix} \quad \text{and}$$

$$\tilde{R} = \frac{1}{4} \begin{bmatrix} 0 & 0 & 0 & 0 \\ 0 & -1 & 1 & 0 \\ 0 & 1 & -1 & 0 \\ 0 & 0 & 0 & 0 \end{bmatrix}.$$

The entries of the matrix L (except the unit diagonal entries) and of U will be stored in the corresponding locations of the original matrix. In practice it is common to store a sparse matrix A in a compact form consisting of three one-dimensional arrays. The first array consists of the entries of A in positions $(i, j) \in S$ (some entries may be zero) stored row-wise (or column-wise). The next array contains the column (row) indices of these entries, and the final array contains the number of stored entries in each row (column). For more details, see Axelsson and Barker (1984), for instance.

The incomplete factorization method can be modified in various ways. For instance, we can add the deleted entries to the diagonal entries in the same row, whereby the innermost (j-)loop of (7.3) takes the form

(7.4)   **if** $(r, j) \in S$ **then**
          begin if $(i, j) \in S$ **then**
                $a(i, j) := a(i, j) - e \times a(r, j)$
                else
                $a(i, i) := a(i, i) - e \times a(r, j)$
          end

As is readily seen, this method preserves the row sums of the original matrix— i.e., $LU\mathbf{e} = A\mathbf{e}$, $\mathbf{e} = (1, 1, \ldots, 1)^T$.

More generally, we shall consider the *"relaxed"* version (see Axelsson and Lindskog, 1986), where the last (modification) step above takes the form

(7.5) $a(i, i) := a(i, i) - \omega \times e \times a(r, j)$, if $(i, j) \notin S$.

Here, ω is a parameter, $\omega \le 1$. For $\omega = 0$ we get the original, unmodified version, (7.3) and for $\omega = 1$ we get the fully modified version (7.4). Still more generally, we can use the components x_j of a positive vector $\mathbf{x}$ as weights and put

(7.6) $a(i, i) := a(i, i) - \omega \times e \times a(r, j)\mathbf{x}(j)/\mathbf{x}(i)$.

For $\omega = 1$ this preserves the generalized row sums $(Ax)_i$ and $LUx = Ax$.

Finally, we can modify the pivot entry by adding a positive (usually small) number to it, when required, before the elimination takes place to guarantee that pivot entries will have sufficient size. For a description and analysis of such an algorithm, see Axelsson and Barker (1984), pp. 344-345. Then the second step of (7.3) takes the form

$$d := (a_{rr}^{(r)} + \alpha_r)^{-1},$$

where α_r is sufficiently large. If we let $\omega = \omega_i$ be variable, this method can be seen to be equivalent to the relaxed method (7.4) for certain ω_i. Gustafsson (1978, 1979) considered similar perturbations when he studied incomplete factorization methods for difference type matrices. See also Axelsson (1972), where such perturbations were used in a generalized SSOR type method.

Incomplete factorization methods can conveniently be described in a recursive two-level formulation that is especially convenient for matrices partitioned in block matrix form, as we shall see in Section 7.2. In the more general case, however, where we permit the use of different approximations of inverses of pivot matrices at each stage r, a formulation similar to the above turns out to be the most convenient one, as we shall see in Section 7.5.

The question remains how to choose the sparsity set S. A common choice is to let

$$S = S_0 = \{i, j; a_{ij} \neq 0\}.$$

Following Gustafsson (1978), we define the sparsity set of order q in the following way. Let $\mathcal{R}(q)$ be the sparsity set defined by the position of the fill-in entries that arise during the incomplete factorization method based on the sparsity set $S(q)$, and let

$$S(q + 1) = S(q) \cup \mathcal{R}(q), \quad q = 0, 1, \ldots,$$

where $S(0) = S_0$. This can be expected to give a reasonable sequence of sparsity sets, at least for not too large values of q. Clearly, it can be combined with the method of incomplete factorization by value—i.e., a fill-in entry position is included in $\mathcal{R}(q)$ only if the corresponding numerical value has a sufficient size satisfying a condition such as in (7.1).

The incomplete factorization algorithm can be shown to exist— i.e., it has $a_{rr}^{(r)} \neq 0$ for some classes of matrices, such as for M-matrices. Note that each new matrix has entries computed as $a_{ij}^{(r)} - a_{ir}^{(r)} a_{rr}^{(r)^{-1}} a_{rj}^{(r)}$ and that these are entries in a Schur complement matrix called $S^{(r+1)}$ here—i.e., $S^{(r+1)}$ is a Z-matrix and $S^{(r+1)}x > 0$ for the same positive vector x for which $Ax > 0$.

Hence, $S^{(r+1)}$ is a M-matrix. The deletion of fill-in entries outside the sparsity pattern S is equivalent to the addition of a nonzero matrix $R^{(r+1)} = [r_{ij}^{(r+1)}]$, where $r_{ij}^{(r+1)} \neq 0$ only where such fill-in has occurred in $S^{(r+1)}$ at stage r. Hence, $A^{(r+1)} = S^{(r+1)} + R^{(r+1)}$, and it is also a M-matrix because $a_{ij}^{(r+1)} = 0$ at such positions and $A^{(r+1)}\mathbf{x} = S^{(r+1)}\mathbf{x} + R^{(r+1)}\mathbf{x} > \mathbf{0}$.

In the relaxed version, we have

$$A^{(r+1)} = S^{(r+1)} + R^{(r+1)} = \omega D^{(r+1)},$$

where $D^{(r+1)}$ is a diagonal matrix, such that

$$D^{(r+1)}\mathbf{x} = R^{(r+1)}\mathbf{x}.$$

Hence,

$$A^{(r+1)}\mathbf{x} = S^{(r+1)}\mathbf{x} + (1 - \omega)R^{(r+1)}\mathbf{x} \geq S^{(r+1)}\mathbf{x} > 0$$

if $\omega \leq 1$ and $A^{(r+1)}$ is a M-matrix.

More general proofs of existence for matrices partitioned in block form will be given in later sections. It should be noted that the incomplete factorization method does not always exist for positive definite matrices. For such matrices, one can use the method of reduction of positive offdiagonal entries and diagonal compensation of them as presented in Chapter 6. The preconditioner is then constructed from the so resulting M-matrix.

For discussions of incomplete factorization methods based on drop tolerance criterions, i.e., neglecting fill-in satisfying (7.1), for instance, see Osterby and Zlatev (1983) and Zlatev (1991).

7.2 Block Incomplete Factorization; Introduction

We shall now present a recursive two-level formulation of incomplete factorization methods. As we shall see, this formulation provides us with a uniform framework for proving existence of the methods. We then let the matrix A be partitioned in a block matrix form,

$$(7.7) \qquad A = \begin{bmatrix} A_{11} & A_{12} & \cdots & A_{1m} & \cdots & A_{1n} \\ A_{21} & A_{22} & \cdots & A_{2m} & \cdots & A_{2n} \\ \vdots & \vdots & \ddots & & \vdots & \\ A_{m1} & A_{m2} & & A_{mm} & \cdots & A_{mn} \end{bmatrix}.$$

Here, A_{ij} has order $p_i \times p_j$, where $1 \le p_i \le m \le n$. Clearly, the presentation includes the case where all or some A_{ij} are scalar entries of A. Note that the matrix blocks occurring in the main diagonal are square matrices. In practical applications, the matrices A_{ij} are frequently sparse and many of the block matrices are zero matrices.

We shall study approximate factorizations of A in block matrix form LU, where L and U are block lower and upper triangular, respectively, and partitioned in blocks consistent with the partitioning of A. L has block order $m \times m$ (i.e., it consists of m block rows and m block columns) and U has block order $m \times n$. In addition, the diagonal blocks of L will be identity matrices.

Let $S^{(r)}$ denote the set of index pairs that at stage r defines the positions where we shall accept nonzero block matrices in positions (i, j), $r + 1 \le i \le m$, $r + 1 \le j \le n$. $S^{(r)}$ will always contain the subset $\{(i, i), r + 1 \le i \le m\}$.

As in the construction of the algorithm in Section 1, we now consider the following approximate factorization algorithm: Starting with $A^{(1)} = A$, at every stage of the elimination process we partition the current matrix, $A^{(r)}$ in two-by-two block form,

$$(7.8) \qquad A^{(r)} = \begin{bmatrix} A_{11}^{(r)} & A_{12}^{(r)} \\ A_{21}^{(r)} & A_{22}^{(r)} \end{bmatrix},$$

where the top diagonal block $A_{11}^{(r)}$ is the current pivot entry, used to eliminate the nonzero blocks of the block matrix column $A_{21}^{(r)}$. (The matrix $A^{(r)}$ considered in Section 7.1 has a larger dimension than the current matrix. For notational simplicity, however, we use the same notations.) Since we are interested in the factored form of $A^{(r)}$, we factor $A^{(r)}$ (approximately) directly, without deriving it via the elimination method. We then get

$$(7.9) \qquad A^{(r)} \simeq \begin{bmatrix} I & 0 \\ A_{21}^{(r)} X^{(r)-1} & I \end{bmatrix} \begin{bmatrix} X^{(r)} & A_{12}^{(r)} \\ 0 & A^{(r+1)} \end{bmatrix}.$$

Note that the right-hand side matrix is

$$\begin{bmatrix} X^{(r)} & A_{12}^{(r)} \\ A_{21}^{(r)} & A^{(r+1)} + A_{21}^{(r)} X^{(r)-1} A_{12}^{(r)} \end{bmatrix}.$$

Here $X^{(r)}$ is nonsingular and an approximation of $A_{11}^{(r)}$ (in practice chosen to enable a fast solution of systems with $X^{(r)}$ as coefficient matrix). $A^{(r+1)}$ is an approximation of the Schur complement type matrix,

$$(7.10) \qquad A_{22}^{(r)} - A_{21}^{(r)} X^{(r)-1} A_{12}^{(r)},$$

chosen as

$$(7.11) \qquad A_{ij}^{(r+1)} = \begin{cases} (S_2^{(r)})_{i,j}, & \text{for } (i,j) \in \mathcal{S}^{(r+1)} \\ 0, & \text{otherwise}, \end{cases}$$

where

$$(7.11') \qquad S_2^{(r)} = A_{22}^{(r)} - A_{21}^{(r)} Y^{(r)} A_{12}^{(r)}.$$

Here, $Y^{(r)}$ is a sparse approximation of $(X^{(r)})^{-1}$, so $S_2^{(r)}$ can be seen as a preliminary approximation of the Schur complement in (7.10). Hence, in general, at every stage we use two or three levels of approximations:

(a) $X^{(r)}$ to approximate $A_{11}^{(r)}$ (frequently $X^{(r)} = A_{11}^{(r)}$).
(b) $Y^{(r)}$ to approximate $(X^{(r)})^{-1}$.
(c) $A^{(r+1)}$ to approximate $S_2^{(r)}$.

In many problems, $A_{11}^{(r)}$ is already a sparse matrix in a form convenient for the solution of systems with it, such as a tridiagonal matrix, and it is then common to let $X^{(r)} = A_{11}^{(r)}$. In general, step (b) is needed when the order of $A_{11}^{(r)}$ is large, because even if $X^{(r)}$ is sparse, its inverse is not (excluding exceptional cases, such as when $X^{(r)}$ is a diagonal matrix). Finally, step (c) is required when fill-in has occurred outside the chosen sparsity structure. For the pointwise factorization method (Section 7.1), we have $x^{(r)} = A_{11}^{(r)} = a_{rr}^{(r)}$, and let $Y^{(r)} = a_{rr}^{(r)-1}$, so that only step (c) takes place. The important topic of how inverses of sparse matrices can be approximated efficiently will be discussed in the next chapter.

The above method is now applied recursively. At every stage we partition the current matrix $A^{(r)}$ in two-by-two block form (7.8) and repeat the above steps. As long as the pivot entries $A_{11}^{(r)}$ are nonsingular, this method can be repeated until we are left with a block matrix, $A^{(m)}$, for which we need only do the first stage of the three levels of approximation, or we may find it most efficient to use another method, such as, for example, a direct solution method.

A computer pseudo-code description of the block incomplete factorization method is found in (7.3) if we exchange the entries a_{ij} with the block matrices A_{ij}, and d with the matrix $Y^{(r)}$. This algorithm is row-oriented. A column-oriented version, including a modification step, can be found in Section 7.5.

The matrix L in the final approximate factorization of A is block lower triangular, has identity matrix diagonal blocks, and the rth column of its lower triangular part is formed by $A_{21}^{(r)} X^{(r)-1}$. (Clearly, the entries of these matrices need not be computed explicitly, because by each action of this matrix on vectors, we compute first the solution of the corresponding system with $X^{(r)}$ and then multiply this solution by matrix $A_{ir}^{(r)}$). The matrix U is block upper triangular, with block diagonal matrix $X = \text{diag}(X^{(1)}, \ldots, X^{(m)})$ and with the rth row of its upper triangular part given by matrices $A_{12}^{(r)}$. The product LU then forms an approximate factorization of A, to be used as a preconditioner in an iterative solution method, such as a conjugate gradient method. At every application of this preconditioner, we need to solve systems with matrices X twice (once in the forward elimination and once in the back elimination). In addition, we need to perform matrix vector multiplications with matrices $A_{21}^{(r)}$ and $A_{12}^{(r)}$.

As is clear from the above, we need to show that each pivot block matrix is nonsingular. This will be shown next for two classes of matrices, M-matrices and positive definite matrices, under certain assumptions of the approximations used for the pivot matrices. For each matrix class, we will show that if matrix $A^{(r)}$ belongs to the class, then so does $A^{(r+1)}$.

7.3 Block Incomplete Factorization of M-Matrices

Assume that A is a M-matrix and let $A^{(1)} = A$, where A has been partitioned as in (7.7). As was shown in Lemma 6.4, this implies that there exists a positive vector $\mathbf{x} = (\mathbf{x}_1^T, \mathbf{x}_2^T, \ldots, \mathbf{x}_n^T)^T$, such that $A^{(1)}\mathbf{x} > 0$. In particular, $A_{11}^{(1)}\mathbf{x}_1 > 0$. Let $\mathbf{x}^{(r+1)} = (\mathbf{x}_{r+1}^T, \ldots, \mathbf{x}_n^T)^T$. We assume that at the rth stage

(7.12)   (a) $X^{(r)}\mathbf{x}_r \geq A_{11}^{(r)}\mathbf{x}_r > 0$ and $X^{(r)}$ is a Z-matrix.

         (b) $O \leq Y^{(r)} \leq X^{(r)^{-1}}$.

Assuming $A^{(r)}$ is a M-matrix, we shall prove that $A^{(r+1)}$ is a M-matrix. For such $A^{(r)}$, (a) shows that $X^{(r)}$ is a M-matrix; so, in particular, it is monotone.

Consider now the intermediate matrix

$$\widetilde{A}^{(r)} = \begin{bmatrix} X^{(r)} & A_{12}^{(r)} \\ A_{21}^{(r)} & A_{22}^{(r)} \end{bmatrix}.$$

Clearly, $\widetilde{A}^{(r)}$ is a Z-matrix, and (7.12a) shows that $\widetilde{A}^{(r)}\mathbf{x}^{(r)} \geq A^{(r)}\mathbf{x}^{(r)}$, the latter being positive by assumption. Let $\widetilde{\mathbf{b}}_r, \mathbf{b}^{(r+1)}$ be the block vector components of $\widetilde{A}^{(r)}\mathbf{x}^{(r)}$, that is, let

$$\tilde{\mathbf{b}}_r = X^{(r)}\mathbf{x}_r + A_{12}^{(r)}\mathbf{x}^{(r+1)}$$
$$\mathbf{b}^{(r+1)} = A_{21}^{(r)}\mathbf{x}_r + A_{22}^{(r)}\mathbf{x}^{(r+1)}.$$

Then $\tilde{\mathbf{b}}_r$ and $\mathbf{b}^{(r+1)}$ are positive and

(7.13) $$\tilde{S}_2\mathbf{x}^{(r+1)} = \mathbf{b}^{(r+1)} - A_{21}X^{(r)^{-1}}\tilde{\mathbf{b}}_r,$$

where

$$\tilde{S}_2 = A_{22}^{(r)} - A_{21}^{(r)}X^{(r)^{-1}}A_{12}^{(r)}$$

is the Schur complement. Since $A_{21}^{(r)} \leq 0$ and $X^{(r)^{-1}} \geq 0$, (7.13) shows that $\tilde{S}_2\mathbf{x}^{(r+1)} \geq \mathbf{b}^{(r+1)}$, which is positive. (7.12b) now shows that

$$A^{(r+1)}\mathbf{x}^{(r+1)} \geq \tilde{S}_2\mathbf{x}^{(r+1)} \geq \mathbf{b}^{(r+1)},$$

so, since $A^{(r+1)}$ is a Z-matrix, $A^{(r+1)}$ is, in fact, a M-matrix.

In particular, Theorem 6.9 shows that $A_{11}^{(r+1)}$ is a M-matrix and, therefore, monotone. The existence of the incomplete factorization method (7.9), applied recursively using approximations $X^{(r)}$ and $Y^{(r)}$ satisfying (7.12) and with possible deletion of fill-in as defined in (7.11), follows now by the induction principle. This important result is stated in the next theorem.

Theorem 7.1 *Consider a M-matrix A partitioned in block matrix form (7.7), where $m = n$. Let $\mathbf{x} > 0$ be such that $A\mathbf{x} > 0$ and partition $\mathbf{x}^T = (\mathbf{x}_1^T, \mathbf{x}_2^T, \ldots, \mathbf{x}_n^T)$, consistent with the partitioning of A. Consider the approximate factorization of A, defined recursively by (7.9), (7.10), (7.11), and (7.11'), where at each stage r, $X^{(r)}$ is a Z-matrix which approximates $A_{11}^{(r)}$, such that $X^{(r)}\mathbf{x}_r \geq A_{11}^{(r)}\mathbf{x}_r$ and $Y^{(r)}$ approximates the inverse of $X^{(r)}$, such that*

$$O \leq Y^{(r)} \leq X^{(r)^{-1}}.$$

Then $X^{(r)}$ is a M-matrix for each r, so the incomplete factorization exists and each intermediate matrix $A^{(r+1)}$ is a M-matrix.

Proof This has already been shown above. $\diamond$

For the application of step (7.12a), we need to know a positive vector for which $A_{11}^{(r)}\mathbf{x}_r$ is positive. Such a vector is the Perron vector of A^{-1}. However, this vector is mostly not known. Often we have $A\mathbf{e} \geq 0$, where $\mathbf{e} =$

$(1, 1, \ldots, 1)^T$. We can then perturb the diagonal of A with arbitrary small positive numbers to make the perturbed matrix $\tilde{A}$ satisfy $\tilde{A}\mathbf{e} > \mathbf{0}$. The incomplete factorization can then be computed from $\tilde{A}$ instead of A. Incidentally, it can be seen that we can permit that $A_{11}^{(r)}$ has positive entries outside the diagonal, if only $A_{11}^{(r)}\mathbf{x}_r > \mathbf{0}$.

7.4 Block Incomplete Factorization of Positive Definite Matrices

We consider now the existence of the incomplete factorization method for positive definite matrices in block tridiagonal form. As we have seen in the previous chapter, any symmetric positive definite matrix A can be preconditioned efficiently by a M-matrix using the method of diagonal compensation of positive entries. This M-matrix may be factored approximately and the resulting matrix can be used as a preconditioner for the original matrix A. However, it can still be of interest to consider a separate existence theory for positive definite matrices. Furthermore, it will be seen that this theory holds for positive definite, but not necessarily symmetric, matrices. Basic for this theory will be a lemma regarding positive definiteness of Schur complements. We recall first that a real-valued matrix A is positive definite if $\mathbf{x}^T A\mathbf{x} > 0$ for all $\mathbf{x} \in \mathbb{R}^n$, $\mathbf{x} \neq \mathbf{0}$ or, equivalently, if its symmetric part, $A^S = 1/2(A + A^T)$ is s.p.d.

For convenience, we recall now some properties of positive definite matrices (cf. Section 3.2).

Lemma 7.2 *If A is positive definite, then A is regular.*

Proof If $A\mathbf{x} = \mathbf{0}$, $\mathbf{x} \neq \mathbf{0}$, then $\mathbf{x}^T A\mathbf{x} = 0$ for some $\mathbf{x} \neq \mathbf{0}$, so A is not positive definite. ◇

Lemma 7.3 *Let A be partitioned in block matrix form, with square diagonal blocks*

$$(7.14) \qquad A = \begin{bmatrix} A_{11} & A_{12} \\ A_{21} & A_{22} \end{bmatrix}.$$

Then the diagonal blocks of A are positive definite if A is positive definite.

Proof Partition $\mathbf{x} = (\mathbf{x}_1, \mathbf{x}_2)$ consistent with the partitioning of A. Then $\mathbf{x}^T A\mathbf{x} = \mathbf{x}_1^T A_{11}\mathbf{x}_1$ and $\mathbf{x}^T A\mathbf{x} = \mathbf{x}_2^T A_{22}\mathbf{x}_2$, if $\mathbf{x} = (\mathbf{x}_1, \mathbf{0})$ and $\mathbf{x} = (\mathbf{0}, \mathbf{x}_2)$, respectively. Hence, for any such vector $\mathbf{x} \neq \mathbf{0}$, $\mathbf{x}_1^T A_{11}\mathbf{x}_1 > 0$ and $\mathbf{x}_2^T A_{22}\mathbf{x}_2 > 0$, respectively. ◇

Lemma 7.4 *The inverse of a positive definite matrix A is positive definite.*

Proof We have $\mathbf{x}^T A\mathbf{x} = \mathbf{y}^T A^{-1}\mathbf{y}$ for $\mathbf{x} = A^{-1}\mathbf{y}$. Hence, $\mathbf{y}^T A^{-1}\mathbf{y} > 0$ for any $\mathbf{y} \neq \mathbf{0}$ is equivalent to $\mathbf{x}^T A\mathbf{x} > 0$ for any $\mathbf{x} \neq \mathbf{0}$.   ◇

Note next that for the inverse of a regular matrix partitioned in the form (7.14), where the A_{ii} are regular, we have

$$(7.15) \qquad \begin{bmatrix} A_{11} & A_{12} \\ A_{21} & A_{22} \end{bmatrix}^{-1} = \begin{bmatrix} S_1^{-1} & * \\ * & S_2^{-1} \end{bmatrix},$$

where S_1 and S_2 are the Schur complements given by

$$(7.16) \qquad S_i = A/A_{jj} \equiv A_{ii} - A_{ij}A_{jj}^{-1}A_{ji}, \quad i \neq j, i, j = 1, 2.$$

Here the entries indicated by ($*$) are irrelevant for the following theorem. This has been shown in equation (3.4).

Theorem 7.5 *Let A be positive definite and partitioned in the two-by-two blockform (7.14). Then its Schur forms,*

$$S_i = A_{ii} - A_{ij}A_{jj}^{-1}A_{ji}, \ i, j = 1, 2, i \neq j,$$

are positive definite.

Proof Lemmas 7.3 and 7.2 show that $A_{ii}, i = 1, 2$ are regular, so $S_i, i = 1, 2$ exists. Therefore the inverse of A can be written in the form (7.15) where S_i are defined by (7.16). Since A is positive definite, so is A^{-1} (by Lemma 7.4), and Lemma 7.3 now shows that the S_i^{-1} are positive definite. Finally, another application of Lemma 7.4 shows that the S_i are positive definite.   ◇

Note that in Theorem 7.5 we have not assumed A to be symmetric, so this theorem generalizes a well-known theorem for symmetric positive definite matrices.

We now consider the incomplete factorization method (7.9), applied recursively. We assume that A is positive definite and consider here only the case where A is block tridiagonal,

$$A = \text{tridiag}\ (A_{i,i-1}, A_{i,i}, A_{i,i+1}).$$

In this case the incomplete factorization method is equivalent to computing matrices $X^{(r)}$, $r = 1, 2, \ldots, n$, and the preconditioner takes the form

$$LD^{-1}U = (X - \widetilde{L})X^{-1}(X - \widetilde{U}),$$

where $\widetilde{L}, \widetilde{U}$ are block strictly lower triangular and block strictly upper triangular, with entries formed by $A_{i,i-1}$ and $A_{i,i+1}$, respectively. X is blockdiagonal,

$$X = \text{blockdiag}\ (X^{(1)}, X^{(2)}, \ldots, X^{(n)}),$$

where the matrices $X^{(r)}$ are defined by the recursion:
Let $A_{11}^{(1)} = A_{11}$, and for $r = 1, 2, \ldots, n - 1$:

(7.17)   (a) Compute $X^{(r)}$ such that $X^{(r)} \geq A_{11}^{(r)}$.

(b) Compute $Y^{(r)}$ such that $Y^{(r)-1} \geq X^{(r)}$,

and let $A_{r+1,r+1}^{(r+1)} = A_{r+1,r+1} - A_{r+1,r} Y^{(r)} A_{r,r+1}$,

where $X^{(r)}$ and $Y^{(r)}$ are sparse matrices as defined before but where the inequalities are now *in the sense of positive semidefiniteness*. For instance, we can let $X^{(r)}$ be a Stieltjes matrix and compute $Y^{(r)}$ by one of the methods to be presented in Chapter 8.

Lemma 7.4 and (7.17a,b) now show that

$$\begin{bmatrix} Y^{(r)-1} & \widetilde{A}_{r,r+1} \\ \widetilde{A}_{r+1,r} & \widetilde{A}_{r+1,r+1} \end{bmatrix} \geq \begin{bmatrix} X^{(r)} & \widetilde{A}_{r,r+1} \\ \widetilde{A}_{r+1,r} & \widetilde{A}_{r+1,r+1} \end{bmatrix} \geq \begin{bmatrix} A_{r,r}^{(r)} & \widetilde{A}_{r,r+1} \\ \widetilde{A}_{r+1,r} & \widetilde{A}_{r+1,r+1} \end{bmatrix},$$

(7.18)

which latter matrix is positive definite. Here, $\widetilde{A}_{r,r+1}$, $\widetilde{A}_{r+1,r}$ and $\widetilde{A}_{r+1,r+1}$ denote the remaining parts of A in block row r, in block column r, and in the lower main block diagonal part of A, of order $(n - r)$, respectively. Since the left-hand matrix in (7.18) is positive definite, so is the Schur complement, $\widetilde{A}_{r+1,r+1} - \widetilde{A}_{r+1,r} Y^{(r)} \widetilde{A}_{r,r+1}$. Since the top diagonal block of this latter matrix is equal to $A_{r+1,r+1}^{(r+1)}$, this matrix is also positive definite, so we can repeat (7.17a,b) with positive definite matrices $X^{(r)}$ until the final step ($r = n$), where it suffices to make the first step in (7.17a).

This result can be stated in a theorem.

Theorem 7.6 *Let A be positive definite (but not necessarily symmetric) and partitioned in block matrix form (7.7), where $m = n$, and assume that A is block tridiagonal. Consider the approximate factorization of A defined by (7.9) and (7.10), where $X^{(r)}$, $Y^{(r)}$, and $A^{(r+1)}$ satisfy (7.17) and the inequalities are in the sense of positive semidefiniteness. Then all pivot block matrices $A_{r,r}^{(r)}$ and the intermediate matrices $A^{(r)}$ are positive definite, and the incomplete factorization exists.*

Proof This proof has already been shown.   ◇

In practice, we can get a matrix $X^{(r)}$ that satisfies condition (7.17a), by using the method of reduction of positive off-diagonal entries and diagonal compensation of them. It is more difficult to find matrices $Y^{(r)}$ that satisfy (7.17b). One possible method is to use projections of vectors to a lower dimension— i.e., in this way reducing the order of the matrices to be inverted and computing the exact inverses of them. Such a method has been used in Chan and Vassilevski (1992).

Let $\{R^{(r)}\}_{r=1}^{n-1}$ be a sequence of restriction matrices that transform a vector of the dimension of $X^{(r)}$ to a lower dimensional vector space—say, of a small fixed size m. We have $R^{(r)}R^{(r)^T} = I_m$, the identity matrix in $\mathbb{R}^m$, and $R^{(r)^T}R^{(r)}$ is a projection matrix. Then let

$$A_{r+1,r+1}^{(r+1)} = A_{r+1,r+1} - A_{r+1,r}R^{(r)^T}\widehat{X}^{(r)^{-1}}R^{(r)}A_{r,r+1},$$

where

$$\widehat{X}^{(r)} = R^{(r)}X^{(r)}R^{(r)^T}.$$

Since the order of $\widehat{X}^{(r)}$ is low, it is feasible to compute its inverse exactly.

In applications to elliptic second-order difference equations, the matrices $A_{r,r+1}$ and $A_{r+1,r}$ are frequently diagonal or bidiagonal, and $A_{r,r}$ are tridiagonal. It is therefore convenient to let $X^{(r)} = A_{r,r}^{(r)}$ and to approximate the inverse of $A_{r,r}^{(r)}$ with a tridiagonal or diagonal matrix $Y^{(r)}$ so that all matrices $A_{r,r}^{(r)}$ during the recursion remain tridiagonal as $A_{r,r}$. However, we can also use a more accurate approximation, such as a pentadiagonal matrix, and then $A_{r,r}^{(r)}$ becomes pentadiagonal also.

Frequently, there is a third step in (7.17)—namely, a modification step in which a diagonal matrix $D^{(r+1)}$ is added. We let

$$A_{r+1,r+1}^{(r+1)} = A_{r+1,r+1} - A_{r+1,r}Y^{(r)}A_{r,r+1} + D^{(r+1)}.$$

Here $D^{(r+1)}$ is commonly computed, such that

(7.19) $D^{(r+1)}\mathbf{e}^{(r+1)} = A_{r+1,r}(Y^{(r)} - A_{r,r}^{(r)^{-1}})A_{r,r+1}\mathbf{e}^{(r+1)},$

where all entries of $\mathbf{e}^{(r+1)}$ are unit numbers. The computation of the right-hand side vector in (7.19) can be performed by matrix vector multiplications or by solving a linear system with matrix $A_{r,r}^{(r)}$. For further related comments, see Axelsson and Polman (1986).

7.5* Incomplete Factorization Methods for Block *H*-Matrices

We shall now consider a more general class of incomplete factorization methods where we can use different approximations of the pivot matrix entries in different columns. This added freedom can have significant benefits in practical applications of the method. The existence of these methods will be shown for the class of block *H*-matrices. To define this class, we recall first some properties of *M*-matrices.

Consider the class of *Z*-matrices, the real square matrices $A = [a_{i,j}]$, for which $a_{i,j} \le 0$, $i \ne j$. Note that any *Z*-matrix can be written in the form $A = \alpha I - B$, where $B \ge 0$ (i.e., B is nonnegative). A matrix A is a *M*-matrix if A can be written in the form $A = \alpha I - B$, with $B \ge 0$ and $\alpha > \rho(B)$, the spectral radius of B. For any $A \in Z$, the following are equivalent (see Ostrowski, 1937, Fiedler and Ptak, 1962, and Berman and Plemmons, 1979):

(a) A is a *M*-matrix.
(b) A is monotone, i.e., $A^{-1} \ge 0$.
(c) There exists a positive vector $\mathbf{x}$, such that $A\mathbf{x} > \mathbf{0}$.
(d) There exists a positive vector $\mathbf{x}$, such that $A^T \mathbf{x} > \mathbf{0}$, or equivalently, such that $\mathbf{x}^T A > \mathbf{0}$.

In order to extend this definition to rectangular matrices, we define here A, of order $m \times n$, to be a *M*-matrix if $A \in \mathbb{Z}$ and there exists a positive vector $\mathbf{x}$, $\mathbf{x} \in \mathbb{R}^m$, such that $\mathbf{x}^T A > \mathbf{0}$ ($\mathbf{x}^T A \in \mathbb{R}^n$), i.e., the weighted column sums of A are positive.

Definition 7.1 Let $\mathcal{M}(A)$, called the *comparison matrix* to A, be defined as

$$\{\mathcal{M}(A)\}_{i,j} = \begin{cases} |a_{i,i}|, & \text{for } i = j \\ -|a_{i,j}|, & \text{for } i \ne j. \end{cases}$$

Then A is said to be a *H-matrix* if $\mathcal{M}(A)$ is a *M*-matrix (Ostrowski, 1956).

Clearly, if A is a *M*-matrix, then $\mathcal{M}(A) = A$, and *M*-matrices form a subclass of *H*-matrices. That it is a proper subclass is shown by Example 7.1. A perhaps better known class of matrices is the class of generalized diagonally dominant matrices. We recall that $A = [a_{i,j}]$ is said to be generalized diagonally dominant if there exists a positive vector **x** such that

$$x_j |a_{j,j}| > \sum_{\substack{i=1 \\ i \neq j}}^{m} x_i |a_{i,j}|, \qquad j = 1, 2, \ldots, n.$$

This means that the weighted column sums of

$$B = [b_{i,j}], \quad b_{i,j} = \begin{cases} |a_{j,j}|, & i = j \\ -|a_{i,j}|, & i \neq j \end{cases}$$

are positive. (Alternatively, we could have defined generalized diagonal dominance using weighted row sums.) As was shown in Lemma 6.4(b), the class of H-matrices and the class of generalized diagonally dominant matrices are, in fact, equivalent.

Lemma 7.7 *A is a H-matrix if and only if A is generalized diagonally dominant.*

Note that a square H-matrix is always nonsingular, but that, in contrast to M-matrices, H-matrices need not be monotone, as shown by the following example.

Example 7.2 Let $A = \begin{bmatrix} 2 & -1 \\ 3 & 2 \end{bmatrix}$. Then it is readily seen that $\mathcal{M}(A)$ is a M-matrix, while $A^{-1} = \frac{1}{7} \begin{bmatrix} 2 & 1 \\ -3 & 2 \end{bmatrix}$. Hence, A^{-1} is not nonnegative.

Consider now a matrix A partitioned in block-matrix form, $A = [A_{i,j}]$, $1 \leq i \leq m$, $1 \leq j \leq n$, where $m \leq n$ and where $A_{i,i}$ are square, nonsingular matrices of order n_i. For such block-matrices where $m = n$, there exist various extensions of the H-matrix concept. Feingold and Varga (1962) define the block comparison matrix $\mathcal{M}_b(A) = [b_{i,j}]$ by

(7.20) $$b_{i,j} = \begin{cases} \|A_{i,i}^{-1}\|^{-1}, & \text{for } i = j \\ -\|A_{i,j}\|, & \text{for } i \neq j \end{cases}$$

and use this for the definition of generalized diagonal dominance. Here,

$$\|A_{i,j}\| = \sup_{\mathbf{x}^{(j)}} \frac{\|A_{i,j}\mathbf{x}^{(j)}\|}{\|\mathbf{x}^{(j)}\|},$$

where the norm is taken in the vector space corresponding to $\mathbf{x}^{(j)}$, the jth block of $\mathbf{x}$.

Definition 7.2 A is called a *block H-matrix* if $\mathcal{M}_b(A)$ is a M-matrix.

Ostrowski (1961) used a similar definition. Robert (1969) calls A a block H-matrix if the diagonal block matrix part, D_A of A, is nonsingular and $\mathcal{M}_b(D_A^{-1}A)$ is a M-matrix. Alternatively, we can call A a block H-matrix if $\mathcal{M}_b(AD_A^{-1})$ is a M-matrix. More generally, Polman (1987) defines a matrix to be a block H-matrix if $\mathcal{M}_b(DAE)$ is a $\mathcal{M}$-matrix, where D and E are nonsingular block diagonal matrices with diagonal blocks of the same orders as $A_{i,i}$, $i = 1, \ldots, m$. For the purpose of this presentation, it suffices to choose D and E identity matrices—i.e., we call A a block H-matrix if D_A is nonsingular and $\mathcal{M}_b(A)$ is a M-matrix. This definition is then valid also for rectangular matrices where $m \neq n$.

Note that the class of block H-matrices can be significantly larger than the class of H-matrices. As an example of a class of block H-matrices that are not H-matrices, see the following.

Example 7.3 *A class of block H-matrices:*

(a) Consider first

$$A = \begin{bmatrix} 660 & 160 & -8/7 & 0 \\ 160 & 40 & 0 & -8/7 \\ -10/9 & 0 & 340 & -80 \\ 0 & -10/9 & -80 & 20 \end{bmatrix} = \begin{bmatrix} A_{11} & A_{12} \\ A_{21} & A_{22} \end{bmatrix},$$

where each A_{ij} has order 2×2, and let $\| \cdot \|$ be the spectral norm. Then, since the A_{ij} are symmetric and definite, we have

$$\|A_{11}^{-1}\|^{-1} = \lambda_{\min}(A_{11}) = \frac{80}{35 + \sqrt{1217}} > 8/7,$$

$$\|A_{22}^{-1}\|^{-1} = \lambda_{\min}(A_{22}) = \frac{20}{9 + \sqrt{80}} > 10/9,$$

$$\|A_{12}\| = 8/9, \quad \|A_{21}\| = 10/9,$$

so $\mathcal{M}_b(A)$ is a M-matrix and A is a block H-matrix. As can be seen, however, A is not generalized diagonally dominant and, hence, is not a H-matrix.

(b) More generally, consider a matrix $A = [A_{i,j}]_{i,j=1}^n$, where each block is square (and, therefore, all matrices $A_{i,j}$ have the same order). Assume that the $A_{i,j}$ are symmetric and negative semidefinite, $i \neq j$, that the $A_{i,i}$ are s.p.d., and that there exists a positive vector $\mathbf{v} \in \mathbb{R}^n$ such that $\sum_{i=1}^n v_j A_{i,j} > 0$, in a positive definite sense. Assume, further, that each $A_{i,j}$, $j = 1, \ldots, n$, has the same eigenvector $\hat{\mathbf{x}}$ for its extreme eigenvalues, $-\|A_{i,j}\| = -\rho(A_{i,j})$, $i \neq j$ and for $\min \lambda(A_{i,i})$, $j = i$. It can be readily seen that the matrix in Example 7.3(a) satisfies these properties.

Then, noting that $\hat{\mathbf{x}}^T A_{i,i} \hat{\mathbf{x}} \geq \min \lambda(A_{i,i}) \hat{\mathbf{x}}^T \hat{\mathbf{x}}$, $\|A_{11}^{-1}\|^{-1} = \min \lambda(A_{i,i})$, and $\mathbf{x}^T \left(\sum_{j=1}^n v_j A_{i,j} \right) \mathbf{x} > 0$ for any $\mathbf{x}$, and in particular for $\hat{\mathbf{x}}$, we see that

$$ v_i \min \lambda(A_{i,i}) > \sum_{j \neq i}^n v_j \|A_{i,j}\|. $$

That is, matrix A is a block H-matrix.

To ascertain whether a matrix is a block H-matrix, we need— among other things—to compute the norms of the $A_{i,i}^{-1}$. If the $A_{i,i}$ are monotone, i.e. $A_{i,i}^{-1} \geq 0$, and in particular if the $A_{i,i}$ are M-matrices, then this can readily be done using the infinity norm and the identity

$$ \|A_{i,i}^{-1}\|_\infty = \|A_{i,i}^{-1} \mathbf{e}\|_\infty, \qquad i = 1, 2, \ldots, m, $$

that is, by solving linear systems with matrices $A_{i,i}$. An interesting method for general matrices for computing the ℓ_1-norm (which is equal to the infinity norm for the transposed matrix, as shown in Appendix A) is discussed in Hager (1984). See also the references in that report. This method gives only approximations of the norm, but they seem to be quite accurate after only few solutions of certain systems with $A_{i,i}$ and $A_{i,i}^T$.

We recall that an incomplete factorization method proceeds as a Gaussian elimination method, but with deletion of entries outside a certain sparsity pattern. This sparsity pattern is defined either before or during the factorization itself. In addition, for matrices partitioned in block form, we may need to approximate the inverses of the pivot block matrix entries. In the previous sections, existence of incomplete factorization methods for M-matrices and for positive definite matrices was shown.

Here we consider block H-matrices that are rectangular and show the existence of relaxed forms of the incomplete block-matrix factorization algorithm. In addition, we show that the approximations of the inverse of the pivot block matrix entries need to satisfy a condition that is even weaker than the condition assumed in the previous section for M-matrices (cf. Axelsson, 1986, and Beauwens and Ben Bouzid, 1987) and, furthermore, that this approximation can vary between matrix block columns. This latter property can have significant advantages in practical applications of the methods. We show that a certain explicit factorization method exists and provides a convergent splitting for the block H-matrix class of matrices.

Consider, then, a matrix $A = [A_{i,j}]$, partitioned in block-matrix form with $m \times n$ blocks and with nonsingular diagonal blocks of order n_i. We shall assume that $m \leq n$. The basic step in any incomplete factorization method is the typical Gaussian elimination step at the rth stage, where we form the Schur complement matrix,

$$(7.21) \qquad A_{i,j}^{(r+1)} := A_{i,j}^{(r)} - A_{i,r}^{(r)} A_{r,r}^{(r)^{-1}} A_{r,j}^{(r)}$$

of the matrix

$$\begin{bmatrix} A_{r,r}^{(r)} & A_{r,j}^{(r)} \\ A_{i,r}^{(r)} & A_{i,j}^{(r)} \end{bmatrix}.$$

As has been shown in Chapter 1, block Gaussian elimination and the factorization of A in lower and upper block triangular (LU) factors are closely related. If $A_{i,j}^{(r)}$ is zero but $A_{i,r}^{(r)} A_{r,r}^{(r)^{-1}} A_{r,j}^{(r)}$ is nonzero, we call the latter a *fill-in* entry. Recall that in an incomplete factorization method, factors L and U will be computed where the product LU only approximates A. We then have a sparsity pattern S that defines the positions where entries will be accepted during the factorization. Hence, the entries $A_{i,j}^{(r+1)}$ in (7.21) will be used only if the index pair (i, j) belongs to the set S. In this way, since not-accepted fill-in entries cannot cause new fill-in entries at later stages, the sparsity pattern of L and U can be controlled, contrary to the case for the full factorization, where there is a tendency of rapid growth of the set of fill-in entries as the factorization proceeds.

The sparsity set S can be defined *a priori*— i.e., before the outset of the factorization, or during the factorization. In the latter case, an entry will be accepted only if

$$\|A_{i,j}^{(r+1)}\|_2 > c\{\|A_{i,i}^{(r+1)}\|_2 \|A_{j,j}^{(r+1)}\|_2\}^{\frac{1}{2}}$$

for some constant c, $0 < c < 1$. (Here $\| \cdot \|_2$ is the Euclidean norm. $c = 0$ gives the full factorization and, for symmetric, positive definite matrices, $c = 1$ gives a diagonal matrix.) It is also possible to combine the two approaches. For definiteness, we consider here only the case where S is chosen *a priori*.

In addition to confining nonzero entries to a sparse set S, when the individual blocks $A_{i,j}^{(r)}$ are sparse themselves, we need to approximate the inverses of the pivot entries, $A_{r,r}^{(r)^{-1}}$, which occur during the factorization. Assume, then, that $D^{(r)}$ is some sparse approximation of $A_{r,r}^{(r)^{-1}}$. In fact, it will be seen that we can use different approximations for the entries in the different columns—i.e., we can use $D_j^{(r)} A_{r,j}^{(r)}$ as an approximation of $A_{r,r}^{(r)^{-1}} A_{r,j}^{(r)}$. This added freedom in the choice of approximations can increase the efficiency of the method significantly, because for blocks $A_{r,j}^{(r)}$ with small entries we frequently can use less accurate—and, hence, less costly—approximations $D_j^{(r)}$ than for blocks with larger entries, without any major deterioration of the quality of the preconditioner.

Finally, instead of simply neglecting the block matrix entries that are deleted during the incomplete factorization (because they fall outside the sparsity pattern S), we can use some information from them by adding certain entries to the diagonal of the diagonal matrix block in the same block matrix column. (In this presentation, we have chosen a column-oriented algorithm and storage of matrices. A row-oriented algorithm would have been equally applicable.) This addition is controlled by the components of a positive vector $\mathbf{x}$ (of order m), such that $\mathbf{x}^T \mathcal{M}_b(A) > \mathbf{0}$, and by the vector $\mathbf{e}_i = (1, 1, \ldots, 1)^T$, of order n_i, so that we compute a vector

$$\mathbf{z}_i^{(j)^T} := x_i x_j^{-1} \mathbf{e}_i^T \left(A_{i,j}^{(r)} - A_{i,r}^{(r)} A_{r,r}^{(r)^{-1}} A_{r,j}^{(r)} \right),$$

if $(i, j) \neq S$, and add the components of the entries of this vector to corresponding positions of the diagonal. Also, we can correct for the errors of the approximations of the inverses in a similar way when we compute the approximations $D_j^{(r)}$. More generally, in the *relaxed method* we add the components of a diagonal matrix $W^{(i)}$ times the above modification vector, where $W^{(i)} = \operatorname{diag}(w_k^{(i)})$, with entries $w_k^{(i)}, 0 \leq w_k^{(i)} \leq 1, k = 1, 2, \ldots$.

Let the sparsity pattern, defining where the block matrix entries will be accepted, be S, and let S^c be the complementary set, i.e.,

$$S^c = \{(i, j); 1 \leq i \leq m, 1 \leq j \leq n\} \setminus S.$$

S will always include at least the set $\{(i, i); 1 \leq i \leq m\}$, i.e., $(i, i) \neq S^c$. In fact, we shall assume for simplicity that S contains at least the set of index pairs defined by the sparsity pattern of A itself. This means that the correction or modification of the diagonal takes place with a vector

$$z_i^{(j)^T} = -x_i x_j^{-1} e_i^T W^{(i)} A_{i,r}^{(r)} A_{r,r}^{(r)^{-1}} A_{r,j}^{(r)}.$$

Remark 7.8 Note that if we let $S^c = \emptyset$, and $D^{(r)} = A_{r,r}^{(r)^{-1}}$, then the algorithm will compute the *exact LU* factorization of A (assuming infinite precision arithmetic). Hence, the full factorization is a special case of the *IBFC* algorithm. Also, we could have used row or column permutations, but that would change the sparsity pattern, so we have not considered that avenue here.

In the following algorithm, we have chosen a column orientation. The proof of existence, to follow, is based on this choice. For a column-oriented factorization, the matrix U will have unit diagonal blocks. The block matrix entries can be stored in the positions originally occupied by the entries of A. (Naturally, for the iterative method, we need to have stored another copy of A.) The entries (except the unit diagonal blocks of U) of the block triangular matrices L and U are found in the lower and upper triangular part of this matrix (array) when the algorithm is completed. If we use a compact storage of the entries of A, i.e., if we store only entries in positions defined by the sparsity pattern of S, the entries of L and U will be found in corresponding positions.

The incomplete factorization method for block matrices (presented in a pseudocode) takes the form shown by the following algorithm.

Algorithm (IBFC)   [C stands for column-oriented]

> **for** $r := 1$ **step** 1 **until** $m - 1$ **do**
>   **for** $j := (r + 1)$ **step** 1 **until** n **do**
>   **begin** compute approximation(s) $D_j^{(r)}$ of $A(r, r)^{-1}$;
>   **if** $(r, j) \in S$ **then**
>   **begin** $F := A(r, j)$; $E := D_j^{(r)} \times F$; $A(r, j) := E$;
>   **for** $i := (r + 1)$ **step** 1 **until** m **do**
>   **if** $(i, r) \in S$ **then**
>   **begin if** $(i, j) \in S$ **then**
> $A(i, j) := A(i, j) - A(i, r) \times E$;
>   **else**
>   diag $(A(j, j)) := $ diag $[A(j, j)] - x_i / x_j e(i)^T W^{(i)}$
> $\times A(i, r) \times A(r, r)^{-1} \times F$

[*Comment:* Here corrections are computed to the diagonal so
that the weighted column sums will be preserved if $w_k^{(i)} = 1$; the
computation is done by vector-matrix multiplications and solution of
a linear system with matrix $A(r, r)^T$. If several fill-in entries
occur in a row, it is advisable to first sum up the i-dependent
part of the correction vectors, so that it suffices with one solution
with $A(r, r)^T$.]

> **end**
> > **end** (loop in i)
> **end**
> **end** (loop in j)
> **end** (loop in r)

The above algorithm adds weighted contributions from deleted entries in each
column to the diagonal entry in the same column. For $w_k^{(i)} = 1$, this insures
"mass balance" (cf. Example 1.4).

Obviously we are interested in knowing if the entries of $A_{r,r}^{(r)}$, which occupy
the main diagonal of L, are nonsingular. We shall show that this is so for the
class of block H-matrices.

Theorem 7.9 *Let A be a block H-matrix and let* **x** *be a positive vector, such
that* $\mathbf{x}^T \mathcal{M}_b(A) > \mathbf{0}$, *where the block diagonal part of A is nonsingular and
$\mathcal{M}_b$ is the comparison matrix, as defined in (7.20). Then the IBFC algorithm
exists—i.e., each pivot entry $A_{r,r}^{(r)}$ is nonsingular if the approximations $D_j^{(r)}$ of
$A_{r,r}^{(r)-1}$, which can differ in different block matrix columns (j), satisfy*

$$(7.22) \qquad \|D_j^{(r)} A_{r,j}^{(r)}\| \le \|A_{r,r}^{(r)-1} A_{r,j}^{(r)}\|, \quad (r, j) \in \mathcal{S}, \ j \ge r + 1.$$

*In addition, at every stage (r) of the method, the remaining part $A^{(r+1)}$ of the
matrix is a block H-matrix.*

Proof (by induction) Let $A^{(1)} = A$,

$$(7.23) \qquad s_j^{(r)} = x_j - \|A_{j,j}^{(r)-1}\| \sum_{i=r, i \ne j}^{m} x_i \|A_{i,j}^{(r)}\|, \quad j = r, \dots, m,$$

and note that $s_j^{(1)}$ is positive, by assumption. First, we need to show that $A_{j,j}^{(r+1)}$

is nonsingular. To this end, using the inverse triangle inequality and noting that $w_k^{(i)} \geq 0$, we find for $j \geq r + 1$,

$$
\begin{aligned}
x_j \|A_{j,j}^{(r+1)}\| &- \sum_{\substack{i=r+1 \\ i \neq j, (i,j) \in S}}^{m} x_i \|A_{i,j}^{(r+1)}\| \\
&\geq x_j \|A_{j,j}^{(r)}\| - x_j \|A_{j,r}^{(r)} D_j^{(r)} A_{rj}^{(r)}\| \\
&\quad - \sum_{\substack{i=r+1 \\ i \neq j, (i,j) \in S}}^{m} \{ x_i \|A_{i,j}^{(r)}\| + x_i \|A_{i,r}^{(r)} D_j^{(r)} A_{r,j}^{(r)}\| \} \\
&\quad - \max_{k,i} w_k^{(i)} \sum_{\substack{i=r+1 \\ (i,j) \in S^c}}^{m} x_i \|A_{i,r}^{(r)} A_{r,r}^{(r)^{-1}} A_{r,j}^{(r)}\| \\
&\geq x_j \|A_{j,j}^{(r)}\| - \sum_{\substack{i=r+1 \\ i \neq j}}^{m} x_i \|A_{i,j}^{(r)}\| \\
&\quad - \|A_{r,r}^{(r)^{-1}} A_{r,j}^{(r)}\| \sum_{i=r+1}^{m} x_i \|A_{i,r}^{(r)}\| + t_j^{(r)},
\end{aligned}
$$

(7.24)

where

$$
t_j^{(r)} = \left(1 - \max_{k,i} w_k^{(i)} \right) \|A_{r,r}^{(r)^{-1}} A_{r,j}^{(r)}\| \sum_{\substack{i=r+1 \\ (i,j) \in S^c}}^{m} x_i \|A_{i,r}^{(r)}\|
$$

and where we have used (7.22). For later use, let

$$
F_{i,j}^{(r)} = A_{i,r}^{(r)} A_{r,r}^{(r)^{-1}} A_{rj}^{(r)}.
$$

Since $\|A\| \geq 1/\|A^{-1}\|$ for any nonsingular matrix A, we find

$$x_j \| A_{j,j}^{(r+1)} \| - \sum_{\substack{i=r+1 \\ i \neq j, (i,j) \in S}}^{m} x_i \| A_{i,j}^{(r+1)} \|$$

$$\geq x_j \| A_{j,j}^{(r)^{-1}} \|^{-1} - \sum_{\substack{i=r \\ i \neq j}}^{m} x_i \| A_{i,j}^{(r)} \| + x_r \| A_{r,j}^{(r)} \|$$

$$- \| A_{r,j}^{(r)} \| \| A_{r,r}^{(r)^{-1}} \| \sum_{i=r+1}^{m} x_i \| A_{i,r}^{(r)} \| + t_j^{(r)}$$

$$= \| A_{j,j}^{(r)^{-1}} \|^{-1} s_j^{(r)} + \| A_{r,j}^{(r)} \| s_r^{(r)} + t_j^{(r)} > 0.$$

Similarly, multiplying all matrix terms by a vector $\mathbf{v} \neq \mathbf{0}$, prior to taking norms in (7.24), yields

$$(7.25) \qquad \| A_{j,j}^{(r+1)} \mathbf{v} \| > \frac{1}{x_j} \sum_{\substack{i=r+1 \\ i \neq j, (i,j) \in S}}^{m} x_i \| A_{i,j}^{(r+1)} \mathbf{v} \| \geq 0,$$

so

$$\inf_{\mathbf{v}; \|\mathbf{v}\|=1} \| A_{j,j}^{(r+1)} \mathbf{v} \| > 0,$$

which shows that $A_{j,j}^{(r+1)}$ is nonsingular, $1 \leq j \leq n$. Using the relation

$$\| (I - B)^{-1} \|^{-1} \geq 1 - \| B \|,$$

which is valid for any matrix with $\| B \| < 1$, we find

$$x_j / \| A_{j,j}^{(r+1)^{-1}} \| = x_j / \left\| \left[A_{j,j}^{(r)} - A_{j,r}^{(r)} D_j^{(r)} A_{rj}^{(r)} - \sum_{\substack{i=r+1 \\ i \neq j, (i,j) \in S^c}}^{m} \operatorname{diag}\left(\frac{x_i}{x_j} (W^{(i)} \mathbf{e}(i))^T F_{i,j}^{(r)} \right) \right]^{-1} \right\|$$

$$\alpha_j \left\| \left[I - A_{j,r}^{(r)} D_j^{(r)} A_{r,j}^{(r)} A_{j,j}^{(r)-1} - \sum_{\substack{i=r+1 \\ i \neq j, (i,j) \in S^c}}^{m} \text{diag}\left(\frac{x_i}{x_j} \mathbf{e}(i)^T W^{(i)} F_{i,j}^{(r)} \right) A_{j,j}^{(r)-1} \right]^{-1} \right\|$$

$$\alpha_j \left(1 - \left[\|A_{j,r}^{(r)}\| \beta_j + \sum_{\substack{i=r+1 \\ i \neq j, (i,j) \in S^c}}^{m} \max_{k,i} w_k^{(i)} \frac{x_i}{x_j} \|A_{i,r}^{(r)}\| \beta_j \right] \|A_{j,j}^{(r)-1}\| \right)$$

$$= \frac{x_j}{\|A_{j,j}^{(r)-1}\|} - \left[x_j \|A_{j,r}^{(r)}\| + \sum_{\substack{i=r+1 \\ i \neq j, (i,j) \in S^c}}^{m} \max_{k,i} w_k^{(i)} x_i \|A_{i,r}^{(r)}\| \right] \|A_{r,r}^{(r)-1} A_{r,j}^{(r)}\|,$$

where $\alpha_j = x_j / \|A_{j,j}^{(r)-1}\|$, $\beta_j = \|A_{r,r}^{(r)-1} A_{r,j}^{(r)}\|$, and a derivation similar to the one in (7.24) reveals that

$$\frac{1}{\|A_{j,j}^{(r+1)-1}\|} s_j^{(r+1)} = \frac{x_j}{\|A_{j,j}^{(r+1)-1}\|} - \sum_{\substack{i=r+1 \\ i \neq j, (i,j) \in S}}^{m} x_i \|A_{i,j}^{(r+1)}\|$$

$$\geq \frac{x_j}{\|A_{j,j}^{(r)-1}\|} - \sum_{\substack{i=r+1 \\ i \neq j}}^{m} x_i \|A_{i,j}^{(r)}\| - \|A_{r,r}^{(r)-1} A_{r,j}^{(r)}\| \sum_{i=r+1}^{m} x_i \|A_{i,r}^{(r)}\| + t_j^{(r)}$$

$$\geq \|A_{j,j}^{(r)-1}\|^{-1} s_j^{(r)} + \|A_{r,j}^{(r)}\| s_r^{(r)} + t_j^{(r)} > 0,$$

which shows that $\mathcal{M}_b(A^{(r+1)})$ is a M-matrix. $\diamond$

Naturally, the above existence theorem is valid also for the exact factorization of a matrix.

Corollary 7.10 *The exact (full) block matrix factorization of a rectangular H-matrix exists. In addition, the diagonal blocks of L are nonsingular.*

Proof $S^c = \emptyset$, the empty set, and $D_i^{(r)} = A_{r,r}^{(r)-1}$ produces the exact factorization. Theorem 7.9 is valid for this choice of S^c and $D_j^{(r)}$ also. $\diamond$

Remark 7.11 (*diagonal modification*) As we saw in the algorithm $IBFC$, we may correct for neglected fill-in entries. If we choose $w_k^{(i)} = 1$ for all k, i then it follows that we have generalized the preservation of (weighted) column sums—i.e., a form of mass-balance.

We can also correct (modify) for the sparse matrix approximation $D_j^{(r)}$ of the inverse $A_{r,r}^{(r)^{-1}}$ of the pivot matrices. To see this, let $G_j^{(r)}$ be an initial approximation of $A_{r,r}^{(r)^{-1}}$, computed, for instance, by one of the methods to be presented in the next section. Let

$$D_r^{(j)} = G_j^{(r)} + \widetilde{D}_j^{(r)},$$

where $\widetilde{D}_j^{(r)}$ is a diagonal matrix, such that

(7.26) $$\left(G_j^{(r)} + \widetilde{D}_j^{(r)}\right)\mathbf{u} = A_{r,r}^{(r)^{-1}}\mathbf{u}$$

and where $\mathbf{u}$ is a positive vector. For instance, if $A_{r,j}^{(r)} \leq 0$ and there is no zero row in $A_{r,j}^{(r)}$, we may choose $\mathbf{u} = \mathbf{u}^{(j)} = -A_{r,j}^{(r)}\mathbf{e}$, where $\mathbf{e} = (1, 1, \ldots, 1)^T$. Then (7.26) shows that

$$D_j^{(r)} A_{r,j}^{(r)}\mathbf{e} = A_{r,r}^{(r)^{-1}} A_{r,j}^{(r)}\mathbf{e},$$

that is,

$$\|D_j^{(r)} A_{r,j}^{(r)}\|_\infty = \|A_{,r,r}^{(r)^{-1}} A_{r,j}^{(r)}\|_\infty,$$

if $D_j^{(r)} \geq 0$ and $A_{r,r}^{(r)}$ is monotone.

Remark 7.12 Clearly, the above includes point-wise versions of the algorithm as well, if the matrix is a H-matrix. The computation of $\mathcal{M}_b(A)$ to check if A is a block H-matrix can be troublesome in practice. Nevertheless, if $A_{j,j}^{-1}$ is nonnegative and each $A_{i,j}$, $i \neq j$ is either nonpositive or nonnegative, then we have

$$\|A_{j,j}^{-1}\|_\infty = \|A_{j,j}^{-1}\mathbf{e}\|_\infty \quad \text{and} \quad \|A_{i,j}\|_\infty = \|A_{i,j}\mathbf{e}\|_\infty,$$

so the norms can be computed simply by solving a linear system with $A_{j,j}$ and using matrix-vector multiplications for $A_{i,j}$, $i \neq j$. Another case occurs if $A_{j,j}$ is symmetric and positive definite. Then we have

$$\|A_{j,j}^{-1}\|_2 = [\lambda_{\min}(A_{j,j})]^{-1}.$$

for the Euclidean norm. Since $\|A\|_2 \leq \{\|A\|_1\|A\|_\infty\}^{\frac{1}{2}}$, we may still use the practically easier norms $\|\cdot\|_1$ and $\|\cdot\|_\infty$ for the estimate of the norms of the offdiagonal blocks.

Finally, it is readily seen from the proof of Theorem 7.9 that, instead of condition (7.22), we could use

$$\|D_j^{(r)}\| \leq \|A_{r,r}^{(r)^{-1}}\|.$$

Remark 7.13 (the case when $\mathcal{M}_b(A)$ is a singular M-matrix) Consider now the case where $\mathcal{M}_b(A)$ is a singular M-matrix. (Note that this does not imply that A itself is singular.) For a singular M-matrix A, there exists no positive vector $\mathbf{x}$ for which $A\mathbf{x} > 0$ (because otherwise A would be a nonsingular M-matrix). Let $\mathbf{x} > 0$ be such that $\mathcal{M}_b(A)\mathbf{x} = 0$. Then it is readily seen that even if we choose $W^{(i)} = I$, (7.25) shows that $A_{i,i}^{(r+1)}$ remains nonsingular as long as there exists at least one nonzero block $A_{i,j}^{(r+1)}$, $i > j$. If the sparsity pattern S equals the sparsity pattern of A, and if there exists at least one nonzero block $A_{i,j}$, $i > j$ for each $i = 1, 2, \ldots, m - 1$, then it is readily seen that $A^{(r+1)}$ also will contain at least one such nonzero block for each $r \leq m - 1$. This is so because the outermost entries in the envelope (skyscraper) structure of A will, in fact, not be changed during the factorization. Hence, the only possible singular diagonal block of L for such a matrix A will occur in the last diagonal block—i.e., only $A_{m,m}^{(m)}$ will be singular.

7.6 Inverse Free Form for Block Tridiagonal Matrices

Following the method used in Section 1.5, we consider now a method to compute a factorization in inverse free form of a block tridiagonal matrix

$$A = \text{tridiag}(A_{i,i-1}, A_{i,i}, A_{i,i+1}), \quad i = 1, 2, \ldots, n,$$

where $A_{i,j}$ has order $n_i \times n_j$. Matrices in this form frequently arise in practice—for instance, for elliptic second-order difference equations. Let

$$A = D_A + L_A + U_A$$

be the splitting of A in its diagonal lower and upper block tridiagonal parts, respectively, and consider the factorization

(7.27) $$C = LDU,$$

where

$$L = D^{-1} + L_A, \quad U = D^{-1} + U_A$$

and where D is to be computed. Since

$$(D^{-1} + L_A)D(D^{-1} + U_A) = D^{-1} + L_A + U_A + L_A D U_A$$
$$= D^{-1} - D_A + L_A D U_A + A,$$

we find that (7.27) shows that

$$(7.28) \qquad D^{-1} = D_A - L_A D U_A.$$

Since L_A and U_A are block bidiagonal, we note that $L_A D U_A$ is block diagonal when D is block diagonal. Hence, (7.28) shows that $D = \text{diag}(D_1, \ldots, D_n)$, where the matrices D_i satisfy the recursion

$$(7.29) \quad D_0 = 0; \quad D_i = (A_{i,i} - A_{i,i-1} D_{i-1} A_{i-1,i})^{-1}, \quad i = 1, 2, \ldots, n.$$

If A is a M-matrix, or more generally a block H-matrix, Corollary 7.10 shows that the D_i exist (i.e., are nonsingular). Also, if A is positive definite, then the D_i are nonsingular. In practice, the D_i will be full matrices even if $A_{i,i}$ and $A_{i,j}, j = i - 1, i + 1$ are sparse matrices, because the inverses occurring in (7.28) are generally full. Therefore, one wants to compute an approximate sparse inverse, which can be done with some of the methods described in the following chapter.

For the solution of systems $C\mathbf{x} = \mathbf{b}$ for a matrix of the form $C = LDU$, such as in (7.27), we need to perform the following steps:

$$(7.30a) \quad L\mathbf{y} = \mathbf{b} \quad \text{or} \quad \mathbf{y} = D\mathbf{b} - DL_A\mathbf{y}, \text{ that is,}$$

$$\mathbf{y}_i \equiv D_i(\mathbf{b}_i - A_{i,i-1}\mathbf{y}_{i-1}), \quad i = 1, 2, \ldots, n,$$

$$(7.30b) \quad DU\mathbf{x} = \mathbf{y} \quad \text{or} \quad \mathbf{x} = \mathbf{y} - DU_A\mathbf{x}, \text{ that is,}$$

$$\mathbf{x}_n \equiv \mathbf{y}_n; \quad \mathbf{x}_i = \mathbf{y}_i - D_i A_{i,i+1}\mathbf{x}_{i+1}, i = n - 1, n - 2, \ldots, 1.$$

The vectors here are partitioned consistently with the partitioning of A.

The computational complexity per block component to compute D_i is one inverse matrix computation (or approximation thereof), two matrix-matrix multiplications, and one matrix addition. To compute the solution using (7.30a,b), we need four matrix vector multiplications and two vector additions. Note that in the forward and back substitutions, we need only multiplications and additions. We still, however, need divisions to actually compute D_i. The forward and back substitutions can be done fairly efficiently on various computer architectures, such as vector and certain parallel computers, because the matrix-vector multiplications can be done in parallel between rows.

An alternative method to compute an inverse free form of factorized matrices is based on computing first the standard factorized form

$$C^{(1)} = (D + L)D^{-1}(D + U) = (D + L)(I + D^{-1}U)$$

and then approximations Y_i of D_i^{-1}, replacing D with Y^{-1} and $C^{(1)}$ with

$$C^{(2)} = (Y^{-1} + L)Y(Y^{-1} + U) = (Y^{-1} + L)(I + YU).$$

This method becomes particularly efficient when we use the method of reduction of block size in the computation as described at the end of Section 7.4. Then we have the following steps:

(a) X_{i-1} approximates D_{i-1} ($X_{i-1} \geq D_{i-1}$ in the positive definite matrix case).

(b) $\widehat{X}_{i-1} = R_{i-1}X_{i-1}R_{i-1}^T$, where R_{i-1} is the reduction matrix at stage i.

(c) $D_i = A_{i,i} - A_{i,i-1}R_{i-1}^T\widehat{X}_{i-1}^{-1}R_{i-1}A_{i-1,i}$.

If we use the Sherman-Morrison formula (Exercise 2.26) to compute the exact inverse of D_i as

$$Y_i = A_{i,i}^{-1} + A_{i,i}^{-1}A_{i,i-1}R_{i-1}^T(\widehat{X}_{i-1} - R_{i-1}A_{i-1,i}A_{i,i}^{-1}A_{i,i-1}R_{i-1}^T)^{-1}R_{i-1}A_{i-1,i}A_{i,i}^{-1},$$

then $Y = D^{-1}$ and $C^{(2)} = C^{(1)}$. However, it is more feasible to approximate $A_{i,i}^{-1}$. Then we have

$$Y_i = B_{i,i} + B_{i,i}A_{i,i-1}R_{i-1}^T(\widehat{X}_{i-1} - R_{i-1}A_{i-1,i}B_{i,i}A_{i,i-1}R_{i-1}^T)^{-1}R_{i-1}A_{i-1,i}B_{i,i},$$

where $B_{i,i}$ is the approximation of $A_{i,i}$. In this case, $C^{(2)}$ is not equal to $C^{(1)}$, in general. Note that the computation of the matrices $B_{i,i}$ and of $R_{i-1}A_{i-1,i}B_{i,i}A_{i,i-1}R_{i-1}^T$ can take place concurrently. Matrix-vector multiplications with Y_i amount to a number of matrix vector multiplications with $B_{i,i}$ and $A_{i,i-1}$, $A_{i-1,i}$ and n solutions of systems with matrices

$$\widehat{X}_{i-1} - R_{i-1}A_{i-1,i}B_{i,i}A_{i,i-1}R_{i-1}^T$$

of order m.

In the symmetric positive definite case where $U = L^T$ and where $B_{i,i}$ are computed as symmetric matrices satisfying $B_{i,i} \geq A_{i,i}^{-1}$ (in a positive semidefinite sense), we readily find that $Y_i \geq D_i^{-1}$. Here we have used

$$R_{i-1}^T \widehat{X}_{i-1}^{-1} R_{i-1} \leq X_{i-1}^{-1}$$

which, as can readily be seen, implies that

$$D_i \geq A_{i,i} - A_{i,i-1} X_{i-1}^{-1} A_{i-1,i}.$$

In addition to the condition $B_{i,i} \geq A_{i,i}^{-1}$, we must assume that $B_{i,i}$ is such that

$$\widehat{X}_{i-1} - R_{i-1} A_{i-1,i} B_{i,i} A_{i,i-1} R_{i-1}^T$$

is s.p.d. Since $\widehat{X}_{i-1} = R_{i-1} X_{i-1} R_{i-1}^T$, this holds if the matrix

$$\begin{bmatrix} X_{i-1} & A_{i-1,i} \\ A_{i,i-1} & B_{i,i}^{-1} \end{bmatrix}$$

is s.p.d. However, this contradicts somewhat the condition $B_{i,i} \geq A_{i,i}^{-1}$. Hence, to make $\widehat{X}_{i-1} - R_{i-1} A_{i-1,i} B_{i,i} A_{i,i-1} R_{i-1}^T$ positive definite, we may have to use a diagonal perturbation of $\widehat{X}_{i-1}$. If the condition holds, then

$$C^{(2)} \geq A,$$

so

$$\lambda_{\min}(C^{(2)^{-1}} A) \geq 1.$$

7.6.1 Modification of Approximate Inverse

The matrix approximation (call it $\widetilde{D}_i$) of the inverse matrix

$$(A_{i,i} - A_{i,i-1} D_{i-1} A_{i-1,i})^{-1}$$

can be modified so that the approximate factorization $C = LDU$ satisfies the generalized row-sum condition

$$C\mathbf{x} = A\mathbf{x}$$

for some positive vector **x**. It will be shown in a later chapter that such modifications can change the spectrum in a way favorable for the iterative acceleration method. By (7.28) this then leads to

$$(7.31) \qquad D^{-1}\mathbf{x} = (D_A - L_A D U_A)\mathbf{x},$$

where $D = \widetilde{D} + \Delta$, and $\Delta = (\Delta_1, \ldots, \Delta_n)$ is a diagonal matrix.

The equality (7.31) then shows that

$$D_i^{-1}\mathbf{x}_i = \mathbf{g}_i,$$

where $\mathbf{g}_i = (A_{i,i} - A_{i,i-1}D_{i-1}A_{i-1,i})\mathbf{x}_i$ or

$$\mathbf{x}_i = \widetilde{D}_i\mathbf{g}_i + \Delta_i\mathbf{g}_i.$$

That is,

$$(7.32) \qquad \Delta_i\mathbf{g}_i = \mathbf{x}_i - \widetilde{D}_i\mathbf{g}_i.$$

We must show that all components of $\mathbf{g}_i$ are nonzero; otherwise we are not able to compute Δ_i from (7.32). We show this in the case when A is a M-matrix and we assume that **x** is a vector for which $A\mathbf{x} > \mathbf{0}$. Consider, then, the approximation $D_1 = \widetilde{D}_1 + \Delta_1$ of A_{11}^{-1}. Here,

$$(7.33) \qquad \Delta_1\mathbf{g}_1 = \mathbf{x}_1 - \widetilde{D}_1\mathbf{g}_1,$$

where $\mathbf{g}_1 = A_{11}\mathbf{x}_1$. It is readily seen that $\mathbf{g}_1 > \mathbf{0}$, and if $0 \leq \widetilde{D}_1 \leq A_{11}^{-1}$ (note that $A_{11}^{-1} > 0$), then we have

$$\mathbf{x}_1 - \widetilde{D}_1\mathbf{g}_1 \geq \mathbf{x}_1 - A_{11}^{-1}\mathbf{g}_1 = \mathbf{0}.$$

Hence, the right-hand side of (7.33) is nonnegative and, since $\mathbf{g}_1 > \mathbf{0}$, we can find a diagonal matrix Δ_1 that satisfies (7.33), where $\Delta_1 \geq 0$.

Assume now that

$$0 \leq \widetilde{D}_i \leq (A_{i,i} - A_{i,i-1}D_{i-1}A_{i-1,i})^{-1}$$

and consider the matrix

$$\begin{bmatrix} D_{i-1}^{-1} & \widehat{A}_{i-1,i} \\ \widehat{A}_{i,i-1} & \widehat{A}_{i,i} \end{bmatrix}, \quad i = 2, 3, \ldots, n.$$

Here $\widehat{A}_{i-1,i}$, $\widehat{A}_{i,i-1}$ and $\widehat{A}_{i,i}$ denotes the remainder of the matrix A from the $(i-1)$th block matrix row and columns, except the pivot block $A_{i-1,i-1}$, which has been replaced by D_{i-1}^{-1}. We shall show first that

(7.34)
$$\begin{bmatrix} f_{i-1} \\ \hat{f}_i \end{bmatrix} = \begin{bmatrix} D_{i-1}^{-1} & \widehat{A}_{i-1,i} \\ \widehat{A}_{i,i-1} & \widehat{A}_{i,i} \end{bmatrix} \begin{bmatrix} x_{i-1} \\ \hat{x}_i \end{bmatrix} > \begin{bmatrix} 0 \\ 0 \end{bmatrix},$$

where the vectors have been partitioned consistently with the partitioning of the matrix and where $\begin{bmatrix} x_{i-1} \\ \hat{x}_i \end{bmatrix}$ is the remainder of the vector x.

Note that (7.34) holds for $i = 2$, because then

$$f_1 = D_1^{-1}x_1 + \widehat{A}_{12}\hat{x}_2 = g_1 + A_{12}x_2 = A_{11}x_1 + A_{12}x_2 = (Ax)_1 > 0,$$

and $\hat{f}_2$ is the remaining part of the vector Ax, which is positive by assumption. Next, note that if (7.34) holds, then

(7.35) $(\widehat{A}_{i,i} - \widehat{A}_{i,i-1}D_{i-1}\widehat{A}_{i-1,i})\hat{x}_i = \hat{f}_i - \widehat{A}_{i,i-1}D_{i-1}f_{i-1} > 0,$

because $\widehat{A}_{i,i-1} \le 0$, $D_{i-1} \ge 0$, $f_{i-1} > 0$ and $\hat{f}_i > 0$. In particular,

$$g_i \equiv (A_{i,i} - A_{i,i-1}D_{i-1}A_{i-1,i})x_i > 0,$$

and, since $D_i^{-1}x_i = g_i$, (7.34) holds for an index "i" if it holds for the preceding index "$i - 1$". Since (7.35) holds, and since the matrix $\widehat{A}_{i,i} - \widehat{A}_{i,i-1}D_{i-1}\widehat{A}_{i-1,i}$ is a Z-matrix, it is, in fact, a M-matrix. By induction it follows that $g_i > 0$, $i = 2, 3, \ldots, n$. Therefore, matrices Δ_i satisfying (7.32) exist.

7.6.2 A Fully Parallelizable Alternative to the Forward and Back Substitution Method: The Euler Expansion Method

Since

$$C = (D^{-1} + L_A)D(D^{-1} + U_A),$$

we have

$$C = (I - \widetilde{L})D^{-1}(I - \widetilde{U}),$$

where $\widetilde{L} = -L_A D$, $\widetilde{U} = -DU_A$.

Hence, solutions of systems with C can be computed as

(7.36) $y = C^{-1}z = (I - \widetilde{U})^{-1}D(I - \widetilde{L})^{-1}z,$

where, for instance, $(I - \tilde{L})^{-1}\mathbf{z}$ can be computed as

(7.37) $$(I - \tilde{L})^{-1}\mathbf{z} = (I + \tilde{L}^{2^s})(I + \tilde{L}^{2^{s-1}}) \ldots (I + \tilde{L})\mathbf{z}.$$

Here $s = \lceil \log_2 n - 1 \rceil$. This is the product expansion method and seems to go back to Euler (see Ostrowski, 1961). In fact, (7.37) follows by a straightforward computation, noting that $\tilde{L}$ is nilpotent, so $\tilde{L}^n = 0$. If we compute the s powers involving $\tilde{L}$, the solution in (7.37) requires only $s + 1$ matrix vector multiplications. Similarly, we can compute the solution of a system with matrix $I - \tilde{U}$. Note that all computations can occur in parallel between rows. Hence, the parallel computational complexity to solve a system, with C as shown in (7.36) using this Euler expansion method, is $O(\log n)$.

7.7 Symmetrization of Preconditioners and the SSOR and ADI Methods

As we have seen, the incomplete factorization methods first require a factorization step. There exists simpler preconditioning methods that require no factorization but have a form similar to the incomplete factorization methods. We shall present two methods of this type. As an introduction, consider first an iterative method of the form

$$M(\mathbf{x}^{l+1} - \mathbf{x}^l) = \mathbf{b} - A\mathbf{x}^l, \quad l = 0, 1, \ldots$$

to solve $A\mathbf{x} = \mathbf{b}$, where A and M are nonsingular. As we saw in Chapter 5, the asymptotic rate of convergence is determined by the spectral radius of the iteration matrix

$$B = I - M^{-1}A.$$

For a method such as the SOR method (which also requires no factorization), with optimal overrelaxation parameter ω (assuming that A has property A^π or A is s.p.d., see Section 6.5), the eigenvalues of the corresponding iteration matrix B are situated on a circle. Again as we saw in Chapter 5, no further acceleration is possible. There is, however, a simple remedy to this, based on taking a step in the forward direction of the chosen ordering, followed by a backward step—i.e., a step in the opposite order of the vector components. This method is said to have its origin in the early days of computers when programs were stored on tapes that had to be rewound before a new forward SOR step could begin. It was found that this otherwise useless computer time for the rewinding could be better used for a backward SOR sweep.

As we shall see, for symmetric and positive definite matrices the combined forward and backward sweeps correspond to a s.p.d. matrix which, contrary to the SOR method, has the advantage that it can be used as a preconditioning matrix in an iterative acceleration method. This method, called the SSOR method, will be defined later. More generally, if A is s.p.d., we consider the symmetrization of an iterative method

$$(7.38) \qquad \mathbf{x}^{l+1} = \mathbf{x}^l + M^{-1}(\mathbf{b} - A\mathbf{x}^l).$$

For the analysis only, we consider the transformed form of (7.38),

$$\mathbf{y}^{l+1} = (I - A^{\frac{1}{2}} M^{-1} A^{\frac{1}{2}})\mathbf{y}^l + \tilde{\mathbf{b}},$$

where

$$\mathbf{y}^l = A^{\frac{1}{2}} \mathbf{x}^l \quad \text{and} \quad \tilde{\mathbf{b}} = A^{\frac{1}{2}} M^{-1} \mathbf{b}.$$

If M is nonsymmetric, the iteration matrix $I - A^{\frac{1}{2}} M^{-1} A^{\frac{1}{2}}$ is also nonsymmetric. We shall now consider a method using M and another preconditioner chosen so that the iteration matrix for the combined method becomes symmetric. We call this the *symmetrization* of the method.

Let M_1, M_2 be two such preconditioning matrices. Let

$$B_i = I - \widetilde{M}_i^{-1}, \quad \widetilde{M}_i = A^{-\frac{1}{2}} M_i A^{-\frac{1}{2}},$$

and consider the combined iteration matrix $B_2 B_1$. As we shall now see, it arises as an iteration matrix for the combined method

$$M_1(\mathbf{x}^{l+\frac{1}{2}} - \mathbf{x}^l) = \mathbf{b} - A\mathbf{x}^l,$$
$$(7.39)$$
$$M_2(\mathbf{x}^{l+1} - \mathbf{x}^{l+\frac{1}{2}}) = \mathbf{b} - A\mathbf{x}^{l+\frac{1}{2}}, \quad l = 0, 1, \ldots.$$

For the analysis only, we transform this to the form

$$\mathbf{y}^{l+\frac{1}{2}} - \mathbf{y}^l = \tilde{\mathbf{b}}^{(1)} - \widetilde{M}_1^{-1} \mathbf{y}^l,$$
$$\mathbf{y}^{l+1} - \mathbf{y}^{l+\frac{1}{2}} = \tilde{\mathbf{b}}^{(2)} - \widetilde{M}_2^{-1} \mathbf{y}^{l+\frac{1}{2}},$$

where

$$\tilde{\mathbf{b}}^{(i)} = A^{\frac{1}{2}} M_i^{-1} \mathbf{b}.$$

This iteration takes the form

$$\mathbf{y}^{l+1} = \tilde{\mathbf{b}}^{(2)} + (I - \tilde{M}_2^{-1})(\tilde{\mathbf{b}}^{(1)} + (I - \tilde{M}_1^{-1})\mathbf{y}^l),$$

that is,

$$\mathbf{y}^{l+1} = \tilde{\mathbf{b}}^{(2)} + (I - \tilde{M}_2^{-1})\tilde{\mathbf{b}}^{(1)} + (I - \tilde{M}_2^{-1})(I - \tilde{M}_1^{-1})\mathbf{y}^l$$

or

(7.40) $$\mathbf{y}^{l+1} = \hat{\mathbf{b}} + B_2 B_1 \mathbf{y}^l, \quad l = 0, 1, \ldots,$$

where

$$\hat{\mathbf{b}} = \tilde{\mathbf{b}}^{(2)} + (I - \tilde{M}_2^{-1})\tilde{\mathbf{b}}^{(1)}.$$

For the following we need a lemma.

Lemma 7.14 *If A, B, and C are Hermitian positive definite and each pair of them commute, then ABC is Hermitian positive definite.*

Proof We have $(ABC)^* = CBA$ and use commutativity to find

$$CBA = BCA = BAC = ABC.$$

Hence, ABC is Hermitian. Next, we show that the product of two s.p.d. matrices that commute is positive definite. We have

$$A^{-\frac{1}{2}}ABA^{\frac{1}{2}} = A^{\frac{1}{2}}BA^{\frac{1}{2}},$$

which is Hermitian positive definite. Hence, by similarity, the eigenvalues of AB are positive and, since

$$(AB)^* = AB,$$

AB is Hermitian positive definite. In the same way, $(AB)C$ is Hermitian positive definite. $\diamond$

Lemma 7.15 *Let A be s.p.d. and assume either of the following additional conditions:*

(a) $M_2^* = M_1$.

(b) M_1, M_2 *are s.p.d.* $\rho(A^{\frac{1}{2}}M_i^{-1}A^{\frac{1}{2}}) < 1$, $i = 1, 2$, *and each pair of matrices* M_1, M_2, A *commutes.*

Then the combined iteration method (7.39) converges if and only if $M_1 + M_2 - A$ is s.p.d.

Proof It is readily seen that $\mathbf{y} = \hat{\mathbf{b}} + B_2 B_1 \mathbf{y}$ (i.e., the iteration method is consistent with $A\mathbf{x} = \mathbf{b}$), where $\mathbf{y} = A^{\frac{1}{2}}\mathbf{x}$ and $\mathbf{x} = A^{-1}\mathbf{b}$. Hence,

$$\mathbf{y} - \mathbf{y}^{l+1} = B_2 B_1 (\mathbf{y} - \mathbf{y}^l),$$

and the iteration method (7.40 and, hence, 7.39) converges for any initial vector if and only if $\rho(B_2 B_1) < 1$, where $\rho(\cdot)$ denotes the spectral radius. But

$$B_2 B_1 = I - \tilde{M}_1^{-1} - \tilde{M}_2^{-1} + \tilde{M}_1^{-1}\tilde{M}_2^{-1}$$

$$= I - A^{\frac{1}{2}} M_1^{-1}(M_1 + M_2 - A)M_2^{-1}A^{\frac{1}{2}}.$$

It is readily seen that under either of the given conditions (a) or (b),

$$M_1^{-1}(M_1 + M_2 - A)M_2^{-1} = M_1^{-1} + M_2^{-1} - M_1^{-1}AM_2^{-1}$$

is symmetric. Further, it is positive definite if and only if $M_1 + M_2 - A$ is positive definite. Hence, $I - B_2 B_1$ is s.p.d. Further, $B_2 B_1 = (I - \tilde{M}_2^{-1})(I - \tilde{M}_1^{-1})$ is symmetric, and a similarity transformation shows that $B_2 B_1$ is similar to $(I - \tilde{M}_2^{-1})^{\frac{1}{2}}(I - \tilde{M}_1^{-1})(I - \tilde{M}_2^{-1})^{\frac{1}{2}}$, which is a congruence transformation of $I - \tilde{M}_1^{-1}$, whose eigenvalues are positive. Hence, $B_2 B_1$ has positive eigenvalues, so the eigenvalues of $B_2 B_1$ are contained in the interval $(0, 1)$ and, in particular, $\rho(B_2 B_1) < 1$. $\diamondsuit$

The proof of Lemma 7.15 shows that $B_2 B_1$ is symmetric, so the combined iteration method is a *symmetrized version* of either of the simple methods.

Let us now consider a special class of symmetrized methods. We let A be split as $A = D + L + U$, where we assume that D is s.p.d., and let

$$(7.41) \qquad V = (1 - \frac{1}{\omega})D + L, \quad H = (1 - \frac{1}{\omega})D + U,$$

$\hat{D} = (\frac{2}{\omega} - 1)D$, where ω is a parameter, $0 < \omega < 2$. (Here L and U are not necessarily the lower and upper triangular parts of A.) Note that

$$(7.42) \qquad \hat{D} + V + H = A,$$

so this is also a splitting of A. As an example of a combined, or symmetrized, iteration method, we consider the preconditioning matrix

$$(7.43) \qquad C = (\hat{D} + V)\hat{D}^{-1}(\hat{D} + H)$$

and show that this leads to a convergent iteration method

$$C(\mathbf{x}^{l+1} - \mathbf{x}^l) = \mathbf{b} - A\mathbf{x}^l, \quad l = 0, 1, \ldots .$$

This corresponds to choosing $M_1 = \widehat{D}^{-\frac{1}{2}}(\widehat{D} + H)$ and $M_2 = (\widehat{D} + V)\widehat{D}^{-\frac{1}{2}}$, and it can be seen that the conditions of Lemma 7.15 hold if the conditions in the next theorem hold.

Theorem 7.16 *Let $A = D + L + U$, where D is s.p.d. Let V, H, $\widehat{D}$ be defined by (7.41), and assume that either (a) or (b) holds, where*

(a) $U = L^*$.

(b) *L, U are s.p.d. and each pair of matrices L, U, D commute.*

Then the eigenvalues λ of the matrix $C^{-1}A$, where C is defined in (7.43), are contained in the interval $0 < \lambda \leq 1$.

Proof This can be shown either by verifying the conditions in Lemma 7.15 or more directly as follows. As in the proof of Lemma 7.15, it follows that C is s.p.d. Hence, the eigenvalues of $C^{-1}A$ are positive. Further,

$$C = \widehat{D} + V + H + V\widehat{D}^{-1}H,$$

so, by (7.42)

$$C = A + V\widehat{D}^{-1}H.$$

Under either condition (a) or (b), $V\widehat{D}^{-1}H$ is symmetric and positive semi-definite.

This shows that $\mathbf{x}^*C\mathbf{x} \geq \mathbf{x}^*A\mathbf{x}$ for all $\mathbf{x}$, so the eigenvalues of $C^{-1}A$ are bounded above by 1. $\diamond$

We shall now show that the matrix C can also efficiently be used as a pre-conditioning matrix, which for a proper value of the parameter ω, and under an additional condition, can even reduce the order of magnitude of the condition number. In this respect, note that when C is used as a preconditioning matrix for the Chebyshev iterative method, it is not necessary to have C scaled so that $\lambda(C^{-1}A) \leq 1$, because it suffices then that $0 < m \leq \lambda(C^{-1}A) \leq M$, for some numbers m, M. Hence, the factor $2/\omega - 1$ in $\widehat{D}^{-1}$ can be neglected.

Theorem 7.17 *Let $A = D + L + U$ be a splitting of A, where A and D are s.p.d. and either (a) $U = L^*$ or (b) L, U are s.p.d. and each pair of D, L, U*

commute. Then, the eigenvalues of matrix $C^{-1}A$, *where*

(7.44)
$$C = (\frac{1}{\omega}D + L)\widehat{D}^{-1}(\frac{1}{\omega}D + U)$$

and $0 < \omega < 2$, $\widehat{D} = (2/\omega - 1)D$, *are contained in the interval*

(7.45)
$$[(2 - \omega)/\{1 + \omega(\frac{1}{\omega} - \frac{1}{2})^2\delta^{-1} + \omega\gamma\}, 1],$$

where

$$\delta = \min_{\mathbf{x} \neq 0} \frac{\mathbf{x}^T A \mathbf{x}}{\mathbf{x}^T D \mathbf{x}}$$

and

$$\gamma = \max_{\mathbf{x} \neq 0} \frac{\mathbf{x}^T(LD^{-1}U - \frac{1}{4}D)\mathbf{x}}{\mathbf{x}^T A \mathbf{x}}.$$

Further, if there exists a vector for which $\mathbf{x}^T(L + U)\mathbf{x} \leq 0$, *then* $\gamma \geq -1/4$, *and if*

$$\rho(\widetilde{L}\widetilde{U}) \leq \frac{1}{4},$$

then $\gamma \leq 0$, *and if*

$$\rho(\widetilde{L}\widetilde{U}) \leq \frac{1}{4} + O(\delta), \text{ then } \gamma \leq O(1), \delta \to 0.$$

Here, $\widetilde{L} = D^{-\frac{1}{2}}LD^{-\frac{1}{2}}$.

Proof It is readily seen that

$$C = \frac{1}{2 - \omega}(\frac{1}{\omega}D + L)(\frac{1}{\omega}D)^{-1}(\frac{1}{\omega}D + U)$$

$$= \frac{1}{2 - \omega}[A + (\frac{1}{\omega} - 1)D + \omega LD^{-1}U]$$

$$= \frac{1}{2 - \omega}[A + \omega(\frac{1}{\omega} - \frac{1}{2})^2 D + \omega(LD^{-1}U - \frac{1}{4}D)].$$

This shows the lower bound in (7.45); the upper bound follows by Theorem 7.16. By choosing a vector for which $\mathbf{x}^T(L + U)\mathbf{x} \le 0$, it follows that

$$\mathbf{x}^T(LD^{-1}U - \frac{1}{4}D)\mathbf{x}/\mathbf{x}^T A\mathbf{x} \ge \mathbf{x}^T(LD^{-1}U\mathbf{x} - \frac{1}{4}D\mathbf{x})/\mathbf{x}^T D\mathbf{x} \ge -\frac{1}{4},$$

which shows $\gamma \ge -1/4$. The remainder of the theorem is immediate. $\diamond$

7.7.1 The Condition Number

Theorem 7.17 shows that the optimal value of ω to minimize the upper bound of the condition number of $C^{-1}A$ is the value that minimizes the real-valued function

$$f(\omega) = \frac{1 + \omega(\frac{1}{\omega} - \frac{1}{2})^2\delta^{-1} + \omega\gamma}{2 - \omega}.$$

It is readily seen (see Axelsson and Barker, 1984) that $f(\omega)$ is minimized for

$$\omega^* = \frac{2}{1 + 2\sqrt{(\frac{1}{2} + \gamma)\delta}}$$

and

$$\min_{\omega} f(\omega) = f(\omega^*) = \sqrt{(\frac{1}{2} + \gamma)\delta^{-1}} + \frac{1}{2}.$$

In general, δ is not known, but we may know that $\delta = O(h^2)$, for some problem parameter, $h \to 0$ (such as for the step length in second-order elliptic problems). Then, if $\gamma = O(1)$, $h \to 0$, we let $\omega = 2/(1 + \zeta h)$ for some $\zeta > 0$, in which case

$$f(\omega) = O(h^{-1}) = O(\sqrt{\delta^{-1}}), \quad h \to 0.$$

This means that $C^{-1}A$ has an order of magnitude smaller condition number than for A itself, which latter is $O(\delta^{-1})$.

We consider now two applications of Theorem 7.17.

7.7.2 The SSOR Method

In the first case, L is the lower triangular part of A, or the lower block triangular part, if A is partitioned in block matrix form and $U = L^*$. Then,

$$C = \frac{1}{2 - \omega}(\frac{1}{\omega}D + L)(\frac{1}{\omega}D)^{-1}(\frac{1}{\omega}D + L^*)$$

is a symmetrized version of the SOR method and is called the SSOR (*symmetric successive overrelaxation*) method.

For elliptic differential equations of second order, it can be seen that the condition $\rho(\widetilde{L}\widetilde{L}^T) \leq 1/4$ holds for problems with Dirichlet boundary conditions and constant coefficients. For extensions of this, see Axelsson and Barker (1984) and Exercises 7.5 and 7.6. For the model difference equation on a square domain with side π, we have

$$\delta = 2(\sin \frac{h}{2})^2, \quad \gamma \leq 0,$$

and we find

(7.46)
$$\omega^* = \frac{2}{1 + 2 \sin \frac{h}{2}} \sim \frac{2}{1 + h},$$

$$f(\omega^*) = \sqrt{\frac{1}{2\delta} + \frac{1}{2}} \sim h^{-1} + \frac{1}{2}, \quad h \to 0.$$

7.7.3 The ADI Method

In the second case, we let L denote the offdiagonal part of the difference operator working in the x-direction and U denote the offdiagonal part of the difference operator in the y-direction. D is its diagonal part. Then the matrix

$$\widehat{C} = (\frac{1}{\omega}D + L)(\frac{1}{\omega}D)^{-1}(\frac{1}{\omega}D + U)$$

is called an *alternating direction* preconditioning matrix and the corresponding iteration method is called the ADI (*alternating direction iteration*) method. In this method, we solve alternately one-dimensional difference equations in x- and y-directions. Much has been written on the ADI-method; see Varga (1962) and Wachspress (1966), for instance.

As we shall see, for the model difference equations we get the same optimal value of ω as in (7.46). The condition $\gamma = O(1)$ may be less restrictive, for the ADI-method, but the condition of commutativity is much more restrictive, as the following lemma shows.

Lemma 7.18 (Varga, 1962) *Let A, B be two Hermitian matrices of order n. Then $AB = BA$ if and only if A and B have a common set of orthonormal eigenvectors.*

Proof If such a common set of eigenvectors $\{v_i\}$ exists, then $Av_i = \sigma_i v_i$, $Bv_i = \tau_i v_i$ and

$$AB v_i = \sigma_i \tau_i v_i = BA v_i, \quad i = 1, 2, \ldots, n.$$

Since the eigenvector space of an Hermitian matrix is complete, we therefore have

$$AB\mathbf{x} = BA\mathbf{x} \quad \text{for all } \mathbf{x} \in \mathbb{C}^n,$$

which shows that $AB = BA$. Conversely, suppose that $AB = BA$. As A is Hermitian, take U to be a unitary matrix that diagonalizes A, that is,

$$\widetilde{A} = UAU^* = \begin{bmatrix} \gamma_1 I_1 & & & 0 \\ & \gamma_2 I_2 & & \\ & & \ddots & \\ 0 & & & \gamma_r I_r \end{bmatrix},$$

where $\gamma_1 < \gamma_2 < \ldots < \gamma_r$ are the distinct eigenvalues of A and I_j is the identity matrix of order n_j, the multiplicity of γ_j. (Here A is possibly permuted accordingly.) Let $\widetilde{B} = UBU^*$ and partition $\widetilde{B}$ corresponding to the partitioning of A, that is,

$$\widetilde{B} = \begin{bmatrix} B_{11} & B_{12} & \ldots & B_{1r} \\ & & \vdots & \\ B_{r1} & B_{r2} & \ldots & B_{rr} \end{bmatrix}.$$

Since $AB = BA$, we have

$$\widetilde{A}\widetilde{B} = UABU^* = UBAU^* = \widetilde{B}\widetilde{A}.$$

Carrying out the block multiplication $\widetilde{A}\widetilde{B} = \widetilde{B}\widetilde{A}$, we find that this, in turn, implies $B_{ij} = 0$, $i \neq j$, since $\gamma_i \neq \gamma_j$, $i \neq j$. Simply stated, a (block) matrix commutes with a (block) diagonal matrix if and only if it is itself (block) diagonal. Hence, $\widetilde{B}$ is block diagonal and each Hermitian submatrix $B_{i,i}$ has n_i orthonormal eigenvectors that are also eigenvectors of the submatrix $\gamma_i I_i$ of $\widetilde{A}$. Since $\sum_{i=1}^{r} n_i = n$ and all eigenvectors are orthonormal, A and B must have the same set of eigenvectors. $\diamond$

For the second-order elliptic difference equation, it turns out that the commutativity of L and U essentially corresponds to the property that the original problem is separable—i.e., that solutions of $\mathcal{L}u = f$ can be written in the form $u = \varphi(x)\psi(y)$. This means that the coefficients $a(x, y)$ and $b(x, y)$ in the differential operator $\frac{\partial}{\partial x}[a(x, y)\frac{\partial u}{\partial x}] + \frac{\partial}{\partial y}[b(x, y)\frac{\partial u}{\partial y}] + c(x, y)u$ must satisfy $a(x, y) = a(x)$, $b(x, y) = b(y)$ and $c(x, y) = c$, a constant. Hence, if $a(x, y) = b(x, y)$, then $a(x, y) = b(x, y) = a$, a constant. Furthermore, the convex closure of the meshpoints must be a rectangle with sides parallel to the coordinate axes (Varga, 1962). If $A = A_1 + A_2$, the ADI-method can be written in the form

(7.47)
$$(I + \tau_1 A_1)\mathbf{x}^{l+\frac{1}{2}} = (I - \tau_1 A_2)\mathbf{x}^l + \tau_1 \mathbf{b}$$

$$(I + \tau_2 A_2)\mathbf{x}^{l+1} = (I - \tau_2 A_1)\mathbf{x}^{l+\frac{1}{2}} + \tau_2 \mathbf{b}, \quad l = 0, 1, \dots.$$

This is the *Peaceman-Rachford* (1955) *iteration method*. The iteration matrix M is similar to

$$(I - \tau_2 A_1)(I + \tau_1 A_1)^{-1}(I - \tau_1 A_2)(I + \tau_2 A_2)^{-1}.$$

When A_1, A_2 are Hermitian positive definite, their eigenvalues $\lambda_i^{(1)}, \lambda_i^{(2)}$ are positive, and

$$\|(I - \tau_2 A_1)(I + \tau_1 A_1)^{-1}\|_2 = \rho((I - \tau_2 A_1)(I + \tau_1 A_1)^{-1}) =$$

$$= \max_i \left| \frac{1 - \tau_2 \lambda_i^{(1)}}{1 + \tau_1 \lambda_i^{(1)}} \right|.$$

Thus,

$$\rho(M) = \rho((I - \tau_2 A_1)(I + \tau_1 A_1)^{-1}(I - \tau_1 A_2)(I + \tau_2 A_2)^{-1})$$

$$\leq \|(I - \tau_2 A_1)(I + \tau_1 A_1)^{-1}(I - \tau_1 A_2)(I + \tau_2 A_2)^{-1}\|_2$$

$$\leq \|(I - \tau_2 A_1)(I + \tau_1 A_1)^{-1}\|_2 \|(I - \tau_1 A_2)(I + \tau_2 A_2)^{-1}\|_2$$

$$= \mu(\tau_1, \tau_2)$$

and

$$\rho(M) \leq \mu(\tau_1, \tau_2) = \max_i \left| \frac{1 - \tau_2 \lambda_i^{(1)}}{1 + \tau_1 \lambda_i^{(1)}} \right| \max_i \left| \frac{1 - \tau_1 \lambda_i^{(2)}}{1 + \tau_2 \lambda_i^{(2)}} \right|.$$

Note that for $\tau_1 = \tau_2 = \tau > 0$, $\mu(\tau_1, \tau_2) = \mu(\tau, \tau) < 1$, so we have $\rho(M) < 1$, that is, convergence for any $\tau > 0$. This holds even if A_1, A_2 do not commute. Note also that when A_1 and A_2 commute, we have

$$\rho(M) = \mu(\tau_1, \tau_2).$$

Let us continue the analyses for the general case where A_1, A_2 do not necessarily commute. We want to compute the optimal values of τ_1 and τ_2 such that $\mu(\tau_1, \tau_2)$ is minimized. For simplicity, we assume that α, β are the same lower and upper bounds of the eigenvalues of A_1 and A_2, that is, $0 < \alpha \le \lambda_i^{(j)} \le \beta$, $j = 1, 2$. We have

$$(7.48) \quad \mu(\tau_1, \tau_2) \le \max\left\{\left|\frac{1 - \tau_2\alpha}{1 + \tau_1\alpha}\right|, \left|\frac{1 - \tau_2\beta}{1 + \tau_1\beta}\right|\right\} \max\left\{\left|\frac{1 - \tau_1\alpha}{1 + \tau_2\alpha}\right|, \left|\frac{1 - \tau_1\beta}{1 + \tau_2\beta}\right|\right\}.$$

We want to choose τ_1, $\tau_2 > 0$ such that this bound is as small as possible. Note, then, that for such values of τ_1, τ_2 we must have $1 - \tau_i\alpha > 0$ and $1 - \tau_i\beta < 0$. Next note that each factor in the bound (7.48) is minimized when

$$\frac{1 - \tau_2\alpha}{1 + \tau_1\alpha} = \frac{\tau_2\beta - 1}{\tau_1\beta + 1} \quad \text{and} \quad \frac{1 - \tau_1\alpha}{1 + \tau_2\alpha} = \frac{\tau_1\beta - 1}{\tau_2\beta + 1},$$

respectively, that is, when

$$\tau_1\tau_2 - \frac{\alpha + \beta}{2\alpha\beta}(\tau_1 - \tau_2) - \frac{1}{\alpha\beta} = 0$$

and

$$\tau_1\tau_2 + \frac{\alpha + \beta}{2\alpha\beta}(\tau_1 - \tau_2) - \frac{1}{\alpha\beta} = 0,$$

respectively. Thus both factors are simultaneously minimized when $\tau_1 = \tau_2$, and then $\tau_1 = \tau_2 = \frac{1}{\sqrt{\alpha\beta}}$.

Theorem 7.19 *Let $A = A_1 + A_2$, where A_1, A_2 are s.p.d., and consider the Peaceman-Rachford ADI method (7.47) to solve $Ax = b$ with $\tau_1 = \tau_2 = 1/\sqrt{\alpha\beta}$. The spectral radius of the corresponding iteration matrix M satisfies*

$$\rho(M) \le \min_{\tau_1, \tau_2} \mu(\tau_1, \tau_2) \le \left(\frac{1 - \sqrt{\frac{\alpha}{\beta}}}{1 + \sqrt{\frac{\alpha}{\beta}}}\right)^2 \sim 1 - 4\sqrt{\frac{\alpha}{\beta}},$$

if $\frac{\alpha}{\beta} \to 0$.

Proof For $\tau_1 = \tau_2 = \frac{1}{\sqrt{\alpha\beta}}$ we have

$$\mu(\tau_1, \tau_2) = \left(\frac{1 - \sqrt{\frac{\alpha}{\beta}}}{1 + \sqrt{\frac{\alpha}{\beta}}}\right)^2. \qquad \diamond$$

Remark 7.20 For a model difference equation for a second-order elliptic differential equation problem on a square with side π, we have with stepsize h,

$$\alpha = \left(\frac{\sin(\frac{h}{2})}{\frac{h}{2}}\right)^2 \sim 1,$$

$$\beta = \left(\frac{\cos(\frac{h}{2})}{\frac{h}{2}}\right)^2 \sim \frac{4}{h^2}, \quad h \to 0.$$

Then,

$$\mu(\tau_1, \tau_2) = \left(\frac{1 - \tan(\frac{h}{2})}{1 + \tan(\frac{h}{2})}\right)^2 = \frac{1 - \sin(h)}{1 + \sin(h)} \sim 1 - 2h, \quad h \to 0.$$

Note that this is just the convergence factor we get for the SOR method with an optimal overrelaxation parameter.

Since $\rho(M) \le \mu(\tau_1, \tau_2)$, this means that the ADI method with parameters (chosen as above) converges at least as fast as the SOR method. Note, however, that in the ADI-method we must solve two systems of equations with tridiagonal coefficient matrices $(I - \tau A_i)$, on each step, while the point-wise SOR method requires no solution of such systems.

7.7.4 The Commutative Case

Assume now that A_1, A_2 commute. Then, as we have seen, M is symmetric and has real eigenvalues, and we can apply the Chebyshev acceleration method. The eigenvalues of the corresponding preconditioned matrix $\widetilde{C}$ are related to the eigenvalues of M by

$$\lambda(\widetilde{C}) = 1 - \lambda(M).$$

Since $-\rho(M) \leq \lambda(M) \leq \rho(M)$, where

$$\rho(M) = \left(\frac{1 - \sqrt{\alpha/\beta}}{1 + \sqrt{\alpha/\beta}}\right)^2 \sim 1 - 4\sqrt{\frac{\alpha}{\beta}},$$

we have

$$4\sqrt{\frac{\alpha}{\beta}} \sim 1 - \rho(M) \leq \lambda(\tilde{C}) = 1 + \rho(M) \sim 2 - 4\sqrt{\frac{\alpha}{\beta}} \sim 2, \quad \alpha/\beta \to 0.$$

The asymptotic rate of convergence of the Chebyshev accelerated method therefore is

$$2\sqrt{\frac{1 - \rho(M)}{1 + \rho(M)}} \sim 2\sqrt{2}\left(\frac{\alpha}{\beta}\right)^{\frac{1}{4}}, \quad \alpha/\beta \to 0.$$

For the model difference equation, we have the asymptotic rate of convergence

$$\sim 2h^{\frac{1}{2}}, h \to 0.$$

7.7.5 The Cyclically Repeated ADI Method

The real power of the ADI method is brought forth when we use a sequence of parameters τ_l. Assume that A_1, A_2 commute; then we choose the parameters τ_l cyclically. With a cycle of q parameters and with the assumption

$$0 < \alpha \leq \lambda_i^{(j)} \leq \beta, \quad j = 1, 2,$$

we get the iteration matrix

$$M^{(q)} = \prod_{p=1}^{q} (I + \tau_p A_2)^{-1}(I - \tau_p A_1)(I + \tau_p A_1)^{-1}(I - \tau_p A_2).$$

The eigenvalues of $M^{(q)}$ are

$$\prod_{p=1}^{q} \frac{1 - \tau_p \lambda_i^{(1)}}{1 + \tau_p \lambda_i^{(1)}} \cdot \frac{1 - \tau_p \lambda_i^{(2)}}{1 + \tau_p \lambda_i^{(2)}}.$$

In the same way as above, $\rho(M^{(q)})$ is minimized when

$$d(\alpha, \beta, q) = \max_{\alpha \leq x \leq \beta} \prod_{p=1}^{q} \left|\frac{1 - \tau_p x}{1 + \tau_p x}\right|$$

is minimized. It can be shown that this mini-max problem has a unique solution $\tau_p^*(\alpha, \beta, q)$. If $q = 2^v$, $v \geq 0$, the following relations hold for the optimal parameters,

$$d(\alpha, \beta, 2q) = d(\sqrt{\alpha\beta}, \frac{\alpha + \beta}{2}, q),$$

$$\tau_p^*(\alpha, \beta, 2q) = 1/[\sqrt{\alpha\beta} \pm \sqrt{\tau_p^*(\sqrt{\alpha\beta}, \frac{\alpha + \beta}{2}, q) - \sqrt{\alpha\beta}}], \quad p = 1, 2, \ldots, q.$$

For instance, starting with

$$\tau_1^*(\alpha, \beta, 1) = \sqrt{\alpha\beta}$$

and

$$d(\alpha, \beta, 1) = \frac{1 - \sqrt{\alpha/\beta}}{1 + \sqrt{\alpha/\beta}},$$

we find for the model difference equation on a square domain with sides of length π,

$$d(\alpha, \beta, 1) = d(1, \frac{4}{h^2}, 1) \sim 1 - h, \quad h \to 0$$

$$d(\alpha, \beta, 2) \sim d(\frac{2}{h}, \frac{2}{h^2}, 1) \sim 1 - 2h^{\frac{1}{2}}, \quad h \to 0$$

$$d(\alpha, \beta, 4) \sim d(\frac{2}{h^{3/2}}, \frac{1}{h^2}, 2) \sim 1 - 2\sqrt{2}h^{\frac{1}{4}}, \quad h \to 0$$

etcetera. The rate of convergence thus increases very fast with increasing q. It can be shown that if

$$c_i = \sqrt{c_{i-1}b_{i-1}}, \quad b_i = \frac{c_{i-1} + b_{i-1}}{2}, \quad i = 1, 2, \ldots$$

and

$$0 < \alpha = c_0 \leq b_0 = \beta$$

then

$$c_{i-1} \leq c_i \leq b_i \leq b_{i-1}$$

and

$$\lim_{i \to \infty} c_i = \lim_{i \to \infty} b_i = \frac{\pi}{2} \beta / \left[\int_0^{\pi/2} \frac{1}{\sqrt{1 - (1 - \frac{\alpha}{\beta})^2 \sin^2 x}} dx \right].$$

A relevant discussion can be found in Wachspress (1966) and Varga (1962). There are also some simple and explicit formulas for choosing a sufficiently good parameter set $\{\tau_p\}$. Two of these can be found in G. Birkhoff, Varga, and Young (1962); see also Rice and de Boor (1963).

7.7.6 Historical Remarks

The idea of an incomplete factorization method goes back to early papers by Buleev (1960), Varga (1962), Oliphant (1962), Dupont, Kendall, and Rachford (1968), Dupont (1968), and Woźnicki (1973), where it was presented for matrices of a type arising from difference approximations of elliptic problems. The first more general form (unmodified methods for point-wise matrices) was studied for M-matrices by Meijerink and van der Vorst (1977). For a review and general formalism for describing such methods, see Axelsson (1977, 1985), Beauwens (1979), and Il'in (1988). For a similar but more involved type of methods for difference matrices, which allowed for variable parameters from one iteration to the next, see Stone (1968).

A modified form of the method, where a certain row sum criterion was imposed, was studied by Gustafsson (1978, 1979). Actually, as is readily seen, the method of Dupont, Kendall, and Rachford (1968) and as further discussed in Axelsson (1972), using a perturbation technique, can be seen as a modified version of the general incomplete factorization method when applied to the five-point elliptic difference matrices, assuming that no fill-in is accepted outside the sparsity structure of A itself and assuming a natural ordering of the grid points. The advantage of modified versions is that they can give condition numbers of the iteration matrices that are of an order of magnitude smaller than for the original matrix.

The incomplete factorization method can be readily generalized to matrices partitioned in block matrix form. This was done first for matrices partitioned in block tridiagonal form in Axelsson, Brinkkemper, and Il'in (1984) and Concus, Golub, and Meurant (1985), the latter being based on earlier work by Underwood (1975). A general form was presented in Axelsson (1986) and Beauwens and Ben Bouzid (1987), where existence of the methods was proved for M-matrices.

The existence—that is, the existence of nonzero pivot entries—of point-wise incomplete factorization methods for M-matrices was first shown by Meijerink and van der Vorst (1977) and, for point-wise H-matrices, by Varga, Saff, and Mehrman (1980). The existence of incomplete factorization methods for M-matrices in block form was shown in Axelsson, Brinkkemper, and Il'in (1984) and Concus, Golub, and Meurant (1985) for block tridiagonal matrices; in Axelsson (1986) and Beauwens and Ben Bouzid (1987) for general block matrices; and in Axelsson and Polman (1986) for relaxed versions of such methods.

For square block H-matrices, Polman (1987) shows the existence of an incomplete block-matrix factorization method (i.e., approximating only the inverses of pivot blocks and with no relaxation). Kolotilina (1989) shows the existence of convergent splittings for block H-matrices, and Axelsson (1991) shows the existence of general incomplete factorizations for block H-matrices.

Exercises

1. Let $A = [a_{ij}]$ be a real symmetric matrix of order n. Establish the identity,

$$\mathbf{x}^T A \mathbf{x} = \sum_{i=1}^{n} (a_{i,i} + \sum_{j \neq i} a_{ij}) x_i^2 + \sum_{j > i} (-a_{ij})(x_i - x_j)^2.$$

2. Let $A = [a_{ij}]$ be a real symmetric matrix of order n. Show that for each $\mathbf{x} \in \mathbb{R}^n$:
   (a) $\mathbf{x}^T A \mathbf{x} = \sum_i a_{i,i} x_i^2 + \sum_i \sum_{j \neq i} a_{ij} x_i x_j.$
   (b) Let $d_1, d_2, \ldots, d_n$ be positive numbers. Show that

$$\frac{d_i}{d_j} |a_{ij}| \left(x_i + \frac{d_j}{d_i} \frac{a_{ij}}{|a_{ij}|} x_j \right)^2 = 2 a_{ij} x_i x_j + \frac{d_i}{d_j} |a_{ij}| x_i^2 + \frac{d_j}{d_i} |a_{ij}| x_j^2$$

and

$$\sum_i \sum_{j \neq i} \frac{d_i}{d_j} |a_{ij}| \left(x_i + \frac{d_j}{d_i} \frac{a_{ij}}{|a_{ij}|} x_j \right)^2 = 2 \sum_i \sum_{j \neq i} a_{ij} x_i x_j$$

$$+ 2 \sum_i \sum_{j \neq i} \frac{d_i}{d_j} |a_{ij}| x_i^2.$$

Note: Here, the convention is that $a_{ij}/|a_{ij}| = 0$ if $a_{ij} = 0$.

(c) Show that

$$\mathbf{x}^T A \mathbf{x} = \sum_i \left(a_{i,i} - \sum_{j \neq i} \frac{d_i}{d_j} |a_{ij}| \right) x_i^2$$

$$+ \frac{1}{2} \sum_i \sum_{j \neq i} \frac{d_i}{d_j} |a_{ij}| \left(x_i + \frac{d_j}{d_i} \frac{a_{ij}}{|a_{ij}|} x_j \right)^2.$$

3. Let $A = [a_{ij}]$ be a real symmetric matrix of order n.

(a) Show that A is positive definite if there exist positive numbers d_i, $1 \leq i \leq n$, such that

$$a_{i,i} - \sum_{j \neq i} \frac{d_i}{d_j} |a_{ij}| > 0, \qquad 1 \leq i \leq n.$$

(b) Let A be irreducible and let

$$a_{i,i} - \sum_{j \neq i} \frac{d_i}{d_j} |a_{ij}| \geq 0, \qquad 1 \leq i \leq n,$$

where strong inequality holds for at least one index i. Show that A is positive definite.

4. Show that the Chebyshev acceleration method, with the SOR matrix $\mathcal{L}_\omega$ as iteration matrix with optimal parameter $\omega = \omega^*$, yields no improvement.

Hint: The eigenvalues of $\mathcal{L}_\omega$ are on a circle of radius $\omega^* - 1$. Hence, the semiaxes a and b satisfy $a = b$ and the Chebyshev method parameter becomes $\alpha_l = 1$, which implies that no gain can be realized by acceleration.

5. Let A be a $n \times n$ symmetric positive definite matrix with the block tridiagonal form

$$A = \begin{bmatrix} D_1 & L_2^T & & \\ L_2 & D_2 & L_3^T & \\ & \ddots & \ddots & \ddots \\ & & L_m & D_m \end{bmatrix}.$$

The block-SSOR preconditioning matrix is defined to be

$$C = \begin{bmatrix} \frac{1}{\omega_1} D_1 & & & \\ L_2 & \frac{1}{\omega_2} D_2 & & \\ & \ddots & \ddots & \\ & & L_m & \frac{1}{\omega_m} D_m \end{bmatrix}$$

$$\times \begin{bmatrix} \left(\frac{1}{\omega_1} D_1 \right)^{-1} & & & \\ & \left(\frac{1}{\omega_2} D_2 \right)^{-1} & & \\ & & \ddots & \\ & & & \left(\frac{1}{\omega_m} D_m \right)^{-1} \end{bmatrix}$$

$$\times \begin{bmatrix} \frac{1}{\omega_1} D_1 & L_2^T & & \\ & \frac{1}{\omega_2} D_2 & L_3^T & \\ & & \ddots & \ddots \\ & & & \frac{1}{\omega_m} D_m \end{bmatrix},$$

where $0 < \omega_1 \leq 1; 0 < \omega_i < 2, i = 2, 3, \ldots, m$.

(a)   Following the proof of Theorem 7.16, show that

$$(2 - \max_{1 \leq i \leq m} \omega_i)^{-1} \mathbf{x}^T C \mathbf{x} \geq \mathbf{x}^T (\widehat{D} + V) \widehat{D}^{-1} (\widehat{D} + V)^T \mathbf{x}$$

$$\geq \mathbf{x}^T A \mathbf{x} \ \forall \mathbf{x} \in \mathbb{R}^n, \ \mathbf{x} = [\mathbf{x}_1^T, \ldots, \mathbf{x}_m^T]^T$$

(**x** is assumed to be partitioned consistently with the blocks of A), where

$$\widehat{D} = \begin{bmatrix} \left(\frac{2}{\omega_1} - 1 \right) D_1 & & & \\ & \left(\frac{2}{\omega_2} - 1 \right) D_2 & & \\ & & \ddots & \\ & & & \left(\frac{2}{\omega_m} - 1 \right) D_m \end{bmatrix},$$

$$V = \begin{bmatrix} \left(1 - \frac{1}{\omega_1} \right) D_1 & & & \\ L_2 & \left(1 - \frac{1}{\omega_2} \right) D_2 & & \\ & \ddots & \ddots & \\ & & L_m & \left(1 - \frac{1}{\omega_m} \right) D_m \end{bmatrix}.$$

(b) Show that $C = A + \tilde{D}$, where

$$\tilde{D} = \text{diag}(\tilde{D}_1, \tilde{D}_2, \ldots, \tilde{D}_m),$$

$$\tilde{D}_1 = (1/\omega_1 - 1)D_1,$$

$$\tilde{D}_i = \frac{1}{4\omega_i}(2 - \omega_i)^2 D_i + \omega_i \left[L_i \left(\frac{\omega_i}{\omega_{i-1}} D_{i-1} \right)^{-1} L_i^T - \frac{1}{4} D_i \right],$$

$$i = 2, 3, \ldots, m$$

(c) Show that the spectral condition number $\kappa(\tilde{A})$ for the preconditioning matrix C, that is, of $C^{-1}A$, satisfies

$$\kappa(\tilde{A}) < (2 - \max_{1 \le i \le m} \omega_i)^{-1}(1 + \gamma\mu + \delta),$$

where

$$\gamma = \max_{2 \le i \le m} [(2 - \omega_i)^2 / 4\omega_i],$$

$$\mu = \max_{\mathbf{x} \ne 0} \{ [\gamma^{-1}(1/\omega_1 - 1)\mathbf{x}_1^T D_1 \mathbf{x}_1 + \mathbf{x}^T D\mathbf{x}] / \mathbf{x}^T A\mathbf{x} \},$$

$$\delta = \max_{\mathbf{x} \ne 0} \frac{\sum_{i=2}^m \omega_i \mathbf{x}_i^T \{ L_i [\omega_i/\omega_{i-1})D_{i-1}]^{-1} L_i^T - \frac{1}{4}D_i \} \mathbf{x}_i}{\mathbf{x}^T A\mathbf{x}},$$

$$D = \text{diag}(D_1, D_2, \ldots, D_m).$$

6. Consider the standard five-point difference matrix with nonzero coefficients $-1, \ldots - 1, 4, -1, \ldots, -1$, on a rectangular domain of meshpoints, where a Neumann boundary condition is imposed on the left edge (the corresponding matrix coefficients on the meshpoints of this edge are $-1/2 \ldots, 0, 2, -1, \ldots, -1/2$). Let the nodes be ordered column-wise, starting at the left edge, where the Neumann boundary condition is imposed. The stiffness matrix A can be partitioned in the block tridiagonal form shown in the preceding exercise, m being the number of columns. We consider the application of block-SSOR preconditioning, where for some positive constant η independent of h

$$\omega_1 = 1/(1 + \eta h), \qquad \omega_i = 2/(1 + \eta h), \qquad i = 2, 3, \ldots, m.$$

(It is assumed that $\eta < h^{-1}$ for the coarsest mesh considered.) Show that $\delta \leq 0$, $\mu = O(h^{-2})$ and, hence, that

$$\kappa(\widetilde{A}) = O(h^{-1}), \qquad h \to 0.$$

Hint:

(a)   Note with regard to δ that

$$\mathbf{x}_2^T \{ L_2[(\omega_2/\omega_1)D_1]^{-1}L_2^T - \frac{1}{4}D_2 \}\mathbf{x}_2 = \mathbf{x}_2^T[\widehat{D}^{-1} - \frac{1}{4}\widehat{D}]\mathbf{x}_2$$

$$\leq (\xi_1^{-1} - \frac{1}{4}\xi_1)\mathbf{x}_2^T\mathbf{x}_2 \quad \forall \mathbf{x}_2,$$

where

$$\widehat{D} = \begin{bmatrix} 4 & -1 & & & \\ -1 & 4 & -1 & & \\ & \ddots & \ddots & \ddots & \\ & & & -1 & 4 \end{bmatrix}$$

and ξ_1 is the smallest eigenvalue of $\widehat{D}$, that is,

$$\xi_1 = 2 + [2\sin(\pi h/2)]^2.$$

(b)   For the analysis of μ, we let $\alpha_m = 2$ and $\alpha_{i-1} = 2 - \alpha_i^{-1}$ for $i = m, m-1, \ldots, 2$. [That is, $\alpha_{m-i} = (i+2)/(i+1)$.] Show that with $D_0 = \frac{1}{2}\widehat{D}$,

$$\mathbf{x}^T A\mathbf{x} = \sum_{i=1}^{m-1} \{ \alpha_{i+1}\mathbf{x}_{i+1}^T D_0\mathbf{x}_{i+1} - 2\mathbf{x}_i^T\mathbf{x}_{i+1} + (2 - \alpha_i)\mathbf{x}_i^T D_0\mathbf{x}_i \}$$

$$+ (\alpha_1 - 1)\mathbf{x}_1^T D_0\mathbf{x}_1.$$

Using the inequality

$$2\mathbf{x}_i^T\mathbf{x}_{i+1} \leq \alpha_{i+1}^{-1}\mathbf{x}_i^T\mathbf{x}_i + \alpha_{i+1}\mathbf{x}_{i+1}^T\mathbf{x}_{i+1}$$

and the positive definite property of $D_0 - I$, show that

$$\mathbf{x}^T A\mathbf{x} \geq (\alpha_1 - 1)\mathbf{x}_1^T D_0\mathbf{x}_1 = m^{-1}\mathbf{x}_1^T D_0\mathbf{x}_1.$$

Hence, defining

$$c(h) = \gamma^{-1}(\omega_1^{-1} - 1), \qquad \mu_0 = \max_{\mathbf{x} \neq 0}(\mathbf{x}^T D \mathbf{x}/\mathbf{x}^T A \mathbf{x}),$$

show that

$$\mu = \max_{\mathbf{x} \neq 0} \frac{c(h)\mathbf{x}_1^T D_1 \mathbf{x}_1 + \mathbf{x}^T D \mathbf{x}}{\mathbf{x}^T A \mathbf{x}} \leq \max_{\mathbf{x} \neq 0} \frac{c(h)\mathbf{x}_1^T D_1 \mathbf{x}_1 + \mathbf{x}^T D \mathbf{x}}{\frac{1}{2}m^{-1}\mathbf{x}_1^T D_0 \mathbf{x}_1 + \frac{1}{2}\mathbf{x}^T A \mathbf{x}}$$

$$\leq 2 \max\{c(h)m, \mu_0\} = O(h^{-2}).$$

7. (J.F. Maître) Let A be s.p.d. and assume that A can be partitioned in the two-by-two block form

$$A = \begin{bmatrix} D_1 & \widetilde{L}^T \\ \widetilde{L} & D_2 \end{bmatrix} \equiv D + L + L^T,$$

where $D = \text{diag}(D_1, D_2)$ and D_i, $i = 1, 2$ are diagonal matrices. Further, let

$$\widehat{A} = D^{-\frac{1}{2}} A D^{-\frac{1}{2}} = \begin{bmatrix} I_1 & F^T \\ F & I_2 \end{bmatrix},$$

where I_1, I_2 are identity matrices. Let

$$\widehat{L} = D^{-1}L, B = I - D^{-1}A[= D^{-1}(L + L^T)],$$

the Jacobi iteration matrix.

(a) Show that $\rho(B)^2 = \rho(F^T F)$.

(b) Using the notations in Theorem 7.17, show the following expressions for γ:

$$\gamma = \max_{\mathbf{y} \neq 0} \frac{\mathbf{y}^T \widehat{L} \widehat{L}^T \mathbf{y} - \frac{1}{4}\mathbf{y}^T \mathbf{y}}{\mathbf{y}^T \widehat{A} \mathbf{y}} =$$

$$= \max\{\lambda \in \mathbb{R}; \ \lambda \widehat{A} - \widehat{L} \widehat{L}^T + \frac{1}{4}I \text{ is singular}\},$$

$$\gamma = \max\{\lambda \in \mathbb{R}; \ (\lambda + \frac{1}{2})^2 F F^T - (\lambda + \frac{1}{4})^2 I_2 \text{ is singular}\},$$

and

$$\gamma = \frac{1}{2}\frac{\rho(B) - \frac{1}{2}}{1 - \rho(B)}.$$

8. Let A be the matrix for the standard five-point elliptic difference equation on a unit square, where a diagonal ordering of the nodepoints has been used and the matrix has been partitioned as in the previous exercise.

   (a) Show that $\rho(B) = \cos \pi h$, $\gamma = \frac{1}{8(\sin \frac{\pi h}{2})^2} - \frac{1}{2}$, $\delta = 2(\sin \frac{\pi h}{2})^2$, and that the value of the parameter ω which minimizes the condition number of the SSOR preconditioned matrix is $\omega^* = 1$.

   (b) Show that the corresponding condition number is

$$\kappa(C^{-1}A) \leq \left(\frac{1}{2 \sin \frac{\pi h}{2}}\right)^2 + \frac{1}{2}.$$

9. Show that every nonsingular triangular matrix is an H-matrix.
   Hint: Use Lemma 7.7 and choose the positive vector $\mathbf{x}$ component-wise (starting with $x_1 = 1$) to satisfy the generalized diagonal dominance.

10. Compute the incomplete block factorization matrix C of the matrix in Example 7.3(a), where A_{11}^{-1} is approximated by $\mathrm{diag}(\frac{1}{20}, 1)$. Compute $\|C^{-1}(A - C)\|$.

11. Consider the matrix in Example 7.3a. Show that it is not a block H-matrix for the partitioning in blocks where A_{11} has order 1×1 and A_{22} has order 3×3.

12. Let A be a nonsingular matrix with a triangular matrix decomposition $A = LU$. Show that

$$|A^{-1}| \leq \mathcal{M}(U)^{-1}\mathcal{M}(L)^{-1}.$$

Here $|B|$, where $B = [b_{ij}]$ denotes the matrix, $B = [|b_{ij}|]$.
   Hint: Use Ostrowski's theorem (Lemma 8.14). Since A is nonsingular, L and U are nonsingular. Using Exercise 9 and Ostrovski's theorem, show

$$|A^{-1}| = |U^{-1}L^{-1}| \leq \mathcal{M}(U)^{-1}\mathcal{M}(L)^{-1}.$$

13. Consider the point-wise incomplete factorization method of a s.p.d. matrix A.

   (a) Show that deletion of fill-in entry $a_{ij}^{(r+1)}$ at stage r in position (i, j), followed by the modification step adding this entry to the diagonal in the same row, is equivalent to the addition to A of the matrix $a_{ij}^{(r+1)} R_{(i,j)}^{(r+1)}$, where

$$
R^{(r+1)}_{(i,j)} = \begin{array}{c} \\ \\ (i) \\ \\ \\ (j) \\ \\ \\ \end{array} \begin{bmatrix} 0 & & & & & & & \\ & \ddots & & & & & & \\ & & 0 & & & & & \\ & & & 1 & \cdots & -1 & & \\ & & & \vdots & & \vdots & & \\ & & & -1 & \cdots & 1 & & \\ & & & & & & 0 & \\ & & & & & & & \ddots \\ & & & & & & & & 0 \end{bmatrix}
$$

(with column labels (i) and (j) over the indicated columns)

where all entries not indicated are zero. Let LU be the incomplete factorization. Show that

$$LU = A + R,$$

where $R = \sum_r \sum_{i,j} a^{(r+1)}_{ij} R^{(r+1)}_{(i,j)}$, the summation takes place over all deleted fill-in entries.

(b) Assume, in addition, that A is a M-matrix. Show that, in this case, all fill-in entries are negative and that the matrix R is negative semidefinite—that is, the eigenvalues of $(LU)^{-1}A$ are bounded below by the value 1.

(c) Consider the following alternate modification method. (Here we do not assume that A is a M-matrix.) If the deleted fill-in entry $a^{(r+1)}_{ij} > 0$, then use the modification step as in (a). However, if $a^{(r+1)}_{ij} < 0$, then use a modification step that corresponds to the addition to A of $-a^{(r+1)}_{ij} Q^{(r+1)}_{(i,j)}$, where

$$
Q^{(r+1)}_{(i,j)} = \begin{array}{c} \\ \\ (i) \\ \\ \\ (j) \\ (j+1) \\ \\ \end{array} \begin{bmatrix} 0 & & & & & & & \\ & \ddots & & & & & & \\ & & 0 & & & & & \\ & & & 1 & \cdots & 1 & -2 & \\ & & & \vdots & & \vdots & \vdots & \\ & & & 1 & \cdots & 1 & -2 & \\ & & & -2 & \cdots & -2 & 4 & \\ & & & & & & & 0 \\ & & & & & & & & \ddots \\ & & & & & & & & & 0 \end{bmatrix}
$$

(with column labels (i), (j), $(j+1)$ over the indicated columns)

Show that $Q^{(r+1)}_{(i,j)}$ is positive semidefinite. Show, in addition, that the eigenvalues of the matrix $\begin{bmatrix} 1 & 1 & -2 \\ 1 & 1 & -2 \\ -2 & -2 & 4 \end{bmatrix}$ are 0, 0, and 6.

(d)   Show that the corresponding LU-factorization matrix satisfies

$$LU = A + Q,$$

where Q is positive semidefinite—that is, the eigenvalues of $(LU)^{-1}A$ are bounded above by the value one, in this case.

(e)   Indicate why the entries in LU may grow rapidly from one stage to the next and why, therefore, the fill-in entries may also grow, when the modification method in (c) is used. However, this method gives always a positive definite matrix LU, even when A is not a M-matrix which does not hold for the method in (a). Note also that the row sums of Q are zero, so $LUe = Ae$ for both methods.

14. Consider the matrix A in Example 7.1. Using step (7.5) of the relaxed incomplete factorization method to compute the matrix C, analyze the condition number of $C^{-1}A$ (possibly using estimates to be presented in Chapter 10).

References

O. Axelsson (1972) A generalized SSOR method. *BIT* **12**, 443–467.

O. Axelsson (1974). On preconditioning and convergence acceleration in sparse matrix problems. CERN Technical Report 74-10, Data Handling Division, Geneva.

O. Axelsson (1977). Solution of linear systems of equations: iterative methods. In *Sparse Matrix Techniques, Lecture Notes in Mathematics #572* (ed. V. A. Barker). Berlin, Heidelberg, New York: Springer-Verlag, pp. 1–50.

O. Axelsson (1985). A survey of preconditioned iterative methods for linear systems of algebraic equations. *BIT* **25**, 166–187.

O. Axelsson (1985). Incomplete block matrix factorization preconditioning methods. The ultimate answer? *J. Comp. Appl. Math.* **12, 13**, 3–18.

O. Axelsson (1986). A general incomplete block-matrix factorization method. *Lin. Alg. Appl.* **74**, 179–190.

O. Axelsson (1991). Preconditioning methods for block *H*-matrices. In *Computer Algorithms for Solving Linear Systems* (ed. E. Spedicato), pp. 169–184. NATO ASI Series vol 77, Berlin, Heidelberg: Springer-Verlag.

O. Axelsson and V. A. Barker (1984). *Finite Element Solution of Boundary Value Problems. Theory and Computation.* Orlando, Fla.: Academic.

O. Axelsson, S. Brinkkemper, and V. P. Il'in (1984). On some versions of incomplete blockmatrix factorization iterative methods. *Lin. Alg. Appl.* **58**, 3–15.

O. Axelsson and G. Lindskog (1986). On the eigenvalue distribution of a class of preconditioning methods. *Numer. Math.* **48**, 479–498.

O. Axelsson and N. Munksgaard (1983). Analysis of incomplete factorizations with fixed storage allocation, in *Preconditioning Methods, Theory and Applications* (ed. D. J. Evans), 219–241. London: Gordon and Breach.

O. Axelsson and B. Polman (1986). On approximate factorization methods for block matrices suitable for vector and parallel processors. *Lin. Alg. Appl.* **77**, 3–26.

R. Beauwens (1979). Factorization iterative methods, M-operators and H-operators. *Numer. Math.* **31**, 335–357.

R. Beauwens and M. Bouzid (1987). On sparse block factorization, iterative methods. *SIAM J. Numer. Anal.* **24**, 1066–1076.

A. Berman and R. J. Plemmons (1979). *Nonnegative Matrices in the Mathematical Sciences.* Orlando, Fla.: Academic.

G. Birkhoff, R. S. Varga, and D. M. Young (1962). Alternating direction implicit methods. In *Advances in Computers*, ed. F. Alt and M. Rubinoff vol. 3, 189–273. Orlando, Fla.: Academic.

N. I. Buleev (1960). A numerical method for the solution of two-dimensional and three-dimensional equations of diffusion. *Math. Sb.* **51**, 227–238. [English transl.: Rep. BNL-TR-551, Brookhaven National Laboratory, Upton, New York, 1973.]

T. F. Chan and P. S. Vassilevski (1992). A framework for block ILU factorizations using block-size reduction. *CAM Report 92-29*, Department of Mathematics, UCLA.

C. Concus, G. H. Golub, and G. Meurant (1985). Block preconditioning for the conjugate gradient method. *SIAM J. Sci. Stat. Comp.* **6**, 220–152.

T. Dupont, R. P. Kendall, and H. H. Rachford, Jr. (1968). An approximate factorization procedure for solving self-adjoint elliptic difference equations. *SIAM J. Numer. Anal.* **5**, 554–573.

T. Dupont (1968). A factorization procedure for the solving of elliptic difference equations. *SIAM J. Numer. Anal.* **5**, 753–782.

D. J. Evans (1968), The use of pre-conditioning in iterative methods for solving linear equations with symmetric positive definite matrices. *J. Inst. Math. Applic.* **4**, 295–314.

M. Fiedler and V. Ptak (1962). On matrices with nonpositive off-diagonal elements and positive principal minors. *Czech. Math. J.* **12**, 382–400.

D. J. Feingold and R. S. Varga (1962). Block diagonally dominant matrices and generalizations of the Gershgorin circle theorem. *Pacific J. Math.* **12**, 1241–1250.

I. Gustafsson (1978). A class of first-order factorization methods. *BIT* **18**, 142–156.

I. Gustafsson (1979). Stability and rate of convergence of modified incomplete Cholesky factorization methods. Report 79.02R, Department of Computer Sciences, Chalmers University of Technology, Goteborg, Sweden.

W. W. Hager (1984). Condition estimates. *SIAM J. Sci. Stat. Comp.* **5**, 311–316.

V. P. Il'in (1988). Incomplete factorization methods. *Sov. J. Numer. Anal. Math. Modelling* **3**, 179–198.

L. Yu. Kolotilina (1989). On approximate inverses of block *H*-matrices, in *Numerical Analysis and Mathematical Modelling.* Moscow (in Russian).

L. Yu. Kolotilina and A. Yu. Yeremin (1986). On a family of two-level preconditionings of the incomplete block factorization type. *Sov. J. Numer. Anal. Math., Modelling* **1**, 293–320.

J. A. Meyerink and H. A. van der Vorst (1977). An iterative solution method for linear systems of which the coefficient matrix is a symmetric *M*-matrix. *Math. Comp.* **31**, 148–162.

T. A. Oliphant (1962). An extrapolation process for solving linear systems. *Quart. Appl. Math.* **20**, 257–267.

O. Østerby and Z. Zlatev (1983). Direct methods for sparse matrices, *in Lecture Notes in Computer Science # 157.* Berlin, Heidelberg, New York: Springer-Verlag.

A. M. Ostrowski (1937). Über die Determinanten mit überwegender Hauptdiagonale. *Comm. Math. Helv.* **10**, 69–96.

A. M. Ostrowski (1956). Determinanten mit überwiegender Hauptdiagonale und die absolute Konvergenz von linearen Iterationsprozessen. *Comm. Math. Helv.* **30**, 175–210.

A. M. Ostrowski (1961). On some metrical properties of operator matrices and matrices partitioned into blocks. *J. Math. Anal.* **2**, 161–209.

D. W. Peaceman and H. H. Rachford, Jr. (1955). The numerical solution of parabolic and elliptic differential equations. *J. Soc. Ind. Appl. Math.* **3**, 28–41.

B. Polman (1987). Incomplete blockwise factorizations of (block) *H*-matrices. *Lin. Alg. Appl.* **90**, 119–132.

J. Rice and C. de Boor (1963). Tchebycheff apoproximation by $\pi[(x - r_j)/(x + r_j)]$ and application to ADI iteration. *J. Soc. Ind. Appl. Math.* **11**, 159–169.

F. Robert (1969). Blocs *H*-matrices et convergence des méthodes iteratives classiques par blocs. *Lin. Alg. Appl.* **2**, 223–265.

J. Sheldon (1955). On the numerical solution of elliptic difference equations. *Math. Tables Aids Comp.* **9**, 101–113.

H. S. Stone (1968). Iterative solution of implicit approximations of multidimensional partial differential equations. *SIAM J. Numer. Anal.* **5**, 530–558.

R. Underwood (1975). An iterative block Lanczos method for the solution of large sparse symmetric eigenproblems. Ph.D. dissertation, Stanford University Computer Science Dept, Report STAN-CS-75-496, Stanford, Calif.

R. S. Varga (1960). Factorizations and normalized iterative methods. In *Boundary Problems in Differential Equations,* ed. R. E. Langer, pp. 121–142. Madison, Wis.: Univ. of Wisconsin Press.

R. S. Varga (1962). *Matrix Iterative Analysis.* Englewood Cliffs, N.J.: Prentice Hall.

R. S. Varga, E. B. Saff, and V. Mehrman (1980). Incomplete factorizations of matrices and connections with *H*-matrices. *SIAM J. Numer. Anal.* **17**, 787–793.

E. L. Wachspress (1966). *Iterative Solution of Elliptic Systems and Applications to the Neutron Diffusion Equations of Reactor Physics.* Englewood Cliffs, N.J.: Prentice Hall.

Z. Woźnički (1973). Two-sweep iterative methods for solving large linear systems and their application to the numerical solution of multi-group multi-dimensional Newton diffusion equation. Report No. 1447/Cytronet/PM/A Dissertation, Institute of Nuclear Research, Warszawa.

Z. Zlatev (1991). *Computational Methods for General Sparse Matrices.* Dordrecht, Boston, London: Kluwer Acad. Pub.

8

Approximate Matrix Inverses and Corresponding Preconditioning Methods

Consider two classes of methods for the construction of preconditioners for a matrix A: (1) *explicit methods*, and (2) *implicit methods*. With explicit methods, we compute an approximation G on explicit form of the inverse A^{-1} of a given nonsingular matrix A, and we use this to form a preconditioning matrix on an explicit form. That is, to solve $A\mathbf{x} = \mathbf{b}$, we consider the system

$$(8.1) \qquad\qquad GA\mathbf{x} = \tilde{\mathbf{b}},$$

where $\tilde{\mathbf{b}} = G\mathbf{b}$, and solve (8.1) by iteration. Since both G and A are explicitly available, each iteration requires only matrix vector multiplications. Alternatively, we can use an explicit right-hand side approximation and solve

$$(8.2) \qquad\qquad AG\mathbf{y} = \mathbf{b}$$

by iteration to yield the vector $\mathbf{y}$ and, finally, compute

$$\mathbf{x} = G\mathbf{y}.$$

In the following, we address only the case of left-hand side preconditioners, but the methods to be presented are equally well applicable for right-hand side preconditioners.

In implicit methods we typically first compute an approximate factorization of A, such as the incomplete LU factorization discussed in the previous chapter, and use this in an implicit way. Namely, given a vector $\mathbf{x}$, each application of this preconditioner requires solving a linear system such as

$$(8.3) \qquad\qquad LU\mathbf{d} = \mathbf{r}, \quad \text{where } \mathbf{r} = A\mathbf{x} - \mathbf{b},$$

for some vector $\mathbf{d}$ which occurs in the iterative method. As we know, (8.3) requires a forward and a back solution. Here the preconditioned matrix is $C^{-1}A$, where $C = LU$.

The matrix G in (8.1) or (8.2) can be viewed as an approximate inverse. An explicit method gives an approximate inverse directly, while an implicit method requires first the factorization of the matrix, which is used to compute approximations $\mathbf{d} = (LU)^{-1}\mathbf{r}$ of the action $A^{-1}\mathbf{r}$ of the inverse in an iterative method, for instance. Hence, in an implicit method, we do not form the approximate inverse explicitly. As we shall see, however, it can also be used to compute an explicit approximate inverse.

One can also form a hybrid method—i.e., combine the two methods, for example, in a two-stage way, where an incomplete factorization (implicit) method like that presented in the previous chapter is used for a matrix partitioned in blocks and a direct method is used to approximate the inverse of the pivot blocks that occur at each stage of the factorization.

We present various methods of computing approximate inverses of matrices. Also discussed is how matrices not known explicitly, but where only the action of the matrix times vectors is available, can be approximated.

A common structure of approximate inverses is the band matrix structure. The rates with which entries of the given matrix decay as their distance to the diagonal increases, is of fundamental importance for the accuracy of such approximations, and this topic is discussed in the final section.

8.1 Two Methods of Computing Approximate Inverses of Block Bandmatrices

To illustrate explicit and implicit approximation methods, we first consider the computation of an approximate inverse of a band matrix. Consider a matrix $B = [b_{ij}]$, where $b_{ij} = 0$ if $j < i - p$ or $j > i + q$. Hence, B is a band matrix with left semi-bandwidth p and right semi-bandwidth q. In fact, we shall consider the more general case, where the entries of B are block matrices. We shall assume that the diagonal blocks $B_{i,i}$ are regular and later make additional assumptions as required. B is said to have left block semi-bandwidth p and right block semi-bandwidth q if $B_{ij} = 0$ if $j < i - p$ or $j > i + q$.

8.1.1 An Explicit Approximation Method

We shall consider a method of computing an approximate inverse G of a band matrix B, partitioned in block matrix form. Let $G = [G_{i,j}]$ be the corresponding partitioning of the approximate inverse of B to be computed and assume that G has block semi-bandwidths p_1 and q_1, respectively, where $p_1 \geq p$ and $q_1 \geq q$. To compute $G_{i,j}$, $i - p_1 \leq j \leq i + q_1$, we let

$$(8.4) \qquad (GB)_{i,j} = \sum_{k=i-p_1}^{i+q_1} G_{i,k} B_{k,j} = \Delta_{i,j}, \quad i - p_1 \leq j \leq i + q_1,$$

where

$$\Delta_{i,j} = \begin{cases} I_i, & \text{for } i = j \\ 0, & \text{otherwise} . \end{cases}$$

Here I_i denotes the identity matrix of the same order as $B_{i,i}$. (8.4) gives $p_1 + q_1 + 1$ block matrix equations that can be used to compute the same number of blocks $G_{i,j}$. However, since $B_{k,j} = 0$ if $k < j - q$ or $k > j + p$, we get

$$(8.5) \qquad \sum_{k=\max\{i-p_1,j-q\}}^{\min\{i+q_1,j+p\}} G_{i,k} B_{k,j} = \Delta_{i,j}, \quad j = i - p_1, \; i - p_1 + 1, \ldots, i + q_1.$$

(Similar, but fewer, equations are valid for $i < 2p_1 + 1$ and $i > m - q_1$.) This shows that if $p_1 + 1 \leq i \leq m - q_1$, where m is the order of B we get,

$$\text{for } j = i - p_1 : \qquad \sum_{k=i-p_1}^{i-p_1+p} G_{i,k} B_{k,i-p_1} = 0,$$

$$\text{for } j = i - p_1 + 1 : \qquad \sum_{k=i-p_1+1}^{i-p_1+p+1} G_{i,k} B_{k,i-p_1+1} = 0,$$

$$\cdots$$

$$\text{for } j = i : \qquad \sum_{\max\{i-p_1,i-q\}}^{\min\{i+q_1,i+p\}} G_{i,k} B_{k,i} = I,$$

$$\cdots$$

$$\text{for } j = i + q_1 : \qquad \sum_{k=i+q_1-p}^{i+q_1} G_{i,k} B_{k,i+q_1} = 0.$$

These equations are solvable under some additional assumptions about B that will be given later. We now illustrate the method for the special, but important, case where $p_1 = p = q_1 = q = 1$, that is, the case where B and G are block triangular, where it suffices for solvability to assume that $B_{i,i}$ and the Schur complement matrix in (8.6) are nonsingular. Then (8.5) shows that

$$G_{i,i-1}B_{i-1,i-1} + G_{i,i}B_{i,i-1} = 0,$$

$$G_{i,i-1}B_{i-1,i} + G_{i,i}B_{i,i} + G_{i,i+1}B_{i+1,i} = I,$$

$$G_{i,i}B_{i,i+1} + G_{i,i+1}B_{i+1,i+1} = 0,$$

so

$$G_{i,i-1} = -G_{i,i}B_{i,i-1}(B_{i-1,i-1})^{-1}$$

$$G_{i,i+1} = -G_{i,i}B_{i,i+1}(B_{i+1,i+1})^{-1}$$

and

$$(8.6) \quad G_{i,i} = [B_{i,i} - B_{i,i-1}(B_{i-1,i-1})^{-1}B_{i-1,i} - B_{i,i+1}(B_{i+1,i+1})^{-1}B_{i+1,i}]^{-1},$$

where we must assume that $B_{i,i}$ are nonsingular and that the inverse of the last (Schur complement) matrix exists. (For instance, if B is a block H-matrix, we shall show later that the conditions for solvability are fulfilled. For $i = 1$ and for $i = n$, one of the terms in the bracket in (8.6) is absent.) It is readily seen that $G_{i,i-1}, G_{i,i}, G_{i,i+1}$ is the ith row of the exact inverse of

$$\widehat{B} = \begin{bmatrix} I & & & & & 0 \\ & \ddots & & & & \\ & & B_{i-1,i-1} & B_{i-1,i} & & \\ & & B_{i,i-1} & B_{i,i} & B_{i,i+1} & \\ & & & B_{i+1,i} & B_{i+1,i+1} & \\ & & & & & \ddots & \\ 0 & & & & & I \end{bmatrix}.$$

This also follows, in fact, from a more general result, to be presented in Remark 8.9.

It is also interesting to note that the matrix in the bracket of (8.6) is the same as the matrix we get (in the ith row) when we make an odd-even reordering of the given block tridiagonal matrix and eliminate the decoupled odd and even blocks, respectively. (Compare with the odd-even reordering method discussed in Chapter 1, and see also Section 8.4.) Note further that G is not symmetric, in general, because the matrices $G_{i,i-1}$ and $G_{i-1,i}^{T}$ are not equal in general, even if B is symmetric. The block entries in every row of G can be computed concurrently.

Finally, it can be seen that GB contains a unit main diagonal and a sub- and super-diagonal at distance two from the main diagonal. Other entries are zero. Hence, it is sparse.

For a scalar tridiagonal matrix we find

$$g_{i,i} = \left(b_{i,i} - \frac{b_{i,i-1}b_{i-1,i}}{b_{i-1,i-1}} - \frac{b_{i,i+1}b_{i+1,i}}{b_{i+1,i+1}} \right)^{-1},$$

$$g_{i,i-1} = -\frac{g_{i,i}b_{i,i-1}}{b_{i-1,i-1}}, \quad g_{i,i+1} = -\frac{g_{i,i}b_{i,i+1}}{b_{i+1,i+1}}, \quad i = 1, 2, \ldots, n.$$

In Section 8.3, a symmetric version of an approximate inverse will be presented.

8.1.2 An Implicit Approximation Method

Consider again a block band matrix B. We shall now present an algorithm to compute the blocks of the exact inverse of B within a band consisting of $p_1 \geq p$ blocks to the left and $q_1 \geq q$ blocks to the right of the diagonal, in each block row. The algorithm was presented in Axelsson et al. (1984) and in Axelsson (1985), and is based on an idea in Takahishi et al. (1973); see also Erisman and Tinney (1975).

The algorithm requires that we factorize B first. (Therefore, this method is practically viable only when the bandwidths p, q of B are small and the entries B_{ij} are scalars or, if block-matrices, have small order.) If B is factored in the form

$$(8.7) \qquad\qquad B = (I - \widetilde{L})D^{-1}(I - \widetilde{U}),$$

where D is block-diagonal and $\widetilde{L}, \widetilde{U}$ are strictly lower and upper block triangular, respectively, then, as we shall see, we do not need to compute any (further) inverses of block-matrices. (Note, however, that the computation of the matrix D in (8.7) requires computations of inverses of pivot block matrices.) A second important advantage of the algorithm is that we need not compute any entries of the inverse outside the required band part of B^{-1}. The algorithm is based on the following identities:

Lemma 8.1 *Let* $B = LD^{-1}U$, *where* $L = I - \widetilde{L}$, $U = I - \widetilde{U}$, *and* $\widetilde{L}$ *and* $\widetilde{U}$ *are strictly lower and upper triangular, respectively. Then:*

(a) $B^{-1} = DL^{-1} + \widetilde{U}B^{-1}$.

(b) $B^{-1} = U^{-1}D + B^{-1}\widetilde{L}$.

Proof We have $B^{-1} = U^{-1}DL^{-1}$, so $(I - \tilde{U})B^{-1} = DL^{-1}$ and $B^{-1}(I - \tilde{L}) = U^{-1}D$, which is (a), respectively (b). $\diamond$

Relations (a) and (b) will be used for the computation of the upper and lower triangular parts of B^{-1}, respectively. Since L^{-1} is lower triangular, with identity matrix diagonal blocks, L^{-1} will not enter in the computation of the upper triangular part of B^{-1}. Correspondingly, U^{-1} will not enter in the computation of the lower triangular part.

The algorithm to compute the bandpart of B^{-1} with block lower and upper semibandwidths p_1 and q_1 therefore takes the following form. We assume that $B = [B_{ij}]$, has block order n, has block lower and upper semibandwidths p and q, and has been factored in the form (8.7).

Algorithm BBI (Band Block Inverse)
For $r = n, n - 1, \ldots, 1$, do

$$(8.8) \qquad (B^{-1})_{r,r} = D_{r,r} + \sum_{s=1}^{\min(q,n-r)} \tilde{U}_{r,r+s}(B^{-1})_{r+s,r}.$$

For $k = 1, 2, \ldots, q_1$, do

$$(8.8a) \qquad (B^{-1})_{r-k,r} = \sum_{s=1}^{\min(q,n-r+k)} \tilde{U}_{r-k,r-k+s}(B^{-1})_{r-k+s,r}.$$

For $k = 1, 2, \ldots, p_1$, do

$$(8.8b) \qquad (B^{-1})_{r,r-k} = \sum_{t=1}^{\min(p,n-r+k)} (B^{-1})_{r,r-k+t}\tilde{L}_{r-k+t,r-k}.$$

Note that only matrix-matrix multiplications occur. Further, if B is symmetric, then only one of equations (8.8a) and (8.8b) is needed.

Remark 8.2 (8.8) shows, in particular, that

$$(B^{-1})_{n,n} = D_{n,n},$$

where $D_{n,n}$ is the last block in D. This will turn out to be a useful property for methods to be presented in Chapter 10.

Remark 8.3 *Envelope sparsity pattern:* The BBI algorithm can readily be extended to the case where B has an envelope sparsity pattern—i.e., there exist p_i, q_i such that $B_{i,j} = 0$ for $i - j > p_i$ and for $j - i > q_i$. In particular, the method can be used to compute an envelope part of the inverse of a quasi-banded matrix, for which $B_{i,j} = 0$, $|i - j| > p$, but $B_{1,n} \neq 0$ and $B_{n,1} \neq 0$. (Such matrices occur, for instance, for elliptic difference equations with periodic boundary value conditions.) In this case, the algorithm takes the form (8.8), but one must add the term $\widetilde{U}_{r-k,n}(B^{-1})_{n,r}$ to (8.8a) and the term $(B^{-1})_{r,n}\widetilde{L}_{n,r-k}$ to (8.8b), because the last row of $\widetilde{L}$ and the last column of $\widetilde{U}$ are, in general, full. Furthermore, for $r = n$ we must compute $(B^{-1})_{n-k,n}$ and $(B^{-1})_{n,n-k}$ for all k, $k = 1, 2, \ldots, n - 1$ in (8.8a) and (8.8b).

8.1.3 Positive Definiteness

The method given above suffers from one disadvantage: Even if B is positive definite, the bandpart $[B^{-1}]^{p_1,q_1}$ of B^{-1} need not be positive definite, as illustrated by the following example.

Example 8.1 Let $G = \begin{bmatrix} 1 & -2 & 1 \\ -2 & 5 & -3 \\ 1 & -3 & 4 \end{bmatrix}$. Then, as is readily seen, G is positive definite, but the $[G]^{1,1}$ bandpart of G, where $[G]^{1,1} = \begin{bmatrix} 1 & -2 & 0 \\ -2 & 5 & -3 \\ 0 & -3 & 4 \end{bmatrix}$, is indefinite. More generally, this is seen from $G^{[p]} = G - R$. Here, $G^{[p]}$ denotes a symmetric bandpart—i.e., with $p = q$, of G. R is indefinite, since it has zero diagonal. Hence, it can occur that the smallest eigenvalue of $G^{[p]}$ will be negative, so $G^{[p]}$ will be indefinite.

However, for an important class of matrices—namely, for monotone matrices B—we can readily modify the matrix $[G]^{(p)}$ by a diagonal compensation for the deleted entries in R, so that the modified matrix becomes positive definite.

Theorem 8.4 (Diagonal compensation) *Let B be monotone, symmetric, and positive definite. Let $G = B^{-1}$ and $R = G - [G]^{(p)}$, where $[G]^{(p)}$ denotes the symmetric bandpart of G with semibandwidth p. Then, $\widetilde{G} = [G]^{(p)} + D$, where D is a diagonal matrix such that $D\mathbf{u} = R\mathbf{u}$, for some positive vector $\mathbf{u}$, is positive definite.*

Proof Since $G = B^{-1} \geq 0$, we have $R \geq 0$. Hence, $D \geq 0$. Let

$$V = \mathrm{diag}(u_1, u_2, \ldots, u_n).$$

Then the matrix $(D - R)V$ is diagonally dominant, so Gershgorin's disc theorem shows that its eigenvalues are nonnegative. The same holds for $V^{-1}(D - R)V$ and, hence, also for $D - R$, because $D - R$ is similarly equivalent to this latter matrix. Hence, $D - R$ is positive semidefinite and the relation

$$\widetilde{G} - G = [G]^{(p)} + D - G = D - R$$

shows that $\widetilde{G} - G$ is positive semidefinite. Hence, $\widetilde{G}$ is positive definite because G is positive definite, being the inverse of a positive definite matrix.

Among possible vectors **u**, one should chose a vector that improves the condition number $G^{(p)}B$ best, if possible. This is similar to the use of a vector for the modification of incomplete factorization methods, as discussed in Chapter 7. It will be seen in Chapter 10 that for certain problems a proper choice of the vector can reduce the order of the condition number.

Historic Remark 8.5 Both the explicit and implicit methods presented above were used in Axelsson, Brinkkemper, and Il'in (1984) to compute tridiagonal approximations of inverses of tridiagonal matrices arising during the block matrix factorization of difference equations for elliptic second-order differential equations in two space dimensions. For such applications, Concus, Golub, and Meurant (1985) similarly used an implicit method to compute the tridiagonal band matrix part of the inverse of a tridiagonal matrix, based on a method using two vectors to generate the matrix first suggested by Asplund (1959). It turns out, however, that this latter method cannot be extended in a stable way to compute the bandpart with semibandwidths ≥ 2, and it therefore is limited in practice to the case $p = q = 1$.

8.2 A Class of Methods for Computing Approximate Inverses of Matrices

We next consider a general framework for a class of methods for constructing approximate inverses. This presentation is largely based on the publications by Kolotilina and Yeremin (1986) and Kolotilina (1989). It will be seen that the previously presented explicit and implicit methods are related—and even equivalent, for symmetric matrices—to two versions of this class of methods.

The basic idea is the following: Given a sparsity pattern, compute an approximate inverse G with this sparsity pattern to a given nonsingular matrix A, of order n, that is a best approximation in some norm. It turns out that the norm based on the traces of the matrix $(I - GA)W(I - GA)^T$, that is, for the square of a weighted Frobenius norm of the error matrix $I - GA$, can give efficient and practical methods for certain choices of the weight matrix W. Hence, consider the functional

$$F_W(G) = \|I - GA\|_W^2 \equiv \text{tr}\{(I - GA)W(I - GA)^T\},$$

where W is a symmetric and positive definite matrix, and assume that G depends on some free parameters, $\alpha_1, \ldots, \alpha_p$. These parameters can be entries of G in locations that are defined by a set of p index pairs $(i, j) \in S$, where S is a subset of the full set, $\{(i, j); \ 1 \le i \le n, \ 1 \le j \le n\}$. Further, $g_{ij} = 0$ for all index pairs outside the sparsity pattern S. Here, n is the order of A.

Hence, S defines the sparsity pattern of $G = [g_{i,j}]$ and $g_{i,j} = 0$ for all $(i, j) \in S^c$, which is the complementary set defined by

$$S^c = \{(i, j) \notin S; \ 1 \le i \le n, \ 1 \le j \le n\}.$$

As before, S contains at least the index pairs $\{(i, i), \ 1 \le i \le n\}$, so the smallest sparsity pattern of G is that of the diagonal matrix. Note that $F_W(G) \ge 0$ and $F_W(G) = 0$ if $G = A^{-1}$. Further, Lemma 8.6 below shows that $\| \cdot \|_W$ is a norm, so $\|I - GA\|_W$ gives a measure of the error when G approximates A^{-1}.

We want to compute the parameters $\{\alpha_i\}$ in order to minimize $F_W(G) = F_W(\alpha_1, \ldots, \alpha_p)$. The solution to this problem must satisfy the stationary relations $\frac{\partial F_W}{\partial \alpha_i} = 0, i = 1, \ldots, p$.

Let $\|B\|_W = \{\text{tr}(BWB^T)\}^{\frac{1}{2}}$. To show that $\| \cdot \|_W$ is a norm, we state the next lemma, where some other useful properties also are derived.

Lemma 8.6 *Let A, B, W be square matrices of order n, and let W be positive definite. Then:*

(a) $\text{tr}(A) = \text{tr}(A^T)$, $\text{tr}(A + B) = \text{tr}(A) + \text{tr}(B)$.

(b) $\text{tr}(A) = \sum_{i=1}^n \lambda_i(A)$.

(c) $\text{tr}(AA^T) = \sum_{i,j=1}^n a_{i,j}^2$, and $\text{tr}(AA^T) \ge \text{tr}(A^2)$.

(d) $\|AB\|_W \le \|A\|_I \|B\|_W$, where $\| \cdot \|_I$ denotes the Frobenius norm, $\|A\|_I \equiv \{\sum_{i,j} a_{i,j}^2\}^{\frac{1}{2}} = \{\text{tr}(AA^T)\}^{\frac{1}{2}}$.

(e) $\|A\|_I \le \|A\|_W$ and $\|AB\|_W \le \|A\|_W \|B\|_W$, if and only if $W - I$ is positive semidefinite.

(f) $\|B\|_W = \{\text{tr}(BWB^T)\}^{\frac{1}{2}}$ is an additive norm.

Proof Part (a) follows from the definition of the trace operator, $\text{tr}(A) = \sum_{i=1}^{n} a_{i,i}$, and (b) has been shown in Chapter 2. The first property of (c) follows by direct computation. To show the inequality, note that

$$\text{tr}(A^2) = \sum_i \sum_j a_{ij}a_{ji} \leq \frac{1}{2} \sum_i \sum_j (a_{ij}^2 + a_{ji}^2) = \sum_i \sum_j a_{ij}^2 = \text{tr}(AA^T).$$

To show (d), note first that $\|AB\|_F \leq \|A\|_F \|B\|_F$ (Appendix A), so

$$\|AB\|_W = \{\text{tr}(ABWB^T A^T)\}^{\frac{1}{2}} = \{\text{tr}(A\widetilde{B}\widetilde{B}^T A^T)\}^{\frac{1}{2}}$$

$$= \|A\widetilde{B}\|_I \leq \|A\|_I \|\widetilde{B}\|_I = \|A\|_I \|B\|_W,$$

where $\widetilde{B} = BW^{\frac{1}{2}}$. Now (e) follows from (d), and

$$\|A\|_I = \{\text{tr}(AA^T)\}^{\frac{1}{2}} = \{\sum \lambda_i(AA^T)\}^{\frac{1}{2}} \leq \{\sum \lambda_i(AWA^T)\}^{\frac{1}{2}} =$$
(8.9)
$$= \{\text{tr}(AWA^T)\}^{\frac{1}{2}} = \|A\|_W,$$

where the inequality holds if and only if $W - I$ is positive semidefinite. Finally,

$$\|A + B\|_W = \{\text{tr}[(AW^{\frac{1}{2}} + BW^{\frac{1}{2}})(AW^{\frac{1}{2}} + BW^{\frac{1}{2}})^T\}^{\frac{1}{2}}$$

$$= \|AW^{\frac{1}{2}} + BW^{\frac{1}{2}}\|_I$$

$$\leq \|AW^{\frac{1}{2}}\|_I + \|BW^{\frac{1}{2}}\|_I \leq \|A\|_W + \|B\|_W. \qquad \diamond$$

Consider now the problem: Find $G \in \mathcal{S}$, such that

$$\|I - GA\|_W \leq \|I - \widetilde{G}A\|_W, \quad \text{for all } \widetilde{G} \in \mathcal{S},$$

i.e., where $\|I - \widetilde{G}A\|_W$ is smallest for $\widetilde{G} = G$ among all matrices with sparsity pattern $\mathcal{S}$. (For notational simplicity, we use here the same notation for the set of matrices having this sparsity pattern as for the corresponding set of indices.) Note that

$$F_W(G) = \text{tr}\{(I - GA)W(I - GA)^T\}$$

$$= \text{tr } W - \text{tr}(GAW) - \text{tr}(GAW^T) + \text{tr}(GAWA^T G^T)$$

$$= \text{tr } W - \sum_{i,j} g_{ij}[(AW)_{j,i} + (AW^T)_{j,i}] + \text{tr}(GAWA^T G^T).$$

Hence, the stationary relations

$$\frac{\partial F_W(G)}{\partial g_{ij}} = 0, \quad (i, j) \in S$$

show that

$$-(AW)_{j,i} - (AW^T)_{j,i} + (AWA^TG^T)_{j,i} + (GAWA^T)_{i,j} = 0,$$

that is,

$$-(W^TA^T)_{i,j} - (WA^T)_{i,j} + (GAW^TA^T)_{i,j} + (GAWA^T)_{i,j} = 0$$

or, since $W = W^T$,

(8.10) $$(GAWA^T)_{i,j} = (WA^T)_{i,j}, \quad (i, j) \in S.$$

(8.10) defines the set of equations that the entries of $G \in S$ must satisfy. Depending on the choice of S and on A, these equations may have or may not have a solution. Note also that (8.10) is a special case of

(8.11) $$(GAV)_{i,j} = V_{i,j}, \quad (i, j) \in S,$$

where $V = WA^T$ in (8.10). It can be seen that (8.11) has a solution that is then unique if all minors of AV, restricted to the set S, are nonsingular. Hence, the problem is to find proper matrices W (or V) and sets S.

Remark 8.7 (Parallel computation) Assuming that V is known in explicit form, the equations for computing the entries of G decouple, because the entries in any row of G are computed independently of the entries in other rows. Hence, the computation of G can be done in parallel between rows. A matrix G defined in this way will, in general, not be symmetric, even if A is symmetric.

We now state the above result in a theorem and consider some particular and practically important choices of the weight matrix W.

Theorem 8.8 *Let A be nonsingular and let $G \in S$ be such that*

$$\|I - GA\|_W \le \|I - \widetilde{G}A\|_W \quad \text{for any } \widetilde{G} \in S.$$

Then G must be a solution of the linear system

$$(GAWA^T)_{i,j} = (WA^T)_{i,j}, \quad (i, j) \in S.$$

For the following choices of W, this system and $F_W(G)$ are as follows:

(a) *For* $W = A^{-1}$, *where A is s.p.d., then*

$$F_W(G) = \operatorname{tr}(A^{-1}) - \operatorname{tr}(G),$$

and G satisfies

$$(GA)_{i,j} = \delta_{i,j}, \quad (i, j) \in \mathcal{S}.$$

(b) *For* $W = I_n$, *then*

$$F_W(G) = n - \operatorname{tr}(GA),$$

and G satisfies

$$(GAA^T)_{i,j} = (A^T)_{i,j}, \quad (i, j) \in \mathcal{S}.$$

(c) *For* $W = A^k$, *where k is a positive integer and A is s.p.d., then*

$$F_W(G) = \operatorname{tr}(A^k - GA^{k+1}),$$

and G satisfies

$$(GA^{k+2})_{i,j} = (A^{k+1})_{i,j}, \quad (i, j) \in \mathcal{S}.$$

(d) *For* $W = (A^T A)^{-1}$, *then*

$$F_W(G) = \operatorname{tr}[(A^{-1} - G)A^{-T}],$$

and G satisfies

$$G_{i,j} = (A^{-1})_{i,j}, \quad (i, j) \in \mathcal{S}.$$

Proof The general part of the statement is (8.10), which has already been shown. For the optimal G, Lemma 8.6(a) shows that

$$F_W(G) = \operatorname{tr} W - \operatorname{tr}(2GAW - GAWA^T G^T)$$

$$= \operatorname{tr} W - 2 \operatorname{tr}(GAW) + \operatorname{tr}(WA^T G^T)$$

$$= \operatorname{tr} W - \operatorname{tr}(GAW) = \operatorname{tr}(W - GAW)$$

$$= \operatorname{tr}((I - GA)W).$$

The statements for the different choices of W now follow readily. $\diamond$

Note that for $W = A^{-1}$ we get the explicit method (8.4) discussed earlier if A is symmetric, and for $W = (A^T A)^{-1}$ we get the implicit method.

For $W = I_n$, then $\|I - GA\|_W = \|I - GA\|_I$, that is, this is the error in the standard Frobenius norm.

Because of the relative simplicity in computing G, the method with $W = A^{-1}$ may be the most important one from a practical point of view when A is s.p.d. However, G generally will not be symmetric.

The functional F_W gives some information about how to choose the sparsity set in an optimal way. Note that $F_W(G) \geq 0$ and, for instance, for $W = A^{-1}$ (if A is s.p.d.) and for $W = I_n$, where $F_W(G) = \text{tr}(A^{-1}) - \text{tr}(G)$ and $F_W(G) = n - \text{tr}(GA)$, respectively, we see that we should try to choose the set S (of some given cardinality) to maximize the traces of G and GA, respectively. It is practically viable to compute $F_W(G)$ only for $W = A^k$, k a nonnegative integer.

Remark 8.9 As noted in Remark 8.7, if the matrix V in (8.11) is available, then the equations to compute G decouple, so that the entries in any row of G can be computed independently of the entries in other rows. In Theorem 8.8, this holds for cases (a), (b), and (c) (for (c), if A^{k+1} is available), but not in case (d). It is readily seen that for the method in (a),

$$(GA)_{i,j} = \delta_{i,j}, \quad (i, j) \in S,$$

not all of the entries of A will actually be used to compute $G \in S$. Since

$$(GA)_{i,j} = \sum_{k,(i,k)\in S} g_{ik} a_{kj} = \delta_{i,j}, \quad (i, j) \in S,$$

we see that for every i only entries a_{kj}, where $(i, k) \in S \wedge (i, j) \in S$, will be used. In the matrix graph for G, this means that for each i, only entries a_{kj} of A will be used, where the vertices k and j are directly connected to the vertex i. For example, if $S = \{(1, 1), (1, 2), (i, i - 1), (i, i), (i, i + 1)$ (for $i = 2, 3, \ldots, n - 1)$, $(n, n - 1)$, $(n, n)\}$, then only entries a_{kj} are required, where

$$k = i - 1, i, i + 1, \quad j = i - 1, i, i + 1.$$

That is, to compute the entries in the ith row of G, $g_{i,i-1}$, $g_{i,i}$, $g_{i,i+1}$, only the entries

$$a_{i-1,i-1}, a_{i-1,i}, a_{i-1,i+1}, a_{i,i-1}, a_{i,i}, a_{i,i+1}, a_{i+1,i-1}, a_{i+1,i}, a_{i+1,i+1}$$

of A will be used. This means that in this case, only the pentadiagonal part of A is used, and other entries of A will have no influence on the approximate inverse G.

More generally, to compute g_{ik}, $(i, k) \in \mathcal{S}$, say for $k = k_1^{(i)}, k_2^{(i)}, \ldots, k_{s_i}^{(i)}$ (if there are s_i index pairs in $\mathcal{S}$ for the ith row) for each i, then only entries a_{kj} with $k = k_1^{(i)}, \ldots, k_{s_i}^{(i)}$, $j = k_1^{(i)}, \ldots, k_{s_i}^{(i)}$ will be used. Naturally, in general we can expect such a method to be accurate only if all other entries of A (outside the corresponding sparsity pattern) are zero or relatively small. The use of diagonal compensation of deleted entries can sometimes improve the approximation significantly.

For case (b) we have

$$\|GA - I\|_W = \|GA - I\|_F^2 = \sum_{i=1}^{n} \|\mathbf{g}_i A - \mathbf{e}_i^T\|_2^2,$$

where $\mathbf{g}_i$ is the ith row of G and $\mathbf{e}_i$ is the ith column of the identity matrix I_n. Hence,

$$\min_{G \in \mathcal{S}} \|GA - I\|_I^2 = \min \sum_{i=1}^{n} \|\mathbf{g}_i A - \mathbf{e}_i^T\|_2^2 = \sum_{i=1}^{n} \min \|\mathbf{g}_i A - \mathbf{e}_i^T\|_2^2.$$

This shows that the computation of G can be carried out independently as a collection of n least squares subproblems, each for a specific row $\mathbf{g}_i$. From what was said about decoupling in Remark 8.7, it follows that only in the case where decoupling does not occur will the approximation G of A preserve symmetry, in general. Hence, even if A is symmetric, the matrices G computed by one of the methods in (a), (b), or (c), will generally not be symmetric.

When the matrix G is applied as a preconditioner in an iterative solution method, we know from Chapter 5 that the spectral condition number can be important for the rate of convergence of the iterative method. Below is shown an estimate of the condition number that is an extension to the case of weighted Frobenius norms of a result that has appeared in Cosgrove, Diaz, and Griewank (1992). Furthermore, it gives a sufficient condition for the positive definiteness of GA.

Theorem 8.10 *Let A be a nonsingular matrix of order n and let G be the approximate inverse obtained by minimizing $\|I_n - GA\|_W$, subject to the given sparsity pattern $\mathcal{S}$, where $W - I_n$ is positive semidefinite. Let G be the optimal solution and let $\varepsilon = \|I_n - GA\|_W$, where we assume that $\varepsilon < 1$. Then*

(a) GA is nonsingular and

$$\kappa(GA) = \|(GA)^{-1}\|_2 \|GA\|_2 \le \frac{1+\varepsilon}{1-\varepsilon}.$$

(b) GA is positive definite.

(Observe that in cases (a), (c), and (d) in Theorem 8.8, we can scale A by a scalar factor to make $W - I_n$ positive semidefinite.)

Proof As shown in Appendix A, $\|B\|_2 \le \|B\|_I$ for any matrix B. Hence, using Lemma 8.6(e), we find

$$\|I_n - GA\|_2 \le \|I_n - GA\|_I \le \|I_n - GA\|_W = \varepsilon,$$

so

$$\big| \|GA\|_2 - \|I_n\|_2 \big| \le \varepsilon$$

or

$$1 - \varepsilon \le \|GA\|_2 \le 1 + \varepsilon.$$

Further, $GA = I_n - (I_n - GA)$, so for any $\mathbf{x} \in \mathbb{R}^n$, we have

$$\mathbf{x}^T GA\mathbf{x} = \mathbf{x}^T\mathbf{x} - \mathbf{x}^T(I_n - GA)\mathbf{x}$$

$$\ge \|\mathbf{x}\|_2^2 - \|\mathbf{x}\|_2^2\|(I_n - GA)\|_2$$

$$\ge (1 - \varepsilon)\|\mathbf{x}\|_2^2,$$

from which we conclude that GA is positive definite when $\varepsilon < 1$. Also, from Appendix A, we have

$$\|(GA)^{-1}\|_2 = \|(I_n - (I_n - GA))^{-1}\|_2 \le \frac{1}{1-\varepsilon},$$

so

$$\kappa(GA) = \|(GA)^{-1}\|_2 \, \|GA\|_2 \le \frac{1+\varepsilon}{1-\varepsilon}.$$

$\diamond$

Note that $\|I_n\|_W^2 = \begin{cases} \operatorname{tr}(A^{-1}), & \text{in case (a)} \\ n, & \text{in case (b)} \\ \operatorname{tr}(A^k),\ k \ge 1, & \text{in case (c)} \\ \operatorname{tr}((A^T A)^{-1}), & \text{in case (d)} \end{cases}$ of Theorem 8.8, and

$\|I_n - GA\|_W$ is, in practice, available only in cases (b) and (c). If an approximate inverse G has been computed (by any method), then the proof of Theorem 8.10 shows that if $\varepsilon' = \|I_n - GA\|_F < 1$, then $\kappa(GA) \leq \frac{1+\varepsilon'}{1-\varepsilon'}$.

The sparsity pattern $\mathcal{S}$ of G does not have to be the same as in the coefficient matrix A. It can even be sparser. This is important for certain integral equations where A is frequently a full matrix, but its inverse can be approximated accurately by a much sparser matrix.

8.2.1 Comparison Results

We now show some comparison results. The first is obvious from what has just been said and from Theorem 8.8(b): If $W = I_n$ and $\mathcal{S}_2 \supseteq \mathcal{S}_1$, then

$$n \geq \text{tr}(G_2 A) \geq \text{tr}(G_1 A),$$

so G_2, corresponding to $\mathcal{S}_2$, is a better approximation to A^{-1} in the Frobenius norm than G_1, corresponding to $\mathcal{S}_1$. Note in passing that if $\mathcal{S} = \mathcal{S}_0 = \{(i, i);\ 1 \leq i \leq n\}$, then Theorem 8.8(b) shows that

$$(GAA^T)_{i,i} = (A^T)_{i,i},$$

where $G = \text{diag}(g_{i,i})$, so

$$g_{i,i} = a_{i,i} / \sum_{j=1}^{n} a_{ij}^2.$$

Hence, $g_{i,i} \leq a_{i,i}^{-1}$ and $G \neq D_A^{-1}$ (unless $A = D_A$), where D_A is the diagonal part of A. Consider next the case $W = A^{-1}$, where A is s.p.d. Then $G = D_A$ if $\mathcal{S} = \mathcal{S}_0$. The following comparison theorem is valid.

Theorem 8.11 *Let* $W = A^{-1}$, *where* $A = [a_{ij}]$ *is s.p.d. Then, if* $\mathcal{S}_2 \supseteq \mathcal{S}_1$, *we have*

(8.12) $$a_{i,i}^{-1} \leq (G_1)_{i,i} \leq (G_2)_{i,i} \leq (A^{-1})_{i,i}.$$

Proof Note first that Theorem 8.8(a) shows that $G = D_A^{-1}$ if $\mathcal{S} = \mathcal{S}_0 = \{(i, i);\ 1 \leq i \leq n\}$. Further, for any $\mathcal{S}$,

$$F_W(G) = \text{tr}(A^{-1} - G) = \sum_{i=1}^{n} ((A^{-1})_{i,i} - g_{i,i}).$$

Since the entries of each row of G are determined independently of the entries in the other rows, we have in fact,

$$F_W(G) = \sum_{i=1}^{n} F_W(G)_i,$$

where $F_W(G)_i = (A^{-1})_{i,i} - g_{i,i}$ and the solution of G is such that each term $F_W(G)_i$ is minimized, independently of the other terms, under the constraint $G \in S$. This shows that each term is nonnegative, that is, $F_W(G)_i \geq 0$, and that $F_W(G_2)_i \leq F_W(G_1)_i$, which is (8.12). The extreme values to the left and right are taken for $S = S_0$ and S, the whole set, respectively. $\Diamond$

Note next that (8.12) does not hold for nonsymmetric positive definite matrices A, for which G satisfies

$$(GA)_{i,j} = \delta_{i,j}, \quad (i, j) \in S.$$

This is shown by the following example.

Example 8.2 Let $A = \begin{bmatrix} 2 & -1 \\ 2 & 2 \end{bmatrix}$. Then $G_1 = D_A^{-1} = \begin{bmatrix} \frac{1}{2} & 0 \\ 0 & \frac{1}{2} \end{bmatrix}$ and $G_2 = A^{-1} = \frac{1}{6}\begin{bmatrix} 2 & 1 \\ -2 & 2 \end{bmatrix}$, so we have $(G_1)_{i,i} > (G_2)_{i,i}$, although $S_1 \subseteq S_2$.

The following lemma will be useful later.

Lemma 8.12 *For $i = 1, 2, \ldots, n$, let $B^{(i)} = \alpha^{(i)} \odot A$, the Hadamard (pairwise entry) product where*

$$\alpha_{kj}^{(i)} = \begin{cases} 1, & \text{for (k, j) such that $(i, k) \in S \wedge (i, j) \in S$} \\ & \text{(all entries in this block are equal to 1)} \\ 1, & \text{for $k = j$} \\ 0, & \text{otherwise.} \end{cases}$$

Assume that $B^{(i)}$ is nonsingular, and consider the method in Theorem 8.8a. Then the nonzero entries of the ith row of G equal the corresponding entries of the exact inverse of $B^{(i)}$.

Proof Let $G^{(i)}$ be the exact inverse of $B^{(i)}$. The equation for the ith row of $G^{(i)}$ is then

$$\sum_{k=1}^{n} G_{i,k}^{(i)} B_{kj}^{(i)} = \delta_{i,j}, \quad j = 1, 2, \ldots, n.$$

By assumption, all entries $B_{kj}^{(i)} = 0$ for $(i, k) \in S^c$ or $(i, j) \in S^c$. Thus, for i fixed and $(i, j) \in S$, we have

$$(G^{(i)} B^{(i)})_{i,j} = \sum_{k,(i,k) \in S} g_{ik}^{(i)} a_{kj} = \delta_{ij},$$

so the statement follows from Remark 8.9. ◇

As an immediate consequence of the definition of a block H-matrix (Chapter 7), we see that if A is a block H-matrix, then $\alpha^{(i)} \odot A$ is also a block H-matrix. Hence, in particular, $B^{(i)}$ is nonsingular if A is a block H-matrix, so $G^{(i)}$ is uniquely determined. This important fact will be restated in Theorem 8.16.

Example 8.3 If S defines the tridiagonal sparsity pattern, as in Remark 8.9, then $G^{(i)}$ is the exact inverse of a matrix $B^{(i)}$ with entries $B_{j,j}^{(i)} = A_{j,j}$ and off-diagonal entries zero, except those (nine in general) in rows, $i - 1, i, i + 1$, which are equal to the entries of A as listed in Remark 8.9. (The remaining diagonal entries are equal to the unit number.)

The following theorem shows that for M-matrices, we may sharpen the comparison result in Theorem 8.11.

Theorem 8.13 *Let A be a M-matrix and let G be determined by $(GA)_{i,j} = \delta_{i,j}$, $(i, j) \in S$.*

> (a) *Then G is nonsingular and $A = G^{-1} - R$ is a weakly regular splitting.*
> (b) *Let $S_2 \supseteq S_1$ and let G_i correspond to S_i, $i = 1, 2$. Then*
>
> $$D_A^{-1} \leq G_1 \leq G_2 \leq A^{-1},$$

that is, monotonicity holds for all *entries, not only for the diagonal, as in Theorem 8.11.*

Proof Lemma 8.12 shows that the nonzero entries of the ith row of G equal the corresponding entries of the ith row of $G^{(i)}$, where $G^{(i)}$ is the exact inverse of $B^{(i)} = \alpha^{(i)} \odot A$ and $\alpha^{(i)}$ is defined as in Lemma 8.12. Since A is a M-matrix, it is readily seen that $B^{(i)}$ is a M-matrix so, in particular, it is nonsingular. Hence, its inverse $G^{(i)}$ is nonnegative, and Lemma 8.12 shows that $G \geq 0$; in addition, G has no zero row. (For instance, Theorem 8.11 shows that $g_{i,i} \geq a_{i,i}^{-1}$.)

Consider now

$$GA = I - GR.$$

We show first that GA is a Z-matrix: For $(i, j) \in S$ we have $(GA)_{i,j} = \delta_{i,j}$, so, in particular, $(GA)_{i,j} = 0$ if $i \neq j$, and for $(i, j) \notin S$ we find that

$$(GA)_{i,j} = \sum_{k,(i,k) \in S} g_{i,k} a_{k,j} = \sum_{\substack{k \\ k \neq j,(i,k) \in S}} g_{i,k} a_{k,j},$$

because $g_{i,j} = 0$ as $(i, j) \notin S$. Hence, for $(i, j) \notin S$ the terms in the above sum are all nonpositive, as $a_{k,j} \leq 0$, $k \neq j$ and $g_{i,k} \geq 0$ (as we have shown above).

Next we show that GA is a M-matrix. Since A is a M-matrix, there exists $\mathbf{x} > \mathbf{0}$, such that $A\mathbf{x} > \mathbf{0}$. Hence, since G is nonnegative and has no zero row, $GA\mathbf{x} > \mathbf{0}$. This, together with the above result, shows that GA is a M-matrix. Since A is a M-matrix, it is, in particular, nonsingular, so G is also nonsingular. This shows that the splitting $A = G^{-1} - R$ is a weakly regular splitting, which is part (a).

To show part (b), note that for $S_2 \supseteq S_1$, the nonzero entries of $(G_2)_i$ (the ith row) equal the corresponding nonzero entries of the exact inverse of $\alpha_2^{(i)} \odot A$, and the nonzero entries of $(G_1)_i$ equal the corresponding nonzero entries of the exact inverse of $\alpha_1^{(i)} \odot A$, where $\alpha_2^{(i)} \geq \alpha_1^{(i)}$ (entrywise). This shows that

$$I - G_2 A \leq I - G_1 A \quad \text{or} \quad G_1 A \leq G_2 A \quad \text{(entrywise)},$$

so, as $A^{-1} \geq 0$, $G_1 \leq G_2$. The lower bound of part (b) follows from Theorem 8.10 and the upper bound follows by letting S_2 be the whole set, for which $G_2 = A^{-1}$. $\diamond$

Corollary 8.14 *Let A be a M-matrix and let G be determined by $(GA)_{i,j} = \delta_{i,j}$, $(i, j) \in S$. Then $\rho(I - GA) < 1$.*

Proof Since A is a M-matrix, we can write $A = D_A - B$, where D_A is the diagonal part of A and $B \geq 0$. This is a regular splitting so, in particular,

$$\rho(I - D_A^{-1} A) = \rho(D_A^{-1} B) < 1.$$

But the proof of Theorem 8.13 shows that $GA \geq D_A^{-1} A$, which—using Perron-Frobenius theorem—shows that $\rho(I - GA) \leq \rho(I - D_A^{-1} A)$, because $I - GA$ is nonnegative. [Note that $(GA)_{i,i} = 1$.] $\diamond$

We now extend this result to the class of H-matrices. First we show a classical comparison result.

Lemma 8.15 (Ostrowski's theorem) *Let A be an H-matrix and let $\mathcal{M}(A)$ be its comparison matrix. Then $|A^{-1}| \leq \mathcal{M}(A)^{-1}$, where $|A|$ denotes the matrix with entries $|a_{i,j}|$.*

Proof Consider the splitting $A = D - B$, where $D = \text{diag}(A)$. Then $\mathcal{M}(A) = |D| - |B|$, and since by assumption $\mathcal{M}(A)$ is a M-matrix, $\rho(|D^{-1}B|) < 1$. Hence, by the proof of Corollary 8.14, $\rho(D^{-1}B) < 1$. Further, $A = D(I - D^{-1}B)$, which shows that

$$A^{-1} = (I - D^{-1}B)^{-1}D^{-1} = [I + D^{-1}B + (D^{-1}B)^2 + \ldots]D^{-1}$$

$$= D^{-1} + D^{-1}BD^{-1} + D^{-1}BD^{-1}BD^{-1} + \ldots.$$

Hence,

$$|A^{-1}| \leq |D|^{-1} + |D|^{-1}|B||D|^{-1} + \ldots$$

$$= (I - |D|^{-1}|B|)^{-1}|D|^{-1} = (|D| - |B|)^{-1} = \mathcal{M}(A)^{-1}. \quad \diamond$$

Theorem 8.16 *Let A be an H-matrix and let G be defined by $(GA)_{i,j} = \delta_{i,j}$, $(i, j) \in \mathcal{S}$. Then G is uniquely determined and $\rho(I - GA) < 1$.*

Proof Similar to the proof of part (b) of Theorem 8.13, we notice first that the ith row of G is the ith row of the exact inverse of $\alpha^{(i)} \odot A$ for a Hadamard matrix product. Clearly, $\alpha^{(i)} \odot A$ is also an H-matrix, so, in particular, its inverse exists. Since this is true for any i, G is uniquely determined.

Let $\mathcal{M}(A)$ be the comparison matrix and let $\widetilde{G} = [\widetilde{g}_{i,j}]$ be the approximate inverse to $\mathcal{M}(A)$ for the set $\mathcal{S}$, that is,

$$\{\widetilde{G}\mathcal{M}(A)\}_{i,j} = \delta_{i,j}, \quad (i, j) \in \mathcal{S}.$$

First Ostrowski's theorem shows that for the exact inverses G_i and $\widetilde{G}_i$ of $\alpha^{(i)} \odot A$ and $\alpha^{(i)} \odot \mathcal{M}(A)$, respectively, we have $|G_i| \leq \widetilde{G}_i$. Thus,

$$|G| \leq \widetilde{G}.$$

We show next that

(8.13) $$|GA| \leq |\widetilde{G}\mathcal{M}(A)|.$$

This inequality is trivial for $(i, j) \in \mathcal{S}$, so we need to consider only $(i, j) \notin \mathcal{S}$.

For such positions, we have

$$|(GA)_{i,j}| = |\sum_{\substack{k:k\neq j \\ (i,k)\in S}} g_{i,k}a_{k,j}| \leq \sum_{\substack{k:k\neq j \\ (i,k)\in S}} |g_{i,k}| \, |\{\mathcal{M}(A)\}_{k,j}|$$

$$\leq \sum_{\substack{k:k\neq j \\ (i,k)\in S}} \tilde{g}_{i,k}[-\{\mathcal{M}(A)\}_{k,j}] = -\{\widetilde{G}\mathcal{M}(A)\}_{i,j}.$$

Hence,

$$|I - GA| \leq |I - \widetilde{G}\mathcal{M}(A)|$$

and, using Perron-Frobenius theorem,

$$\rho(I - GA) \leq \rho(|I - GA|) \leq \rho(|I - \widetilde{G}\mathcal{M}(A)|)$$

$$= \rho(I - \widetilde{G}\mathcal{M}(A)) < 1,$$

where the last inequality follows from Corollary 8.14.   ◇

Remark 8.17 (Block H-matrices) Taking norms instead of absolute values, we find that Theorem 8.16 holds also for block H-matrices and that

(8.14) $\|G_{i,j}\| \leq \tilde{g}_{i,j}$ and $\|(I - GA)_{i,j}\| \leq |(I - \widetilde{G}\mathcal{M}_b(A))_{i,j}|,$

where $\tilde{g}_{i,j}$ are the entries of the matrix $\widetilde{G}$ satisfying $\{\widetilde{G}\mathcal{M}_b(A)\}_{i,j} = \delta_{i,j}$, $(i,j) \in \mathcal{S}$, and $\mathcal{M}_b(A)$ is the block matrix comparison matrix.

Corollary 8.18 *Let A be an H-matrix and $\mathcal{M}(A)$ its comparison matrix, and let $\mathbf{x} > 0$ be such that $\mathcal{M}(A)\mathbf{x} > 0$. Then $\mathcal{M}(GA)\mathbf{x} > 0$, where G is defined as in Theorem 8.16—i.e., the generalized diagonal dominance is preserved.*

Proof By (8.13) we have $|GA| \leq |\widetilde{G}\mathcal{M}(A)|$. Since $\mathcal{M}(A)\mathbf{x} > 0$ for some positive $\mathbf{x}$ and $\widetilde{G} \geq 0$, we have $\widetilde{G}\mathcal{M}(A)\mathbf{x} > 0$. Noting that the diagonal entries of GA are unit numbers and that $(GA)_{i,j} \leq 0$, $i \neq j$ (see the proof of Theorem 8.13), we find that

$$\mathcal{M}(GA)\mathbf{x} = \mathbf{x} - (I - GA)\mathbf{x} \geq \mathbf{x} - |I - GA|\mathbf{x}$$

$$\geq \mathbf{x} - |I - \widetilde{G}\mathcal{M}(A)|\mathbf{x} = \mathbf{x} - [I - \widetilde{G}\mathcal{M}(A)]\mathbf{x}$$

$$= \widetilde{G}\mathcal{M}(A)\mathbf{x} > 0,$$

where we have used the property that $\widetilde{G}\mathcal{M}(A)$ is a Z-matrix.   ◇

Historical Remark Explicit preconditioning methods that are best approximations in the Frobenius norm were first proposed by Benson (1973), Frederickson (1975), Benson and Frederickson (1982), and Ong (1984). They were later extended with careful analysis to best approximations in generalized Frobenius norms by Kolotilina and Yeremin (1986) and Kolotilina (1989). Two versions of the latter turned out to be identical to two methods presented earlier in Axelsson, Brinkkemper, and Il'in (1984). For another method of studying optimal preconditioners, see Greenbaum and Rodrique (1989).

8.3 A Symmetric and Positive Definite Approximate Inverse

It is readily seen that the approximate inverses in Theorem 8.8 suffer from the disadvantage that they do not preserve symmetry and/or positive definiteness of A in general. When using an iterative method, such as the Chebyshev method, it is of interest to preserve symmetry and positive definiteness, (i.e., to use a symmetric and positive definite preconditioning matrix) when A is symmetric and positive definite. The approximate inverse in Theorem 8.8(d) will clearly be symmetric if S is a symmetric sparsity set, but it may be indefinite, even if A is positive definite. Furthermore, it is practically viable only when S has a special structure, such as for a band matrix. In order to compute a s.p.d. preconditioner to a s.p.d. matrix A, we can modify the approximate inverse by using the method of diagonal compensation (Theorem 8.4) if A is monotone. Otherwise, we can proceed as follows.

Assume first that we have factored A as $A = L_A D^{-1} L_A^T$, where D is diagonal and where $L_A = I - \tilde{L}$ and $\tilde{L}$ is strictly lower triangular. We then compute an approximate inverse L_G of L_A. This can be done using formulas corresponding to (8.10), where we choose the weight matrix properly. For instance, we get

$$(8.15a) \qquad (L_G L_A)_{i,j} = \delta_{i,j}, \qquad (i, j) \in S_L$$

and

$$(8.15b) \qquad (L_G)_{i,j} = (L_A^{-1})_{i,j}, \qquad (i, j) \in S_L$$

for $W = (L_A^T)^{-1}$ and $W = (L_A^T L_A)^{-1}$, respectively, where S_L denotes the "lower triangular part" of S i.e. $S_L = \{(i, j) \in S, j \leq i\}$. The preconditioner

$$G = L_G^T D L_G$$

to be used in a multiplicative preconditioner is now symmetric and positive definite. If a factorization of A is not known (for instance, might be too expensive to compute) then we can use the following method, again based on minimizing the approximation error in a (weighted) Frobenius norm.

To derive this method, we let $A = L_A D^{-1} L_A^T$. It is our intention to find a method where L_A (that is, $\widetilde{L}$) actually is not needed. We want to minimize

$$F_W(L_G) = \operatorname{tr}\{(I - L_G L_A) W (I - L_G L_A)^T\}$$

for all L_G with sparsity pattern S_L. Then for $W = (L_A^T L_A)^{-1}$, (8.10) shows that

$$[L_G L_A (L_A^T L_A)^{-1} L_A^T]_{i,j} = [(L_A^T L_A)^{-1} L_A^T]_{i,j}, \qquad (i, j) \in S_L$$

or

$$(L_G)_{i,j} = (L_A^{-1})_{i,j} \qquad (i, j) \in S_L,$$

which is (8.15b). Hence, this choice of W requires that L_A has been computed so that it is unpractical, but we see that the above characterizes the solution of (8.15b) as an optimal solution in the corresponding norm.

We now try $W = D^{-1}$. Then (8.10) shows that

$$(L_G L_A D^{-1} L_A^T)_{i,j} = (D^{-1} L_A^T)_{i,j}, \qquad (i, j) \in S_L,$$

or

(8.16a) $$(L_G A)_{i,j} = \{(L_A D^{-1})^T\}_{i,j}, \qquad (i, j) \in S_L.$$

Since S_L is a lower triangular set but the right matrix is upper triangular, (8.16a) takes the form

(8.16b) $$(L_G A)_{i,j} = \begin{cases} 0, & i \neq j \\ D_i^{-1}, & i = j \end{cases}.$$

Hence, for this choice of weight matrix W, we do not need to know L_A!

If D is also not known, Kolotilina and Yeremin (1990) suggested the use of diagonal scaling. Compute first $\widehat{L}_G$ such that

$$(\widehat{L}_G A)_{i,j} = \delta_{i,j}, \qquad (i, j) \in S_L.$$

Then set $L_G = \widehat{D}^{\frac{1}{2}} \widehat{L}_G$, where $\widehat{D}$ is a diagonal matrix to be defined below, and let $G = L_G^T L_G = \widehat{L}_G^T \widehat{D} \widehat{L}_G$ be the preconditioner. $\widehat{D}$ will be chosen so that

(8.17) $$(L_G A L_G^T)_{i,i} = 1,$$

that is,

$$(\widehat{D}^{\frac{1}{2}}\widehat{L}_G A \widehat{L}_G^T \widehat{D}^{\frac{1}{2}})_{i,i} = 1.$$

With $\widehat{D} = \mathrm{diag}(\hat{d}_1, \ldots, \hat{d}_n)$, we find then

$$\sum_{j=1}^{n} \hat{d}_i (\widehat{L}_G A)_{i,j} (\widehat{L}_G^T)_{i,j} = 1.$$

Since $(\widehat{L}_G A)_{i,j} = \delta_{i,j}$, $(i, j) \in \mathcal{S}_L$ and $(\widehat{L}_G^T)_{i,j} = 0$, $(i, j) \in \mathcal{S}_L$ if $i > j$, we find

$$\hat{d}_i (\widehat{L}_G)_{i,i} = 1,$$

that is,

$$\widehat{D} = [\mathrm{diag}(\widehat{L}_G)]^{-1}.$$

Hence, the preconditioner is

$$L_G^T L_G = \widehat{L}_G^T \widehat{D} \widehat{L}_G = \widehat{L}_G^T [\mathrm{diag}(\widehat{L}_G)]^{-1} \widehat{L}_G,$$

which is symmetric and positive definite, and the method to compute it takes the following form:

(a) Compute $\widehat{L}_G$ with sparsity pattern $\mathcal{S}_L$ such that

$$(\widehat{L}_G A)_{ij} = \delta_{i,j}, \quad (i, j) \in \mathcal{S}_L.$$

(b) Let $\widehat{D} = (\mathrm{diag}(\widehat{L}_G))^{-1}$ and $L_G = \widehat{D}^{\frac{1}{2}} \widehat{L}_G$.

The diagonal entries of $\widehat{L}_G$ are positive; thus, in particular, $\widehat{D}$ exists. This follows because Lemma 8.12 shows that its entries in the ith row are the entries of the lower triangular part of the ith row of the exact inverse of $\alpha^{(i)} \odot L_A$, where $\alpha^{(i)}$ is defined as in Lemma 8.12 (but for the sparsity pattern $\mathcal{S}_L$) and L_A is the Cholesky factor of A. We recall further from the construction of $\widehat{D}$ that

$$\widehat{D}^{-1} = \mathrm{diag}(\widehat{L}_G) = \mathrm{diag}(\widehat{L}_G A \widehat{L}_G^T),$$

so

$$\mathrm{diag}(\widehat{D}^{\frac{1}{2}} \widehat{L}_G A \widehat{L}_G^T \widehat{D}^{\frac{1}{2}}) = I.$$

Hence, $\widehat{D}^{\frac{1}{2}}$ scales $\widehat{L}_G A \widehat{L}_G^T$ to have a unit diagonal. ($\widehat{D}^{\frac{1}{2}} \widehat{L}_G A \widehat{L}_G^T \widehat{D}^{\frac{1}{2}} - I$ generally is indefinite.)

Example 8.4 Let $A = \text{tridiag}(-1, 4, -1)$ and the index set

$$S = (i - 1, i, i + 1)_{i=1}^n.$$

Find L_G in the corresponding approximate inverse $L_G L_G^T$. We have $S_L = (i - 1, i)_{i=1}^n$ and $(\widehat{L}_G A)_{i,j} = \delta_{i,j}$, $(i, j) \in S_L$. This shows that the ith row of $\widehat{L}_G$ is equal to the ith row of the inverse of $\text{diag}(1, \ldots, 1, \begin{bmatrix} 4 & -1 \\ -1 & 4 \end{bmatrix}$, $1, \ldots, 1)$, that is, $(\widehat{L}_G)_{i,i-1} = \frac{1}{15}$, $(\widehat{L}_G)_{i,i} = \frac{4}{15}$, except the first entry, which is $(\widehat{L}_G)_{1,1} = \frac{1}{4}$. Hence,

$$\widehat{L}_G = \begin{bmatrix} \frac{1}{4} & & & & 0 \\ \frac{1}{15} & \frac{4}{15} & & & \\ & \ddots & \ddots & & \\ 0 & & & \frac{1}{15} & \frac{4}{15} \end{bmatrix}.$$

Note that

$$\widehat{L}_G A = \begin{bmatrix} 1 & -\frac{1}{4} & & & 0 \\ 0 & 1 & -\frac{4}{15} & & \\ -\frac{1}{15} & 0 & 1 & -\frac{4}{15} & \\ & \ddots & \ddots & \ddots & \\ 0 & & -\frac{1}{15} & 0 & 1 \end{bmatrix}.$$

We have $\widehat{D}^{\frac{1}{2}} = (\text{diag}\,\widehat{L}_G)^{-\frac{1}{2}} = \text{diag}(2, \frac{1}{2}\sqrt{15}, \ldots, \frac{1}{2}\sqrt{15})$, $L_G = \widehat{D}^{\frac{1}{2}} \widehat{L}_G =$

$$\begin{bmatrix} \frac{1}{2} & & & \\ \frac{1}{2\sqrt{15}} & \frac{2}{\sqrt{15}} & & \\ & \ddots & \ddots & \\ 0 & & \frac{1}{2\sqrt{15}} & \frac{2}{\sqrt{15}} \end{bmatrix}, \text{ and we find}$$

$$L_G A L_G^T = \begin{bmatrix} 1 & 0 & -\frac{1}{4\sqrt{15}} & & \cdots & 0 \\ 0 & 1 & -\frac{1}{60} & -\frac{1}{15} & & \\ -\frac{1}{4\sqrt{15}} & -\frac{1}{60} & 1 & -\frac{1}{60} & -\frac{1}{15} & \\ & \ddots & \ddots & \ddots & \ddots & \\ 0 & & & & -\frac{1}{15} & -\frac{1}{60} & 1 \end{bmatrix}.$$

The above choice of scaling is motivated by the following well-known results (Van der Sluis, 1969; Forsythe and Strauss, 1955): Let A be s.p.d. and $D = \text{diag}(A)$. Then:

(a) $\kappa(D^{-\frac{1}{2}} A D^{-\frac{1}{2}}) \leq m \, \kappa(\widetilde{D} A \widetilde{D})$ for any positive definite diagonal matrix $\widetilde{D}$, where m is the maximum number of nonzero entries in any row of A and $\kappa(B)$ denotes

$$\kappa(B) = \lambda_{\max}(B)/\lambda_{\min}(B),$$

i.e., the spectral condition number.

(b) If, in addition, A is block-tridiagonal, then

$$\kappa(D^{-\frac{1}{2}} A D^{-\frac{1}{2}}) \leq \kappa(\widetilde{D} A \widetilde{D}).$$

Furthermore, the following minimization property holds (see Kolotilina and Yeremin, 1990).

Theorem 8.19 *Let A be a s.p.d. and consider matrices L with sparsity pattern $\mathcal{S}_L$. Then:*

(a) *The matrix L_G computed by*
  (i) $(\widehat{L}_G A)_{i,j} = \delta_{ij}$, $(i, j) \in \mathcal{S}_L$ *and*
  (ii) $L_G = (\text{diag}(\widehat{L}_G))^{-\frac{1}{2}} \widehat{L}_G$
  minimizes the functional

$$\frac{1}{n} \text{tr}(LAL^T)/\det(LAL^T)^{\frac{1}{n}}$$

  for all $L \in \mathcal{S}_L$.

(b) *The minimal value is*

$$\frac{1}{n} \text{tr}(L_G A L_G^T)/\det(L_G A L_G^T)^{\frac{1}{n}} = \left[\frac{\prod_{i=1}^{n}[(I - \widetilde{L}_G)A(I - \widetilde{L}_G^T)]_{i,i}}{\det(A)} \right]^{\frac{1}{n}},$$

  where $\widetilde{L}_G = I - \text{diag}(L_G)^{-1} L_G$.

Proof For any nonsingular matrix X we have

$$\text{tr}(XAX^T) = \sum_{i=1}^{n} \lambda_i(XAX^T) = \sum_{i=1}^{n}(XAX^T)_{i,i}$$

and

$$\det(XAX^T) = \det(X)^2 \det(A).$$

Hence,

$$\frac{\frac{1}{n}\operatorname{tr}(XAX^T)}{\det(XAX^T)^{\frac{1}{n}}} = \frac{\frac{1}{n}\sum_{i=1}^{n}(XAX^T)_{i,i}}{\det(X)^{\frac{2}{n}}\det(A)^{\frac{1}{n}}}.$$

For $X = L = D(I - \widetilde{L})$, where $D = \operatorname{diag}(L) = \operatorname{diag}(d_1, d_2, \ldots, d_n)$ and where $\widetilde{L} = L - D^{-1}L$ (which is strictly lower triangular), we find

$$\frac{\frac{1}{n}\operatorname{tr}(LAL^T)}{\det(LAL^T)^{\frac{1}{n}}} = \frac{\frac{1}{n}\sum_{1}^{n}(LAL^T)_{i,i}^{\frac{1}{n}}}{\det(L)^{\frac{2}{n}}\det(A)^{\frac{1}{n}}} = \frac{\frac{1}{n}\sum_{i=1}^{n}d_i^2[(I-\widetilde{L})A(I-\widetilde{L}^T)]_{i,i}}{(\prod_{1}^{n}d_i^2)^{\frac{1}{n}}\det(A)^{\frac{1}{n}}} =$$

$$= \frac{\frac{1}{n}\sum_{1}^{n}\alpha_i^2}{(\prod_{1}^{n}\alpha_i^2)^{\frac{1}{n}}} \cdot \frac{\{\prod_{1}^{n}[(I-\widetilde{L})A(I-\widetilde{L}^T)]_{i,i}\}^{\frac{1}{n}}}{\det(A)^{\frac{1}{n}}},$$

where

$$\alpha_i^2 = d_i^2[(I-\widetilde{L})A(I-\widetilde{L}^T)]_{i,i}.$$

Using the inequality between the arithmetic and geometric means, we find

$$\frac{1}{n}\sum_{1}^{n}\alpha_i^2 / (\prod_{1}^{n}\alpha_i^2)^{\frac{1}{2}} \geq 1,$$

where equality holds if and only if

$$\alpha_1^2 = \alpha_2^2 = \ldots = \alpha_n^2,$$

that is, when $D = \operatorname{diag}(L)$ is chosen such that

$$(LAL^T)_{i,i} = \text{constant}, \quad i = 1, 2, \ldots, n.$$

The second factor,

$$\prod_{i=1}^{n}[(I-\widetilde{L})A(I-\widetilde{L}^T)]_{i,i},$$

can be minimized by minimizing each of the factors,

$$[(I-\widetilde{L})A(I-\widetilde{L}^T)]_{i,i}, \quad i = 1, 2, \ldots, n$$

separately. However, this problem is of the same form as the minimization problem in Theorem 8.8 (with $G = \widetilde{L}$, $A = I_n$ and weight matrix $W = A$ and $S = S_L$). Hence, its optimal solution satisfies

$$(\widetilde{L}A)_{i,j} = A_{i,j}, \quad (i, j) \in \widetilde{S}_L$$

or

$$[(I - \widetilde{L})A]_{i,j} = 0, \quad (i, j) \in \widetilde{S}_L,$$

and its solution has the same form as the correspondingly scaled solution in (8.16b). The above shows that with such a matrix, $L = D(I - \widetilde{L})$, where D is chosen to make $(LAL^T)_{i,i} = \text{constant} (= 1, \text{say})$, $i = 1, 2, \ldots, n$, condition (8.17) holds, and we have $L = L_G$. Hence, as has been shown above, L_G minimizes the functional

$$\frac{1}{n} \text{tr}(LAL^T)/\det(LAL^T)^{\frac{1}{n}} \quad \text{for all } L \in S_L.$$

The minimum takes the value

$$\frac{1}{n} \text{tr}(L_G A L_G^T)/\det(L_G A L_G^T)^{\frac{1}{n}} = \{\prod_1^n [(I - \widetilde{L}_G)A(I - \widetilde{L}_G^T)]_{i,i}\}^{\frac{1}{n}} \det(A)^{-\frac{1}{n}},$$

where $\widetilde{L}_G = \widehat{D}(I - \widehat{L}_G)$. $\diamond$

The quantity $[\frac{1}{n} \text{tr}(B)]^n/\det(B)$ is a (nonstandard) condition number of B, and Theorem 8.19 will turn out to be useful when analyzing the rate of convergence of the conjugate gradient method (see Chapter 13).

Note that when L is diagonal (that is, $S_L = \{\emptyset\}$) and L scales A as in (8.17), then we have

$$\left[\frac{1}{n} \text{tr}(LAL^T)\right]^n / \det(LAL^T) = \frac{\prod_1^n a_{i,i}}{\det(A)},$$

while $\left[\frac{1}{n} \text{tr}(A)\right]^n / \det(A) = \left[\frac{1}{n} \sum_1^n a_{i,i}\right]^n / \det(A)$. Hence, diagonal scaling can only improve the condition number significantly if the diagonal entries of A vary significantly, so that the geometric average of $\{a_{i,i}\}_{i=1}^n$ is much smaller than the arithmetic average.

8.4 Combinations of Explicit and Implicit Methods

It turns out that while some explicit preconditioners are readily implemented in an iterative solution method, even on parallel computer architectures, they frequently require $|S|$, the cardinality of S, to be relatively large to get an accurate approximation of the inverse. In particular, this is the case when A has a large condition number, as we shall see in Section 8.6. On the other hand, implicit preconditioners can be accurate approximations of A if A is sparse, even when the corresponding $|S|$ is relatively small.

However, combinations of explicit and implicit preconditioners can be most efficient. We consider here two such combinations, one of multiplicative (i.e., explicit) form and another using the recursive two-level structure of implicit incomplete factorization of matrices partitioned in block form.

8.4.1 Incomplete Factorization of an Explicitly Preconditioned Matrix

Consider first the multiplicative form. Here we first compute an explicit preconditioner G and then an implicit preconditioner C to $I - E$, where $E = I - GA$. The resulting preconditioner (of multiplicative form) is then $C^{-1}G$. We shall prove the existence of this for the class of block H-matrices. To this end we shall use the next lemma.

Lemma 8.20 *Let A be a M-matrix and let G be defined by $(GA)_{i,j} = \delta_{i,j}$, $(i, j) \in S$. Then $E = I - GA \geq 0$ (entrywise), and there exists a positive vector $\mathbf{x}$ such that for each i we have either $\{(I - GA)\mathbf{x}\}_i > 0$ or $\{(I - GA)\mathbf{x}\}_i = 0$ for some i. In the latter case, the ith row of E is zero.*

Proof The proof of Theorem 8.13 shows that GA is a M-matrix, so, in particular, since $(GA)_{i,i} = 1$, $E = I - GA \geq 0$. Hence, $E\mathbf{x} \geq \mathbf{0}$. If $(E\mathbf{x})_i = 0$ for some i, then $(GA\mathbf{x})_i = x_i$, so $x_i = (GA\mathbf{x})_i = \sum_j (GA)_{i,j}x_j = x_i + \sum_{j \neq i}(GA)_{i,j}x_j$, which together with $(GA)_{i,j} \geq 0$ for $i \neq j$ shows that $(GA)_{i,j} = 0$, $j \neq i$, that is, the ith row of E is zero.   ◇

Lemma 8.20 shows, in particular, that $I - E$ is a M-matrix. We extend now this result to block H-matrices.

Theorem 8.21 *Let A be a block H-matrix, and let G be defined by $(GA)_{i,j} = \delta_{i,j}$, $(i, j) \in S$. Then $I - E$ is a block H-matrix, where $E = I - GA$. Further, the incomplete block-matrix factorization C of $I - E$ exists, and therefore, the combined preconditioner $C^{-1}G$ exists.*

Proof (8.13) and (8.14) show that for $i \neq j$,

$$\|E_{i,j}\| = \|(I - GA)_{i,j}\| \leq |\{1 - \tilde{G}\mathcal{M}_b(A)\}_{i,j}| = |\tilde{E}_{i,j}|,$$

where $\mathcal{M}_b(A)$ is the block comparison matrix to A, where $\{\tilde{G}\mathcal{M}_b(A)\}_{i,j} = \delta_{i,j}$, $(i, j) \in \mathcal{S}$, and where $\tilde{E} = I - \tilde{G}\mathcal{M}_b(A)$. Note that $E_{i,j}$ are block matrix entries and $\tilde{E}_{i,j}$ are scalar entries. But Lemma 8.20 shows that $\tilde{E} \geq 0$, so $\|E_{i,j}\| \leq \tilde{E}_{i,j}$. Hence,

$$x_i - \sum_{j \neq i} \|E_{i,j}\| x_j \geq x_i - \sum_{j \neq i} \tilde{E}_{i,j} x_j > 0,$$

where $\mathbf{x}$ is a positive vector such that $(I - \tilde{E})\mathbf{x} > \mathbf{0}$. This shows that $I - E$ is a block H-matrix. The remainder of the theorem follows by Theorem 7.9. ◇

Since the implementation of the explicit preconditioner G costs relatively little, the hybrid preconditioner $C^{-1}G$ can be more efficient than an implicit preconditioner C alone, in particular when E is sparser than A. However, to be practically viable, G must have a simple sparsity pattern such as a diagonal or tridiagonal structure. This method takes the following form:

(a) Compute an explicit preconditioner G such that

$$(GA)_{i,j} = \delta_{i,j}, \quad (i, j) \in \mathcal{S}.$$

Let $E = I - GA$.

(b) Compute an implicit preconditioner to $GA = I - E$, for instance, using a block incomplete factorization of $I - E$.

(c) Solve a given system $A\mathbf{x} = \mathbf{b}$, or

$$GA\mathbf{x} = \tilde{\mathbf{b}} \equiv G\mathbf{b},$$

using an iterative method with the preconditioner for $I - E$.

8.4.2 Incomplete Factorization of Block Tridiagonal Matrices

Consider next the combination of explicit and implicit preconditioners in the block matrix incomplete factorization method. This will be done for matrices on block tridiagonal form. Such matrices occur frequently in practice. Here, as is shown in Chapter 7, we approximate the inverses $A_{r,r}^{(r)^{-1}}$ of the pivot block matrices that occur at every stage (r) of the algorithm by an explicit method.

This approximation can be modified, as shown in Chapter 7, or computed as a symmetric and positive definite approximate inverse, as shown in Section 8.3. Such methods have already been extensively studied for M-matrices in block tridiagonal form by Axelsson, Brinkkemper, and Il'in (1984) and by Concus, Golub, and Meurant (1985) and for matrices with a general sparsity pattern by Axelsson (1985) and Axelsson and Polman (1986). The idea of combining explicit and implicit approximation methods goes back to Underwood (1976). See also Axelsson (1985), Kolotilina and Yeremin (1986), and Diaz and Macedo (1989).

For matrices A in block tridiagonal form, i.e.,

$$A = \text{block tridiag}(A_{i,i-1}, A_{i,i}, A_{i,i+1})_{i=1}^{n},$$

we have seen in Chapter 1 that, assuming D_i, $i = 1, 2, \ldots, n$ are nonsingular, A can be factored in two forms:

(a) $A = (D + L)D^{-1}(D + U) = (D + L)(I + D^{-1}U)$.
(b) $A = (\widetilde{D}^{-1} + L)\widetilde{D}(\widetilde{D}^{-1} + U) = (\widetilde{D}^{-1} + L)(I + \widetilde{D}U)$.

The second form was called inverse free form (meaning that linear systems could be solved using just matrix-vector products). Here $A = \widehat{D} + L + U$, where $\widehat{D}, L$, and U are the block-diagonal and the lower and upper block triangular parts of A, respectively. The matrices $D, \widetilde{D}$ satisfy the recursions

$$D_1 = A_{11}, \quad D_i = A_{i,i} - A_{i,i-1}D_{i-1}^{-1}A_{i-1,i}, \qquad i = 2, \ldots, n$$

and

$$\widetilde{D}_0 = 0, \quad \widetilde{D}_i = (A_{i,i} - A_{i,i-1}\widetilde{D}_{i-1}A_{i-1,i})^{-1}, \qquad i = 1, 2, \ldots, n,$$

respectively. As we saw in Chapter 7, the factorization exists, that is, D_i (and hence $\widetilde{D}_i$) are nonsingular if either A is s.p.d. or A is a block H-matrix (in particular, if A is a M-matrix).

In general, the matrices $D_i, \widetilde{D}_i$ are full, even if $A_{i,j}$, $j = i - 1, i, i + 1$ are sparse, because the inverse of a sparse matrix is full, in general (excluding special matrices). Therefore, to compute a sparse matrix approximate factorization of A, we can use sparse approximations of the inverses of the form discussed earlier in this chapter, and we may even use further sparse approximations of the matrix product terms that occur during the factorization. Hence, to compute a sparse approximate factorization of the form

$$C = (D + L)D^{-1}(D + U)$$

or

$$C = (\widetilde{D}^{-1} + L)\widetilde{D}(\widetilde{D}^{-1} + U),$$

we let

$$D_1 = A_{1,1}, \quad D_i = A_{i,i} - Y[A_{i,i-1}X(D_{i-1}^{-1})A_{i-1,i}], \quad i = 2, \ldots, n$$

and

$$\widetilde{D}_0 = 0, \quad \widetilde{D}_i = X[(A_{i,i} - Y[A_{i,i-1}\widetilde{D}_{i-1}A_{i-1,i}])^{-1}], \quad i = 1, 2, \ldots, n.$$

Here $X[B^{-1}]$ denotes a sparse approximation to B^{-1}, computed by one of the previously presented methods for instance, and $Y[B]$ denotes a sparse part of B. (Sometimes $Y[B] = B$, if B is already sufficiently sparse.) In both cases, we need to specify the sparsity pattern used. In many problems, such as for elliptic difference equations, a band-matrix structure is most convenient. This follows because frequently the entries of the inverse of a band matrix decay as fast as their distance to the diagonal increases. In particular, this is the case for diagonally dominant matrices.

As shown in Chapter 7, instead of completely neglecting entries outside the chosen sparsity pattern, we may use these entries in the following way. Here we modify the diagonal so that for the value 1 of the method relaxation parameter ω, the action of the approximating matrices D_i, $\widetilde{D}_i$ on a certain vector is the same as the given exact matrix. This implies that the matrix C has the same action as A for a corresponding block vector. This modification is, in fact, a special form of a method of probing that will be studied in more detail in the following section.

Let $\mathbf{v} = (\mathbf{v}_1, \mathbf{v}_2, \ldots, \mathbf{v}_n)$ be a block vector with nonzero components, partitioned consistently with the partitioning of A. For M-matrices or for block H-matrices, the vector $\mathbf{v}$ is positive such that for M-matrices $A\mathbf{v} > 0$ and, correspondingly, $\mathcal{M}_b(A)\tilde{\mathbf{v}} > 0$ for block H-matrices, where $\tilde{\mathbf{v}}$ is a certain n-dimensional vector.

The relaxed version of the incomplete factorization method takes the form

$$D_1 = A_{1,1}, \quad D_i = A_{i,i} - Y[A_{i,i-1}X[D_{i-1}^{-1}]A_{i-1,i}] - \omega\Delta_i, \quad i = 2, \ldots, n,$$

where Δ_i is a diagonal matrix chosen to satisfy

(8.18) $\Delta_i \mathbf{v}_i = A_{i,i-1} D_{i-1}^{-1} A_{i-1,i} \mathbf{v}_i - Y[A_{i,i-1} X[D_{i-1}^{-1}] A_{i-1,i}] \mathbf{v}_i.$

Here we usually choose the method parameter ω in the inverval $0 \le \omega \le 1$, but in some cases even a negative value can be necessary for getting sufficiently large diagonal entries. Since

$$C = D + L + U + LD^{-1}U$$

$$= A + D - (\widehat{D} - LD^{-1}U)$$

and $(LD^{-1}U)_{i,i} = A_{i,i-1} D_{i-1}^{-1} A_{i-1,i}$, it is readily seen that for $\omega = 1$ we have

$$C\mathbf{v} = A\mathbf{v},$$

i.e., the approximate factorization matrix C has the same action as A for the vector $\mathbf{v}$.

It can be shown that the components of Δ_i are positive for the case of M-matrices (see, for instance, Axelsson and Polman 1986).

Note that the computation of the entries of Δ_i is inexpensive if D_{i-1} has a band structure, for instance. The right-hand side of (8.18) is computed using matrix vector multiplications with sparse matrices and a solution of a linear system with matrix D_{i-1}. The latter can be done cheaply if D_{i-1} is a band matrix. The choice $\omega = 1$ and $\mathbf{v} = \mathbf{e} = (1, 1, \ldots, 1)^T$ leads to so-called modified factorization methods of generalized SSOR type.

For the standard difference equation for second-order elliptic problems with constant coefficients on a unit square, the condition numbers are reduced dramatically using the above preconditioner. The condition number of the corresponding difference matrix A_h can be shown to be $\kappa(A_h) = (2/\pi)^2 h^{-2} - 2/3 + O(h^2)$, $h \to 0$, while for the preconditioned matrix $C_h^{-1} A_h$ with $\omega = 1$ it is

$$\kappa(C_h^{-1} A_h) \simeq 0.08 h^{-1} + 0.25 + 3h + O(h^2), \quad h \to 0,$$

when we use tridiagonal approximations $X[B^{-1}]$ of the so arising inverses of tridiagonal matrices B. (We have used here a natural ordering of the meshpoints. The condition number $\kappa(C_h^{-1} A_h)$ has been computed numerically.) As the asymptotic convergence factor of the preconditioned Chebyshev iteration method (or of the preconditioned conjugate gradient method) is

$(1 - \sqrt{1/\kappa})/(1 + \sqrt{1/\kappa})$, it can be seen to be less than $1/2$ if $h \geq 10^{-2}$, that is, a fairly fast decay.

Consider now the second, inverse free form. The modified version to recursively compute the matrix sequence $\{\tilde{D}_i\}$ takes the form

$$\tilde{D}_0 = 0, \quad B_i = A_{i,i} - Y[A_{i,i-1}\tilde{D}_{i-1}A_{i-1,i}],$$

$$\tilde{D}_i = X[B_i^{-1}] + \omega\Delta_i, \quad i = 1, 2, \ldots, n,$$

where Δ_i is a diagonal matrix determined by

(8.19) $$\Delta_i B_i \mathbf{v}_i = (I - X[B_i^{-1}]B_i)\mathbf{v}_i.$$

Here, $\mathbf{v} = (\mathbf{v}_1, \ldots, \mathbf{v}_n) > \mathbf{0}$. For $\omega = 1$ we then have

$$\tilde{D}_i B_i \mathbf{v}_i = \mathbf{v}_i,$$

from which it follows as before that

$$C\mathbf{v} = A\mathbf{v}$$

in that case. Note that for the inverse free form, the matrix Δ_i can be computed directly from matrix vector products and there is no need to solve a linear system of equations as in (8.18).

Naturally, we must show that each component of $B_i\mathbf{v}_i$ is nonzero. This can be done for M-matrices, for instance (see Axelsson and Polman, 1986). As has been discussed earlier in this chapter, a sparse approximation—such as on band matrix form—of the inverse of a positive definite, matrix is not always positive definite, but may be indefinite. However, preservation of positive definiteness for Stieltjes matrices has been shown to hold in the above reference if we let $\omega = 1$ or take ω sufficiently close to 1. Here $X[B_i^{-1}]$ is assumed to be a symmetric approximation of B_i^{-1}. For $\omega = 1$, it follows in fact from the diagonal compensation Theorem 8.4.

8.4.3 The Odd-Even Reordering Method

As has been shown in Chapter 1, the odd-even reordering method, when applied recursively, can be efficient for the solution of tridiagonal systems of equations, in particular for parallel computers. This method is applicable also for block tridiagonal matrices and, as remarked in Section 8.1, has some similarities with the explicit approximation method to compute an approximate inverse. Consider, then,

$$A = \text{block tridiag}(A_{i,i-1}, A_{i,i}, A_{i,i+1}), \quad i = 1, \ldots, n,$$

which we reorder, using the odd-even reordering, to

$$\widehat{A}^{(0)} = \begin{bmatrix} \widehat{A}_{11}^{(0)} & \widehat{A}_{12}^{(0)} \\ \widehat{A}_{21}^{(0)} & \widehat{A}_{22}^{(0)} \end{bmatrix}.$$

Then $\widehat{A}_{11}^{(1)}$, $\widehat{A}_{22}^{(2)}$ becomes block diagonal and $\widehat{A}_{12}$ and $\widehat{A}_{21}$ become block bidiagonal matrices. It is readily seen that each matrix (or all matrices within a block row) of the reduced system

$$A^{(1)} = \widehat{A}_{22}^{(0)} - \widehat{A}_{21}^{(0)} \widehat{A}_{11}^{(0)^{-1}} \widehat{A}_{12}^{(0)}$$

can be reordered in the same way, and we get

$$\widehat{A}^{(1)} = \begin{bmatrix} \widehat{A}_{11}^{(1)} & \widehat{A}_{12}^{(1)} \\ \widehat{A}_{21}^{(1)} & \widehat{A}_{22}^{(1)} \end{bmatrix},$$

a matrix of the order of $\widehat{A}_{22}^{(0)}$ that has about only half the number of block rows as in $\widehat{A}^{(0)}$. This can be reduced in its turn, reordered and reduced etcetera, until after at most $\lfloor \log_2 n \rfloor$ steps we are left with just a single block matrix of the order of corresponding original diagonal matrix— say, A_{i_0,i_0}. The method can also be used to factorize the block matrix, and it will then give an exact factorization.

As for the block matrix factorization for the lexicographic ordering (i.e., no reordering) of a block tridiagonal matrix, the inverses of the block matrices will be full matrices, in general, even if the individual block matrices $A_{i,j}$ are sparse. Hence, as before, for practical reasons we must approximate these by some sparse matrices. We consider, then,

$$\widehat{A}^{(0)} = \begin{bmatrix} D^{(0)} & E^{(0)} \\ F^{(0)} & G^{(0)} \end{bmatrix},$$

where

$$D_{i,i}^{(0)} = A_{i,i}^{(0)}, \quad i = 1, 2, \ldots, \left\lfloor \frac{n+1}{2} \right\rfloor, \quad D_{i,j}^{(0)} = 0, \quad i \neq j,$$

$$E_{i,i}^{(0)} = A_{i,i+\lfloor \frac{n+1}{2} \rfloor}, \quad i = 1, 2, \ldots, \left\lfloor \frac{n-1}{2} \right\rfloor$$

$$E_{i,i-1}^{(0)} = A_{i,i+\lfloor \frac{n-1}{2} \rfloor}, \quad i = 2, \ldots, \left\lfloor \frac{n+1}{2} \right\rfloor, \quad E_{i,j}^{(0)} = 0, \text{ otherwise,}$$

$$F_{i,i}^{(0)} = A_{i,i-\lfloor \frac{n+1}{2} \rfloor}, \quad i = \left\lfloor \frac{n+3}{2} \right\rfloor, \ldots, n,$$

$$F_{i,i+1}^{(0} = A_{i,i-\lfloor \frac{n-1}{2} \rfloor}, \quad i = \left\lfloor \frac{n+3}{2} \right\rfloor, \ldots, n, \quad F_{i,j}^{(0)} = 0, \text{ otherwise,}$$

$$G_{i,i}^{(0)} = A_{i,i}^{(0)}, \quad i = \left\lfloor \frac{n+3}{2} \right\rfloor, \ldots, n, \quad G_{i,j}^{(0)} = 0, \quad i \neq j.$$

We now factorize $\widehat{A}^{(0)}$ as $\widehat{L}^{(1)}\widehat{U}^{(1)}$, where

$$\widehat{L}^{(1)} = \begin{bmatrix} X^{(1)-1} & 0 \\ F^{(0)} & I \end{bmatrix}, \quad \widehat{U}^{(1)} = \begin{bmatrix} I & X^{(1)}E^{(0)} \\ 0 & A^{(1)} \end{bmatrix},$$

$$X_{i,i}^{(1)} = X\left[(A_{i,i}^{(0)})^{-1}\right], \quad i = 1, 2, \ldots, \left\lfloor \frac{n+1}{2} \right\rfloor, \quad X_{i,j}^{(1)} = 0, \quad i \neq j,$$

and

$$A^{(1)} = G^{(0)} - F^{(0)}X^{(1)}E^{(0)}.$$

Note that $A^{(1)}$ is block tridiagonal. Here, as before, $X[B^{-1}]$ denotes a sparse approximation of the inverse of B.

For $A^{(1)}$ we use the same type of odd-even reordering. After factorization of the permuted matrix, denoted by $\widehat{A}^{(1)}$, we have $\widehat{A}^{(1)} = L^{(2)}U^{(2)}$, where

$$\widehat{A}^{(1)} = \begin{bmatrix} D^{(1)} & E^{(1)} \\ F^{(1)} & G^{(1)} \end{bmatrix} \quad L^{(2)} = \begin{bmatrix} X^{(2)-1} & 0 \\ F^{(1)} & I \end{bmatrix}, \quad \text{and } U^{(2)} = \begin{bmatrix} I & X^{(2)}E^{(1)} \\ 0 & A^{(2)} \end{bmatrix}.$$

Here, $X^{(2)} = X[(D^{(1)})^{-1}]$ and $A^{(2)} = G^{(1)} - F^{(1)}X^{(2)}E^{(1)}$. For $A^{(2)}$ we use the same process, and this is repeated recursively at most $\lfloor \log_2 n \rfloor$ times.

The resulting matrix (which is not explicitly calculated) will then be used as a sparse preconditioner for an iterative method as described earlier. It can be seen that when the original diagonal block matrices are diagonally dominant, this property remains valid for the matrices $D_{i,i}^{(k)}$, $k = 1, 2, \ldots, \lfloor \log_2 n \rfloor$. For a model Laplacian problem on a unit square with Dirichlet boundary conditions and discretized by the usual five-point difference method, it is readily seen that the row sums of $D_{i,i}^{(k)}$ are bounded below by 2^{1-k}, $k = 0, 1, \ldots, \lceil \log_2 n \rceil$. As we shall see in the final section of this chapter, this implies that the accuracy of the approximate inverses decreases with the stage (level) number k. To balance this decreasing accuracy, it can be advisable to increase the cardinality of the sparsity set $S = S^{(k)}$. If $S^{(k)}$ defines a band matrix, we can increase the half bandwidth p, like $p = 2^k$, for instance.

8.5 Methods of Matrix Action

In some problems, the entries of the matrix we intend to approximate are too costly to compute and therefore not known explicitly. Nevertheless, the computation of the action of the matrix on vectors may still be viable. Such a case occurs for the Schur complement,

$$S = A_{22} - A_{21}A_{11}^{-1}A_{12}$$

of a matrix A partitioned in two-by-two blockform

$$A = \begin{bmatrix} A_{11} & A_{12} \\ A_{21} & A_{22} \end{bmatrix},$$

where the order of A_{11} is so large that the computation of its inverse would be too costly. If the order, say m, of A_{22} is small, it can still be efficient to compute the Schur complement explicitly by simply computing

$$(8.20) \qquad Se_i = (A_{22} - A_{21}A_{11}^{-1}A_{12})e_i, \quad i = 1, 2, \ldots, m$$

for the unit coordinate vectors, e_i. Since Se_i equals the ith column of S, the application of (8.20) will produce all the m columns of S. To compute the action of the Schur complement on these vectors, we need to perform matrix-vector multiplications with matrices A_{12}, A_{21}, and A_{22}, and, in addition, we need to solve m linear systems with the matrix A_{11}. In some problems, where we use proper substructurings or domain decompositions, A_{11} corresponds frequently to some substructures for which efficient computer software is available. At any rate, we shall assume here that linear systems with A_{11} can be solved efficiently.

If m is small (say $m \leq 10$), such a method can be efficient. When the Schur complement has been computed, we solve the system(s) with S using a direct solution method (see Chapter 1). If m is not particularly small, it can be more efficient to use an iterative solution method, which also requires only the computation of the action of the Schur complement on some vectors—namely, in this case, on the iteration vectors. In order to speed up the rate of convergence, we need a good preconditioner. Two methods of matrix actions to construct a preconditioner are (1) the method of probing, and (2) the method of action on consistency vectors.

8.5.1 The Method of Probing

The method of probing has been extensively used, particularly in numerical optimization methods. Suppose we want to compute a band matrix B that has the same action as a matrix S, where S is not generally known in explicit form, but the action of S on vectors can be computed. Furthermore, if S is close to a band matrix of the same form and order as B, then we want B to be close to the bandpart of S. (By S being close to a band matrix, we mean that the absolute values of entries outside the band are small relative to the dominant entries in the band.) Such a matrix B can be computed by "probing" S on a particular set of vectors. This set consists of vectors with components equal to zero except at certain periodically appearing positions, where the entry is the unit number. If the semibandwidth of B is p, the period will be equal to the bandwidth, $2p + 1$. To be specific, let $p = 2$, that is, we want to compute a pentadiagonal matrix B. Let

$$\hat{\mathbf{e}}_i = (\delta_{1,i}, \delta_{2,i}, \delta_{3,i}, \delta_{4,i}, \delta_{5,i}, \delta_{1,i}, \delta_{2,i}, \delta_{3,i}, \delta_{4,i}, \delta_{5,i}, \ldots)^T, \quad i = 1, 2, \ldots, 5,$$

where $\delta_{i,j}$, is the Kronecker symbol, and note that for any pentadiagonal matrix B,

$$B\hat{\mathbf{e}}_1 = (b_{1,1}, b_{2,1}, b_{3,1}, b_{4,6}, b_{5,6}, b_{6,6}, b_{7,6}, b_{8,6}, \ldots)^T$$

$$B\hat{\mathbf{e}}_2 = (b_{1,2}, b_{2,2}, b_{3,2}, b_{4,2}, b_{5,7}, b_{6,7}, b_{7,7}, b_{8,7}, \ldots)^T$$

$$B\hat{\mathbf{e}}_3 = (b_{1,3}, b_{2,3}, b_{3,3}, b_{4,3}, b_{5,3}, b_{6,8}, b_{7,8}, b_{8,8}, \ldots)^T$$

$$B\hat{\mathbf{e}}_4 = (0, b_{2,4}, b_{3,4}, b_{4,4}, b_{5,4}, b_{6,4}, b_{7,9}, b_{8,9}, \ldots)^T$$

$$B\hat{\mathbf{e}}_5 = (0, 0, b_{3,5}, b_{4,5}, b_{5,5}, b_{6,5}, b_{7,5}, b_{8,10}, \ldots)^T,$$

Hence, the nonzero entries of any row of B can be found in the above vectors, and B can be computed to have the same action as S on these vectors $\hat{\mathbf{e}}_i$,

$i = 1, 2, \ldots, 5$ by equating the entries of $B\hat{e}_i$ with the corresponding entries of the vectors Se_i. (Some entries in S will be neglected.) From $B\hat{e}_1$ we find the entries in the 1st, 6th, 11th, ... columns of B; from $B\hat{e}_2$, the entries in the 2nd, 7th, 12th, ...; from $B\hat{e}_3$, the entries in the 3rd, 8th, 13th, ... ; from $B\hat{e}_4$, the entries in the 4th, 9th, 14th; and, finally, from $B\hat{e}_5$, we find the entries in the 5th, 10th, 15th, etcetera columns of B.

As can be seen, the method is easy to apply, and the matrix-vector multiplications can occur in parallel. A disadvantage with this method is that we can loose positive definiteness—that is, B can become indefinite even if S is definite, as the following example shows.

Example 8.5 Let $S = \begin{bmatrix} 2 & 0 & 0 & -3 \\ 0 & 1 & 0 & 0 \\ 0 & 0 & 1 & 0 \\ -3 & 0 & 0 & 5 \end{bmatrix}$ and compute a tridiagonal matrix B having the same actions as S on the vectors $e_i = (\delta_{1,i}, \delta_{2,i}, \delta_{3,i}, \delta_{1,i}, \delta_{2,i}, \delta_{3,i}, \ldots)^T$, $i = 1, 2, 3$. Note that S is positive definite. Then we get

$$B = \begin{bmatrix} -1 & 0 & 0 & 0 \\ 0 & 1 & 0 & 0 \\ 0 & 0 & 1 & 0 \\ 0 & 0 & 0 & 2 \end{bmatrix},$$

which is indefinite.

8.5.2 Method of Action on Consistency Vectors

Let us now consider an alternative method, where positive definiteness will be preserved. This method will be presented for symmetric and tridiagonal matrices B only. The idea, again, is to construct an approximation B of G that is exact on subspaces spanned by some particular vectors. Here, these vectors will be chosen to guarantee that B becomes s.p.d. when G is s.p.d. The method was first presented in Axelsson and Polman (1988).

Lemma 8.22 *Let G be some possibly implicitly defined symmetric matrix of order m, and let $\mathbf{e} = (1, 1, \ldots, 1)^T$ and $\mathbf{v} = (1, 2, \ldots, m)^T$. There then exists a unique symmetric, tridiagonal matrix $B = \text{tridiag}(b_{r,r-1}, b_{r,r}, b_{r,r+1})$, $r = 1, 2, \ldots, m$, such that $B\mathbf{e} = G\mathbf{e}$ and $B\mathbf{v} = G\mathbf{v}$. Furthermore, let $\mathbf{b} = (b_{2,1}, b_{3,2}, \ldots, b_{m,m-1})^T$ and*

$$L = \begin{bmatrix} 1 & 0 & \cdots & & 0 \\ 1 & 1 & \cdots & & 0 \\ \vdots & & \ddots & & \\ 1 & 1 & \cdots & 1 & 0 \end{bmatrix}_{(m-1) \times m}$$

Then

(8.21) $$\mathbf{b} = L(GD - DG)\mathbf{e},$$

where $D = \text{diag}(0, 1, \ldots, m - 1)$. *Also*

$$b_{r,r} = (G\mathbf{e})_r - b_{r,r-1} - b_{r,r+1}.$$

Proof Note first that $\mathbf{v} = (I + D)\mathbf{e}$, and

$$\{B(\mathbf{v} - r\mathbf{e})\}_r = \{G(\mathbf{v} - r\mathbf{e})\}_r, \quad r = 1, 2, \ldots, m.$$

It is readily seen that

$$\{G(\mathbf{v} - r\mathbf{e})\}_r = \{(GD - DG)\mathbf{e}\}_r$$

and, similarly, with $b_{1,0} = 0$,

$$\{B(\mathbf{v} - r\mathbf{e})\}_r = \{(BD - DB)\mathbf{e}\}_r$$

$$= -b_{r,r-1} + b_{r,r+1}$$

$$= -b_{r,r-1} + b_{r+1,r}, \quad r = 1, 2, \ldots, m - 1.$$

Hence,

$$\begin{bmatrix} 1 & & & 0 \\ -1 & 1 & & \\ & & \ddots & \\ 0 & & -1 & 1 \end{bmatrix} \mathbf{b} = \{(GD - DG)\mathbf{e}\}_1^{m-1}$$

or

$$\mathbf{b} = \begin{bmatrix} 1 & & & 0 \\ 1 & 1 & & \\ \vdots & & \ddots & \\ 1 & 1 & \cdots & 1 \end{bmatrix} \{(GD - DG)\mathbf{e}\}_1^{m-1},$$

which shows (8.21). Finally, $(B\mathbf{e})_r = (G\mathbf{e})_r$ shows the expression for $b_{r,r}$. ◇

The vectors **e** and **v** are called consistency vectors because, for difference or finite element matrices corresponding to second- or higher-order elliptic differential equations, they belong to the nullspace of the matrices (if we neglect boundary condition effects). As Lemma 8.22 shows, the algorithm for the computation of the tridiagonal matrix, using the action on these vectors, takes the following form:

Algorithm (Computation of a symmetric tridiagonal matrix, using matrix action on consistency vectors)

$$b_{1,0} := 0, \quad \mathbf{h}^{(1)} := G\mathbf{e}, \quad \mathbf{h}^{(2)} := G\mathbf{v};$$

for $r = 1, 2, \ldots, m - 1$, do

$$b_{r+1,r} := b_{r,r+1} := b_{r,r-1} + h_r^{(2)} - r h_r^{(1)},$$

$$b_{rr} := h_r^{(1)} - b_{r,r-1} - b_{r,r+1}.$$

We next show that B is at least as positive definite as G if the entries of G are nonpositive outside its tridiagonal part.

Theorem 8.23 *Let G be symmetric with entries $g_{i,j}$ satisfying $g_{i,j} \leq 0$, $j \geq i + 2$. Then the quadratic form of B defined in Lemma 8.20 is bounded below by the quadratic form of G, that is,*

$$(B\mathbf{x}, \mathbf{x}) \geq (G\mathbf{x}, \mathbf{x}) \quad \forall \mathbf{x} \in \mathbb{R}^m.$$

Proof A straightforward computation shows that

$$(G\mathbf{x}, \mathbf{x}) = \sum_{i=1}^m \left(\sum_{j=1}^m g_{i,j} \right) x_i^2 - \sum_{i=1}^{m-1} \sum_{j=i+1}^m g_{i,j}(x_i - x_j)^2.$$

Hence, since $G\mathbf{e} = B\mathbf{e}$, we have

$$(G\mathbf{x}, \mathbf{x}) = \sum_{i=1}^m \left(\sum_{j=1}^m b_{i,j} \right) x_i^2 + \sum_{i=1}^{m-1} \sum_{j=i+1}^m (-g_{i,j})(x_i - x_j)^2,$$

and it remains to show that for all $\mathbf{x} \in \mathbb{R}^m$,

$$(8.22) \qquad \sum_{i=1}^{m-1} \sum_{j=i+1}^m (-g_{i,j})(x_i - x_j)^2 \leq \sum_{i=1}^{m-1} (-b_{i,i+1})(x_i - x_{i+1})^2.$$

We have

$$(x_i - x_j)^2 \le (j - i) \sum_{k=i}^{j-1} (x_k - x_{k+1})^2, \quad j \ge i + 2.$$

Hence,

$$\sum_{i=1}^{m-1} \sum_{j=i+1}^{m} (-g_{i,j})(x_i - x_j)^2 = \sum_{i=1}^{m-1} (-g_{i,i+1})(x_i - x_{i+1})^2$$

$$+ \sum_{i=1}^{m-2} \sum_{j=i+2}^{m} (-g_{i,j})(x_i - x_j)^2$$

$$\le \sum_{i=1}^{m-1} (-g_{i,i+1})(x_i - x_{i+1})^2$$

$$+ \sum_{i=1}^{m-2} \sum_{j=i+2}^{m} (-g_{i,j})(j - i) \sum_{k=i}^{j-1} (x_k - x_{k+1})^2$$

$$= \sum_{i=1}^{m-1} \sum_{j=i+1}^{m} (-g_{i,j})(j - i) \sum_{k=i}^{j-1} (x_k - x_{k+1})^2$$

$$= \sum_{k=1}^{m} (x_k - x_{k+1})^2 \sum_{i=1}^{k} \sum_{j=k+1}^{m} (j - i)(-g_{i,j})$$

$$= \sum_{k=1}^{m-1} (x_k - x_{k+1})^2 (-b_{k,k+1}),$$

which is (8.22). $\diamond$

Theorem 8.23 shows that the matrix B, computed as shown in Lemma 8.20, is always positive definite when G has the sign pattern $g_{i,j} \le 0$ for $|i - j| \ge 2$. Such sign patterns occur especially when G is a Schur complement matrix, $G = A_{22} - A_{21} A_{11}^{-1} A_{12}$, where the entries of A_{21} and A_{12} have the same sign, A_{11} is monotone, and the entries of A_{22} have the required sign pattern—i.e., $(A_{22})_{i,j} \le 0$, if $(i - j) \ge 2$. Note that if G is tridiagonal, then B will equal G. Hence, if G is almost tridiagonal, we can expect that B becomes a close approximation of G. The method can be extended to the case where $g_{i,j}$ are block matrices of the same order and we compute a block tridiagonal matrix B with block entries, $b_{r+1,r} = b_{r,r+1}$, defined by (8.21), and

$$b_{r,r} = \text{diag}(G\mathbf{e})_r - b_{r,r-1} - b_{r,r+1} + g_{r,r} - \text{diag}(g_{r,r}).$$

The last formula modifies the diagonal of $b_{r,r}$ so that the row sum criterion $B\mathbf{e} = G\mathbf{e}$ becomes satisfied. The disadvantage with the method as compared with some of the explicit approximate inverse methods of Section 8.2 is that it requires a recursive computation of the entries—i.e., they cannot be computed in parallel. However, clearly the vectors $G\mathbf{e}$ and $G\mathbf{v}$ can be computed in parallel, and the computation of the entries of B can be done efficiently on vector computers. A further disadvantage is that there is a tendency of increasing entries $b_{r,r-1}, b_{r,r}, b_{r,r+1}$ in matrix B as r increases.

A similar method of probing using vectors $\mathbf{e}_r^{(s)} = \sin(\nu_s r\pi h)$, $s = 1, 2$, and $r = 1, 2, \ldots, n$, $h = 1/(n + 1)$ corresponding to the first eigenvectors of the Laplacian operator has been used by Wittum (1991) to obtain a so-called smoothing method for multigrid iteration methods.

8.6 Decay Rates of (Block-) Entries of Inverses of (Block-) Tridiagonal s.p.d. Matrices

As is readily understood from the construction of the methods discussed in the previous sections, the rate of decay with which (block) entries of inverses decrease as their distance to the main diagonal increase is of importance when we want to estimate the accuracy of the explicit approximations and to decide which bandwidth to use for the approximation of the inverses. This topic will be addressed in this section.

The first technique uses the Chebyshev theory of best polynomial approximations to show how the entries of A^{-1} decay away from the main diagonal as a function of $q = (1 - \sqrt{a/b})/(1 + \sqrt{a/b})$, where b/a is the condition number of A. It is valid for a general s.p.d. (block) matrix. The other technique we will use (purely algebraic) is based on certain Schur complement matrices and follows the presentation in Vassilevski (1990). It assumes that A has a (block) tridiagonal form.

8.6.1 Decay Rate Estimates Based on a Chebyshev Polynomial Expansion

Consider a symmetric positive definite band matrix A. We want to estimate the rate of decay of entries of A^{-1} as a function of their distance to the main diagonal.

As every band matrix can be partitioned in a block-tridiagonal matrix, it suffices to consider matrices that are block tridiagonal. As we shall see, the rate of decay can be estimated using the condition number of A. To this end we consider the Chebyshev polynomial expansion of the scalar function x^{-1}, where $0 < a \leq x \leq b$. We have (see Meinardus, 1964)

$$x^{-1} = \frac{1}{2}c_0 + \sum_{k=1}^{\infty} c_k T_k(z),$$

where $z = \frac{2}{b-a}[x - \frac{a+b}{2}]$, and T_k are the Chebyshev polynomials

$$T_0(z) = 1, \quad T_1(z) = z, \quad T_k(z) = 2zT_{k-1}(z) - T_{k-2}(z), \quad k = 2, 3, \ldots.$$

For $-1 \leq z \leq 1$, we have $T_k(z) = \cos k\varphi$, $\varphi = \cos^{-1} z$. Noting that $x^{-1} = \frac{2}{b-a} \cdot \frac{1}{z+\beta}$, $\beta = \frac{b+a}{b-a}$, and using the orthogonality property $\int_0^\pi \cos k\varphi \cos j\varphi d\varphi = 0, k \neq j$ and $\int_0^\pi (\cos k\varphi)^2 d\varphi = \frac{\pi}{2}$, one finds

$$c_k = \frac{4}{(b-a)\pi} \int_0^\pi \frac{\cos k\varphi}{\cos\varphi + \beta} d\varphi,$$

which can be computed exactly to give

$$c_k = \frac{2}{\sqrt{ab}}(-q)^k, \quad q = \frac{1 - \sqrt{a/b}}{1 + \sqrt{a/b}}.$$

Similarly, if A is a matrix with spectrum in $[a, b]$, where $a > 0$, we find

$$A^{-1} = \frac{c_0}{2}I + \sum_{k=1}^{\infty} c_k T_k(Z),$$

where

$$Z = \frac{2}{b-a}[A - \frac{a+b}{2}I].$$

Note that the spectrum of Z is contained in $[-1, 1]$.

We want to estimate the maximum norm of entries outside a symmetric bandpart of A^{-1} as a function of the bandwith. To this end, we need the following lemmas. This presentation follows closely the one in Demko, Moss, and Smith (1984).

Lemma 8.24 *Let B be a block tridiagonal matrix, where each block has order $p \times p$. Then, B^k is a block banded matrix with, at most, $2k + 1$ symmetrically* *located nonzero blocks in each block row. Furthermore,*

$$\|B^k\|_\infty \le \sqrt{p(2k+1)}\,\|B^k\|_2.$$

Proof The first statement follows readily by induction. Since the block matrices of B have order p, it follows also that there are at most $p(2k+1)$ nonzero (scalar) entries in a row of B^k. Now,

$$\|B^k\|_\infty = \sup_{\mathbf{x}}\sup_i\{|(B^k\mathbf{x})_i|;\ \|\mathbf{x}\|_\infty = 1,\ x_j \ne 0 \text{ for at most } p(2k+1) \text{ entries}\}.$$

Since $\|B^k\mathbf{x}\|_\infty \le \|B^k\mathbf{x}\|_2$ and $\|\mathbf{x}\|_2 \le \sqrt{p(2k+1)}\|\mathbf{x}\|_\infty$, and the inequalities are sharp, we find

$$\|B^k\|_\infty \le \sup_{\mathbf{x}}\{\|B^k\mathbf{x}\|_2;\ \|\mathbf{x}\|_\infty = 1,\ x_j \ne 0 \text{ for at most } p(2k+1) \text{ entries}\}$$

$$\le \sqrt{p(2k+1)}\,\sup_{\mathbf{x}}\{\|B^k\mathbf{x}\|_2;\ \|\mathbf{x}\|_2 = 1\}$$

$$= \sqrt{p(2k+1)}\,\|B^k\|_2. \qquad\qquad \Diamond$$

Lemma 8.25 *Let* $q = (1 - \sqrt{a/b})/(1 + \sqrt{a/b})$. *Then,*

$$\sum_{k=r}^{\infty} q^k \sqrt{k} \le q^r\left[\frac{1}{4}\left(\frac{\pi}{2}\right)^{\frac{1}{2}}\left(\frac{b}{a}\right)^{\frac{3}{4}} + \frac{1}{2}\sqrt{r}\left(\left(\frac{b}{a}\right)^{\frac{1}{2}} + 1\right) + g\right],$$

where $g = \left(\frac{b}{a}\right)^{\frac{1}{4}}/(2\sqrt{e})$.

Proof Let $\alpha = \ln q^{-1}$ and note that $q^k = e^{-\alpha k}$. It is also readily seen that $\alpha \ge 2\sqrt{\frac{a}{b}}$. We have

$$\sum_{k=r}^{\infty} q^k \sqrt{k} = q^r \sum_{k=0}^{\infty} q^k \sqrt{k+r} \le q^r \sum_{k=0}^{\infty} q^k(\sqrt{k} + \sqrt{r}).$$

From an elementary computation using the graph of the function $f(x) = \sqrt{x}e^{-\alpha x}$, it follows that for the first term above

$$\sum_{k=0}^{\infty} q^k \sqrt{k} \le \int_0^{\infty} \sqrt{x}e^{-\alpha x}dx + g.$$

Note: f attains a maximum at $x = \frac{1}{2\alpha}$ and $f(\frac{1}{2\alpha}) \leq g$. Using the substitution $t = \alpha x$ and subsequently $t = y^2$, we find

$$\sum_{k=0}^{\infty} q^k \sqrt{k} \leq \alpha^{-\frac{3}{2}} \int_0^{\infty} \sqrt{t} e^{-t} dt + g$$

$$= \alpha^{-\frac{3}{2}} \int_0^{\infty} 2y^2 e^{-y^2} dy + g.$$

By partial integration we find

$$\int_0^{\infty} 2y^2 e^{-y^2} dy = \int_0^{\infty} d(-e^{-y^2}) y = \int_0^{\infty} e^{-y^2} dy$$

which, as is well known, takes the value $\frac{1}{2}\sqrt{\pi}$. For the second term, we find

$$\sum_{k=0}^{\infty} \sqrt{r} q^k = \sqrt{r} \frac{1}{1-q} = \frac{1}{2}\sqrt{r}[(\frac{b}{a})^{\frac{1}{2}} + 1],$$

which completes the proof. $\diamond$

Note that g is a lower-order term that can be neglected when $b \gg a$.

Let $X^{(s)}[A^{-1}]$ denote the symmetric bandpart of A^{-1}, which consists of $2s + 1$ blocks, and let the remainder of $X^{(s)}$ contain zero matrix blocks. We want to estimate the size of the entries outside the band when $s = r - 1$. To this end, we consider $\|A^{-1} - X^{(r-1)}[A^{-1}]\|_\infty$. The following theorem holds.

Theorem 8.26 *Let A be s.p.d., with spectrum contained in the interval $[a, b]$, $a > 0$. Then,*

$$\|A^{-1} - X^{(r-1)}[A^{-1}]\|_\infty \leq \sqrt{2p+1} q^r \left[(\frac{b}{a})^{\frac{1}{4}} a^{-1} \sqrt{\frac{\pi}{2}} + \sqrt{2r}\, a^{-1} + \tau \right],$$

where $q = (1 - \sqrt{a/b})/(1 + \sqrt{a/b})$ and τ contains lower-order forms.

Proof Since all terms of A^{-1} up to the $(r-1)$st are contained in the block band part $X^{(r-1)}[A^{-1}]$, Lemma 8.24 shows that

$$\|A^{-1} - X^{(r-1)}[A^{-1}]\|_\infty \leq \sum_{k=r}^{\infty} |c_k| \|T_k(Z)\|_\infty$$

$$\leq \sum_{k=r}^{\infty} \frac{2}{\sqrt{ab}} q^k \sqrt{p(2k+1)} \|T_k(Z)\|_2.$$

Using $p(2k + 1) \leq \frac{3}{2}k(2p + 1)$ and $\|T_k(Z)\|_2 = 1$ (because the eigenvalues of Z are contained in $[-1, 1]$), we find that

$$\|A^{-1} - X^{(r-1)}[A^{-1}]\|_\infty \leq \sqrt{3p + \frac{3}{2}\frac{2}{\sqrt{ab}}\sum_{k=r}^\infty q^k\sqrt{k}},$$

and Lemma 8.25 shows that

$$\|A^{-1} - X^{(r-1)}[A^{-1}]\|_\infty \leq \sqrt{3p + \frac{3}{2}q^r}\left[\sqrt{\frac{\pi}{2}}(\frac{b}{a})^{\frac{1}{4}}a^{-1} + \sqrt{2r}a^{-1} + \tau\right]. \quad \Diamond$$

If a and b are the extreme eigenvalues of A, then $a = 1/\|A^{-1}\|_2$ and $b = \|A\|_2$.

Remark 8.27 If $r = 0$, Theorem 8.26 shows that

$$\|A^{-1}\|_\infty \leq \sqrt{3p + \frac{3}{2}(\frac{b}{a})^{\frac{1}{4}}}\left[\sqrt{\frac{\pi}{2}}a^{-1} + \frac{1}{2\sqrt{e}}\right],$$

where $a = 1/\|A^{-1}\|_2$ and $b = \|A\|_2$. The theorem shows that the entries in A^{-1} within matrix blocks a distance r or more from the main block diagonal decay with increasing r at least as fast as const$\times q^r$ for $r \leq \sqrt{b/a}$ and as $\sqrt{r}q^r$ for $r > \sqrt{b/a}$. Clearly, if the condition number is very large—i.e., when the value of q is close to 1—the decay of entries can be very slow. An illustration of this point follows.

Example 8.6 Let

$$A = \begin{bmatrix} 1 & -1 & & & 0 \\ -1 & 2 & -1 & & \\ & \ddots & \ddots & \ddots & \\ 0 & & & -1 & 2 \end{bmatrix}_{n \times n}, \quad \text{then}$$

$$A^{-1} = \begin{bmatrix} n & (n-1) & (n-2) & \cdots & 1 \\ (n-1) & (n-1) & (n-2) & \cdots & 1 \\ \vdots & & & & \\ 2 & 2 & 2 & \cdots & 1 \\ 1 & 1 & 1 & \cdots & 1 \end{bmatrix}.$$

On the other hand, if A is diagonally dominant, the decay can be rapid.

Let A be diagonally scaled and write A as $A = I + B$. For diagonally dominant matrices, we then have $\|B\| < 1$ and

$$\lambda_{\min} \geq a = 1 - \|B\|, \quad \lambda_{\max} \leq b = 1 + \|B\|.$$

So, when the diagonal dominance is pronounced, b/a is not large and the entries decay fast away from the main diagonal.

8.6.2 Polynomial Preconditioning

The results above can also be used to construct a polynomial matrix preconditioner to A when A is s.p.d. It can be seen that

$$G = \frac{c_0}{2} I + \sum_{k=0}^{r-1} c_k T_k(Z)$$

is s.p.d. and, hence, GA has positive eigenvalues that can be estimated using Theorem 8.26. Thus, we can apply the Chebyshev acceleration method to $GAx = Gb$ in order to solve $Ax = b$. The condition number of GA is bounded by b/a, where

$$b = 1 + \|B_r\|, \quad a = 1 - \|B_r\|,$$

and

$$\|B_r\| = \|\sum_{k=r}^{\infty} c_k T_k(Z) A\| \leq \sum_{k=r}^{\infty} |c_k| \, \|A\|,$$

so $\|B_r\| < 1$ if r is sufficiently large and $\|B_r\| \to 0$ as $O(\sqrt{r}q^r), r \to \infty$.

Note in passing that this preconditioner can be implemented efficiently. For every iteration step, we need to compute

$$G(Ax^l - b), \quad l = 0, 1, \ldots.$$

The terms in the matrix vector product $G \cdot v$, where $v = Ax^l - b$, can be computed efficiently by recursion as follows:

$$T_0(Z)v = v, \quad T_1(Z)v = Zv,$$

$$T_k(Z)v = 2Z(T_{k-1}(Z)v) - T_{k-2}(Z)v, \quad k = 2, 3, \ldots, r - 1.$$

Hence, it can be computed as a sum of vectors where each term, $c_k T_k(Z)v$, involves only one new matrix vector product and some vector operations. This can be done efficiently, especially on vector and parallel computers. Note that the complexity grows linearly with r. The condition number of GA decays (at best) as $(1 + cq^r)/(1 - cq^r)$, for some constant c ($c = c_r < 1$). Thus, the Chebyshev iteration error decays as

$$1/T_l \left([1 + \frac{1 + cq^r}{1 - cq^r}]/[1 - \frac{1 + cq^r}{1 - cq^r}]\right),$$

with the iteration number l. It can be seen that as r increases, eventually the reduction of the number of iterations does not compensate for the increased cost, $O(r)$, per iteration step. However, the method can be efficient to reduce the work for some $r > 1$ that are not too large.

Actually, we see that the best preconditioning polynomial $P_{r-1}(A)$ is the one that minimizes the condition number of $P_{r-1}(A)A$, that is, the polynomial of degree $r - 1$ for which

$$\max_{a \leq x \leq b} P_{r-1}(x)x / \min_{a \leq x \leq b} P_{r-1}(x)x$$

is least. (Note that $P_{r-1}(x)x$ must be positive in $[a, b]$). This polynomial is unique only up to a constant factor. It can be seen that the polynomial is

$$P_{r-1}(x) = \frac{1}{x}\left[T_r(\frac{b+a}{b-a}) - T_r(\frac{b+a-2x}{b-a})\right],$$

for which the condition number of $P_{r-1}(A)A$ becomes

$$\frac{T_r(\frac{b+a}{b-a}) + 1}{T_r(\frac{b+a}{b-a}) - 1}.$$

8.6.3 An Algebraic, Schur Complement Matrix Technique to Estimate the Decay of Entries of the Inverse of a Matrix

The following method to estimate the decay rate is purely algebraic. Consider, then, an s.p.d. matrix A partitioned in block matrix form

(8.23)
$$A = \begin{bmatrix} A_{11} & A_{12} \\ A_{21} & A_{22} \end{bmatrix}.$$

Let $\mathbf{v} = (\mathbf{v}_1^T, \mathbf{v}_2^T)^T$ be a partitioning of vectors $\mathbf{v}$ in the domain of definition dom(A) of A and let

$$V_1 = \left\{ \begin{bmatrix} \mathbf{v}_1 \\ 0 \end{bmatrix} \right\}, \quad V_2 = \left\{ \begin{bmatrix} 0 \\ \mathbf{v}_2 \end{bmatrix} \right\}, \quad \mathbf{v} \in \text{dom}(A)$$

be the corresponding vector spaces.

Lemma 8.28 *Let A be an s.p.d. matrix partitioned as in (8.23). Then there exists a positive constant $\gamma < 1$ such that*

$$\gamma \geq \sup_{\substack{\mathbf{v} \in V_1 \\ \mathbf{u} \in V_2}} |\mathbf{v}^T A \mathbf{u}| / \{\mathbf{v}_1^T A_{11} \mathbf{v}_1 \cdot \mathbf{v}_2^T A_{22} \mathbf{v}_2\}^{\frac{1}{2}},$$

where $\mathbf{u} = \begin{bmatrix} 0 \\ \mathbf{v}_2 \end{bmatrix}$, $\mathbf{v} = \begin{bmatrix} \mathbf{v}_1 \\ 0 \end{bmatrix}$. Here $\mathbf{v}^T A \mathbf{u} = \mathbf{v}_1^T A_{12} \mathbf{v}_2$.

Proof Let $\mathbf{v}^T = (\mathbf{v}_1^T, \mathbf{v}_2^T) \neq \mathbf{0}$. Then

$$0 < \mathbf{v}^T A \mathbf{v} = (\mathbf{v}_1^T, \mathbf{v}_2^T) A (\mathbf{v}_1^T, \mathbf{v}_2^T)^T = \mathbf{v}_1^T A_{11} \mathbf{v}_1 + 2\mathbf{v}_1^T A_{12} \mathbf{v}_2 + \mathbf{v}_2^T A_{22} \mathbf{v}_2,$$

so

$$(8.24) \qquad |\mathbf{v}_1^T A_{12} \mathbf{v}_2| < \frac{1}{2}(\mathbf{v}_1^T A_{11} \mathbf{v}_1 + \mathbf{v}_2^T A_{22} \mathbf{v}_2).$$

Let $\mathbf{w}_i = A_{i,i}^{-\frac{1}{2}} \mathbf{v}_i$. Then (8.24) shows that

$$|\mathbf{w}_1^T \widetilde{A}_{12} \mathbf{w}_2| < \frac{1}{2}[(\mathbf{w}_1^T \mathbf{w}_1) + (\mathbf{w}_2^T \mathbf{w}_2)],$$

where $\widetilde{A}_{12} = A_{11}^{-\frac{1}{2}} A_{12} A_{22}^{-\frac{1}{2}}$, so

$$(8.25) \qquad \sup_{\mathbf{w}_1, \mathbf{w}_2} \frac{|\mathbf{w}_1^T \widetilde{A}_{12} \mathbf{w}_2|}{\|\mathbf{w}_1\|^2 + \|\mathbf{w}_2\|^2} = \frac{1}{2}\widetilde{\gamma}, \quad \mathbf{w}_i = A_{ii}^{-\frac{1}{2}} \mathbf{v}_i,$$

for some $\widetilde{\gamma} < 1$. Here $\widetilde{\gamma}$ must be strictly less than one, because the vector spaces are finite dimensional and (8.24) holds for any vectors $\mathbf{v}_1, \mathbf{v}_2$. Note that by symmetry arguments, the supremum in (8.25) must be taken for some $\mathbf{w}_1$, $\mathbf{w}_2$, for which $\|\mathbf{w}_1\| = \|\mathbf{w}_2\|$. Hence,

$$\widetilde{\gamma} = \sup_{\substack{\mathbf{w}_1, \mathbf{w}_2 \\ \|\mathbf{w}_1\| = \|\mathbf{w}_2\|}} \frac{2|\mathbf{w}_1^T \widetilde{A}_{12} \mathbf{w}_2|}{\|\mathbf{w}_1\|^2 + \|\mathbf{w}_2\|^2} = \sup_{\substack{\mathbf{w}_1, \mathbf{w}_2 \\ \|\mathbf{w}_1\| = \|\mathbf{w}_2\|}} \frac{|\mathbf{w}_1^T \widetilde{A}_{12} \mathbf{w}_2|}{\|\mathbf{w}_1\| \, \|\mathbf{w}_2\|}$$

$$= \sup_{\mathbf{w}_1, \mathbf{w}_2} \frac{|\mathbf{w}_1^T \widetilde{A}_{12} \mathbf{w}_2|}{\|\mathbf{w}_1\| \, \|\mathbf{w}_2\|} = \sup_{\mathbf{v}_1, \mathbf{v}_2} \frac{|\mathbf{v}_1^T A_{12} \mathbf{v}_2|}{\{\mathbf{v}_1^T A_{11} \mathbf{v}_1 \cdot \mathbf{v}_2^T A_{22} \mathbf{v}_2\}^{\frac{1}{2}}}. \qquad \diamond$$

Remark 8.29 The inequality,

$$|v_1^T A_{12} v_2| \leq \gamma \{v_1^T A_{11} v_1\}^{\frac{1}{2}} \{v_2^T A_{22} v_2\}^{\frac{1}{2}},$$

valid for all $v = (v_1, v_2)$, where $0 < \gamma < 1$ is called the *strengthened Cauchy-Bunyakowski-Schwarz (C.B.S.) inequality* (strengthened because $\gamma < 1$). (See Appendix A, Example 1.) In the vector space with inner product $\langle u, v \rangle = u^T A v$ and norm $\|u\| = \{u^T A u\}^{1/2}$, γ equals the cosine of the smallest angle between any pair of vectors in V_1, V_2. As we shall see, this result will turn out to be useful for estimation of the decay rates. We need also an algorithm for computation of entries of the inverse of a block tridiagonal matrix.

8.6.4 Recursive Computation of Entries of the Inverse of a Block-Tridiagonal Matrix

Consider a block-tridiagonal s.p.d. matrix A,

$$A = \begin{bmatrix} A_{1,1} & A_{1,2} & & & & 0 \\ A_{2,1} & A_{2,2} & A_{2,3} & & & \\ & \cdot & \cdot & \cdot & & \\ & & \cdot & \cdot & \cdot & \\ & & & \cdot & \cdot & \\ & & & A_{n-1,n-2} & A_{n-1,n-1} & A_{n-1,n} \\ 0 & & & & A_{n,n-1} & A_{n,n} \end{bmatrix},$$

(8.26)

and let block matrices D_i be computed recursively by

(8.27) $D_1 = A_{1,1}^{-1}, \quad D_i = (A_{i,i} - A_{i,i-1} D_{i-1} A_{i-1,i})^{-1}, \quad i = 2, \ldots, n.$

As A is s.p.d., each Schur complement which occurs during the factorization is s.p.d., so the $\{D_i\}$, being inverses of the leading diagonal blocks of these Schur complements, are s.p.d., and, in particular, nonsingular.

A modification of the BBI algorithm (8.8), with $p = q = 1$ and $p_1 = q_1 = n - 1$, can now be applied to compute the entries of $B = A^{-1}$. Note that $\widetilde{U} = I - U$, $\widetilde{L} = I - L$ and $L_{i,i-1} = D_i A_{i,i-1}$, $U_{i,i+1} = A_{i,i+1} D_{i+1}$. The recursion takes the form:

(8.28)   begin $B_{n,n} = D_n$;
            for $i = n - 1, n - 2, \ldots 1$ do
                for $k = 1, 2, \ldots, n - i$ do
                    $B_{i,i+k} = -D_i A_{i,i+1} B_{i+1,i+k}$;
                    $B_{i+k,i} = -B_{i+k,i+1} A_{i+1,i} D_i$
                end (of k loop)
                $B_{i,i} = D_i + D_i A_{i,i+1} B_{i+1,i+1} A_{i+1,i} D_i$
            end (of i loop)
            end (of algorithm)

8.6.5 Decay Rate

The decay rate can now be estimated as follows. For any i, $1 \leq i \leq n - 1$ we partition A as follows,

(8.29)
$$A = \begin{bmatrix} A_i & C_i \\ C_i^T & A_i' \end{bmatrix},$$

where

$$A_i = \begin{bmatrix} A_{11} & A_{12} & & & 0 \\ A_{21} & A_{22} & A_{23} & & \\ & \cdot & \cdot & \cdot & \\ & & \cdot & \cdot & \cdot \\ & & & \cdot & \cdot \\ 0 & & & A_{i,i-1} & A_{i,i} \end{bmatrix},$$

$$A_i' = \begin{bmatrix} A_{i+1,i+1} & A_{i+1,i+2} & & & 0 \\ A_{i+2,i+1} & A_{i+2,i+2} & A_{i+2,i+3} & & \\ & \cdot & \cdot & \cdot & \\ & & \cdot & \cdot & \cdot \\ & & & \cdot & \cdot \\ 0 & & & A_{n,n-1} & A_{n,n} \end{bmatrix},$$

and C_i is a rectangular block matrix of order $(i \times (n - i) + 1)$ with zero blocks, except at the lower left corner, where the block is $A_{i,i+1}$. Let γ_i be the constant in the strengthened C.B.S. inequality corresponding to the partitioning (8.28)—that is, let γ_i be the smallest constant in

(8.30)

$$\begin{bmatrix} \mathbf{v}_1 \\ \vdots \\ \mathbf{v}_i \end{bmatrix}^T C_i^T \begin{bmatrix} \mathbf{v}_{i+1} \\ \vdots \\ \mathbf{v}_n \end{bmatrix}$$

$$\leq \gamma_i \left\{ \begin{bmatrix} \mathbf{v}_1 \\ \vdots \\ \mathbf{v}_i \end{bmatrix}^T A_i \begin{bmatrix} \mathbf{v}_1 \\ \vdots \\ \mathbf{v}_i \end{bmatrix} \right\}^{\frac{1}{2}} \left\{ \begin{bmatrix} \mathbf{v}_{i+1} \\ \vdots \\ \mathbf{v}_n \end{bmatrix}^T A_i' \begin{bmatrix} \mathbf{v}_{i+1} \\ \vdots \\ \mathbf{v}_n \end{bmatrix} \right\}^{\frac{1}{2}}$$

for all $\mathbf{v}_1, \mathbf{v}_2, \ldots, \mathbf{v}_n$. As the left-hand side equals $\mathbf{v}_i^T A_{i,i+1}\mathbf{v}_{i+1}$, and therefore is independent of $\mathbf{v}_1, \ldots, \mathbf{v}_{i-1}, \mathbf{v}_{i+2}, \ldots, \mathbf{v}_n$, (8.30) shows that

(8.31)

$$\mathbf{v}_i^T A_{i,i+1}\mathbf{v}_{i+1} \leq \gamma_i \inf_{\begin{bmatrix} \mathbf{v}_1 \\ \vdots \\ \mathbf{v}_{i-1} \end{bmatrix}} \left\{ \begin{bmatrix} \mathbf{v}_1 \\ \vdots \\ \mathbf{v}_i \end{bmatrix}^T A_i \begin{bmatrix} \mathbf{v}_1 \\ \vdots \\ \mathbf{v}_i \end{bmatrix} \right\}^{\frac{1}{2}} \inf_{\begin{bmatrix} \mathbf{v}_{i+2} \\ \vdots \\ \mathbf{v}_n \end{bmatrix}} \left\{ \begin{bmatrix} \mathbf{v}_{i+1} \\ \vdots \\ \mathbf{v}_n \end{bmatrix}^T A_i' \begin{bmatrix} \mathbf{v}_{i+1} \\ \vdots \\ \mathbf{v}_n \end{bmatrix} \right\}^{\frac{1}{2}}$$

$$= \gamma_i \{\mathbf{v}_i^T S_i \mathbf{v}_i\}^{\frac{1}{2}} \{\mathbf{v}_{i+1}^T S_i' \mathbf{v}_{i+1}\}^{\frac{1}{2}}, \quad \text{for all } \mathbf{v}_i, \mathbf{v}_{i+1}.$$

where S_i and S_i' are the following Schur complements of A_i and A_i', respectively:

(8.32)
$$S_i = A_{i,i} - [0, \ldots, 0, A_{i,i-1}] A_{i-1}^{-1} \begin{bmatrix} 0 \\ \vdots \\ 0 \\ A_{i-1,i} \end{bmatrix}$$

and

(8.33) $$S_i' = A_{i+1,i+1} - [A_{i+1,i+2}, 0 \ldots, 0] (A_{i+1}')^{-1} \begin{bmatrix} A_{i+2,i+1} \\ 0 \\ \vdots \\ 0 \end{bmatrix}.$$

In (8.31) we have used Theorem 3.8.

This result, stated in the following lemma, is important in its own right.

Lemma 8.30 *Let the block tridiagonal matrix A in (8.26) be partitioned as in (8.29), and let the Schur complements S_i and S_i' be defined by (8.32) and (8.33), respectively. Then*

$$\mathbf{v}_i^T A_{i,i+1}\mathbf{v}_{i+1} \le \gamma_i\{\mathbf{v}_i^T S_i\mathbf{v}_i\}^{\frac{1}{2}}\{\mathbf{v}_{i+1}^T S_i'\mathbf{v}_{i+1}\}^{\frac{1}{2}}$$

for all $\mathbf{v}_i$, $\mathbf{v}_{i+1}$, where γ_i is the smallest constant in (8.31). Next, we give a relation between the Schur complements and the block matrix entries of the inverse of A.

Lemma 8.31 Let S_i, S_i' be defined as above, and let $\{B_{ij}\}$ be the block-matrix entries of the inverse of A defined by (8.26). Then

$$\begin{bmatrix} S_i & A_{i,i+1} \\ A_{i+1,i} & S_i' \end{bmatrix}^{-1} = \begin{bmatrix} B_{i,i} & B_{i,i+1} \\ B_{i+1,i} & B_{i+1,i+1} \end{bmatrix}.$$

Proof Note that (8.27) shows that $S_i = D_i^{-1}$ and $S_i' = D_{i+1}'^{-1}$, where D_j' is computed as D_i in (8.27) but backwards, $j = n, n-1, \ldots$. Hence,

$$\begin{bmatrix} S_i & A_{i,i+1} \\ A_{i+1,i} & S_i' \end{bmatrix}\begin{bmatrix} B_{i,i} & B_{i,i+1} \\ B_{i+1,i} & B_{i+1,i+1} \end{bmatrix}$$

$$= \begin{bmatrix} D_i^{-1}B_{i,i} + A_{i,i+1}B_{i+1,i} & D_i^{-1}B_{i,i+1} + A_{i,i+1}B_{i+1,i+1} \\ A_{i+1,i}B_{i,i} + D_{i+1}'^{-1}B_{i+1,i} & A_{i+1,i}B_{i,i+1} + D_{i+1}'^{-1}B_{i+1,i+1} \end{bmatrix},$$

and (8.28) shows that this matrix is equal to the identity matrix. ◇

Lemma 8.32 Let S_i, S_i' and $B_{i,i}$ be defined as above. Then

$$\frac{\mathbf{v}_i^T B_{i,i}\mathbf{v}_i}{\mathbf{v}_i^T S_i^{-1}\mathbf{v}_i} \le \frac{1}{1 - \gamma_i^2},$$

where γ_i is the smallest constant in Lemma 8.30.

Proof Lemma 8.31 shows that

$$\begin{bmatrix} I & S_i^{-\frac{1}{2}}A_{i,i+1}(S_i')^{-\frac{1}{2}} \\ (S_i')^{-\frac{1}{2}}A_{i+1,i}S_i^{-\frac{1}{2}} & I \end{bmatrix}^{-1}$$

$$= \begin{bmatrix} S_i^{\frac{1}{2}}B_{i,i}S_i^{\frac{1}{2}} & S_i^{\frac{1}{2}}B_{i,i+1}(S_i')^{\frac{1}{2}} \\ (S_i')^{\frac{1}{2}}B_{i+1,i}S_i^{\frac{1}{2}} & (S_i')^{\frac{1}{2}}B_{i+1,i+1}(S_i')^{\frac{1}{2}} \end{bmatrix},$$

so, in particular, $S_i^{\frac{1}{2}} B_{i,i} S_i^{\frac{1}{2}} = [I - (S_i')^{-\frac{1}{2}} A_{i+1,i} S_i^{-1} A_{i,i+1} (S_i')^{-\frac{1}{2}}]^{-1}$. But Lemma 8.30 shows that

$$\mathbf{v}_i^T \left(S_i^{\frac{1}{2}} B_{i,i} S_i^{\frac{1}{2}} \right)^{-1} \mathbf{v}_i = \mathbf{v}_i^T (I - (S_i)^{-\frac{1}{2}} A_{i+1,i} S_i'^{-1} A_{i,i+1} (S_i)^{-\frac{1}{2}}) \mathbf{v}_i$$

$$\geq (1 - \gamma_i^2) \mathbf{v}_i^T \mathbf{v}_i$$

for any $\mathbf{v}_i$, which proves the lemma.   ◇

Now we are in a position to formulate the main result concerning the decay rate of the block-entries of $A^{-1} = [B_{i,j}]_{i,j=1}^n$ (see Vassilevski, 1990).

Theorem 8.33 *Let A be a block-tridiagonal s.p.d. matrix, and let γ_i be the smallest constant in (8.30), $i = 1, \ldots, n - 1$. Then the following estimates are valid for the block matrix entries $B_{i,j}$ of A^{-1}:*

$$\| B_{i+k,i} \| \leq \left(\prod_{s=0}^{k-1} \gamma_{i+s} \right) \left\{ \| B_{i,i} \| \, \| B_{i+k,i+k} \| \right\}^{\frac{1}{2}},$$

where

$$\| B_{i,j} \| = \sup_{\mathbf{v}_j \neq 0} \left\{ \| B_{i,j} \mathbf{v}_j \| / \| \mathbf{v}_j \| \right\}.$$

Note: For notational simplicity, we have omitted the index notation for the norms.

Proof Algorithm (8.27) shows that

(8.34) $$B_{i+k,i} = B_{i+k,i+1}(-A_{i+1,i} D_i),$$

where $D_i = S_i^{-1}$ and S_i is defined in (8.32). By recursion, (8.34) shows that

$$B_{i+k,i} = B_{i+k,i+k} \prod_{s=k}^{1} (-A_{i+s,i+s-1} D_{i+s-1})$$

$$= B_{i+k,i+k}^{\frac{1}{2}} \prod_{s=k}^{1} (-B_{i+s,i+s}^{\frac{1}{2}} A_{i+s,i+s-1} D_{i+s-1} B_{i+s-1,i+s-1}^{-\frac{1}{2}}) B_{i,i}^{\frac{1}{2}}.$$

Hence,

$$\|B_{i+k,i}\|$$

(8.35)
$$\leq \|B_{i+k,i+k}^{\frac{1}{2}}\| \ \|B_{i,i}^{\frac{1}{2}}\| \prod_{s=1}^{k} \|B_{i+s,i+s}^{\frac{1}{2}} A_{i+s,i+s-1} D_{i+s-1} B_{i+s-1,i+s-1}^{-\frac{1}{2}}\|.$$

Now we shall estimate the factors in this product. We have

$$g_j = \|B_{j+1,j+1}^{\frac{1}{2}} A_{j+1,j} D_j B_{j,j}^{-\frac{1}{2}}\|^2$$

(8.36)
$$= \sup_{\mathbf{v}_j} \frac{\mathbf{v}_j^T B_{j,j}^{-\frac{1}{2}} D_j A_{j,j+1} B_{j+1,j+1} A_{j+1,j} D_j B_{j,j}^{-\frac{1}{2}} \mathbf{v}_j}{\mathbf{v}_j^T \mathbf{v}_j}$$

$$= \sup_{\mathbf{v}_j} \frac{\mathbf{v}_j^T D_j A_{j,j+1} B_{j+1,j+1} A_{j+1,j} D_j \mathbf{v}_j}{\mathbf{v}_j^T B_{j,j} \mathbf{v}_j}.$$

(8.28) shows that

(8.37)
$$g_j = \sup_{\mathbf{v}_j} \frac{\mathbf{v}_j^T (B_{j,j} - D_j) \mathbf{v}_j}{\mathbf{v}_j^T B_{j,j} \mathbf{v}_j} = 1 - \inf_{\mathbf{v}_j} \frac{\mathbf{v}_j^T D_j \mathbf{v}_j}{\mathbf{v}_j^T B_{j,j} \mathbf{v}_j},$$

and (8.35) and (8.36) show that

(8.38)
$$\|B_{i+k,i}\| \leq \|B_{i+k,i+k}^{\frac{1}{2}}\| \ \|B_{i,i}^{\frac{1}{2}}\| \prod_{s=0}^{k-1} g_{i+s}.$$

It remains to estimate

$$\inf_{\mathbf{v}_j} \frac{\mathbf{v}_j^T D_j \mathbf{v}_j}{\mathbf{v}_j^T B_{j,j} \mathbf{v}_j}.$$

But since $D_j = S_j^{-1}$, Lemma (8.27) shows that

$$\inf_{\mathbf{v}_j} \frac{\mathbf{v}_j^T D_j \mathbf{v}_j}{\mathbf{v}_j^T B_{i,j} \mathbf{v}_i} \geq 1 - \gamma_j^2$$

and (8.37) then shows that

$$g_j \leq 1 - (1 - \gamma_j^2) = \gamma_j^2.$$

This, together with (8.38), completes the proof, since $\| B_{i,i}^{\frac{1}{2}} \| = \| B_{i,i} \|^{\frac{1}{2}}$ because $B_{i,i}$ is s.p.d. $\Diamond$

Remark 8.34 The above estimate is sharp for scalar tridiagonal matrices A because all inequalities in the above proofs are then equalities. The theorem shows that the norm of an offdiagonal block of the inverse matrix is bounded by a product of CBS-numbers γ_{i+s} and the geometric average of the norm of the corresponding two diagonal blocks. Hence, the decay rate depends on certain products of the values of γ_{i+s}, all of which are numbers in the interval $(0, 1)$.

References

E. Asplund (1959). Inverses of matrices $\{a_{ij}\}$ which satisfy $a_{ij} = 0$ for $j > i + p$. *Mathematica Scandinavia* **7**, 57–60.

O. Axelsson (1985). A survey of preconditioned iterative methods for linear systems of algebraic equations. *BIT* **25**, 166–167.

O. Axelsson (1985). Incomplete block-matrix factorization preconditioning methods. The ultimate answer? *J. Comp. Appl. Math.* **12, 13**, 3–18.

O. Axelsson (1986). A general incomplete block-matrix factorization method. *Linear Algebra Appl.* **74**, 179–190.

O. Axelsson (1991). Preconditioning methods for block H-matrices, in *Computer Algorithms for Solving Linear Systems* (ed. E. Spedicato), NATO ASI Series vol. 77, 169–184. Berlin, Heidelberg: Springer-Verlag.

O. Axelsson, S. Brinkkemper, and V. P. Il'in (1984). On some versions of incomplete block–matrix factorization iterative methods. *Linear Algebra Appl.* **58**, 3–15.

O. Axelsson and B. Polman (1986). On approximate factorization methods for block matrices suitable for vector and parallel processors. *Lin. Alg. Appl.* **77**, 3–26.

O. Axelsson and B. Polman (1988). A robust preconditioner based on algebraic substructuring and two-level grids. In *Robust Multi-Grid Methods* ed. W. Hackbusch, pp. 1–26. *Notes on Numerical Fluid Mechanics*, Vol. 23, Wieweg, Braunschweig, 1988.

R. Beauwens and M. Ben Bouzid (1987). On sparse block factorization, iterative methods. *SIAM J. Numer. Anal.* **24**, 1066–1076.

M. W. Benson (1973). Iterative solution of large scale linear systems. Thesis, Lakehead University, Thunder Bay, Canada.

M. W. Benson and P. O. Frederickson (1982). Iterative solution of large sparse linear systems arising in certain multidimensional approximation problems. *Utilitas Math.* **22**, 127–140.

J. D. F. Cosgrove, J. C. Diaz, and A. Griewank (1992). Approximate inverse preconditionings for sparse linear systems. *Inter. J. Comp. Math.*, to appear.

P. Concus, G. H. Golub, and G. Meurant (1985). Block preconditioning for the conjugate gradient method. *SIAM J. Sci. Stat. Comp.* **6**, 220–252.

S. Demko (1986). Spectral bounds for $|A^{-1}|_\infty$. *J. Approx. Theory* **48**, 207–212.

S. Demko, W. F. Moss, and P. W. Smith (1984). Decay rates for inverses of band matrices. *Math. Comp.* **43**, 491–499.

J. C. Diaz and C. G. Macedo, Jr. (1989). Fully vectorizable block preconditionings with approximate inverses for non-symmetric systems of equations. *Int. J. Numer. Meth. Eng.* **27**, 501–522.

A. M. Erisman and W. F. Tinney (1975). On computing certain elements of the inverse of a sparse matrix. *Comm. ACM* **18**, 177–179.

P. O. Frederickson (1975). Fast approximate inversion of large sparse linear systems. *Math. Report* **7**, Lakehead University, Thunder Bay, Canada.

G. E. Forsythe and E. G. Strauss (1955). On the best conditioned matrices, *Proc. Amer. Math. Soc.* **6**, 340–345.

A. Greenbaum and G. H. Rodrique (1989). Optimal preconditioners of a given sparsity pattern. *BIT* **29**, 610–634.

W. W. Hager (1984). Condition estimates. *SIAM J. Sci. Stat. Comp.* **5**, 311–316.

L. Yu. Kolotilina (1989). On approximate inverses of block H-matrices. In *Numerical Analysis and Mathematical Modelling*, Moscow (in Russian).

L. Yu. Kolotilina and A.Yu. Yeremin (1986). On a family of two-level preconditionings of the incomplete block factorization type. *Sov. J. Numer. Anal. Math. Modelling* **1**, 292–320.

L.Yu. Kolotilina and A.Yu. Yeremin (1990). Factorized sparse approximate inverse preconditionings. Submitted to *SIMAX*.

G. Meinardus (1964). *Approximation von Funktionen und ihre numerische Behandlung*. Heidelberg: Springer-Verlag.

Hoon Liong Ong (1984). Fast approximate solution of large-scale sparse linear systems. *J. Comp. Appl. Math.* **10**, 45–54.

A. M. Ostrowski (1937). Über die Determinanten mit überwiegender Hauptdiagonale. *Comm. Math. Helv.* **10**, 69–96.

A. van der Sluis (1969). Condition numbers and equilibration of matrices. *Numer. Math.* **14**, 14–23.

K. Takahishi, J. Fagan, and M. S. Chen (1973). Formation of a sparse bus impedance matrix and its application to short circuit study. Proc. 8th PICA Conference, Minneapolis, Minn., 63–69.

R. R. Underwood (1976). An approximate factorization procedure based on the block Cholesky decomposition and its use with the conjugate gradient method. Report NEDO-11386, General Electric Co., Nuclear Energy Div., San Jose, Calif.

P. S. Vassilevski (1990). On some ways of approximating inverses of banded matrices in connection with deriving preconditioners based on incomplete block factorizations. *Computing* **43**, 277–296.

G. Wittum (1991). An ILU-based smoothing correction scheme. In *Parallel Algorithms for Partial Differential Equations*, ed. W. Hackbusch. *Proceedings at Sixth GAMM-Seminar*, Kiel, Jan. 19–21, 1990 *Notes on Numerical Fluid Mechanics*, Vol. 31, Vieweg, Braunschweig, Germany.

9

Block Diagonal and Schur Complement Preconditionings

In many applications, there is a natural partitioning of the given matrix in 2×2 blocks,

$$(9.1) \qquad A = \begin{bmatrix} A_{11} & A_{12} \\ A_{21} & A_{22} \end{bmatrix}$$

Such cases arise, for instance, when the set of unknowns has been partitioned into two sets, which we call here the "red" and the "black" sets, and the matrix has been permuted accordingly, so that A_{11}, A_{22} correspond to the equations for the red and black unknowns, respectively, and A_{12}, A_{21} define the coupling between the two sets.

As an application, consider a finite difference equation for an elliptic partial differential equation. Then such a partitioning arises when we use the familiar red-black (or checkerboard) ordering of the meshpoints, for instance. Another case where such a partitioning arises is that of domain decomposition methods, where the red points correspond to the interior points in the subdomains and the black points correspond to the points on the interior subdomain boundaries. In linearly equality constrained quadratic optimization problems, symmetric matrices of the form (9.1) arise, where $A_{22} = 0$ and A_{21} corresponds to the constraints. Naturally, it is of importance to examine if the corresponding system

$$(9.2) \qquad \begin{bmatrix} A_{11} & A_{12} \\ A_{21} & A_{22} \end{bmatrix} \begin{bmatrix} \mathbf{x}_1 \\ \mathbf{x}_2 \end{bmatrix} = \begin{bmatrix} \mathbf{b}_1 \\ \mathbf{b}_2 \end{bmatrix},$$

can be solved more efficiently by utilizing this partitioning. In this respect, note that subsystems with matrix A_{11} can be solved efficiently in the aforementioned applications for difference methods. Here, for the red-black ordering, A_{11} is a diagonal matrix for certain, five-point difference matrices, and

A_{11} is a block diagonal matrix for the domain-decomposition method. In both cases, the sub-systems in A_{11} can be solved concurrently, if a computer with parallel processors is available.

In this chapter, various methods to solve (9.2) will be examined—in particular, using the Schur complement matrix

$$S = A_{22} - A_{21} A_{11}^{-1} A_{12}$$

or approximations of this matrix. The Schur complement arises both when we use a block factorization of (9.1) and when we reduce system (9.2) to a system for the black unknowns alone. In the first case, we have

$$(9.3) \qquad A \begin{bmatrix} \mathbf{x}_1 \\ \mathbf{x}_2 \end{bmatrix} = \begin{bmatrix} I_1 & 0 \\ A_{21} A_{11}^{-1} & I_2 \end{bmatrix} \begin{bmatrix} A_{11} & A_{12} \\ 0 & S \end{bmatrix} \begin{bmatrix} \mathbf{x}_1 \\ \mathbf{x}_2 \end{bmatrix} = \begin{bmatrix} \mathbf{b}_1 \\ \mathbf{b}_2 \end{bmatrix},$$

where I_1, I_2, denote identity matrices of orders equal to the number of unknowns in the red and black sets, respectively. This system can be solved by the forward and back block-matrix substitution method.

In the second case, we solve

$$(9.4) \qquad\qquad\qquad S\mathbf{x}_2 = \mathbf{b}_2 - A_{21} A_{11}^{-1} \mathbf{b}_1.$$

If a direct solution method is used, both methods require the formation of the Schur complement matrix prior to solving systems (9.3) and (9.4). In addition, they each require the solution of two systems with matrix A_{11} and one with matrix S, plus some matrix-vector multiplications. The first system with matrix A_{11} occurs for (9.4) when we form the right-hand side, and the second system occurs when, after computing $\mathbf{x}_2$, we want to compute $\mathbf{x}_1$. It can, in fact, be seen that when we use direct solution methods, the two methods are equivalent and require the same computational effort. Only when it suffices to compute the solution in the black points is the second method computationally more efficient.

Actually, when we use direct solution methods for (9.2), we can as well use a factored form of the inverse of (9.1) and compute $A^{-1} \begin{bmatrix} \mathbf{b}_1 \\ \mathbf{b}_2 \end{bmatrix}$. Using the factorization of A, we find

$$A = \begin{bmatrix} I_1 & 0 \\ A_{21} A_{11}^{-1} & I_2 \end{bmatrix} \begin{bmatrix} A_{11} & 0 \\ 0 & S \end{bmatrix} \begin{bmatrix} I_1 & A_{11}^{-1} A_{12} \\ 0 & I_2 \end{bmatrix},$$

which shows that

$$A^{-1} = \begin{bmatrix} I_1 & -A_{11}^{-1} A_{12} \\ 0 & I_2 \end{bmatrix} \begin{bmatrix} A_{11}^{-1} & 0 \\ 0 & S^{-1} \end{bmatrix} \begin{bmatrix} I_1 & 0 \\ -A_{21} A_{11}^{-1} & I_2 \end{bmatrix},$$

or

$$A^{-1} = \begin{bmatrix} I_1 & -A_{11}^{-1}A_{12} \\ 0 & I_2 \end{bmatrix} \begin{bmatrix} A_{11}^{-1} & 0 \\ -S^{-1}A_{21}A_{11}^{-1} & S^{-1} \end{bmatrix},$$

so

$$\begin{bmatrix} \mathbf{x}_1 \\ \mathbf{x}_2 \end{bmatrix} = A^{-1} \begin{bmatrix} \mathbf{b}_1 \\ \mathbf{b}_2 \end{bmatrix} = \begin{bmatrix} -A_{11}^{-1} \left\{ A_{12}S^{-1}(\mathbf{b}_2 - A_{21}A_{11}^{-1}\mathbf{b}_1) \right\} + A_{11}^{-1}\mathbf{b}_1 \\ S^{-1}(\mathbf{b}_2 - A_{21}A_{11}^{-1}\mathbf{b}_1) \end{bmatrix}.$$

To compute the solution vector $\begin{bmatrix} \mathbf{x}_1 \\ \mathbf{x}_2 \end{bmatrix}$, we therefore need to perform the following steps:

1. Solve $A_{11}\mathbf{v}_1 = \mathbf{b}_1$.
2. $\mathbf{v}_2 := \mathbf{b}_2 - A_{21}\mathbf{v}_1$.
3. Solve $S\mathbf{x}_2 = \mathbf{v}_2$.
4. $\mathbf{y}_1 := A_{12}\mathbf{x}_2$.
5. Solve $A_{11}\mathbf{z}_1 = \mathbf{y}_1$.
6. $\mathbf{x}_1 := \mathbf{v}_1 - \mathbf{z}_1$.

For large-scale problems, iterative solution methods are usually more efficient than direct solution methods. The most important aspect of the iterative solution method is the choice of the preconditioning matrix. Various methods to construct preconditioners will be considered and the corresponding condition numbers will be analyzed. It will be seen that iterative preconditioned methods based on the reduced system (9.4) can be more efficient than iterative methods based on the global matrix in (9.2). The following constant will be seen to play a basic role in the estimation of condition numbers.

9.1 The C.B.S. Constant

Let V_1, V_2 be finite dimensional spaces, $W = V_1 \times V_2$, and let

$$A = \begin{bmatrix} A_{11} & A_{12} \\ A_{21} & A_{22} \end{bmatrix}$$

be partitioned consistently with V_1, V_2. We assume that A is symmetric and positive semidefinite and that A_{11} is positive definite. The following parameter will be of crucial importance for the analysis of condition numbers of certain preconditioned forms of A, as we shall see. Let

$$(9.5) \qquad \gamma = \sup_{\mathbf{w}_1 \in W_1, \mathbf{w}_2 \in W_2} \frac{\mathbf{w}_1^T A \mathbf{w}_2}{\{\mathbf{w}_1^T A \mathbf{w}_1 \mathbf{w}_2^T A \mathbf{w}_2\}^{\frac{1}{2}}},$$

where the subspaces W_1 and W_2 are defined by

$$W_1 = \left\{ v = \begin{bmatrix} v_1 \\ 0 \end{bmatrix}, \ v_1 \in V_1 \right\}, \quad W_2 = \left\{ v = \begin{bmatrix} 0 \\ v_2 \end{bmatrix}, \ v_2 \in V_2 \right\}.$$

Thus, γ is the cosine of the angle between the two subspaces W_1, W_2, with inner products defined by A. Hence, $\gamma \leq 1$. Note that $W_1 \cap W_2 = \begin{bmatrix} 0 \\ 0 \end{bmatrix}$, that is, the intersection of W_1, W_2, contains only the trivial element. Therefore $\gamma < 1$ if A is positive definite. When A is singular, however, we must examine the nullspace of A and of $A_{i,i}$, $i = 1, 2$. (9.5) shows that

(9.6) $|w_1^T A w_2| \leq \gamma \{ w_1^T A w_1 \cdot w_2^T A w_2 \}^{\frac{1}{2}}$ for all $w_1 \in W_1$, $w_2 \in W_2$,

and γ is the smallest possible such constant. Inequality (9.6) is referred to as the strengthened Cauchy-Schwarz-Bunyakowski (C.B.S.) inequality. The inequality is an extension of the Schwarz inequality (see Exercise 3.10b).

Using the block matrices in the partitioning of A, we find that an equivalent form of (9.5) is

(9.7) $$\gamma = \sup_{v_1 \in V_1, v_2 \in V_2} \frac{v_1^T A_{12} v_2}{\{ v_1^T A_{11} v_1 \cdot v_2^T A_{22} v_2 \}^{\frac{1}{2}}}.$$

Lemma 9.1 *Let γ be defined by (9.5) or (9.7). Then:*

(a) $\gamma \leq 1$ *and* $\gamma = 1$ *if there exists a vector* $\begin{bmatrix} v_1 \\ v_2 \end{bmatrix}$ *in* $\mathcal{N}(A)$ *for which*
    $v_2 \neq \mathcal{N}(A_{22})$.

(b) $\gamma < 1$ *if for any vector* $\begin{bmatrix} v_1 \\ v_2 \end{bmatrix}$ *in the nullspace of A it holds that*
    $A_{22} v_2 = 0$, *i.e.* $v_2 \in \mathcal{N}(A_{22})$.

(c) *Under the assumption of (b),*

$$\gamma = \sup_{\substack{v_i \in V_i \backslash \mathcal{N}(A_{ii}) \\ i=1,2}} \frac{v_1^T A_{12} v_2}{\{ v_1^T A_{11} v_1 \cdot v_2^T A_{22} v_2 \}^{\frac{1}{2}}}.$$

[*By assumption* $\mathcal{N}(A_{11}) = \{0\}$.]

Proof Since A is symmetric and positive semidefinite, it holds that $w^T A w \geq 0$, $w = \begin{bmatrix} v_1 \\ v_2 \end{bmatrix}$, or

(9.8) $v_1^T A_{11} v_1 + v_2^T A_{22} v_2 + 2 v_1^T A_{12} v_2 \geq 0$, for all $v_1 \in V_1$, $v_2 \in V_2$.

Let $\gamma(\mathbf{v}_1, \mathbf{v}_2) = \mathbf{v}_1^T A_{12} \mathbf{v}_2 / \{\mathbf{v}_1^T A_{11} \mathbf{v}_1 \cdot \mathbf{v}_2^T A_{22} \mathbf{v}_2\}^{\frac{1}{2}}$ and note that $\gamma(\alpha \mathbf{v}_1, \beta \mathbf{v}_2) = \gamma(\mathbf{v}_1, \mathbf{v}_2)$ for any $\alpha, \beta \neq 0$. Hence, in analyzing γ, it suffices to consider nonzero vectors $\mathbf{v}_1, \mathbf{v}_2$. Since A_{11} is positive definite for any $\mathbf{v}_2$ such that $\mathbf{v}_2^T A_{22} \mathbf{v}_2 \neq 0$, we can then take $\mathbf{v}_1$ such that $\mathbf{v}_1^T A_{11} \mathbf{v}_1 = \mathbf{v}_2^T A_{22} \mathbf{v}_2$. For such vectors, (9.8) shows that

$$|\mathbf{v}_1^T A_{12} \mathbf{v}_2| \leq \mathbf{v}_1^T A_{11} \mathbf{v}_1,$$

so (9.7) shows that $\gamma \leq 1$. If $\mathbf{v}_2$ is such that $\mathbf{v}_2^T A_{22} \mathbf{v}_2 = 0$, we take $\mathbf{v}_1 = \tau \hat{\mathbf{v}}_1$ for some $\hat{\mathbf{v}}_1$, where $\tau > 0$. Then (9.8) shows that

$$(9.9) \qquad \tau \hat{\mathbf{v}}_1^T A_{11} \hat{\mathbf{v}}_1 + 2 \hat{\mathbf{v}}_1^T A_{12} \mathbf{v}_2 \geq 0.$$

As this holds for any $\tau > 0$, we must have $\hat{\mathbf{v}}_1^T A_{12} \mathbf{v}_2 \geq 0$. If $A_{12} \mathbf{v}_2 \neq \mathbf{0}$, we can take $\hat{\mathbf{v}}_1^T = -A_{12} \mathbf{v}_2$, so $-\|A_{12} \mathbf{v}_2\| \geq 0$, which shows that $A_{12} \mathbf{v}_2 = \mathbf{0}$. This shows that $\mathbf{v}_2^T A_{22} \mathbf{v}_2 = 0 \Rightarrow A_{12} \mathbf{v}_2 = \mathbf{0}$. If $A\mathbf{w} \neq \mathbf{0}$, then $\mathbf{w}^T A \mathbf{w} > 0$ and (9.8) holds with inequality and shows that

$$|\mathbf{v}_1^T A_{12} \mathbf{v}_2| < \mathbf{v}_1^T A_{11} \mathbf{v}_1,$$

so

$$\frac{|\mathbf{v}_1^T A_{12} \mathbf{v}_2|}{\{\mathbf{v}_1^T A_{11} \mathbf{v}_1 \cdot \mathbf{v}_2^T A_{22} \mathbf{v}_2\}^{\frac{1}{2}}} < 1$$

for such vectors. If, on the other hand, $A\mathbf{w} = \mathbf{0}$, $\mathbf{w} = \begin{bmatrix} \mathbf{v}_1 \\ \mathbf{v}_2 \end{bmatrix}$, but $\mathbf{v}_2^T A_{22} \mathbf{v}_2 \neq 0$ and $\mathbf{v}_1 \neq \mathbf{0}$. Then, taking $\mathbf{v}_1$ such that $\mathbf{v}_1^T A_{11} \mathbf{v}_1 = \mathbf{v}_2^T A_{22} \mathbf{v}_2$, we find $\mathbf{v}_1^T A_{12} \mathbf{v}_2 = -\mathbf{v}_1^T A_{11} \mathbf{v}_1$, so $\gamma = 1$. This completes the proof of part (a).

If $A\mathbf{w} = \mathbf{0}$ and $A_{22} \mathbf{v}_2 = \mathbf{0}$, then $\mathbf{w}^T A \mathbf{w} = 0$, $\mathbf{v}_2^T A_{22} \mathbf{v}_2 = 0$. Then (9.9) shows that $A_{12} \mathbf{v}_2 = \mathbf{0}$ and, hence, $A_{11} \mathbf{v}_1 = \mathbf{0}$, that is, because A_{11} is s.p.d., $\mathbf{v}_1 = \mathbf{0}$. Thus, under the assumption of part (b), the value γ in (9.7) is not assumed for vectors $\mathbf{w} = \begin{bmatrix} \mathbf{v}_1 \\ \mathbf{v}_2 \end{bmatrix}$, where $A\mathbf{w} = \mathbf{0}$, $A_{22} \mathbf{v}_2 = \mathbf{0}$. This proves parts (b) and (c). $\diamondsuit$

We shall now derive some other useful relations involving the constant γ.

Lemma 9.2 *Let A be symmetric and positive semidefinite, with A_{11} positive definite, such that $A\mathbf{w} = \mathbf{0} \Rightarrow A_{22} \mathbf{v}_2 = \mathbf{0}$, where $\mathbf{w} = \begin{bmatrix} \mathbf{v}_1 \\ \mathbf{v}_2 \end{bmatrix}$. Then:*

(a) $\gamma^2 = \sup_{v_2 \in V_2 \backslash \mathcal{N}(A_{22})} \dfrac{v_2^T A_{21} A_{11}^{-1} A_{12} v_2}{v_2^T A_{22} v_2}$.

(b) $1 - \gamma^2 \le \dfrac{v_2^T S v_2}{v_2^T A_{22} v_2} \le 1$ *for all* v_2, *where* $S = A_{22} - A_{21} A_{11}^{-1} A_{12}$ *and where the left-hand side inequality is sharp and the right-hand side inequality is sharp if* $\mathcal{N}(A_{12})$ *is nontrivial.*

Proof Lemma 9.1 (c) shows that

$$\gamma = \sup_{v_i \in V_i \backslash \mathcal{N}(A_{ii})} \frac{v_1^T A_{12} v_2}{\{v_1^T A_{11} v_1 \cdot v_2^T A_{22} v_2\}^{\frac{1}{2}}}$$

$$= \sup_{\substack{v_1 \ne 0 \\ v_2 \in V_2 \backslash \mathcal{N}(A_{22})}} \frac{v_1^T A_{11}^{-\frac{1}{2}} A_{12} v_2}{\{v_1^T v_1 \cdot v_2^T A_{22} v_2\}^{\frac{1}{2}}},$$

where the supremum is taken for $v_1 = A_{11}^{-\frac{1}{2}} A_{12} v_2$, so

$$\gamma = \sup_{v_2 \in V_2 \backslash \mathcal{N}(A_{22})} \left\{ \frac{v_2^T A_{12}^T A_{11}^{-1} A_{12} v_2}{v_2^T A_{22} v_2} \right\}^{\frac{1}{2}}.$$

This proves part (a). Part (b) follows directly from (a). ◇

The above discussion of the C.B.S. constant has been based on presentations in Axelsson and Gustafsson (1983) and Axelsson (1982). For a similar presentation, see Eijkhout and Vassilevski (1991).

It turns out that in many applications, the constant γ does not depend on the dimension of the problem. As has been shown in Bank and Dupont (1980), Axelsson and Gustafsson (1983), Axelsson (1982), and Maitre and Musy (1982), this situation occurs, for instance, in the variational, finite element solution of elliptic boundary value problems. In such problems, γ can be computed by taking the maximum value of constants γ_i computed locally for individual elements. Therefore, γ does not depend on the number of elements. For an algebraic presentation of such results, see Axelsson and Kolotilina (1992). Lemma 9.1 (or Definition 9.5) and Lemma 9.2 give two alternative ways to compute γ; in the latter case, the computation can be done by computing an extreme eigenvalue of a generalized eigenvalue problem involving a Schur complement. We shall now analyze the condition numbers of block-diagonal, Schur complement, and full block-matrix factorization preconditioners.

It turns out that in many cases, the condition number depends only on γ. In such instances, when γ does not depend on the dimension, the rate of convergence of the corresponding preconditioned Chebyshev and conjugate gradient iteration methods does not depend on the dimension of the problem. The cost of each iteration depends mainly on the costs of solving subproblems associated with A_{11} and the Schur complement of A with respect to A_{11}. For certain multilevel extensions of the methods (not to be presented here), the cost of each iteration is typically proportional to the dimension. This shows that the cost of solving such problems is proportional to the dimension. Clearly, this is an optimal computational complexity result, which is important when very large systems of equations must be solved.

9.2 Block-Diagonal Preconditioning

In the sequel, we assume that A is symmetric and positive semidefinite, with A_{11} positive definite, and that $A\mathbf{w} = \mathbf{0} \Rightarrow A_{22}\mathbf{v}_2 = \mathbf{0}$, where $\mathbf{w} = \begin{bmatrix} \mathbf{v}_1 \\ \mathbf{v}_2 \end{bmatrix}$. Consider now preconditioning of A by a block diagonal matrix

$$B = \begin{bmatrix} B_{11} & 0 \\ 0 & B_{22} \end{bmatrix},$$

where B_{ii} are symmetric, and assume that

(9.10)
$$\alpha_1 A_{11} \le B_{11} \le \alpha_0 A_{11},$$
$$\beta_1 A_{22} \le B_{22} \le \beta_0 A_{22}$$

for some constants α_i, β_i. We assume that $\alpha_0 \ge \beta_0$, which can be achieved by scaling B_{11} properly. The inequalities are in a positive semidefinite sense—that is, $\alpha_1 \mathbf{x}_1^T A_{11} \mathbf{x}_1 \le \mathbf{x}_1^T B_{11} \mathbf{x}_1$, etcetera. Hence, B_{11} is s.p.d. and $\mathcal{N}(B_{22}) = \mathcal{N}(A_{22})$. We want to estimate the condition number of $B^\dagger A$, where $B^\dagger$ is the Moore-Penrose generalized inverse—i.e., we want to find the ratio of the extreme eigenvalues of the generalized eigenvalue problem

(9.11)
$$\lambda B\mathbf{x} = A\mathbf{x}, \quad \mathbf{x} \ne \mathbf{0}.$$

Theorem 9.3 *Let A be symmetric and positive semidefinite, with A_{11} positive definite and such that $A\mathbf{w} = \mathbf{0} \Rightarrow A_{22}\mathbf{v}_2 = \mathbf{0}$, $\mathbf{w} = \begin{bmatrix} \mathbf{v}_1 \\ \mathbf{v}_2 \end{bmatrix}$. Further, let*

$B = \mathrm{diag}(B_{11}, B_{22})$, where B_{ii} satisfies (9.10) with $\alpha_0 \geq \beta_0$. Then the condition number $\mathcal{K}$ of $B^\dagger A$ satisfies

$$\mathcal{K} \leq \frac{\alpha_0}{\alpha_1(1-\gamma^2)} \left\{ \frac{1}{2}\left(1 + \frac{\alpha_1}{\beta_1}\right) + \left[\left(\frac{1}{2}\left(1 - \frac{\alpha_1}{\beta_1}\right)\right)^2 + \frac{\alpha_1}{\beta_1}\gamma^2\right]^{\frac{1}{2}} \right\} \times$$

$$\left\{ \frac{1}{2}\left(1 + \frac{\beta_0}{\alpha_0}\right) + \left[\left(\frac{1}{2}\left(1 - \frac{\beta_0}{\alpha_0}\right)\right)^2 + \frac{\beta_0}{\alpha_0}\gamma^2\right]^{\frac{1}{2}} \right\},$$

where γ is the C.B.S. constant in (9.7) for A corresponding to the vector spaces V_1, V_2.

Proof The extreme eigenvalues of (9.11) are the extreme values of

(9.12)
$$\frac{\mathbf{x}^T A \mathbf{x}}{\mathbf{x}^T B \mathbf{x}} = \frac{\mathbf{x}_1^T A_{11}\mathbf{x}_1 + 2\mathbf{x}_1^T A_{12}\mathbf{x}_2 + \mathbf{x}_2^T A_{22}\mathbf{x}_2}{\mathbf{x}_1^T B_{11}\mathbf{x}_1 + \mathbf{x}_2^T B_{22}\mathbf{x}_2}.$$

Using (9.6) and the arithmetic-geometric inequality $\sqrt{ab} \leq \frac{1}{2}(\zeta a + \zeta^{-1}b)$, we find

$$|\mathbf{x}_1^T A_{12}\mathbf{x}_2| \leq \gamma \left\{ \mathbf{x}_1^T A_{11}\mathbf{x}_1 \cdot \mathbf{x}_2^T A_{22}\mathbf{x}_2 \right\}^{\frac{1}{2}}$$

$$\leq \frac{1}{2}\left[\gamma\zeta\mathbf{x}_1^T A_{11}\mathbf{x}_1 + \gamma\zeta^{-1}\mathbf{x}_2^T A_{22}\mathbf{x}_2 \right],$$

where ζ is a positive parameter. (9.12) then shows

$$\frac{\mathbf{x}^T A \mathbf{x}}{\mathbf{x}^T B \mathbf{x}} \leq \frac{(1+\zeta\gamma)\mathbf{x}_1^T A_{11}\mathbf{x}_1 + (1+\zeta^{-1}\gamma)\mathbf{x}_2^T A_{22}\mathbf{x}_2}{\mathbf{x}_1^T B_{11}\mathbf{x}_1 + \mathbf{x}_2^T B_{22}\mathbf{x}_2}.$$

Applying the relations (9.10), we find that

(9.13)
$$\frac{\mathbf{x}^T A \mathbf{x}}{\mathbf{x}^T B \mathbf{x}} \leq \max\left\{ \frac{1+\zeta\gamma}{\alpha_1}, \frac{1+\zeta^{-1}\gamma}{\beta_1} \right\}.$$

The minimum of this upper bound is taken for ζ, such that

$$1 + \zeta\gamma = \frac{\alpha_1}{\beta_1}(1 + \zeta^{-1}\gamma),$$

that is,

$$\zeta \gamma = \frac{1}{2} \left(\frac{\alpha_1}{\beta_1} - 1 \right) + \left[\left(\frac{1}{2} \left(\frac{\alpha_1}{\beta_1} - 1 \right) \right)^2 + \frac{\alpha_1}{\beta_1} \gamma^2 \right]^{\frac{1}{2}}.$$

Hence, (9.13) shows that, for all $\mathbf{x}$,

$$(9.14) \qquad \frac{\mathbf{x}^T A \mathbf{x}}{\mathbf{x}^T B \mathbf{x}} \leq \frac{1}{\alpha_1} \left\{ \frac{1}{2} \left(\frac{\alpha_1}{\beta_1} + 1 \right) + \left[\left(\frac{1}{2} \left(1 - \frac{\alpha_1}{\beta_1} \right) \right)^2 + \frac{\alpha_1}{\beta_1} \gamma^2 \right]^{\frac{1}{2}} \right\}.$$

In a similar way, we derive a lower bound and, if $\gamma < \zeta < \gamma^{-1}$, we get

$$\frac{\mathbf{x}^T A \mathbf{x}}{\mathbf{x}^T B \mathbf{x}} \geq \min \left\{ \frac{1 - \zeta \gamma}{\alpha_0}, \frac{1 - \zeta^{-1} \gamma}{\beta_0} \right\}.$$

We now choose ζ, such that

$$1 - \zeta^{-1} \gamma = \frac{\beta_0}{\alpha_0} (1 - \zeta \gamma),$$

that is,

$$(9.15) \qquad \zeta \gamma = \frac{1}{2} \left(1 - \frac{\alpha_0}{\beta_0} \right) + \left[\left(\frac{1}{2} \left(1 - \frac{\alpha_0}{\beta_0} \right) \right)^2 + \frac{\alpha_0}{\beta_0} \gamma^2 \right]^{\frac{1}{2}}.$$

(Note that since $\alpha_0 \geq \beta_0$ and $\gamma^2 \leq 1$, we have $\gamma^2 \leq \zeta \gamma \leq 1$, so $\gamma \leq \zeta \leq \gamma^{-1}$.)

Hence, for all $\mathbf{x}$,

$$\frac{\mathbf{x}^T A \mathbf{x}}{\mathbf{x}^T B \mathbf{x}} \geq \frac{1 - \gamma^2}{\alpha_0} \left\{ \frac{1}{2} \left(1 + \frac{\beta_0}{\alpha_0} \right) + \left[\left(\frac{1}{2} \left(1 - \frac{\beta_0}{\alpha_0} \right) \right)^2 + \frac{\beta_0}{\alpha_0} \gamma^2 \right]^{\frac{1}{2}} \right\}^{-1}.$$

(9.14) and (9.15) establish the bound for $\mathcal{K}$. $\qquad \diamond$

If the constants $\alpha_i, \beta_i, i = 1, 2$, and γ do not depend on the dimension of A, then Theorem 9.3 shows that the condition number of $B^\dagger A$ does not depend on the dimension. We now consider a special choice of B_{ii} where the expression in the upper bound simplifies.

Corollary 9.4 *If $B_{ii} = A_{ii}, i = 1, 2$, then*

$$\operatorname{cond}(B^\dagger A) = \frac{1 + \gamma}{1 - \gamma}.$$

Proof In this case, $\alpha_i = \beta_i = 1$ and Theorem 9.3 shows that

$$\text{cond}(B^{\dagger}A) \leq \frac{(1+\gamma)^2}{1-\gamma^2} = \frac{1+\gamma}{1-\gamma}.$$

Since γ is the best possible constant in (9.7), it can be seen that this bound is sharp. ◇

9.3 Schur Complement Preconditioning

Consider now solving system (9.4) by preconditioned iteration methods. Note, first, that the system can be written in the form

$$(9.16) \qquad \mathbf{r}(\mathbf{x}_2) = A_{22}\mathbf{x}_2 - \mathbf{b}_2 - A_{21}A_{11}^{-1}(A_{12}\mathbf{x}_2 - \mathbf{b}_1) = 0.$$

When residuals are computed in this way, we save the initial computation of the right-hand side vector

$$\tilde{\mathbf{b}}_2 = \mathbf{b}_2 - A_{11}^{-1}A_{12}\mathbf{b}_1,$$

which would be required if the residuals were computed by

$$(9.17) \qquad \mathbf{r}(\mathbf{x}_2) = S\mathbf{x}_2 - \tilde{\mathbf{b}}_2.$$

Hence, form (9.16) can be efficient when the cost of solving systems with A_{11} is comparatively large. Let A_{22} be the preconditioner for S. Then Lemma 9.2(b) shows that the condition number is

$$\mathcal{K}(A_{22}^{\dagger}S) = \frac{1}{1-\gamma^2}.$$

This result and Corollary 9.4 show that the ratio of the condition number for the block-diagonal, or block Jacobi preconditioning $\text{diag}(A_{11}, A_{22})$ and this condition number, is

$$\mathcal{K}(B^{\dagger}A)/\mathcal{K}(A_{22}^{\dagger}S) = \frac{(1-\gamma^2)(1+\gamma)}{1-\gamma} = (1+\gamma)^2.$$

When γ is close to 1, this ratio is close to 4, and preconditioned iterative methods such as the Chebyshev or the conjugate gradient method can be expected to converge with about $1/(1+\gamma) \sim 1/2$ iterations less for the reduced Schur complement system than for the global system.

Also note that in an iterative solution method, there is no need to form the Schur complement system explicitly. In addition, all vector operations take place for vectors with a shorter length (equal to the dimension of V_2). Note also that when the form (9.16) is used, the block vector component x_1 of the solution vector is already available, as $x_1 = A_{11}^{-1}(b_1 - A_{12}x_2)$, and this is part of the computation of $r(x_2)$. Therefore, by using the form (9.16) instead of (9.17), we in fact save *two* solutions of systems with A_{11}.

The residual is then computed in the following steps:

1. $y_1 = b_1 - A_{12}x_2$.
2. Solve $A_{11}x_1 = y_1$.
3. $y_2 = A_{22}x_2 - b_2$ (can be computed concurrently with steps 1 and 2).
4. $r(x_2) = y_2 + A_{21}x_1$.

As has been shown in Axelsson and Gustafsson (1983), there exists a C.B.S. constant γ independent of the dimension of the problem when one uses so-called *hierarchical basis function* finite element methods. In the iterative method, however, it is frequently more efficient to use standard finite element nodal basis function methods, because the corresponding matrix is more sparse than the hierarchical basis function method.

If $A = \begin{bmatrix} A_{11} & A_{12} \\ A_{21} & A_{22} \end{bmatrix}$ is the standard basis function matrix, it turns out that there exists a transformation matrix of the form

$$J = \begin{bmatrix} I_1 & K \\ 0 & I_2 \end{bmatrix},$$

which takes A into the hierarchical basis function matrix,

$$\widetilde{A} = J^T A J = \begin{bmatrix} \widetilde{A}_{11} & \widetilde{A}_{12} \\ \widetilde{A}_{21} & \widetilde{A}_{22} \end{bmatrix},$$

where $\widetilde{A}_{11} = A_{11}$, $\widetilde{A}_{12} = A_{11}K + A_{12}$, $\widetilde{A}_{21} = K^T A_{11} + A_{21}$, and

$$\widetilde{A}_{22} = A_{22} + K^T A_{12} + A_{21}K + K^T A_{11}K.$$

An elementary computation now shows that the two Schur complement matrices,

$$S(\widetilde{A}) = \widetilde{A}_{22} - \widetilde{A}_{21}\widetilde{A}_{11}^{-1}\widetilde{A}_{12}$$

and

$$S(A) = A_{22} - A_{21}A_{11}^{-1}A_{12},$$

are identical. (This observation was made by Vassilevski in 1989. See also Axelsson and Vassilevski, 1989.) Thus, the iterative method can use $S(A)$, but the computation of its rate of convergence can be based on the C.B.S. constant γ computed from $\widetilde{A}$. (The latter computation is frequently more accessible.) However, *the preconditioner must be* $\widetilde{A}_{22}$ for $\mathrm{cond}(\widetilde{A}_{22}^{\dagger}S) = 1/(1 - \gamma^2)$ to hold. In the context of finite element matrices, this matrix corresponds to the *standard finite element basis functions* on the coarser mesh.

The computation of a preconditioner to S that is more efficient on parallel computers can be done using a combination of the method to compute symmetric and positive definite preconditioners on the form $L_G L_G^T$, as presented in Chapter 8, and the method of probing to first compute approximations of the entries of S that will be needed in the computation of L_G using $(\widehat{L_G S})_{i,j} = \delta_{i,j}$, $(i, j) \in S_L$, where S_L is the chosen sparsity pattern of L_G and $\widehat{S}$ is the approximation of S that arises from the method of probing. This method can be accurate only if the condition number of S is moderate.

For the solution of second-order elliptic difference equations using so-called domain decomposition methods, where the domain of definition has been split into two parts, it will be seen in Chapter 10 that the corresponding C.B.S. constant $\gamma = 1 - |O(h)|$ and, hence, $\mathrm{cond}(B^{\dagger}S) = O(h^{-1})$ (and $\mathrm{cond}(S) = O(h^{-1})$), $h \to 0$, where h is the standard mesh size parameter and V_2 is the vector space corresponding to the points on the intersecting line.

The block diagonal and Schur complement preconditioners can be extended in various ways to multilevel versions (see Axelsson and Gustafsson, 1983, and Vassilevski, 1989).

9.4 Full Block-Matrix Factorization Methods

In the previous section, we considered a preconditioning method for the Schur complement system requiring the solution of a system with matrix A_{11} at each iteration step. Similarly, if we use $B_{11} = A_{11}$ in the block-diagonal preconditioning method, we must solve such a system at each iteration step. If the cost associated with A_{11} is large, it therefore can be more efficient to use another choice of B_{11}. In the context of finite element matrices, it turns out that A_{11} is often well-conditioned uniformly in the parameter h, enabling one to compute efficient, sparse approximations of A_{11} (or even of A_{11}^{-1}). (See Axelsson and Gustafsson, 1983, for instance.) Furthermore, it can be efficient to use a more accurate preconditioner of A, namely, by using a full block-matrix factorization method.

Consider a preconditioner of A in the factorized form,

$$(9.18) \quad C = \begin{bmatrix} C_{11} & C_{12} \\ C_{21} & C_{22} \end{bmatrix} = \begin{bmatrix} I_1 & 0 \\ A_{21}B_{11}^{-1} & I_2 \end{bmatrix} \begin{bmatrix} B_{11} & A_{12} \\ 0 & B_{22} \end{bmatrix},$$

where we assume that systems with B_{11} (which occur twice) and B_{22} can be solved with relatively little computational effort. This preconditioner may be written as

$$\begin{bmatrix} C_{11} & C_{12} \\ C_{21} & C_{22} \end{bmatrix} = \begin{bmatrix} B_{11} & A_{12} \\ A_{21} & B_{22} + A_{21}B_{11}^{-1}A_{12} \end{bmatrix}.$$

Here, in particular, we shall consider cases where B_{11} has a sparsity structure such that its inverse can be formed readily. For instance, B_{11} may be a diagonal matrix or, more generally, B_{11} can be the *inverse* of a sparse matrix, such as a bandmatrix. A convenient choice of B_{22} then is

$$B_{22} = \widetilde{B}_{22} - A_{21}B_{11}^{-1}A_{12}.$$

This matrix can be formed explicitly. Here $\widetilde{B}_{22}$ is an approximation of A_{22}. Hence,

$$C = \begin{bmatrix} B_{11} & A_{12} \\ A_{21} & \widetilde{B}_{22} \end{bmatrix}.$$

In practice, the approximation $\widetilde{B}_{22}$ should be such that the computational effort in solving the systems with B_{22}, which occur during each iteration step, typically correspond to the cost for computing some matrix-vector products with A, at most.

To estimate the condition number of $C^{-1}A$ or, in the singular matrix case, of $C^{\dagger}A$, we shall assume a spectral relation between B_{11} and A_{11} and between $\widetilde{B}_{22} = B_{22} + A_{21}B_{11}^{-1}A_{12}$ and A_{22}. Let

$$\alpha_1 A_{11} \le B_{11} \le \alpha_0 A_{11}$$

$$(9.19)$$

$$\beta_1 A_{22} \le \widetilde{B}_{22} \le \beta_0 A_{22},$$

where we assume that $\alpha_0 \ge 1 \ge \alpha_1 > \gamma^2$, $\beta_0 \ge 1 \ge \beta_1 > \gamma^2$.

Theorem 9.5 *Let* $A = \begin{bmatrix} A_{11} & A_{12} \\ A_{21} & A_{22} \end{bmatrix}$ *be symmetric and positive semidefinite where* A_{11} *is positive definite. In addition, assume that if* $A \begin{bmatrix} v_1 \\ v_2 \end{bmatrix} = 0$, *then* $A_{22}v_2 = 0$. *Let*

$$C = \begin{bmatrix} B_{11} & A_{12} \\ A_{21} & \widetilde{B}_{22} \end{bmatrix}, \quad \widetilde{B}_{22} = B_{22} + A_{21}B_{11}^{-1}A_{12},$$

where B_{11}, $\widetilde{B}_{22}$ *satisfy (9.19). Then, the extreme eigenvalues of* $C^\dagger A$ *are bounded by*

$$\lambda_{\min}(C^\dagger A) \geq \left\{ 1 + \frac{\max(\alpha_0, \beta_0) - 1}{1 - \gamma^2} \left[\frac{1 + r_0}{2} + \sqrt{\left(\frac{1 - r_0}{2} \right)^2 + r_0\gamma^2} \right] \right\}^{-1},$$

where $r_0 = \min\left\{ \frac{\alpha_0 - 1}{\beta_0 - 1}, \frac{\beta_0 - 1}{\alpha_0 - 1} \right\}$, $\alpha_0 > 1$ *and/or* $\beta_0 > 1$, *and*

$$\lambda_{\max}(C^\dagger A) \leq \left\{ 1 - \frac{1 - \min(\alpha_1, \beta_1)}{1 - \gamma^2} \left[\frac{1 + r_1}{2} + \sqrt{\left(\frac{1 - r_1}{2} \right)^2 + r_1\gamma^2} \right] \right\}^{-1},$$

where $r_1 = \min\left\{ \frac{1 - \alpha_1}{1 - \beta_1}, \frac{1 - \beta_1}{1 - \alpha_1} \right\}$, $\alpha_1 < 1$, *and/or* $\beta_1 < 1$.

Proof With $\mathbf{x}^T = (\mathbf{x}_1^T, \mathbf{x}_2^T)$ partitioned consistently with A, we have

$$\frac{\mathbf{x}^T(C - A)\mathbf{x}}{\mathbf{x}^T A \mathbf{x}} = \frac{\mathbf{x}_1^T(B_{11} - A_{11})\mathbf{x}_1 + \mathbf{x}_2^T(\widetilde{B}_{22} - A_{22})\mathbf{x}_2}{\mathbf{x}_1^T A_{11}\mathbf{x}_1 + 2\mathbf{x}_1^T A_{12}\mathbf{x}_2 + \mathbf{x}_2^T A_{22}\mathbf{x}_2},$$

and the strengthened C.B.S. inequality shows as in the proof of Theorem 9.3,

(9.19′)
$$\frac{\mathbf{x}^T(C - A)\mathbf{x}}{\mathbf{x}^T A \mathbf{x}} \leq \max\left\{ \frac{\alpha_0 - 1}{1 - \zeta_0\gamma}, \frac{\beta_0 - 1}{1 - \zeta_0^{-1}\gamma} \right\},$$

$$\gamma < \zeta_0 < \gamma^{-1}, \quad \mathbf{x} \notin \mathcal{N}(A),$$

and

$$\frac{\mathbf{x}^T(C - A)\mathbf{x}}{\mathbf{x}^T A \mathbf{x}} \geq -\max\left\{ \frac{1 - \alpha_1}{1 - \zeta_1\gamma}, \frac{1 - \beta_1}{1 - \zeta_1^{-1}\gamma} \right\},$$

$$\gamma < \zeta_1 < \gamma^{-1}, \quad \mathbf{x} \notin \mathcal{N}(A).$$

Here, we choose ζ_0 such that if $\beta_0 \geq \alpha_0$,

$$\frac{1 - \zeta_0\gamma}{1 - \zeta_0^{-1}\gamma} = r_0 = \frac{\alpha_0 - 1}{\beta_0 - 1}.$$

Then,

$$\zeta_0\gamma = \frac{1 - r_0}{2} + \sqrt{\left(\frac{1 - r_0}{2} \right)^2 + r_0\gamma^2}$$

and

$$\frac{\alpha_0 - 1}{1 - \zeta_0 \gamma} = \frac{(\alpha_0 - 1)\left[\frac{1+r_0}{2} + \sqrt{\left(\frac{1-r_0}{2}\right)^2 + r_0\gamma^2}\right]}{r_0(1 - \gamma^2)}$$

$$= \frac{(\beta_0 - 1)\left[\frac{1+r_0}{2} + \sqrt{\left(\frac{1-r_0}{2}\right)^2 + r_0\gamma^2}\right]}{1 - \gamma^2}$$

If $\alpha_0 \geq \beta_0$, the symmetric roles played by α_0 and β_0 show that we shall just exchange α_0, β_0 in the above expression. (9.19′) shows that the reciprocal of $1 + (\alpha_0 - 1)/(1 - \zeta_0 \gamma)$ becomes the lower bound of $\lambda_{\min}(C^\dagger A)$. Similarly, the upper bound of $\lambda_{\max}(C^\dagger A)$ follows. $\diamond$

9.4.1 Some Particular Choices of B_{11}, B_{22}

Depending on the type of problem, some of the following choices of B_{11}, B_{22} can be practically viable. For instance, if solutions of systems with A_{11} are relatively inexpensive, we can let $B_{11} = A_{11}$. Similarly, if systems with A_{22} can be solved readily and if it is not efficient to form a more accurate approximation $\widetilde{B}_{22}$ of A_{22} than

$$\widetilde{B}_{22} = A_{22} + A_{21}B_{11}^{-1}A_{12},$$

then we can let $B_{22} = A_{22}$. More generally, we can approximate the inverse of A_{11}^{-1} by D to let

$$B_{22} = A_{22} - A_{21}DA_{12}$$

be an approximation of S. If D is a sufficiently sparse matrix, such as a diagonal matrix, then systems with B_{22} may still be sufficiently simple to solve. Here we shall consider four different classes of approximations.

Case 9.1 $B_{11} = A_{11}$. Here, $\alpha_1 = \alpha_0 = 1$, so $r_0 = r_1 = 0$ (if $\beta_1 < 1$, $\beta_0 > 1$) and

(9.20)

$$\lambda_{\min}(C^\dagger A) \geq \left\{1 + \frac{\beta_0 - 1}{1 - \gamma^2}\right\}^{-1} = \frac{1 - \gamma^2}{\beta_0 - \gamma^2}$$

$$\lambda_{\max}(C^\dagger A) \leq \left\{1 - \frac{1 - \beta_1}{1 - \gamma^2}\right\}^{-1} = \frac{1 - \gamma^2}{\beta_1 - \gamma^2}.$$

Hence, Theorem 9.5 shows that the condition number of $C^\dagger A$ is

$$(9.21) \qquad \mathcal{K}(C^\dagger A) = \frac{\beta_0 - \gamma^2}{\beta_1 - \gamma^2}.$$

Consider preconditioning of the Schur complement $S(A) = A_{22} - A_{21}A_{11}^{-1}A_{12}$ with the Schur complement $S(C) = \widetilde{B}_{22} - A_{21}A_{11}^{-1}A_{12}$ of C. As the next lemma shows, for this case we have the same extreme eigenvalues of the generalized eigenvalue problems $\lambda A\mathbf{x} = C\mathbf{x}$ and $\lambda S(A)\mathbf{x}_2 = S(C)\mathbf{x}_2$.

Lemma 9.6 *Let* $A = \begin{bmatrix} A_{11} & A_{12} \\ A_{21} & A_{22} \end{bmatrix}$, *where* A *is symmetric and positive semidefinite,* A_{11} *is positive definite, and* $A_{22}\mathbf{v}_2 = \mathbf{0}$ *if* $A\mathbf{v} = \mathbf{0}$, $\mathbf{v} = \begin{bmatrix} \mathbf{v}_1 \\ \mathbf{v}_2 \end{bmatrix}$. *Let* $C = \begin{bmatrix} A_{11} & A_{12} \\ A_{21} & \widetilde{B}_{22} \end{bmatrix}$, *where we assume that* $\widetilde{B}_{22}$ *satisfies*

$$\beta_1 A_{22} \le \widetilde{B}_{22} \le \beta_0 A_{22}$$

and $\gamma^2 < \beta_1 \le 1 \le \beta_0$. *Then (with inequalities in a positive semidefinite sense),*

$$\lambda_1 A \le C \le \lambda_0 A$$

and

$$\lambda_1 S(A) \le S(C) \le \lambda_0 S(A),$$

where

$$S(A) = A_{22} - A_{21}A_{11}^{-1}A_{12}, \quad S(C) = B_{22} \equiv \widetilde{B}_{22} - A_{21}A_{11}^{-1}A_{12}$$

and

$$\lambda_1 = \frac{\beta_1 - \gamma^2}{1 - \gamma^2}, \quad \lambda_0 = \frac{\beta_0 - \gamma^2}{1 - \gamma^2}.$$

Proof We have

$$S(C) = S(A) + [S(C) - S(A)]$$

$$= S(A) + (\widetilde{B}_{22} - A_{22}),$$

so

$$\frac{\mathbf{x}^T S(C)\mathbf{x}}{\mathbf{x}^T S(A)\mathbf{x}} = 1 + \frac{\mathbf{x}^T (\widetilde{B}_{22} - A_{22})\mathbf{x}}{\mathbf{x}^T S(A)\mathbf{x}} \le 1 + (\beta_0 - 1)\frac{\mathbf{x}^T A_{22}\mathbf{x}}{\mathbf{x}^T S(A)\mathbf{x}},$$

and Lemma 9.2 (b) shows that

$$\frac{\mathbf{x}^T S(C)\mathbf{x}}{\mathbf{x}^T S(A)\mathbf{x}} \leq 1 + \frac{\beta_0 - 1}{1 - \gamma^2} = \frac{\beta_0 - \gamma^2}{1 - \gamma^2} = \lambda_0.$$

Similarly, we find

$$\frac{\mathbf{x}^T S(C)\mathbf{x}}{\mathbf{x}^T S(A)\mathbf{x}} \geq 1 - \frac{1 - \beta_1}{1 - \gamma^2} = \lambda_1.$$

(9.20) shows that the same extreme eigenvalues hold for the generalized eigenvalue problem $\lambda A\mathbf{x} = C\mathbf{x}$. ◇

Case 9.1(a) Let $B_{11} = A_{11}$ and assume, in addition, that $B_{22} = A_{22}$. Then $S(C) = A_{22}$ and $\tilde{B}_{22} = A_{22} + A_{21}A_{11}^{-1}A_{12}$, so $\beta_1 = 1$, $\beta_0 = 1 + \gamma^2$. Hence,

$$S(A) \leq A_{22} = S(C) \leq \frac{1}{1 - \gamma^2} S(A),$$

which is the already familiar result from Section 9.2. In this case the full block factorization preconditioning matrix can be written in the form

$$C = \begin{bmatrix} I_1 & 0 \\ A_{21}A_{11}^{-1} & I_2 \end{bmatrix} \begin{bmatrix} A_{11} & 0 \\ 0 & A_{22} \end{bmatrix} \begin{bmatrix} I_1 & A_{11}^{-1}A_{12} \\ 0 & I_2 \end{bmatrix},$$

which corresponds to a symmetric block Gauss-Seidel method. For its implementation, we use the form

$$C = \begin{bmatrix} I_1 & 0 \\ A_{21}A_{11}^{-1} & I_2 \end{bmatrix} \begin{bmatrix} A_{11} & A_{12} \\ 0 & A_{22} \end{bmatrix}.$$

This requires two solutions of systems with matrix A_{11} at each iteration step.

Table 9.1 is a comparison of the condition numbers and arithmetic work of the block Jacobi, the symmetric block Gauss-Seidel method, and the Schur complement reduced system method. Here, n_1 and n_2 are the orders of A_{11} and A_{22}, respectively.

For domain decomposition problems for second-order elliptic difference equations, A_{11} corresponds to the interior of the subdomain problems, A_{22} corresponds to the equations on the intersecting lines (planes), and, as we shall see in Chapter 10, $\gamma = 1 - O(h)$, where h is a meshwidth parameter. Hence, when $B_{22} = A_{22}$, we have

$$\mathcal{K}(C^\dagger A) = O(h^{-1}), \quad h \to 0.$$

Table 9.1.   *Condition numbers and arithmetic work of three typical methods*
for matrices partitioned in 2 × 2 blocks

Method	Condition Number	Solutions of Systems per Iteration with A_{11}	with A_{22}	Vector Length
Block Jacobi	$(1 + \gamma)/(1 - \gamma)$	1	1	$n_1 + n_2$
Symmetric block Gauss-Seidel	$1/(1 - \gamma^2)$	2	1	$n_1 + n_2$
Schur complement	$1/(1 - \gamma^2)$	1	1	n_2

Case 9.2 $B_{22} = A_{22}$, and $B_{11} \geq A_{11}$. In this case, $\tilde{B}_{22} = A_{22} + A_{12}B_{11}^{-1}A_{12} \geq A_{22}$ and $\beta_1 = 1$, $\beta_0 = 1 + \gamma^2$. Furthermore, $\alpha_1 = 1$. Hence, Theorem 9.5 shows that

$$\lambda_{\max}(C^{\dagger}A) \leq 1$$

and

$$\lambda_{\min}(C^{\dagger}A) \geq \left\{ 1 + \frac{\max(\alpha_0, \beta_0) - 1}{1 - \gamma^2} \left[\frac{1 + r_0}{2} + \sqrt{\left(\frac{1 - r_0}{2} \right)^2 + r_0\gamma^2} \right] \right\}^{-1},$$

where $r_0 = \min \left\{ \frac{\alpha_0 - 1}{\beta_0 - 1}, \frac{\beta_0 - 1}{\alpha_0 - 1} \right\}$.
If $1 \leq \alpha_0 \leq 1 + \gamma^2$, then

$$K(C^{\dagger}A) \leq 1 + \frac{\gamma^2}{1 - \gamma^2} \left[\frac{1 + r_0}{2} + \sqrt{\left(\frac{1 - r_0}{2} \right)^2 + r_0\gamma^2} \right], \ r_0 = \frac{\alpha_0 - 1}{\gamma^2},$$

and if $\alpha_0 \geq 1 + \gamma^2$, then

$$K(C^{\dagger}A) \leq 1 + \frac{\alpha_0 - 1}{1 - \gamma^2} \left[\frac{1 + r_0}{2} + \sqrt{\left(\frac{1 - r_0}{2} \right)^2 + r_0\gamma^2} \right], \ r_0 = \frac{\gamma^2}{\alpha_0 - 1}.$$

Since $\gamma^2 < 1$, this is bounded in both cases by

$$K(C^{\dagger}A) \leq \begin{cases} 1 + \dfrac{\gamma^2}{1 - \gamma^2}(1 + r_0) \\[4mm] 1 + \dfrac{\alpha_0 - 1}{1 - \gamma^2}(1 + r_0) \end{cases} = \frac{\alpha_0}{1 - \gamma^2}.$$

The above choice, $B_{22} = A_{22}$, is practically viable if systems with A_{22} are readily solvable but systems with A_{11} are not. Such a case occurs for domain decomposition problems with few domains, for example. In this case, B_{11} might be chosen as an incomplete factorization of A_{11}, scaled to make $B_{11} \geq A_{11}$.

Case 9.3 $\widetilde{B}_{22} = A_{22}$ and $B_{11} \leq A_{11}$. Here $B_{22} = A_{22} - A_{21}B_{11}^{-1}A_{12}$, so generally this choice is practically viable only when B_{11} is a matrix for which B_{11}^{-1} can be formed explicitly—for instance, when B_{11} is the inverse of a sparse matrix, such as a bandmatrix. In the simplest case, B_{11} is diagonal. Here, $\beta_1 = \beta_0 = 1$ and $\alpha_0 = 1$. Hence, Theorem 9.5 shows that

$$\lambda_{\min}(C^{\dagger}A) \geq 1,$$

$$\lambda_{\max}(C^{\dagger}A) \leq \left\{1 - \frac{1-\alpha_1}{1-\gamma^2}\right\}^{-1} = \frac{1-\gamma^2}{\alpha_1 - \gamma^2}$$

and

$$\mathcal{K}(C^{\dagger}A) \leq \frac{1-\gamma^2}{\alpha_1 - \gamma^2}.$$

An example of a problem where this situation occurs is the red-black ordering of a finite difference mesh for a nine-point difference approximation. Here, the nine-point difference approximation is first modified to a five-point difference approximation in the set of red meshpoints. This corresponds to approximating A_{11} with a diagonal matrix B_{11}. Then the corresponding matrix B_{22} can be formed explicitly and used in the preconditioning matrix C. (For details, see Axelsson and Eijkhout, 1991.)

Case 9.4 General case: Both A_{11} and S are approximated. Consider now the more general case, where we let

$$B_{22} = A_{22} - A_{21}DA_{12}$$

for some symmetric positive definite matrix D, which in practice is sparse. Let B_{11} be a symmetric positive definite approximation of A_{11} and assume that the following relations hold between D, B_{11}, and A_{11}:

(a) $$0 \leq B_{11} - A_{11} \leq \xi A_{11}$$

and

(b) $$0 \leq B_{11}^{-1} - D \leq \eta A_{11}^{-1},$$

for some positive numbers ξ, η. Here and in the remainder of this section, all inequalities are in a positive semidefinite sense. Note that (a) and (b) imply that

$$(9.22) \qquad\qquad A_{11} \le B_{11} \le D^{-1}.$$

To analyze the condition number of the corresponding preconditioner

$$C = \begin{bmatrix} B_{11} & A_{12} \\ A_{21} & \tilde{B}_{22} \end{bmatrix}$$

of A, where $\tilde{B}_{22} = A_{22} + A_{21}(B_{11}^{-1} - D)A_{12}$, we note first that

$$C - A = \begin{bmatrix} (B_{11} - A_{11}) & 0 \\ 0 & A_{21}(B_{11}^{-1} - D)A_{12} \end{bmatrix},$$

so

$$\frac{\mathbf{x}^T(C - A)\mathbf{x}}{\mathbf{x}^T A \mathbf{x}} = \frac{\mathbf{x}_1^T(B_{11} - A_{11})\mathbf{x}_1 + \mathbf{x}_2^T A_{21}(B_{11}^{-1} - D)A_{12}\mathbf{x}_2}{\mathbf{x}_1^T A_{11}\mathbf{x}_1 + 2\mathbf{x}_1^T A_{12}\mathbf{x}_2 + \mathbf{x}_2^T A_{22}\mathbf{x}_2}.$$

This and (9.22) show that

$$\mathbf{x}^T(C - A)\mathbf{x} \ge 0.$$

Next, as in the derivation of Theorem 9.5, we find that for any ζ, $\gamma < \zeta < \gamma^{-1}$, we have

$$(9.23) \qquad \frac{\mathbf{x}^T(C - A)\mathbf{x}}{\mathbf{x}^T A \mathbf{x}} \le \max\left\{ \frac{\xi}{1 - \zeta\gamma}, \frac{\eta\gamma^2}{1 - \zeta^{-1}\gamma} \right\}.$$

For this bound to be small, both ξ and η must be small. However, for any given matrix D, ξ becomes small if B_{11} is close to A_{11}, and η becomes small if B_{11} is close to D^{-1}. To balance these two extremes, we consider now the following choice of B_{11}. Given D, let B_{11} be defined by

$$B_{11}^{-1} = (I - P_\nu(DA_{11}))A_{11}^{-1},$$

where P_ν is a polynomial of degree ν such that

$$(c) \qquad\qquad 1 > P_\nu(t) \ge 0,$$

and

$$(d) \qquad\qquad 1 - t - P_\nu(t) \ge 0.$$

for any $t \in [\lambda_{min}(DA_{11}), \lambda_{max}(DA_{11})]$. Then,

$$B_{11}^{-1} \leq A_{11}^{-1},$$

so $A_{11} \leq B_{11}$ and

$$[I - DA_{11} - P_\nu(DA_{11})]A_{11}^{-1} \geq 0.$$

That is,

$$B_{11}^{-1} - D \geq 0.$$

Further,

(9.24)
$$B_{11} - A_{11} = A_{11}\left\{[I - P_\nu(DA_{11})]^{-1} - I\right\}$$

$$= A_{11}[I - P_\nu(DA_{11})]^{-1} P_\nu(DA_{11}).$$

(9.22) shows that the eigenvalues of DA_{11} are found in the interval $(0, 1]$, so (9.23) implies that

(9.25)
$$\xi = \max_{t_0 \leq t \leq t_1} \frac{P_\nu(t)}{1 - P_\nu(t)},$$

where

$$0 < t_0 = \lambda_{min}(DA_{11}), \quad t_1 = \lambda_{max}(DA_{11}) \leq 1.$$

In practice, we let D and B_{11} be such that $t_1 = 1$. We then have

$$B_{11}^{-1} - D = (I - P_\nu(DA_{11}) - DA_{11})A_{11}^{-1},$$

so

(9.26)
$$\eta = \max_{t_0 \leq t \leq t_1} (1 - t - P_\nu(t)).$$

To make ξ small, according to (9.25), $P_\nu(t)$ should be small in the interval $[t_0, t_1]$, while (9.26) shows that to make η small, $P_\nu(t)$ should be close to $1 - t$ in this interval. A polynomial that satisfies conditions (c, d) above and is a compromise between these two conflicting conditions is $P_\nu(t) = (1 - t)^\nu$. Then, the best choice of ν to minimize the bound of the condition number of $C^{-1}A$ is the value of ν, which minimizes

$$\max\{\xi, \eta\gamma^2\} = \max\left\{\frac{(1 - t_0)^\nu}{1 - (1 - t_0)^\nu}, \gamma^2 \max_{t_0 \leq t \leq t_1} [1 - t - (1 - t)^\nu]\right\}.$$

Obviously, if

$$t_0 \leq 1 - \left(\frac{1}{\nu}\right)^{\frac{1}{\nu-1}},$$

then

$$\max\{\xi, \eta\gamma^2\} = \max\left\{\frac{(1-t_0)^\nu}{1-(1-t_0)^\nu}, \gamma^2 \cdot \left(\frac{1}{\nu}\right)^{\frac{1}{\nu-1}} \frac{\nu-1}{\nu}\right\}.$$

For $\nu = 2$, we then get, if $t_0 \leq \frac{1}{2}$,

$$\max\{\xi, \eta\gamma^2\} = \max\left\{\frac{(1-t_0)^2}{1-(1-t_0)^2}, \left(\frac{\gamma}{2}\right)^2\right\} = \frac{(1-t_0)^2}{1-(1-t_0)^2}.$$

This holds if we let $\zeta = 1$ in (9.23) and the bound of the condition number becomes

$$\frac{1}{1-\gamma} \frac{(1-t_0)^2}{1-(1-t_0)^2}, \text{ if } t_0 \leq \frac{1}{2}.$$

However, we can let ζ be determined by (9.15), with $\beta_0/\alpha_0 = \eta\gamma^2/\xi$. This gives a somewhat smaller condition number. The upper bound is never smaller than $\frac{1}{1-\gamma^2} \frac{(1-t_0)^2}{1-(1-t_0)^2}$ for this choice of the polynomial P_ν.

9.4.2 Conclusions

It has been shown that if the computational effort in solving systems with A_{11} is not too large, then the most efficient method for solving systems of the form (9.2) is to use a preconditioned form of the Schur complement system. Here, the residuals should be computed by use of (9.16). The preconditioner could be some matrix of the form

$$B_{22} = A_{22} - A_{21}DA_{12},$$

as discussed in Section 9.4, or some modification of it using the method of matrix actions as described in Chapter 8. If the expense in solving systems with A_{11} is large compared with other work involved, then some version of the full block matrix incomplete factorization method will usually be most efficient.

The methods studied above for matrices partitioned in two-by-two block form correspond to two-level versions of more general multilevel iteration methods. In the two-level method for finite element spaces, the space is split into only two subspaces, necessitating the solution with the finite element matrix on the coarser level. However, both the block diagonal and the full block factorization preconditioning methods can be extended into a multilevel form of preconditioner. This can be done by simply recursively continuing the partitioning of each new matrix that arises (in the lower right corner) in two-by-two block form and preconditioning this in the same way as was done for the original matrix. For instance, in the block diagonal case, the three-level preconditioning matrix takes the form

$$
\begin{bmatrix}
A_{11}^{(l)} & 0 & 0 \\
0 & A_{11}^{(l-1)} & 0 \\
0 & 0 & A_{22}^{(l-1)}
\end{bmatrix},
$$

where

$$
A^{(l)} = \begin{bmatrix}
A_{11}^{(l)} & A_{12}^{(l)} \\
A_{21}^{(l)} & A_{22}^{(l)}
\end{bmatrix}
$$

is the matrix on the original level and $A_{22}^{(l)} = A^{(l-1)}$ and $A_{22}^{(l-1)} = A^{(l-2)}$ are the matrices on the next two levels. In the context of finite element matrices, the matrices involved should here correspond to the hierarchical basis functions, so $A^{(l-1)}$ and $A^{(l-2)}$ correspond to the standard finite element matrices on the coarser levels.

Using certain polynomials in a similar way as was done in Case 9.4, one can improve the multilevel methods so that the condition number becomes independent on the level number l, while the arithmetic cost of implementing the preconditioner is proportional to the number of unknowns on that level. (For details concerning the latter, see, for instance, Axelsson and Vassilevski, 1989, 1990, and Axelsson and Neytcheva, 1992.)

9.5 Indefinite Systems

In linear equality constrained (optimization) problems, there occur systems of the form

(9.27)
$$
\begin{bmatrix}
M & A^T \\
A & 0
\end{bmatrix}
\begin{bmatrix}
y \\
x
\end{bmatrix}
=
\begin{bmatrix}
b \\
a
\end{bmatrix},
$$

where M is symmetric. M can be indefinite in general. A is $n \times m$, $n < m$, rank $(A) = n$, and we assume M is positive definite on $\mathcal{N}(A)$, which implies that

$$\varepsilon M + A^T A, \; 0 < \varepsilon < \varepsilon_0$$

is positive definite for some sufficiently small ε_0.

It is readily seen that the matrix in (9.27) is indefinite—i.e., it has both positive and negative eigenvalues (except in the trivial case when A is a zero matrix and M is semidefinite). Solving indefinite matrix problems by iteration requires special care, because the iterative method may diverge or a breakdown (such as division by zero) or a near breakdown (division by a number which is small) can occur. (Division by small numbers usually causes large relative round-off errors.) For more details, see Chapter 11. Here, we shall instead consider methods based on Schur complements.

Assume first that M is positive definite. Then we can eliminate the vector block component **y** to get

(9.28) $$A M^{-1} A^T \mathbf{x} = A M^{-1} \mathbf{b} - \mathbf{a}.$$

Here $A M^{-1} A^T$ is symmetric. Clearly, in addition, it is positive semidefinite. Since A has full row rank, it is, in fact, positive definite. (9.28) can frequently be solved efficiently by some iterative method; then there is no need to form the matrix $A M^{-1} A^T$ explicitly. In particular, conjugate gradient solution methods (see Chapter 11) are convenient, because in such methods there is no method parameter that must be estimated. A method such as the Chebyshev iteration method, on the other hand, requires that the extreme eigenvalues of the iteration matrix $A M^{-1} A^T$ be estimated. To estimate these eigenvalues can be costly, because the matrix usually is not explicitly available.

In an iterative method, we need to perform only matrix-vector multiplications in addition to some vector computations. In the present context, the computation of $A M^{-1} A^T \mathbf{v}$ for some vector **v** involves a matrix-vector multiplication with A^T, the solution of a linear system with matrix M, and a matrix-vector multiplication with A. Usually, the most expensive part is the solution of systems with matrix M. This can also be done by iteration. As these computations occur once at every iteration step, it can be important to use an efficient preconditioner. For special problems in the numerical solution of partial differential equations, it has been shown how such preconditioners can be constructed. For references to such methods and for the analysis of coupled inner-outer iteration methods to solve systems of the above type, see Axelsson and Vassilevski (1991). In a coupled inner-outer iteration method, both the (outer) system with $AM^{-1}A^T$ and the (inner) systems with M are solved by iteration.

Consider now the general case where M is indefinite. Note first that (9.27) is equivalent to

$$(9.29) \quad \begin{bmatrix} (M + \frac{1}{\varepsilon}A^T A) & A^T \\ A & 0 \end{bmatrix} \begin{bmatrix} \mathbf{y} \\ \mathbf{x} \end{bmatrix} = \begin{bmatrix} \mathbf{b} + \frac{1}{\varepsilon}A^T \mathbf{a} \\ \mathbf{a} \end{bmatrix}$$

in the sense that both systems have the same solution. (By the assumptions made, a solution exists.) Similar to the derivation above, this shows that for $0 < \varepsilon < \varepsilon_0$,

$$A(M + \frac{1}{\varepsilon}A^T A)^{-1}A^T \mathbf{x} = A(M + \frac{1}{\varepsilon}A^T A)^{-1}(\mathbf{b} + \frac{1}{\varepsilon}A^T \mathbf{a}) - \mathbf{a}$$

or

$$(9.30) \quad A(\varepsilon M + A^T A)^{-1}A^T \mathbf{x} = A(\varepsilon M + A^T A)^{-1}(\mathbf{b} + \frac{1}{\varepsilon}A^T \mathbf{a}) - \frac{1}{\varepsilon}\mathbf{a}.$$

Before analyzing solution methods for (9.30), we shall first show that (9.30) reduces to (9.28) as $\varepsilon \to \infty$, if M is positive definite. We then need the following lemma.

Lemma 9.7 *Let M be positive definite and $\varepsilon > 0$. Then:*

> *(a)* $A(\varepsilon M + A^T A)^{-1} = (\varepsilon I + AM^{-1}A^T)^{-1}AM^{-1}$.
> *(b)* $A(\varepsilon M + A^T A)^{-1}A^T = I - (I + \frac{1}{\varepsilon}AM^{-1}A^T)^{-1}$.

Proof See Exercise 3.16.   ◇

Using Lemma 9.7(b), (9.30) can be rewritten as

$$A(M + \frac{1}{\varepsilon}A^T A)^{-1}A^T \mathbf{x} = A(M + \frac{1}{\varepsilon}A^T A)^{-1}\mathbf{b} - (I + \frac{1}{\varepsilon}AM^{-1}A^T)^{-1}\mathbf{a},$$

and, as $\varepsilon \to +\infty$, this takes the form (9.28).

It can happen that for a finite value of ε, (9.30) can be solved more efficiently than (9.28), even when M is positive definite. Before we analyze this possibility, we first show that the system (9.29) arises when we use the augmented Lagrangian method to solve linearly constrained optimization problems.

9.5.1 Augmented Lagrangian Method

Consider the optimization problem,

$$\min_{\mathbf{y}} \left\{ \frac{1}{2}\mathbf{y}^T M \mathbf{y} - \mathbf{b}^T \mathbf{y} \right\},$$

subject to the constraint $A\mathbf{y} = \mathbf{a}$. If $\mathbf{x}$ is the Lagrangian multiplier, the augmented Lagrangian optimization problem takes the form

$$\inf_{\mathbf{y}} \sup_{\mathbf{x}} \left\{ \frac{1}{2}\mathbf{y}^T M \mathbf{y} - \mathbf{b}^T \mathbf{y} + \mathbf{x}^T (A\mathbf{y} - \mathbf{a}) + \frac{1}{2\varepsilon}(A\mathbf{y} - \mathbf{a})^T(A\mathbf{y} - \mathbf{a}) \right\},$$

where $\varepsilon > 0$. It is readily seen that the stationary equations for this problem are the equations in (9.29). These equations, in fact, determine the saddle point for the Lagrangian functional. Let

$$\mathbf{v}_\varepsilon = A(\varepsilon M + A^T A)^{-1}(\mathbf{b} + \frac{1}{\varepsilon}A^T \mathbf{a}) - \frac{1}{\varepsilon}\mathbf{a}, \quad M_\varepsilon = \varepsilon M + A^T A.$$

Then (9.30) can be written as

(9.31) $$A M_\varepsilon^{-1} A^T \mathbf{x} = \mathbf{v}_\varepsilon.$$

This system can be solved by iteration in the same way as (9.28). The rate of convergence depends on the extreme, positive eigenvalues of the matrix $A M_\varepsilon^{-1} A^T$. To analyze this, we assume first that M is positive definite.

Theorem 9.8 *Let M be positive definite and let $M_\varepsilon = \varepsilon M + A^T A$, $\varepsilon > 0$. Then the spectrum of $A M_\varepsilon^{-1} A^T$ is contained in the interval $[\lambda_1/(\varepsilon + \lambda_1), 1)$, where λ_1 is the smallest eigenvalue of $A M^{-1} A^T$.*

Proof Lemma 9.7(b) shows that

$$A M_\varepsilon^{-1} A^T = I - \varepsilon(\varepsilon I + A M^{-1} A^T)^{-1},$$

where the eigenvalues of $(\varepsilon I + A M^{-1} A^T)^{-1}$ are contained in $[0, (\varepsilon + \lambda_1)^{-1}]$.
$\diamond$

Theorem 9.8 shows that, under the stated assumptions, we can make the matrix $A M_\varepsilon^{-1} A^T$ arbitrarily well-conditioned by choosing ε sufficiently small. Hence, we can get an arbitrary fast rate of convergence, typically leading to just a few—say, two iteration steps—when solving (9.31) by an iterative method such as the conjugate gradient method. However, each iteration step requires the solution of a system with matrix M_ε. As ε decreases, M_ε becomes increasingly ill-conditioned, which may require more computational efforts for solving systems with M_ε.

The optimal choice of ε for a smallest computational complexity must, thus, be a balance between the rate of convergence of the global iterative method to solve (9.31) and the cost of solving systems with M_ε, possibly using (inner) iterations.

Remark 9.9 (Elasticity equations) The matrix M_ε is the negative Schur complement of the perturbed matrix

$$A_\varepsilon = \begin{bmatrix} M & A^T \\ A & -\varepsilon I \end{bmatrix} = A - \begin{bmatrix} 0 & 0 \\ 0 & \varepsilon I \end{bmatrix}.$$

Matrices of this form arise in many applications. An important example is the discretized elasticity equation. Elasticity equations, which describe the displacement of an elastic body, consist of a system of three partial differential equations. Their discretization leads to a matrix in the form A_ε, which depends on two parameters, the stepsize h and $\varepsilon = 1 - 2\nu$. Here, $\nu, 0 < \nu < 1/2$ is called the Poisson ratio. $\nu = 1/2$ corresponds to (volume-wise) incompressible materials. For almost incompressible materials (i.e. $\nu \to 1/2$, or $\varepsilon \to 0$), the matrix becomes increasingly ill-conditioned. On the other hand, for common materials (such as steel), $\nu \le 1/2 - \delta$ for some fixed δ and the condition number of A_ε is $O(h^{-2})$, essentially independent of ε. For further details and references, see Axelsson (1979).

We now consider the case where M is positive definite, but A is rank-deficient. This means that $A M_\varepsilon^{-1} A^T$ is singular. It will be seen in Chapter 11 that for the conjugate gradient method, the rate of convergence depends only on the positive eigenvalues. Hence, the method is applicable also to consistent singular matrices. The following extension of the previous theorem holds.

Theorem 9.10 *Let M be positive definite, let λ_1 be the smallest positive eigenvalue of $A M^{-1} A^T$, and let $M_\varepsilon = \varepsilon M + A^T A$, $\varepsilon > 0$. Then the nonzero part of the spectrum of $A M_\varepsilon^{-1} A^T$ is contained in $[\lambda_1/(\varepsilon + \lambda_1), 1]$.*

Proof This follows from the proof of Theorem 9.8, since the eigenvalues of $(\varepsilon I + A M^{-1} A^T)^{-1}$ are positive and

$$\sup_{\substack{x^T x=1 \\ x \in R(A) = \mathcal{N}(A^T)^{\perp}}} \mathbf{x}^T (\varepsilon I + AM^{-1}A^T)^{-1}\mathbf{x}$$

$$= 1 / \inf_{\substack{x^T x=1, \\ x \in R(A)}} \mathbf{x}^T (\varepsilon I + AM^{-1}A^T)\mathbf{x} = 1/(\varepsilon + \lambda_1)$$

$\diamond$

The above method, which uses M_ε, is an algebraic formulation of an algorithm known as the Uzawa algorithm, which has been used to solve constrained problems for elliptic partial differential equations. (See, for instance, Fortin and Glowinski, 1983, and Axelsson, 1979.)

One can, alternatively, solve the original problem for the indefinite matrix A. The spectrum of A can be seen to consist of two intervals, one to the left and one to the right of the origin. If accurate bounds are known for the endpoints of these intervals, one can use a matrix polynomial transformation as shown in Section 5.4 and then use Chebyshev iteration. Since, in general, such bounds are not known, one can instead use a minimum residual conjugate gradient method or Lanczos iteration method (see Chapter 11). The rate of convergence depends on the ratios of the extreme eigenvalues for the two intervals, as shown in Section 5.4. The rate of convergence of such methods for the unreduced problem and of the Schur complement methods can be improved by preconditioning. It is essential to use a symmetric form of the preconditioned matrix; otherwise, it may have complex eigenvalues.

Let $\tilde{B}$ be the preconditioned matrix, where

$$\tilde{B} = \begin{bmatrix} L^{-1} & 0 \\ 0 & S^{-1} \end{bmatrix} \begin{bmatrix} M & A^T \\ A & 0 \end{bmatrix} \begin{bmatrix} L^{-T} & 0 \\ 0 & S^{-T} \end{bmatrix}.$$

Then an elementary computation shows that

$$\tilde{B} = \begin{bmatrix} L^{-1}ML^{-T} & L^{-1}A^T S^{-T} \\ S^{-1}AL^{-T} & 0 \end{bmatrix}.$$

The negative Schur complement of $\tilde{B}$ equals

$$S^{-1}AL^{-T}(L^{-1}ML^{-T})^{-1}L^{-1}A^T S^{-T} = S^{-1}AM^{-1}A^T S^{-T},$$

which, hence, does not depend on L. Therefore, we choose S, independent of L, so that the outer system matrix $S^{-1}(AM^{-1}A^T)S^{-T}$ is well-conditioned. Further, we can choose L independently, so that $L^{-1}ML^{-T}$ is well-conditioned, and the inner systems therefore can be solved with few iterations. Naturally, for practical reasons, the cost of solving systems with the (triangular) factors $S,\ S^T$ and $L,\ L^T$ must be relatively modest. Note that the matrix $\widetilde{B}$ corresponds to the constrained minimization problem

$$\min\left\{\frac{1}{2}\mathbf{z}^T\widetilde{M}^{-1}\mathbf{z} - \mathbf{z}^T L^{-1}\mathbf{b}\right\},$$

subject to $\widetilde{A}\mathbf{z} = S^{-1}\mathbf{a}$, where $\widetilde{M} = L^{-1}ML^{-T}$, $\widetilde{A} = S^{-1}AL^{-T}$.

Remark 9.11 Underdetermined systems: Consider the constrained minimization problem

$$\min_{\mathbf{y}}\frac{1}{2}\mathbf{y}^T M\mathbf{y},$$

subject to $A\mathbf{y} = \mathbf{a}$, where we assume that M is s.p.d. This is equivalent to

$$\begin{bmatrix} M & A^T \\ A & 0 \end{bmatrix}\begin{bmatrix} \mathbf{y} \\ \mathbf{x} \end{bmatrix} = \begin{bmatrix} 0 \\ \mathbf{a} \end{bmatrix},$$

and its solution satisfies

$$AM^{-1}A^T\mathbf{x} = -\mathbf{a},$$

$$M\mathbf{y} = -A^T\mathbf{x}.$$

The solution $\mathbf{y}$ found will be a generalized solution to $A\mathbf{y} = \mathbf{a}$, with smallest $M^{\frac{1}{2}}$-norm, $\|\mathbf{y}\|_{M^{\frac{1}{2}}} = \{\mathbf{y}^T M\mathbf{y}\}^{\frac{1}{2}}$. Hence, if only an underdetermined system $A\mathbf{y} = \mathbf{a}$ is given, then we can choose M for convenience to get a solution that best satisfies certain additional requirements. The underdetermined system has infinitely many solutions, but for each M we get a particular solution that satisfies some additional properties. For instance, M can be chosen to smooth out highly oscillatory components of the solution. This process can be useful when solving certain ill-posed problems arising in so-called inverse problems, typically of integral equation type. If we let $M = I$, then the solution found will be the mininum (Euclidean) norm solution, where all components have equal weight.

References

O. Axelsson (1979). Preconditioning of indefinite problems by regularization. *SIAM J. Numer. Anal.* **16**, 58–69.

O. Axelsson (1982). On multigrid methods of two-level type. In *Multigrid Methods* ed. W. Hackbusch and U. Trottenburg), LNiM vol. 960, pp. 352–367. Berlin Heidelberg, New York: Springer-Verlag.

O. Axelsson and V. A. Barker (1984). *Finite Element Solution of Boundary Value Problems. Theory and Computation.* Orlando, Fla.: Academic.

O. Axelsson and V. Eijkhout (1991). The nested recursive two-level factorization method for nine-point difference matrices. *SIAM J. Scientific and Statistical Computations* **12**, 1373–1400.

O. Axelsson and I. Gustafsson (1983). Preconditioning and two-level multigrid methods of arbitrary degree of approximation. *Math. Comp.* **40**, 219–242.

O. Axelsson and L. Kolotilina (1992). Diagonally compensated reduction and related preconditioning methods. *Numer. Lin. Alg. Appl.*, to appear.

O. Axelsson and M. Neytcheva (1992). Algebraic multilevel iteration methods for Stieltjes matrices. *Numer. Lin. Alg. Appl.*, to appear.

O. Axelsson and P. S. Vassilevski (1989). Algebraic multilevel preconditioning methods I. *Numer. Math.* **56**, 157–177.

O. Axelsson and P. S. Vassilevski (1990). Algebraic multilevel preconditioning methods II. *SIAM J. Numer. Anal.* **27**, 1569–1590.

O. Axelsson and P. S. Vassilevski (1991). A black box generalized conjugate gradient solver with inner iterations and variable-step preconditioning. *SIAM J. Matrix. Anal. Appl.* **12**, 625–644.

R. E. Bank and T. F. Dupont (1980). Analysis of a two-level scheme for solving finite element equations. Techn. Rep. CNA-159, Center for Numerical Analysis, University of Texas at Austin.

V. Eijkhout and P. S. Vassilevski (1991). The role of the strengthened Cauchy-Buniakowskii-Schwarz inequality in multi-level methods. *SIAM Review* **33**, 405–419.

M. Fortin and R. Glowinski (1983). Augmented Lagrangian Methods: Application to the Numerical Solution of Boundary-Value Problems. In *Studies in Mathematics and Its Applications*, Vol. 15. Amsterdam: North-Holland.

J. Mandel (1990). On block diagonal and Schur complement preconditioning. *Numer. Math.* **58**, 79–93.

J. F. Maitre and F. Musy (1982). The contraction number of a class of two-level methods; an exact evaluation for some finite element subspaces and model problems. In *Multigrid Methods*, ed. W. Hackbusch and U. Trottenberg. LNiM Vol. 960, pp. 535–544. Berlin, Heidelberg, New York: Springer-Verlag.

P. S. Vassilevski (1989). Nearly optimal iterative methods for solving finite element equations based on multilevel splitting of the matrix. Report 1989–09, Institute for Scientific Computation, Univ. of Wyoming, Laramie.

10

Estimates of Eigenvalues and Condition Numbers for Preconditioned Matrices

The rate of convergence of preconditioned iterative methods such as the Chebyshev iterative method and (generalized) conjugate gradient methods can be estimated when the condition number of the preconditioned matrix is known. The Chebyshev method requires even that the extreme eigenvalues are known, or estimated from below and above, respectively. As will be seen in Chapter 13, the rate of convergence of the conjugate gradient method depends, in fact, more precisely on the distribution of the eigenvalues.

Therefore, eigenvalue and condition number estimates for preconditioned iteration matrices provide the information required to estimate the parameters involved in methods such as the Chebyshev iteration methods and the rate of convergence of iterative methods, as preconditioned conjugate gradient methods. Such estimates will be derived for symmetric positive (semi)definite matrices. As was already shown in Theorem 8.10 for explicit preconditioners, if $\|I - GA\|_W \leq \varepsilon < 1$ and $W - I$ is positive semidefinite, then

$$\kappa(GA) = \|GA\|_2 \|(GA)^{-1}\|_2 \leq \frac{1 + \varepsilon}{1 - \varepsilon}.$$

In Section 10.2, upper eigenvalue bounds for implicit preconditioners in incomplete point-wise or block-wise form are derived. This includes two-sided estimates of individual eigenvalues. A new upper bound $2m$ for the largest eigenvalue of matrices partitioned in $m \times m$ blocks is shown to hold for M-matrices.

A frequently used alternative technique in the construction of preconditioners is the use of small perturbations of the given matrix during the factorization, when the preconditioner must satisfy certain conditions for the upper eigenvalue bounds to hold. We use some graph theory to estimate the perturbation parameters and the lower eigenvalue bounds resulting from them.

For many problems, the condition number depends on some problem parameters. For difference equations for elliptic equations, for example, this parameter is usually the meshwidth (h) of the difference grid, and we want to estimate the order of the condition number w.r.t. this parameter (as $h \to 0$). For certain preconditioners, one can reduce the condition number by an order of magnitude for difference equations for second-order problems from $O(h^{-2})$ to $O(h^{-1})$. By deriving a lower bound of the condition number, it will be seen that this order is also the best possible, if the sparsity of the matrices involved in the preconditioner is of the same order as for the given matrix.

The following notations are used in this chapter: Unless otherwise stated, $A \geq B$ means that $A - B$ is positive semidefinite. $\lambda_i(A)$ denotes the ith eigenvalue of a symmetric matrix A, where the eigenvalues are numbered in a nondecreasing order. $\lambda_{\max}(A)$ denotes the maximal eigenvalue of A, and $\lambda_{\min}^{+}(A)$ denotes the smallest positive eigenvalue. A is partitioned in block matrices (which may be scalars) and split as

$$A = D_A + L_A + L_A^T,$$

where D_A, L_A is the block diagonal part and lower block triangular part of A, respectively.

10.1 Upper Eigenvalue Bounds

Let A be symmetric and positive semidefinite, and consider a preconditioner C in the factorized form

(10.1) $$C = (X + L)X^{-1}(X + L^T),$$

where X is a (block) diagonal s.p.d. matrix and L is a (block) lower triangular matrix. X and L are partitioned in blocks consistent with the partitioning of A. For example, C can be a point-wise or block matrix incomplete factorization of A, as described in Chapter 7. This type of preconditioner is called implicit because application of it requires solving a linear system, here using forward and back substitutions.

To estimate the rate of convergence of (generalized) conjugate gradient iteration methods in detail, we need to know the distribution of eigenvalues of the preconditioned matrix. Hence, we need to estimate individual eigenvalues. Naturally, this problem can be more difficult than just estimating the condition number, but it will be shown that, under certain conditions, quite accurate lower and upper eigenvalue bounds can be derived and that these provide the information needed to compare, for example, modified and unmodified incomplete factorization methods. This estimate provides, in particular, an estimate of the largest eigenvalue of $C^{-1}A$. Three alternate estimates of this type will be given, one of which gives a general bound $2m$ for M-matrices partitioned in $m \times m$ blocks.

In the theorems to follow in this section, unless stated otherwise, we assume only that A is symmetric and positive semidefinite—i.e., we do not assume that A is a M-matrix, for instance.

The following result for estimating upper and lower bounds for eigenvalues of the matrix $C^{-1}A$ relates these eigenvalues to the eigenvalues of A. It is a slight extension of Theorem 3.16 and can be proved as in the proof of that theorem.

Theorem 10.1 *Let A be symmetric positive semidefinite and let C be s.p.d. Let the eigenvalues of A and $C^{-1}A$ be numbered in a nondecreasing order— i.e., $\lambda_i(A) \leq \lambda_{i+1}(A)$, $\lambda_i(C^{-1}A) \leq \lambda_{i+1}(C^{-1}A)$, $i = 1, 2, \ldots$, and let μ_1, μ_2 be sufficiently large positive numbers such that*

$$(10.2a) \qquad \lambda_{\max}(\mu_1 C - A) \geq 0$$

and

$$(10.2b) \qquad \lambda_{\min}(\mu_2 C - A) \geq 0$$

hold. Then, for all positive eigenvalues of A, the following lower and upper bounds of the eigenvalues of $C^{-1}A$ hold:

$$\frac{\mu_1 \lambda_i(A)}{\lambda_i(A) + \lambda_{\max}(\mu_1 C - A)} \leq \lambda_i(C^{-1}A) \leq \frac{\mu_2 \lambda_i(A)}{\lambda_i(A) + \lambda_{\min}(\mu_2 C - A)}.$$

We shall now give conditions for which (10.2a,b) hold, with the objective of finding the best values of μ_1, μ_2 in the sense of finding the largest possible lower eigenvalue bound and the smallest possible upper eigenvalue bound. Clearly, any μ_2 such that $\lambda_{\min}(\mu_2 C - A) \geq 0$ is an upper eigenvalue bound for all eigenvalues of $C^{-1}A$, and any μ_1 such that $\lambda_{\max}(\mu_1 C - A) \leq 0$ is a lower eigenvalue bound for all eigenvalues of $C^{-1}A$. As we shall see later, the analysis can also be used to develop strategies for modifying the incomplete factorization to satisfy an a priori chosen upper bound.

Consider, then, a preconditioning matrix C of the form (10.1). The next theorem gives a sufficient and readily checked condition for (10.2b) to hold. We first prove a lemma.

Lemma 10.2 *Let A be symmetric positive semidefinite. Let C have the form (10.1), where X is s.p.d., and let $K = A - L - L^T$. Then, $\lambda_{\min}(\mu C - A) \geq 0$ if there exists a positive number μ such that*

$$\lambda_{\min}\left((2 - \frac{1}{\mu})X - K\right) \geq 0.$$

Proof Let $V = (1 - 1/\mu)X + L$ and note that

$$V + V^T = 2(1 - \frac{1}{\mu})X + A - K$$

and

$$\mu C = (V + \frac{1}{\mu}X)(\frac{1}{\mu}X)^{-1}(V^T + \frac{1}{\mu}X)$$

$$= \mu V X^{-1} V^T + V + V^T + \frac{1}{\mu}X.$$

Hence,

$$\mu C - A = \mu V X^{-1} V^T + (2 - \frac{1}{\mu})X - K,$$

and since $\mu V X^{-1} V^T$ is positive semidefinite for any positive number μ, it follows that

$$\lambda_{\min}(\mu C - A) \geq \lambda_{\min}\left((2 - \frac{1}{\mu})X - K\right) \geq 0 \qquad \diamond$$

In some cases, we have $L = L_A$, and then K becomes (block) diagonal, $K = D_A$. Lemma 10.2 shows that we can replace the condition $\lambda_{\min}(\mu C - A) \geq 0$ with the corresponding condition on $\lambda_{\min}((2 - \frac{1}{\mu})X - K)$, which—as follows from the next theorem—can be easier to check.

Theorem 10.3 *Let A be a symmetric positive semidefinite matrix, let D_A denote the (block) diagonal part of A, and let $\mu_0 > 1/2$ be given. Assume that for any $\mu \geq \mu_0$:*

(a) *X and $(2 - 1/\mu)X - D_A$ are Z-matrices,*
(b) *The offdiagonal block entries of $L + L^T$ are not larger than the corresponding entries of A,*

(c)

$$Xv > 0 \quad \text{and} \quad \left[(2 - \frac{1}{\mu_0})X - K\right]v \geq 0,$$

for some positive vector **v**, *where* $K = A - L - L^T$. *Then* X *is a Stieltjes matrix and*

$$\lambda_{\min}(\mu C - A) \geq \lambda_{\min}\left((2 - \frac{1}{\mu})X - K\right) \geq 0.$$

Proof We have

$$(10.3) \qquad K = A - L - L^T = D_A + L_A + L_A^T - (L + L^T),$$

where L_A denotes the lower (block) triangular part of A. In the block matrix case, we assume that A and $L + L^T$ are partitioned in the same way. Hence, by assumption (b), the offdiagonal block entries of K are nonnegative and this, together with assumption (a), shows that

$$(2 - \frac{1}{\mu})X - K = (2 - \frac{1}{\mu})X - D_A + L + L^T - (L_A + L_A^T)$$

is a Z-matrix. By assumption (c), it now follows that the matrix

$$(2 - \frac{1}{\mu})X - K$$

is positive semidefinite—i.e., for any $\mu \geq \mu_0$

$$(10.4) \qquad \lambda_{\min}\left((2 - \frac{1}{\mu})X - K\right) \geq 0.$$

Since, by assumption, X is a Z-matrix and $Xv > 0$, X is then positive definite, so C in (10.1) exists. This, with Lemma 10.2, completes the proof.   ◇

Corollary 10.4 *Let the conditions in the previous theorem hold. Then, for any eigenvalue* $\lambda_i(C^{-1}A)$, *such that* $\lambda_i(A) > 0$, *it holds*

$$\lambda_i(C^{-1}A) \leq \frac{\mu}{1 + \lambda_{\min}(\mu C - A)/\lambda_i(A)},$$

where $\mu \geq \mu_0$. *In particular,*

$$\lambda_i(C^{-1}A) \leq \mu.$$

for any μ sufficiently large to make $\lambda_{\min}(\mu C - A) \geq 0$, that is, in particular, to make $\lambda_{\min}((2 - 1/\mu)X - K) \geq 0$.

Proof This proof follows from Theorems 10.1 and 10.3. $\diamond$

10.1.1 Essentially Point-wise Incomplete Factorization Method

The next theorem gives more detailed estimates of $\lambda_i(C^{-1}A)$ for diagonally scaled matrices. By a diagonally scaled matrix, we mean the matrix $D_A^{-\frac{1}{2}} A D_A^{-\frac{1}{2}}$, and the incomplete factorization is applied to this matrix. In this case, the block diagonal part is I, the block identity matrix. We shall, however, let this diagonal matrix be $d \cdot I$ for some positive scalar d, in order to see better the influence of the diagonal. Clearly, diagonal scaling is practically viable only if the orders of the diagonal blocks in A are small. Hence, the next result is, in practice, applicable only to point-wise incomplete factorization methods or block-wise with a small order of the blocks.

Theorem 10.5 *Let A be positive semidefinite and diagonally scaled such that $A = d \cdot I + L_A + L_A^T$ for some positive scalar d. Assume that X is a Stieltjes matrix and that $L = L_A$. Then the following hold:*

(a) $K = d \cdot I$ and

$$\lambda_{\min}\left[(2 - \frac{1}{\mu})X - K\right] = \lambda_{\min}\left[(2 - \frac{1}{\mu})X - D_A\right] = (2 - \frac{1}{\mu})x - d,$$

where $x = \lambda_{\min}(X)$, the smallest eigenvalue of X, and $\mu \geq x/(2x - d)$.

(b) If, in addition, $2x - d > 0$, then

$$\lambda_i(C^{-1}A) \leq \begin{cases} \dfrac{4x\lambda_i(A)}{[2x-d+\lambda_i(A)]^2}, & \text{if } \lambda_i(A) \leq 2x - d \\[2mm] \dfrac{x}{2x-d}, & \text{if } \lambda_i(A) \geq 2x - d \end{cases}.$$

Proof Part (a) follows from

$$K = A - L - L^T = A - L_A - L_A^T = d \cdot I.$$

Now Corollary 10.4 and Theorem 10.3 show that

$$(10.5) \qquad \lambda_i(C^{-1}A) \leq \mu\lambda_i(A)/[\lambda_i(A) + (2 - \frac{1}{\mu})x - d]$$

for any μ, such that

(10.6) $$\left(2 - \frac{1}{\mu}\right)x - d \geq 0.$$

Differentiation of the right-hand side of (10.5), with respect to μ, shows the stationary point

(10.7) $$\mu_i = \frac{2x}{2x - d + \lambda_i(A)}; \quad \lambda_i(C^{-1}A) \leq \bar{\lambda}_i \equiv \frac{4x\lambda_i(A)}{[2x - d + \lambda_i(A)]^2}.$$

However, by (10.6) this holds only for eigenvalues $\lambda_i(A)$, such that

$$\left(2 - \frac{1}{\mu_i}\right)x - d \geq 0,$$

that is,

$$\mu_i \geq \frac{x}{2x - d}.$$

Therefore, (10.7) gives a minimum in the interval

$$\frac{x}{2x - d} \leq \mu < \infty$$

only if

$$\mu_i \geq \frac{x}{2x - d}.$$

That is, by (10.7), we have

$$2x - d + \lambda_i(A) \leq 2(2x - d),$$

which holds for any sufficiently small eigenvalue, such that

$$\lambda_i(A) \leq 2x - d.$$

In this case, $\lambda_i(C^{-1}A) \leq \bar{\lambda}_i$, which is the first part of (b). On the other hand, if

$$\lambda_i(A) \geq 2x - d,$$

then (10.5) shows that

$$\lambda_i(C^{-1}A) = \frac{\mu}{1 + [(2 - \frac{1}{\mu})x - d]/\lambda_i(A)} \leq \frac{\mu}{1 + ((2 - \frac{1}{\mu})x - d)/(2x - d)}.$$

Hence, with $\mu = x/(2x - d)$,

$$\lambda_i(C^{-1}A) \le \frac{x}{2x - d}.$$

Note, finally, that the last bound holds for any eigenvalue if $2x - d > 0$. $\diamond$

The previous results yield, in particular, upper bounds for the largest eigenvalue. The next theorem gives an alternate bound, but here we exclude the case where X is block diagonal. Furthermore, we assume that L is nonpositive.

Theorem 10.6 (X diagonal; Beauwens [1984]) *Let A be symmetric and positive semidefinite, and let $X + L$ be a M-matrix, where X is nonsingular and diagonal and L is lower triangular. Assume that there exists a positive vector $\mathbf{v}$, such that $A\mathbf{v} \ge \mathbf{0}$, and that:*

(a) the off-diagonal entries of $L + L^T$ are not larger than the corresponding entries of A.
(b) Let $C = (X + L)X^{-1}(X + L^T)$ and assume that

$$C\mathbf{v} \ge (1 - \tau_0)A\mathbf{v}, \quad \text{for some } \tau_0, \ 0 < \tau_0 < 1.$$

(c) $\tau_1 \equiv \max_i\{-(X^{-1}L^T\mathbf{v})_i/\mathbf{v}_i\} < 1$.

Then,

$$\lambda_{\max}(C^{-1}A) \le \frac{1}{1 - \tau},$$

where $\tau = \max(\tau_0, \tau_1)$.

Proof Let $B = (X_0 + L)X^{-1}(X_0 + L^T)$, where X_0 is a diagonal matrix such that $X_0\mathbf{v} = -L^T\mathbf{v}$. Since L is nonpositive, we have $X_0 \ge 0$. Let $A = D_A + L_A + L_A^T$, where D_A is the diagonal and L_A the lower triangular part of A. Then

$$C - (1 - \tau)A - B = X + L + L^T + LX^{-1}L^T$$

$$- (1 - \tau)A - X_0X^{-1}X_0 - X_0X^{-1}L^T - LX^{-1}X_0 - LX^{-1}L^T$$

$$= X - X_0X^{-1}X_0 - (1 - \tau)D_A$$

$$+ (1 - \tau)(L + L^T - L_A - L_A^T)$$

$$+ L(\tau I - X^{-1}X_0) + (\tau I - X_0X^{-1})L^T.$$

Note next that $\tau \mathbf{v} - X^{-1} X_0 \mathbf{v} = \tau \mathbf{v} + X^{-1} L^T \mathbf{v} \geq \tau_1 \mathbf{v} + X^{-1} L^T \mathbf{v} \geq \mathbf{0}$. Hence, $\tau \mathbf{v} - X^{-1} X_0 \mathbf{v} \geq \mathbf{0}$, and since $\tau I - X^{-1} X_0$ is a diagonal matrix, $\tau I - X^{-1} X_0 \geq 0$. Similarly, $\tau I - X_0 X^{-1} \geq 0$, which, together with assumption (a), shows that the offdiagonal entries of $C - (1 - \tau) A - B$ are nonpositive. Since $B \mathbf{v} = \mathbf{0}$ and $A \mathbf{v} \geq \mathbf{0}$, (b) shows that $[C - (1 - \tau) A - B] \mathbf{v} \geq C \mathbf{v} - (1 - \tau_0) A \mathbf{v} \geq \mathbf{0}$, that is, $C - (1 - \tau) A - B$ is positive semidefinite. Since B is positive semidefinite, $C - (1 - \tau) A$ is positive semidefinite.   ◇

In the next theorem, the condition $[(2 - 1/\mu) X - K] \mathbf{v} \geq \mathbf{0}$ in Theorem 10.3 is replaced by the weaker condition $(2X - K) \mathbf{v} \geq \mathbf{0}$.

Theorem 10.7 (X [block] diagonal) *Let A be symmetric and positive semidefinite, with positive diagonal entries. Assume that the entries of $L + L^T$, where L is a block lower triangular matrix, are not larger than the corresponding entries of A. In addition, assume that X and $2X - D_A$ are symmetric Z-matrices and that there exists a positive vector $\mathbf{v}$ such that*

(10.8) $$X \mathbf{v} > \mathbf{0} \quad \text{and} \quad (2X - K) \mathbf{v} \geq \mathbf{0},$$

where $K = A - L - L^T$. Then X is a Stieltjes matrix and

$$\lambda_{\max}(C^{-1} A) \leq \rho(M).$$

Here,

$$C = (X + L) X^{-1} (X + L^T),$$

$$M = (I + X^{-\frac{1}{2}} L X^{-\frac{1}{2}})^{-1} + (I + X^{-\frac{1}{2}} L^T X^{-\frac{1}{2}})^{-1}$$

and $\rho(M)$ denotes the spectral radius of M.

Proof Since $K = A - L - L^T$, we know—as shown in the proof of Theorem 10.3—that the offdiagonal (block) entries of K are nonnegative. Since, by assumption, $2X - D_A$ is a Z-matrix, $2X - K$ is also a Z-matrix. By (10.8), $X \mathbf{v} > \mathbf{0}$, and since X is a Z-matrix, X is positive definite—i.e., a Stieltjes matrix. Thus, X is nonsingular, so C exists and is positive definite.

Next, note that

(10.9) $$A = (K - 2X) + (X + L) + (X + L^T).$$

Since $2X - K$ is a Z-matrix, (10.8) shows that $2X - K$ is positive semidefinite—that is, $K - 2X$ is negative semidefinite.

A similarity transformation of $C^{-1}A$ and (10.9) now shows that

$$X^{-\frac{1}{2}}(X + L^T)C^{-1}A(X + L^T)^{-1}X^{\frac{1}{2}}$$

$$= X^{\frac{1}{2}}(X + L)^{-1}A(X + L^T)^{-1}X^{\frac{1}{2}}$$

$$= X^{\frac{1}{2}}(X + L)^{-1}(K - 2X)(X + L^T)^{-1}X^{\frac{1}{2}}$$

$$+ X^{\frac{1}{2}}(X + L^T)^{-1}X^{\frac{1}{2}} + X^{\frac{1}{2}}(X + L)^{-1}X^{\frac{1}{2}}$$

$$\leq X^{\frac{1}{2}}(X + L^T)^{-1}X^{\frac{1}{2}} + X^{\frac{1}{2}}(X + L)^{-1}X^{\frac{1}{2}} = M.$$

This is a symmetric matrix, and the above shows that the maximal eigenvalue of $C^{-1}A$ is bounded by the maximal eigenvalue of M, so

$$\lambda_{\max}(C^{-1}A) \leq \rho(M), \qquad \diamond$$

As we shall see, Theorems 10.3 and 10.5 will enable us to compute perturbations to get a preconditioner that satisfies an a priori chosen upper bound. On the other hand, the last theorem cannot readily be used for this purpose, but the next corollary shows that it can be used to estimate the maximal eigenvalue.

Corollary 10.8 *Let A be partitioned in $m \times m$ blocks and assume that the conditions of Theorem 10.7 hold and that $\|\widetilde{L}\|_2 \leq 1 + \frac{c}{m}$, where $\widetilde{L} = X^{-\frac{1}{2}}LX^{-\frac{1}{2}}$, for some constant c that does not depend on m. Then*

$$\lambda_{\max}(C^{-1}A) \leq 2me^c.$$

Proof We have

$$\|X^{\frac{1}{2}}(X + L^T)^{-1}X^{\frac{1}{2}}\|_2 = \|(I + \widetilde{L})^{-1}\|_2$$

$$= \|I + \widetilde{L} + \ldots + \widetilde{L}^{m-1}\|_2 \leq$$

$$\leq 1 + \|\widetilde{L}\|_2 + \ldots + \|\widetilde{L}\|_2^{m-1} \leq me^c,$$

so

$$\rho(M) \leq 2me^c. \qquad \diamond$$

Remark 10.9 (General upper bound) Under the assumptions of Theorem 10.7, we have $2X + L + L^T = (2X - K) + A$, which is a sum of two positive semidefinite matrices, so $2I + \widetilde{L} + \widetilde{L}^T$ is positive semidefinite. Hence, as will be shown below, the numerical radius of $\widetilde{L}$ is bounded by 1 if L is nonpositive. This does not imply that $\|\widetilde{L}\|_2 \leq 1$, in general. Note, however, that if $\|\widetilde{L}\|_2 \leq 1$, then $\lambda_{\max}(C^{-1}A) \leq 2m$. This latter bound will now be shown to hold if we assume, in addition, that L is nonpositive.

Theorem 10.10 *Let A be partitioned in $m \times m$ blocks and assume that the conditions of Theorem 10.7 hold. Assume, in addition, that L is nonpositive. Then*

$$\lambda_{\max}(C^{-1}A) \leq 2m.$$

Proof Theorem 10.7 shows that $\lambda_{\max}(C^{-1}A) \leq \rho(M)$, where $M = (I - \widetilde{L})^{-1} + (I - \widetilde{L}^T)^{-1}$ and $\widetilde{L} = -X^{-\frac{1}{2}}LX^{-\frac{1}{2}}$. Since X is a Stieltjes matrix, Theorem 6.15 shows that $X^{\frac{1}{2}}$ is a Stieltjes matrix, so $X^{-\frac{1}{2}}$ is nonnegative. Then, $\widetilde{L}$ is nonnegative because L is nonpositive. Note also that $\widetilde{L}$ is strictly block lower triangular. Hence,

$$(I - \widetilde{L})^{-1} = \sum_{k=0}^{m-1} \widetilde{L}^k,$$

and

$$M = 2I + \sum_{k=1}^{m-1} (\widetilde{L}^k + \widetilde{L}^{k^T}).$$

Consider now the numerical radius,

$$r(B) = \sup_{\substack{x^*x=1 \\ x \in \mathbb{C}^n}} |\mathbf{x}^* B\mathbf{x}|,$$

where B is a square matrix. As has been shown in Theorem 4.15, this is additive and $\rho(B) \leq r(B)$. Hence,

$$(10.10) \quad \rho(M) \leq 2 \left[1 + \sum_{k=1}^{m-1} r(\tfrac{1}{2}(\widetilde{L}^k + \widetilde{L}^{k^T})) \right] = 2 \left[1 + \sum_{k=1}^{m-1} r(\widetilde{L}^k) \right].$$

Next, note that

$$2X + L + L^T = (2X - K) + A,$$

which is a sum of two positive semidefinite matrices. Thus, $2I - \widetilde{L} - \widetilde{L}^T$ is positive semidefinite and

$$\max_{\mathbf{x}^T\mathbf{x}=1} \mathbf{x}^T \frac{1}{2}(\widetilde{L} + \widetilde{L}^T)\mathbf{x} \leq 1.$$

Corollary 4.20 shows that

$$r(B) = \max_{\substack{\mathbf{x}^T\mathbf{x}=1 \\ \mathbf{x}\in\mathbb{R}^n}} \mathbf{x}^T \frac{1}{2}(B + B^T)\mathbf{x}$$

for any nonnegative square matrix B. Since $\widetilde{L}$ is nonnegative, the above shows that

$$r(\widetilde{L}) \leq 1.$$

Also, it is shown in Theorem 4.19 that if $r(B) \leq 1$ for some matrix B, then also $r(B^k) \leq 1$, $k \geq 1$. The latter and (10.10) show that

$$\rho(M) \leq 2m. \qquad \diamond$$

Block matrix incomplete factorization methods can be efficient preconditioning methods for certain parallel computers when parallelism is utilized on micro-level—i.e., in the solution of systems with matrices X_i and in matrix-vector multiplications with block matrices. For efficiency reasons, the block size should not be too small, and a large block size implies a small value of m, which can be expected to be good for the fast rate of convergence of the iterative solution method.

Finally we remark that knowing an (accurate) upper eigenvalue bound is important when one uses a Chebyshev iterative acceleration method.

10.2 Perturbation Methods

Perturbation methods are convenient and easily implementable techniques to control the largest eigenvalue of the preconditioned matrix $C^{-1}A$. In such methods, the given matrix A (usually only its diagonal) is perturbed in order to satisfy certain bounds. The perturbations can take place á priori—i.e., before the factorization—or dynamically—i.e., during the factorization. The first use of perturbations occurred for second-order elliptic boundary value problems in Axelsson (1972) and was later extended in Gustafsson (1979). Here, a relative amount c_0h^2 was added to the diagonal of A (or even c_1h, where the coefficients made jumps). The parameter h is an average step size in the finite difference mesh. Hence, $\widetilde{A} = A + \Delta$, where Δ is a diagonal matrix with

$$\Delta_{ii} = \begin{cases} c_0 h^2, & \text{for all points except where the matrix coefficients} \\ & \text{in the diagonal made jumps} \\ c_1 h, & \text{otherwise,} \end{cases}$$

where c_0, c_1 are positive numbers. The preconditioning matrix was computed using $\widetilde{A}$ instead of A, so that

$$C\mathbf{v} = \widetilde{A}\mathbf{v}$$

(called the modification step) for some given positive vector $\mathbf{v}$. The commonly used vector was

$$\mathbf{v} = \mathbf{e} = (1, 1, \ldots, 1)^T,$$

and in this case the modification step corresponded to a row sum criterion: The sum of entries in C equals the sum of entries in A row-wise.

Under certain conditions, one could show an upper bound $O(h^{-1})$ for the largest eigenvalue. For Stieltjes matrices, it is readily seen that, due to the modification step, the lower eigenvalue of $C^{-1}\widetilde{A}$ is 1. However, we want a lower bound of the eigenvalues of $C^{-1}A$. It turned out that these could be estimated also, and, under certain conditions, they were $O(1)$, $h \to 0$ (see Axelsson and Barker, 1984, for instance). A thorough investigation was also undertaken to show when a condition number $O(h^{-1})$ could be achieved without use of perturbations (see Beauwens, 1985, and Kutcherov and Makarov, 1992, for details).

The conditions

(10.11a)
$$[(2 - \frac{1}{\mu})X - K]\mathbf{v} \geq \mathbf{0}$$

or

(10.11b)
$$(2X - K)\mathbf{v} \geq \mathbf{0},$$

as used in Theorems 10.3 and 10.7, respectively, or the conditions

$$C\mathbf{v} \geq (1 - \tau_0)A\mathbf{v}, \quad 0 < \tau_0 < 1$$

(10.11c)
$$\tau_1 \equiv \max_i \{(-X^{-1}L^T\mathbf{v})_i / v_i\} < 1$$

$$\tau = \max(\tau_0, \tau_1)$$

in Theorem 10.6 lend themselves readily to a dynamic procedure to determine perturbations. In these, the perturbations will take place only when needed and their size will be appropriate to satisfy any of the stated conditions. They can, therefore, be expected to provide preconditioners with—for the rate of convergence of the preconditioned iterative method—better distributions of eigenvalues than when a priori perturbations are used.

Having chosen any of these conditions for an appropriate value of μ or τ in (a) and (c), respectively, one checks the conditions when computing the entry X_i in the ith (block) row. If they are violated, a positive entry of sufficient size is added to (the diagonal of) X_i and, formally, also to the corresponding entrie(s) of A. Due to the factors $[2 - 1/\mu]$, 2 and $[1/(1 - \tau_0)]$ of X_i, which are all bigger than the factor one of $(D_A)_i$ in these conditions, such perturbations exist.

Theorems 10.3 (or Corollary 10.4) and 10.6 then show that, under the other stated conditions, μ, $2m$ and $1/(1 - \tau)$, respectively, become upper bounds for the largest eigenvalue, $\lambda_{\max}(C^{-1}A)$. These bounds can be chosen by the user. In elliptic second-order boundary value problems, they are typically $\mu = O(h^{-1})$ and $\tau = 1 - O(h)$.

If the conditions in Corollary 10.4 hold, then this Corollary and the general bound in Theorem 10.7 show that for M-matrices A and X,

$$\lambda_{\max}(C^{-1}A) \leq \min\{2m, \mu\}.$$

This holds because $[(2 - \frac{1}{\mu})X - K]\mathbf{v} \geq \mathbf{0}$ implies, in particular, $(2X - K)\mathbf{v} \geq \mathbf{0}$.

We next consider implementations of algorithms to satisfy the conditions required in Theorem 10.3 (Corollary 10.4).

10.2.1 Implementations

Assume that A is a Stieltjes matrix. Having defined a nonnegative matrix L we define $X = \text{blockdiag}(X_1, X_2, \ldots, X_m)$, where

$$X_1 = Z[(D_A)_{1,1}] - D'_1,$$

$$X_i = Z[(D_A)_{i,i} - (LY[X^{-1}]L^T)_{i,i}] - D'_i, \quad i = 2, 3, \ldots, m.$$

We can also define the matrices $L_{i,j}$ recursively during the computation of X_i. Here $Y[B]$ denotes a sparse approximation of B, which is nonnegative, such that $Y[B] \leq B$. Similarly, $Z[M]$ denotes a sparse approximation in the form of a Z-matrix such that $Z[M] \geq M$, where M is a Z-matrix. In particular, $Z[M]$ leaves the diagonal of M unaffected. Finally, the diagonal compensation modification matrix D_i' is determined, such that

$$(10.12) \qquad X_i \mathbf{v}_i = (D_A)_{i,i} \mathbf{v}_i - (LX^{-1}L^T)_{i,i} \mathbf{v}_i$$

holds, where $\mathbf{v}_i$ is the ith block of a positive vector $\mathbf{v}$. Hence,

$$(10.13) \qquad \begin{aligned} D_i' \mathbf{v}_i &= Z[(D_A)_{i,i} - (LY[X^{-1}]L^T)_{i,i}] \mathbf{v}_i \\ &\quad - \{(D_A)_{i,i} - (LX^{-1}L^T)_{i,i}\} \mathbf{v}_i. \end{aligned}$$

Note that because all entries of L have the same sign, the definitions of the sparsity operators $Z[\cdot]$ and $Y[\cdot]$ show that D_i' is nonnegative. It can also be seen that X_i is a Stieltjes matrix. Hence, given a symmetric and positive semidefinite matrix A, we consider approximate factorizations of the form

$$C = (X + L)X^{-1}(X + L^T),$$

where X is a blockdiagonal matrix and L is block lower triangular. There are various ways such a preconditioner can be computed. Here, we choose a strategy to get the preconditioner to satisfy a given upper eigenvalue bound of $C^{-1}A$. Corollary 10.4 shows that

$$\lambda_{\max}(C^{-1}A) \leq \mu$$

for any a priori chosen upper bound μ ($\mu > 1/2$), for which

$$(10.14) \qquad \lambda_{\min}(\mu C - A) \geq 0.$$

As has been shown in Theorem 10.3, (10.14) holds if:

  (a) X and $(2 - 1/\mu)X - D_A$ are Z-matrices,
  (b) the block off-diagonal entries of $L + L^T$ are not larger than the corresponding entries of A, and
  (c) $X\mathbf{v} > \mathbf{0}$ and $[(2 - \frac{1}{\mu})X - K]\mathbf{v} \geq \mathbf{0}$ for some positive vector $\mathbf{v}$, where $K = A - L - L^T$.

Condition (b) holds always when $L = L_A$ and condition (a) holds, in particular, if X and D_A are diagonal matrices. In more general cases, we simply let

$$L_{i,j} = A_{i,j}$$

to make (b) hold, if an entry (of) $L_{i,j}$ computed by a (block) matrix incomplete factorization, for instance, happens to be larger than the corresponding entry (of) $A_{i,j}$. Similarly, we let the offdiagonal entries of X be modified, if required, for (a) to hold.

It remains to make $(2 - 1/\mu)X - K$ positive semidefinite, and, as Theorem 10.3 shows, this holds if (c) is valid.

We now consider the condition

$$(10.15) \qquad [(2 - \frac{1}{\mu})X - K]\mathbf{v} \geq \mathbf{0}$$

in Theorem 10.3 for some positive μ. By constructing X_i as in (10.12), there is in general no guarantee that (10.15) holds. One way to overcome this problem is to use the *method of perturbations:* During the incomplete factorization— i.e., during the computation of the sequence X_i in (10.12)—we add a nonnegative diagonal matrix to $(D_A)_{i,i}$ to make $X_i \mathbf{v}_i$ bigger, so that

$$(2 - \frac{1}{\mu})\widetilde{X}_i\mathbf{v}_i = (\widetilde{K})_{i,i}\mathbf{v}_i$$

holds for some positive μ where $\widetilde{X}_i$, $\widetilde{K}$ are the so perturbed matrices. Note that (10.13) shows that the modification matrix D_i' does not change when we use such perturbations. If we use diagonal approximations $Z[M]$ of $M = (D_A)_{i,i} - [LY(X^{-1})L^T]_{i,i}$, then the above method resembles a point-wise incomplete factorization method for which we have also the general upper bound $2m$, where m is the number of blockmatrices in D_A.

In the above method, the diagonal entries of X_i will be changed twice: once by subtracting the nonnegative matrix D_i' (due to neglected fill-in) and, if condition (10.16) fails, by adding positive entries. The reason we have chosen this form is that the lower bound of the eigenvalues of $C^{-1}A$ equals one if no perturbations are applied. Perturbations change the lower eigenvalue bound, but these changes can be estimated by a separate technique to be presented in the next section.

In practice, it may be more convenient to change the diagonal of X_i just once. This can be done if we use a method of relaxation, which means that we use a factor ω_i, $\omega_i < 1$ of the amount in (10.13) that results from deletion of entries (fill-in)—that is, we let D_i' in X_i equal ω_i times D_i' in (10.13). We then let $\omega_i = 1 - \xi_i$, $\xi_i > 0$ and choose the smallest value of ξ_i for which the condition

$$(10.16) \qquad \{(2 - \frac{1}{\mu})X - K)\mathbf{v}\}_i = \mathbf{0}$$

(and, similarly, for the condition 10.11c) holds. An implementation of this method for a point-wise incomplete factorization method has been presented in Section 7.1.

In the relaxed incomplete factorization method (RIC) presented by Axelsson and Lindskog (1986a), ω was chosen fixed. For $\omega = 1$, one gets the so-called modified incomplete factorization method (MIC), and for $\omega = 0$, the (unmodified) incomplete factorization method (IC). Nevertheless, as explained above, it is generally most efficient to let ω depend on i and to choose it so that condition (10.16) holds exactly.

10.3 Lower Eigenvalue Bounds for M-Matrices

Both upper and lower eigenvalue bounds of the eigenvalues of $C^{-1}A$ were derived in Theorem 10.1. In particular, if A is s.p.d. and if

$$\lambda_{\max}(\mu_1 C - A) = 0$$

for some positive (sufficiently small) number μ_1, then this theorem shows that μ_1 becomes a lower bound. For modified methods, where $C\mathbf{v} = A\mathbf{v}$ for some positive vector $\mathbf{v}$ and where we assume that A is a M-matrix and $L = L_A$, the value 1 becomes a lower bound. This is discussed next.

Let $A = D_A + L_A + L_A^T$ be a standard splitting of A and assume that A is a M-matrix. Let

$$C = (X + L)X^{-1}(X + L^T),$$

where $L = L_A$, and let X be diagonal—or, more generally, let X be such that the offdiagonal entries of $D_A - X$ are nonpositive. Then,

$$A - C = D_A - X - LX^{-1}L^T.$$

By the assumptions made, $A - C$ is a Z-matrix, and this, together with $(A - C)\mathbf{v} = \mathbf{0}$, shows that $A - C$ is positive semidefinite. Hence,

(10.17) $$\lambda_{\min}(C^{-1}A) \geq 1.$$

This means that the condition $\lambda_{\max}(\mu_1 C - A) \geq 0$ for the lower bound in Theorem 10.1 holds, with $\mu_1 = 1$ in this case.

When the method of perturbations has been used, the effect of the perturbations on the lower eigenvalue bound must also be estimated.

10.3.1 Lower Eigenvalue Bounds for the Method of Perturbations

We saw in the previous section that, in some cases, perturbations by positive numbers of the diagonal part of a given matrix A can help in the construction of the preconditioner and in getting upper eigenvalue bounds. Here, we want to estimate the change of the lower eigenvalue bounds due to these perturbations. This will be done for M-matrices.

We assume that a method of diagonal compensation has been used to get a matrix that is a M-matrix. Hence, given A, such that $A\mathbf{v} > \mathbf{0}$ for some positive vector $\mathbf{v}$, all positive offdiagonal entries of A are reduced and compensated for in the main diagonal to form a matrix A_1, where $A_1\mathbf{v} = A\mathbf{v}$. Next, A_1 may be perturbed by a diagonal matrix D' with nonnegative entries to form $A_2 = A_1 + D'$. Then,

$$A_2 - A = A_1 - A + D',$$

which is positive semidefinite because its offdiagonal entries are nonpositive and $(A_2 - A)\mathbf{v} = D'\mathbf{v} \geq \mathbf{0}$, so $D' \geq 0$. As has been shown in the previous section, the perturbations can be computed dynamically during the factorization. For the analysis, however, it is easier to assume that they have been added a priori before the factorization.

Let now C be a preconditioner that is constructed from A_2 and assume that

$$\lambda_{\max}(C^{-1}A_2) \leq \mu \text{ and } \lambda_{\min}(C^{-1}A_2) \geq \alpha.$$

Then,

$$\lambda_{\max}(C^{-1}A) \leq \mu\lambda_{\max}(A_2^{-1}A) \leq \mu$$

and

$$\lambda_{\min}(C^{-1}A) \geq \alpha\lambda_{\min}(A_2^{-1}A) \geq \alpha\lambda_{\min}(A_2^{-1}A_1)\lambda_{\min}(A_1^{-1}A).$$

Assuming that $\lambda_{\min}(A_1^{-1}A)$ has been estimated (by the method of diagonal compensation), it remains to estimate $\lambda_{\min}(A_2^{-1}A_1)$, where $A_2 = A_1 + D'$ and A_1 is a M-matrix. To simplify the notations, we now let A be a given M-matrix, and $\widetilde{A} = A + D'$, where D' is nonnegative and we want to estimate $\lambda_{\min}(\widetilde{A}^{-1}A)$. To simplify the presentation further, we assume that the vector $\mathbf{v} = \mathbf{e} = (1, 1, \ldots, 1)^T$. Then, $A\mathbf{e} \geq \mathbf{0}$, that is, the row sums of A are nonnegative. [If we let $V = \text{diag}(v_1, \ldots, v_n)$, then the scaled (and symmetric) matrix VAV has nonnegative row sums. Hence, if $A\mathbf{v} \geq \mathbf{0}$ but $\mathbf{v} \neq \mathbf{e}$, then we consider VAV instead of A.]

We shall use the following lemma later.

Lemma 10.11 *Given a Stieltjes matrix A of order n, then*

$$\mathbf{x}^T A \mathbf{x} \geq \sum_{i=1}^{n} \sum_{j>i} (-a_{i,j})(x_i - x_j)^2, \quad \text{for all } \mathbf{x} \in \mathbb{R}^n.$$

Proof For any symmetric matrix A, we have the identity

$$\mathbf{x}^T A \mathbf{x} = \sum_{i=1}^{n} \left\{ \left(a_{i,i} + \sum_{j \neq i} a_{i,j} \right) x_i^2 + \sum_{j>i} (-a_{i,j})(x_i - x_j)^2 \right\},$$

which is readily shown by computations. Note now that the row sums of entries of A are nonnegative, so the first part of the r.h.s. is nonnegative. Further, since $a_{ij} \leq 0$, the second part is nonnegative.   ◇

To estimate $\lambda_{\min}(\widetilde{A}^{-1} A)$, we shall use matrix graphs. We recall that a graph G is defined as an ordered pair (V, E), where V is a set of vertices (or nodes) and E is a set of pairs of elements of V, that represent edges in the graph. For the graph $G(A)$ induced by a square matrix $A = [a_{i,j}]$ of order n, we have $V = \{1, 2, \ldots, n\}$, where a vertex "i" in V corresponds to row "i" in A. An edge $(i, j) \in E$ if and only if $a_{i,j} \neq 0$. Self edges (i, i), that is, edges with both endpoints at a single vertex, are not included in the graph. The so-constructed graph forms the *directed graph* of A.

Given a subset $\mathcal{N}$ of V, we construct a *directed set of paths* $\{P(i), i \in \mathcal{N}\}$ from every node of $\mathcal{N}$, where every path starts in node $i \in \mathcal{N}$ and passes consecutive nodes

$$j_r(i), \quad r = 1, 2, \ldots, q_i,$$

terminating in node $j_{q_i}(i)$. Hence,

$$P(i) = \{j_0(i) = i, \ j_1(i), j_2(i), \ldots, j_{q_i}(i)\}.$$

The number q_i is the *length* of $P(i)$. We construct only one path originating from each node $i \in \mathcal{N}$, but different paths can pass through one and the same node. (In practice, we will try to limit the number of such occurrences as much as possible, but, on the other hand, we want the path lengths to be sufficiently long, as we shall see.)

In our application, the set $\mathcal{N}$ will be the set of nodes where perturbations have taken place in the corresponding diagonal entries of A. To the nodes in a given path $P(i)$, we associate a positive number $a^{(i)}$ and a diagonal matrix $D^{(i)}$ with entries $D^{(i)}_{k,k} = a^{(i)}$, $k = j_r(i)$, $r = 0, 1, \ldots, q_i$; Otherwise, $D^{(i)}_{k,k} = 0$. We let A be such that for each k there is at least one path passing node k. Hence,

$$
\text{(10.18)} \qquad \sum a^{(i)} \begin{cases} i \in \mathcal{N}, \\ j_r(i) = k, \quad 0 \le r \le q_i \end{cases} \le a_{k,k}
$$

and

$$
\text{(10.19)} \qquad \sum_{i \in \mathcal{N}} \mathbf{x}^T D^{(i)} \mathbf{x} \le \mathbf{x}^T D \mathbf{x}, \quad \text{for all } \mathbf{x} \in R^n,
$$

where $D = \mathrm{diag}(A)$. This means that if a particular path passes, only through nodes where no other path passes, we can let $a^{(i)} = \min a_{k,k}$, $k = j_r(i)$, $r = 0, 1, \ldots, q_i$. However, if there are two or more paths through a particular node k, we must satisfy (10.18). The simplest choice is to let $a^{(i)} = 1/v_k a_{k,k}$, where v_k is the number of paths through k.

Given positive perturbations $\delta_i a_{i,i}$, $i \in \mathcal{N}$ to the diagonal of A, we want to estimate $\sum_{i \in \mathcal{N}} \delta_i a_{i,i} x_i^2$ from above. To this end, following the approach taken in Gustafsson (1978) and Axelsson and Barker (1984), we shall first estimate $a_{i,i} x_i^2$ from above. Since $x_i^2 = x_{j_r(i)}^2 + (x_i^2 - x_{j_r(i)}^2)$, we note that

$$
\sum_{r=1}^{q_i} a_{i,i} x_i^2 = \sum_{r=1}^{q_i} a_{i,i} x_{j_r(i)}^2 + \sum_{r=1}^{q_i} a_{i,i} (x_i^2 - x_{j_r(i)}^2)
$$

or

$$
\text{(10.20)} \qquad q_i a_{i,i} x_i^2 = \sum_{r=1}^{q_i} a_{i,i} x_{j_r(i)}^2 + \sum_{r=1}^{q_i} a_{i,i} (x_i^2 - x_{j_r(i)}^2).
$$

Here, by the definition of $D^{(i)}$,

$$
\sum_{r=1}^{q_i} a_{i,i} x_{j_r(i)}^2 \le \frac{a_{i,i}}{a^{(i)}} \mathbf{x}^T D^{(i)} \mathbf{x},
$$

so, by (10.20),

$$\sum_{i \in \mathcal{N}} \delta_i a_{i,i} x_i^2 \le \max_{i \in \mathcal{N}} \left(\frac{\delta_i}{q_i} \right) \sum_{i \in \mathcal{N}} q_i a_{i,i} x_i^2$$

$$(10.21) \qquad \le \max_{i \in \mathcal{N}} \left(\frac{\delta_i}{q_i} \right) \left\{ a_0 \sum_{i \in \mathcal{N}} \mathbf{x}^T D^{(i)} \mathbf{x} + \sum_{i \in \mathcal{N}} \sum_{r=1}^{q_i} a_{i,i} (x_i^2 - x_{j_r(i)}^2) \right\},$$

where

$$a_0 = \max_{i \in \mathcal{N}} a_{i,i} / a^{(i)}.$$

Consider now the second term of the right-hand side of (10.20). The relation

$$x_i^2 - x_{j_r(i)}^2 = \sum_{s=0}^{r-1} (x_{j_s(i)}^2 - x_{j_{s+1}(i)}^2)$$

and the Cauchy-Schwarz inequality imply

$$|x_i^2 - x_{j_r(i)}^2| \le \sum_{s=0}^{r-1} |x_{j_s(i)} - x_{j_{s+1}(i)}| \, |x_{j_s(i)} + x_{j_{s+1}(i)}|$$

$$\le \left\{ \sum_{s=0}^{r-1} (x_{j_s(i)} - x_{j_{s+1}(i)})^2 \right\}^{\frac{1}{2}} \cdot \left\{ \sum_{s=0}^{r-1} (x_{j_s(i)} + x_{j_{s+1}(i)})^2 \right\}^{\frac{1}{2}}.$$

Multiplying by $a_{i,i}$, summing with respect to i, and using the Cauchy-Schwarz inequality again, we obtain

$$(10.22) \qquad \begin{aligned} |\sum_{i \in \mathcal{N}} a_{i,i} (x_i^2 - x_{j_r(i)}^2)| &\le \left\{ \sum_{i \in \mathcal{N}} a_{i,i} \sum_{s=0}^{r-1} (x_{j_s(i)} - x_{j_{s+1}(i)})^2 \right\}^{\frac{1}{2}} \\ &\times \left\{ \sum_{i \in \mathcal{N}} a_{i,i} \sum_{s=0}^{r-1} (x_{j_s(i)} + x_{j_{s+1}(i)})^2 \right\}^{\frac{1}{2}}. \end{aligned}$$

Now, for $r = 1, 2, \ldots, q_i$, we have

$$\sum_{i \in \mathcal{N}} a_{i,i} \sum_{s=0}^{r-1} (x_{j_s(i)} - x_{j_{s+1}(i)})^2 \le \sum_{i \in \mathcal{N}} a_{i,i} \sum_{s=0}^{q_i-1} (x_{j_s(i)} - x_{j_{s+1}(i)})^2$$

$$\le a_1^{-1} \sum_{i \in \mathcal{N}} \sum_{s=0}^{q_i-1} |a_{j_s(i), j_{s+1}(i)}| (x_{j_s(i)} - x_{j_{s+1}(i)})^2,$$

where

$$a_1 = \min_{i \in \mathcal{N}} \min_{0 \le s \le q_i - 1} |a_{j_s(i), j_{s+1}(i)} / a_{i,i}|.$$

Lemma 10.11 now shows that

$$(10.23) \quad \sum_{i \in \mathcal{N}} a_{i,i} \sum_{s=0}^{r-1} (x_{j_x(i)} - x_{j_{s+1}(i)})^2 \le a_1^{-1} \nu \mathbf{x}^T A \mathbf{x}, \quad \text{for all } \mathbf{x} \in \mathbb{R}^n,$$

where $\nu = \max_k \nu_k$ is the maximum number of times any node is passed by different paths. For the second factor of (10.23) we find

$$\sum_{i \in \mathcal{N}} a_{i,i} \sum_{s=0}^{r-1} (x_{j_s(i)} + x_{j_{s+1}(i)})^2 \le 2 \sum_{i \in \mathcal{N}} a_{i,i} \sum_{s=0}^{q_i - 1} (x_{j_s(i)}^2 + x_{j_{s+1}(i)}^2)$$

$$(10.24) \qquad\qquad \le 4 \sum_{i \in \mathcal{N}} a_{i,i} \sum_{s=0}^{q_i} x_{j_s(i)}^2 \le 4 a_0 \sum_{i \in \mathcal{N}} a^{(i)} \sum_{s=0}^{q_i} x_{j_s(i)}^2$$

$$\le 4 a_0 \sum_{i \in \mathcal{N}} \mathbf{x}^T D^{(i)} \mathbf{x}, \quad \text{for all } \mathbf{x} \in \mathbb{R}^N.$$

(10.19) and (10.21) to (10.24) now imply

$$\sum_{i \in \mathcal{N}} \delta_i a_{i,i} x_i^2 \le \max_{i \in \mathcal{N}} \left(\frac{\delta_i}{q_i} \right) \left\{ a_0 \mathbf{x}^T D \mathbf{x} + 2\overline{q} [\nu a_1^{-1} \mathbf{x}^T A \mathbf{x}]^{\frac{1}{2}} [a_0 \mathbf{x}^T D \mathbf{x}]^{\frac{1}{2}} \right\},$$

where $\overline{q} = \max_{i \in \mathcal{N}} q_i$.
Hence,

$$\frac{1}{\mathbf{x}^T A \mathbf{x}} \sum_{i \in \mathcal{N}} \delta_i a_{i,i} x_i^2 \le \max_{i \in \mathcal{N}} \left(\frac{\delta_i}{q_i} \right) \left(a_0 \frac{\mathbf{x}^T D \mathbf{x}}{\mathbf{x}^T A \mathbf{x}} \right)^{\frac{1}{2}} \left\{ \left(a_0 \frac{\mathbf{x}^T D \mathbf{x}}{\mathbf{x}^T A \mathbf{x}} \right)^{\frac{1}{2}} + 2\overline{q} \left(\frac{\nu}{a_1} \right)^{\frac{1}{2}} \right\}.$$

We can now state the main result in this section.

Theorem 10.12 *Let A be a Stieltjes matrix with nonnegative row sums and let $\tilde{A} = A + D'$, where D' is a diagonal perturbation matrix,*

$$(D')_{i,i} = \delta_i a_{i,i}, \quad i \in \mathcal{N} \subset V = \{1, \ldots, n\}$$

$$(D')_{i,i} = 0, \quad i \in V \setminus \mathcal{N}.$$

Then,

$$(10.25) \quad \frac{\mathbf{x}^T \tilde{A} \mathbf{x}}{\mathbf{x}^T A \mathbf{x}} \le 1 + \max_i \left(\frac{\delta_i}{q_i} \right) (a_0 \lambda_0)^{\frac{1}{2}} \left[(a_0 \lambda_0)^{\frac{1}{2}} + 2\bar{q} \left(\frac{v}{a_1} \right)^{\frac{1}{2}} \right],$$

where

$$\lambda_0 = \max_{\mathbf{x} \ne 0} \frac{\mathbf{x}^T D \mathbf{x}}{\mathbf{x}^T A \mathbf{x}}.$$

Here q_i is the length of the path in the matrix graph representation of A, originating in node i, $\bar{q} = \max_{i \in \mathcal{N}} q_i$, and v is the maximal number of paths passing one and the same node. ◇

We want to estimate the order of the eigenvalue bound as a function of some problem parameter, which we call h and which converges to zero. In practice, λ_0 grows with some power of h^{-1}. Frequently one uses two or more order of magnitudes—i.e., $O(h)$ or $O(h^2)$, for the perturbations with respect to this parameter. Consider, then, perturbations $\delta_i^{(1)}$, $i \in \mathcal{N}_1$ and $\delta_i^{(2)}$, $i \in \mathcal{N}_2$, where $\mathcal{N}_1 \cap \mathcal{N}_2 = \emptyset$. We let $v = v^{(1)}$ for paths originating in $\mathcal{N}_1$ and $v = v^{(2)}$ for paths originating in $\mathcal{N}_2$. The estimate in Theorem 10.12 takes the form

$$\frac{\mathbf{x}^T \tilde{A} \mathbf{x}}{\mathbf{x}^T A \mathbf{x}} \le 1 + \max_{i \in \mathcal{N}_1} \left(\frac{\delta_i^{(1)}}{q_i^{(1)}} \right) \left(a_0^{(1)} \lambda_0 \right)^{\frac{1}{2}} \left\{ \left(a_0^{(1)} \lambda_0 \right)^{\frac{1}{2}} + 2\bar{q}^{(1)} \left(\frac{v^{(1)}}{a_1^{(1)}} \right)^{\frac{1}{2}} \right\}$$

$$+ \max_{i \in \mathcal{N}_2} \left(\frac{\delta_i^{(2)}}{q_i^{(2)}} \right) \left(a_0^{(2)} \lambda_0 \right)^{\frac{1}{2}} \left\{ \left(a_0^{(2)} \lambda_0 \right)^{\frac{1}{2}} + 2\bar{q}^{(2)} \left(\frac{v^{(2)}}{a_1^{(2)}} \right)^{\frac{1}{2}} \right\}.$$

(10.18) shows that typically $a_0 = O(v)$, while $a_1 = O(1)$, $h \to 0$. To get a smallest possible order of the upper bound, we shall take q_i large enough to balance the large eigenvalue λ_0. At the same time, however, v should be kept sufficiently small. To balance the terms in the brackets, we choose $v^{(i)}$ and $\bar{q}^{(i)}$, such that $\bar{q}^{(i)} = O(\lambda_0^{\frac{1}{2}})$, $h \to 0$, and to balance the two terms we let

$$\frac{\delta^{(1)}}{\bar{q}^{(1)}} a_0^{(1)} = O \left(\frac{\delta^{(2)}}{\bar{q}^{(2)}} a_0^{(2)} \right),$$

that is,

$$\delta^{(1)} a_0^{(1)} = O(\delta^{(2)} a_0^{(2)})$$

or

$$\delta^{(1)} v^{(1)} = O(\delta^{(2)} v^{(2)}), \ h \to 0.$$

Example 10.1 Consider a difference mesh for a second-order partial differential equation. Here, h is the meshwidth and $\lambda_0 = O(h^{-2})$. If $\delta^{(1)} = O(h)$ and $\delta^{(2)} = O(1)$, then we construct paths of length

$$q^{(i)} = O(\lambda_0)^{\frac{1}{2}} = O(h^{-1})$$

and choose $v^{(1)}$, $v^{(2)}$, such that

$$\delta^{(1)} v^{(1)} = \delta^{(2)} v^{(2)}.$$

That is, with $v^{(2)} = 1$ say, we can let

$$v^{(1)} = O(h^{-1}).$$

The upper bound then becomes

$$\frac{\mathbf{x}^T \tilde{A} \mathbf{x}}{\mathbf{x}^T A \mathbf{x}} \le O(h^{-1}), \ h \to 0.$$

If $\delta^{(1)} = O(h^2)$ and $\delta^{(2)} = O(h)$, then, similarly we let $q^{(i)} = O(h^{-1})$, and we can choose $v^{(1)} = O(h^{-1})$, $v^{(2)} = O(1)$ to get

$$\frac{\mathbf{x}^T \tilde{A} \mathbf{x}}{\mathbf{x}^T A \mathbf{x}} = O(1), \ h \to 0.$$

In this case, it is also possible to let $q^{(1)} = O(1)$ and $v^{(1)} = O(1)$ to get the upper bound $O(1)$.

Clearly, the larger the sets $\mathcal{N}_1, \mathcal{N}_2$ are, the more difficult it becomes to choose paths of sufficiently large lengths without having to permit too many paths to pass through the same node. However, even if $\mathcal{N}_1$ is the whole set of node points, then—as is readily seen—we can still find (one-directional) paths from each nodepoint of length $O(h^{-1})$, where at most $O(h^{-1})$ different paths pass a single node.

10.4 Upper and Lower Bounds of Condition Numbers

Using the upper and lower bounds of the condition number in Sections 10.1 and 10.3, we find an upper estimate of the spectral condition number. Assuming that A is a M-matrix and C satisfies the conditions stated in Section 10.3, and if the perturbations used are sufficiently small and/or sufficiently infrequent so that the lowest eigenvalue becomes $O(1)$, then the condition number is bounded by

$$\kappa(C^{-1}A) \leq C \min\{2m, \mu\},$$

where μ is the number in Corollary 10.4. It is interesting to note that the upper bound becomes $2m$ for sufficiently large values of μ. For an elliptic difference equation, the number of blocks m is typically $O(h^{-1})$ for a two-dimensional domain. If the domain is an oblong rectangular domain, let the number of nodepoints be $N_1 \times N_2$ where, say, $N_1 > N_2$. If we partition the domain such that each meshline corresponds to a matrix block, and if we number the points in such an order that the dimension of the blocks is $N_1 \times N_1$, then there are $m = N_2$ blocks in the main diagonal and the above bound shows that

$$\lambda_{\max}(C^{-1}A) \leq 2N_2,$$

which, hence, does not depend on N_1. This is an interesting observation, in particular, if $N_1 \gg N_2$.

It will be illustrative to consider now a somewhat simpler type of incomplete factorization method, namely, the block SSOR method.

10.4.1 The Block SSOR Method

Assume that A is block tridiagonal. In the block SSOR method (see Section 7.5), we let $L = L_A$ and $X_i = 1/\omega_i(D_A)_{i,i}$, $i = 1, \ldots, m$, where $1 \leq \omega_i < 2$ and $\omega_1 = 1$. Assume that D_A is a Z-matrix, $D_A\mathbf{v} > \mathbf{0}$ for some positive $\mathbf{v}$, and that L_A is nonpositive. The sequence ω_i must be determined such that the conditions for the lower eigenvalue bound ($=1$) and upper eigenvalue bound ($= \mu$) hold. In this case, the two conditions to be satisfied are

(a) $\{(A - C)\mathbf{v}\}_i = (1 - \dfrac{1}{\omega_i})(D_A)_{i,i}\mathbf{v}_i - \omega_{i-1}(LD_A^{-1}L^T)_{i,i}\mathbf{v}_i \geq 0, \quad i \geq 2,$

and

(b) $[(2 - \dfrac{1}{\mu})X - D_A]_{i,i}\mathbf{v}_i = [(2 - \dfrac{1}{\mu})\omega_i^{-1} - 1](D_A)_{i,i}\mathbf{v}_i \geq 0.$

The latter is satisfied with $\mu = 1/(2 - \max_i \omega_i)$, when $\max_i \omega_i \leq 2 - 1/\mu$. As has been shown in Axelsson and Barker (1984) (see also Exercise 7.5), for certain matrices A it is possible to choose the sequence ω_i such that the first condition (a) holds. Otherwise, if it does not hold, one can use a method of perturbations of D_A as presented previously. A sufficient condition for (a) to hold is

$$1 - \frac{1}{\omega_i} - \omega_{i-1}\rho([\tilde{L}\tilde{L}^T]_{i,i}) \geq 0, \quad i = 2, \ldots, m,$$

where $\rho(\cdot)$ denotes the spectral radius and $\tilde{L} = D_A^{-\frac{1}{2}} L D_A^{-\frac{1}{2}}$.

Case 10.1 Assume, first, that

$$\rho([\tilde{L}\tilde{L}^T]_{i,i}) \leq \frac{1}{4}.$$

Then let

$$\omega_1 = 1, \quad 1 - \frac{1}{\omega_i} - \omega_{i-1}/4 = 0,$$

that is, let $\omega_i = \frac{2i}{i+1}$, $i = 1, 2, \ldots, m$. Then,

$$(10.26) \qquad \lambda_{\max}(C^{-1}A) \leq \mu = 1/(2 - \omega_m) = \frac{m+1}{2}.$$

Case 10.2 Let $\rho_i = \rho([\tilde{L}\tilde{L}^T]_{i,i})$ and assume next that there exists a positive number ε such that

$$\varepsilon \leq \rho_{i-1} \leq \rho_i \leq \rho_0,$$

where $\rho_0 \leq (\frac{1+\varepsilon}{2})^2$. Then let

$$(10.27) \qquad \omega_1 = 1, \quad \omega_i^{-1} = 1 + \varepsilon - \rho_i\omega_{i-1}, \quad i = 2, 3, \ldots.$$

This choice corresponds to perturbing D_A by εD_A. Then it can be seen that

$$1 \leq \omega_i \leq \omega_{i+1} \leq \omega_0 = 2/[1 + \varepsilon + \sqrt{(1 + \varepsilon)^2 - 4\rho_0}], \quad i = 2, 3, \ldots.$$

Hence, in this case,

$$\mu = \frac{1}{2} + \frac{1}{2}\left[\varepsilon + \sqrt{(1 + \varepsilon)^2 - 4\rho_0}\right]^{-1},$$

and Corollary 10.4 and Theorem 10.7 show that

$$(10.28) \qquad \lambda_{\max}(C^{-1}A) \leq \min(\mu, 2m).$$

However, with the above choice (10.27) of the sequence ω_i, condition (a) is not satisfied and we have only

$$\{(\widetilde{A} - C)\mathbf{v}\}_i \geq \mathbf{0}, \quad i \geq 2,$$

when $\widetilde{A} = A + \varepsilon D_A$.

Then,

$$(10.29) \qquad \lambda_{\min}(C^{-1}A) \geq \lambda_{\min}(\widetilde{A}^{-1}A) = \frac{1}{1 + \varepsilon\mu_0},$$

where $\mu_0 = \lambda_{\max}(A^{-1}D_A)$. Note that μ_0 is the order of the condition number of the Jacobi (block-diagonal) preconditioned matrix and that, in practical (interesting) applications, μ_0 is a large number. Hence, ε must be small to balance μ_0. Now, (10.27) to (10.29) show that

$$\kappa(C^{-1}A) \leq \min\{(1 + \varepsilon\mu_0)2m, \frac{1}{2}(1 + \varepsilon\mu_0)[1 + \left[\varepsilon + \sqrt{(1 + \varepsilon)^2 - 4\rho_0}\right]^{-1}]\}.$$

Assume next that $\rho_0 = 1/4$. Even in this case, we can sometimes reduce the condition number by using perturbations of proper order. Then (10.26) and (10.28) show that

$$\kappa(C^{-1}A)$$

$$(10.30a)$$
$$\leq \min\{\frac{m + 1}{2}, \frac{1}{2}(1 + \varepsilon\mu_0)[1 + \left[\varepsilon + \sqrt{(1 + \varepsilon)^2 - 4\rho_0}\right]^{-1}]\},$$

and in this case we can choose the perturbation parameter ε to minimize the upper bound. It can be seen that as $\mu_0 \to \infty$, the (asymptotically) optimal value of ε should minimize $(1 + \varepsilon\mu_0)1/\sqrt{2\varepsilon}$, that is, $\varepsilon = \mu_0^{-1}$. The upper bound in (10.30a) then takes (asymptotically) the form

$$(10.30b) \qquad \kappa(C^{-1}A) \leq \min\{\frac{m + 1}{2}, 1 + \sqrt{\frac{\mu_0}{2}}\}.$$

For the standard five-point second-order elliptic difference equation, one finds $\mu_0 = 2/(\pi h)^2$, so

$$\kappa(C^{-1}A) \leq \min\left\{\frac{m + 1}{2}, 1 + \frac{1}{\pi h}\right\}.$$

As we have seen, upper bounds of condition numbers for elliptic difference equations preconditioned by modified (and slightly perturbed, if necessary) methods are $O(h^{-1})$, $h \to 0$, if certain additional conditions hold. A natural question to ask is if this order is sharp or if it can be improved by some other choices of X_i. By deriving lower bounds on the condition number, we show that the order of the bound is sharp if X_i has a fixed sparsity pattern. This does not exclude that choices of X_i with other sparsity patterns and, in particular, a sparsity pattern that grows with i and h^{-1}, can reduce the order of the condition number. In the extreme, if we let $X_i = A_{i,i} - A_{i,i-1}X_{i-1}^{-1}A_{i-1,i}$, then C becomes the exact factorization of A and $\text{cond}(C^{-1}A) = 1$. However, here X_i are full matrices.

The lower bounds of the condition number will be derived using Schur complements.

10.4.2 Lower Bounds of Condition Numbers Based on Schur Complements

In order to find a lower bound of the condition number, we present first an elementary but useful lemma for quadratic forms of blocktridiagonal matrices.

Lemma 10.13 *Let* $A = \text{blocktridiag}(A_{i,i-1}, A_{i,i}, A_{i,i+1})$, *where* $A_{i,i}$ *are square matrices, be symmetric and positive definite. Then,*

$$(10.31) \quad x^T A x = \sum_{i=1}^{n-1} (x_i + S_i^{-1} A_{i,i+1} x_{i+1})^T S_i (x_i + S_i^{-1} A_{i,i+1} x_{i+1}) + x_n^T S_n x_n,$$

where $x = (x_1, x_2, \ldots, x_n)$ *is partitioned consistently with the partitioning of* A *and where* S_i *are the Schur complements defined by the recursion*

$$S_1 = A_{11}, \quad S_{i+1} = A_{i+1,i+1} - A_{i+1,i} S_i^{-1} A_{i,i+1}, \quad i = 1, 2, \ldots, n-1.$$

Proof Consider, first, a partitioning of A in block two-by-two form, $A = \begin{bmatrix} A_{11} & \widehat{A}_{12} \\ \widehat{A}_{21} & \widehat{A}_{22} \end{bmatrix}$, and note that A_{11} is positive definite. Then, for $x = (x_1, \hat{x}_2)$, $\hat{x}_2 = (x_2, x_3, \ldots, x_n)$, we have

$$x^T A x = x_1^T A_{11} x_1 + x_1^T \widehat{A}_{12} \hat{x}_2 + \hat{x}_2^T \widehat{A}_{21} x_1 + \hat{x}_2^T \widehat{A}_{22} \hat{x}_2$$

or, due to the special form of matrices $\widehat{A}_{12} = \widehat{A}_{21}^T$, we have $x_1^T \widehat{A}_{12} \hat{x}_2 = x_1^T A_{12} x_2$ and $\hat{x}_2^T \widehat{A}_{21} x_1 = x_2^T A_{21} x_1$. Hence,

$$x^T A x = (x_1 + A_{11}^{-1} A_{12} x_2)^T A_{11} (x_1 + A_{11}^{-1} A_{12} x_2) + \hat{x}_2^T (\widehat{A}_{22} - \widehat{A}_{21} A_{11}^{-1} \widehat{A}_{12}) \hat{x}_2.$$

Also, due to the block tridiagonal form of A,

$$\widehat{A}_{22} - \widehat{A}_{21}A_{11}^{-1}\widehat{A}_{12} = \begin{bmatrix} S_2 & \widehat{A}_{23} \\ \widehat{A}_{32} & \widehat{A}_{33} \end{bmatrix}.$$

Repeated use of the above now shows (10.31).   ◇

Similarly, for the matrix

(10.32) $$C = (D - L)D^{-1}(D - L^T),$$

where $D = \text{diag}(D_1, D_2, \ldots, D_n)$ and D_i are symmetric and positive definite, we have

$$C = \begin{bmatrix} D_1 & A_{12} & & \\ A_{21} & (D_2 + A_{21}D_1^{-1}A_{12}) & A_{23} & 0 \\ & \cdot & \cdot & \cdot \\ 0 & & A_{n,n-1} & (D_n + A_{n,n-1}D_{n-1}^{-1}A_{n-1,n}) \end{bmatrix},$$

whence D_i are the corresponding Schur complements for C and

(10.33)
$$\mathbf{x}^T C \mathbf{x} = \sum_{i=1}^{n-1}(\mathbf{x}_i + D_i^{-1}A_{i,i+1}\mathbf{x}_{i+1})^T D_i(\mathbf{x}_i + D_i^{-1}A_{i,i+1}\mathbf{x}_{i+1})$$
$$+ \mathbf{x}_n^T D_n \mathbf{x}_n.$$

We can now give an upper bound for the smallest eigenvalue and a lower bound for the largest eigenvalue of $C^{-1}A$.

Lemma 10.14 *Let A be defined as in Lemma 10.13 and let C have the form (10.32). Then, for $r \geq 2$,*

(a) $\min_{\mathbf{x}\neq 0} \dfrac{\mathbf{x}^T A \mathbf{x}}{\mathbf{x}^T C \mathbf{x}} \leq \dfrac{\hat{\mathbf{x}}_r^T S_r \hat{\mathbf{x}}_r}{\hat{\mathbf{x}}_r^T D_r \hat{\mathbf{x}}_r + f_r(\hat{\mathbf{x}}_2,\ldots,\hat{\mathbf{x}}_r)}$ *and*

(b) $\max_{\mathbf{x}\neq 0} \dfrac{\mathbf{x}^T A \mathbf{x}}{\mathbf{x}^T C \mathbf{x}} \geq \dfrac{\hat{\mathbf{y}}_r^T S_r \hat{\mathbf{y}}_r + g_r(\hat{\mathbf{y}}_2,\ldots,\hat{\mathbf{y}}_r)}{\hat{\mathbf{y}}_r^T D_r \hat{\mathbf{y}}_r}$,

where

$$f_r(\mathbf{x}_2, \mathbf{x}_3, \ldots, \mathbf{x}_r) = \sum_{i=1}^{r-1}[(D_i^{-1} - S_i^{-1})A_{i,i+1}\mathbf{x}_{i+1}]^T D_i[(D_i^{-1} - S_i^{-1})A_{i,i+1}\mathbf{x}_{i+1}],$$

$$g_r(\mathbf{y}_2, \mathbf{y}_3, \ldots, \mathbf{y}_r) = \sum_{i=1}^{r-1}[(D_i^{-1} - S_i^{-1})A_{i,i+1}\mathbf{y}_{i+1}]^T S_i[(D_i^{-1} - S_i^{-1})A_{i,i+1}\mathbf{y}_{i+1}],$$

and $\hat{\mathbf{x}}$ *and* $\hat{\mathbf{y}}$ *are defined by*

$$\hat{\mathbf{x}}_i = -S_i^{-1} A_{i,i+1} \hat{\mathbf{x}}_{i+1},$$

$$i = r-1, n-2, \ldots, 1, \ \hat{\mathbf{x}}_r \ \text{arbitrary}, \ \hat{\mathbf{x}}_{r+1} = \ldots = \hat{\mathbf{x}}_n = \mathbf{0},$$

$$\hat{\mathbf{y}}_i = -D_i^{-1} A_{i,i+1} \hat{\mathbf{y}}_{i+1},$$

$$i = r-1, n-2, \ldots, 1, \ \hat{\mathbf{y}}_r \ \text{arbitrary}, \ \hat{\mathbf{y}}_{r+1} = \ldots = \hat{\mathbf{y}}_n = \mathbf{0},$$

respectively.

Proof By direct substitution of $\hat{\mathbf{x}}$ and $\hat{\mathbf{y}}$, respectively, in (10.31) and (10.33).

$\diamond$

Consider next the spectral condition number $\kappa(C^{-1}A)$.

Corollary 10.15 $\kappa(C^{-1}A) \geq \max_{1 \leq r \leq n} \kappa(D_r^{-1}S_r)$.

Proof Let $\hat{\mathbf{x}}_r$ be the eigenvector for the smallest and $\hat{\mathbf{y}}_r$ the eigenvector for the largest eigenvalue of $D_r^{-1}S_r$, respectively. Then, since $f_r \geq 0$ and $g_r \geq 0$ (because the Schur complements S_i and D_i are positive definite, as A and C are positive definite), Lemma 10.14 shows that

$$\min_{\mathbf{x} \neq 0} \frac{\mathbf{x}^T A \mathbf{x}}{\mathbf{x}^T C \mathbf{x}} \leq \frac{\hat{\mathbf{x}}_r^T S_r \hat{\mathbf{x}}_r}{\hat{\mathbf{x}}_r^T D_r \hat{\mathbf{x}}_r} = \lambda_{\min}(D_r^{-1}S_r)$$

and

$$\max_{\mathbf{x} \neq 0} \frac{\mathbf{x}^T A \mathbf{x}}{\mathbf{x}^T C \mathbf{x}} \geq \frac{\hat{\mathbf{y}}_r^T S_r \hat{\mathbf{y}}_r}{\hat{\mathbf{y}}_r^T D_r \hat{\mathbf{y}}_r} = \lambda_{\max}(D_r^{-1}S_r). \qquad \diamond$$

Corollary 10.15 shows that $\text{cond}(C^{-1}A) \geq \text{cond}(D_n^{-1}S_n)$. Note that this lower bound holds independently of the choice of D_i, $i = 1, 2, \ldots, n-1$. In particular, it holds even if $D_i = S_i$, $i = 1, 2 \ldots, n-1$. (In this case the bound is sharp.) To estimate the order of $\text{cond}(D_n^{-1}S_n)$, we shall first relate S_n to $A_{n,n}$. Let $S_1 = A_{11}$, let A_i, A'_{i+1} be defined by (8.29). Then,

(10.34) $$S_i = A_{i,i} - \widehat{A}_{i,i-1}(A_{i-1})^{-1}\widehat{A}_{i-1,i}, \ 1 < i < n$$

and

$$S'_n = A_{n,n},$$

(10.35) $\quad S'_{i+1} = A_{i+1,i+1} - \widehat{A}_{i+1,i+2}(A'_{i+2})^{-1}\widehat{A}_{i+2,i+1}, \; 0 < i < n-1.$

It follows from Theorem 3.8 that

$$\hat{\mathbf{v}}_i^T A_i \hat{\mathbf{v}}_i \geq \mathbf{v}_i^T S_i \mathbf{v}_i \; \text{ for any } \hat{\mathbf{v}}_i$$

and

$$\hat{\mathbf{u}}_{i+1}^T A'_{i+1} \hat{\mathbf{u}}_{i+1} \geq \mathbf{v}_{i+1}^T S'_{i+1} \mathbf{v}_{i+1} \; \text{ for any } \hat{\mathbf{u}}_{i+1}.$$

Further,

(10.36) $$\inf_{\hat{\mathbf{v}}_{i-1}} \hat{\mathbf{v}}_i^T A_i \hat{\mathbf{v}}_i = \mathbf{v}_i^T S_i \mathbf{v}_i$$

and

(10.37) $$\inf_{\hat{\mathbf{u}}_{i+2}} \hat{\mathbf{u}}_{i+1}^T A'_{i+1} \hat{\mathbf{u}}_{i+1} = \mathbf{v}_{i+1}^T S'_{i+1} \mathbf{v}_{i+1}.$$

We recall the following relations discussed in Chapter 8.

Lemma 10.16 *Let A, A_i, A'_{i+1} be defined by (8.26), (8.29) and let S_i, S'_{i+1} be defined by (10.34), (10.35). Then,*

(10.38) $$\mathbf{v}_i^T A_{i,i+1} \mathbf{v}_{i+1} \leq \gamma_i \{\mathbf{v}_i^T S_i \mathbf{v}_i\}^{\frac{1}{2}} \{\mathbf{v}_{i+1}^T S'_{i+1} \mathbf{v}_{i+1}\}^{\frac{1}{2}},$$

and this bound is sharp.

Proof The upper bound follows from (8.31). Here, γ_i is the best constant because it is the best constant in (8.31). $\diamond$

Note that (10.38) is applicable, in particular, for $i = n - 1$, where $S'_{i+1} = S'_n = A_{n,n}$. We shall use Lemma 10.16 to prove a relation between the Schur complement S_n and $A_{n,n}$. To this end, recall first such a relation for a general matrix, partitioned in two-by-two block form (see Lemma 9.2). Let $A = \begin{bmatrix} A_{1,1} & A_{1,2} \\ A_{2,1} & A_{2,2} \end{bmatrix}$ be symmetric and positive definite. Then,

(10.39a)
$$\gamma^2 \equiv \sup_{\mathbf{x}_1, \mathbf{x}_2} \frac{(\mathbf{x}_1^T A_{1,2} \mathbf{x}_2)^2}{(\mathbf{x}_1^T A_{1,1} \mathbf{x}_1)(\mathbf{x}_2^T A_{2,2} \mathbf{x}_2)}$$
$$= \sup_{\mathbf{x}_2} \frac{\mathbf{x}_2^T A_{2,1} A_{1,1}^{-1} A_{1,2} \mathbf{x}_2}{\mathbf{x}_2^T A_{2,2} \mathbf{x}_2}$$

(10.39b)
$$1 - \gamma^2 \leq \frac{x_2^T S_2 x_2}{x_2^T A_{2,2} x_2} \leq 1 \quad \text{for all } x_2, \text{ where}$$

$$S_2 = A_{2,2} - A_{2,1} A_{1,1}^{-1} A_{1,2}.$$

Further, the lower bound is sharp and the upper bound is sharp if the nullspace of $A_{1,2}$ or of $A_{2,1}$ is nontrivial.

Corollary 10.17 *Let* $S_n = A_{n,n} - A_{n,n-1} S_{n-1}^{-1} A_{n-1,n}.$ *Then,*

(10.40)
$$1 - \gamma_{n-1}^2 \leq \frac{v_n^T S_n v_n}{v_n^T A_{n,n} v_n} \leq 1, \quad v_n \in \mathbb{R}^{\dim(A_{n,n})}$$

where γ_{n-1} *is defined in (10.38), with* $i = n - 1$, *and where the lower bound is sharp and the upper bound is sharp if* $A_{n,n-1} S_{n-1}^{-1} A_{n-1,n}$ *has a nontrivial nullspace.*

Proof Let $i = n - 1$ in (10.38). For the matrix $\begin{bmatrix} S_{n-1} & A_{n-1,n} \\ A_{n,n-1} & A_{n,n} \end{bmatrix}$, which is symmetric and positive definite, (10.38) shows that $\gamma^2 = \gamma_{n-1}^2$ and (10.40) follows then by (10.39b). $\diamond$

10.5 Asymptotic Estimates of Condition Numbers for Second-Order Elliptic Problems

We have seen that the asymptotic estimate for second-order elliptic difference equation problems, preconditioned by incomplete factorization methods, is $O(h^{-1})$, $h \to 0$, if certain conditions are satisfied. In this section, some further results related to asymptotic estimates will be presented.

We show, first, that the relaxed method [cf. (7.5)] with a fixed parameter $\omega < 1$ (independent of h), i.e., the essentially unmodified incomplete factorization method, cannot improve the order of the condition number of the given matrix, which is $O(h^{-2})$, $h \to 0$.

Furthermore, it will be shown that the smallest eigenvalues are distributed essentially as for the given matrix. This shows that the method can be used as an efficient smoother in connection with multigrid methods, for instance. For a model type problem, it will also be shown that using a modification step based on the eigenvector corresponding to the eigenvalue of A, the constant c in the asymptotic estimate

$$\kappa(C^{-1}A) \sim ch^{-1}$$

will be smaller than for the commonly used vector

$$\mathbf{e} = (1, 1, \ldots, 1)^T.$$

Finally, it will be shown that the lower bound of the condition number of block incomplete factorization methods is $O(h^{-1})$ when a fixed sparsity pattern of the last diagonal block matrix (X_n) is chosen. This holds even if the factorization is exact up to (and including) the next-to-final block.

As the first application of the eigenvalue estimates, we consider the finite difference approximation of the problem

$$-\delta u_{xx} - u_{yy} = f \text{ in } [0, 1]^2,$$

where $\delta > 0$, with Dirichlet boundary conditions, using a uniform mesh. Using a natural ordering of meshpoints, one finds a symmetric matrix A with nonzero entries,

$$a_{i,i-n} = -1, \ a_{i,i-1} = -\delta, \ a_{i,i} = d, \ a_{i,i+1} = -\delta, \ a_{i,i+n} = -1,$$

where $d = 2(1 + \delta)$, and the mesh width is $h = 1/(n + 1)$.

For simplicity we consider a point-wise factorization, where the sparsity pattern $S = \{(i, j); a_{i,j} \neq 0\}$, and let $X = \mathrm{diag}(x_1, x_2, \ldots)$ where we let x_i be defined by the recursion (method of relaxation)

$$(10.41) \qquad x_1 = a_{1,1}, x_i = a_{i,i} - \sum_{i<j} \ell_{i,j} x_j^{-1} \ell_{j,i} - \omega(\tilde{R}\mathbf{e})_i, \ i = 2, 3, \ldots,$$

where $\tilde{R}$ contains the deleted (fill-in) entries, or

$$x_i = 2(1 + \delta) - \delta^2 x_{i-1}^{-1} - x_{i-n}^{-1} - \omega\delta(x_{i-n}^{-1} + x_{i-1}^{-1})$$

(apart from obvious corrections at points next to the boundary). Here, $\omega < 1$ is a relaxation parameter, and this will be chosen so that $2x - d > 0$, where $x = \min_i x_i$. We see readily that as $i \to \infty$ and $h \to 0$, x_i converges to a lower bound x, where

$$x = 2(1 + \delta) - (1 + 2\omega\delta + \delta^2)x^{-1}$$

or

$$(10.42) \qquad x = 1 + \delta + \{2\delta(1 - \omega)\}^{\frac{1}{2}}.$$

It also follows readily that the nonzero entries of $R = C - A$ converge to

$$r_{i,i\pm(n-1)} = \delta x^{-1}, \ r_{i,i} = -2\omega\delta x^{-1}.$$

Thus,

(10.43) $\lambda_{max}(R) \leq 2\delta(1-\omega)x^{-1}$ and $\lambda_{min}(R) \geq -2\delta(1+\omega)x^{-1}$.

Note also that

(10.44) $$2x - d = 2\{2\delta(1-\omega)\}^{\frac{1}{2}} > 0.$$

Hence, Theorem 10.5 shows that

(10.45) $$\mu = \frac{x}{2x-d} = \frac{1}{2} + \frac{1+\delta}{2\{2\delta(1-\omega)\}^{\frac{1}{2}}}$$

is an upper bound of the eigenvalues of $C^{-1}A$. Furthermore, Theorems 10.1 and 10.5 show that

(10.46) $$\lambda_i(C^{-1}A) \geq \underline{\lambda}_i = \frac{\lambda_i(A)}{\lambda_i(A) + \lambda_{max}(C-A)}$$

[By Theorem 10.1 with $\mu_1 = 1$, it suffices that $\lambda_{max}(C-A) = \lambda_{max}(R) \geq 0$ for this to hold] and

(10.47) $$\lambda_i(C^{-1}A) \leq \overline{\lambda}_i = \frac{4x\lambda_i(A)}{[2x - d + \lambda_i(A)]^2},$$

where the latter holds for all $\lambda_i(A) \leq 2x - d$.

Hence, for the difference between these upper and lower bounds using (10.49) and (10.46), (10.47), we find, if $\lambda_i(A) \leq 2\{2\delta(1-\omega)\}^{\frac{1}{2}}$,

(10.48)
$$\overline{\lambda}_i - \underline{\lambda}_i \leq \left(\frac{4x}{[2x-d+\lambda_i(A)]^2} - \frac{1}{\lambda_i(A) + 2\delta(1-\omega)x^{-1}} \right) \lambda_i(A)$$
$$= \frac{[2d - \lambda_i(A)](\lambda_i(A))^2}{[\lambda_i(A) + 2\{2\delta(1-\omega)\}^{\frac{1}{2}}]^2[\lambda_i(A) + 2\delta(1-\omega)x^{-1}]}.$$

If $\lambda_i(A) = O(h^\alpha)$, $h \to 0$, then for some $\alpha > 0$, this shows that

(10.49) $$\lambda_i(C^{-1}A) = \frac{x}{2\delta(1-\omega)}\lambda_i(A) + O(\lambda_i(A)^2), \quad h \to 0.$$

Therefore, for any fixed value of ω, $\omega < 1$ (independent of the problem parameter h), the eigenvalues λ_i of $C^{-1}A$ are very close to the factor

$$x/[2\delta(1-\omega)] = \frac{1 + \delta + \{2\delta(1-\omega)\}^{1/2}}{2\delta(1-\omega)}$$

times the corresponding eigenvalues $\lambda_i(A)$ for all small eigenvalues $\lambda_i(A) = O(h^\alpha)$, $h \to 0$.

In fact, a closer asymptotic analysis, which we omit, shows that

$$\lambda_i(C^{-1}A) = \frac{x}{2\delta(1-\omega)}\lambda_i(A)\left(1 + O\left(\frac{\lambda_i(A)}{1-\omega}\right)\right)$$

$$= \frac{x}{2\delta(1-\omega)}\lambda_i(A)(1 + o(1)) \quad h \to 0,$$

for all eigenvalues $\lambda_i(A) = O(h^\alpha)$, if

$$\omega = 1 - O(h^\beta), \quad h \to 0, \quad 0 \le \beta < \alpha.$$

The above can be used to derive an optimal value of ω to minimize the condition number of $C^{-1}A$, and this will be done in Example 10.2.

This shows that the essentially unmodified incomplete factorization method changes the smallest eigenvalues in the main only by a constant factor. This factor increases as ω is taken closer to 1. The consequence for the rate of convergence of the corresponding preconditioned conjugate gradient method was discussed in Axelsson (1992). It will be seen in Chapter 13 that for well-separated small eigenvalues, the rate of convergence can be much faster than what the condition number alone would predict. If $\delta \le 1$ and $\omega \le 0$, then (10.42) and (10.48) show that the gap ratio between successive small eigenvalues is larger by at least the factor $1 + 1/\sqrt{2}$ for the preconditioned method compared with the unpreconditioned. Anyhow, as was shown in Axelsson (1992) and Chapter 13, when δ and ω do not depend on h, this increase of the gap ratio does not improve the asymptotic order of the rate of convergence.

The above shows that the incomplete factorization method can be used as a good smoother for multigrid methods, possibly in combination with a Chebyshev or conjugate gradient iterative method. In a multigrid method, one or few smoothing steps take place on the fine (given) difference mesh and are followed by a correction on a coarse mesh. The method can be repeated as an iterative method. It can also be extended by use of a sequence of meshes where smoothing takes place, followed by a correction on the coarsest mesh. For a detailed analysis of this method, see Hackbusch (1985), for instance. For an application of incomplete factorization smoothers in the multigrid method, see Kettler (1982).

There is an alternative analysis of condition numbers possible for the case $\omega = 0$ for the above-considered problem. Since C is nonsingular and $B = C^{-1}R$ is nonnegative, the splitting $A = C - R$ is nonnegative (in fact, even regular). Hence, Theorem 6.16 shows that

$$\rho(C^{-1}R) \le \frac{\rho(A^{-1}R)}{1 + \rho(A^{-1}R)}.$$

Since $C^{-1}A = I - C^{-1}R$, we find

$$\lambda_1(C^{-1}A) = 1 - \rho[C^{-1}R] \ge 1/(1 + \rho(A^{-1}R))]$$

and

$$\lambda_n(C^{-1}A) \le 1 + \rho(C^{-1}R) \le \frac{1 + 2\rho(A^{-1}R)}{1 + \rho(A^{-1}R)}.$$

To estimate $\rho(A^{-1}R)$, we use

$$\rho(A^{-1}R) \le \|A^{-1}R\| \le \|A^{-1}\|\|R\| \le 2\delta x^{-1}/\lambda_1(A).$$

This shows

$$\lambda_1(C^{-1}A) \ge \frac{\lambda_1(A)}{\lambda_1(A) + 2\delta/x},$$

which is the same as (10.46), and

$$\lambda_n(C^{-1}A) \lesssim 2.$$

The latter estimate is worse than the bound (10.45), except when $(1 + \delta)/\sqrt{2\delta} \ge 3$, that is, when $\delta \lesssim 1/16$, in which case the bound 2 is better.

Example 10.2 (Condition number of the relaxed method) Consider the above difference approximation matrix A_h preconditioned by the point-wise relaxed incomplete factorization method, with $\omega < 1$ and with preconditioning matrix

$$C = (X - L)X^{-1}(X - L^T),$$

where $X = \text{diag}(x_1, x_2, \ldots, x_n)$ and $\{x_i\}$ are defined by the recursion (10.41). The extreme eigenvalues of A are

$$\lambda_1(A) = 4(1 + \delta)(\sin \frac{\pi h}{2})^2 \text{ and } \lambda_n(A) = 4(1 + \delta) - |O(h^2)|.$$

To find the condition number of $C^{-1}A$, we use first (10.49), which shows that the lower eigenvalue satisfies

$$\lambda_1(C^{-1}A) = \frac{x}{2\delta(1 - \omega)}\lambda_1(A) + [O(\lambda_1(A)]^2, \; h \to 0,$$

where x is defined by (10.42). Hence, if ω is fixed (independent of h),

$$\lambda_1(C^{-1}A) = \frac{1 + \delta + \{2\delta(1-\omega)\}^{\frac{1}{2}}}{2\delta(1-\omega)}(1+\delta)(\pi h)^2 + O(h^4), \quad h \to 0.$$

To find the upper eigenvalue, we use (10.45) to find

$$\lambda_n(C^{-1}A) \leq \frac{1}{2} + \frac{1+\delta}{2\{2\delta(1-\omega)\}^{\frac{1}{2}}}.$$

[For $\delta = 1$ and $\omega = 0$, we find $\lambda_n \leq \frac{1}{2}(1+\sqrt{2}) \simeq 1.207$.] Hence, the condition number of $C^{-1}A$ is

$$\kappa(C^{-1}A) = \lambda_n(C^{-1}A)/\lambda_1(C^{-1}A) = \frac{\{2\delta(1-\omega)\}^{\frac{1}{2}}}{2(1+\delta)}(\pi h)^{-2} + O(1), \quad h \to 0.$$

The condition number of A is

$$\kappa(A) = 4(\pi h)^{-2} + O(1), \quad h \to 0,$$

so the ratio of the condition numbers is

$$\kappa(C^{-1}A)/\kappa(A) = \frac{1}{8(1+\delta)}\{2\delta(1-\omega)\}^{\frac{1}{2}} + O(h^2), \quad h \to 0,$$

which takes the value $\sqrt{2}/16 + O(h^2)$ for $\delta = 1$ and $\omega = 0$.

The above analysis assumes that $\omega < 1$ and does not depend on h but indicates that values of ω close to 1 give the smallest condition numbers. To ascertain this, we use (10.46), which shows the lower bound

$$\lambda_1(C^{-1}A) \geq \frac{\lambda_1(A)}{\lambda_1(A) + 2\delta(1-\omega)/x}.$$

Using this, together with the upper bound $x/(2x - d)$, we find

$$\kappa(C^{-1}A) \leq \frac{\bar{\lambda}_n}{\bar{\lambda}_1} \leq \frac{1}{2}\left(\frac{x}{\{2\delta(1-\omega)\}^{\frac{1}{2}}} + \frac{\{2\delta(1-\omega)\}^{\frac{1}{2}}}{\lambda_1(A)}\right)$$

This upper bound is minimized for ω, such that

$$2\delta(1-\omega) = x\lambda_1(A)$$

or

$$\omega = \omega_{opt} = 1 - \frac{x\lambda_1(A)}{2\delta},$$

and for this value of ω we find

$$\kappa(C^{-1}A) \le \sqrt{\frac{x}{\lambda_1(A)}} \sim \sqrt{\frac{1+\delta}{\lambda_1(A)}} \sim \frac{1}{\pi h}, \quad h \to 0.$$

This is an order of magnitude improvement compared to the condition number of A, which is

$$\kappa(A) = \lambda_n(A)/\lambda_1(A) \sim 4(1+\delta)/\lambda_1(A) \sim (\frac{2}{\pi h})^2, \quad h \to 0.$$

We next consider incomplete factorization for a block matrix. Let A be a block tridiagonal matrix,

$$A = \text{blocktridiag}(A_{i,i-1}, A_{i,i}, A_{i,i+1}), \quad i = 1, 2, \ldots, n,$$

which is assumed, in addition, to be symmetric and positive definite. Let A be split as

$$A = D_A + L + L^T,$$

where D_A is the blockdiagonal part and L is the lower block triangular part of A. Consider the special incomplete factorization method used in forming the generalized SSOR preconditioned matrix

(10.50)
$$C = (X + L)X^{-1}(X + L^T),$$

where X is nonsingular and diagonal or block diagonal, $X = \text{diag}(X_1, \ldots, X_n)$, partitioned as D_A. In practice, X_i will be sparse matrices, such as bandmatrices. We shall derive upper and lower asymptotic bounds of the condition number of $C^{-1}A$.

As we saw in Chapter 7, a common method of computing the matrices X_i is the following:

(10.51) $\quad X_i = A_{i,i} - A_{i,i-1}D_{i-1}A_{i-1,i} + D_i', \quad i = 1, 2, \ldots, n.$

Here, $D_0 = 0$ and $D_1' = 0$ (i.e., $X_1 = A_{11}$) and D_{i-1}, $i \ge 2$ is a sparse non-negative approximation to X_{i-1}^{-1}, such as a bandmatrix, and D_i' is a diagonal matrix chosen such that

(10.52) $\quad\quad\quad X_i \mathbf{v} = (A_{i,i} - A_{i,i-1}X_{i-1}^{-1}A_{i-1,i})\mathbf{v}$

for some positive vector $\mathbf{v}$. (10.51) and (10.52) show that

$$D_i'\mathbf{v} = A_{i,i-1}(D_{i-1} - X_{i-1}^{-1})A_{i-1,i}\mathbf{v}.$$

Hence, D_i' compensates for the error of the approximation D_{i-1} of X_{i-1}^{-1} in such a way that $C\mathbf{v} = A\mathbf{v}$, that is, C and A have the same action on the vector $\mathbf{v}$. (10.51) can be generalized to allow perturbations of A, taking the form

$$X_i = (A_{i,i} + \Delta_i) - A_{i,i-1}D_{i-1}A_{i-1,i} + D_i', \quad i = 1, 2, \ldots, n,$$

where Δ_i is a diagonal matrix that contains small (nonnegative) perturbations of $\mathrm{diag}(A_{i,i})$. Alternatively, we may use the relaxed method.

Example 10.3 An important application of generalized SSOR methods is to second-order elliptic difference equations. Here the problem parameter h is the meshwidth. Consider, then, the model equation $au_{xx} + bu_{yy} = f$ on a unit square domain, where a and b are positive numbers, and assume that $b \leq a$. Using orderings along the x-axis, the matrices

$$A_{i,i} = \mathrm{tridiag}[-a, 2(a+b), -a] \text{ and}$$

(10.53)

$$A_{i,i-1} = A_{i-1,i} = b \, \mathrm{diag}(1, 1, \ldots, 1).$$

Then the smallest eigenvalue of $A_{i,i}$ is $\lambda_1 = 2b + a(2\sin \pi h/2)^2$. *If $\mathbf{v}_1$ is the corresponding eigenvector, then (10.52) with $\mathbf{v} = \mathbf{v}_1$ shows that this becomes an eigenvector of X_i also*, and the corresponding eigenvalues of X_i become

$$\lambda_1^{(1)} = \lambda_1, \quad \lambda_1^{(i)} = \lambda_1 - b^2 \lambda_1^{(i-1)^{-1}}, \quad i = 2, 3, \ldots, n.$$

Thus, $\lambda_1^{(i)}$ converges monotonically to the lower bound,

$$\lambda_1^{(i)} \to b + 2a(\sin \frac{\pi h}{2})^2 + 2\sin \frac{\pi h}{2}\sqrt{ab + (a\sin \frac{\pi h}{2})^2}.$$

Hence,

$$\lambda_1^{(n)} \geq b + \pi h\sqrt{ab} + 2a(\sin \frac{\pi h}{2})^2.$$

For the choice $\mathbf{v} = \mathbf{e} = (1, 1, \ldots, 1)^T$, it can be seen that the sequence $\{\lambda_1^{(i)}\}_{i \geq 1}$ is bounded below by $\{\hat{\lambda}_1^{(i)}\}_{i \geq 1}$, where

$$\hat{\lambda}_1^{(1)} = 2b, \quad \hat{\lambda}_1^{(i)} = 2b - b^2 \hat{\lambda}_1^{(i-1)^{-1}}, \quad i \geq 2,$$

that is, $\hat{\lambda}_1^{(i)} = b(1 + \frac{1}{i})$. Note that this bound does not depend on a. Corollary 10.4 shows that for any μ such that

$$\left(2 - \frac{1}{\mu}\right)\lambda_1^{(n)} - \lambda_1 \geq 0,$$

we have $\lambda_{\max}(C^{-1}A) \leq \mu$. Hence, we can let

$$(10.54) \quad \mu = \frac{\lambda_1^{(n)}}{2\lambda_1^{(n)} - \lambda_1} \simeq \begin{cases} \frac{1}{2}(1 + \frac{\sqrt{\frac{b}{a}}}{\pi h}), & h \to 0, \text{ for the vector } \mathbf{v} \\ & \text{(eigenvector)} \\ \frac{1}{2h}, & h \to 0, \text{ for the vector } \mathbf{e} \end{cases}$$

[we have $h = 1/(n + 1)$]. Since the lower bound of $\lambda(C^{-1}A)$ is 1, we have

$$\kappa(C^{-1}A) \leq \mu,$$

that is, we have a condition number bounded by $O(h^{-1})$, $h \to 0$. It is seen from (10.54) that the constant in the asymptotic estimate is about a factor $\sqrt{b/a}/\pi$ smaller if the modification is done with smallest eigenvector than with the vector $\mathbf{e}$.

For later comparison, we state here a result regarding the unmodified block incomplete factorization matrix. This has the form (10.50), but with $D_i' = 0$ in (10.51). (In a relaxed version, we would add $\omega D_i'$. In other words, the unmodified factorization corresponds to the relaxed version with $\omega = 0$.) If we let D_{i-1} be a tridiagonal matrix approximation of X_{i-1}, computed as described in (8.8) with $p = q = 1$, then the corresponding error,

$$\|R\| \leq \sup_i \|A_{i,i-1}(D_{i-1} - X_{i-1}^{-1})A_{i-1,i}\|$$

turns out to be $\|R\| \simeq 0.123$ for $\delta = 1$. (For details, see Axelsson and Eijkhout, 1987.) Further, this matrix is a regular splitting, so

$$\lambda_1(C^{-1}A) \geq 1/(1 + \rho(A^{-1}R)) \sim \lambda_1(A)/0.123.$$

Since $\lambda_n(C^{-1}A) \leq 2$, we find

$$\kappa(C^{-1}A) \leq 0.246\lambda_1(A)^{-1} \sim 0.123(\pi h)^{-2}.$$

For an extension of the above limit matrix analysis to elliptic difference equations in three space dimensions and to more general incomplete factorization methods, see Axelsson and Eijkhout (1987, 1989).

Example 10.4 (Condition numbers of modified methods) Consider modified or relaxed preconditioned methods for the central difference approximation of the equation $au_{xx} + bu_{yy} = f$ on a unit square with Dirichlet boundary conditions. We want to compare the condition numbers of the modified block SSOR preconditioned method in (10.54) with the perturbed version of this method analyzed in (10.27) to (10.30a,b) and with the relaxed point-wise incomplete factorization method with relaxation parameter $\omega \to 1$ ($h \to 0$) defined in (10.41).

First, (10.30b) shows that the condition number for the perturbed method with perturbed diagonal matrix $(1 + \varepsilon)D_A$, satisfies

$$\kappa(C^{-1}A) \le \max\left\{\frac{1}{2h}, 1 + \sqrt{\frac{\mu_0}{2}}\right\},$$

with $\varepsilon = \mu_0^{-1} = 1/\lambda_{\max}(A^{-1}D_A)$. Since $\mu_0^{-1} = (a + b)(2\sin\frac{\pi h}{2})^2/2b$, we find

$$\kappa(C^{-1}A) \le \max\left\{\frac{1}{2h}, 1 + \sqrt{\frac{b}{b + a}}(\pi h)^{-1}\right\}.$$

For the point-wise method, (10.45), (10.43), and (10.46) show that

$$\kappa(C^{-1}A) \le \frac{x}{2x - d}\left[1 + \frac{2\delta(1 - \omega)}{x\lambda_1(A)}\right],$$

where $2x - d = 2\{2\delta(1 - \omega)\}^{\frac{1}{2}}$. Hence,

$$\kappa(C^{-1}A) \le \frac{x}{2}\left[\{2\delta(1 - \omega)\}^{-\frac{1}{2}} + \frac{\{2\delta(1 - \omega)\}^{\frac{1}{2}}}{x\lambda_1(A)}\right],$$

which is minimized (asymptotically, neglecting lower-order terms) for

$$2\delta(1 - \omega) = x\lambda_1(A),$$

or, letting $x = 1 + \delta$, for

$$\omega = \omega_{\text{opt}} = 1 - \frac{x\lambda_1(A)}{2\delta}.$$

Hence,

$$\kappa(C^{-1}A) \le (1 + \delta)\{x\lambda_1(A)\}^{-\frac{1}{2}} \sim (\pi h)^{-1}, \quad h \to 0,$$

Table 10.1. *Asymptotic condition numbers for modified, perturbed, and relaxed incomplete factorization methods ($\delta = a/b$.)*

Method	RIC (ω indep. of h)	RIC ω_{opt}	Block MIC $\mathbf{v} = \mathbf{e}$	Block MIC $\mathbf{v} = \mathbf{v}_1$	Perturbed Block MIC
$\kappa(C^{-1}A)$	$\frac{\{2\delta(1-\omega)\}^{\frac{1}{2}}}{2(1+\delta)}(\pi h)^{-2}$	$(\pi h)^{-1}$	$(2h)^{-1}$	$\sqrt{b/a}(2\pi h)^{-1}$	$\max\{(2h)^{-1}, \sqrt{\frac{b}{b+a}}(\pi h)^{-1}\}$

and

$$\omega = \omega_{opt} = 1 - \frac{(1+\delta)^2}{2\delta}(\pi h)^2.$$

We collect these results and the result in (10.54) in the Table 10.1.

It is seen that the perturbed block MIC method has the smallest condition number when $b/(b+a) < b/(4a)$, that is, when $b > 3a$; otherwise, it is smallest for the block MIC with $\mathbf{v} = \mathbf{v}_1$. The point-wise method has a condition number at least $\sqrt{2}$, or even twice as large as the best block-wise methods, if $b \le a$.

Although the above analysis has been done for a model type problem, it turns out that similar gains are often achieved also for more general problems. The relaxed methods turn out to be most robust with respect to various problem parameters.

We next consider the estimate of the condition number of the Schur complements in (10.40) based on estimates of the constant in the strengthened C.B.S. inequality, when the matrix is the finite element matrix for a second-order self-adjoint elliptic problem with anisotropy, using piece-wise linear finite element approximations. This discretization actually leads to almost the same matrix as in (10.53). However, here we consider Neuman boundary conditions on one boundary line, so the final block is different.

Hence, consider

(10.55) $$au_{xx} + bu_{yy} = f,$$

where a and b are positive numbers on a rectangular domain Ω, with homogeneous Dirichlet boundary conditions on all sides except at the right side, Γ_N, where we assume a homogeneous Neuman boundary condition. For the discretization, we use piece-wise linear basis functions on an isosceles right-angled triangular mesh.

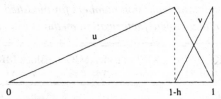

Figure 10.1.   Functions u, v for which (10.57) takes its supremum.

When we order the vertical lines from left to right, the global matrix takes a block tridiagonal form and the system reduced to the last two lines to the right (including Γ_N) takes the form in (10.34), with $i = n - 1$.

The variational formulation of (10.55) is

$$a(u, v) = \int_\Omega (au_x v_x + bu_y v_y)dxdy = \int_\Omega fvdxdy,$$

for all $u, v \in H^1(\Omega)(= \{v \in H^1(\Omega), v = 0 \text{ at } \Gamma \setminus \Gamma_N\})$. As is readily seen, (10.40) (with $i = n - 1$) can be written in the form

(10.56) $$a(u, v) \le \gamma_{n-1} a(u, u)^{\frac{1}{2}} a(v, v)^{\frac{1}{2}}$$

for all functions u, v, where u is piece-wise linear on Ω and zero on Γ_N and where v is piece-wise linear on the strip $\Omega_2 : 1 - h \le x \le 1$ and zero on the line $x = 1 - h$ and in $\Omega_1 = \Omega \setminus \Omega_2$. Hence, to compute γ_{n-1} we may use

(10.57) $$\gamma_{n-1} = \sup_{u,v} \frac{a(u, v)}{\{a(u, u)a(v, v)\}^{\frac{1}{2}}},$$

where the supremum is taken on the set of such functions u, v.

It is readily seen that because of the boundary condition at Γ_N, the supremum is taken for such functions where $a(u, v)$ contains a term with factor a but no term with factor b. That is why, for anisotropic problems, γ_{n-1} takes its largest value when $b = 0$ (but the dependence on b is a second-order term anyway). Therefore, we shall compute γ_{n-1} for $b = 0$. Then $a(u, v) = \int_0^1 au_x v_x dx$. It can be seen that the supremum in (10.57) is taken for the functions u, v in Figure 10.1.

Here, $u = x/(1 - h)$, $0 < x < 1 - h$, $u = (1 - x)/h$, $1 - h \le x \le 1$ and $v = 0$, $0 < x < 1 - h$, $v = 1 - (1 - x)/h$, $1 - h \le x \le 1$. An elementary computation now shows that

$$\gamma_{n-1} = \sqrt{1 - h}.$$

For positive values of b (but $b \le a$) it turns out that this estimate is correct in its order—that is, $1 - \gamma_{n-1}^2 = O(h)$. Corollaries 10.15 and 10.17 show that

$$(10.58) \qquad \kappa(C^{-1}A) \geq \kappa(D_n^{-1}S_n) = \kappa(A_{n,n}^{-1}S_n) = \frac{1}{1 - \gamma_{n-1}^2} = \frac{1}{h},$$

if $D_n = A_{n,n}$.

For constant coefficient problems, we can actually derive the asymptotically correct constant in the above estimate. If we let $a = 1$, we then have

$$(10.59) \qquad A = \begin{bmatrix} 2G & -I & & 0 \\ -I & 2G & -I & \\ & \ddots & \ddots & \\ 0 & & -I & G \end{bmatrix},$$

where $G = I + b/2 \, \text{tridiag}(-1, 2, -1)$. As has been shown in Axelsson and Polman (1988), the eigenvalues λ_i of $A_{n,n}^{-1}S_n = G^{-1}S_n$ can be computed using Chebyshev polynomials of the second kind,

$$U_\ell(x) = \frac{1}{2}[(x + \sqrt{x^2 - 1})^{\ell+1} - (x - \sqrt{x^2 - 1})^{\ell+1}](x^2 - 1)^{-\frac{1}{2}}.$$

One finds

$$S_\ell = U_\ell(G)U_{\ell-1}(G)^{-1}, \quad \ell = 1, 2, \ldots, n - 1 \; (U_0 = I)$$

and

$$(10.60) \qquad S_n = 2G - S_{n-1}^{-1}.$$

The eigenvalues of G are

$$\mu_i = 1 + 2b(\sin i \frac{\pi}{2(m + 1)})^2, \quad i = 1, 2, \ldots, m,$$

and for the smallest eigenvalue, λ_1, in the limit case $b = 0$, one finds

$$\lambda_1 = 1 - \frac{n - 1}{n}, \quad \text{or } \lambda_1 = \frac{1}{n},$$

which is in accordance with the result above using the C.B.S. constant γ.

Above we considered $D_n = G$. Even if we use another choice of D_n, it can be seen that we cannot improve the order $O(h^{-1})$ of the condition number of S_n, as long as D_n is a bandmatrix (with fixed bandwidth independent of h). The fundamental property of S_n, which explains the difficulty in preconditioning it by matrices with a fixed sparsity pattern, is that S_n is a full matrix where entries decay fairly slowly away from the main diagonal. For the model problem matrix (10.59), the limit form of S_n can actually be seen from (10.60) to be

$$S_n \sim (G^2 - I)^{\frac{1}{2}}, \ n \to \infty$$

(for details, see Axelsson and Polman, 1988), which illustrates this slow decay of entries.

Note that the Schur complement is one part of the reduced system one gets for the line of intersection when a domain Ω is divided into two subdomains Ω_i, $i = 1, 2$. The part arising from the second subdomain has properties similar to the one from the first part. Let

$$\begin{bmatrix} A_{11} & 0 & A_{13} \\ 0 & A_{22} & A_{23} \\ A_{31} & A_{32} & A_{33} \end{bmatrix}$$

be the matrix that arises from such a domain decomposition, where $A_{i,i}$ is the matrix that corresponding to subdomain Ω_i. Let $A_{33} = A_{33}^{(1)} + A_{33}^{(2)}$, where $A_{33}^{(i)}$ is the part corresponding to the part of the coupling of the nodes on the intersecting lines Γ arising from the domain Ω_i and $A_{i,3}$ is the matrix containing entries arising from the coupling of the nodes on Γ with nodes in Ω_i. Then the Schur complement consists of two terms,

$$S = S_1 + S_2,$$

where

$$S_i = A_{33}^{(i)} - A_{3,i} A_{i,i}^{-1} A_{i,3}.$$

Note that S_i equals the final diagonal block in the matrix block factorization of $\begin{bmatrix} A_{i,i} & A_{i,3} \\ A_{3,i} & A_{33}^{(i)} \end{bmatrix}$. As we saw in (10.58), the condition number of S_i has order $O(h^{-1})$. This holds also for the preconditioned matrices $A_{33}^{(i)^{-1}} S_i$ and $A_{33}^{-1}(S_1 + S_2)$. A more efficient preconditioned form is

(10.61) $$(S_1^{-1} + S_2^{-1})(S_1 + S_2).$$

It can be seen that under certain conditions, this matrix has condition number $O(1)$, $h \to 0$. In an iterative solution method, we need to solve systems with the interior domain matrices $A_{i,i}$ and with S_i, $i = 1, 2$. Since the iterations with matrix (10.61) take place in the subspace corresponding to the vector components for the nodes on Γ, we need only the part of the solution vectors for the above matrices corresponding to this subspace. For further details, see Axelsson and Polman (1988).

The elementary derivation of the condition number of the Schur complement used above was based on the C.B.S. constant. For proofs based on Steklov-Poincaré operators and inverse inequalities in certain Sobolev spaces, see Marchuk, Kuznetsov, and Matsokin (1986). See also Bjørstad and Widlund (1986), Bramble, Pasciak, and Schatz (1986), and M. Práger (1991).

10.5.1 Historical Remarks

A modified incomplete factorization method for second-order elliptic boundary-value problems was first presented in Dupont, Kendall, and Rachford (1968) and later analyzed with perturbations in Axelsson (1972). As was shown in the latter paper, the method corresponds to a generalized SSOR method with variable parameters ω_i. Modified and perturbed methods for more general incomplete factorizations were presented by Gustafsson (1978, 1979). Relaxed versions of these were presented in Axelsson and Lindskog (1986a,b). Relaxed methods can be seen as perturbation methods with variable perturbations. A dynamic method for implementing such methods was analyzed in Axelsson and Barker (1984).

Beauwens and his students have considered the possibility of getting a condition number of $O(h^{-1})$ without use of perturbations (see Beauwens, 1985 and 1990, for instance). See also Kutcherov and Makarov (1992). It seems, however, that the condition numbers and eigenvalue distributions of properly relaxed methods are not worse and frequently better than the condition numbers for purely modified methods (satisfying a generalized row sum criterion).

References

O. Axelsson (1972). A generalized SSOR method. *BIT* **13**, 443–467.

O. Axelsson (1992). Bounds of eigenvalues of preconditioned matrices. *SIAM J. Matrix Anal. Appl.* **13**, 847–862.

O. Axelsson and V. A. Barker (1984). *Finite Element Solution of Boundary Value Problems, Theory and Computation*. Orlando, Fla.: Academic.

O. Axelsson and V. Eijkhout (1987). Robust vectorizable preconditioners for three-dimensional elliptic difference equations with anisotropy. In *Algorithms and Applications on Vector and Parallel Computers*, ed. H. J. J. te Riele, et al. Amsterdam: North-Holland.

O. Axelsson and V. Eijkhout (1989). Vectorizable preconditioners for elliptic difference equations in three space dimensions. *J. Comp. Appl. Math.* **27**, 299–321.

O. Axelsson and G. Lindskog (1986a). On the eigenvalue distribution of a class of preconditioning methods. *Numer. Math.* **48**, 479–498.

O. Axelsson and G. Lindskog (1986b). On the rate of convergence of the preconditioned conjugate gradient method. *Numer. Math.* **48**, 499–523.

O. Axelsson and B. Polman (1988). Block preconditioning and domain decomposition methods II. *J. Comp. Appl. Math.* **24**, 55–72.

R. Beauwens (1985). On Axelsson's perturbations. *Lin. Alg. Appl.* **68**, 221–242.

R. Beauwens (1984). Upper eigenvalue bounds for pencils of matrices. *Lin. Alg. Appl.* **62**, 87–104.

R. Beauwens (1990), Modified incomplete factorization strategies. In *Preconditioned Conjugate Gradient Methods*, ed. O. Axelsson and L. Kolotilina, pp. 1–16. *Lecture Notes in Mathematics*, No. 1457. Berlin, Heidelberg, New York: Springer-Verlag.

P. Bjørstad and O. Widlund (1986). Iterative methods for the solution of elliptic problems on regions partitioned into substructures. *SIAM J. Numer. Anal.* **23**, 1097–1120.

J. Bramble, J. Pasciak, and A. Schatz (1986). An iterative method for elliptic problems on regions partitioned in substructures. *Math. Comp.* **46**, 361–369.

T. Dupont, R. Kendall, and H. Rachford (1968). An approximate factorization procedure for solving self-adjoint elliptic difference equations. *SIAM J. Numer. Anal.* **5**, 559–573.

I. Gustafsson (1978). A class of first order factorization methods. *BIT* **18**, 142–156.

I. Gustafsson (1979). Stability and rate of convergence of modified incomplete Cholesky factorization methods. Thesis, Department of Computer Sciences, Göteborg, Sweden.

W. Hackbusch (1985). *Multigrid Methods and Applications*. New York: Springer-Verlag.

R. Kettler (1982). Analysis and comparison of relaxed schemes in robust multigrid and preconditioned conjugate methods. In Multigrid Methods, ed. W. Hackbusch and U. Trottenberg. Lecture Notes in Mathematics, Vol. 960. Berlin, Heidelberg, New York: Springer-Verlag.

A. B. Kutcherov and M. M. Makarov (1992). An approximate factorization method for solving discrete elliptic problems on stretched domains. *J. Lin. Alg. Appl.* **1** 1–26.

M. M. Magolu and Y. Notay (1991). On the conditioning analysis of block approximate factorization methods. *Lin. Alg. Appl.* **154–156**, 583–599.

G. I. Marchuk, Y. A. Kuznetsov, and A. M. Matsokin (1986). Fictitious domain and domain decomposition methods. *Sov. J. Numer. Anal. Math. Modelling* **1**, 3–35.

Y. Notay (1990). Solving positive (semi) definite linear systems by preconditioned iterative methods. In *Preconditioned Conjugate Gradient Methods*, ed. by O. Axelsson and L. Yu. Kolotilina, pp. 105–125. Lecture Notes in Mathematics 1457. Berlin, Heidelberg, New York: Springer-Verlag.

Y. Notay (1991). Conditioning analysis of modified block incomplete factorizations. *Lin. Alg. Appl.* **154–156**, 711–722.

M. Práger (1991). An iterative method of alternating type for systems with special block matrices. *Appl. Math.* **36**, 72–78.

11

Conjugate Gradient and Lanczos-Type Methods

The conjugate gradient method—or, rather, the conjugate gradient methods, because there exists a plethora of such methods—can be seen as iterative solution methods to solve linear systems of equations $A\mathbf{x} = \mathbf{b}$ by minimizing quadratic functionals, such as $f(\mathbf{x}) = (\frac{1}{2}\mathbf{x}^T A\mathbf{x}) - \mathbf{b}^T\mathbf{x}$, if A is symmetric and positive definite, or the residual functional $f(\mathbf{x}) = (A\mathbf{x} - \mathbf{b})^T(A\mathbf{x} - \mathbf{b})$ in the general case. The minimization takes place over certain vector spaces called *Krylov spaces*.

If A is nonsingular (and s.p.d. as in the first case above), both minimization problems can be readily seen to be equivalent to solving the linear system $A\mathbf{x} = \mathbf{b}$. This is obvious for the second case. For the first, we simply note that

$$\frac{1}{2}(A\mathbf{x} - \mathbf{b})^T A^{-1}(A\mathbf{x} - \mathbf{b}) = \frac{1}{2}\mathbf{x}^T A\mathbf{x} - \mathbf{b}^T\mathbf{x} + \frac{1}{2}\mathbf{b}^T A^{-1}\mathbf{b},$$

where the last term is constant. Since A is positive definite, there is a unique minimum of this function, attained by the solution of the stationary equations $A\mathbf{x} = \mathbf{b}$. If we define the norm

$$\|\mathbf{u}\|_{A^{-\frac{1}{2}}} = \|A^{-\frac{1}{2}}\mathbf{u}\| = \{\mathbf{u}^T A^{-1}\mathbf{u}\}^{\frac{1}{2}},$$

this is equivalent to minimizing the function $\|A\mathbf{x} - \mathbf{b}\|_{A^{-\frac{1}{2}}}$.

The minimization takes place on a sequence of subspaces V_k of increasing dimension that are constructed recursively by adding a new basis vector $A^k\mathbf{r}^0$ to those of the previous subspace V_{k-1}, that is,

$$V_k = V_{k-1} \oplus \{A^k\mathbf{r}^0\}, \quad k = 1, 2, 3, \ldots,$$

where $V_0 = \{\mathbf{r}^0\}$. Here, $\mathbf{r}^0 = A\mathbf{x}^0 - \mathbf{b}$ is the residual for the initial vector $\mathbf{x}^0$ (which can be an arbitrary approximation of $\mathbf{x}$). Hence, V_k is spanned by the vectors $r^0, Ar^0, \ldots, A^k r^0$ and is called the Krylov space associated with A and r^0 (Faddeev and Faddeeva, 1963).

For reasons of efficiency, it is expedient to construct a sequence of orthogonal, or conjugate, orthogonal vectors (*conjugate* stands for *orthogonal* with respect to an inner product with a weight matrix, such as $(x, y) = x^T Ay$, if A is s.p.d.). As we shall see, the actual construction of these vectors can occur in a recursion involving only a few vectors if A is selfadjoint w.r.t. the inner product. It will also be seen that because the Krylov vectors eventually span the whole space (or the space spanned by the eigenvectors of A, represented in r^0), the methods will give the exact solution after at most n steps, where n is the order of A. Therefore, the method can be seen as a direct solution method. However, in the presence of roundoff errors, the generated vectors will not be exactly (conjugately) orthogonal and the method may need more iterations to reach machine number precision. More importantly, we shall see that the conjugate gradient method can generate approximations to the solution vector x that are sufficiently accurate after many fewer steps than n. The method is then used as an iterative solution method.

As follows from the above, conjugate gradient methods can be seen as (generalized) least square methods where the minimization takes place on a particular vector subspace, the Krylov space. More generally, one can consider such minimization methods where the subspaces are constructed in a different way. They are referred to as *subspace iteration methods*.

Since a matrix-vector multiplication with A is needed to compute the new residual (the latter being required if we want to estimate the accuracy of the current approximation to the solution vector), it will be seen that the Krylov space can be generated at no extra cost, and Krylov subspace methods are therefore most frequently used. (However, if one computes and checks the residual less frequently, this argument is less relevant.) Compared with other types of iterative methods such as SOR, it turns out that the conjugate gradient method can converge with a faster rate—at least, when a proper preconditioning is used. In addition, conjugate gradient methods are parameter-free. Hence, although many iterative methods, such as SOR, require little work per iteration and little storage, they are generally not so efficient as conjugate gradient methods.

We first describe a version where the vectors can be computed from a three-term form involving residual vectors similar to the Chebyshev iterative method presented in Chapter 5. Next, the more standard form is presented. This form uses a set of search vectors which, again, can be computed from a recursion containing only a few vectors (two if A is symmetric).

Preconditioned versions of the algorithms with s.p.d. preconditioners are presented by simply incorporating the preconditioners in the inner product as a weight matrix and by considering the so-called pseudoresiduals.

Finally, we describe Lanczos-type methods to generate A-orthogonal (i.e., conjugately orthogonal) vectors, their application to compute solutions of linear systems and extreme eigenvalues of A, and modifications of the three-term recurrence method to solve nonsymmetric systems. When A is indefinite, the standard conjugate gradient method can break down. Similarly, the Lanczos method may fail to produce linearly independent search vectors. Some recently proposed remedies to avoid such terminations of the method are also discussed.

Historical Remarks

A three-term form of the classical conjugate gradient method for s.p.d. systems was first presented by Lanczos (1950, 1952). See also references to papers by Rutishauser in Engeli et al. (1959). The standard form of the conjugate gradient method was first developed in the late 1940s and early 1950s. See Hestenes and Stiefel (1952) and the interesting papers on the history of conjugate gradient method by Golub and O'Leary (1989) and by Hestenes (1990). See also Engeli et al. (1959). (Incidentally, in Birman [1950], an extension of the steepest descent method using s-inner steps with a Krylov subspace is discussed and the standard error estimate involving Chebyshev polynomials is derived. Here, the vectors were made orthogonal by a Gram-Schmidt orthogonalization method.) However, the method did not come into widespread use until the early 1970s; see Reid (1971), Axelsson (1972, 1974, 1976), Concus, Golub, and O'Leary (1976), and Meijerink and van der Vorst (1977). The popularity of the method is due to many factors:

(a) It has an optimality property over the relevant solution space, which usually means convergence to an acceptable accuracy with far fewer steps than the number required for the finite termination property— i.e., it has a relatively high rate of convergence.

(b) The rate of convergence can be much improved with various preconditioning techniques.

(c) The method is parameter-free—i.e., the user is not required to estimate any method parameters.

(d) The short recurrence relation makes the execution time per iteration and the memory requirements acceptable.

(e) The roundoff error properties are acceptable.

By the late 1970s, an extensive search was under way for generalized CG methods that could be applied to nonsymmetric and/or indefinite problems and that possessed all, or most, of these appealing properties. This search led to methods such as the Orthomin method of Vinsome (1976), the Concus, Golub (1976) and Widlund (1978) method, and several methods developed by Axelsson (1978, 1979, 1980), for example. For an earlier extension of the p-step steepest descent method to nonsymmetric problems, see Marchuk and Kuznetsov (1968).

By the early 1980s there was a plethora of new versions of generalized conjugate gradient methods, as can be seen by surveys of Young and Jea (1980), Saad and Schultz (1983, 1985), and Dennis and Turner (1988), for instance. However, none of the methods for general nonsymmetric problems worked in practice as well as the CG method did for s.p.d. problems, in that for some problems either a very large number of iterations were required or the method did not even converge. In addition, as explained by the results in papers by Voevodin (1983), Faber and Manteuffel (1984), Joubert and Young (1986), and Axelsson (1987), the short recurrence relation is essentially valid only for H-normal matrices.

As has been shown in papers by Elman (1982), Axelsson et al. (1987), Axelsson and Vassilevski (1991), and Vassilevski (1992), generalized conjugate gradient methods can still work very satisfactorily even for nonsymmetric problems if they are properly preconditioned. Hence, it is a combination of preconditioning with the optimality property of some generalized conjugate gradient methods that seems to show the greatest robustness.

The generalized conjugate gradient methods will be presented in the following chapter. The corresponding references given above can be found in that chapter.

11.1 The Three-Term Recurrence Form of the Conjugate Gradient Method

Let $(\mathbf{x}, \mathbf{y})$ be an inner product in $\mathbb{R}^n$, defined by $(\mathbf{x}, \mathbf{y}) = \mathbf{x}^T W \mathbf{y}$ for some symmetric and positive definite matrix W. We assume that $(\mathbf{x}, A\mathbf{y}) = (A\mathbf{x}, \mathbf{y})$, that is, that A is selfadjoint with respect to this inner product. To solve $A\mathbf{x} = \mathbf{b}$, consider the following method, which generates a sequence of orthogonal residual vectors with respect to the inner product: Let $\mathbf{x}^0$ be a given initial vector and let

$$(11.1a) \qquad\qquad \mathbf{x}^1 = \mathbf{x}^0 - \beta_0 \mathbf{r}^0.$$

Multiplying (11.1a) by A and subtracting the vector $\mathbf{b}$ from both sides, we obtain

$$\mathbf{r}^1 = \mathbf{r}^0 - \beta_0 A \mathbf{r}^0,$$

where $\mathbf{r}^k = A\mathbf{x}^k - \mathbf{b}$ denotes the kth residual. Assuming that $\mathbf{r}^0 \neq \mathbf{0}$ and using the required orthogonality $(\mathbf{r}^1, \mathbf{r}^0) = 0$, we find

(11.1b)
$$\beta_0 = \frac{(\mathbf{r}^0, \mathbf{r}^0)}{(\mathbf{r}^0, A\mathbf{r}^0)}.$$

After the initialization steps (11.1a,b), the general recursion takes the form (also valid for $k = 0$, if we let $\alpha_0 = 1$)

(11.2)
$$\mathbf{x}^{k+1} = \alpha_k \mathbf{x}^k + (1 - \alpha_k)\mathbf{x}^{k-1} - \beta_k \mathbf{r}^k, \quad k = 1, 2, \ldots$$

Multiplying (11.2) by A and subtracting the vector $\mathbf{b}$, we see that the residuals can also be computed by recursion,

(11.3)
$$\mathbf{r}^{k+1} = \alpha_k \mathbf{r}^k + (1 - \alpha_k)\mathbf{r}^{k-1} - \beta_k A\mathbf{r}^k, \quad k = 1, 2, \ldots.$$

Assuming that $(\mathbf{r}^k, \mathbf{r}^{k-j}) = 0$, $j = 1, 2, \ldots, k$ (which by the initialization step is valid for $k = 1$) and that $\mathbf{r}^k \neq \mathbf{0}$, we shall show that the parameters α_k, β_k can be uniquely determined from $(\mathbf{r}^{k+1}, \mathbf{r}^k) = 0$, $(\mathbf{r}^{k+1}, \mathbf{r}^{k-1}) = 0$ and that, furthermore, for this choice of parameters, $(\mathbf{r}^{k+1}, \mathbf{r}^{k-j}) = 0$, $j = 0, 1, \ldots, k$, that is, the new residual is orthogonal to all previous residuals (which already form an orthogonal set among themselves). It will also be seen later that $\beta_k > 0$ if $\mathbf{r}_k \neq \mathbf{0}$. Note that since A is positive definite, (11.1b) shows that $\beta_0 > 0$ when $\mathbf{r}^0 \neq \mathbf{0}$.

Assuming, by induction, that $\beta_j \neq 0$ and $(\mathbf{r}^k, \mathbf{r}^{k-1-j}) = 0$, $j = 0, 1, \ldots, k - 1$, then (11.3) and $(\mathbf{r}^{k+1}, \mathbf{r}^{k-j}) = 0$, $j = 0, 1$ show that

(11.4a)
$$\alpha_k = \beta_k \mu_k$$

and

(11.4b)
$$\alpha_k \delta_{k-1} + \beta_k (\mathbf{r}^{k-1}, A\mathbf{r}^k) = \delta_{k-1},$$

where

$$\mu_k = (\mathbf{r}^k, A\mathbf{r}^k)/(\mathbf{r}^k, \mathbf{r}^k)$$

and

$$\delta_k = (\mathbf{r}^k, \mathbf{r}^k).$$

Since A is selfadjoint, it follows by (11.3) (with k replaced by $k - 1$) that

$$(\mathbf{r}^{k-1}, A\mathbf{r}^k) = (A\mathbf{r}^{k-1}, \mathbf{r}^k)$$

$$= \frac{1}{\beta_{k-1}}(\alpha_{k-1}\mathbf{r}^{k-1} + (1 - \alpha_{k-1})\mathbf{r}^{k-2} - \mathbf{r}^k, \mathbf{r}^k) = -\frac{\delta_k}{\beta_{k-1}}.$$

This and (11.4a,b) show that

$$\beta_k \mu_k \delta_{k-1} - \beta_k \delta_k \beta_{k-1}^{-1} = \delta_{k-1}$$

or

$$(11.5) \qquad \beta_k^{-1} = \mu_k - \frac{\delta_k}{\delta_{k-1}}\beta_{k-1}^{-1}, \quad k = 1, 2, \dots .$$

This recursion and (11.4a) define β_k, α_k if it can be shown, in addition, that the right-hand side of (11.5) is nonzero. In fact, it will be seen that if $\mathbf{r}^k \neq \mathbf{0}$, then $\alpha_k > 1$ and $\beta_k > 0$. The next theorem shows this and some further results, such as the optimality of the method—i.e., the residual computed by (11.3), (11.5), and (11.4a) is the smallest possible, even if we let $\mathbf{x}^{k+1}$ be a linear combination of all previously generated vectors.

Theorem 11.1 *Let A be selfadjoint and positive definite w.r.t. $(\cdot, \cdot)$, and consider the iterative method defined by*

$$\mu_k = (\mathbf{r}^k, A\mathbf{r}^k)/(\mathbf{r}^k, \mathbf{r}^k), \quad \delta_k = (\mathbf{r}^k, \mathbf{r}^k), \quad \alpha_k = \beta_k \mu_k$$

and (11.2), (11.3). Let $\Delta\mathbf{x}^l \equiv \mathbf{x}^l - \mathbf{x}^{l-1}$, $\Delta\mathbf{r}^l \equiv \mathbf{r}^l - \mathbf{r}^{l-1}$, $\tilde{\alpha}_l \equiv \alpha_l - 1$. Then, for $k \geq 0$,

(a) $\Delta\mathbf{x}^{k+1} = \tilde{\alpha}_k \Delta\mathbf{x}^k - \beta_k \mathbf{r}^k$, *and*

$$(11.6) \qquad \Delta\mathbf{r}^{k+1} = \tilde{\alpha}_k \Delta\mathbf{r}^k - \beta_k A\mathbf{r}^k, \quad k = 1, 2, \dots .$$

(b) $(\mathbf{r}^{k+1}, \mathbf{r}^{k-j}) = 0$, $j = 0, 1, \dots, k$, *and* $(\mathbf{r}^{k+1}, A^{-1}\Delta\mathbf{r}^{k-j+1}) = 0$, $j = 0, 1, \dots, k$.

(c) β_k, α_k *computed by (11.5) and (11.4a) satisfy also*

$$(11.7) \qquad \begin{bmatrix} (\mathbf{r}^k, A\mathbf{r}^k) & -(\mathbf{r}^k, \mathbf{r}^k) \\ -(\mathbf{r}^k, \mathbf{r}^k) & (\Delta\mathbf{r}^k, A^{-1}\Delta\mathbf{r}^k) \end{bmatrix} \begin{bmatrix} \beta_k \\ \tilde{\alpha}_k \end{bmatrix} = \begin{bmatrix} (\mathbf{r}^k, \mathbf{r}^k) \\ 0 \end{bmatrix},$$

which, unless $\mathbf{r}^k = \mathbf{0}$, shows that $\beta_k > 0$, $\tilde{\alpha}_k > 0$ and $\alpha_k > 1$. This shows the existence of the sequence defined in (11.5).

(d) Among all vectors

$$\tilde{\mathbf{x}}^{k+1} = \sum_{j=0}^{k} \xi_j \mathbf{x}^j - \sum_{j=0}^{k} \eta_j \mathbf{r}^j,$$

the vector with $\xi_k = \alpha_k$, $\xi_{k-1} = 1 - \alpha_k$, $\xi_j = 0$, $0 \leq j < k - 1$ and $\eta_k = \beta_k$, $\eta_j = 0$, $0 \leq j \leq k - 1$, that is, $\tilde{\mathbf{x}}^{k+1} = \mathbf{x}^{k+1}$ gives the smallest value of the functional

(11.8)
$$f(\xi_0, \dots, \xi_k, \eta_0, \dots, \eta_k) \equiv (A\tilde{\mathbf{x}}^{k+1} - \mathbf{b}, A^{-1}(A\tilde{\mathbf{x}}^{k+1} - \mathbf{b}))$$

$$= (\tilde{\mathbf{r}}^{k+1}, A^{-1}\tilde{\mathbf{r}}^{k+1}),$$

where $\tilde{\mathbf{r}}^{k+1} = A\tilde{\mathbf{x}}^{k+1} - \mathbf{b}$.

Proof Part (a) follows from (11.2) and (11.3). To show part (b), assume, by induction, that $(\mathbf{r}^k, \mathbf{r}^{k-1-j}) = 0$, $j = 0, 1, \dots, k - 1$. Since $(\mathbf{r}^{k+1}, \mathbf{r}^{k-j}) = 0$, $j = 0, 1$ by the choice of α_k, β_k, and for $j \geq 2$,

$$(\mathbf{r}^{k+1}, \mathbf{r}^{k-j}) = (\alpha_k \mathbf{r}^k + (1 - \alpha_k)\mathbf{r}^{k-1} - \beta_k A\mathbf{r}^k, \mathbf{r}^{k-j})$$

$$= -\beta_k(A\mathbf{r}^k, \mathbf{r}^{k-j}) = -\beta_k(\mathbf{r}^k, A\mathbf{r}^{k-j}) = 0,$$

because $A\mathbf{r}^{k-j} \in V_{k-j+1} \subset V_{k-1}$, then by the induction assumption, $\mathbf{r}^k$ is orthogonal to any vector in V_{k-1}. Further, for $j \geq 0$,

$$(\mathbf{r}^{k+1}, A^{-1}\Delta\mathbf{r}^{k-j+1}) = (\mathbf{r}^{k+1}, \tilde{\alpha}_{k-j}A^{-1}\Delta\mathbf{r}^{k-j} - \beta_{k-j}\mathbf{r}^{k-j})$$

$$= \tilde{\alpha}_{k-j}(\mathbf{r}^{k+1}, A^{-1}\Delta\mathbf{r}^{k-j}) = \cdots$$

$$= \prod_{s=j}^{k-1} \tilde{\alpha}_{k-s}(\mathbf{r}^{k+1}, A^{-1}\Delta\mathbf{r}^1)$$

$$= -\beta_0 \prod \tilde{\alpha}_{k-s}(\mathbf{r}^{k+1}, \mathbf{r}^0) = 0.$$

To show (c), use $(\mathbf{r}^k, \mathbf{r}^{k-j}) = 0$, $j = 1, 2, \dots, k$ and $(\mathbf{r}^k, A^{-1}\Delta\mathbf{r}^{k-j}) = 0$, $j = 0, \dots, k - 1$, and multiply (11.6) by $\mathbf{r}^k$ and $A^{-1}\Delta\mathbf{r}^k$, respectively. Note, then, that $(A^{-1}\Delta\mathbf{r}^k, \Delta\mathbf{r}^{k+1}) = 0$ by (b) and that $(A\mathbf{r}^k, A^{-1}\Delta\mathbf{r}^k) = (\mathbf{r}^k, \Delta\mathbf{r}^k) = (\mathbf{r}^k, \mathbf{r}^k)$, which shows (11.7).

Consider now the determinant of the matrix in (11.7),

$$d \equiv (\mathbf{r}^k, A\mathbf{r}^k)(\Delta\mathbf{r}^k, A^{-1}\Delta\mathbf{r}^k) - (\mathbf{r}^k, \mathbf{r}^k)^2$$

$$= (\mathbf{r}^k, A\mathbf{r}^k)(\Delta\mathbf{r}^k, A^{-1}\Delta\mathbf{r}^k) - (\Delta\mathbf{r}^k, \mathbf{r}^k)^2$$

$$= (\mathbf{r}^k, A\mathbf{r}^k)(\Delta\mathbf{r}^k, A^{-1}\Delta\mathbf{r}^k) - [(B^{-T}W^{\frac{1}{2}}\Delta\mathbf{r}^k)^T BW^{\frac{1}{2}}\mathbf{r}^k]^2,$$

where B is such that

$$W^{1/2}B^T BW^{1/2} = WA.$$

Such a matrix B exists because A is symmetric and positive definite w.r.t. the inner product $(\mathbf{x}, \mathbf{y}) = \mathbf{x}^T W\mathbf{y}$, that is, WA is s.p.d. By the Cauchy-Bunyakowski-Schwarz inequality, it follows that $d \geq 0$, and $d = 0$ if and only if $B^{-T}W^{\frac{1}{2}}\Delta\mathbf{r}^k$ and $BW^{\frac{1}{2}}\mathbf{r}^k$ are linearly dependent. If they are linearly dependent, then for some constants c_1, c_2,

$$c_1 B^{-T}W^{\frac{1}{2}}\Delta\mathbf{r}^k = c_2 BW^{\frac{1}{2}}\mathbf{r}^k,$$

which shows that

$$c_1 W\Delta\mathbf{r}^k = c_2 WA\mathbf{r}^k$$

or

(11.9) $$c_1 A^{-1}\Delta\mathbf{r}^k = c_2\mathbf{r}^k.$$

But then, by part (b),

$$0 = c_1(\mathbf{r}^k, A^{-1}\Delta\mathbf{r}^k) = c_2(\mathbf{r}^k, \mathbf{r}^k).$$

Hence, either $c_2 = 0$ or $(\mathbf{r}^k, \mathbf{r}^k) = 0$. If $c_2 = 0$, then (11.9) shows that $\Delta\mathbf{r}^k = \mathbf{0}$. But $(\mathbf{r}^k, \mathbf{r}^k) = (\mathbf{r}^k, \Delta\mathbf{r}^k)$, so then $\mathbf{r}^k = \mathbf{0}$. Thus, in either case, $\mathbf{r}^k = \mathbf{0}$, that is, a solution has already been found. If $\mathbf{r}^k \neq \mathbf{0}$, then $d > 0$, and (11.7) implies that

$$\begin{bmatrix} \beta_k \\ \tilde{\alpha}_k \end{bmatrix} = \frac{1}{d}\begin{bmatrix} (\Delta\mathbf{r}^k, A^{-1}\Delta\mathbf{r}^k) & (\mathbf{r}^k, \mathbf{r}^k) \\ (\mathbf{r}^k, \mathbf{r}^k) & (\mathbf{r}^k, A\mathbf{r}^k) \end{bmatrix}\begin{bmatrix} (\mathbf{r}^k, \mathbf{r}^k) \\ 0 \end{bmatrix}$$

$$= \frac{1}{d}\begin{bmatrix} (\Delta\mathbf{r}^k, A^{-1}\Delta\mathbf{r}^k)\cdot(\mathbf{r}^k, \mathbf{r}^k) \\ (\mathbf{r}^k, \mathbf{r}^k)^2 \end{bmatrix},$$

which is positive, since $\mathbf{r}^k \neq \mathbf{0}$ (which implies, $\Delta\mathbf{r}^k \neq \mathbf{0}$). Hence, $\beta_k > 0$, $\tilde{\alpha}_k > 0$. It remains to prove part (d), the optimality property. A necessary and sufficient condition for f in (11.8) to take and extreme (minimal) value is that all its partial derivatives are zero, that is,

$$\frac{\partial f}{\partial \xi_j} = 0, \, j = 0, 1, \ldots, k, \, \frac{\partial f}{\partial \eta_j} = 0, \, j = 0, 1, \ldots, k.$$

This shows that

$$(\mathbf{r}^j, A^{-1}\tilde{\mathbf{r}}^{k+1}) = 0, \, j = 0, 1, \ldots, k$$

and

$$(A\mathbf{r}^j, A^{-1}\tilde{\mathbf{r}}^{k+1}) = 0, \, j = 0, 1, \ldots, k,$$

that is

$$(\tilde{\mathbf{r}}^{k+1}, A^{-1}\Delta\mathbf{r}^j) = 0, \, j = 1, 2, \ldots, k$$

and

$$(\tilde{\mathbf{r}}^{k+1}, \mathbf{r}^j) = 0, \, j = 0, 1, \ldots, k.$$

Hence, in particular, we have

$$(\tilde{\mathbf{r}}^{k+1}, \mathbf{r}^k) = 0 \text{ and } (\tilde{\mathbf{r}}^{k+1}, \mathbf{r}^{k-1}) = 0,$$

and, as we have seen, these conditions alone determine $\xi_k = \alpha_k, \xi_{k-1} = 1 - \alpha_k, \eta_k = \beta_k$. Part (b) shows that the remaining conditions are then automatically satisfied—that is, the remaining coefficients ξ_j, η_j are then zero.   ◇

The property $\alpha_k > 1$ indicates a slight instability of the recursion for round-off errors, but this will not be further discussed here. It will now be seen that one can save some computational effort rewriting the algorithm using the "Δ" quantities introduced in Theorem 11.1.

11.1.1 Computational Considerations

The algorithm (11.2), (11.3), (11.4a), (11.5) is computationally feasible if the inner products $(\mathbf{r}^k, \mathbf{r}^k)$ and $(\mathbf{r}^k, A\mathbf{r}^k)$ are feasible—i.e., do not involve solving a system with A. This holds, for example, if A is s.p.d. and $(\mathbf{u}, \mathbf{v}) = \mathbf{u}^T A^s \mathbf{v}$ for some integer $s \geq 0$, but we later also consider an inner product defined by another s.p.d. matrix.

As we have seen, the algorithm can be written in the computationally some-what more efficient form

$$\delta^k = (\mathbf{r}^k, \mathbf{r}^k), \ \mu_k = (\mathbf{r}^k, A\mathbf{r}^k)/\delta_k,$$

$$\beta_k^{-1} = \mu_k - \frac{\delta_k}{\delta_{k-1}} \beta_{k-1}^{-1}, \ \tilde{\alpha}_k = \beta_k \mu_k - 1,$$

(11.10)

$$\Delta\mathbf{x}^{k+1} = \tilde{\alpha}_k \Delta\mathbf{x}^k - \beta_k \mathbf{r}^k, \ \mathbf{x}^{k+1} = \mathbf{x}^k + \Delta\mathbf{x}^{k+1},$$

$$\Delta\mathbf{r}^{k+1} = \tilde{\alpha}_k \Delta\mathbf{r}^k - \beta_k A\mathbf{r}^k, \ \mathbf{r}^{k+1} = \mathbf{r}^k + \Delta\mathbf{r}^{k+1}.$$

The computational complexity involves the following:

(a) a matrix vector multiplication with A,
(b) two inner products, and
(c) eight vector operations (scalar times vector and vector additions), which equals 4 "flops."

We need to store five vectors,

$$\mathbf{x}^k, \ \Delta\mathbf{x}^k, \ \mathbf{r}^k, \ \Delta\mathbf{r}^k, \ A\mathbf{r}^k.$$

Instead of using the recurrence relation (11.3) to compute $\mathbf{r}^{k+1}$, in some cases it can be more efficient to compute $\mathbf{r}^{k+1} = A\mathbf{x}^{k+1} - \mathbf{b}$. Then we save three vector operations, but we need one more matrix vector computation.

11.1.2 The Inner Product and the Preconditioned Variant

The inner product $(\cdot, \cdot)$ can be simply $(\mathbf{x}, \mathbf{y}) = \mathbf{x}^T\mathbf{y}$ or, more generally, may be defined by a symmetric, positive definite matrix C, that is, $(\mathbf{x}, \mathbf{y}) = \mathbf{x}^T C\mathbf{y}$. Since we require that A is selfadjoint w.r.t. the inner product, we must have that

(11.11) $$CA = A^T C$$

that is, $CA = AC$, when A is symmetric. This means that A must commute with C, which is a quite restrictive assumption. However, when we use a preconditioning of A with the matrix C that is symmetric and positive definite, we may consider the *pseudoresiduals*

$$\tilde{\mathbf{r}}^k = C^{-1}(A\mathbf{x}^k - \mathbf{b})$$

and then replace A with $B = C^{-1}A$ in the algorithm (11.10), requiring instead that B is selfadjoint w.r.t. the inner product, $(\mathbf{x}, \mathbf{y}) = \mathbf{x}^T C\mathbf{y}$. By (11.11) this means $CB = B^T C$, that is, $CC^{-1}A = A^T C^{-T} C$ or $A = A^T$, since C is symmetric. *Hence, in this case B becomes selfadjoint w.r.t. the inner product* $(\mathbf{x}, \mathbf{y}) = \mathbf{x}^T C \mathbf{y}$ if A is symmetric. The algorithm (11.10), where A is replaced by $B = C^{-1}A$ and $\mathbf{r}^k$ by the pseudoresiduals $\tilde{\mathbf{r}}^k$, is then the optimal algorithm to solve $C^{-1}A\mathbf{x} = C^{-1}\mathbf{b}$ over the Krylov spaces *generated by the matrix B*. Note that in this case

$$\mu_k = (\tilde{\mathbf{r}}^k, C^{-1}A\tilde{\mathbf{r}}^k) = \tilde{\mathbf{r}}^{k^T}A\tilde{\mathbf{r}}^k$$

and

$$\delta_k = (\tilde{\mathbf{r}}^k, \tilde{\mathbf{r}}^k) = \tilde{\mathbf{r}}^{k^T}\mathbf{r}^k,$$

where $\mathbf{r}^k = A\mathbf{x}^k - \mathbf{b}$. Theorem 11.1 shows that $\mathbf{r}^{k+1^T}C^{-1}\mathbf{r}^{k-j} = 0$, $j = 0, 1, \ldots, k$ and that the same functional $\mathbf{r}^{k+1^T}A^{-1}\mathbf{r}^{k+1}$ is minimized as in the unpreconditioned version. Such a three-term recurrence preconditioned method was first used in Axelsson (1972); see also Axelsson (1974). More generally, if $(\mathbf{x}, \mathbf{y}) = \mathbf{x}^T W \mathbf{y}$ for some s.p.d. matrix W, then

$$WB = B^T W,$$

so we must have

$$WC^{-1}A = A^T C^{-T} W.$$

For instance, if $W = A^T A$, then we must have

$$AC^{-1} = C^{-T}A^T,$$

that is, AC^{-1} must be symmetric. For related discussions, see Hageman and Young (1981).

11.2 The Standard Conjugate Gradient Method

To solve $A\mathbf{x} = \mathbf{b}$, $\mathbf{x} \in \mathbb{R}^n$, where A is s.p.d., we describe now a more standard form of conjugate gradient methods that can be seen as a minimization method to find the minimizer of a functional

(11.12) $$f(\mathbf{x}) = \frac{1}{2}(\mathbf{r}, A^{-1}\mathbf{r}),$$

where $\mathbf{r} = A\mathbf{x} - \mathbf{b}$ and $(\cdot, \cdot)$ is an inner product. It is assumed here that A is selfadjoint and positive definite w.r.t. this inner product. Clearly, the minimizer of f is the solution $\mathbf{x} = A^{-1}\mathbf{b}$ of $A\mathbf{x} = \mathbf{b}$, since A^{-1} is positive definite and selfadjoint w.r.t. the same inner product. For the standard inner product, it is readily seen that $f(\mathbf{x}) = 1/2\mathbf{x}^T A\mathbf{x} - \mathbf{b}^T\mathbf{x} + 1/2\mathbf{b}^T A^{-1}\mathbf{b}$. The *gradient* of f at $\mathbf{x}$ is the vector

$$\mathbf{g}(\mathbf{x}) = \left(\frac{\partial f}{\partial x_1}, \ldots, \frac{\partial f}{\partial x_n}\right)^T.$$

Let $\mathbf{d}$ be any nonzero vector in $\mathbb{R}^n$. Then $\mathbf{g}$ can be computed in the following way: We have

$$(\mathbf{g}, \mathbf{d}) = \lim_{\tau \to 0} \frac{1}{\tau}[f(\mathbf{x} + \tau\mathbf{d}) - f(\mathbf{x})],$$

and an elementary computation, using the selfadjointness of A, reveals that

$$f(\mathbf{x} + \tau\mathbf{d}) - f(\mathbf{x}) = \frac{1}{2}(\mathbf{r} + \tau A\mathbf{d}, A^{-1}(\mathbf{r} + \tau A\mathbf{d})) - \frac{1}{2}(\mathbf{r}, A^{-1}\mathbf{r})$$
(11.13)
$$= \tau(\mathbf{r}, \mathbf{d}) + \frac{1}{2}\tau^2(\mathbf{d}, A\mathbf{d}),$$

so

$$(\mathbf{g}, \mathbf{d}) = (\mathbf{r}, \mathbf{d}).$$

Since this is valid for any $\mathbf{d}$, we have $\mathbf{g} = \mathbf{r}$, that is, the gradient of the functional equals the residual.

11.2.1 Minimization

As was seen in the introduction, if A is positive definite, then there is a unique minimizer of f. To find the minimizer of f, we will use an iterative method where at each stage we construct a new *search direction* $\mathbf{d}^k$ (which will be conjugately orthogonal to the previous search directions). We compute the local minimizer along this search direction—i.e., given $\mathbf{x}^k$, the approximation at stage k, we compute $\tau = \tau_k$ such that

$$f(\mathbf{x}^k + \tau\mathbf{d}^k), \quad -\infty < \tau < \infty$$

is minimized by τ_k and then let

(11.14)
$$\mathbf{x}^{k+1} = \mathbf{x}^k + \tau_k\mathbf{d}^k$$

be the new approximation.

Let $\mathbf{r}^k = A\mathbf{x}^k - \mathbf{b}$ and let $h(\tau) \equiv f(\mathbf{x}^k + \tau\mathbf{d}^k) - f(\mathbf{x}^k)$. Note that (11.13) implies that

$$h(\tau) = \tau(\mathbf{r}^k, \mathbf{d}^k) + \frac{1}{2}\tau^2(\mathbf{d}^k, A\mathbf{d}^k).$$

This is quadratic, and differentiation w.r.t. τ shows that h takes its smallest value when

$$(\mathbf{r}^k, \mathbf{d}^k) + \tau(\mathbf{d}^k, A\mathbf{d}^k) = 0,$$

that is, when

(11.15) $$\tau = \tau_k \equiv -(\mathbf{r}^k, \mathbf{d}^k)/(\mathbf{d}^k, A\mathbf{d}^k).$$

Since, by (11.14),

(11.16) $$\mathbf{r}^{k+1} = \mathbf{r}^k + \tau_k A\mathbf{d}^k,$$

then (11.15) shows that

(11.17) $$(\mathbf{r}^{k+1}, \mathbf{d}^k) = 0,$$

that is, the gradient becomes orthogonal to the search direction.

For the next iteration step, we need a new search direction, $\mathbf{d}^{k+1}$, and this will be computed so that

(11.18) $$(\mathbf{d}^{k+1}, A\mathbf{d}^j) = 0, \ 0 \le j \le k,$$

that is, in such a way that the search directions become mutually conjugate orthogonal w.r.t. A and the inner product.

The above holds for many sets of search directions. Let us now consider the following practically important choice.

11.2.2 The Krylov Space of Search Directions

Let $V_k = \text{span }\{\mathbf{d}^0, \mathbf{d}^1, \ldots, \mathbf{d}^k\}$, that is, let the vector space V_k be spanned by $\mathbf{d}^0, \mathbf{d}^1, \ldots, \mathbf{d}^k$, which we assume to be mutually conjugate orthogonal. Assume also that

(11.19) $$(\mathbf{r}^k, \mathbf{d}^j) = 0, \ 0 \le j \le k - 1,$$

where $k \ge 1$. Then (11.16) implies that

$$(\mathbf{r}^{k+1}, \mathbf{d}^j) = (\mathbf{r}^k, \mathbf{d}^j) + \tau_k(A\mathbf{d}^k, \mathbf{d}^j),$$

and (11.18), (11.19) and (11.17) show that

(11.20) $$(\mathbf{r}^{k+1}, \mathbf{d}^j) = 0, \ 0 \le j \le k.$$

Hence, by induction, it follows that when the search directions are conjugately orthogonal, the residuals (or gradients) become orthogonal to the previous search directions. As we shall see, this property implies that the method computes the best approximation $\mathbf{x}^{k+1} = \mathbf{x}^k + \mathbf{d}$ of all vectors $\mathbf{d}$ in V_k. It remains, therefore, only to compute the search directions in an efficient way to make them mutually A-orthogonal. To this end, let

(11.21) $$\mathbf{d}^{k+1} = -\mathbf{r}^{k+1} + \beta_k \mathbf{d}^k, \ k = 0, 1, 2, \ldots,$$

where initially we let $\mathbf{d}^0 = -\mathbf{r}^0$ and $\beta_0, \beta_1, \ldots$ remain to be determined. (Other choices of $\mathbf{d}^0$ are conceivable.) The relation $(\mathbf{d}^{k+1}, A\mathbf{d}^k) = 0$ shows that

(11.22) $$\beta_k = \frac{(\mathbf{r}^{k+1}, A\mathbf{d}^k)}{(\mathbf{d}^k, A\mathbf{d}^k)}.$$

Next note that (11.21) shows that

$$(\mathbf{d}^{k+1}, A\mathbf{d}^j) = -(\mathbf{r}^{k+1}, A\mathbf{d}^j) + \beta_k(\mathbf{d}^k, A\mathbf{d}^j).$$

Now, by the induction assumption and (11.21) for $j \le k - 1$, the last term is zero. The choice $\mathbf{d}^0 = -\mathbf{r}^0$ allows us to define V_k alternatively by

$$V_k = \text{span}\{\mathbf{r}^0, \mathbf{r}^1, \ldots, \mathbf{r}^k\}.$$

Hence, $A\mathbf{d}^j \in V_{j+1}$. Since (11.20) shows that $\mathbf{r}^{k+1}$ is orthogonal to any vector in V_k, then

(11.22a) $$(\mathbf{r}^{k+1}, A\mathbf{d}^j) = 0, \ 0 \le j \le k - 1.$$

Hence, we have shown by induction and by the choice (11.22) of β_k in (11.21) that

$$(\mathbf{d}^{k+1}, A\mathbf{d}^j) = 0, \ 0 \le j \le k,$$

which is (11.18).

Remark 11.2 The method whereby the search directions are computed by (11.21, 11.22) is called the *conjugate gradient method*. There is a somewhat simpler, but less efficient, method of computing the minimizer of f, called the *steepest descent* method. In this one, let $\beta_k = 0$ in (11.21), that is, the search of a local minimum takes place along the current gradient vector. For details, see Axelsson and Barker (1984), for instance. On the other hand, in the conjugate gradient method we move along a plane spanned by the gradient at the most recent point and the most recent search direction.

As remarked in the introduction, we can also use other sets of orthogonal search directions. For example, if we let $\mathbf{d}^k = \mathbf{e}^k$, where $\mathbf{e}^k = (0, \ldots, 1, \ldots 0)^T$, that is, $\mathbf{e}^k$ is the kth unit coordinate vector with zero entries except the unit entry at the kth position, then minimizing $f(\mathbf{x}^k + \tau \mathbf{d}^k)$ gives $\tau = \tau_k = -(\mathbf{r}^k, \mathbf{e}^k)/(\mathbf{e}^k, A\mathbf{e}^k) = -r_k^k/a_{kk}$. (Here r_k^k denotes the kth component of $\mathbf{r}^k$.) Hence, $\mathbf{x}^{k+1} = \mathbf{x}^k + \tau_k \mathbf{e}^k$, or $x_j^{k+1} = x_j^k, j \neq k, x_k^{k+1} = x_k^k - r_k^k/a_{kk}, k = 1, 2, \ldots, n$, which is nothing but one step of the Gauss-Seidel iterative method, usually a slow process. Here, a_{kk} are the diagonal entries of A.

Similarly, it has been shown in Hestenes (1980) that Gaussian elimination can be described as an optimization method using also the coordinate vectors for the generation of search directions. For a more general presentation of such relations, see Abaffy and Spedicato (1989).

11.2.3 The Finite Termination Property

Since if $\mathbf{d}^k = \mathbf{0}$, τ_k is not defined by (11.15), it is important to investigate the situation that arises when (11.21) produces a zero search vector. Replacing $(k + 1)$ by k in (11.21) and taking the product with $\mathbf{r}^k$, we obtain $(\mathbf{r}^k, \mathbf{d}^k) = -(\mathbf{r}^k, \mathbf{r}^k) + \beta_{k-1}(\mathbf{r}^k, \mathbf{d}^{k-1})$. But by (11.20), $(\mathbf{r}^k, \mathbf{d}^{k-1}) = 0$, so

$$(11.22b) \qquad\qquad (\mathbf{r}^k, \mathbf{d}^k) = -(\mathbf{r}^k, \mathbf{r}^k)$$

and, if $\mathbf{d}^k = \mathbf{0}$, then $(\mathbf{r}^k, \mathbf{r}^k) = 0$, implying $\mathbf{r}^k = A\mathbf{x}^k - \mathbf{b} = \mathbf{0}$ and $\mathbf{x}^k = \mathbf{x}$, the solution of $A\mathbf{x} = \mathbf{b}$. Thus, a zero search direction can be produced only after the minimizer has already been found, at which stage we stop the iterations.

On the other hand, if $\mathbf{x}^k \neq \mathbf{x}$, then $\mathbf{r}_k \neq \mathbf{0}$, for $\mathbf{x}^k \neq \mathbf{x}$ implies $\|\mathbf{r}^k\| \neq 0$ and (11.22b) shows that $(\mathbf{r}^k, \mathbf{d}^k) \neq 0$ and, hence, $\tau_k \neq 0$. These remarks make it clear that *regardless of the choice* of $\beta_0, \beta_1, \ldots$, the iterative process defined by (11.14), (11.15), and (11.21) will either be nonterminating, with $\mathbf{x}^k \neq \mathbf{x}$, $\mathbf{d}^k \neq \mathbf{0}$, and $\tau_k \neq 0$ for all values of k, or there will be some integer m such that $\mathbf{x}^k \neq \mathbf{x}$, $\mathbf{d}^k \neq \mathbf{0}$, and $\tau_k \neq 0$ for $k = 0, 1, \ldots, m - 1$, and $\mathbf{x}^m = \mathbf{x}$. However, *the particular choice* of $\beta_0, \beta_1, \ldots$, making the set of search directions conjugately orthogonal, must stop with $m \leq n$, because we can generate at most n such mutually orthogonal vectors in $\mathbb{R}^n$. Hence, the conjugate gradient method where the search directions are computed by (11.21, 11.22) has a finite termination property. This holds, at least in the absence of rounding errors.

Note, finally, that if A is indefinite, even in exact arithmetic the method can break down with $(\mathbf{d}^k, A\mathbf{d}^k) = 0$ but $\mathbf{d}^k \neq \mathbf{0}$.

11.2.4 The Optimization Property

As we have seen, the residuals $\mathbf{r}^{k+1}$ at stage $k+1$ are elements of V_{k+1}. Furthermore, (11.20) and (11.21) show that

$$(11.23) \qquad (\mathbf{r}^{k+1}, \mathbf{r}^j) = 0, \quad 0 \le j \le k.$$

Hence, we see that $(\mathbf{r}^{k+1}, A^{-1}A\mathbf{r}^j) = (\mathbf{r}^{k+1}, A^{-1}\mathbf{h}) = 0$ for all $\mathbf{h} \in S_{k+1}$, where

$$S_{k+1} = \operatorname{span}\{A\mathbf{r}^0, A^2\mathbf{r}^0, \ldots, A^{k+1}\mathbf{r}^0\}.$$

Letting $\mathbf{h}^{k+1} = \mathbf{r}^{k+1} - \mathbf{r}^0$, this can be written in the form

$$(11.24) \qquad (\mathbf{r}^0 + \mathbf{h}^{k+1}, A^{-1}\mathbf{h}) = 0 \text{ for all } \mathbf{h} \in S_{k+1}$$

and can be illustrated as in Figure 11.1.

This orthogonality property shows, hence, that $(\mathbf{r}^0 + \mathbf{h}, A^{-1}(\mathbf{r}^0 + \mathbf{h}))$ is smallest among all $\mathbf{h} \in S_{k+1}$ if and only if $\mathbf{h} = \mathbf{h}^{k+1}$, that is, $\mathbf{r}^0 + \mathbf{h} = \mathbf{r}^0 + \mathbf{h}^{k+1} = \mathbf{r}^{k+1}$, where $\mathbf{r}^{k+1}$ has been computed by the conjugate gradient method. The residual is, therefore, the smallest possible among any residual $\mathbf{r} = \mathbf{r}^0 + \mathbf{h}$, where $\mathbf{h}$ is taken from this subspace.

The following example gives an alternative presentation of the optimality and termination properties that hold for the conjugate gradient method. Note, first, that the graphs of $f(\mathbf{x}) = 1/2(\mathbf{r}, A^{-1}\mathbf{r}) = \text{constant}$, where $\mathbf{r} = A(\mathbf{x} - \hat{\mathbf{x}})$, $\hat{\mathbf{x}} = A^{-1}\mathbf{b}$ and $\mathbf{x}$ is variable, are ellipses. (This holds for any s.p.d. matrix A.)

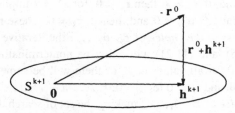

Figure 11.1.   A geometrical interpretation of the best approximation to $-\mathbf{r}^0$ in the subspace S_{k+1}.

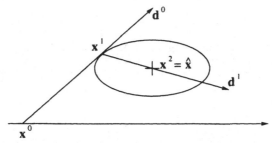

Figure 11.2. An alternative geometric presentation of the best approximation and termination properties.

Note next that the point $\mathbf{x}^1$ taken in the direction $\mathbf{d}^0$ is closest to the solution point in the norm defined by the inner product. Hence, $\mathbf{d}^0$ is tangent to the corresponding ellipse. Finally, at $\mathbf{x}^1$, a search direction $\mathbf{d}^1$ is found such that the line $\mathbf{x}^1 + \tau\mathbf{d}^1$, $\tau > 0$, passes the solution point.

Example 11.1 Let

$$A = \begin{bmatrix} 1 & 0 \\ 0 & 2 \end{bmatrix}, \quad \mathbf{b} = \begin{bmatrix} 2 \\ 2 \end{bmatrix}$$

(that is, $\hat{\mathbf{x}} = A^{-1}\mathbf{b} = \begin{bmatrix} 2 \\ 1 \end{bmatrix}$), let $(\mathbf{u}, \mathbf{v}) = \mathbf{u}^T\mathbf{v}$, and let $\mathbf{x}^0 = \begin{bmatrix} 0 \\ 0 \end{bmatrix}$. Then,

$$\mathbf{d}^0 = -\mathbf{r}^0 = \begin{bmatrix} 2 \\ 2 \end{bmatrix}, \quad \tau_0 = \frac{2}{3}, \quad \mathbf{x}^1 = \frac{4}{3}\begin{bmatrix} 1 \\ 1 \end{bmatrix},$$

$$\mathbf{r}^1 = \frac{2}{3}\begin{bmatrix} -1 \\ 1 \end{bmatrix}, \quad \beta_0 = \frac{1}{9}, \quad \mathbf{d}^1 = \frac{4}{9}\begin{bmatrix} 2 \\ -1 \end{bmatrix}$$

and

$$\mathbf{x}^1 + \tau\mathbf{d}^1 = \hat{\mathbf{x}} \quad \text{for } \tau = \frac{3}{4}.$$

The next example illustrates that breakdown of the conjugate gradient algorithm can occur if A is indefinite.

Example 11.2 Let

$$A = \begin{bmatrix} -1 & 0 \\ 0 & 4 \end{bmatrix}, \quad \mathbf{b} = \begin{bmatrix} 2 \\ 1 \end{bmatrix}, \quad (\mathbf{u}, \mathbf{v}) = \mathbf{u}^T\mathbf{v},$$

and let $x^0 = \begin{bmatrix} 0 \\ 0 \end{bmatrix}$. Then $d^0 = \begin{bmatrix} 2 \\ 1 \end{bmatrix}$, $Ad^0 = \begin{bmatrix} -2 \\ 4 \end{bmatrix}$, and $(d^0, Ad^0) = 0$. Hence, the algorithm fails in computing the length along the search direction d^0. However, with $x^0 = \begin{bmatrix} -1 \\ 0 \end{bmatrix}$, for instance, we find $d^0 = -r^0 = \begin{bmatrix} 1 \\ 1 \end{bmatrix}$, $Ad^0 = \begin{bmatrix} -1 \\ 4 \end{bmatrix}$, and $(d^0, Ad^0) = 3$. Thus, $\tau_0 = \frac{2}{3}$, $x^1 = \frac{1}{3}\begin{bmatrix} -1 \\ 2 \end{bmatrix}$

$$r^1 = \frac{5}{3}\begin{bmatrix} -1 \\ 1 \end{bmatrix}, \quad \beta_0 = \frac{25}{9},$$

$$d^1 = \frac{10}{9}\begin{bmatrix} 4 \\ 1 \end{bmatrix}, \quad Ad^1 = \frac{40}{9}\begin{bmatrix} -1 \\ 1 \end{bmatrix}, \quad \tau_1 = -\frac{3}{8},$$

and $x^2 = \begin{bmatrix} -2 \\ \frac{1}{4} \end{bmatrix}$, which is the solution to $Ax = b$.

We collect the results found for the conjugate gradient algorithm in the following theorem.

Theorem 11.3 *Let A be selfadjoint and positive definite w.r.t. the inner product* $(\cdot, \cdot)$. *Let* x^0 *be an arbitrary vector and let* $d^0 = -r^0$, *where* $r^0 = Ax^0 - b$. *For* $k = 0, 1, \ldots,$ *let*

$$x^{k+1} = x^k + \tau_k d^k$$

$$r^{k+1} = r^k + \tau_k Ad^k$$

$$d^{k+1} = -r^{k+1} + \beta_k d^k,$$

where τ_k *and* β_k *are computed by (11.15) and (11.22), respectively. Then:*

(a) $\qquad (r^{k+1}, A^{-1}r^{k+1}) \le (r, A^{-1}r), \ r = r^0 + h$

for all $h \in S_{k+1} = \text{span} \{Ar^0, \ldots, A^{k+1}r^0\}$, *that is,*

the residual r^{k+1} *is smallest possible in norm* $\|r\| = (r, A^{-1}r)^{1/2}$ *among all residuals of the form* $r^0 + h$, *where* h *is taken from the subspace* S_{k+1}.

(b) *The following orthogonality or conjugate orthogonality properties hold:*

$$(\mathbf{r}^{k+1}, \mathbf{r}^j) = 0, \quad 0 \le j \le k,$$

$$(\mathbf{r}^{k+1}, \mathbf{d}^j) = 0, \quad 0 \le j \le k,$$

$$(\mathbf{d}^{k+1}, A\mathbf{d}^j) = 0, \quad 0 \le j \le k,$$

$$(\mathbf{r}^{k+1}, A\mathbf{d}^j) = 0, \quad 0 \le j \le k - 1,$$

$$(\mathbf{r}^i, \mathbf{d}^j) = (\mathbf{r}^0, \mathbf{d}^j), \quad 0 \le i < j.$$

(c) *If the inner product* $(\mathbf{u}, \mathbf{v}) = \mathbf{u}^T \mathbf{v}$, *then the conjugate gradient method minimizes*

$$(\mathbf{r}^{k+1})^T A^{-1} \mathbf{r}^{k+1} = (\mathbf{e}^{k+1})^T A \mathbf{e}^{k+1},$$

where $\mathbf{e}^{k+1} = \mathbf{x}^{k+1} - \hat{\mathbf{x}}$, $\hat{\mathbf{x}} = A^{-1}\mathbf{b}$ *is the iteration error. Here,*

$$(\mathbf{d}^{k+1})^T A\mathbf{d}^j = 0, 0 \le j \le k, \text{ and}$$

$$(\mathbf{r}^{k+1})^T \mathbf{r}^j = 0, 0 \le j \le k.$$

If $(\mathbf{u}, \mathbf{v}) = \mathbf{u}^T A\mathbf{v}$, *then the conjugate gradient method gives the best least square* residual *solution— i.e., at every iteration step, the norm of the current residual,*

$$\|\mathbf{r}^{k+1}\|_2 = \{(\mathbf{r}^{k+1})^T \mathbf{r}^{k+1}\}^{\frac{1}{2}},$$

is minimized over the corresponding Krylov space. Here,

$$(A\mathbf{d}^{k+1})^T A\mathbf{d}^j = 0, 0 \le j \le k, \text{ and}$$

$$(\mathbf{r}^{k+1})^T A\mathbf{r}^j = 0, 0 \le j \le k.$$

If $(\mathbf{u}, \mathbf{v}) = \mathbf{u}^T A^{-1}\mathbf{v}$, *then the conjugate gradient method gives the least square iterative* solution *error, that is,* $\|\mathbf{e}^{k+1}\|$ *is minimized. Here,* $(\mathbf{d}^{k+1})^T \mathbf{d}^j = 0$, $0 \le j \le k$, *and* $(\mathbf{r}^{k+1})^T A^{-1}\mathbf{r}^j = 0$, $0 \le j \le k$. *(This version is not viable in practice, because the computation of inner products like* $(\mathbf{r}^k, \mathbf{r}^k)$ *requires the solution of a linear system with A.)*

Proof The properties have already been shown by (11.17), (11.20), (11.23), and (11.22a). It remains only to show the final property in part (b). To this end, note that the recurrence form implies that

$$\mathbf{r}^i = \mathbf{r}^0 + Aq_{i-1}(A)\mathbf{r}^0$$

$$= \mathbf{r}^0 - Aq_{i-1}(A)\mathbf{d}^0$$

for some polynomial q_{i-1} of degree $i - 1$. Hence, for $i < j$,

$$(\mathbf{r}^i, \mathbf{d}^j) = (\mathbf{r}^0 - Aq_{i-1}(A)\mathbf{d}^0, \mathbf{d}^j) = (\mathbf{r}^0, \mathbf{d}^j),$$

because $q_{i-1}(A)\mathbf{d}^0 \in V_{i-1}$ and the conjugate orthogonality shows that

$$(Aq_{i-1}(A)\mathbf{d}^0, \mathbf{d}^j) = 0. \qquad \qquad \diamond$$

An accurate estimate of the decay of $(\mathbf{r}^{k+1}, \mathbf{r}^{k+1})$ is presented in a more general context in Theorem 12.5. It follows that $(\mathbf{r}^{k+1}, \mathbf{r}^{k+1})$ can be written as a continued fraction.

Remark 11.4 (Short recurrence form) Part (a) of Theorem 11.2 shows that even if we let $\mathbf{x}^{k+1}$ and/or $\mathbf{d}^{k+1}$ contain additional terms involving some of the previous search directions to $\mathbf{d}^k$, we cannot get better approximations than by (11.14). The short recurrence relation is one reason for the efficiency of the conjugate gradient method. Note that it corresponds to the three-term recurrence relation for the algorithm in the previous section. In fact, due to the uniqueness of the optimal solution, *both algorithms generate identical approximations* $\mathbf{x}^k$ if we use the same inner product and neglect the influence of round-off errors (see also the formula below, which gives a relation between the coefficients in the recursions used in the two methods).

Because of the conjugate orthogonality property $(\mathbf{d}^i, A\mathbf{d}^j) = 0, i \neq j$, the method was named the *conjugate gradient method* (see Hestenes and Stiefel, 1952). Since it is actually the search directions that are conjugate orthogonal, a more proper name might be the *conjugate direction method*. However, for the inner product $(\mathbf{x}, \mathbf{y}) = \mathbf{x}^T A\mathbf{y}$, the residuals (or gradients) indeed become conjugate orthogonal w.r.t. the matrix A. The relation between the coefficients in the conjugate gradient method and in the three-term recurrence form presented in Section 11.1 can readily be found in the following way: We have

$$\mathbf{x}^{k+1} = \mathbf{x}^k + \tau_k \mathbf{d}^k = \mathbf{x}^k + \tau_k(-\mathbf{r}^k + \beta_{k-1}\mathbf{d}^{k-1}),$$

and substitution of

$$\mathbf{d}^{k-1} = \tau_{k-1}^{-1}(\mathbf{x}^k - \mathbf{x}^{k-1})$$

shows that

$$\mathbf{x}^{k+1} = (1 + \frac{\tau_k}{\tau_{k-1}}\beta_{k-1})\mathbf{x}^k - \frac{\tau_k}{\tau_{k-1}}\beta_{k-1}\mathbf{x}^{k-1} - \tau_k\mathbf{r}^k.$$

Comparing with (11.2), where we write the coefficient as $\hat{\beta}_k$ to distinguish it from the coefficient β_k in (11.22), we find

$$\mathbf{x}^{k+1} = \alpha_k\mathbf{x}^k + (1 - \alpha_k)\mathbf{x}^{k-1} - \hat{\beta}_k\mathbf{r}^k,$$

where

$$\alpha_k = 1 + \frac{\tau_k}{\tau_{k-1}}\beta_{k-1}, \quad \hat{\beta}_k = \tau_k.$$

It will be seen below that $\beta_k > 0$ and that $\tau_k > 0$ when A is s.p.d., so these relations give an alternative proof of the properties $\alpha_k > 1$ and $\hat{\beta}_k > 0$ (cf. Theorem 11.1(c)).

11.2.5 Computational Considerations

Various identities allow a number of formulations of the conjugate gradient methods. For example, the gradient can be computed directly by the formula $\mathbf{r}^k = A\mathbf{x}^k - \mathbf{b}$ instead of the recursion in (11.16). There are also other expressions for β_k and τ_k. Using $\mathbf{d}^k = -\mathbf{r}^k + \beta_{k-1}\mathbf{d}^{k-1}$, $A\mathbf{d}^k = 1/\tau_k(\mathbf{r}^{k+1} - \mathbf{r}^k)$ and orthogonality properties, we find

$$-(\mathbf{r}^k, \mathbf{d}^k) = (\mathbf{r}^k, \mathbf{r}^k),$$

$$(\mathbf{r}^{k+1}, A\mathbf{d}^k) = \frac{1}{\tau_k}(\mathbf{r}^{k+1}, \mathbf{r}^{k+1} - \mathbf{r}^k) = \frac{1}{\tau_k}(\mathbf{r}^{k+1}, \mathbf{r}^{k+1})$$

and

$$\tau_k(\mathbf{d}^k, A\mathbf{d}^k) = (\mathbf{d}^k, \mathbf{r}^{k+1} - \mathbf{r}^k) = -(\mathbf{d}^k, \mathbf{r}^k) = (\mathbf{r}^k, \mathbf{r}^k).$$

Hence, by (11.15) and (11.22),

$$\tau_k = (\mathbf{r}^k, \mathbf{r}^k)/(\mathbf{d}^k, A\mathbf{d}^k),$$

$$\beta_k = (\mathbf{r}^{k+1}, \mathbf{r}^{k+1})/(\mathbf{r}^k, \mathbf{r}^k).$$

Note that with the latter formulas for τ_k and β_k, there is no need to compute the inner products $(\mathbf{r}^k, \mathbf{d}^k)$ and $(\mathbf{r}^{k+1}, A\mathbf{d}^k)$. Furthermore, the inner product

$(\mathbf{r}^{k+1}, \mathbf{r}^{k+1})$ can be used to test when to stop the iterations. Then the conjugate gradient method takes the form

$$\tau_k = (\mathbf{r}^k, \mathbf{r}^k)/(\mathbf{d}^k, A\mathbf{d}^k),$$

$$\mathbf{x}^{k+1} = \mathbf{x}^k + \tau_k \mathbf{d}^k,$$

(11.25) $$\mathbf{r}^{k+1} = \mathbf{r}^k + \tau_k A\mathbf{d}^k,$$

$$\beta_k = (\mathbf{r}^{k+1}, \mathbf{r}^{k+1})/(\mathbf{r}^k, \mathbf{r}^k),$$

$$\mathbf{d}^{k+1} = -\mathbf{r}^{k+1} + \beta_k \mathbf{d}^k,$$

$k = 0, 1, \dots$ until $(\mathbf{r}^{k+1}, \mathbf{r}^{k+1}) \leq \varepsilon$. Initially we choose $\mathbf{x}^0$, calculate $\mathbf{r}^0 = A\mathbf{x}^0 - \mathbf{b}$, and put $\mathbf{d}^0 = -\mathbf{r}^0$. The computer implementation of this algorithm is clarified as follows:

```
     x := x⁰; r := b;
     r := Ax − r; δ0 := (r, r);
     IF δ0 ≤ ε THEN STOP;
     d := −r;
R:   h := Ad;
     τ := δ0/(d, h);
     x := x + τd;
     r := r + τh; δ1 := (r, r);
     IF δ1 ≤ ε THEN STOP;
     β := δ1/ δ0; δ0 := δ1;
     d := −r + βd;
     GOTO R;
```

Note: $\mathbf{x}_0$ and $\mathbf{b}$ in the first line of the code are input vectors.

Regarding the storage requirement, n-dimensional arrays are needed for $\mathbf{h}, \mathbf{x}, \mathbf{r},$ and $\mathbf{d}$. The storage requirement for A depends on the number of nonzero entries and the chosen data structure. In our problems, A is usually large and sparse, and the simplest data structure—an $n \times n$ array—is much too wasteful of storage. See Axelsson and Barker (1984), for instance, where a commonly used storage scheme suitable for a general sparse symmetric matrix is presented.

Most of the computational labor per iteration is usually expended in the computation of $A\mathbf{d}$. There are, in addition, two inner products and three recursion formulas (of so-called linked triad type) per iterative step, requiring about $5n$ multiplications and $5n$ additions for the standard inner product. This compares somewhat more favorably with the three-term recurrence form in the previous section, where we required about $6n$ multiplications and $6n$ additions in addition to the matrix vector multiplication with A.

As we have seen, the conjugate gradient method has the property of finite termination in the absence of round-off errors. However, it turns out that in practice, there is a tendency toward increasing round-off error and loss of orthogonality in the method. On the other hand, one frequently has a sufficiently accurate solution after a number of steps much smaller than n, and therefore the test above, $(\mathbf{r}^{k+1}, \mathbf{r}^{k+1}) \leq \varepsilon$, may stop the iterations after a relatively small number of steps. As we shall see in Chapter 13, the rate of convergence depends on the distribution of eigenvalues. With a proper preconditioning of the matrix, this distribution can be such that the method converges much faster with preconditioning than without.

The conjugate gradient method was originally proposed by Hestenes and Stiefel (1952). It can be derived from the Lanczos algorithm (to be presented in the next section). It can, in fact, be derived in many equivalent forms; see, for example, Young, Hayes, and Jea (1981).

11.2.6 The Preconditioned Conjugate Gradient Method

As has been seen in previous chapters, the use of preconditioners can increase the rate of convergence of iterative solution methods considerably. For the conjugate gradient method, this will be further analyzed in a later chapter. We consider now the computational aspects of the preconditioned version of the standard conjugate gradient method. This will be derived using an inner product defined by the preconditioning matrix.

If C is s.p.d. (in practice an approximation to A), we can let the inner product be defined by C, that is,

$$(\mathbf{x}, \mathbf{y}) = \mathbf{x}^T C \mathbf{y},$$

and consider the pseudoresiduals $\mathbf{h}^k = C^{-1}(A\mathbf{x}^k - \mathbf{b})$. Note that $C^{-1}A$ is s.p.d. with respect to this inner product. Then, replacing A in (11.25) and in the Krylov subspace by $C^{-1}A$, the preconditioned conjugate gradient method takes the form

$$\tau_k = \mathbf{h}^{k^T} C \mathbf{h}^k / \mathbf{d}^{k^T} A \mathbf{d}^k,$$

$$\mathbf{x}^{k+1} = \mathbf{x}^k + \tau_k \mathbf{d}^k,$$

(11.26) $$\mathbf{h}^{k+1} = \mathbf{h}^k + \tau_k C^{-1} A \mathbf{d}^k,$$

$$\beta_k = \mathbf{h}^{k+1^T} C \mathbf{h}^{k+1} / \mathbf{h}^{k^T} C \mathbf{h}^k,$$

$$\mathbf{d}^{k+1} = -\mathbf{h}^{k+1} + \beta_k \mathbf{d}^k,$$

and a pseudocode giving a partial description of the computer implementation can take the form:

$$\mathbf{x} := \mathbf{x}^0; \ \mathbf{r} := A\mathbf{x} - \mathbf{b};$$
$$\mathbf{h} := C^{-1}\mathbf{r};$$
$$\mathbf{d} := -\mathbf{h}; \ \delta 0 := \mathbf{r}^T \ \mathbf{h};$$
IF $\delta 0 \le \varepsilon$ THEN STOP;
$R:$ $\mathbf{h} := \ A\mathbf{d};$
$$\tau := \delta 0 / \ \mathbf{d}^T \ \mathbf{h};$$
$$\mathbf{x} := \mathbf{x} + \tau \mathbf{d};$$
$$\mathbf{r} := \mathbf{r} + \tau \mathbf{h};$$
$$\mathbf{h} := C^{-1}\mathbf{r}; \ \delta 1 := \mathbf{r}^T \ \mathbf{h};$$
IF $\delta 1 \le \varepsilon$ THEN STOP;
$$\beta := \delta 1 / \ \delta 0; \ \delta 0 := \delta 1;$$
$$\mathbf{d} := -\mathbf{h} + \beta \mathbf{d};$$
GOTO $R;$

The statement $\mathbf{h} := C^{-1}\mathbf{r}$ is to be interpreted as solving the system $C\mathbf{h} = \mathbf{r}$. The vectors $\mathbf{x}^0$ and $\mathbf{b}$ are input, or, alternatively, $\mathbf{b}$ is input and $\mathbf{x}^0 = C^{-1}\mathbf{b}$. Note that array $\mathbf{h}$ is used to store both $A\mathbf{d}^k$ and $\mathbf{h}^{k+1}$.

Since $f(\mathbf{x})$ in (11.12) now takes the form

$$f(\mathbf{x}) = \frac{1}{2}(\mathbf{h}, (C^{-1}A)^{-1}\mathbf{h}) = \frac{1}{2}(C^{-1}\mathbf{r})^T C (C^{-1}A)^{-1} C^{-1}\mathbf{r}$$

$$= \frac{1}{2}\mathbf{r}^T A^{-1}\mathbf{r},$$

we see that the above preconditioned algorithm *minimizes the same functional*, $\mathbf{r}^T A^{-1}\mathbf{r}$, where $\mathbf{r} = A\mathbf{x} - \mathbf{b}$ as for the unpreconditioned method but on the Krylov subspace $\{\mathbf{r}^0, AC^{-1}\mathbf{r}^0, \ldots, (AC^{-1})^k\mathbf{r}^0\}$. With an appropriate preconditioner, this Krylov subspace can generate vectors to minimize $f(\mathbf{x})$ much faster than for the unpreconditioned subspace. We can also consider a minimum residual algorithm where we minimize $\mathbf{r}^T\mathbf{r}$, and this will be presented in a more general context in the following chapter.

Note also that

$$\mathbf{d}^{k^T}A\mathbf{d}^j = (\mathbf{d}^k, C^{-1}A\mathbf{d}^j) = 0, \quad 0 \le j < k,$$

which is zero by Theorem 11.3(b), with A replaced by $C^{-1}A$. We collect the results for the preconditioned conjugate gradient algorithm in the next theorem.

Theorem 11.5 *Let A and C be symmetric and positive definite. Let $\mathbf{x}^0$ be an arbitrary vector and let $\mathbf{h}^0 = C^{-1}\mathbf{r}^0$ and $\mathbf{d}^0 = -\mathbf{h}^0$, where $\mathbf{r}^0 = A\mathbf{x}^0 - \mathbf{b}$, and for $k = 0, 1, \ldots$ let*

$$\tau_k = \mathbf{h}^{k^T}\mathbf{r}^k/\mathbf{d}^{k^T}A\mathbf{d}^k,$$

$$\mathbf{x}^{k+1} = \mathbf{x}^k + \tau_k\mathbf{d}^k,$$

$$\mathbf{r}^{k+1} = \mathbf{r}^k + \tau_k A\mathbf{d}^k,$$

$$\mathbf{h}^{k+1} = C^{-1}\mathbf{r}^{k+1},$$

$$\beta_k = \mathbf{h}^{k+1^T}\mathbf{r}^{k+1}/\mathbf{h}^{k^T}\mathbf{r}^k,$$

$$\mathbf{d}^{k+1} = -\mathbf{h}^{k+1} + \beta_k\mathbf{d}^k.$$

Then:

(a) $\mathbf{r}^{k+1^T}A^{-1}\mathbf{r}^{k+1} \le \mathbf{r}^T A^{-1}\mathbf{r}$, $\mathbf{r} = \mathbf{r}^0 + \mathbf{h}$ *for all* $\mathbf{h} \in \text{span}\{AC^{-1}\mathbf{r}^0, \ldots,$
$(AC^{-1})^{k+1}\mathbf{r}^0\}$, *where* $\mathbf{r}^{k+1} = A\mathbf{x}^{k+1} - \mathbf{b}$,
(b) $\mathbf{r}^{k+1^T}C^{-1}\mathbf{r}^j = 0$, $0 \le j \le k$, $\mathbf{r}^{k+1^T}\mathbf{d}^j = 0$, $0 \le j \le k$, $\mathbf{d}^{k+1^T}A\mathbf{d}^j = 0$, $0 \le j \le k$.

Proof This proof follows from Theorem 11.3, with $(\mathbf{u}, \mathbf{v}) = \mathbf{u}^T C\mathbf{v}$ and A replaced by $C^{-1}A$. $\diamond$

Note that the quantity that is minimized, $\mathbf{r}^{k+1^T}A^{-1}\mathbf{r}^{k+1}$, is not available in algorithm (11.26), so it cannot be used for a termination criterion. Instead, the algorithm uses $\mathbf{r}^{k+1^T}C^{-1}\mathbf{r}^{k+1} \le \varepsilon$, which is available. Alternatively, one could use $\mathbf{r}^{k+1^T}\mathbf{r}^{k+1} \le \varepsilon$, but this would require the computation of an extra inner product.

11.2.7 Efficient Implementation

For some preconditioners C, one finds that the error matrix $R = A - C$ is sparser than A, in which case it can be more efficient to modify the algorithm so that R and C are used instead of A and C. An implementation of such a modification is shown below. It assumes that C has the form

$$C = LDL^T,$$

where L is nonsingular, D is symmetric, and multiplications with D are inexpensive (typically D is diagonal or blockdiagonal). An efficient implementation can then be based on the following observations:

(a) $\mathbf{h}^{k^T}C\mathbf{h}^k = \mathbf{r}^{k^T}C^{-1}\mathbf{r}^k = (D^{-1}L^{-1}\mathbf{r}^k)^T D(D^{-1}L^{-1}\mathbf{r}^k)$.

(b) $-\mathbf{d}^{k^T}\mathbf{r}^k = -(C\mathbf{d}^k)^T\mathbf{h}^k = -(\mathbf{d}^k, \mathbf{h}^k) = (\mathbf{h}^k, \mathbf{h}^k) = \mathbf{h}^{k^T}C\mathbf{h}^k$, where (11.22b) has been used, whereby the pseudoresidual $\mathbf{h}^k$ has replaced $\mathbf{r}^k$.

(c) $\mathbf{d}^{k^T}A\mathbf{d}^k = (A\mathbf{d}^k)^T\mathbf{d}^k = (C^{-1}A\mathbf{d}^k)^T C\mathbf{d}^k$

$$= (C^{-1}A\mathbf{d}^k)^T(-\mathbf{r}^k + \beta_{k-1}C\mathbf{d}^{k-1})$$

$$= -(C^{-1}A\mathbf{d}^k)^T\mathbf{r}^k = -\mathbf{d}^{k^T}\mathbf{r}^k - (C^{-1}R\mathbf{d}^k)^T\mathbf{r}^k$$

$$= \mathbf{h}^{k^T}C\mathbf{h}^k - (L^{-1}R\mathbf{d}^k)^T(D^{-1}L^{-1}\mathbf{r}^k), \text{ where (b) has been}$$

used.

(d) $\tilde{\mathbf{x}}^{k+1} \equiv L^T\mathbf{x}^{k+1} = L^T\mathbf{x}^k + \tau_k L^T\mathbf{d}^k$.

(e) $\mathbf{r}^{k+1} = \mathbf{r}^k + \tau_k(C + R)\mathbf{d}^k$ or $\tilde{\mathbf{r}}^{k+1} \equiv D^{-1}L^{-1}\mathbf{r}^{k+1} = D^{-1}L^{-1}\mathbf{r}^k + \tau_k(L^T\mathbf{d}^k + D^{-1}L^{-1}R\mathbf{d}^k)$.

(f) $\tilde{\mathbf{d}}^{k+1} \equiv L^T\mathbf{d}^{k+1} = -L^T\mathbf{h}^{k+1} + \beta_k L^T\mathbf{d}^k = -D^{-1}L^{-1}\mathbf{r}^{k+1} + \beta_k(L^T\mathbf{d}^k) = -\tilde{\mathbf{r}}^{k+1} + \beta_k\tilde{\mathbf{d}}^k$.

(g) $D^{-1}L^{-1}R\mathbf{d}^k = D^{-1}L^{-1}RL^{-T}(L^T\mathbf{d}^k) = D^{-1}L^{-1}RL^{-T}\tilde{\mathbf{d}}^k = \tilde{R}\tilde{\mathbf{d}}^k$, where $\tilde{R} = D^{-1}L^{-1}RL^{-T}$.

The algorithm (11.26) now takes the following form:

$$\tau_k = \tilde{\mathbf{r}}^{k^T} D \tilde{\mathbf{r}}^k / [\tilde{\mathbf{r}}^{k^T} D \tilde{\mathbf{r}}^k - (\tilde{R}\tilde{\mathbf{d}}^k)^T D \tilde{\mathbf{r}}^k],$$

$$\tilde{\mathbf{x}}^{k+1} = \tilde{\mathbf{x}}^k + \tau_k \tilde{\mathbf{d}}^k,$$

(11.26a) $\qquad \tilde{\mathbf{r}}^{k+1} = \tilde{\mathbf{r}}^k + \tau_k (\tilde{\mathbf{d}}^k + \tilde{R}\tilde{\mathbf{d}}^k),$

$$\beta_k = \tilde{\mathbf{r}}^{k+1^T} D \tilde{\mathbf{r}}^{k+1} / \tilde{\mathbf{r}}^{k^T} D \tilde{\mathbf{r}}^k,$$

$$\tilde{\mathbf{d}}^{k+1} = -\tilde{\mathbf{r}}^{k+1} + \beta_k \tilde{\mathbf{d}}^k,$$

which can be implemented in a way similar to (11.25) and (11.26).

Note that $\tilde{R}\tilde{\mathbf{d}}^k$ and $D\tilde{\mathbf{r}}^k$ need to be computed only once. If L has the form $L = D^{-1} + \tilde{L}$, where $\tilde{L}$ is strictly lower triangular, then

$$\tilde{R}\tilde{\mathbf{d}}^k = (I + \tilde{L}D)^{-1} R(I + D\tilde{L}^T)^{-1} D\tilde{\mathbf{d}}^k,$$

which can be computed using matrix-vector multiplications with D and R and using back and forward substitution recursions with matrices $I + D\tilde{L}^T$ and $I + \tilde{L}D$, respectively. No multiplication with A occurs. Algorithm (11.26a) involves two inner products, three linked triads (i.e., computations of the form vector + scalar times vector), multiplications with D and $\tilde{R}$, and a vector addition. Note that the multiplication with $\tilde{R}$ involves the same steps as required for the multiplication with C^{-1} (i.e., solving the corresponding system) plus a multiplication with the matrix R.

Hence, comparing with the work involved in the computer implementation of algorithm (11.26), we see that the multiplication with A in the latter has been replaced with a multiplication with R, a multiplication with D, and an extra vector addition. Unless R is much sparser than A, this means that we may not have gained much in reducing the computational effort. However, note that the call for $\tilde{L}$ (and $\tilde{L}^T$), together with the call for R, occurs in only one place, which can save overhead (administrative) computer time if this takes place from an outer storage or from a subroutine. An additional advantage is that the algorithm (11.26a) is applicable both for point-wise and block-wise incomplete factorization methods.

Some initialization computations are needed. When the iterations are stopped, one computes $\mathbf{x}^{k+1} = L^{-1}\tilde{\mathbf{x}}^{k+1} = (I + D\tilde{L}^T)^{-1} D\tilde{\mathbf{x}}^{k+1}$. Variations of similar efficient implementation methods can be found in previous publications by Conrad and Wallach (1979), Eisenstat (1981), Axelsson and Barker (1984), Bank and Douglas (1985), and Ortega (1988). Not all of these are applicable for block-wise factorization methods.

11.2.8 The Conjugate Gradient Method for Singular and Nearly Singular Matrices

Assume that A is singular but positive semidefinite. Recall that $\mathcal{R}(A)$ and $\mathcal{N}(A)$ denote the range and nullspace of A, respectively. In the solution of a singular system $A\mathbf{x} = \mathbf{b}$, we must consider two effects of the singularity:

(a) Consistency: The system may not be consistent, i.e., $\mathbf{b} \notin \mathcal{R}(A)$, in which case there exists no solution. Lemma 1.7(a) shows that the system is consistent if and only if $\mathbf{b} \in \mathcal{N}(A^T)^\perp$. Recall that, by Remark 1.8, $\mathcal{R}(A) = \mathcal{N}(A^T)^\perp$.

(b) Nonuniqueness: If the system is consistent, there exist infinitely many solutions. All solutions have the form $\mathbf{x} = \hat{\mathbf{x}} + \mathbf{s}$, where $A\hat{\mathbf{x}} = \mathbf{b}$ and $\mathbf{s}$ is an arbitrary vector in $\mathcal{N}(A)$.

Example 11.3 Let $A = \begin{bmatrix} 1 & -1 \\ -2 & 2 \end{bmatrix}$ and $\mathbf{b} = \begin{bmatrix} -1 \\ 2 \end{bmatrix}$. Then $\mathcal{N}(A^T) = \text{span}\{\mathbf{v}\}$, $\mathbf{v} = \begin{bmatrix} 1 \\ \frac{1}{2} \end{bmatrix}$, and $\mathbf{b}^T\mathbf{v} = 0$, so $\mathbf{b} \in \mathcal{N}(A^T)^\perp$ and $A\mathbf{x} = \mathbf{b}$ is consistent. Its solution is $\mathbf{x} = c\mathbf{s} + \begin{bmatrix} 1 \\ 2 \end{bmatrix}$, where $\mathcal{N}(A) = \text{span}\{\mathbf{s}\}$, $\mathbf{s} = \begin{bmatrix} 1 \\ 1 \end{bmatrix}$ and c is an arbitrary constant.

Consider now a system $A\mathbf{x} = \mathbf{b}$, where A is singular and positive semidefinite. If $\mathbf{b} \notin \mathcal{R}(A)$, that is, the system is inconsistent, then we can solve the normal equations, $A^T A\mathbf{x} = \tilde{\mathbf{b}}$, $\tilde{\mathbf{b}} \equiv A^T\mathbf{b}$, which is consistent. However, its condition number $\lambda_{\max}(A^T A)/\lambda_{\min}^+(A^T A)$, where $\lambda_{\min}^+(A^T A)$ denotes the smallest positive eigenvalue, can be very large. It is the square of the corresponding condition number of A.

Alternatively, if vectors $\{\mathbf{v}_i\}$ which span $\mathcal{N}(A^T)$ are known, then we can proceed as follows. Assume for simplicity that $\mathcal{N}(A^T) = \text{span}\{\mathbf{v}\}$, i.e., $\dim \mathcal{N}(A^T) = 1$. Then we let

$$\tilde{\mathbf{b}} = \mathbf{b} - \frac{(\mathbf{b}, \mathbf{v})}{(\mathbf{v}, \mathbf{v})}\mathbf{v}.$$

Here $(\tilde{\mathbf{b}}, \mathbf{v}) = 0$, so $A\mathbf{x} = \tilde{\mathbf{b}}$ is a consistent system, which we then solve. Its solution is accepted as a solution of the given system $A\mathbf{x} = \mathbf{b}$.

Assume now that $A\mathbf{x} = \mathbf{b}$ is consistent. Let the initial vector $\mathbf{d}^0 = -A^T\mathbf{r}^0$, $\mathbf{r}^0 = A\mathbf{x}^0 - \mathbf{b}$, where $\mathbf{x}^0$ is an initial vector. Then $\mathbf{d}^0 \in \mathcal{N}(A)^\perp$. Let $\mathbf{x}^0 = \hat{\mathbf{x}}^0 + \mathbf{s}^0$, where $\hat{\mathbf{x}}^0 \in \mathcal{N}(A)^\perp$, $\mathbf{s}^0 \in \mathcal{N}(A)$. Then, due to the construction of the vector $\mathbf{x}^k$ in the conjugate gradient method, all vectors $\mathbf{x}^k$ will contain the same component s^0 in $\mathcal{N}(A)$ as the initial vector. Hence, the limit vector contains this component also. If we let $x^0 = 0$ or, if we make x^0 orthogonal to $\mathcal{N}(A)$ (assuming the vectors which span $\mathcal{N}(A)$ are known), then this component will be zero. If $Ax = b$ is consistent and A is symmetric, then $b \in \mathcal{R}(A) = \mathcal{N}(A^T)^\perp = \mathcal{N}(A)^\perp$, in which case we can let $d^0 = -r^0$, so $d^0 \in \mathcal{N}(A)^\perp$ and x^k will still contain the same component in $\mathcal{N}(A)$ as x^0. It will be seen in Chapter 13 that the rate of convergence of the conjugate gradient method to solve a consistent system with positive semidefinite matrix is determined by the positive part of the spectrum.

In many problems nearly singular systems occur where there are one or a few eigenvalues very close to zero. The remainder of the spectrum can contain eigenvalues in a wider interval. For notational simplicity, take the case where there is just one such eigenvalue, λ_1, and let v_1 be the corresponding eigenvector. The solution $\hat{x}$ of $Ax = b$, where A is s.p.d., satisfies

$$\hat{x} = \frac{b^T v_1}{\lambda_1} v_1 + \hat{x}_d,$$

where

$$\hat{x}_d = \sum_{i=2}^{n} \frac{b^T v_i}{\lambda_i} v_i$$

and λ_i, v_i, $i = 1, 2, \ldots, n$ are the eigensolutions of A where v_i are normalized. (v_i are not known, in general, so the above cannot be used to actually compute x.)

Frequently, the first component of the solution vector is dominating the solution, but in some applications it may be of interest to compute $\hat{x}_d$ accurately also. The conjugate gradient method applied directly on A will, in general, give poor approximations of $\hat{x}$ and $\hat{x}_d$. If v_1 (and hence also λ_1) is known, we can compute the first component, $(\lambda_1^{-1} b^T v_1) v_1$ of $\hat{x}$ and $\hat{x}_d$, separately. $\hat{x}_d$ can then be computed using an initial vector orthogonal to v_1 as mentioned above. If v_1 is not known, we can compute an (accurate) approximation of it, for instance, using the Lanczos method described in Section 11.3. Alternatively, the following method can be used.

11.2.9 The Augmented System Method

The system with A is bordered with a row and column to make the new matrix singular. Assume that we can find a vector v in $\mathbb{R}^n$ such that Av is nonnnegative, nonzero, and $\mathbf{v}^T \mathbf{v}_1 \neq 0$. Frequently $\mathbf{v} = \mathbf{e} = (1, 1, \ldots, 1)^T$ is a good choice. Then

$$\tilde{A} = \begin{bmatrix} A & -A\mathbf{v} \\ -\mathbf{v}^T A & \mathbf{v}^T A\mathbf{v} \end{bmatrix}$$

is singular for the vector $\begin{bmatrix} \mathbf{v} \\ 1 \end{bmatrix}$. Furthermore, the same vector is in $\mathcal{N}(\tilde{A}^T)$ and the right-hand side vector is orthogonal to it. Hence, the above bordered or augmented system is consistent. Its solution is $(\mathbf{x}^T, \mathbf{0})^T$, where $A\mathbf{x} = \mathbf{b}$. It can be seen that $\tilde{A}$ is, in general, better conditioned than A. If $\mathbf{v} = \mathbf{v}_1$, then $\begin{bmatrix} \mathbf{v} \\ 1 \end{bmatrix}$ is an eigenvector to $\tilde{A}$ with eigenvalue $\tilde{\lambda}_0 = 0$ and the remaining eigenvalues are then not perturbed. This follows because the corresponding eigenvectors are orthogonal to $\mathbf{v}$ and, therefore, $\mathbf{v}^T A\mathbf{v}_i = 0$, $i = 2, \ldots, n$. Besides these and the zero eigenvalue, there is an extra eigenvalue $\lambda_1(1 + \mathbf{v}_1^T \mathbf{v}_1)$, which replaces the smallest eigenvalue λ_1 of A and has the eigenvector $\begin{bmatrix} \mathbf{v}_1 \\ -\mathbf{v}_1^T \mathbf{v}_1 \end{bmatrix}$.

Example 11.4 Let

$$A = \begin{bmatrix} (1+\varepsilon) & -1 & & 0 \\ -1 & (2+\varepsilon) & -1 & \\ & \ddots & \ddots & \ddots \\ 0 & & -1 & (1+\varepsilon) \end{bmatrix}$$

have order n and consider the bordered matrix

$$\tilde{A} = \begin{bmatrix} (1+\varepsilon) & -1 & & & -\varepsilon \\ -1 & (2+\varepsilon) & -1 & & -\varepsilon \\ & & \ddots & & \vdots \\ 0 & 0 & \cdots & -1 & (1+\varepsilon) & -\varepsilon \\ -\varepsilon & -\varepsilon & \cdots & -\varepsilon & -\varepsilon & n\varepsilon \end{bmatrix}.$$

Let $\{\lambda_i\}_{i=1}^n$ and $\{\tilde{\lambda}_i\}_{i=0}^n$ be the eigenvalues of A and $\tilde{A}$, respectively. Then $\mathbf{v}_1 = (1, 1, \ldots, 1)^T$ is an eigenvector of A for $\lambda_1 = \varepsilon$. The remaining eigenvalues of A satisfy $\lambda_i > \varepsilon$, $i \geq 2$, and the eigenvalues of $\tilde{A}$ are $\tilde{\lambda}_0 = 0$, $\tilde{\lambda}_i = \lambda_i$, $i = 2, \ldots, n$ and $\tilde{\lambda}_1 = (1 + \mathbf{v}_1^T \mathbf{v}_1)\lambda_1 = (n+1)\varepsilon$. For the case where there are several nearly zero eigenvalues, one can border A with the corresponding number of columns and rows. The eigenvalues are then perturbed as stated in the next theorem.

Theorem 11.6 *Let A be of order $n \times n$ and V of order $n \times m$. Consider the augmented matrix*

$$\tilde{A} = \begin{bmatrix} A & -AV \\ -V^T A & V^T AV \end{bmatrix}.$$

(a) *$\tilde{A}$ has at least m zero eigenvalues. There are n eigenvalues $\tilde{\lambda}_i$ that equal those of $(I + VV^T)A$.*

(b) *If A is s.p.d., then to every eigenvalue λ_i of A there is a $\tilde{\lambda}_i$, such that $\tilde{\lambda}_i \geq \lambda_i$.*

(c) *If A is nonsingular and symmetric and $V = [\alpha_1 \mathbf{v}_1, \ldots, \alpha_m \mathbf{v}_m]$, where $\mathbf{v}_i$, $i = 1, 2, \ldots, m$ are the normalized eigenvectors of A, corresponding to λ_i, $i = 1, 2, \ldots, m$, then the nonzero eigenvalues of $\tilde{A}$ equal $\tilde{\lambda}_i = (1 + \alpha_i^2)\lambda_i$, $i = 1, 2, \ldots, m$, $\tilde{\lambda}_i = \lambda_i$, $i = m + 1, \ldots, n$.*

Proof We have

$$\begin{bmatrix} I_n & 0 \\ -V^T & I_m \end{bmatrix} \begin{bmatrix} A & 0 \\ 0 & 0 \end{bmatrix} \begin{bmatrix} I_n & -V \\ 0 & I_m \end{bmatrix} = \begin{bmatrix} A & -AV \\ -V^T A & V^T AV \end{bmatrix},$$

so $\tilde{A}$ is similar to

$$\begin{bmatrix} I_n & -V \\ 0 & I_m \end{bmatrix} \begin{bmatrix} I_n & 0 \\ -V^T & I_m \end{bmatrix} \begin{bmatrix} A & 0 \\ 0 & 0 \end{bmatrix} = \begin{bmatrix} (I + VV^T)A & 0 \\ -V^T A & 0 \end{bmatrix}.$$

This shows part (a). Part (b) follows directly from (a). If $A\mathbf{v}_i = \lambda_i \mathbf{v}_i$, $i = 1, 2, \ldots, n$, then

$$(I + VV^T)A\mathbf{v}_i = \lambda_i(I + VV^T)\mathbf{v}_i = \begin{cases} \lambda_i(1 + \alpha_i^2 \mathbf{v}_i^T \mathbf{v}_i)\mathbf{v}_i, i = 1, \ldots, m, \\ \lambda_i \mathbf{v}_i, \quad i = m + 1, \ldots, n, \end{cases}$$

where we have used the orthogonality of the eigenvalues. This proves part (c).
◇

If A is s.p.d., Theorem 11.6(c) shows that in order to increase the rate of convergence as much as possible when solving the bordered system with the conjugate gradient method, one generally lets λ_i, $i = 1, \ldots, m$ be the smallest eigenvalues and one chooses α_i such that $\tilde{\lambda}_i = \lambda_n$, $i = 1, \ldots, m$, where $\lambda_n = \max \lambda_i$, that is, $\alpha_i = (\lambda_n/\lambda_i - 1)^{\frac{1}{2}}$.

It is also seen from Theorem 11.6 that if instead of iterating on the augmented system, one iterates on the preconditioned system $B\mathbf{x} = \hat{\mathbf{b}}$, where

$B = (I + VV^T)A$ and $\hat{\mathbf{b}} = (I + VV^T)\mathbf{b}$, the rate of convergence of the methods will be the same. In the latter method, the lengths of the iteration vectors will be slightly smaller.

Although the method works well in case of just one vector, however, in general there can be practical difficulties in choosing appropriate vectors to border A with. On the other hand, the case, where there is just a "rank-one" nearly singularity, occurs in many important applications, such as in continuation methods to compute bifurcation points (see Keller, 1977), for instance.

It can be shown that we can let the preconditioner to A in the corresponding preconditioned conjugate gradient method also be singular, if $\mathcal{N}(C) \subset \mathcal{N}(A)$, where C is the preconditioning matrix. Naturally, we can let C be nonsingular, but for certain problems the preconditioner might be more efficient if we permit it to be singular when A is singular. For a further discussion of incomplete factorization methods for singular matrices, see Notay (1989).

11.3 The Lanczos Method for Generating A-Orthogonal Vectors

We have shown in the previous section a method to generate A-orthogonal vectors $\mathbf{d}^k$, that is, vectors that satisfy $(\mathbf{d}^k, A\mathbf{d}^j) = 0$ for all $j \neq k$, when A is selfadjoint w.r.t. the inner product. We present here another method for this purpose, attributed to Lanczos (1950, 1952). We show why this method can be efficient when we want to estimate the extreme eigenvalues of A, and we present a variant of the method that solves nonsymmetric linear systems.

Given $\mathbf{d}^0 \neq \mathbf{0}$, for $k = 0, 1, 2, \ldots$, let

$$(11.27) \qquad \mathbf{d}^{k+1} = A\mathbf{d}^k - r_k\mathbf{d}^k - s_{k-1}\mathbf{d}^{k-1},$$

where $s_{-1} = 0$ and

$$r_k = \frac{(A\mathbf{d}^k, A\mathbf{d}^k)}{(\mathbf{d}^k, A\mathbf{d}^k)}, \quad s_{k-1} = \frac{(A\mathbf{d}^k, A\mathbf{d}^{k-1})}{(\mathbf{d}^{k-1}, A\mathbf{d}^{k-1})}.$$

One can show that the vector sequence $\{\mathbf{d}^j\}$ is mutually $A-$ orthogonal and that the expression for s_{k-1} can be simplified as in the following lemma.

Lemma 11.7

   (a) For any $k \neq j$, we have $(\mathbf{d}^k, A\mathbf{d}^j) = 0$.

   (b) $s_{k-1} = \frac{(\mathbf{d}^k, A\mathbf{d}^k)}{(\mathbf{d}^{k-1}, A\mathbf{d}^{k-1})}$, $k \geq 1$.

Proof Note, first, that for $k = 0$,

$$(\mathbf{d}^1, A\mathbf{d}^0) = (A\mathbf{d}^0, A\mathbf{d}^0) - r_0(\mathbf{d}^0, A\mathbf{d}^0) = 0.$$

Assume that $(\mathbf{d}^k, A\mathbf{d}^j) = 0$ for all $j < k$. Then, by induction,

$$(11.28) \quad (\mathbf{d}^{k+1}, A\mathbf{d}^j) = (A\mathbf{d}^k, A\mathbf{d}^j) - r_k(\mathbf{d}^k, A\mathbf{d}^j) - s_{k-1}(\mathbf{d}^{k-1}, A\mathbf{d}^j),$$

and for $j = k$,

$$(\mathbf{d}^{k+1}, A\mathbf{d}^k) = -s_{k-1}(\mathbf{d}^{k-1}, A\mathbf{d}^k) = -s_{k-1}(A\mathbf{d}^{k-1}, \mathbf{d}^k) = 0$$

by the choice of r_k, the symmetry of A, and the induction assumption. For $j = k - 1$, $(\mathbf{d}^{k+1}, A\mathbf{d}^{k-1}) = 0$ by the choice of s_{k-1} and the induction assumption. For $j \le k - 2$, we use the induction assumption and

$$A\mathbf{d}^j = \mathbf{d}^{j+1} + r_j\mathbf{d}^j + s_{j-1}\mathbf{d}^{j-1}.$$

Together with (11.28), this shows that

$$(\mathbf{d}^{k+1}, A\mathbf{d}^j) = (A\mathbf{d}^k, \mathbf{d}^{j+1} + r_j\mathbf{d}^j + s_{j-1}\mathbf{d}^{j-1}).$$

The above and the symmetry of A imply that $(\mathbf{d}^{k+1}, A\mathbf{d}^j) = 0$, which completes the proof of part (a). Part (b) follows by

$$(A\mathbf{d}^k, A\mathbf{d}^{k-1}) = (A\mathbf{d}^k, \mathbf{d}^k + r_{k-1}\mathbf{d}^{k-1} + s_{k-2}\mathbf{d}^{k-2}) = (A\mathbf{d}^k, \mathbf{d}^k)$$

and the formula for s_{k-1} below (11.27). $\diamond$

If we assume the spectral representation of $\mathbf{d}^0$ includes all eigenvectors of A (which span $\mathbb{R}^n$ because A is selfadjoint), then $\text{span}\{\mathbf{d}^0, \mathbf{d}^1, \ldots, \mathbf{d}^{n-1}\} = \mathbb{R}^n$.

11.3.1 The Preconditioned Version

If the matrix has the form $B = C^{-1}A$, where C is s.p.d. and A is symmetric, then we can choose an inner product

$$(\mathbf{u}, \mathbf{v}) = \mathbf{u}^T C \mathbf{v},$$

and we get

$$(\mathbf{u}, B\mathbf{v}) = \mathbf{u}^T A\mathbf{v} = (C^{-1}A\mathbf{u})^T C\mathbf{v} = (B\mathbf{u}, \mathbf{v}),$$

that is, B is selfadjoint w.r.t. the inner product. The recursion (11.27) takes the form

$$\mathbf{d}^{k+1} = C^{-1}A\mathbf{d}^k - r_k\mathbf{d}^k - s_{k-1}\mathbf{d}^{k-1},$$

where

$$r_k = \frac{(C^{-1}A\mathbf{d}^k)^T A\mathbf{d}^k}{\mathbf{d}^{k^T}A\mathbf{d}^k}, \quad s_{k-1} = \frac{\mathbf{d}^{k^T}A\mathbf{d}^k}{\mathbf{d}^{k-1^T}A\mathbf{d}^{k-1}}.$$

We need to store $\mathbf{d}^{k-1}$, $\mathbf{d}^k$, $A\mathbf{d}^k$ and $C^{-1}A\mathbf{d}^k$ and to compute two inner products per iteration step, in addition to computing $A\mathbf{d}^k$ and $C^{-1}(A\mathbf{d}^k)$. Observe that for the above inner product, the vectors $\mathbf{d}^k$ are A-orthogonal w.r.t. the standard inner product, i.e.,

$$\mathbf{d}^{j^T}A\mathbf{d}^i = 0, \quad i \neq j,$$

because $\mathbf{d}^{j^T}A\mathbf{d}^i = (\mathbf{d}^j, B\mathbf{d}^i) = 0, \quad i \neq j.$

11.3.2 The C-Orthogonal Version

If $B = C^{-1}A$, A is s.p.d. and C is symmetric and nonsingular, then we can make the alternative choice of the inner product, namely,

$$(\mathbf{u}, \mathbf{v}) = \mathbf{u}^T C A^{-1} C \mathbf{v}.$$

In this case, we find

$$(B\mathbf{u}, B\mathbf{v}) = \mathbf{u}^T (C^{-1}A)^T (CA^{-1}C)(C^{-1}A\mathbf{v}) = \mathbf{u}^T A\mathbf{v},$$

and

$$(\mathbf{u}, B\mathbf{v}) = \mathbf{u}^T (CA^{-1}C)C^{-1}A\mathbf{v} = \mathbf{u}^T C\mathbf{v}.$$

Then the recurrence coefficients are found to be

$$r_k = \frac{\mathbf{d}^{k^T}A\mathbf{d}^k}{\mathbf{d}^{k^T}C\mathbf{d}^k}, \quad s_{k-1} = \frac{\mathbf{d}^{k^T}C\mathbf{d}^k}{\mathbf{d}^{k-1^T}C\mathbf{d}^{k-1}},$$

and the vectors $\{\mathbf{d}^k\}$ form a C-orthogonal set, i.e.,

$$\mathbf{d}^{j^T} C \mathbf{d}^i = 0, \quad i \neq j.$$

Clearly, this version is applicable also when $C = I$, the identity matrix, in which case the vectors $\{\mathbf{d}^k\}$ form an orthogonal set. When A and C are s.p.d., the behavior of the two versions, the A-orthogonal or the C-orthogonal, is similar. If A is indefinite, then the first method may break down, because the denominator, $\mathbf{d}^{k^T} A \mathbf{d}^k$, may become zero. Similarly, the second method can break down when C is indefinite.

If C is s.p.d. and A is symmetric but possibly indefinite, then the C-orthogonal version (in particular, the orthogonal direction vector version, where C is the identity matrix) is applicable. It cannot break down by a division by zero, but it stalls, i.e., fails to generate new direction vectors when $A\mathbf{d}^k = \mathbf{0}$. However, as suggested in Faddeev and Faddeeva (1963), in this case we can take an arbitrary vector, $\mathbf{y}$ and compute a new (nonzero) initial direction vector, $\mathbf{d}^{1,0} = \mathbf{y} - \sum_{i=0}^{k} c_i \mathbf{d}^i$, to be C-orthogonal to the previous vectors, i.e., to let

$$c_i = \frac{\mathbf{y}^T C \mathbf{d}^i}{\mathbf{d}^{i^T} C \mathbf{d}^i}, \quad i = 0, 1, \ldots, k.$$

The new direction vectors $\mathbf{d}^{1,i}$, $i = 1, 2 \ldots$ are computed with the C-orthogonal Lanzcos method and, because $\mathbf{d}^{1,0}$ is in the C-orthogonal complement of span $(\mathbf{d}^0, \mathbf{d}^1, \ldots, \mathbf{d}^k)$, it can be seen that these vectors will also be in this orthogonal complement. If $A\mathbf{d}^{1,k_1} = \mathbf{0}$ for some $k_1 \geq 1$, then we can repeat the above, and in this way the vector space is broken down into the direct sum of several pair-wise C-orthogonal subspaces, each spanned by a sequence of direction vectors computed until the sequence possibly stalls with $A\mathbf{d}^{i,k_i} = \mathbf{0}$ for some $k_i \geq 1$.

There are two major applications of the A-orthogonal (or C-orthogonal) set of vectors: (a) to solve linear systems and (b) to compute eigensolutions.

11.3.3 Using A-Orthogonal (or C-Orthogonal) Vectors to Solve Linear Systems

Let $\mathbf{x} = \sum_{j=0}^{n-1} \alpha_j \mathbf{d}^j$, where n is the order of A. Then the A-orthogonality of the vector set $\{\mathbf{d}^j\}$ and $(\mathbf{d}^j, A\mathbf{x} - \mathbf{b}) = 0$ show that

$$(11.29) \qquad \alpha_j = (\mathbf{d}^j, \mathbf{b})/(\mathbf{d}^j, A\mathbf{d}^j), \quad j = 0, 1, \ldots, n - 1.$$

Example 11.5 Let $C = \begin{bmatrix} \frac{1}{2} & 0 & 0 \\ 0 & 1 & 0 \\ 0 & 0 & \frac{1}{2} \end{bmatrix}$, $A = \begin{bmatrix} 2 & -1 & 0 \\ -1 & 2 & -1 \\ 0 & -1 & 1 \end{bmatrix}$, and $\mathbf{d}^0 = [0, 1, -1]^T$. Then one finds

$$\mathbf{d}^{0^T} A \mathbf{d}^0 = 5, \quad C^{-1} A \mathbf{d}^0 = [-2, 3, -4]^T, \quad r_0 = \frac{19}{5},$$

$$\mathbf{d}^1 = \frac{1}{5}[-10, -4, -1]^T,$$

$$\mathbf{d}^{1^T} A \mathbf{d}^1 = \frac{29}{5}, \quad C^{-1} A \mathbf{d}^1 = \frac{1}{5}[-32, 3, 6]^T, \quad r_1 = \frac{539}{145}, \quad s_0 = \frac{29}{25},$$

$$\mathbf{d}^2 = \frac{10}{29}[3, 7, 9]^T,$$

$$\mathbf{d}^{2^T} A \mathbf{d}^2 = \frac{100}{29}, \quad C^{-1} A \mathbf{d}^2 = \frac{10}{29}[-2, 2, 4]^T, \quad r_2 = \frac{14}{29}, \quad s_1 = \frac{500}{29^2}.$$

For later use, note that the matrix $H = \text{tridiag}(1, r_i, s_i)$,

$$H = \begin{bmatrix} \frac{19}{5} & \frac{29}{25} & 0 \\ 1 & \frac{539}{145} & \frac{500}{29^2} \\ 0 & 1 & \frac{14}{29} \end{bmatrix},$$

has the characteristic polynomial

$$\det(A - \lambda I) = -\lambda^3 + 8\lambda^2 - 16\lambda + 4,$$

which is identical to the characteristic polynomial of $C^{-1}A$. To solve the system $C^{-1}A\mathbf{x} = \mathbf{b} = [2, 0, 0]^T$, $\mathbf{x} = \sum_{j=0}^{2} \alpha_j \mathbf{d}^j$, one finds

$$C^{-1} A \mathbf{x} = \sum_{j=0}^{2} \alpha_j C^{-1} A \mathbf{d}^j = \mathbf{b}.$$

Taking inner products with $C\mathbf{d}^j$, we get

$$\alpha_0 = 0, \quad \alpha_1 = -\frac{5}{29}, \quad \alpha_2 = \frac{3}{10}$$

and

$$\mathbf{x} = \sum_{j=0}^{2} \alpha_j \mathbf{d}^j = [1, 1, 1]^T.$$

When n is large, there can be a significant loss of orthogonality among the vectors $\mathbf{d}^j$ because of rounding errors. If one attempts to compute an approximation to $\mathbf{x}$ using only a few of the vectors, then this approximation cannot be expected to be accurate unless the condition number of A is moderate and $\mathbf{d}^0$ is related to the system, such as $\mathbf{d}^0 = \mathbf{r}^0 = A\mathbf{x}^0 - \mathbf{b}$, where $\mathbf{x}^0$ is an initial vector ($\mathbf{x}^0 = \mathbf{0}$ is a possible choice).

11.3.4 Using A-Orthogonal (or C-Orthogonal) Vectors to Compute Eigensolutions

Another application of the A-orthogonal set of vectors is to compute the eigenvalues of A. Note, first, that when $(\mathbf{d}^j, A\mathbf{d}^j) \neq 0$, $0 \leq j \leq n - 1$, the vector set $\{\mathbf{d}^j\}_{j=0}^{n-1}$ is linearly independent, so the matrix

$$Q = [\mathbf{d}^0, \mathbf{d}^1, \ldots, \mathbf{d}^{n-1}]$$

is nonsingular. Furthermore, the recursion (11.27) shows that

$$A[\mathbf{d}^0, \mathbf{d}^1, \ldots, \mathbf{d}^{n-1}] = [\mathbf{d}^0, \mathbf{d}^1, \ldots, \mathbf{d}^{n-1}] \begin{bmatrix} r_0 & s_0 & & O \\ 1 & r_1 & s_1 & \\ & \ddots & \ddots & \ddots \\ O & & 1 & r_{n-1} \end{bmatrix},$$

that is,

$$AQ = QH,$$

where $H = \mathrm{tridiag}(1, r_i, s_i)$. Therefore, $H = Q^{-1}AQ$ and this similarity transformation shows that H has the same eigenvalues as A (compare Example 11.5). *Hence Q reduces A to tridiagonal form.*

Householder similarity transformation matrices (discussed in Exercise 2.27) can also be used to reduce A to a tridiagonal form. However, this approach is impractical if A large and sparse, because Householder transformations tend to destroy sparsity and, as a result, unacceptably large, dense matrices arise during the reduction.

11.3.5 Normalized Form of the Preconditioned Version

If the vectors $\{\mathbf{d}^k\}$ are normalized, the matrix Q becomes C-orthonormal, that is, $Q^T C Q = I$, in case the normalization takes place w.r.t. the matrix C. As is readily seen, the preconditioned form of the recursion (11.27), takes now the following form:

$$\tilde{s}_k \tilde{\mathbf{d}}^{k+1} := \mathbf{d}^{k+1} = C^{-1} A \tilde{\mathbf{d}}^k - \tilde{r}_k \tilde{\mathbf{d}}^k - \tilde{s}_{k-1} \tilde{\mathbf{d}}^{k-1},$$

where $\tilde{r}_k = \tilde{\mathbf{d}}^{k^T} A \tilde{\mathbf{d}}^k$ and $\tilde{s}_k = \tilde{\mathbf{d}}^{k+1^T} C \tilde{\mathbf{d}}^{k+1}$. The vectors $\{\tilde{\mathbf{d}}^k\}$ now satisfy $\tilde{\mathbf{d}}^{j^T} C \tilde{\mathbf{d}}^i = \delta_{ij}$ and the matrix H becomes symmetric, $H = \text{tridiag}(\tilde{s}_{i-1}, \tilde{r}_i, \tilde{s}_i)$.

As we have seen, the eigenvalues of A can be computed by computing the eigenvalues of H. In practice, however, the entries of H are perturbed by round-off errors and the eigenvalues will only be approximations of the exact eigenvalues of A. What makes the above method particularly interesting is that, as it turns out, one can often get accurate estimates of some of the eigenvalues of A—in particular, of the extreme eigenvalues—from the portion of the matrix H obtained at the kth step,

$$H_k = \begin{bmatrix} r_0 & s_0 & & 0 \\ 1 & r_1 & s_1 & \\ & \ddots & \ddots & \ddots \\ 0 & & 1 & r_k \end{bmatrix}$$

for k considerably smaller than $n - 1$. For instance, as can be seen from the next theorem, the smallest eigenvalue of A will be approximated accurately by the smallest eigenvalue of H_k if the condition number of A is small and if the first two disjoint eigenvalues of A are well separated. The theorem is presented here for the case where the smallest eigenvalue has multiplicity $r - 1$.

Theorem 11.8 (Lanczos, 1950, Kaniel, 1966, and Paige, 1971) *Let A be symmetric, with eigenvalues $\lambda_1 = \lambda_2 = \ldots = \lambda_{r-1} < \lambda_r \leq \lambda_{r+1} \leq \ldots \leq \lambda_n$, where $r \geq 2$, and let Z_1 be the vector space spanned by the eigenvectors corresponding to $\lambda_1, \ldots, \lambda_{r-1}$. Further, let μ_1 be the smallest eigenvalue of H_k for some $k \geq 1$. Then, the following estimate holds:*

$$0 \leq \mu_1 - \lambda_1 \leq (\lambda_n - \lambda_1) \left[\frac{\tan \phi_1}{T_k \left(\frac{\kappa_r + 1 - 2\lambda_1/\lambda_r}{\kappa_r - 1} \right)} \right]^2,$$

where $\kappa_r = \lambda_n/\lambda_r$, ϕ_1 is the angle between the vector $\mathbf{d}^0$ and Z_1, that is,

$$\cos \phi_1 = \max_{\mathbf{z}_1 \in Z_1} |(\mathbf{d}^0, \mathbf{z}_1)| / \{(\mathbf{d}^0, \mathbf{d}^0)(\mathbf{z}_1, \mathbf{z}_1)\}^{\frac{1}{2}}$$

and T_k is the Chebyshev polynomial of degree k.

Proof Let $\mathbf{v}_i$ be the normalized eigenvectors of A, that is, let $A\mathbf{v}_i = \lambda_i \mathbf{v}_i$, $\mathbf{v}_i \neq \mathbf{0}$, $i = 1, 2, \ldots, n$ and $(\mathbf{v}_i, \mathbf{v}_j) = \delta_{ij}$, and let

$$(11.30) \qquad \mathbf{d}^0 = \sum_{i=1}^{n} \alpha_i \mathbf{v}_i.$$

Further, let

$$V_k = \text{span}\{\mathbf{d}^0, \mathbf{d}^1, \ldots, \mathbf{d}^k\}.$$

This is the Krylov space for $\mathbf{d}^0$ and A. The Courant-Fischer theorem shows that

$$\mu_1 = \min_{\substack{\mathbf{x} \neq 0 \\ \mathbf{x} \in \mathbb{R}^{k+1}}} \frac{\mathbf{x}^T H_k \mathbf{x}}{\mathbf{x}^T \mathbf{x}} = \min_{\substack{\mathbf{y} \neq 0 \\ \mathbf{y} \in V_k}} \frac{\mathbf{y}^T A \mathbf{y}}{\mathbf{y}^T \mathbf{y}}$$

and that

$$\lambda_1 = \min_{\substack{\mathbf{x} \neq 0 \\ \mathbf{x} \in \mathbb{R}^n}} \frac{\mathbf{x}^T A \mathbf{x}}{\mathbf{x}^T \mathbf{x}} = \min_{\substack{\mathbf{y} \neq 0 \\ \mathbf{y} \in V_{n-1}}} \frac{\mathbf{y}^T A \mathbf{y}}{\mathbf{y}^T \mathbf{y}}.$$

Let P_k denote the set of polynomials of degree k at most. The Lanczos method to generate the vector sequence $\{\mathbf{d}^k\}$ shows that any vector $\mathbf{y} \in V_k$ can be represented by a polynomial $p \in P_k$, such that $\mathbf{y} = p(A)\mathbf{d}^0$. Hence,

$$\mu_1 = \min_{p \in P_k} \frac{\mathbf{d}^{0T} p(A^T) A p(A) \mathbf{d}^0}{\mathbf{d}^{0T} p(A^T) p(A) \mathbf{d}^0}.$$

Using (11.30), we find

$$\mu_1 = \min_{p \in P_k} \frac{\sum_{i=1}^{n} \alpha_i^2 \lambda_i p(\lambda_i)^2}{\sum_{i=1}^{n} \alpha_i^2 p(\lambda_i)^2}.$$

Hence, for any polynomial,

$$\mu_1 \leq \lambda_1 + \sum_r^n \alpha_i^2 (\lambda_i - \lambda_1) p(\lambda_i)^2 / \sum_1^n \alpha_i^2 p(\lambda_i)^2$$

$$\leq \lambda_1 + (\lambda_n - \lambda_1) \sum_r^n \alpha_i^2 p(\lambda_i)^2 / \sum_1^n \alpha_i^2 p(\lambda_i)^2.$$

$$= \lambda_1 + (\lambda_n - \lambda_1) \frac{\sum_r^n \alpha_i^2 p(\lambda_i)^2}{\left(\sum_1^{r-1} \alpha_i^2\right) p(\lambda_1)^2 + \sum_r^n \alpha_i^2 p(\lambda_i)^2}.$$

To get a smallest upper bound for any sequence of numbers $\alpha_1, \ldots, \alpha_n$, we take

$$p(\lambda) = T_k \left(\frac{\lambda_n + \lambda_r - 2\lambda}{\lambda_n - \lambda_r} \right).$$

Then, $|p(\lambda_n)| = 1$, $|p(\lambda_i)| \leq 1, i = r, r+1, \ldots, n-1$, while $|p(\lambda_1)|$ is the largest possible of all so normalized polynomials (see Appendix B). For this choice, we find

$$\mu_1 \leq \lambda_1 + (\lambda_n - \lambda_1) \frac{\sum_r^n \alpha_i^2}{p(\lambda_1)^2 \sum_1^{r-1} \alpha_i^2 + \sum_r^n \alpha_i^2}$$

$$\leq \lambda_1 + (\lambda_n - \lambda_1) \frac{\sum_r^n \alpha_i^2 / \sum_1^{r-1} \alpha_i^2}{p(\lambda_1)^2}.$$

Finally, note that

$$\sum_r^n \alpha_i^2 / \sum_1^{r-1} \alpha_i^2 = [\tan(\phi_1)]^2$$

and

$$p(\lambda_1) = T_k \left(\frac{\kappa_r + 1 - 2\frac{\lambda_1}{\lambda_r}}{\kappa_r - 1} \right). \qquad \diamond$$

Remark 11.9 A similar estimate holds for the largest eigenvalue, μ_{k+1} of H_k. For related estimates, see Parlett (1980), Saad (1980), and Golub and Van Loan (1989).

Theorem 11.8 shows that when A is well-conditioned, for instance, when $A := M^{-1}A$ is a preconditioned matrix with an accurate preconditioner, k can be chosen quite small and we would still have accurate estimates. In fact, if the condition number κ of A (or of $M^{-1}A$) does not depend on the order of A (i.e., if M is a spectrally equivalent preconditioner to A) and if, in addition, the gap between λ_r and λ_1 is uniform in n (that is, $\lambda_r - \lambda_1 \geq \delta\lambda_1$, where $\delta > 0$ does not depend on n and $\tan\phi_1 \leq$ const, where const does not depend on n), then we can choose k to be independent of n also. For example, suppose that we want to compute μ_1 with a relative accuracy ε, such that

$$0 \leq \mu_1 - \lambda_1 \leq \varepsilon\lambda_1.$$

Then, noting that $\kappa_r \leq \kappa$, where $\kappa = \lambda_n/\lambda_1$, Theorem 11.7 shows that it suffices to choose k, such that

$$(\mu_1 - \lambda_1)/\lambda_1 \leq (\kappa - 1)\left[\frac{\gamma}{T_k(\frac{\kappa+\frac{\lambda_r}{\lambda_1}-2}{\kappa-1})}\right]^2 \leq \varepsilon.$$

That is, we need to find the smallest k, such that

$$T_k(\frac{\kappa + \frac{\lambda_r}{\lambda_1} - 2}{\kappa - 1}) \geq \gamma(\frac{\kappa - 1}{\varepsilon})^{1/2},$$

where $\gamma = \tan\phi_1$. But such a k is smaller than the smallest k, such that

$$T_k(1 + \frac{\delta}{\kappa - 1}) \geq \gamma(\frac{\kappa - 1}{\varepsilon})^{1/2},$$

and this k does not grow with n, because by assumption, δ, κ, and γ do not depend on n.

We see also that for $k \leq n - 1$, the following matrix equality holds:

$$(11.31) \qquad AQ_k = Q_kH_k + [\mathbf{0}, \ldots, \mathbf{0}, \mathbf{d}^{k+1}],$$

where $Q_k = [\mathbf{d}^0, \ldots, \mathbf{d}^k]$, a matrix of order $n \times (k + 1)$.

The computation of eigenvalues $\tilde{\lambda}_j$ of the tridiagonal matrix H_k can be done with many algorithms, such as bisection algorithms, (see Wilkinson, 1961). In more recent years, some variants of a cyclic reduction algorithm have become popular (see Cuppen, 1981, for instance).

The corresponding approximations, $\mathbf{v}^j$ of the eigenvectors can be found from the eigenvectors $\mathbf{w}^j$ of H_k as $\mathbf{v}^j = Q_k\mathbf{w}^j$, because if $H_k\mathbf{w}^j = \tilde{\lambda}_j\mathbf{w}^j$, then (11.31) shows that $A\mathbf{v}^j = AQ_k\mathbf{w}^j = Q_kH_k\mathbf{w}^j + \mathbf{w}^j_{k+1}\mathbf{d}^{k+1} = \tilde{\lambda}_k\mathbf{v}^j + \mathbf{w}^j_{k+1}\mathbf{d}^{k+1}$. Hence,

$$\text{(11.32)} \qquad \|A\mathbf{v}^j - \tilde{\lambda}_k \mathbf{v}^j\| = |w_{k+1}^j| \|\mathbf{d}^{k+1}\|,$$

which can be used as an inexpensive way to check the convergence of the eigenpair solutions. As is shown in Appendix A, the error in the eigenvalue is of the order of the square of the residual norm in (11.32), while the angle between the exact eigenvector and the approximate one is of the same order as the residual norm.

If we let $\mathbf{d}^0 = \mathbf{r}^0$, then the Lanczos method with A-orthogonal vectors with $(\mathbf{d}^j, A\mathbf{d}^j) \neq 0$ generates the same Krylov space $V_k = \text{span}\{\mathbf{r}^0, A\mathbf{r}^0 \dots, A^k\mathbf{r}^0\}$ as the conjugate gradient method. Furthermore, if we let $\mathbf{x}^{k+1}$ be defined as the approximation for which the residual $\mathbf{r}^{k+1} = A\mathbf{x}^{k+1} - \mathbf{b}$ becomes orthogonal to the Krylov space, i.e.,

$$(\mathbf{r}^{k+1}, \mathbf{g}) = 0 \text{ for all } \mathbf{g} \in V_k,$$

then $\mathbf{x}^{k+1}$ is identical to the corresponding approximation in the conjugate gradient method. For a derivation of the conjugate gradient method based on the Lanczos method, see Cullum and Willoughby (1977, 1985) and Chandra (1978).

Clearly, the Lanczos method is even more closely related to the three-term recurrence form in Section 11.1. Note, however, that there we constructed orthogonal (i.e., not A-orthogonal) residual vectors, but with a proper choice of inner products and with $\mathbf{d}^0 = \mathbf{r}^0$, the methods become identical. This can be seen if we write (11.3) in the form

$$\beta_k^{-1} \mathbf{r}^{k+1} = \mu_k \mathbf{r}^k + (\beta_k^{-1} - \mu_k)\mathbf{r}^{k-1} - A\mathbf{r}^k.$$

Hence, the vectors $\mathbf{r}^k$ and $\mathbf{d}^k$ differ only in a scaling factor.

11.3.6 Several Right-Hand Sides

The two uses of A-orthogonal vectors can be combined—i.e., during the generation of the vector sequence $\{\mathbf{d}^k\}$ in (11.27) or $\{\mathbf{r}^k\}$ in (11.3), one can compute the condition number of the current matrix H_k. Using (5.32), this condition number estimate can be used to estimate the number of iterations (k_{iter}) that are needed for a given relative precision (ε), assuming that we let the initial vector $\mathbf{d}^0 = -\mathbf{r}^0$, where $\mathbf{r}_0$ is the initial residual. It will be shown in Chapter 13 that the expression in (5.32) is an upper bound also for the number of iterations in the conjugate gradient and, hence, the Lanczos method, if a proper norm of the residuals is chosen.

During the generation of the sequence $\mathbf{d}^k$, one computes the coordinates $\alpha_k = \frac{(\mathbf{b}, \mathbf{d}^k)}{(\mathbf{d}^k, A\mathbf{d}^k)}$ of the solution vector, $\mathbf{x} = \sum_{k=1}^{k_{\text{iter}}} \alpha_k \mathbf{d}^k$, and updates the current value of the approximate solution vector accordingly. This will save storage of the vectors $\mathbf{x}^{k-1}$, $\mathbf{r}^{k-1}$, $\mathbf{r}^k$, which are required if (11.2), (11.3) are used. To solve several systems with the same matrix, all can be computed in this way using the same vectors $\mathbf{d}^k$ if the different right-hand sides differ only little. If they differ much, we must recompute the search directions for each new right-hand side, but this computation can take place in parallel for the different right-hand sides.

11.3.7 Nonsymmetric Matrices

If A is nonsymmetric, then it can be seen that the three-term recursion in (11.3) or in the Lanczos method (11.27) will not generate an orthogonal, or conjugately orthogonal, set of vectors in general. Let us now consider a version of the Lanczos method for solving nonsymmetric problems, *the Lanczos biorthogonal vector algorithm*, where biorthogonal sets of vectors are generated and the three-term recursion form is still maintained.

Instead of considering only the solution of the system

$$(11.33a) \qquad\qquad A\mathbf{x} = \mathbf{b},$$

we also consider an auxiliary system,

$$(11.33b) \qquad\qquad A^*\tilde{\mathbf{x}} = \tilde{\mathbf{b}},$$

where A^* is the adjoint of A *w.r.t. the inner product*. In some problems, one is interested in solving both systems. If, however, one wants only the solution of $A\mathbf{x} = \mathbf{b}$, there is no need to choose $\tilde{\mathbf{b}}$, because we need only an initial residual $\tilde{\mathbf{r}}^0 = A^*\tilde{\mathbf{x}}^0 - \tilde{\mathbf{b}}$, which can be chosen independently. Also, we do not need to compute approximations of $\tilde{\mathbf{x}}$. The three-term recurrence relations in (11.2) and (11.3) for (11.33a) and (11.33b) are then

$$(11.34) \qquad \begin{cases} \mathbf{x}^{k+1} = \alpha_k \mathbf{x}^k + (1 - \alpha_k)\mathbf{x}^{k-1} - \beta_k \mathbf{r}^k, \\[2mm] \mathbf{r}^{k+1} = \alpha_k \mathbf{r}^k + (1 - \alpha_k)\mathbf{r}^{k-1} - \beta_k A\mathbf{r}^k, \\[2mm] \tilde{\mathbf{r}}^{k+1} = \alpha_k \tilde{\mathbf{r}}^k + (1 - \alpha_k)\tilde{\mathbf{r}}^{k-1} - \beta_k A^*\tilde{\mathbf{r}}^k, \end{cases}$$

$k = 0, 1, \ldots$, where $\mathbf{x}^0$ is arbitrary, $\mathbf{r}^0 = A\mathbf{x}^0 - \mathbf{b}$, $\alpha_0 = 1$, and $\tilde{\mathbf{r}}^0$ is arbitrary. As we shall see, however, there are some restrictions on $\tilde{\mathbf{r}}^0$ to avoid a breakdown of the algorithm. In the first place, as will be seen below, $\tilde{\mathbf{r}}^0$ must not be orthogonal to $\mathbf{r}^0$. A good choice is $\tilde{\mathbf{r}}^0 = \mathbf{r}^0$, at least if A is almost symmetric. We now seek to choose the parameters α_k, β_k, such that the sets $\mathbf{r}^0, \mathbf{r}^1, \ldots$ and $\tilde{\mathbf{r}}^0, \tilde{\mathbf{r}}^1, \ldots$ become biorthogonal in the sense that for $i \neq j$, we have

$$(11.35) \qquad\qquad (\mathbf{r}^i, \tilde{\mathbf{r}}^j) = 0.$$

Assume by induction that (11.35) holds for $i \neq j$, $0 \leq i \leq k$, $0 \leq j \leq k$. Then, $(\mathbf{r}^{k+1}, \tilde{\mathbf{r}}^k) = 0$ shows that (cf. Section 11.1)

$$\alpha_k = \mu_k \beta_k,$$

where

$$(11.36) \qquad\qquad \mu_k = (A\mathbf{r}^k, \tilde{\mathbf{r}}^k)/(\mathbf{r}^k, \tilde{\mathbf{r}}^k),$$

and $(\mathbf{r}^{k+1}, \tilde{\mathbf{r}}^{k-1}) = 0$ shows that

$$(11.37) \qquad\qquad (1 - \alpha_k)(\mathbf{r}^{k-1}, \tilde{\mathbf{r}}^{k-1}) = \beta_k(A\mathbf{r}^k, \tilde{\mathbf{r}}^{k-1})$$

or

$$(\beta_k^{-1} - \mu_k)(\mathbf{r}^{k-1}, \tilde{\mathbf{r}}^{k-1}) = (\mathbf{r}^k, A^*\tilde{\mathbf{r}}^{k-1}) = \beta_{k-1}^{-1}(\mathbf{r}^k, \tilde{\mathbf{r}}^k).$$

Hence,

$$\beta_0 = \mu_0^{-1} = (\mathbf{r}^0, \tilde{\mathbf{r}}^0)/(A\mathbf{r}^0, \tilde{\mathbf{r}}^0)$$

and

$$\beta_k^{-1} = \mu_k - \frac{\delta_k}{\delta_{k-1}} \beta_{k-1}^{-1}, k = 1, 2, \ldots,$$

where

$$(11.38) \qquad\qquad \delta_k = (\mathbf{r}^k, \tilde{\mathbf{r}}^k).$$

Note that the latter recurrence relation is the same as in (11.5). In fact, if $A^* = A$ and $\tilde{\mathbf{r}}^0 = \mathbf{r}^0$, then the recurrence (11.34) is identical to the recurrence (11.2), (11.3). The equality (11.37) shows that the recurrence relations require that $(\mathbf{r}^{k-1}, \tilde{\mathbf{r}}^{k-1}) \neq 0$ and that $\alpha_k \neq 1$. If $\alpha_k = 1$, then (11.37) and (11.34) imply

$$0 = \beta_k(A\mathbf{r}^k, \tilde{\mathbf{r}}^{k-1}) = -\beta_k \beta_{k-1}^{-1}(\mathbf{r}^k, \tilde{\mathbf{r}}^k).$$

Since $\beta_k \neq 0$ (because $\alpha_k = 1$), we must then have $(\mathbf{r}^k, \tilde{\mathbf{r}}^k) = 0$. However, if $\mathbf{r}^k = \mathbf{0}$, then we have already found a solution to $A\mathbf{x} = \mathbf{b}$, and if $\tilde{\mathbf{r}}^k = \mathbf{0}$, then we have found a solution to the auxiliary system. The case where $(\mathbf{r}^k, \tilde{\mathbf{r}}^k) = 0$, although neither $\mathbf{r}^k = \mathbf{0}$ nor $\tilde{\mathbf{r}}^k = \mathbf{0}$, means a breakdown of algorithm (11.34). (It has been called a *serious* breakdown by Wilkinson, 1961, p. 389.)

Note that a breakdown is the result of the choice of $\tilde{\mathbf{r}}_0$, not of some defect of the matrix A. For any matrix, it is possible to generate breakdowns by choosing $\tilde{\mathbf{r}}^0$ appropriately related to $\mathbf{r}_0$; in particular, the algorithm breaks down at the first step if $\tilde{\mathbf{r}}_0$ and $\mathbf{r}_0$ are orthogonal. If $(\tilde{\mathbf{r}}^0, \mathbf{r}^0) \neq 0$, it can be seen that such a breakdown cannot occur if A is nondegenerate—i.e., it contains a complete eigenvector space and $\tilde{\mathbf{r}}^0$ has a spectral representation with nonzero components in each of the vectors. Hence, if none of the numbers $(\mathbf{r}^0, \tilde{\mathbf{r}}^0)$, $(\mathbf{r}^1, \tilde{\mathbf{r}}^1)$, ... vanishes, then the process will converge in at most n iterations, because the vectors $\mathbf{r}^0, \mathbf{r}^1, \ldots$ are linearly independent and the condition $(\mathbf{r}^n, \tilde{\mathbf{r}}^i) = 0$, $i = 0, 1, \ldots, n-1$ implies that $\mathbf{r}^n = \mathbf{0}$. (It can, in fact, be seen that the method converges in a number of steps equal to the degree of the minimal polynomial of the restriction of A to the maximum Krylov space generated by A from $\mathbf{r}^0$, cf. Chapter 12.) If these conditions are violated, however, then the method can break down. Furthermore, near breakdowns (i.e., divisions with small positive numbers) can occur, and this can cause cancellation of significant digits and a loss of orthogonality among the vectors. This result implies that the conditions for the three-term recursion and for termination do not hold.

Also, for nonsymmetric and indefinite problems, there is in general no minimization property associated with the Lanczos (or the three-term recurrence or the classical conjugate gradient) method and, therefore, there is *no monotone convergence* of the residuals.

Some improvements of the above algorithm will be presented below. An alternative presentation (see Young, Hayes, and Jea, 1981) of the above biorthogonal Lanczos method is possible if we consider the extended matrix $\begin{bmatrix} A & O \\ O & A^* \end{bmatrix}$ and the indefinite inner product with the matrix $\begin{bmatrix} O & I \\ I & O \end{bmatrix}$. Then it can be seen that the three-term recurrence or the conjugate gradient method give the same approximations as in the biorthogonal methods.

Remark 11.10 The Lanczos biorthogonal vector algorithm to solve a linear system $A\mathbf{x} = \mathbf{b}$ can also be applied with preconditioning matrices. We replace A in the algorithm with $B = C^{-1}A$, where C is the preconditioner. Alternatively, if C is s.p.d.—for instance, if C is the symmetric part of A and this is positive definite—then we can let the innerproduct, $(\cdot, \cdot)$ in (11.36) and (11.38) be defined by C, and the algorithm simplifies in a similar way as discussed below (11.11).

Instead of using the three-term recurrence relations (for A and for A^*), we can use the recurrence relations corresponding to the standard conjugate gradient method (for A and for A^*). This method, called the *biconjugate gradient method* (BCG) was presented by Fletcher (1976); see also Lanczos (1952). It takes the following form [cf. (11.25)]:

$$\tau_k = (\tilde{\mathbf{r}}^k, \mathbf{r}^k)/(\tilde{\mathbf{d}}^k, A\mathbf{d}^k);$$

$$\mathbf{x}^{k+1} = \mathbf{x}^k + \tau_k \mathbf{d}^k;$$

$$\mathbf{r}^{k+1} = \mathbf{r}^k + \tau_k A\mathbf{d}^k;$$

(11.39) $$\tilde{\mathbf{r}}^{k+1} = \tilde{\mathbf{r}}^k + \tau_k A^* \tilde{\mathbf{d}}^k;$$

$$\beta_k = (\tilde{\mathbf{r}}^{k+1}, \mathbf{r}^{k+1})/(\tilde{\mathbf{r}}^k, \mathbf{r}^k);$$

$$\mathbf{d}^{k+1} = -\mathbf{r}^{k+1} + \beta_k \mathbf{d}^k;$$

$$\tilde{\mathbf{d}}^{k+1} = -\tilde{\mathbf{r}}^{k+1} + \beta_k \tilde{\mathbf{d}}^k;$$

$k = 0, 1, \ldots$ until $(\mathbf{r}^{k+1}, \mathbf{r}^{k+1}) \leq \varepsilon$ or $(\tilde{\mathbf{r}}^{k+1}, \tilde{\mathbf{r}}^{k+1}) \leq \varepsilon$.

When near breakdown occurs—that is, $(\tilde{\mathbf{r}}^k, \mathbf{r}^k)$ is close to zero—then one observes an erratic convergence behavior of $(\mathbf{r}^k, \mathbf{r}^k)$ with spikes.

11.3.8 The Classical Lanczos Method

When the main purpose of using the Lanczos algorithm is to compute a set of biorthogonal vectors, there is no need to use the consistent form in the recursion (11.34), where the first two coefficients are related (their sum is one). The more general algorithm then takes the following form (cf. Freund, Golub, and Nachtigal, 1991):

Choose $\tilde{\mathbf{v}}^1, \tilde{\mathbf{w}}^1 \in \mathbb{C}^n$, with $(\tilde{\mathbf{w}}^1, \tilde{\mathbf{v}}^1) \neq 0$, and set $\mathbf{v}^0 = \mathbf{w}^0 = \mathbf{0}$. For $k = 1, 2, \ldots$ do:

(a) Compute $\delta_k = (\tilde{\mathbf{w}}^k, \tilde{\mathbf{v}}^k)$; if $\delta_k = 0$ stop (breakdown).
(b) Otherwise, choose $\beta_k, \gamma_k \in \mathbb{C}$, with $\beta_k \gamma_k = \delta_k$; set $\mathbf{v}^k = \frac{1}{\gamma_k} \tilde{\mathbf{v}}^k$ and $\mathbf{w}^k = \frac{1}{\beta_k} \tilde{\mathbf{w}}^k$.
(c) Compute

$$\alpha_k = (\mathbf{w}^k, A\mathbf{v}^k),$$

(11.40)
$$\tilde{\mathbf{v}}^{k+1} = A\mathbf{v}^k - \alpha_k\mathbf{v}^k - \beta_k\mathbf{v}^{k-1},$$

$$\tilde{\mathbf{w}}^{k+1} = A^*\mathbf{w}^k - \overline{\alpha}_k\mathbf{w}^k - \overline{\gamma}_k\mathbf{w}^{k-1}.$$

If $\tilde{\mathbf{v}}^{k+1} = \mathbf{0}$ or $\tilde{\mathbf{w}}^{k+1} = \mathbf{0}$, stop (termination).

It is readily seen that the biorthogonality relation $(\mathbf{w}^i, \mathbf{v}^j) = 0$, $i \neq j$, $1 \leq i$, $j \leq k + 1$ holds. The degree of freedom left in choosing β_k or γ_k can, for instance, be used to scale the vectors $\mathbf{v}^k$, $\mathbf{w}^k$, so that $\|\mathbf{v}^k\| = \|\mathbf{w}^k\|$.

As in the earlier derivation for the case of a selfadjoint matrix, we find with

$$V_k = [\mathbf{v}^1, \mathbf{v}^2, \ldots, \mathbf{v}^k], \quad W_k = [\mathbf{w}^1, \mathbf{w}^2, \ldots, \mathbf{w}^k],$$

$$H_k = \begin{bmatrix} \alpha_1 & \beta_2 & & 0 \\ \gamma_2 & \alpha_2 & \beta_3 & \\ \ddots & \ddots & & \\ 0 & & \gamma_k & \alpha_k \end{bmatrix}$$

that

$$AV_k = V_kH_k + [0, \ldots, 0, \tilde{\mathbf{v}}^{k+1}],$$

$$A^*W_k = W_kH_k^* + [0, \ldots, 0, \tilde{\mathbf{w}}^{k+1}],$$

while the biorthogonality relation can be written as

$$W_k^*V_k = I_k,$$

where I_k is the identity matrix of order k.

Remark 11.11 There is a variant of the Lanczos method that uses the transpose A^T instead of the adjoint matrix A^* in the recursion (11.40) and no complex conjugates of the recurrence coefficients. (For details, see Freund et al., 1991.)

The Lanczos method reduces to only one recursion in case $A = A^*$ or $A = A^T$ (A is called complex symmetric) if we choose $\tilde{\mathbf{w}}^1 = \tilde{\mathbf{v}}^1$ and $\beta_k = \gamma_k$. The first method is the *Hermitian Lanczos method* and the second is the *complex symmetric Lanczos process*. The latter method can break down even if A is positive definite (see Cullum and Willoughby, 1985).

11.3.9 Transpose-Free Methods

As we have seen, the Lanczos algorithm and the similar algorithms presented above for nonsymmetric problems involve matrix-vector products with both A and A^T (or with A and A^* in the complex matrix case). This can be a disadvantage, because A^T is not always readily available. It is possible, however, to construct Lanczos-based algorithms that do not involve the transpose or adjoint of A. For simplicity, we consider below only the case of real matrices.

Note, first, that for the transpose of A, we have $A^T = SAS^{-1}$ for some nonsingular matrix S because A and A^T are similar. (See Chapter 2. This relation can also be seen from the Jordan normal forms of A and A^T.) If we choose $\mathbf{w}^1 = \|S\mathbf{v}^1\|S\mathbf{v}^1$, then it can readily be seen that the Lanczos algorithm generates vectors that satisfy

$$(11.41) \qquad \mathbf{w}^k = \frac{1}{\|S\mathbf{v}^k\|} S\mathbf{v}^k, \qquad k = 1, 2, \dots .$$

Hence, instead of updating the Lanczos vectors $\mathbf{w}^k$ using the recursion (11.40), we can compute them by (11.41). Clearly, this is a viable approach only if we can find such a matrix S, and for which matrix-vector products can be computed at an expense comparable with those for A, or cheaper.

This approach holds if A is symmetric with respect to either its diagonal (i.e., the trivial case where $A = A^T$) or its antidiagonal (such matrices are called *centrosymmetric;* so-called Toeplitz matrices are examples of centrosymmetric matrices), in which case $A^T = SAS^{-1}$, with

$$S = \begin{bmatrix} 0 & \dots & 0 & 1 \\ 0 & \dots & 1 & 0 \\ \cdot & \cdot & \cdot & \cdot \\ 1 & \dots & 0 & 0 \end{bmatrix}.$$

Example 11.6 Let A be a Jordan block matrix (here shown of order 3), $A = \begin{bmatrix} \lambda & 1 & 0 \\ 0 & \lambda & 1 \\ 0 & 0 & \lambda \end{bmatrix}$. This is centrosymmetric (in fact, a Toeplitz matrix). Then we factor A as

$$A = \begin{bmatrix} 0 & 1 & \lambda \\ 1 & \lambda & 0 \\ \lambda & 0 & 0 \end{bmatrix} \begin{bmatrix} 0 & 0 & 1 \\ 0 & 1 & 0 \\ 1 & 0 & 0 \end{bmatrix},$$

where both factors are symmetric, and we have $SAS^{-1} = A^T$. If A is a symmetric matrix, we can let M be another symmetric matrix and consider the (preconditioned) matrix $B = MA$. For this, we have with $S = M$,

$$SB^T S^{-1} = MA = B \text{ and } B^T = AM.$$

Hence, we can apply the Lanczos algorithm to B, with no need to compute matrix-vector products with B^T. Note that we do not require that M or A be positive definite.

Incidentally, it can be seen from the Jordan canonical form that any matrix can be written as a product of two symmetric matrices—namely, if $A = S^{-1}JS$, where J is the Jordan normal form, then $A = (S^{-1} \Lambda S^{-T})(S^T C S)$, where $J = \Lambda C$ is factored as in Example 11.6. However, this property is, in general, practically useless, because it presupposes computation of the Jordan normal form.

Consider now transpose-free variants of the Lanczos type algorithms. Such a variant was first presented by Sonneveld (1984, 1989) for the BCG method and was called the *squared biconjugate gradient method* (BCGS). It was based on the observation that the residuals $\mathbf{r}^k$ in any Krylov subspace method are related to the initial residual $\mathbf{r}^0$ via a matrix polynomial $P_k(A)$, that is, $\mathbf{r}^k = P_k(A)\mathbf{r}^0$, where the polynomial coefficients are defined by the coefficients in the recurrence relations of the Lanczos type method used.

The trick to avoid the recurrence for the vector sequence $\tilde{\mathbf{r}}^k$ in (11.34), for instance, and the use of the adjoint (transpose) matrix, is based on the observation that both $\mathbf{r}^k$ and $\tilde{\mathbf{r}}^k$ are generated by the same polynomial $P_k(\cdot)$ (defined by the coefficients α_i, β_i, $i = 0, 1, \ldots, k - 1$), that is,

$$\mathbf{r}^k = P_k(A)\mathbf{r}^0, \quad \tilde{\mathbf{r}}^k = P_k(A^*)\tilde{\mathbf{r}}^0.$$

Hence, we see that the inner products occurring in (11.36) and (11.37) can be computed as follows:

(11.42)
$$(\mathbf{r}^k, \tilde{\mathbf{r}}^k) = (P_k(A)\mathbf{r}^0, P_k(A^*)\tilde{\mathbf{r}}^0)$$
$$= (P_k^2(A)\mathbf{r}^0, \tilde{\mathbf{r}}^0) = (P_k(A)\mathbf{r}^k, \tilde{\mathbf{r}}^0) = (\hat{\mathbf{r}}^k, \tilde{\mathbf{r}}^0),$$

where $\hat{\mathbf{r}}^k = P_k(A)\mathbf{r}^k$ and

$$(A\mathbf{r}^k, \tilde{\mathbf{r}}^k) = (AP_k(A)^2\mathbf{r}^0, \tilde{\mathbf{r}}^0) = (AP_k(A)\mathbf{r}^k, \tilde{\mathbf{r}}^0) = (A\hat{\mathbf{r}}^k, \tilde{\mathbf{r}}^0).$$

Note that

(11.43)
$$\hat{\mathbf{r}}^k = [P_k(A)]^2 \mathbf{r}^0$$

and $\hat{\mathbf{r}}^k$ can be generated by the same recurrence, starting with $\mathbf{r}^k$, as $\mathbf{r}^k$ is generated starting with $\mathbf{r}^0$. Further, it is readily seen that the recurrence steps in the biconjugate gradient (or Lanczos) method to generate the sequence $\mathbf{r}^{k+1}$ commute—i.e., we can permute the order in which the steps take place. Using this property, it can be seen that the computation of the sequence $\hat{\mathbf{r}}^k$ can take place in an order as illustrated below. Here $\xrightarrow{i}$ denotes the recursion steps that take place at the ith step:

$$\mathbf{r}^0 \xrightarrow{1} \mathbf{r}^1 \xrightarrow{2} \mathbf{r}^2 \xrightarrow{3} \mathbf{r}^3$$

$$\mathbf{r}^0 \xrightarrow{1} \vdots \xrightarrow{1} \hat{\mathbf{r}}^1 \xrightarrow{2} \vdots \xrightarrow{2} \hat{\mathbf{r}}^2 \xrightarrow{3} \vdots \xrightarrow{3} \hat{\mathbf{r}}^3.$$

Hence, the sequence $\hat{\mathbf{r}}^k$ can be generated taking two steps of the recurrence used for $\mathbf{r}^k$ at every iteration. Note that we need $\mathbf{r}^k$ when we use the recurrence relation for $\mathbf{x}^{k+1}$ in (11.34).

Besides vector updates and inner products, this method requires three matrix-vector multiplications with A within each complete step, while the BCG (or Lanczos) method requires one multiplication with A and one with A^T. On the other hand, we get a residual $\hat{\mathbf{r}}^k$, which is the "square" of $\mathbf{r}^k$ [in the sense $\hat{\mathbf{r}}^k = P_k(A)^2\tilde{\mathbf{r}}^0$, $\mathbf{r}^k = P_k(A)\mathbf{r}^0$]. Hence, if $\tilde{\mathbf{r}}^0 = \mathbf{r}^0$ and if $\mathbf{r}^k$ has been reduced by $P_k(A)$, $\hat{\mathbf{r}}^k$ will be squared as much reduced, and we can expect that the BCGS method will stop with a sufficiently small residual at an earlier stage than the BCG method. On the other hand, if there is an erratic convergence behavior in the BCG method, the behavior of the BCGS method will be squared so erratic.

Furthermore, if we do not want the intermediate BCG-approximations $\mathbf{x}^{k+1}$ (and $\mathbf{r}^{k+1}$), we can neglect the recursions for $\mathbf{x}^{k+1}$, $\mathbf{r}^{k+1}$, and $\tilde{\mathbf{r}}^{k+1}$ and only compute the "double-step" recursions needed for $\hat{\mathbf{r}}^k = [P_k(A)]^2\mathbf{r}^0$. The coefficients needed for this recursion will then be computed using the relations

$$(\mathbf{r}^k, \tilde{\mathbf{r}}^k) = (\hat{\mathbf{r}}^k, \tilde{\mathbf{r}}^0), \ (A\mathbf{r}^k, \tilde{\mathbf{r}}^k) = (A\hat{\mathbf{r}}^k, \tilde{\mathbf{r}}^0).$$

In this case it suffices with two matrix-vector multiplications with A at every cycle of the double-step recursion. For further details, see Sonneveld (1989).

It can be seen that the BCGS algorithm can be implemented in the following form. Let $\mathbf{r}^0 = A\mathbf{x}^0 - \mathbf{b}$, $\hat{\mathbf{r}}^0 = \mathbf{r}^0$, let $\tilde{\mathbf{r}}^0$ be an arbitrary vector such that $(\mathbf{r}^0, \tilde{\mathbf{r}}^0) \neq 0$, and let $\mathbf{s} = \mathbf{d}^0 = -\mathbf{r}^0$. Then, for $k = 0, 1, \ldots$ perform the following steps:

$$\tau_k = (\hat{\mathbf{r}}^k, \tilde{\mathbf{r}}^0)/(A\hat{\mathbf{r}}^k, \tilde{\mathbf{r}}^0);$$

$$\mathbf{d}^{k+1} = \mathbf{s} + \tau_k A\hat{\mathbf{r}}^k;$$

$$\mathbf{x}^{k+1} = \mathbf{x}^k + \tau_k(\mathbf{s} + \mathbf{d}^{k+1});$$

$$\mathbf{r}^{k+1} = \mathbf{r}^k + \tau_k A(\mathbf{s} + \mathbf{d}^{k+1});$$

$$\beta = (\mathbf{r}^{k+1}, \tilde{\mathbf{r}}^0)/(\mathbf{r}^k, \tilde{\mathbf{r}}^0);$$

$$\mathbf{s} = -\mathbf{r}^{k+1} + \beta\mathbf{d}^k;$$

$$\hat{\mathbf{r}}^{k+1} = -\mathbf{s} + \beta(-\mathbf{d}^k + \beta\hat{\mathbf{r}}^k).$$

11.3.10 Hybrid Methods

Lanczos type methods are not based on any minimization property and often exhibit a rather erratic convergence behavior. As is clear from (11.41), these effects are magnified in the BCGS method, because this method tends to accelerate convergence as well as divergence of the BCG method. Moreover, there are cases for which BCGS diverges, while BCG still converges.

As will be shown in the following chapter, there exist monotonically converging methods, of minimal residual type. It will be seen that the cost of these methods increases quadratically with the number of steps (the storage required increases linearly), so they can become too costly when we need to perform many iteration steps. Therefore, they must in practice be truncated to a form where every minimization takes place on a space spanned by a small set of vectors, or one must use the method with restarts.

Alternatively one can combine minimal residual methods with Lanczos methods to form a *hybrid method*, where within each step of the minimal residual method one computes a number (say s) of Lanczos type steps. If at stage k, the latter correspond to a polynomial $P_s^{(k)}(A)$, the global polynomial after k minimal residual steps will have degree ks. If s is chosen appropriately, one can expect that after few steps (k), a sufficiently small residual has been found. The inner iteration (Lanczos) steps can be seen as a preconditioner to A. This case therefore corresponds to a variable-step preconditioner. A theory for convergence of variable-step preconditioners will be found in Chapter 12.

Such or similar methods to stabilize the Lanczos type methods have been used in practice. Based on an earlier proposal by Sonneveld (1984), van der Vorst (1992) proposed a hybrid biconjugate gradient and mimimal residual method where every step of the biconjugate gradient method is followed by a step of a minimum residual method. This corresponds to multiplying the reduction polynomial by an extra factor $(1 - \xi_{k+1}\lambda)$. The roots ξ_{k+1}^{-1} can be computed by a (line search) minimization of the residual. More generally, one can compute several minimal residual steps after each cycle of the Lanczos algorithm. Here ξ_{k+1} is real. Gutknecht (1991) extended this idea, to make it more efficient in the case of complex eigenvalues, by use of two extra factors, $(1 - \xi_{k+1}\lambda)(1 - \xi_{k+2}\lambda)$. Here, ξ_{k+1}, ξ_{k+2} are either real or complex conjugate numbers.

Schönauer (1987) has used smoothing methods also based on one-dimensional minimization methods. The zeros need not be computed; instead one constructs a recursive method by minimizing the residual locally—i.e., within a two-dimensional subspace. For details, see Gutknecht (1993). For further results, see Weiss (1994). It has also been observed (see Axelsson and Makarov, 1993), that the behavior of the methods can differ significantly when the error is measured in different norms, such as the Euclidean or the maximum norm.

11.3.11 Look-Ahead Lanczos

We have seen that the Lanczos type algorithms break down if the vectors $\mathbf{v}^k, \mathbf{w}^k$ become orthogonal, $(\mathbf{v}^k, \mathbf{w}^k) = 0$, without either $\mathbf{v}^k$ nor $\mathbf{w}^k$ being zero. This prevents the computation of vectors $\tilde{\mathbf{v}}^{k+1}, \tilde{\mathbf{w}}^{k+1}$, which are orthogonal to the previous vectors. However, it could happen that the biorthogonality condition can be fulfilled for a pair of vectors corresponding to some higher power of A and A^T. Such procedures, which skip over a breakdown by looking ahead of the process, were considered first by Parlett, Taylor, and Liu (1985) and later in a different form by Freund, Gutknecht, and Nachtigal (1992) and by Freund and Nachtigal (1990). They will not be considered further here.

In the next chapter, we consider a method with monotone convergence even for nonsymmetric problems, under the assumption that the symmetric part of the matrix is positive definite. For this method, it will be seen that we must store more vectors and compute more inner products than for the methods considered in the present chapter. However, this disadvantage can be partly offset when we use accurate preconditioners, for which convergence to a desired accuracy takes place in few iteration steps.

To summarize, bi-conjugate gradient type algorithms terminate in at most n steps or, more precisely, in a number of steps equal to the maximal dimension of the Krylov space generated by A from the initial residual $\mathbf{r}^0$. In practice, this property is often irrelevant, because this number may be very large and the bi-orthogonality property is partly lost due to round-off errors. What is important in practice is the small memory requirement of the methods. In contrast to the conjugate gradient method for s.p.d. matrices, however, the methods in this class do not minimize the residual in any (fixed) norm. This can show up in an erratic behavior of the residuals. Furthermore, the methods can break down before the iterates have converged, either by division by zero, or by the division by too small numbers. The latter causes some local peaks in the convergence of the residuals, and the corresponding corrections to the current approximate solution vector may cause cancellation of significant digits. The stabilized or hybrid versions of the algorithms referred to above frequently behave significantly better.

References

J. Abaffy and E. Spedicato (1989). *ABS Projection Algorithms: Mathematical Techniques for Linear and Nonlinear Equations*. Ellis Horwood Ltd, Chichester.

O. Axelsson (1972). A generalized SSOR method. *BIT*, 443– 467.

O. Axelsson (1974). On preconditioning and convergence acceleration in sparse matrix problems. CERN Technical Report 74-10, Data Handling Division, Geneva.

O. Axelsson (Copenhagen, 1976). Solution of linear systems of equations: iterative methods. In *Sparse Matrix Techniques*, ed. V. A. Barker. Berlin: Springer-Verlag.

O. Axelsson and V.A. Barker (1984). *Finite Element Solution of Boundary Value Problems. Theory and Computation*. Orlando, Fla.: Academic.

O. Axelsson and M. Makarov (1993). The choice of search directions in the iterative solution of nonsymmetric indefinite linear systems. Technical Report, Faculty of Mathematics and Informatics, University of Nijmegen, The Netherlands.

R. Bank and C. Douglas (1985). An efficient implementation for SSOR and incomplete factorization preconditionings. *Appl. Numer. Math.* **1**, 489–492.

M. S. Birman (1950). Some estimates for the method of steepest descent (in Russian). *Uspechi Mat. Nauk.* **5**, 152–155.

R. Chandra (1978). Conjugate gradient methods for partial differential equations. Research Report #128, Yale University, Department of Computer Science.

P. Concus, G. H. Golub, and D. P. O'Leary (1976). A generalized conjugate gradient method for the numerical solution of elliptic partial differential equations. In *Sparse Matrix Computations*, eds. J. R. Bunch and D. J. Rose, pp. 309–332. Orlando, Fla.: Academic.

V. Conrad and Y. Wallach (1979). Alternating methods for sets of linear equations. *Numer. Math.* **27**, 371–372.

J. K. Cullum and R. A. Willoughby (1977). The equivalence of the Lanczos and the conjugate gradient algorithms. Research Report, RC 6903 12/20/77, Mathematical

Sciences Department, IBM T. J. Watson Research Center, Yorktown Heights, New York.

J. K. Cullum and R. A. Willoughby (1985). *Lanczos Algorithms for Large Symmetric Eigenvalue Computations*, Vols. I and II. Basel: Birkhäuser.

J. J. J. Cuppen (1981). A divide and conquer method for the symmetric tridiagonal eigenproblem. *Numerische Mathematik* **36**, 177–195.

S. C. Eisenstat (1981). Efficient implementation of a class of conjugate gradient methods. *SIAM J. Sci. Stat. Comp.* **2**, 1–4.

M. Engeli, Th. Ginsburg, H. Rutishauser, and E. Stiefel (1959). *Refined Iterative Methods for Computation of the Solution and the Eigenvalues of Self-Adjoint Boundary Value Problems*. Mitt. Inst. f. Angew. Math. ETH, Zürich, Nr. 8. Basel: Birkhäuser.

D. K. Faddeev and V. N. Faddeeva (1963). *Computational Methods of Linear Algebra*. San Francisco, Calif.: Freeman.

R. Fletcher (1976). Conjugate gradient methods for indefinite systems. *Lecture Notes in Mathematics*, No. 506, pp. 73–89. Heidelberg: Springer-Verlag.

R. W. Freund, G. H. Golub, and N. M. Nachtigal (1991). Iterative solution of linear systems. RIACS Technical Report 92-21, Moffett Field, Calif. *Acta Numerica*, to appear.

R. Freund, M. Gutknecht, and N. Nachtigal (1992). An implementation of the look-ahead Lanczos algorithm for non-Hermitian matrices. *SIAM J. Sci. Stat. Comp.*, to appear.

R. W. Freund and N. M. Nachtigal (1990). A quasi-minimal residual method for non-Hermitian linear systems. RIACS Technical Report 90.51, Moffett Field, Calif.

G. H. Golub and D. P. O'Leary (1989). Some history of the conjugate gradient and Lanczos algorithms: 1948–1976. *SIAM Review* **31**, 50–102.

G. Golub and C. van Loan (1989). *Matrix Computations*, 2nd ed. Baltimore: Johns Hopkins Univ.

M. G. Gutknecht (1991). Variants of BICGSTAB for matrices with complex spectrum. IPS Research Report No. 91.14, ETH-Zentrum, Zürich.

M.G. Gutknecht (1993). Changing the norm in conjugate gradient type algorithms. *SIAM J. Numer. Anal.* **30**, 40–56.

L. A. Hageman and D. M. Young (1981). *Applied Iterative Methods*. Orlando, Fla.: Academic.

M. R. Hestenes (1990). Conjugacy and gradients. In *A History of Scientific Computing*, ed. S. G. Nash, pp. 167–179. Reading, Mass.: Addison-Wesley.

M. R. Hestenes (1980). *Conjugate Direction Methods in Optimization*. New York, Heidelberg, Berlin: Springer-Verlag.

M. R. Hestenes and E. Stiefel (1952). Methods of conjugate gradients for solving linear systems. *J. Res. Nat. Bur. Standards Sect. B* **49**, 409–436.

S. Kaniel (1966). Estimates for some computational techniques in linear algebra. *Math. Comp.* **20**, 369–378.

B. Keller (1977). Numerical solution of bifurcation and nonlinear eigenvalue problems. In *Applications of Bifurcation Theory*, ed. P. Rabinowitz, pp. 359–384. Orlando, Fla.: Academic.

C. Lanczos (1950). An iteration method for the solution of the eigenvalue problem of linear differential and integral operators. *J. Res. Nat. Bur. Standards* **45**, 255–282.

C. Lanczos (1952). Solutions of systems of linear equations by minimized iterations. *J. Res. Nat. Bur. Standards, Sect. B* **49**, 33–53.

D. G. Luenberger (1969). Hyperbolic pairs in the method of conjugate gradients. *SIAM J. Appl. Math.* **17**, 1263–1267.

G. I. Marchuk and Yu. A. Kuznetsov (1968). On optimal iteration processes. *Sov. Math. Dokl.* **9**, 1041–1045.

J. A. Meijerink and H. A. van der Vorst (1977). An iterative method for linear systems of which the coefficient matrix is a symmetric M-matrix. *Math. Comp.* **31**, 148–162.

Y. Notay (1989). Incomplete factorizations of singular linear systems. *BIT* **29**, 577–582.

J. M. Ortega (1988). Efficient implementation of certain iterative methods. *SIAM J. Sci. Stat. Comp.* **9**, 882–891.

C. C. Paige (1971). The computation of eigenvalues and eigenvectors of very large sparse matrices. Ph.D. thesis, London University.

B. N. Parlett (1980). *The Symmetric Eigenvalue Problem.* Englewood Cliffs, N.J.: Prentice Hall.

B. N. Parlett, D. R. Taylor, and Z. A. Liu (1985). A look-ahead Lanczos algorithm for unsymmetric matrices. *Math. Comp.*, **44**, pp. 105–124.

J. Reid (1971). On the method of conjugate gradients for the solution of large sparse systems of linear equations, In *Large Sparse Sets of Linear Equations*, pp. 231–254. Orlando, Fla.: Academic.

Y. Saad (1980). On the rates of convergence of the Lanczos and the block Lanczos methods. *SIAM J. Num. Anal.* **17**, 687–706.

W. Schönauer (1987). *Scientific Computing on Vector Computers.* Amsterdam: North-Holland.

P. Sonneveld (1984). CGS, a fast Lanczos-type solver for nonsymmetric linear systems. Report 84-16, Department of Mathematics and Informatics, Delft University of Technology, Delft, The Netherlands.

P. Sonneveld (1989). CGS, a fast Lanczos-type solver for nonsymmetric linear systems. *SIAM J. Sci. Statist. Comput.* **10**, 36–52.

H. van der Vorst (1992). Bi-CGSTAB: A fast and smoothly converging variant of Bi-CG for the solution of non-symmetric linear systems. *SIAM J. Sci. Statist. Comput.* **13**, 631–644.

R. Weiss (1994). Properties of generalized conjugate gradient methods. *Num. Lin. Alg. with Appl.* **1**, 45–63.

J. Wilkinson (1961). *The Algebraic Eigenvalue Problem*, Oxford, Clarendon Press.

D. M. Young, L. J. Hayes, and K. C. Jea (1981). Generalized conjugate gradient acceleration of iterative methods, I: The symmetrizable case. Report CNA-162, Center for Numerical Analysis, The Univ. of Texas at Austin.

12
Generalized Conjugate Gradient Methods

As we saw in the previous chapter, there exist two types of methods to solve systems of equations with short recurrence relations: (1) minimization algorithms, such as the conjugate gradient type algorithms, and (2) Lanczos-type algorithms. The short recurrence and the minimization properties have been shown to hold for the conjugate gradient methods for matrices that are selfadjoint and positive definite w.r.t. to the inner product used in the algorithm. The short recurrence holds for the biconjugate gradient-type Lanczos algorithms, also for nonsymmetric matrices, but these algorithms can break down when the matrix is indefinite.

In this chapter, it will first be shown that such short recurrence relations for the conjugate gradient minimization type algorithms exist for a broader class of matrices, the H-normal class w.r.t. the initial vector. This extends the applicability of these methods. However, many matrices occurring in practice belong to still more general classes. On the other hand, the Lanczos-type algorithms do not have a minimization property as a rule and, as just said, can even break down.

We shall also analyze a general class of methods based on minimizing the least square norm of the residual (w.r.t. an inner product) and using a set of search directions (orthogonal w.r.t. another inner product, in general). Such methods are referred to as minimal residual, or least squares, methods. It turns out that in this case, we do not obtain a short recurrence relation normally, but on the other hand, we can show monotone convergence for all matrices for which a certain stagnation does not occur and even convergence with at least a geometric rate for matrices with a positive definite symmetric part.

Since the computational cost per iteration now increases with every iteration, it is of interest also to consider truncated or restarted versions of these algorithms. There also exist similar methods, based on making the current iteration error orthogonal (w.r.t. some inner product) to the subspace spanned by the previous search direction vectors. Such methods will also be presented and analyzed.

The problem of increasing cost per iteration is not detrimental if we need only a few iterations for convergence to a sufficient accuracy. Therefore, efficient preconditioners are particularly important for these methods. Preconditioners have been presented in Chapters 7 to 10.

Even if A is symmetric and positive definite but has a very large condition number, it can be advisable to use a generalized conjugate gradient method as referred to above, or at least a truncated form of it, because for matrices with large condition numbers, the loss of orthogonality among the search directions due to round-off errors is more pronounced.

We consider, first, the generalized conjugate gradient least squares method, analyze its convergence, and determine the conditions under which it truncates to a short recurrence form. Next, a similar analysis will be given for orthogonal error type methods.

In many contexts there arise matrices, such as Schur complement matrices of the form $S = A_{22} - A_{21}A_{11}^{-1}A_{12}$, where the inefficiency of forming the matrix explicitly forces us to use iterative solution methods where only the action of S on vectors is required. Moreover, the inner systems with matrix A_{11} are then usually solved most efficiently also with iterations, which we call inner iterations. In some cases, we have eigenvalue information about A_{11} available that enables us to choose a polynomial—such as a Chebyshev polynomial—approximation of A_{11}^{-1}, and this can be used as a preconditioner for the inner iterations. In more general problems, however, such information is not present, and we need to use a conjugate gradient type method also for the inner iterations. Moreover, it can be efficient to let the number of inner iterations vary for the different outer iterations. The use of a conjugate gradient method as inner iteration method corresponds to the use of a variable (or nonlinear) preconditioner which can change from one iteration to the next. In Section 12.3, we analyze the case when such methods are used as preconditioners for a generalized conjugate gradient method.

The use of Schur complements S can be particularly efficient when we have a matrix of the form $A = \begin{bmatrix} A_{11} & A_{12} \\ A_{21} & A_{22} \end{bmatrix}$, where A is symmetric, A_{11} is positive definite, and A_{22} is negative semidefinite. Then, $-S$ becomes positive definite. Problems of this type occur in numerical optimization, for instance; see discussion in Chapter 9.

12.1 Generalized Conjugate Gradient, Least Squares Methods

To solve $A\tilde{\mathbf{x}} = \tilde{\mathbf{a}}$, $\tilde{\mathbf{x}} \in \mathbb{R}^n$, where A is real but in general nonsymmetric and/or indefinite or even rectangular (or order $m \times n$), we consider methods based on minimizing the quadratic functional

$$f(\mathbf{x}) = \frac{1}{2}(\mathbf{r}, \mathbf{r})_0 = \frac{1}{2}(B\mathbf{x} - \mathbf{b}, B\mathbf{x} - \mathbf{b})_0.$$

Here, $B\mathbf{x} = \mathbf{b}$ is a transformation of $A\tilde{\mathbf{x}} = \tilde{\mathbf{a}}$, for instance, by a left preconditioner Q (of order $n \times m$) if $n \le m$ and $B = QA$, $\mathbf{x} = \tilde{\mathbf{x}}$, $\mathbf{b} = Q\tilde{\mathbf{a}}$. If $m < n$, we use a right preconditioner Q (of order $n \times m$) and let $B = AQ$, $\tilde{\mathbf{x}} = Q\mathbf{x}$, $\mathbf{b} = \tilde{\mathbf{a}}$. Further, $(\cdot, \cdot)_0$ denotes an inner product in $\mathbb{R}^n$ and in $\mathbb{R}^m$, respectively. If A is square, we can let $B = C^{-1}A$, where C is a nonsingular approximation (preconditioner) to A. There is an important special choice of the matrix Q that can be efficient to use in certain situations: $Q = A^T$. Then,

$$QA\tilde{\mathbf{x}} = A^T A\tilde{\mathbf{x}} = A^T\tilde{\mathbf{a}}$$

becomes the normal equations corresponding to $A\tilde{\mathbf{x}} = \tilde{\mathbf{a}}$, and

$$AQ\mathbf{x} = AA^T\mathbf{x} = \tilde{\mathbf{a}}, \quad \tilde{\mathbf{x}} = Q\mathbf{x}$$

is the system of equations used in Craig's (1955) method.

The corresponding conjugate gradient methods become the minimal residual and minimal error methods, and the Krylov vector space is generated by matrices $A^T A$ and $A A^T$, respectively. Naturally, there is no need to form these matrices explicitly, which is an important observation when A is sparse. It can be seen that $A^T A$ and $A A^T$ can be almost full even if A is very sparse. In some special cases, such as when A is close to a unitary matrix, the normal equation matrix and the minimal error equation matrix are close to the identity matrix and the corresponding iteration method can converge very fast. In general, however, when A is square, the spectral condition number of $A^T A$ and $A A^T$ is much larger than the condition number of A itself (the square of it, if A is s.p.d.). To keep the notations consistent, we assume that

$$B\mathbf{x} = \mathbf{b}, \quad \mathbf{x}, \mathbf{b} \in \mathbb{R}^n$$

and that B is a square matrix (of order n). The inner product is defined by a matrix M_0, that is,

$$(\mathbf{x}, \mathbf{y})_0 = \mathbf{x}^T M_0 \mathbf{y},$$

where M_0 is s.p.d. The corresponding norm is

$$\|\mathbf{x}\|_0 = (\mathbf{x}, \mathbf{x})_0^{\frac{1}{2}}.$$

A vector $\hat{\mathbf{x}}$, such that

$$f(\hat{\mathbf{x}}) = \inf_{\mathbf{x} \in \mathbb{R}^n} f(\mathbf{x}),$$

is called a *minimizer* of f. The existence of such a minimizer is evident, because $\mathbb{R}^n$ is a finite dimensional vector space. If $\mathbf{b} \in \mathbb{R}(B)$, the range of B, then $f(\hat{\mathbf{x}}) = 0$; otherwise $\hat{\mathbf{x}}$ is the least squares solution of $B\mathbf{x} = \mathbf{b}$.

For the numerical computation of the minimizer, we use a set of recursively computed search directions $\{\mathbf{d}^j\}$ such that $B\mathbf{d}^k$ becomes linearly independent of the set of vectors $\{B\mathbf{d}^j\}_{k-s_k}^{k-1}$, $1 \le s_k \le k$. Here, $s_k = \min\{s_{k-1} + 1, s\}$ and $(s + 1)$, where $s \ge 0$, is the maximal length of the vector recursion used to update the current approximation to $\mathbf{x}$, as we shall see. With the choice of search directions, we then determine $s_k + 1$ parameters $\alpha_j^{(k)} \in (-\infty, \infty)$, $k - s_j \le j \le k$, such that $f(\mathbf{x}^{k+1})$ is minimized, where we let

$$(12.1) \qquad \mathbf{x}^{k+1} = \mathbf{x}^k + \sum_{j=k-s_k}^{k} \alpha_j^{(k)} \mathbf{d}^j.$$

A necessary condition for f to take a minimum is $\partial f / \partial \alpha_j^{(k)} = 0$, that is,

$$(12.2) \qquad (B\mathbf{x}^{k+1} - \mathbf{b}, B\mathbf{d}^i)_0 = 0, \quad k - s_k \le i \le k,$$

or

$$(12.3) \qquad \sum_{j=k-s_k}^{k} \alpha_j^{(k)}(B\mathbf{d}^j, B\mathbf{d}^i)_0 = -(\mathbf{r}^k, B\mathbf{d}^i)_0, \quad k - s_k \le i \le k, \ k - s_k \le j \le k.$$

Here,

$$(12.4a) \qquad \mathbf{r}^k = B\mathbf{x}^k - \mathbf{b}.$$

As for the standard conjugate gradient methods, we will sometimes find it convenient to use the recursion formula, which follows from (12.1),

$$(12.4b) \qquad \mathbf{r}^{k+1} = \mathbf{r}^k + \sum_{j=k-s_k}^{k} \alpha_j^{(k)} B\mathbf{d}^j,$$

instead of computing $\mathbf{r}^k$ by the defining expression (12.4a). This can be more efficient, at least when s is small. Note that by (12.2), the following orthogonality condition is satisfied,

$$(12.5) \qquad (\mathbf{r}^{k+1}, B\mathbf{d}^i)_0 = 0, \quad k - s_k \le i \le k.$$

Further, (12.3) can be written as

$$(12.6) \qquad \Lambda^{(k)} \underline{\alpha}^{(k)} = \underline{\gamma}^{(k)},$$

where

$$\Lambda^{(k)} = [(B\mathbf{d}^{k+1-i}, B\mathbf{d}^{k+1-j})_0], \quad 1 \le i \le s_k + 1, 1 \le j \le s_k + 1$$

$$(12.7) \quad (\underline{\alpha}^{(k)})_j = \alpha_{k+1-j}^{(k)}, \quad 1 \le j \le s_k + 1$$

$$(\underline{\gamma}^{(k)})_1 = -(\mathbf{r}^k, B\mathbf{d}^k)_0, \quad (\underline{\gamma}^{(k)})_j = 0, \quad 2 \le j \le s_k + 1.$$

Here $\Lambda^{(k)}$ has order $s_k + 1$ and is symmetric and positive semidefinite. It is nonsingular if and only if the vectors $\{B\mathbf{d}^j\}_{k-s_k}^k$ are linearly independent. (Or, equivalently, if B is nonsingular, then $\Lambda^{(k)}$ is nonsingular if and only if $\{\mathbf{d}^j\}_{k-s_k}^k$ are linearly independent.) If $s_k = s_{k-1} + 1$, the matrix $\Lambda^{(k)}$ equals $\Lambda^{(k-1)}$, augmented with a row and a column.

We shall now consider the construction of search directions such that this vector set is linearly independent. At the kth step, we compute a new search direction as a linear combination of the residual $\mathbf{r}^{k+1}$, corresponding to the most recent approximation $\mathbf{x}^{k+1}$ and the $s_k + 1$ previous search directions, i.e.,

$$(12.8) \qquad \mathbf{d}^{k+1} = -\mathbf{r}^{k+1} + \sum_{k-s_k}^k \beta_j^{(k)} \mathbf{d}^j,$$

where the parameters $\beta_j^{(k)}$, $k - s_k \le j \le k$ are determined by the conjugate orthogonality condition

$$(12.9) \qquad (B\mathbf{d}^{k+1}, B\mathbf{d}^j)_1 = 0, \quad k - s_k \le j \le k.$$

Here, $(\cdot, \cdot)_1$ is an inner product in $\mathbb{R}^n$, defined by $(\mathbf{x}, \mathbf{y})_1 = \mathbf{x}^T M_1 \mathbf{y}$, where M_1 is s.p.d. The new search direction $\mathbf{d}^{k+1}$ is added to the set of current search directions and, if $s_k = s_{k-1}$, the oldest direction is deleted from the set. By induction, it follows that the orthogonality (12.9) holds for

$$(B\mathbf{d}^p, B\mathbf{d}^q)_1 = 0, \quad p \ne q, k - s_k \le p, q \le k + 1,$$

because $s_k \le s_{k-1} + 1$. Hence, (12.8) and (12.9) show that

$$(12.10) \qquad \beta_j^{(k)} = (B\mathbf{r}^{k+1}, B\mathbf{d}^j)_1 / (B\mathbf{d}^j, B\mathbf{d}^j)_1, \quad k - s_k \le j \le k.$$

If we prescribe $\mathbf{x}^0$ and $\mathbf{d}^0$, then (12.1), (12.4), and (12.8), where $\alpha_j^{(k)}$ and $\beta_j^{(k)}$ are computed from (12.7) and (12.10), respectively, completely define the algorithm, which we call the *generalized truncated conjugate gradient, least squares*, GCG-LS(s) algorithm. If $s = n - 1$, we get the untruncated version of the algorithm, which we call GCG-LS. $\mathbf{x}^0$ can be chosen arbitrarily, and frequently one puts $\mathbf{d}^0 = -\mathbf{r}^0$.

The only possibility of a breakdown of the algorithm is when $\Lambda^{(k)}$ becomes singular. If B is nonsingular, we shall show that this can only happen if $\mathbf{x}^k$ is already a solution. Nevertheless, a stagnation occurs if $(\mathbf{r}^k, B\mathbf{d}^k)_0 = 0$, because then (12.7) shows that the vectors $\underline{\alpha}^{(k)}$ become zero, so $\mathbf{x}^{k+1} = \mathbf{x}^k$.

Note that for the algorithm to be practically viable, the computation of B times a vector and the computation of the inner products $(\cdot, \cdot)_0$ and $(\cdot, \cdot)_1$ must be performed with little expense, as compared with solving the system $B\mathbf{x} = \mathbf{b}$ by some direct method, for instance. Note, also, that if the two inner products are identical and if B is selfadjoint with respect to the inner product, then the algorithm reduces to a minimum residual form of the standard conjugate gradient method. This will follow also from a more general result about automatic truncation to be presented later.

Lemma 12.1 *If $\Lambda^{(j)}$ is nonsingular, $j = 0, 1, \ldots, k$, then:*

(a) $(\mathbf{r}^{k+1}, B\mathbf{d}^i)_0 = 0$, $k - s_k \leq i \leq k$.

(b) $(\mathbf{r}^{k+1}, B\mathbf{r}^i)_0 = 0$, $s_{i-1} + k - s_k + 1 \leq i \leq k$.

(c) *If $s_j = j$, $j = 0, 1, \ldots$, then* $(\mathbf{r}^{k+1}, B\mathbf{r}^i)_0 = 0$, $0 \leq i \leq k$.

(d) $(\mathbf{r}^k, B\mathbf{d}^k)_0 = -(\mathbf{r}^k, B\mathbf{r}^k)_0$.

Proof Part (a) is (12.5). To prove part (b), note that by (12.8), with $k + 1$ replaced by i, we get

$$(\mathbf{r}^{k+1}, B\mathbf{r}^i)_0 = (\mathbf{r}^{k+1}, -B\mathbf{d}^i + \sum_{j=i-1-s_{i-1}}^{i-1} \beta_j^{(i-1)} B\mathbf{d}^j)_0 = 0,$$

if $i \leq k$ and $k - s_k \leq i - 1 - s_{i-1}$, that is, if $k \geq i \geq s_{i-1} + k - s_k + 1$.

If $s_j = j$, then $s_{i-1} = i - 1$, and part (c) follows from part (b). Finally, (12.8) shows that

$$(12.11) \quad -(\mathbf{r}^k, B\mathbf{d}^k)_0 = (\mathbf{r}^k, B\mathbf{r}^k - \sum_{j=k-1-s_{k-1}}^{k-1} \beta_j^{(k-1)} B\mathbf{d}^j)_0 = (\mathbf{r}^k, B\mathbf{r}^k)_0,$$

because (a) shows that $(\mathbf{r}^k, B\mathbf{d}^j)_0 = 0$, if $k - 1 \geq j \geq k - 1 - s_{k-1}$. $\diamond$

Theorem 12.2 *Assume that $M_0 B + B^T M_0$ is positive definite. If $\mathbf{r}^k \neq \mathbf{0}$, then $\Lambda^{(k)}$, defined by (12.7) is nonsingular.*

Proof $\Lambda^{(k)}$ is singular if and only if the set $\{B\mathbf{d}^j\}_{k-s_k}^k$ is linearly dependent. Assume now, by induction, that $\Lambda^{(k-1)}$ is nonsingular and that this set is linearly dependent. Then, $\{B\mathbf{d}^j\}_{k-s_k}^{k-1}$ is linearly independent. (Otherwise $\Lambda^{(k-1)}$ would be singular.) Then, for some $\lambda_j^{(k)}$,

$$B\mathbf{d}^k = \sum_{j=k-s_k}^{k-1} \lambda_j^{(k)} B\mathbf{d}^j,$$

and (12.8) (with $k + 1$ replaced by k) shows that, for some $\tilde{\lambda}_j^{(k)}$,

$$B\mathbf{r}^k = \sum_{j=\min(k-1-s_{k-1}, k-s_k)}^{k-1} \tilde{\lambda}_j^{(k)} B\mathbf{d}^j.$$

But $s_k \leq s_{k-1} + 1$, so Lemma 12.1(a) shows that

$$(\mathbf{r}^k, B\mathbf{r}^k) = (\mathbf{r}^k, \sum_{j=k-1-s_{k-1}}^{k-1} \tilde{\lambda}_j^{(k)} B\mathbf{d}^j) = 0.$$

Hence, $\mathbf{r}^{k^T} M_0 B\mathbf{r}^k = 0$ or

$$\mathbf{r}^{k^T}(M_0 B + B^T M_0)\mathbf{r}^k = 0.$$

But, by assumption, $M_0 B + B^T M_0$ is positive definite, showing that $\mathbf{r}^k = \mathbf{0}$, and this contradiction shows that $\{B\mathbf{d}^j\}_{k-s_k}^k$ is linearly independent. $\diamond$

Remark 12.3 Theorem 3.4 shows that the assumption that $K = M_0 B + B^T M_0$ is positive definite implies that B is positive stable—i.e., its eigenvalues have positive real parts. As was seen in the discussion of Theorem 3.4, it suffices that K is positive definite on the subspace spanned by the eigenvectors of B to conclude that B is positive stable. Similarly, if $\mathbf{d}^0$ is taken from this space, then if $\mathbf{d}^0 \neq \mathbf{0}$, it can be seen from the proof of Theorem 12.2 that it suffices that K is positive definite on this subspace for $\Lambda^{(k)}$ to be nonsingular.

We shall now present a monotone convergence result of the method. To this end, we first state the next lemma.

Lemma 12.4 (Meinardus, 1964) *Let* $\mathbf{f} \in \mathbb{R}^n$ *be given and let* $\{\mathbf{g}^j\}_{j=0}^k$, $k \leq n -$
1, *where* $\mathbf{g}^0 = \mathbf{f}$, *be linearly independent vectors in* $\mathbb{R}^n$. *Let* $\Lambda^{(k)} = [(\mathbf{g}^i, \mathbf{g}^j)]$,
$0 \leq i, j \leq k$ *be the associated* Gramian *matrix of order* $k + 1$, *let* $\Lambda_0^{(k)}$ *be the*
first principal minor of $\Lambda^{(k)}$, *and let* $V_k = span\{\mathbf{g}^1, \ldots, \mathbf{g}^k\}$. *Then,*

$$\frac{\det(\Lambda^{(k)})}{\det(\Lambda_0^{(k)})} = \min_{\mathbf{h} \in V_k} \|\mathbf{f} - \mathbf{h}\|^2,$$

where $\| \cdot \| = (\cdot, \cdot)^{\frac{1}{2}}$.

Proof Let $\mathbf{g} = \sum_{j=1}^k \lambda_j \mathbf{g}^j$ be the best least square approximation to $\mathbf{f}$, that is,
let $\mathbf{g}$ be defined by

$$(12.12a) \qquad\qquad (\mathbf{f} - \mathbf{g}, \mathbf{h}) = 0, \quad \forall \mathbf{h} \in V_k.$$

Then, with $\mathbf{h} = \mathbf{g}^i$ we get

$$(12.12b) \qquad\qquad (\mathbf{f}, \mathbf{g}^i) = (\mathbf{g}, \mathbf{g}^i),$$

or

$$(12.12c) \qquad (\mathbf{f}, \mathbf{g}^i) - \sum_{j=1}^k \lambda_j (\mathbf{g}^j, \mathbf{g}^i) = 0, \quad i = 1, 2, \ldots, k.$$

Hence,

$$(\mathbf{f}, \mathbf{g}) - \sum_{j=1}^k \lambda_j (\mathbf{g}^j, \mathbf{g}) = 0.$$

By the orthogonality property (12.12a),

$$\|\mathbf{f} - \mathbf{g}\|^2 = (\mathbf{f} - \mathbf{g}, \mathbf{f} - \mathbf{g}) = (\mathbf{f} - \mathbf{g}, \mathbf{f}) = (\mathbf{f}, \mathbf{f}) - (\mathbf{g}, \mathbf{f})$$

or, combining the last two results, we find

$$(12.13) \qquad\qquad (\mathbf{f}, \mathbf{f}) - \sum_{j=1}^k \lambda_j (\mathbf{g}^j, \mathbf{g}) = \|\mathbf{f} - \mathbf{g}\|^2.$$

Let $\lambda_0 = -1$, $\mathbf{g}^0 = \mathbf{f}$. Then

$$\sum_{j=0}^k \lambda_j (\mathbf{g}^j, \mathbf{g}^i) = \begin{cases} -\|\mathbf{f} - \mathbf{g}\|^2 & , i = 0 \text{ [by (12.13)]} \\ 0 & , i = 1, \ldots, k \text{ [by (12.12c)]}. \end{cases}$$

Cramer's rule then shows that

$$\lambda_0 = \frac{\det(\Lambda_0^{(k)})}{\det(\Lambda^{(k)})}(-\|\mathbf{f} - \mathbf{g}\|^2),$$

or

$$\frac{\det(\Lambda^{(k)})}{\det(\Lambda_0^{(k)})} = \|\mathbf{f} - \mathbf{g}\|^2 = \min_{\mathbf{h} \in V_k} \|\mathbf{f} - \mathbf{h}\|^2. \qquad \diamond$$

Theorem 12.5 *Assume that $M_0 B + B^T M_0$ is s.p.d. Then, for the GCG-LS(s) method, we have:*

(a) *$\alpha_k^{(k)} = (\det \Lambda^{(k)})^{-1} \det(\Lambda_{k,0}^{(k)})(\mathbf{r}^k, B\mathbf{r}^k)_0$, where $\Lambda_{k,0}^{(k)}$ is the first principal minor of $\Lambda^{(k)}$. If $\mathbf{r}^k \neq \mathbf{0}$, then $\alpha_k^{(k)} > 0$.*

(b) *Unless $\mathbf{r}^k = \mathbf{0}$, the method converges monotonically, $f(\mathbf{x}^{k+1}) < f(\mathbf{x}^k)$, $k = 0, 1, \ldots$ and with a rate defined by*

$$(\mathbf{r}^{k+1}, \mathbf{r}^{k+1})_0 = (\mathbf{r}^k, \mathbf{r}^k)_0 - \det(\Lambda^{(k)})^{-1} \det(\Lambda_{k,0}^{(k)})(\mathbf{r}^k, B\mathbf{r}^k)_0^2,$$

$$k = 0, 1, 2, \ldots.$$

(c) *Furthermore,*

$$(\mathbf{r}^{k+1}, \mathbf{r}^{k+1})_0 = (\mathbf{r}^k, \mathbf{r}^k)_0 - (\mathbf{r}^k, B\mathbf{r}^k)_0^2 / \min_{\mathbf{g} \in W_{k-1}} \|B\mathbf{r}^k - \mathbf{g}\|_0^2$$

$$\leq (1 - \xi)(\mathbf{r}^k, \mathbf{r}^k)_0, \quad \text{if } s_k \geq 0,$$

where

$$\xi = \lambda_{\min}(\widetilde{B}^S) \cdot \lambda_{\min}((\widetilde{B}^{-1})^S),$$

$$\widetilde{B} = M_0^{\frac{1}{2}} B M_0^{-\frac{1}{2}}, \quad \widetilde{B}^S = \frac{1}{2}(\widetilde{B} + \widetilde{B}^T),$$

$$W_{k-1} = \text{span}\{B\mathbf{d}^{k-1}, \ldots, B\mathbf{d}^{k-s_k}\},$$

and $\lambda_{\min}(\widetilde{B})$ denotes the smallest eigenvalue of $\widetilde{B}$. Hence, as $k \to \infty$, $(\mathbf{r}^k, \mathbf{r}^k)_0$ converges monotonically to zero for any $s \geq 0$.

(d)

$$(\mathbf{r}^{k+1}, \mathbf{r}^{k+1})_0 = (\mathbf{r}^k, \mathbf{r}^k)_0 - \frac{(\mathbf{r}^k, B\mathbf{r}^k)_0^2}{(B\mathbf{r}^k, B\mathbf{r}^k)_0 - \frac{(B\mathbf{r}^k, B^2\mathbf{r}^k)_0^2}{\min_{g \in W_{k-1} \ominus B\mathbf{d}^{k-1}} \|B^2\mathbf{r}^k - g\|_0^2}},$$

if $s_k \geq 1$, and, if $\xi < \frac{1}{2}$,

$$(\mathbf{r}^{k+1}, \mathbf{r}^{k+1})_0 \leq (1 - \frac{\xi}{1-\xi})(\mathbf{r}^k, \mathbf{r}^k)_0.$$

Proof To prove part (a), note that by Cramer's rule, it follows from (12.7) that

$$\alpha_k^{(k)} = \det(\Lambda^{(k)})^{-1} \sum_{j=0}^{s_k} (-1)^j \det(\Lambda_{k,j}^{(k)})(\underline{\gamma}^k)_j,$$

where $\Lambda_{k,j}^{(k)}$ denotes the comatrix of the corresponding entry of $\Lambda^{(k)}$ for the first row. (12.7) now shows that

$$\alpha_k^{(k)} = -\det(\Lambda^{(k)})^{-1} \det(\Lambda_{k,0}^{(k)})(\mathbf{r}^k, B\mathbf{d}^k)_0,$$

so Lemma 12.1(d) shows that

$$\alpha_k^{(k)} = \det(\Lambda^{(k)})^{-1} \det(\Lambda_{k,0}^{(k)})(\mathbf{r}^k, B\mathbf{r}^k)_0.$$

Since $\Lambda^{(k)}$ is s.p.d., it follows that $\det(\Lambda^{(k)}) > 0$. The same holds for the principal minors. Hence, (12.10) and

$$(\mathbf{r}^k, B\mathbf{r}^k)_0 = \mathbf{r}^{k^T}(M_0 B + B^T M_0)\mathbf{r}^k > 0$$

(because $M_0 B + B^T M_0$ is positive definite) show that $\alpha_k^{(k)} > 0$ if $\mathbf{r}^k \neq \mathbf{0}$. To show part (b), use (12.4b) and Lemma 12.1(b), to find

$$(\mathbf{r}^{k+1}, \mathbf{r}^{k+1})_0 = (\mathbf{r}^{k+1}, \mathbf{r}^k + \sum_{j=k-s_k}^{k} \alpha_j^{(k)} B\mathbf{d}^j)_0$$

$$= (\mathbf{r}^{k+1}, \mathbf{r}^k)_0 = (\mathbf{r}^k + \sum_{k-s_k}^{k} \alpha_j^{(k)} B\mathbf{d}^j, \mathbf{r}^k)_0$$

$$= (\mathbf{r}^k, \mathbf{r}^k)_0 + \alpha_k^{(k)} (B\mathbf{d}^k, \mathbf{r}^k)_0$$

(here noting that $k - s_k \geq k - 1 - s_{k-1}$).

Hence, Lemma 12.1(d) shows that

$$(\mathbf{r}^{k+1}, \mathbf{r}^{k+1})_0 = (\mathbf{r}^k, \mathbf{r}^k)_0 - \alpha_k^{(k)} (B\mathbf{r}^k, \mathbf{r}^k)_0,$$

and part (a) completes the proof of part (b). To prove part (c), note that $\Lambda^{(k)}$ is a Gramian matrix,

$$\Lambda^{(k)} = [\lambda_{ij}^{(k)}], \quad \text{where} \quad \lambda_{i,j}^{(k)} = (B\mathbf{d}^i, B\mathbf{d}^j)_0.$$

Lemma 12.4 then shows that

$$\frac{\det(\Lambda_{k,0}^{(k)})}{\det(\Lambda^{(k)})} = 1/\min_{\mathbf{g} \in W_{k-1}} \|B\mathbf{r}^k - \mathbf{g}\|_0^2,$$

which together with part (b) shows the first part of (c). If we take $\mathbf{g} = 0$ (which corresponds to $s = 0$), then

$$(\mathbf{r}^{k+1}, \mathbf{r}^{k+1})_0 = (\mathbf{r}^k, \mathbf{r}^k)_0 - (\mathbf{r}^k, B\mathbf{r}^k)_0^2 / \min_{\mathbf{g} \in W_{k-1}} \|B\mathbf{r}^k - \mathbf{g}\|_0^2$$

(12.14)

$$\leq (\mathbf{r}^k, \mathbf{r}^k)_0 - (\mathbf{r}^k, B\mathbf{r}^k)_0^2 / (B\mathbf{r}^k, B\mathbf{r}^k)_0.$$

Note now that, with $\tilde{\mathbf{r}}^k = M_0^{\frac{1}{2}} \mathbf{r}^k$,

$$\frac{(\mathbf{r}^k, B\mathbf{r}^k)_0^2}{(\mathbf{r}^k, \mathbf{r}^k)_0 (B\mathbf{r}^k, B\mathbf{r}^k)_0} = \frac{\tilde{\mathbf{r}}^{k^T} \tilde{B} \tilde{\mathbf{r}}^k}{\tilde{\mathbf{r}}^{k^T} \tilde{\mathbf{r}}^k} \frac{(\tilde{B}\tilde{\mathbf{r}}^k)^T \tilde{B}^{-T}(\tilde{B}\tilde{\mathbf{r}}^k)}{(\tilde{B}\tilde{\mathbf{r}}^k)^T (\tilde{B}\tilde{\mathbf{r}}^k)} \geq \lambda_{\min}(\tilde{B}^S) \lambda_{\min}([\tilde{B}^{-1}]^S).$$

The relation (12.14) shows that

$$(\mathbf{r}^{k+1}, \mathbf{r}^{k+1})_0 \leq (1 - \xi)(\mathbf{r}^k, \mathbf{r}^k)_0.$$

To prove part (d), note first that

$$(\mathbf{r}^{k+1}, \mathbf{r}^{k+1})_0 = \min \|B\mathbf{x} - \mathbf{b}\|_0^2, \quad \mathbf{x} \in \mathbf{x}^0 \oplus W_k.$$

Hence, using (12.4b) we find

$$(\mathbf{r}^{k+1}, \mathbf{r}^{k+1})_0 = \min \|\mathbf{r}^k - \mathbf{w}\|_0^2, \quad \mathbf{w} \in W_{k-1} \oplus B\mathbf{d}^k.$$

As in the derivation of (12.14), for $s_k \geq 1$, we find

$$\min_{\mathbf{g} \in W_{k-1}} \|B\mathbf{r}^k - \mathbf{g}\|_0^2 =$$

(12.15)

$$(B\mathbf{r}^k, B\mathbf{r}^k)_0 - \frac{(B\mathbf{r}^k, B^2\mathbf{r}^k)_0^2}{\min_{\mathbf{h} \in W_{k-1} \ominus B\mathbf{d}^{k-1}} \|B^2\mathbf{r}^k - \mathbf{h}\|_0^2}.$$

Choosing $\mathbf{h} = \mathbf{0}$ shows

$$\min_{g \in W_{k-1}} \| B\mathbf{r}^k - \mathbf{g} \|_0^2 = (B\mathbf{r}^k, B\mathbf{r}^k)_0 (1 - \frac{(B\mathbf{r}^k, B^2\mathbf{r}^k)_0^2}{(B\mathbf{r}^k, B\mathbf{r}^k)_0 (B^2\mathbf{r}^k, B^2\mathbf{r}^k)_0})$$

$$\leq (1 - \xi)(B\mathbf{r}^k, B\mathbf{r}^k)_0,$$

which, together with the first part of (12.14), completes the proof of (d). $\diamond$

Remark 12.6 (Continued fraction) Repeated use of Theorem 12.5(c,d) yields a continued fraction expression if we continue to write $\| B^2\mathbf{r}^k - \mathbf{h} \|_0^2$, etcetera in a form similar to (12.15). This gives an alternative illustration of the optimality property of the generalized conjugate gradient method. It can be seen that a similar continued fraction expression for the residuals holds even if we use subspaces different from the Krylov space.

Remark 12.7 The expression for the number ξ in Theorem 12.5(c) shows that ξ can be seen as the inverse of a condition number for the matrix B. If B is s.p.d., then

$$\xi \geq \lambda_{\min}(\widetilde{B}^S)\lambda_{\min}([\widetilde{B}^{-1}]^S) = \lambda_{\min}(B)/\lambda_{\max}(B),$$

that is, the inverse of the spectral condition number. We have

$$(\widetilde{B}^{-1})^S = \frac{1}{2}(\widetilde{B}^{-1} + \widetilde{B}^{-T}) = \widetilde{B}^{-1} \left[\frac{1}{2}(\widetilde{B} + \widetilde{B}^T) \right] \widetilde{B}^{-T}$$

$$= \widetilde{B}^{-1} \widetilde{B}^S \widetilde{B}^{-T}.$$

Hence, $\lambda_1 = \lambda_{\min}\left([\widetilde{B}^{-1}]^S\right)$ is equal to the smallest eigenvalue of the spectrally equivalent matrix

$$\widetilde{B}^{-T}(\widetilde{B}^{-1})^S\widetilde{B}^T = (\widetilde{B}\widetilde{B}^T)^{-1}\widetilde{B}^S$$

or

$$\widetilde{B}^S\mathbf{v}_1 = \lambda_1 \widetilde{B}\widetilde{B}^T\mathbf{v}_1,$$

where $\mathbf{v}_1$ is the corresponding eigenvector. This shows that

$$\lambda_1 = \frac{\mathbf{v}_1^T\widetilde{B}^S\mathbf{v}_1}{\|\widetilde{B}^T\mathbf{v}_1\|_2^2} \geq \frac{\lambda_{\min}(\widetilde{B}^S)}{\|\widetilde{B}\|_2^2}.$$

Hence,

$$\xi \geq \lambda_{\min}(\widetilde{B}^S)\lambda_{\min}\left([\widetilde{B}^{-1}]^S\right) \geq \left(\frac{\lambda_{\min}(\widetilde{B}^S)}{\|\widetilde{B}\|_2}\right)^2.$$

The latter lower bound of the rate of convergence in Theorem 12.5(c) was derived earlier in Elman (1982); see also Dennis and Turner (1988). We see that our bound on ξ is sharper. Furthermore, note that our sharp bound in (12.14), with $\mathbf{g} \neq \mathbf{0}$ (which appeared originally in Axelsson, 1987), shows that this lower bound can be quite pessimistic. The upper bound in Theorem 12.5(c) is independent of s. This indicates that it must be inaccurate for large values of s.

Theorem 12.5(b) shows that the monotone convergence holds also for truncated versions of the algorithm for any length $s + 1$, $s \geq 0$ of the vector recursion.

Theorem 12.5(c,d) holds even if $M_0 B + B^T M_0$ is indefinite. Hence, as long as $\mathbf{r}^k$ is such that $(\mathbf{r}^k, B\mathbf{r}^k)_0 \neq 0$, the method converges monotonically! On the other hand, if $(\mathbf{r}^k, B\mathbf{r}^k)_0 = 0$, the method stagnates. The rate of convergence of the conjugate gradient methods will be further discussed in Chapter 13.

Note, finally, that Theorem 12.5 is applicable also for the restarted GCG-LS(s) method, where after, say, s_0 steps ($s_0 \geq s$), one restarts the method with the latest found approximation, $\mathbf{x}^{(s_0)}$, as an initial approximation. This procedure can be repeated an arbitrary number of times. The truncated method with $s = 0$ and the restarted method with $s_0 = 0$ are identical.

12.1.1 Termination

We now consider termination properties of the full (untruncated) GCG-LS algorithm—i.e., assuming exact arithmetic, we want to find a number of steps sufficient to get a zero residual. Let $\mathbf{d}^0 = -\mathbf{r}^0$. By the construction of $\mathbf{d}^k$ and $\mathbf{r}^k$, it follows that $\mathbf{d}^k$ is a linear combination of the vectors $B^j \mathbf{r}^0$, $0 \leq j \leq k$ that form the Krylov set, denoted by

$$V_k = V_k(\mathbf{r}^0) = \text{SPAN}\{\mathbf{r}^0, B\mathbf{r}^0, \ldots, B^k \mathbf{r}^0\}.$$

Similarly, $\mathbf{r}^k$ is a direct sum

$$\mathbf{r}^k = \mathbf{r}^0 \oplus B V_{k-1}$$

of $\mathbf{r}^0$ and a linear combination of the vectors in $B V_{k-1}$. Hence,

$$\mathbf{r}^k = [I + p_k(B)]\mathbf{r}^0, \qquad k = 0, 1, 2, \ldots$$

for some polynomial p_k satisfying $p_k(0) = 0$. This yields

$$f(\mathbf{x}^k) = \frac{1}{2}(\mathbf{r}^k, \mathbf{r}^k)_0 = \frac{1}{2}\|\mathbf{r}^0 + p_k(B)\mathbf{r}^0\|_0^2.$$

We may regard $-p_k(B)\mathbf{r}^0$ as an approximation of $\mathbf{r}^0$ and $\|\mathbf{r}^0 + p_k(B)\mathbf{r}^0\|_0$ as the corresponding error. Since all previous search directions are used at each stage, then

(12.16) $$(\mathbf{r}^k, \mathbf{r}^k)_0 = \min_{p_k \in \pi_k^0} \|[I + p_k(B)]\mathbf{r}^0\|_0^2,$$

where π_k^0 denotes the set of polynomials of degree k at most, such that $p_k(0) = 0$. This is true whatever choice of $\beta_j^{(k)}$ has been made, as long as $\{Bd^j\}$, $k - s_k \le j \le k$ is a linearly independent set. As follows by Theorem 12.5, the choice recommended in (12.9) implies that this set is linearly independent.

To solve (12.16) is a pure approximation problem: The rate of convergence $\|\mathbf{r}^k\|_0 \to 0$ is determined by the best polynomial approximation. Recall the following from Chapter 2:

(a) A polynomial $q(\lambda)$ of degree ≥ 1 is called an *annihilating polynomial* of A if $q(A) = 0$. We shall consider here only real polynomials. Note that $q(\lambda) = \det(A - \lambda I)$ is an annihilating real polynomial of degree n if A is a real matrix of order n (Cayley-Hamilton theorem).

(b) An annihilating polynomial of A of minimal degree is called a *minimal polynomial* of A. Its degree is denoted by $m(A)$.

It follows that $m(A) \le n$. If A is nonsingular and if the lowest order coefficient of the minimal polynomial, denoted by $\hat{p}_m(\lambda)$, is zero, then $A^{-1}\hat{p}_m(A) = 0$ and $p_{m-1}(\lambda) = \lambda^{-1}\hat{p}_m(\lambda)$ would be an annihilating polynomial of degree $m - 1$. Hence, the lowest-order coefficient is nonzero. We shall normalize the minimal polynomial corresponding to the nonsingular matrix B so that this coefficient is equal to 1. The minimal polynomial then is unique. Now let us generalize this property.

Definition 12.1 Let A be a square matrix. Any polynomial $q(\lambda)$ of minimal degree denoted by $m(A, \mathbf{r}^0)$, such that $q(A)\mathbf{r}^0 = \mathbf{0}$ is called a *minimal polynomial* of A and $\mathbf{r}^0$. Clearly, $m(A, \mathbf{r}^0) \le m(A)$.

For a nonsingular matrix A, we normalize the minimal polynomial of A and $\mathbf{r}_0$ as described above.

Theorem 12.8 *Unless stagnation, i.e.,* $(\mathbf{r}^k, B\mathbf{r}^k)_0 = 0$ *occurs at some step with* $\mathbf{r}^k \neq 0$, *the full generalized conjugate gradient least square method terminates with residual* $= 0$ *after* $m(B, \mathbf{r}^0)$ *steps.*

Proof By Definition 12.1, the minimal polynomial q_m to B and $\mathbf{r}^0$ satisfies

$$(12.17) \qquad\qquad q_m(B)\mathbf{r}^0 = \mathbf{0}.$$

Note that by the normalization of the minimal polynomial, $q_m(0) = 1$. It follows from (12.17) that for some polynomial $\hat{p}_m$ of degree m, such that $\hat{p}_m(0) = 0$, we have $\mathbf{r}^0 + \hat{p}_m(B)\mathbf{r}^0 = \mathbf{0}$. Hence, if no stagnation occurs, the GCG-LS method will generate this polynomial, and by (12.16)

$$0 \leq (\mathbf{r}^m, \mathbf{r}^m)_0 = \min_{p_m \in \pi_m^0} \|[I + p_m(B)]\mathbf{r}^0\|_0^2 = 0,$$

where the minimal value is taken for $p_m = \hat{p}_m$. Thus, $\mathbf{r}^m = \mathbf{0}$ (and m is the smallest number for which this can occur), which means that the algorithm terminates after $m = m(B, \mathbf{r}^0)$ steps. $\diamond$

Corollary 12.9 *Assume that* $\mathbf{r}^0$ *can be written as a linear combination of some eigenvectors of* B. *Let* ν *be the number of distinct values of the corresponding eigenvalues of* B. *Then the full version of the generalized conjugate gradient least square method terminates in, at most,* ν *steps.*

Proof By assumption, we have $[I + p_k(B)]\mathbf{r}^0 = \sum_{i=1}^{\nu'} a_i[1 + p_k(\lambda_i)]\mathbf{v}_i$ for some scalars a_i and eigensolutions $(\lambda_i, \mathbf{v}_i)$, $i = 1, \ldots, \nu'$ of B, where $\nu' \geq \nu$, and ν' is the number of eigenvalues (counting multiplicities) corresponding to the eigenvalue expansion of $\mathbf{r}^0$. Here p_k is a real polynomial.

Since B is a real matrix, its eigenvalues are real or conjugate complex. Hence, if $k = \nu$, we may choose $p_k = \hat{p}_k$, such that $1 + \hat{p}_k(\lambda_i) = 0$, $i = 1, \ldots, \nu$, where $\hat{p}_k \in \pi_k^0$ is real. Then, $[I + \hat{p}_k(B)]\mathbf{r}^0 = \mathbf{0}$ if $k = \nu$, that is, the minimal polynomial to B and $\mathbf{r}^0$ has degree, at most, ν. By virtue of Theorem 12.8, the algorithm terminates in, at most, ν steps. $\diamond$

It can be seen from a result to be derived in Chapter 13 that for a degenerate matrix B, the method terminates in, at most, $\sum s_i$ steps, where the summation takes place over the distinct eigenvalues of B and where s_i is the maximal order of any Jordan box for B corresponding to an eigenvalue λ_i. (Here $s_i = 1$ if the corresponding eigenvector space to λ_i is complete.)

12.1.2 Efficient Combinations of Inner Products and Computational Complexity

Consider the case where $B = C^{-1}A$ and C is a preconditioner of A. In a straightforward application of the GCG-LS(s) method, we must reserve computer storage for $k + 1$ vectors $\mathbf{d}^j$, $k + 1$ vectors $B\mathbf{d}^j$, $0 \leq j \leq k$ (if $k \leq s$), and for $\mathbf{x}^{k+1}$, $\mathbf{r}^{k+1}$, $B\mathbf{r}^{k+1}$: in total, $2k + 5$ vectors. However, when the inner products $(\cdot, \cdot)_0$ and $(\cdot, \cdot)_1$ are equal, it suffices to store only one of the sets $\{\mathbf{d}^j\}$ and $\{B\mathbf{d}^j\}$, $0 \leq j \leq k$, and the total demand of storage is reduced to $k + 5$ vectors. The reason for this is that when the inner products are equal, the conjugate orthogonality (12.9) among the vectors $\{\mathbf{d}^j\}$ implies that the matrix $\Lambda^{(k)}$ in (12.7) becomes diagonal. Therefore, in addition to the parameters β_j, $0 \leq j \leq k$, we need to compute only the parameter

$$\alpha_k^{(k)} = (\mathbf{r}^k, B\mathbf{r}^k)_0/(B\mathbf{d}^k, B\mathbf{d}^k)_0.$$

Let us consider first two versions of such methods.

Algorithm 1 Here we let $(\cdot, \cdot)_0 = (\cdot, \cdot)_1$ be the standard inner product—i.e., we let $M_0 = M_1 = I$. We must store $\{\mathbf{d}^j\}$, $\{B\mathbf{d}^j\}$, $0 \leq j \leq k$ ($k < s$), $\mathbf{x}^{k+1}$, $\mathbf{r}^{k+1}$, and $B\mathbf{r}^{k+1}$: in total, $2k + 5$ vectors. After proper initialization steps, it takes the form:

Compute $(B\mathbf{d}^k)^T B\mathbf{d}^k$;
$\alpha_k^{(k)} := (B\mathbf{r}^k)^T \mathbf{r}^k/(B\mathbf{d}^k)^T B\mathbf{d}^k$;
$\mathbf{x}^{k+1} := \mathbf{x}^k + \alpha_k^{(k)}\mathbf{d}^k$;
$\mathbf{r}^{k+1} := \mathbf{r}^k + \alpha_k^{(k)} B\mathbf{d}^k$;
compute $B\mathbf{r}^{k+1}$;
$\beta_j^{(k)} := (B\mathbf{r}^{k+1})^T B\mathbf{d}^j/(B\mathbf{d}^j)^T B\mathbf{d}^j$, $k - s_k \leq j \leq k$;
$\mathbf{d}^{k+1} := -\mathbf{r}^{k+1} + \sum_{k-s_k}^k \beta_j^{(k)}\mathbf{d}^j$;
$B\mathbf{d}^{k+1} := -B\mathbf{r}^{k+1} + \sum_{k-s_k}^k \beta_j^{(k)} B\mathbf{d}^j$;

Hence, per iteration step there are $s_k + 3$ inner products, $2s_k + 4$ linked triadic vector operations, one matrix-vector multiplication with $B = C^{-1}A$, which includes a matrix-vector multiplication with A and a solution with the preconditioning matrix C. If we compute $B\mathbf{d}^{k+1}$ as $C^{-1}A\mathbf{d}^{k+1}$, we save $s_k + 1$ linked triadic vector operations, but at the expense of an extra multiplication with A and a solution with C.

Algorithm 1 computes the minimum pseudoresidual solution—that is, a solution for which $r^{k+1^T} r^{k+1}$, where $r^{k+1} = Bx^{k+1} - b$ is minimized at each step.

Remark 12.10 The special case of the generalized conjugate gradient method, where $M_0 = M_1 = I$ (the identity matrix), was presented in Vinsome (1976) and was called the ORTHOMIN method. An earlier presentation of minimal residual methods can be found in Khabaza (1963).

Algorithm 2 Here we let $M_0 = M_1 = C^T C$ (note that this matrix is s.p.d.). We need to store d^j, $0 \le j \le k$, x^{k+1}, r^{k+1}, $\tilde{r}^{k+1}$ and h^{k+1}; the latter vectors will be defined below. By the above choice of inner products, this reduction of storage can be achieved without increasing the computational cost.

Notice first that the old vectors Bd^j, $0 \le j \le k$, are required only for the computation of the numerators in (12.10), because the values of the denominators can be stored from previous steps. Now we rewrite the numerator as

$$(Br^{k+1}, Bd^j)_1 = (h^{k+1})^T d^j,$$

where

(12.18) $$h^{k+1} = B^T M_1 Br^{k+1}.$$

By use of (12.18) we can now compute $\beta_j^{(k)}$ in (12.10), with no need to store the vectors Bd^j. However, to compute h^{k+1} in this form would require matrix-vector multiplications with B, M_1, and B^T. To reduce the number of matrix-vector multiplications, we shall consider the inner product where $M_0 = C^T C$. Then (12.18) takes the form

$$h^{k+1} = A^T C^{-T} C^T C (C^{-1} A) r^{k+1},$$

that is,

$$h^{k+1} = A^T A r^{k+1}.$$

In practical applications, the cost of a matrix-vector multiplication with the given matrix A, or with its transpose A^T, is usually smaller than the cost of solving a linear system with the preconditioning matrix C. *Hence it is cost-efficient to have just one application of the preconditioner per iteration step, and this occurs now only in the computation of the pseudoresidual* r^{k+1}. At the same time, we can reduce the number of linked triadic operations per step to $s_k + 3$, but at the expense of an extra multiplication with A and one with A^T, as we shall see. Furthermore, with this choice of inner product, the remaining computations of the coefficients $\beta_j^{(k)}$ and $\alpha_j^{(k)}$ also simplify. We have, namely,

$$(B\mathbf{d}^k, B\mathbf{d}^k)_1 = (B\mathbf{d}^k)^T C^T C (B\mathbf{d}^k) = (A\mathbf{d}^k)^T A\mathbf{d}^k,$$

$$(\mathbf{r}^k, B\mathbf{d}^k)_0 = \mathbf{r}^{k^T} C^T C B\mathbf{d}^k = (C\mathbf{r}^k)^T A\mathbf{d}^k.$$

We collect these results in the next theorem.

Theorem 12.11 *Let* $(\cdot, \cdot)_0 = (\cdot, \cdot)_1$ *and* $M_0 = M_1 = C^T C$. *Then, for the* GCG-LS(s) *method with* $B = C^{-1}A$,

$$\alpha_k^{(k)} = \frac{(\mathbf{r}^k, B\mathbf{r}^k)_0}{(B\mathbf{d}^k, B\mathbf{d}^k)_0} = \frac{(C\mathbf{r}^k)^T A\mathbf{d}^k}{(A\mathbf{d}^k)^T A\mathbf{d}^k}, \quad \alpha_j^{(k)} = 0, \; k - s_k \leq j \leq k - 1,$$

$$\beta_j^{(k)} = \frac{\mathbf{h}^{k+1^T} \mathbf{d}^j}{(A\mathbf{d}^j)^T A\mathbf{d}^j}, \quad k - s_k \leq j \leq k,$$

where $\mathbf{h}^{k+1} = A^T A\mathbf{r}^{k+1}$. *This algorithm computes the minimum residual solution (w.r.t. the true residual), that is,* $(A\mathbf{x}^k - \mathbf{b})^T (A\mathbf{x}^k - \mathbf{b})$ *is minimized over all solutions* $\mathbf{x}^k = \mathbf{x}^0 \oplus B V_{k-1}(\mathbf{r}^0)$, *where*

$$V_{k-1}(\mathbf{r}^0) = \text{span}\{\mathbf{r}^0, B\mathbf{r}^0, \dots, B^{k-1}\mathbf{r}^0\}.$$

The algorithm takes the following form:

Algorithm 2(a)

    Compute $A\mathbf{d}^k$ (can be stored in the same space as $\mathbf{r}^k$);
    compute $(A\mathbf{d}^k)^T A\mathbf{d}^k$;
    compute $\tilde{\mathbf{r}}^{k^T} A\,\mathbf{d}^k$;
    $\alpha_k^{(k)} := -\tilde{\mathbf{r}}^{k^T} A\mathbf{d}^k / (A\mathbf{d}^k)^T A\mathbf{d}^k$;
    $\mathbf{x}^{k+1} := \mathbf{x}^k + \alpha_k^{(k)} \mathbf{d}^k$;
    $\tilde{\mathbf{r}}^{k+1} := \tilde{\mathbf{r}}^k + \alpha_k^{(k)} A\mathbf{d}^k$; (here $\tilde{\mathbf{r}}^{k+1} = A\mathbf{x}^k - \mathbf{b}$)
    solve $C\mathbf{r}^{k+1} = \tilde{\mathbf{r}}^{k+1}$;
    $\mathbf{h}^{k+1} := A^T A\,\mathbf{r}^{k+1}$;
    $\beta_j^{(k)} := (\mathbf{h}^{k+1})^T \mathbf{d}^j / (A\mathbf{d}^j)^T A\mathbf{d}^j, \quad k - s_k \leq j \leq k$;
    $\mathbf{d}^{k+1} := -\mathbf{r}^{k+1} + \sum_{k-s_k}^{k} \beta_j^{(k)} \mathbf{d}^j$;

Hence, per iteration step, there are $s_k + 3$ inner products, $s_k + 3$ linked triadic vector operations, two matrix-vector multiplications with A and one with A^T, and one solution with the preconditioning matrix.

Algorithm 3 In the above algorithm, we need to compute the action of A twice and that of A^T once. In some applications, such as for reduced Schur complement matrix systems, for instance, arising from certain domain-decomposed problems, the cost of computing the action of A can be much larger than the remaining computations in the GCG-LS method, including the cost of the preconditioner. We now consider a variant of the method where only one action of A (and still only one action of C^{-1}) is required, but we need to store two sets of vectors, $\{\mathbf{d}^k, \mathbf{d}^{k-1}, \ldots, \mathbf{d}^{k-s_k}\}$ and $\{A\mathbf{d}^k, A\mathbf{d}^{k-1}, \ldots, A\mathbf{d}^{k-s_k}\}$, in addition to the vectors $\mathbf{x}^{k+1}, \mathbf{r}^{k+1}, \tilde{\mathbf{r}}^{k+1}$. *Hence, this version is efficient mostly in cases where we have reduced the order of the vectors significantly by solving a reduced Schur complement system* instead of the original system. (The Schur complement system is usually not formed explicitly.) Here it is expected that the major cost is the application of the action of A, while the remaining computations are relatively minor because they involve only vectors of the order of the reduced system.

In this variant, we let the inner products be defined by

$$M_0 = C^T C \quad \text{and} \quad M_1 = (BB^T)^{-1}.$$

Then the inner products involved in the method take the form

$$(B\mathbf{r}^{k+1}, B\mathbf{d}^j)_1 = (\mathbf{r}^{k+1})^T \mathbf{d}^j, \quad k - s_k \le j \le k,$$

$$(B\mathbf{d}^k, B\mathbf{d}^k)_1 = \mathbf{d}^{k^T} \mathbf{d}^k,$$

$$(\mathbf{r}^k, B\mathbf{d}^k)_0 = (C\mathbf{r}^k)^T A\mathbf{d}^k,$$

and

$$(B\mathbf{d}^i, B\mathbf{d}^j)_0 = (A\mathbf{d}^i)^T A\mathbf{d}^j, \quad k - s_k \le i, j \le k.$$

In this algorithm, we will also compute *the best minimal (true) residual solution*. It takes the following form:

Compute $A\mathbf{d}^k$;
compute $(A\mathbf{d}^k)^T A\mathbf{d}^j, \quad k - s_k \le j \le k$;
compute $(\tilde{\mathbf{r}}^k)^T A\mathbf{d}^k$;
solve $\Lambda^{(k)} \underline{\alpha}^{(k)} = \underline{\gamma}^{(k)}$;

compute $\mathbf{x}^{k+1} = \mathbf{x}^k + \sum_{j=k-s_k}^{k} \alpha_j^{(k)} \mathbf{d}^j$;

compute $\tilde{\mathbf{r}}^{k+1} = \tilde{\mathbf{r}}^k + \sum_{j=k-s_k}^{k} \alpha_j^{(k)} A\mathbf{d}^j$; (here $\tilde{\mathbf{r}}^{k+1} = A\mathbf{x}^{k+1} - \mathbf{b}$).

Solve $C\mathbf{r}^{k+1} = \tilde{\mathbf{r}}^{k+1}$;

compute $\beta_j^{(k)} = (\mathbf{r}^{k+1})^T \mathbf{d}^j / {\mathbf{d}^j}^T \mathbf{d}^j$, $k - s_k \leq j \leq k$;

compute $\mathbf{d}^{k+1} = -\mathbf{r}^{k+1} + \sum_{j=k-s_k}^{k} \beta_j^{(k)} \mathbf{d}^j$.

The computational cost of this algorithm consists of $2s_k + 4$ inner products, $3s_k + 3$ linked triadic vector operations, and one action of A and one of C^{-1}. Therefore, the computational cost might be too large when k grows, so we need to limit k to a small number, either by choosing s (the maximal number of search directions used) small or by *restarting* the algorithm after, say, every s iteration.

12.1.3 Automatic Truncation of the Generalized Conjugate Gradient Method

We now consider conditions for which the recurrence relations in the generalized conjugate gradient method with $\mathbf{d}^0 = -\mathbf{r}^0$ truncate to short-term recurrence relations. Note, first, that by the construction of $\mathbf{d}^k$ and $\mathbf{r}^k$ in (12.8) and (12.4b), respectively, it follows that $\mathbf{d}^k$ is a linear combination of the vectors $B^j \mathbf{r}^0, 0 \leq j \leq k$, which form the Krylov set,

$$V_k = V_k(\mathbf{r}^0, B) = \text{span}\{\mathbf{r}^0, B\mathbf{r}^0, \ldots, B^k \mathbf{r}^0\}.$$

Similarly, $\mathbf{r}^k$ is a direct sum

$$\mathbf{r}^k = \mathbf{r}^0 \oplus B V_{k-1}$$

of $\mathbf{r}^0$ and a linear combination of the vectors in $B V_{k-1}$. Hence, there is a relation

$$\mathbf{r}^k = [I - Bq_{k-1}(B)]\mathbf{r}^0, \quad k = 1, 2, \ldots$$

for some polynomial q_{k-1}, of degree $k - 1$. Also,

$$f(\mathbf{x}^k) = \frac{1}{2}(\mathbf{r}^k, \mathbf{r}^k)_0 = \frac{1}{2}\|\mathbf{r}^0 - Bq_{k-1}(B)\mathbf{r}^0\|_0^2.$$

We may view $Bq_{k-1}(B)\mathbf{r}^0$ as an approximation of $\mathbf{r}^0$ and $\|\mathbf{r}^0 - Bq_{k-1}(B)\mathbf{r}^0\|_0$ as the corresponding error. Furthermore, $q_{k-1}(B)$ can be considered a matrix polynomial approximating B^{-1}. However, this approximation is biased, because it is based only on that part of the spectral information of B involved in the spectral representation of $\mathbf{r}^0$ in the eigenvectors of B. If B is degenerate, then $\mathbf{r}^0$ may have components not represented by the eigenvectors of B, and then the situation becomes more involved. (For related comments see Chapter 13.) The matrix polynomial q_{k-1} depends not only on B and $\mathbf{r}^0$ but also on the inner products chosen for the minimization of $f(\mathbf{x})$.

Let us now consider conditions for the automatic truncation of the recurrence relations occurring in the generalized conjugate gradient algorithm. We recall (Chapter 2) that given a Hermitian positive definite matrix H, the matrix $B' = H^{-1}B^*H$ is called the H-adjoint to B. The following are also equivalent:

(a) B' commutes with B for some H.

(b) B is diagonalizable by a similarity transformation for some H.

(c) There exists a polynomial p and an H such that $B' = p(B)$.

It is readily seen that if $\langle \mathbf{u}, \mathbf{v} \rangle = \mathbf{u}^*H\mathbf{v}$, the inner product defined by the Hermitian positive definite matrix H, then $\langle B'\mathbf{u}, \mathbf{v} \rangle = \langle \mathbf{u}, B\mathbf{v} \rangle$, that is, B' is the adjoint to B with respect to this inner product. Note also that $(B')' = B$.

If B commutes with its H-adjoint, we say that B is H-normal. If B is H-normal, then the polynomial $\hat{p}$ of smallest degree, $n(B, H)$ such that $B' = \hat{p}(B)$, is called the H-normal polynomial to B and $n(B, H)$ is called the H-normal degree. We now extend this definition.

Definition 12.2 A polynomial $\hat{p}$ of smallest degree, $n(B, H, \mathbf{r}^0)$ [possibly $n(B, H, \mathbf{r}^0) = +\infty$], such that

$$B'q(B)\mathbf{r}^0 = \hat{p}(B)q(B)\mathbf{r}^0$$

for any polynomial q, is called the *H-normal polynomial to B with respect to* $\mathbf{r}^0$, and $n(B, H, \mathbf{r}^0)$ is called the *H-normal degree w.r.t.* $\mathbf{r}^0$.

Clearly, if B is H-normal and of order N, then it is H-normal w.r.t. any vector $\mathbf{r}^0$ of dimension N, so $n(B, H, \mathbf{r}^0) \leq n(B, H)$.

Example 12.1

(a) Let B have the form

$$B = \begin{bmatrix} B_{11} & 0 \\ 0 & B_{22} \end{bmatrix},$$

where B_{ii} are square, and assume that B_{11} has order m. Then B has the same (I-normal) degree w.r.t. any vector $\mathbf{v} = \begin{bmatrix} \mathbf{v}_1 \\ 0 \end{bmatrix}$, where $\mathbf{v}_1$ has dimension m, as B_{11} has w.r.t. $\mathbf{v}_1$.

(b) Let B be the shift matrix of order n,

$$B = \begin{bmatrix} 0 & 1 & 0 & \cdots & 0 \\ 0 & 0 & 1 & \cdots & 0 \\ \cdot & \cdot & \cdot & & \\ 0 & 0 & 0 & \cdots & 1 \\ 1 & 0 & 0 & \cdots & 0 \end{bmatrix}.$$

Let $\mathbf{e}_1 = (1, 1, \ldots, 1)^T$. Then $B\mathbf{e} = \mathbf{e}$, $B^T\mathbf{e} = \mathbf{e}$, and B has degree 1 w.r.t. $\mathbf{e}$. [Here $\hat{p}(B) = B$.] However, with $\mathbf{v} = \mathbf{e}_i$, the ith coordinate vector, then $B\mathbf{e}_i = \mathbf{e}_{i-1}$, $B^T\mathbf{e}_i = \mathbf{e}_{i+1}$ (where $\mathbf{e}_0 = \mathbf{e}_n$ and $\mathbf{e}_{n+1} = \mathbf{e}_1$) and B has degree $n - 1$ w.r.t. such a vector $\mathbf{v}$. [Here $\hat{p}(B) = B^{n-1}$.]

We can now prove the main theorem for automatic truncation. In stating this theorem, we assume exact arithmetic operations—i.e., that no round-off errors occur. We consider only the case where $M_0 = M_1$.

Theorem 12.12 *Let $M_0 = M_1$ be Hermitian and positive definite and let B be M_0-normal w.r.t. $\mathbf{r}^0$ of M_0-normal degree w.r.t. $\mathbf{r}^0$, $n = n(B, M_0, \mathbf{r}^0)$. Assume that $M_0 B + B^T M_0$ is positive definite. Then, the truncated generalized conjugate gradient least square GCG-LS(s) algorithm is identical to the full (i.e. untruncated) algorithm if and only if $s = n(B, M_0, \mathbf{r}^0) - 1$.*

Proof Consider the full version, where $\beta_j^{(k)}$ is determined by (12.10). By assumption, we have $(\mathbf{u}, \mathbf{v})_0 = (\mathbf{u}, \mathbf{v})_1 = \mathbf{u}^* M_0 \mathbf{v}$. Since $\mathbf{d}^j \in V_j$, the Krylov degree j, with $B' = M_0^{-1} B^T M_0$ we have

(12.19) $\quad (B\mathbf{r}^{k+1}, B\mathbf{d}^j)_1 = (\mathbf{r}^{k+1}, B'B\mathbf{d}^j)_1 = (\mathbf{r}^{k+1}, B'Bp_j(B)\mathbf{r}^0)_1$

for some polynomial p_j of degree j, $0 \le j \le k$. By assumption, B is M_0-normal w.r.t. $\mathbf{r}^0$. Hence,

(12.20) $\quad\quad\quad (B\mathbf{r}^{k+1}, B\mathbf{d}^j)_0 = (\mathbf{r}^{k+1}, \hat{p}_n(B)Bp_j(B)\mathbf{r}^0)_0,$

where $\hat{p}_n$ is the M_0-normal polynomial w.r.t. B and $\mathbf{r}^0$, and has degree $n = n(B, M_0, \mathbf{r}^0)$.

The degree of the polynomial $\hat{p}_n(B)p_j(B)$ is $n + j$. On the other hand, by Lemma 12.1(b) we have

(12.21) $\quad\quad\quad\quad (\mathbf{r}^{k+1}, Bp_k(B)\mathbf{r}^0)_0 = 0$

for any polynomial p_k of degree k or less. Hence, since $(\cdot, \cdot)_0 = (\cdot, \cdot)_1$, (12.19) and (12.20) show that

(12.22) $(Br^{k+1}, Bd^j)_1 = 0$,   if $n + j \leq k$ or $j \leq k - n$.

By (12.10), this shows that, for $k \geq n$, $\beta_j^{(k)} = 0$, $0 \leq j \leq k - n$, so the recurrence (12.8) can be written

$$d^{k+1} = -r^{k+1} + \sum_{j=k-s}^{k} \beta_j^{(k)} d^j, \quad \text{for any } s \geq n - 1.$$

Further, (12.7) and $(\cdot, \cdot)_0 = (\cdot, \cdot)_1$ show that $\Lambda^{(k)}$ is a diagonal matrix and $\alpha_j^{(k)} = 0$, $0 \leq j \leq k - 1$. Thus, the truncated algorithm GCG-LS(s), with $s \geq n - 1$, is identical to the full version.

To show the converse statement, assume now that the truncated version is identical to the full version for some $s < n - 1$. Then, $\beta_j^{(k)} = 0$, $0 \leq j \leq k - (s + 1)$, and (12.19) and (12.21) show that

$$(r^{k+1}, B'Bp_j(B)r^0 - Bp_k(B)r^0)_0 = 0, \quad 0 \leq j \leq k - (s + 1),$$

or

$$(r^{k+1}, (p_n(B)p_j(B) - p_k(B))Br^0)_0 = 0, \quad 0 \leq j \leq k - (s + 1).$$

Here, for $j = k - n + 1$, $k \geq n$, the degree of the matrix polynomial $p_n(B)\, p_j(B) - p_k(B)$ is $n + j = k + 1$, so, for some polynomial $\tilde{q}_{k+1}$ of degree $k + 1$ (exactly),

$$(r^{k+1}, B\tilde{q}_{k+1}(B)r^0)_0 = 0,$$

or, by the orthogonality (12.5),

$$(r^{k+1}, \gamma Br^{k+1})_0 = 0$$

for some scalar $\gamma \neq 0$. Since, by assumption, $M_0B + B^T M_0$ is positive definite, this shows that $r^{k+1} = 0$, that is, the solution ($x = x^{k+1}$) has already been found. Hence, the algorithm will either terminate at some step ($k + 1$), $k \leq n - 1$ or the GCG-LS(s) truncates with $s = n - 1$.   ◇

Remark 12.13 If we assume that the GCG-LS(s) algorithm truncates for some s for *any* initial vector, it follows that

$$(p_{k+1}(B)r^0, (B'p_j(B) - p_k(B))Br^0)_0 = 0, \quad 0 \leq j \leq k - (s + 1)$$

for any $\mathbf{r}^0$ where $B' = M_0^{-1} B^* M_0$. It can be seen that this can only hold if $B' p_j(B) - p_k(B)$, $0 \le j \le k - (s + 1)$, is a matrix polynomial in B, implying that B' is a matrix polynomial in B, that is, B is M_0-normal. For further discussions related to this topic, see Faber and Manteuffel (1984) and Joubert and Young (1987). In practice, however, we are interested only in the existence of an H-normal matrix w.r.t. the particular initial residual $\mathbf{r}^0$, which corresponds to the chosen initial approximation $\mathbf{x}^0$. In many cases, we can choose this initial approximation so that $\mathbf{r}^0$ will be a vector in a proper subspace of $\mathbb{R}^N$, where N is the order of the matrix B. For vectors from this subspace, it may hold that the M_0-normal degree is small.

We now present two important examples of M_0-normal matrices B.

Example 12.2 Let $B = C^{-1}A$, where C and A are s.p.d., and let $M_0 = CA^{-1}C$. Then, for the M_0-adjoint matrix B' to B, we have

$$B' = M_0^{-1} B^T M_0 = C^{-1}AC^{-1}(AC^{-1})CA^{-1}C = C^{-1}A = B.$$

Hence, B' is selfadjoint w.r.t. the inner product defined by M_0, and B is M_0-normal with degree 1. Theorem 12.12 shows that the GCG-LS(0) method is equivalent to the full GCG-LS method. Moreover, after some manipulations, one finds that the method reduces to the standard preconditioned conjugate gradient method (see Chapter 11), with preconditioning matrix C, and the algorithm computes the best approximation with respect to the norm $\{\mathbf{r}^T A^{-1} \mathbf{r}\}^{\frac{1}{2}}$, where $\mathbf{r} = A\mathbf{x} - \mathbf{b}$.

Example 12.3 Let $A = M + N$, where $M = 1/2(A + A^T)$, $N = 1/2(A - A^T)$, that is, consider the splitting of A in its symmetric and skewsymmetric parts. Assume that M is s.p.d., and use M as a preconditioner to A, that is, let $B = M^{-1}A = I + M^{-1}N$. Further, let $M_0 = M$. We then have $B' = M^{-1}B^T M = M^{-1}(I - NM^{-1})M = I - M^{-1}N = 2I - B$. Hence, $B' = p_1(B)$, where $p_1(\lambda) = 2 - \lambda$ and B is of M-normal degree 1, and Theorem 12.12 shows that the corresponding GCG-LS(0) method is identical to the full version. The algorithm is similar to the method by Concus and Golub (1976) and Widlund (1978), the so called CGW-method.

Relevant to Example 12.3, note that any system $A\mathbf{x} = \mathbf{a}$ with a nonsingular matrix A can be transformed into a system, $B\mathbf{x} = (\nu A^T A + A)\mathbf{x} = \nu A^T \mathbf{a} + A\mathbf{x} = \mathbf{b}$, where $\nu > 0$ is large enough to satisfy the condition that $B + B^T = 2\nu A^T A + A^T + A$ is positive definite. Here we choose ν small but large enough so that this condition is fulfilled. Naturally, if ν must be taken too large, we might as well consider the normal equations $A^T A\mathbf{x} = A^T \mathbf{a}$. We hope that v can be taken so small that the condition number of $vA^T A + A$ is much smaller than that for $A^T A$.

Examples 12.2 and 12.3 above show cases where the M_0-normal degree of a matrix B is equal to 1 and the GCG-LS(0) method is identical to the full (untruncated) version. The following disappointing result shows that unless the M_0-normal degree is 1, it will, in general, be a large number. The following considerations follow from Theorem 2.23.

Let H be a Hermitian positive definite matrix and assume that B is H-normal. Let $n(B, H) > 1$ be the H-normal degree of B, and let $m(B)$ be its minimal polynomial degree. Then,

$$n(B, H) \geq m(B)^{\frac{1}{2}}.$$

Theorem 12.12 shows that unless $n(B, H) = 1$, we must, in general, choose $s \geq n(B, H) - 1 \geq m(B)^{\frac{1}{2}} - 1$ to get a truncated algorithm identical to the full version for an arbitrary initial vector. As a general rule, $m(B)$ is very large. We hope, therefore, that the normal degree $n(B, H, \mathbf{r}^0)$ is much smaller than $n(B, H)$ [if $n(B, H) > 1$] for the particular vector $\mathbf{r}^0 = B\mathbf{x}^0 - \mathbf{b}$.

The following corollary is also of interest.

Corollary 12.14 *If B is M_0-normal w.r.t. $\mathbf{r}^0$ or M_0-normal of degree $n = n(B, M_0, \mathbf{r}^0)$, then the GCG-LS$(n - 1)$ method terminates with residual $= 0$ after $m(B, \mathbf{r}^0)$ steps.*

Proof This proof follows from Theorems 12.12 and 2.23. $\diamond$

Remark 12.15 The case of automatic truncation of the GCG-LS method for an arbitrary vector $\mathbf{r}^0$ has been addressed by Voevodin (1983) and by Faber and Manteuffel (1984). As we have seen above (except for exceptional matrices), $n = 1$ is a necessary condition for an $(s + 1)$-term expansion—i.e., that the full-term expansion is identical to the GCG-LS(0) algorithm.

Let $M_0 = M_1 = I$. If a matrix has normal degree $n = 1$, then $B^* = p(B)$, where p is of first degree. Let $p(B) = \alpha_0 + i\beta_0 + (\alpha_1 + i\beta_1)B$ for some real scalars α_k, β_k, $k = 0, 1$. Then, if $\lambda + i\mu$ is an eigenvalue of B, $B^* = p(B)$ shows that

$$\lambda - i\mu = \alpha_0 + i\beta_0 + (\alpha_1 + i\beta_1)(\lambda + i\mu)$$

or

$$(1 - \alpha_1)\lambda + \beta_1\mu = \alpha_0,$$

$$\beta_1\lambda + (1 + \alpha_1)\mu = -\beta_0.$$

Hence, if the corresponding determinant, $1 - \alpha_1^2 - \beta_1^2 \neq 0$, the equation can only be satisfied for a single eigenvalue $(\lambda + i\mu)$. However, if $\alpha_1^2 + \beta_1^2 = 1$, then (letting $\alpha_1 \neq 1$) we have $\lambda + \frac{\beta_1}{1-\alpha_1}\mu = \frac{\alpha_0}{1-\alpha_1}$ for any eigenvalue $\lambda + i\mu$. Thus, the eigenvalues of B must be situated along a straight line in the complex plane. As also pointed out by Voevodin (1983) and Faber and Manteuffel (1984), this means that we are essentially back to the classical case where the eigenvalues of B are real (and positive).

12.1.4 General Subspace Iteration Methods

The primary purpose of the generalized conjugate gradient least squares method is to compute approximations $\mathbf{x}^{k+1}$ to the solution of $B\mathbf{x} = \mathbf{b}$, where $\mathbf{x}^{k+1} - \mathbf{x}^k$ is a linear combination of some search direction vectors, such that the residual $\mathbf{r}^{k+1} = B\mathbf{x}^{k+1} - \mathbf{b}$ is minimized. This has been analyzed above for the case where the search directions $\{\mathbf{d}^j\}$ are computed from the Krylov set of vectors. A natural extension of the method is to let the search directions be taken from a more general vector subspace than the Krylov subspace.

Assume that the vectors $\mathbf{d}^0, \mathbf{d}^1, \dots, \mathbf{d}^k$ have been generated such that $B\mathbf{d}^0, B\mathbf{d}^1, \dots, B\mathbf{d}^k$ are linearly independent. As in (12.1)-(12.3), vectors $\mathbf{x}^{k+1}$ are computed to minimize

$$f(\mathbf{x}) = \frac{1}{2}(B\mathbf{x} - \mathbf{b}, B\mathbf{x} - \mathbf{b})_0.$$

Hence, the vector $\underline{\alpha}^{(k)}$ of coefficients $\alpha_j^{(k)}$ in (12.1) must satisfy (12.7). Since $\{B\mathbf{d}^j\}_{k-s_k}^k$ form a linearly independent set of vectors, the matrix $\Lambda^{(k)}$ in (12.7) is nonsingular, and there exists a unique set of parameters $\alpha_j^{(k)}$ which minimize $f(\mathbf{x})$.

In general, we must store both of the sets of vectors $\mathbf{d}^j$ and $B\mathbf{d}^j$, $k - s_k \leq j \leq k$. There are $s_k + 2$ inner products and $2s_k + 2$ linked triadic vector computations. If we compute $\mathbf{r}^{k+1} = B\mathbf{x}^{k+1} - \mathbf{b}$ instead of using the recurrence (12.4b), then we save $s_k + 1$ of the latter computations, but at the expense of an extra computation with matrix B.

If we only want to generate the Krylov subspace, we can simply choose the search directions as

$$\mathbf{d}^{k+1} = -\mathbf{r}^{k+1}$$

or as

$$\mathbf{d}^{k+1} = -\mathbf{r}^{k+1} + \beta_k^{(k)} \mathbf{d}^k,$$

where, in the latter case, $\beta_k^{(k)}$ is chosen to make $(B\mathbf{d}^{k+1}, B\mathbf{d}^k)_0 = 0$, for instance. However, the search direction vector $\mathbf{d}^{k+1}$ is generally not orthogonal to the previous vectors. Least square methods of this type was first presented in Axelsson (1980). Convergence results similar to those in Theorem 12.5 (see also Remark 12.6) can be shown to hold also for the general subspace iteration methods.

12.2 Orthogonal Error Methods

There exist two classes of generalized conjugate gradient methods: (1) *minimum residual* or *least squares methods*, as presented in the previous section, and (2) *orthogonal error* or *orthogonal projection methods*, where the iteration error is made to be orthogonal with respect to a given bilinear form to the current Krylov, or another, vector subspace.

The first type of methods are also known as *Ritz* methods, while the latter are also known as *Galerkin* methods. If the orthogonal errors are computed with respect to a symmetric, positive definite form, it turns out that the corresponding Galerkin method is, in fact, a Ritz method. As we saw in the previous section, the convergence of Ritz methods is monotone and the corresponding approximations are best approximations in the corresponding norm. If the bilinear form is not symmetric, then the Galerkin approximations generally do not converge monotonically and the approximations are not optimal. It will be shown in Chapter 13 that they are quasi-optimal, in a certain sense, if the bilinear form is positive definite. The presentation here mainly follows the presentation in Axelsson (1979). For special choices of the bilinear form, it will be seen that the orthogonal error method reduces to methods known as the CGW-method and to a method similar to the popular GMRES method.

Let $a(\mathbf{u}, \mathbf{v})$ be a bilinear form, defined on $V \times V$, where V is equal to $\mathbb{R}^n$ or a subspace of $\mathbb{R}^n$. We assume that $a(\cdot, \cdot)$ is such that

$$a(\mathbf{u}, \mathbf{v}) = (F\mathbf{u}, \mathbf{v}), \quad \mathbf{u}, \mathbf{v} \in V$$

for some real matrix F of order $n \times n$, which is positive definite w.r.t. an inner product $(\mathbf{u}, \mathbf{v})$. This means that $a(\cdot, \cdot)$ is *positive definite*, that is, there exists a positive number ρ such that

$$(12.23a) \qquad a(\mathbf{u}, \mathbf{u}) \geq \rho \|\mathbf{u}\|^2, \quad \text{for all } \mathbf{u} \in V,$$

and *bounded*, that is, there exists a constant K such that

$$(12.23b) \qquad |a(\mathbf{u}, \mathbf{v})| \leq K \|\mathbf{u}\| \, \|\mathbf{v}\|, \quad \text{for all } \mathbf{u}, \mathbf{v} \in V.$$

Here, $\|\mathbf{u}\| = (\mathbf{u}, \mathbf{u})^{\frac{1}{2}}$, where $(\cdot, \cdot)$ is an inner product on V, which is defined by some given positive definite symmetric matrix M on V, that is,

$$(\mathbf{u}, \mathbf{v}) = \mathbf{u}^T M \mathbf{v}.$$

Let B be a given matrix on $V \to V$. We shall consider two bilinear forms,

$$(12.24a) \qquad a_1(\mathbf{u}, \mathbf{v}) = (B\mathbf{u}, \mathbf{v})$$

and

$$(12.24b) \qquad a_2(\mathbf{u}, \mathbf{v}) = (B\mathbf{u}, B\mathbf{v}).$$

In the first case, we assume that the M-symmetric part of B, $B^T M + M B$ is positive definite. In our applications:

   (a) $B = CA$,
   (b) $B = AC$,

or

   (c) $B = C^{-1}A$,

if, in the latter case, C is nonsingular, where C is a preconditioning matrix to A. If A is rectangular of order $m \times n$, then we must choose C to be rectangular of order $n \times m$ to make B a square matrix. The iteration method to be presented will be used to solve a linear system $B\mathbf{x} = \mathbf{b}$. Here, in the three above cases,

   $\mathbf{b} = C\mathbf{a}$, $A\mathbf{x} = \mathbf{a}$ is the given system [case (a)],
   $\mathbf{b} = \mathbf{a}$, $A\mathbf{y} = \mathbf{a}$ is the given system and $\mathbf{y} = C\mathbf{x}$ [case (b)],
   $\mathbf{b} = C^{-1}\mathbf{a}$ and $A\mathbf{x} = \mathbf{a}$ [case (c)].

Usually we try to choose C such that the M-symmetric part of B is positive definite and, in addition, we try to make the eigenvalues of B more clustered—i.e., contained in one or a few short intervals or small-size domains (such as ellipses). The goal is to make the iteration methods converge faster for this choice than for (most) other choices, while still keeping the cost of applying C—i.e., multiplications with C in (a) and (b) and solutions of systems in (c)—not much larger than the cost of a matrix-vector multiplication with A. The rate of convergence of generalized conjugate gradient methods is discussed in Chapter 13.

The bilinear form $a_1(\cdot, \cdot)$ need not be symmetric, that is, $a_1(\mathbf{u}, \mathbf{v}) \neq a_1(\mathbf{v}, \mathbf{u})$ in general. It is, however, symmetric—for instance, in the following two cases: (1) B is selfadjoint and commutes with M, and (2) $B = C^{-1}A$, C is selfadjoint and positive definite, $M = C$, and A is selfadjoint. In these cases and in all cases when the bilinear form $a_2(\cdot, \cdot)$ is used, the corresponding orthogonal error method reduces to a minimal residual method.

To define the orthogonal error iteration method, we let $\mathbf{d}^j$, $j = 0, 1, \ldots$ be a sequence of vectors, search directions, and $\mathbf{x}^j$, $j = 0, 1, 2, \ldots$ successive approximations of a solution $\hat{\mathbf{x}} \in V$ of $B\mathbf{x} = \mathbf{b}$. We consider two cases:

(A) The search directions have already been computed by some separate method, such as a Lanczos method, to be $a(\cdot, \cdot)$ orthogonal, that is,

$$a(\mathbf{d}^k, \mathbf{d}^j) = 0, \quad 0 \le j \le k - 1,$$

or $a(\cdot, B\cdot)$ orthogonal, that is,

$$a(\mathbf{d}^k, B\mathbf{d}^j) = 0, \quad 0 \le j \le k - 1.$$

(B) The search directions are computed, together with the successive approximations. In case (A), we let

$$\mathbf{x}^{k+1} = \mathbf{x}^k + \sum_{j=0}^{k} \alpha_j^{(k)} \mathbf{d}^j, \quad k = 0, 1, \ldots$$

and $\alpha_j^{(k)}$ is computed such that

(12.25) $a(\mathbf{e}^{k+1}, \mathbf{d}^j) = 0, \quad 0 \le j \le k$

and

(12.26) $a(\mathbf{r}^{k+1}, \mathbf{d}^j) = 0, \quad 0 \le j \le k,$

respectively, where $\mathbf{e}^{k+1} = \hat{\mathbf{x}} - \mathbf{x}^{k+1}$ is the new iteration error and $\mathbf{r}^{k+1} =$

$B\mathbf{x}^{k+1} - \mathbf{b}$ is the new residual. Note that even though $\mathbf{e}^{k+1}$ is not available, $a(\mathbf{e}^{k+1}, \mathbf{d}^j)$ is computable when $a(\cdot, \cdot)$ is defined by (12.24a) or (12.24b), because $B\mathbf{e}^{k+1} = -\mathbf{r}^{k+1}$, which is available.

In case (B), we follow the method proposed for the general subspace iteration method, presented at the end of the previous section, and let

$$(12.27) \qquad \mathbf{d}^k = -\mathbf{r}^k + \beta_{k-1}\mathbf{d}^{k-1}$$

$$(12.28) \qquad \mathbf{x}^{k+1} = \mathbf{x}^k + \sum_{j=0}^{k} \alpha_j^{(k)}\mathbf{d}^j, \quad k = 0, 1, \dots.$$

Here, $\beta_{-1} = 0$, and $\{\alpha_j^{(k)}\}_{j=0}^k$ and β_k, $k = 0, 1, \dots$ will be determined below.

Remark 12.16 There exist various alternative forms of the above methods. As in the GCG-LS methods, we can let the vector $\mathbf{d}^{k+1}$ be computed from all previous vectors $\mathbf{d}^j$, $0 \le j \le k$ and, possibly, made orthogonal with respect to another bilinear form or inner product. Also, we can let $\mathbf{x}^{k+1}$ be determined by a truncated expression—i.e., use only a set of the more recent vectors in the recursion. However, in the interest of keeping the presentation clear, we consider here only the above-mentioned cases (A) and (B).

Now, (12.28) and $B\hat{\mathbf{x}} = \mathbf{b}$ show that

$$(12.29) \qquad \mathbf{e}^{k+1} = \mathbf{e}^k - \sum_{j=0}^{k} \alpha_j^{(k)}\mathbf{d}^j,$$

$$(12.30) \qquad \mathbf{r}^{k+1} = \mathbf{r}^k + \sum_{j=0}^{k} \alpha_j^{(k)} B\mathbf{d}^j.$$

Let

$$S_k = \text{span}\{\mathbf{r}^0, B\mathbf{r}^0, \dots, B^k\mathbf{r}^0\}.$$

Then we see that, by (12.27), $\mathbf{d}^k \in S_k$, $\mathbf{e}^k - \mathbf{e}^0 \in S_{k-1}$ and $\mathbf{r}^k - \mathbf{r}^0 \in BS_{k-1}$, so $\mathbf{r}^k \in S_k$.

As above, we determine $\alpha_j^{(k)}$, $j = 0, 1, \dots, k$ by the *Galerkin* or *orthogonal (error) projection method*:

$$(12.31) \qquad a(\mathbf{e}^{k+1}, \mathbf{v}) = 0, \quad \text{for all } \mathbf{v} \in S_k.$$

Let $\mathbf{v} = \mathbf{d}^l$, $l = k, k-1, \dots, 0$. Then, (12.29) and (12.31) show that

$$(12.32) \qquad \sum_{j=0}^{k} \alpha_j^{(k)} a(\mathbf{d}^j, \mathbf{d}^l) = a(\mathbf{e}^k, \mathbf{d}^l), \quad l = k, k-1, \ldots, 0.$$

Similarly, we can determine the coefficients $\alpha_j^{(k)}$ by the orthogonal residual condition

$$a(\mathbf{v}, \mathbf{r}^{k+1}) = 0, \quad \text{for all } \mathbf{v} \in S_k.$$

However, since $\mathbf{r}^{k+1} = -B\mathbf{e}^{k+1}$, we see that any orthogonal error method for a bilinear form $a(\mathbf{u}, \mathbf{v}) = a_2(\mathbf{u}, \mathbf{v}) = (B\mathbf{u}, B\mathbf{v})$ corresponds to an orthogonal residual method for the bilinear form $a_1(\mathbf{u}, \mathbf{v}) = (B\mathbf{u}, \mathbf{v})$. Hence, there is no loss of generality by considering (12.32) only.

Lemma 12.17 *If $\mathbf{d}^0, \mathbf{d}^1, \ldots, \mathbf{d}^k$ are linearly independent, then the matrix $\Lambda^{(k)}$, where $\Lambda_{i,j}^{(k)} = a(\mathbf{d}^j, \mathbf{d}^i), \; 0 \le i, j \le k,$ is positive definite and there exists a unique solution of (12.32). Further, the leading coefficients satisfy*

$$\alpha_k^{(k)} = \frac{\det(\Lambda^{(k-1)})}{\det(\Lambda^{(k)})} a(\mathbf{e}^k, \mathbf{d}^k), \quad k = 0, 1, \ldots.$$

Proof Let $\mathbf{u} = \sum_{j=0}^{k} x_j \mathbf{d}^j$. Then, with $\mathbf{x} = [x_0, x_1, \ldots, x_k]$, we have

$$\mathbf{x}^T \Lambda^{(k)} \mathbf{x} = a(\mathbf{u}, \mathbf{u}),$$

and (12.22a) shows that

$$\mathbf{x}^T \Lambda^{(k)} \mathbf{x} \ge \rho \|\mathbf{u}\|^2 > 0, \quad \text{for all } \mathbf{u} \ne \mathbf{0}$$

and

$$\mathbf{x}^T \Lambda^{(k)} \mathbf{x} = 0, \quad \text{if and only if } \mathbf{u} = \mathbf{0}.$$

But since $\{\mathbf{d}^j\}_0^k$ is a linearly independent set, this shows that $\mathbf{x}^T \Lambda^{(k)} \mathbf{x} = 0$ if and only if $\mathbf{x} = \mathbf{0}$. Hence, $\Lambda^{(k)}$ is positive definite. Further, making use of $a(\mathbf{e}^k, \mathbf{d}^j) = 0, 0 \le j \le k-1$, Cramer's rule shows that

$$(12.33) \qquad \alpha_k^{(k)} = \frac{\det(\Lambda^{(k-1)})}{\det(\Lambda^{(k)})} a(\mathbf{e}^k, \mathbf{d}^k). \qquad \qquad \diamond$$

We shall discuss two choices of the coefficients β_{k-1}:

(C) $\beta_{k-1} = \frac{a(\mathbf{d}^{k-1}, \mathbf{r}^k)}{a(\mathbf{d}^{k-1}, \mathbf{d}^{k-1})}$, $k = 1, 2, \ldots$, $\beta_{-1} = 0$, which makes

(12.34) $$a(\mathbf{d}^{k-1}, \mathbf{d}^k) = 0.$$

(D) $\beta_{k-1} = 0$, $k = 0, 1, \ldots$.

It will be seen that for each choice, the matrix $\Lambda^{(k)}$ takes a particular form.

We now consider the bilinear form $a_1(\cdot, \cdot)$ and also comment briefly on the bilinear form $a_2(\cdot, \cdot)$. For the bilinear form $a(\cdot, \cdot) = a_1(\cdot, \cdot)$ the following relations hold.

Lemma 12.18 *Let $a(\mathbf{u}, \mathbf{v}) = (B\mathbf{u}, \mathbf{v})$ and let (12.27), (12.28), and (12.31) hold. Then:*

(a) *$(\mathbf{r}^{k+1}, \mathbf{v}) = 0$ for all $\mathbf{v} \in S_k = \mathrm{span}\{\mathbf{r}^0, B\mathbf{r}^0, \ldots, B^k\mathbf{r}^0\}$.*
(b) *$(B\mathbf{d}^j, \mathbf{r}^k) = 0$, $0 \le j \le k - 2$.*
(c) *$(B\mathbf{d}^j, \mathbf{d}^k) = \left(\prod_{s=j+1}^{k-1} \beta_s\right)(B\mathbf{d}^j, \mathbf{d}^{j+1})$, $0 \le j \le k - 2$.*
(d) *$(B\mathbf{d}^{k-1}, \mathbf{d}^k) = -(B\mathbf{d}^{k-1}, \mathbf{r}^k) + \beta_{k-1}(B\mathbf{d}^{k-1}, \mathbf{d}^{k-1})$.*

Proof $B\mathbf{e}^{k+1} = -\mathbf{r}^{k+1}$ and (12.31) show that

$$(\mathbf{r}^{k+1}, \mathbf{v}) = 0 \quad \text{for all } \mathbf{v} \in S_k.$$

Further, for $0 \le j \le k - 2$, we have

$$a(\mathbf{d}^j, \mathbf{r}^k) = (B\mathbf{d}^j, \mathbf{r}^k) = (\mathbf{r}^k, B\mathbf{d}^j) = -(B\mathbf{e}^k, B\mathbf{d}^j) = -a(\mathbf{e}^k, B\mathbf{d}^j) = 0,$$

because $B\mathbf{d}^j \in S_{j+1} \subset S_{k-1}$. Hence,

$$a(\mathbf{d}^j, \mathbf{d}^k) = a(\mathbf{d}^j, -\mathbf{r}^k + \beta_{k-1}\mathbf{d}^{k-1})$$

$$= \begin{cases} -a(\mathbf{d}^{k-1}, \mathbf{r}^k) + \beta_{k-1}a(\mathbf{d}^{k-1}, \mathbf{d}^{k-1}), & j = k - 1 \\ \beta_{k-1}a(\mathbf{d}^j, \mathbf{d}^{k-1}), & 0 \le j \le k - 2 \end{cases}$$

and, by induction,

$$a(\mathbf{d}^j, \mathbf{d}^k) = \prod_{s=j+1}^{k-1} \beta_s a(\mathbf{d}^j, \mathbf{d}^{j+1}), \ 0 \le j \le k - 2. \qquad \diamond$$

We now consider the bilinear form $a_1(\cdot, \cdot)$ for cases (C) and (D).

The Bilinear Form $a_1(\cdot, \cdot)$ for Case (C)

Here, by (12.34), $a(\mathbf{d}^{k-1}, \mathbf{d}^k) = 0$, and Lemma 12.18(c) shows that

$$(12.35) \qquad\qquad a(\mathbf{d}^j, \mathbf{d}^k) = 0, \ \ 0 \le j \le k - 1.$$

Hence, $\Lambda^{(k)}$ is a *lower triangular* matrix. Since $\Lambda^{(k)}$ is positive definite, the diagonal entries of $\Lambda^{(k)}$ are positive. Further, Lemma 12.17, (12.35), and (12.27) show that

$$\alpha_k^{(k)} = \frac{1}{a(\mathbf{d}^k, \mathbf{d}^k)} a(\mathbf{e}^k, \mathbf{d}^k) = \frac{-a(\mathbf{e}^k, \mathbf{r}^k)}{a(\mathbf{d}^k, \mathbf{d}^k)},$$

so, using $B\mathbf{e}^k = -\mathbf{r}^k$, we find

$$(12.36) \qquad\qquad \alpha_k^{(k)} = \frac{(\mathbf{r}^k, \mathbf{r}^k)}{(B\mathbf{d}^k, \mathbf{d}^k)}.$$

The remaining entries $\alpha_j^{(k)}, 0 \le j \le k - 1$ are computed by (12.32). We collect these and some additional results in a theorem.

Theorem 12.19 *Let $a(\mathbf{u}, \mathbf{v}) = (B\mathbf{u}, \mathbf{v})$, where B is positive definite, and consider case (C). Then:*

(a) *$\Lambda^{(k)} = [a(\mathbf{d}^j, \mathbf{d}^i)], 0 \le i, j \le k$ is lower triangular with positive diagonal entries, and the leading coefficient $\alpha_k^{(k)}$ in the expansion of $\mathbf{x}^{k+1}$ satisfies $\alpha_k^{(k)} = \frac{(\mathbf{r}^k, \mathbf{r}^k)}{(B\mathbf{d}^k, \mathbf{d}^k)}$, so $\alpha_k^{(k)} \ge 0$. Further, $(B\mathbf{d}^k, \mathbf{d}^k) > 0$ and $\alpha_k^{(k)} > 0$, unless $\mathbf{x}^k$ is already a solution to $B\mathbf{x} = \mathbf{b}$, which means that the method does not break down before the solution has been computed.*

(b) *The residual vectors are mutually orthogonal, i.e., $(\mathbf{r}^j, \mathbf{r}^k) = 0$, $j \ne k$.*

(c) *$\beta_{k-1} = \frac{(\mathbf{r}^k, \mathbf{r}^k)}{(\mathbf{r}^{k-1}, \mathbf{r}^{k-1})}$.*

Proof Of part (a) it remains to show that if $\alpha_k^{(k)} = 0$ or if $(B\mathbf{d}^k, \mathbf{d}^k) = 0$, then $\mathbf{r}^k = \mathbf{0}$. Clearly, (12.36) shows that if $(B\mathbf{d}^k, \mathbf{d}^k) \ne 0$ but $\alpha_k^{(k)} = 0$, then $\mathbf{r}^k = \mathbf{0}$. If $(B\mathbf{d}^k, \mathbf{d}^k) = 0$, then the positivity assumption (12.32a) shows that $\mathbf{d}^k = \mathbf{0}$, in which case (12.27) shows that $\mathbf{r}^k = \beta_{k-1}\mathbf{d}^{k-1}$. Then, by Lemma 12.18(a), $\mathbf{r}^k = \mathbf{0}$. Since, by (12.30), $\mathbf{r}^j$ is a linear combination of $\mathbf{r}^{j-1}$ and $B\mathbf{d}^t$, $0 \le t \le j - 1$, (12.35), and Lemma 12.18(a) show that (b) holds. To prove part (c), note that when $\alpha_{k-1}^{(k-1)} > 0$ (i.e., when $\mathbf{r}^{k-1} \neq \mathbf{0}$), (12.28), with k exchanged with $k - 1$, shows that

$$B\mathbf{d}^{k-1} = \frac{\mathbf{r}^k - \mathbf{r}^{k-1}}{\alpha_{k-1}^{(k-1)}} - \sum_{j=0}^{k-2} \frac{\alpha_j^{(k-1)}}{\alpha_{k-1}^{(k-1)}} B\mathbf{d}^j.$$

Hence, by (C) and using Lemma 12.18(b) and part (a) of the present theorem,

$$\beta_{k-1} = \frac{a(\mathbf{d}^{k-1}, \mathbf{r}^k)}{a(\mathbf{d}^{k-1}, \mathbf{d}^{k-1})} = \frac{(B\mathbf{d}^{k-1}, \mathbf{r}^k)}{(B\mathbf{d}^{k-1}, \mathbf{d}^{k-1})} = \frac{(\mathbf{r}^k - \mathbf{r}^{k-1}, \mathbf{r}^k)}{(\mathbf{r}^k - \mathbf{r}^{k-1}, \mathbf{d}^{k-1})},$$

which, by Lemma 12.18(a) and (12.27), simplifies to

$$\beta_{k-1} = \frac{-(\mathbf{r}^k, \mathbf{r}^k)}{(\mathbf{r}^{k-1}, \mathbf{d}^{k-1})} = \frac{(\mathbf{r}^k, \mathbf{r}^k)}{(\mathbf{r}^{k-1}, \mathbf{r}^{k-1})}. \qquad \diamond$$

Remark 12.20 If a_1 is a symmetric bilinear form, i.e., if B is selfadjoint w.r.t. the inner product $(\cdot, \cdot)$, then $\Lambda^{(k)}$ becomes symmetric. Hence, since $\Lambda^{(k)}$ is, in addition, triangular, $\Lambda^{(k)}$ must be diagonal. Then (12.32) and (12.31) show that $\alpha_j^{(k)} = 0$, $0 \leq j \leq k - 1$, and the above method reduces to the classical conjugate gradient (or direction) method in Section 11.2. The property $a(\mathbf{d}^j, \mathbf{d}^k) = (B\mathbf{d}^j, \mathbf{d}^k)$, $0 \leq j \leq j \leq k - 1$, then corresponds to the conjugate orthogonality among the search directions, because when a is symmetric, we have

$$a(\mathbf{d}^k, \mathbf{d}^j) = 0, \quad \text{for all } j \neq k.$$

The Bilinear Form $a_1(\cdot, \cdot)$ for Case (D) $(\beta_{k-1} = 0)$

In this case, $\mathbf{r}^k = -\mathbf{d}^k$, so Lemma 12.18(b) shows that $(B\mathbf{d}^j, \mathbf{d}^k) = 0$, $0 \leq j \leq k - 2$. Hence, the matrix $\Lambda^{(k)}$, corresponding to the linear system (12.32), has a *lower Hessenberg* form. Again, it can readily be seen that the method cannot break down before $\mathbf{r}^k = \mathbf{0}$, when B is positive definite.

The following coefficients of $\Lambda^{(k)}$ must be computed at step k:

$$(B\mathbf{d}^k, \mathbf{d}^j), \ 0 \leq j \leq k, \ (B\mathbf{d}^k, \mathbf{d}^{k-1}), \ \text{and } (\mathbf{r}^k, \mathbf{d}^k) = -(\mathbf{r}^k, \mathbf{r}^k).$$

The coefficients $\alpha_j^{(k)}$, $j = 0, 1, \ldots, k$ are solutions of the linear system

$$
\begin{bmatrix}
(Bd^k, d^k) & (Bd^{k-1}, d^k) & 0 & \cdots & 0 \\
(Bd^k, d^{k-1}) & (Bd^{k-1}, d^{k-1}) & (Bd^{k-2}, d^{k-1}) & \cdots & 0 \\
\cdots & & & & \\
(Bd^k, d^1) & (Bd^{k-1}, d^1) & (Bd^{k-2}, d^1) & \cdots & (Bd^0, d^1) \\
(Bd^k, d^0) & (Bd^{k-1}, d^0) & (Bd^{k-2}, d^0) & \cdots & (Bd^0, d^0)
\end{bmatrix}
\begin{bmatrix}
\alpha_k^{(k)} \\
\alpha_{k-1}^{(k)} \\
\vdots \\
\alpha_0^{(k)}
\end{bmatrix}
=
\begin{bmatrix}
(r^k, r^k) \\
0 \\
\vdots \\
0 \\
0
\end{bmatrix}.
$$

(12.37)

This Hessenberg matrix problem can be solved, for instance, by QR factorization, where Q is orthogonal and R is lower triangular. If $\Lambda_k = Q_k R_k$, then

$$
\Lambda_k^T \Lambda_k = R_k^T Q_k^T Q_k R_k = R_k^T R_k.
$$

Therefore, the coefficient vector $\underline{\alpha}^{(k)}$ can be computed from

$$
\Lambda_k^T \Lambda_k \underline{\alpha}^{(k)} = \|r^k\|^2 \Lambda_k^T e_1, \quad e_1 = [1, 0, \ldots, 0]^T
$$

or from the Cholesky factorization,

$$
R_k^T R_k \underline{\alpha}^{(k)} = \|r^k\|^2 \Lambda_k^T e_1.
$$

As in Theorem 12.19, it is readily seen that the following holds for version (D).

Theorem 12.21 *Let* $a(u, v) = (Bu, v)$ *and consider case (D). Then:*

(a) $\Lambda^{(k)} = [a(d^j, d^i)]$, $0 \le i, j \le k$ *is lower Hessenberg.*
(b) $(d^j, d^k) = 0$, $0 \le j \le k - 2$.

For small values of k relative to the order of the system, the cost of solving the systems $\Lambda^{(k)} \underline{\alpha}^{(k)} = \underline{\gamma}^{(k)}$ can be neglected. The main cost is associated with forming the matrix $\Lambda^{(k)}$ from (12.32) and with updating the vectors d^k (if $\beta_{k-1} \ne 0$), x^{k+1} and r^{k+1} in (12.27), (12.28), and (12.30), respectively. These costs are equal in cases (C) and (D), and since the solution of the linear system $\Lambda^{(k)} \underline{\alpha}^{(k)} = \underline{\gamma}^{(k)}$ associated with case (C) is somewhat simpler, this should be recommended.

12.2.1 An Efficient Implementation

There is a substantial saving of computational work possible for the above methods, because it is not required to update the vectors x^{k+1} at each step. Instead, we compute just the vectors d^k and r^{k+1}, starting with $d^0 = -r^0 =$

$\mathbf{b} - A\mathbf{x}^0$, where $\mathbf{x}^0$ is an initial approximation. When $(\mathbf{r}^{k+1}, \mathbf{r}^{k+1})$ is sufficiently small for some value $k = k_0$, we can compute the corresponding vector $\mathbf{x}^{k_0+1}$, using

$$(12.38) \qquad \mathbf{x}^{k_0+1} = \mathbf{x}^0 + \sum_{j=0}^{k_0} a_j^{(k_0)} \mathbf{r}^j,$$

where the scalars $a_j^{(k_0)}$ are computed in the following way, described here for the orthogonal residual method defined in Theorem 12.19.

We note first that as $B\mathbf{d}^j \in V_k$, $0 \le j \le k - 1$, we can rewrite (12.30) as

$$\mathbf{r}^{k+1} = \alpha_k^{(k)} B\mathbf{d}^k + \sum_{j=0}^{k} \xi_j^{(k)} \mathbf{r}^j,$$

where $\alpha_k^{(k)} = (\mathbf{r}^k, \mathbf{r}^k)/(B\mathbf{d}^k, \mathbf{d}^k)$. In addition, we recall from Theorem 12.19 that

$$\mathbf{d}^{k+1} = -\mathbf{r}^{k+1} + \beta_k \mathbf{d}^k,$$

where $\beta_k = (\mathbf{r}^{k+1}, \mathbf{r}^{k+1})/(\mathbf{r}^k, \mathbf{r}^k)$. Further, the orthogonality of the set $\mathbf{r}^j$ implies

$$\xi_l^{(k)} = -\alpha_k^{(k)} \frac{(B\mathbf{d}^k, \mathbf{r}^l)}{(\mathbf{r}^l, \mathbf{r}^l)}, \quad 0 \le l \le k.$$

If, for some $k = k_0$, $(\mathbf{r}^{k_0+1}, \mathbf{r}^{k_0+1}) \le \varepsilon$, for predetermined ε, then we can compute the corresponding approximation $\mathbf{x}^{k+1}$ on the form (12.38) as follows. Making use of (12.38), we can write

$$\mathbf{r}^{k_0+1} = \mathbf{r}^0 + \sum_{0}^{k_0} a_j^{(k_0)} B\mathbf{r}^j.$$

Theorem 12.19(b) shows that

$$\sum_{0}^{k_0} a_j^{(k_0)} (B\mathbf{r}^j, \mathbf{r}^0) = -(\mathbf{r}^0, \mathbf{r}^0),$$

$$\sum_{j=l-1}^{k_0} a_j^{(k_0)} (B\mathbf{r}^j, \mathbf{r}^l) = 0, \quad l = 1, \dots, k_0.$$

Now, because of $\mathbf{r}^j = -\mathbf{d}^j + \beta_{j-1}\mathbf{d}^{j-1}$, we obtain

$$(B\mathbf{r}^j, \mathbf{r}^l) = -(B\mathbf{d}^j, \mathbf{r}^l) + \beta_{j-1}(B\mathbf{d}^{j-1}, \mathbf{r}^l).$$

This shows, in particular, that

$$(B\mathbf{r}^{l-1}, \mathbf{r}^l) = -(B\mathbf{d}^{l-1}, \mathbf{r}^l).$$

For each j, the numbers $(B\mathbf{d}^j, \mathbf{r}^l)$, $0 \le l \le j + 1$ are computed and stored. Then, $(B\mathbf{r}^j, \mathbf{r}^l)$, $l = 0, 1, \ldots, j + 1$ can be computed with no further inner products, using the above and these numbers. The matrix with coefficients $(B\mathbf{r}^j, \mathbf{r}^l)$, $0 \le j$, $l \le k_0$, which appears in the system for $a_j^{(k_0)}$, has Hessenberg form, and the vector $\{a_j^{(k_0)}\}$, $j = 0, 1, \ldots, k_0$ can readily be computed using a QR factorization of the matrix, for instance. The corresponding algorithm takes the form:

Algorithm 4 (GCG-OE)

Let $\mathbf{x}^0$, $\mathbf{b}$ be input vectors;
compute $\mathbf{r}^0 = B\mathbf{x}^0 - \mathbf{b}$, and let $\mathbf{d}^0 = -\mathbf{r}^0$;
compute $(\mathbf{r}^0, \mathbf{r}^0)$; If$(\mathbf{r}^0, \mathbf{r}^0) \le \varepsilon$ then stop;
for $k = 0, 1, \ldots$ until convergence
compute $B\mathbf{d}^k$;
compute (and store) $(B\mathbf{d}^k, \mathbf{d}^k)$ and $(B\mathbf{d}^k, \mathbf{r}^l)$, $0 \le l \le k$;
If $(B\mathbf{d}^k, \mathbf{d}^k) \ne 0$ compute
$\quad \alpha_k := (\mathbf{r}^k, \mathbf{r}^k)/(B\mathbf{d}^k, \mathbf{d}^k)$;
for $l = 0, 1, \ldots, k$ compute
$\quad \xi_l := -\alpha_k(B\mathbf{d}^k, \mathbf{r}^l)/(\mathbf{r}^l, \mathbf{r}^l)$;
$\quad \mathbf{r}^{k+1} := \alpha_k B\mathbf{d}^k + \sum_0^k \xi_l \mathbf{r}^l$;
compute (and store) $(\mathbf{r}^{k+1}, \mathbf{r}^{k+1})$;
if $(\mathbf{r}^{k+1}, \mathbf{r}^{k+1}) \le \varepsilon$ go to R;
$\quad \beta_k := (\mathbf{r}^{k+1}, \mathbf{r}^{k+1})/(\mathbf{r}^k, \mathbf{r}^k)$;
$\quad \mathbf{d}^{k+1} := -\mathbf{r}^{k+1} + \beta_k \mathbf{d}^k$;
compute (and store) $(B\mathbf{d}^k, \mathbf{r}^{k+1})$;
R: for $l = 0, 1, \ldots, k$; for $j = l - 1, l, \ldots, k$;
compute
$(B\mathbf{r}^j, \mathbf{r}^l) = -(B\mathbf{d}^j, \mathbf{r}^l) + \beta_{j-1}(B\mathbf{d}^{j-1}, \mathbf{r}^l)$
[comment: Here$(B\mathbf{r}^{l-1}, \mathbf{r}^l) = -(B\mathbf{d}^{l-1}, \mathbf{r}^l)$.]

Solve for $a_j^{(k)}$ [using a QR-factorization] and let

$$\mathbf{x}^{k+1} := \mathbf{x}^0 + \sum_{j=0}^k a_j^{(k)} \mathbf{r}^j;$$

Remarks: Storage of $\mathbf{x}^0, \mathbf{d}^k, B\mathbf{d}^k, \mathbf{r}^0, \mathbf{r}^1, \ldots, \mathbf{r}^k$ is required. Note that there occurs only one action of B and only one vector update of increasing length $(k+1)$ per iteration. Note also that the computational effort in computing the numbers ξ_l, in solving the system for $a_j^{(k)}$, and in computing the numbers $(B\mathbf{r}^j, \mathbf{r}^l)$, is negligible when k is much smaller than the order of the system. There is a similar method, known as the GMRES method, with the same savings in the computational effort (see Saad, Schultz, 1986). It is based on a Lanczos-type process to compute orthogonal vectors, in this case, referred to as the Arnoldi method (Arnoldi, 1951).

Algorithm Arnoldi   Choose $\mathbf{d}^0 = \mathbf{r}^0/(\mathbf{r}^0, \mathbf{r}^0)^{1/2}$. For $k = 1, 2, \ldots$, do:

(a) For $j = 1, 2, \ldots, n$ compute

$$\lambda_{j,k} = \mathbf{d}^{jT} B\mathbf{d}^k.$$

(b) Set

$$\tilde{\mathbf{d}}^{k+1} = B\mathbf{d}^k - \sum_{j=0}^{k} \lambda_{j,k}\mathbf{d}^j.$$

(c) Compute

$$\lambda_{k+1,k} = \left\{ (\tilde{\mathbf{d}}^{k+1}, \tilde{\mathbf{d}}^{k+1}) \right\}^{1/2}.$$

(d) If $\lambda_{k+1,k} = 0$ stop; otherwise, set

$$\mathbf{d}^{k+1} = \frac{1}{\lambda_{k+1,k}}\tilde{\mathbf{d}}^{k+1}$$

and repeat.

It can be seen that the approximate solution $\mathbf{x}^{k+1}$ can be computed as

$$\mathbf{x}^{k+1} = \mathbf{x}^0 + \sum_{j=0}^{k} a_j\mathbf{d}^j,$$

where the vector $\mathbf{a}^{(k)} = [a_0, \ldots, a_k]^T$ is computed either as a solution to a minimal residual problem involving a matrix of order $(k+2) \times (k+1)$ of Hessenberg form or solving a $(k+1) \times (k+1)$ linear system with a Hessenberg matrix. Only when $\mathbf{x}^{k+1}$ has been computed can one compute $\mathbf{r}^{k+1}$

to check if $(\mathbf{r}^{k+1}, \mathbf{r}^{k+1}) < \varepsilon$ for some chosen accuracy number ε. If this test fails, one must compute more vectors using the Arnoldi method. For details, see Saad, Schultz (1986) and Saad (1981).

In the orthogonal error GCG-OE algorithm presented above, we avoided working with Hessenberg matrices. Furthermore, the norm of the residual is directly available *at each step* during the construction of the vectors $\mathbf{d}^k$. The arithmetic costs of the algorithms are, otherwise, the same.

12.2.2 A Special Generalized Conjugate Gradient Method

Consider now the special case when $B = I + M^{-1}N$, where M is s.p.d. and N is skewsymmetric. Further, let the inner product be $(\mathbf{u}, \mathbf{v}) = \mathbf{u}^T M \mathbf{v}$. This can be seen as the preconditioned form of $A = M + N$, where M is the symmetric part of A (which we assume to be positive definite), and N is the skewsymmetric part. In this case,

$$(B\mathbf{d}^k, \mathbf{d}^j) = \left([I + M^{-1}N]\mathbf{d}^k\right)^T M \mathbf{d}^j$$

$$= \mathbf{d}^{k^T}(M - N)\mathbf{d}^j = (\mathbf{d}^k, [I - M^{-1}N]\mathbf{d}^j).$$

Since $(\mathbf{d}^k, B\mathbf{d}^{k-1}) = 0$ and since $M^{-1}N\mathbf{d}^j = (B - I)\mathbf{d}^j$, we have by Lemma 12.18(b,c),

$$(B\mathbf{d}^k, \mathbf{d}^j) = 0, \ 0 \le j \le k - 1,$$

which shows that the matrix $\Lambda^{(k)}$ is diagonal, and algorithm GCG-OE simplifies (truncates) accordingly. In particular, $\alpha_j^{(k)} = 0, \ 0 \le j \le k - 1$. The resulting algorithm is similar to the CGW method. See Example 12.3 for further comments on this method.

12.3 Generalized Conjugate Gradient Methods and Variable (Nonlinear) Preconditioners

As will be shown in Chapter 13, the standard analysis of the rate of convergence of conjugate gradient methods uses properties of best approximation polynomials based on the Krylov set of vectors and the eigenvalue distribution of the matrix $B \ (= C^{-1}A)$, which generates the Krylov set. Hence, this analysis is based on the use of a fixed matrix and, in particular, of a fixed preconditioner.

There are situations, however, where it can be more efficient to use a variable preconditioner (or even a variable matrix A). For instance, during the first iterations, we may find it efficient to use a less costly preconditioner than during the final iterations. The preconditioner may even be updated during the course of iterations, using information of the spectrum gathered during the previous iterations. (For adaptively updated SSOR preconditioners, see Hageman and Young, 1981.) If the matrix A is a Schur complement matrix, we may need to use inner iterations to compute the action of A on vectors, and the number of such inner iterations may also vary from one iteration to the next. That is why we need a convergence theory for variable preconditioners and even for variable matrices.

We first present the typical case of a matrix partitioned in two-by-two block form, where we will use inner-outer iterations. When we solve systems with the inner matrix (A_{11}), we may not have sufficient information of its spectrum to be able to construct a fixed matrix polynomial to approximate the inverse of this matrix. Hence, we must replace the action of the inverse with an approximate action, and this approximation can vary from one step to the next. Also, we explain why it is advisable to use a generalized conjugate gradient method as an acceleration method, even if the given matrix is symmetric and positive definite.

We then present the generalized conjugate gradient method with variable preconditioner. Assuming a positivity (coercivity) and boundedness condition, we show the convergence of the method. For the case of matrices partitioned in two-by-two block form, we verify these positivity and boundedness conditions when we use sufficiently accurate inner procedures to compute the action of the matrices occurring in the inner processes.

This presentation is based on Axelsson and Vassilevski (1991). Other methods using inner-outer iterations can be found in Axelsson (1973), Golub and Overton (1988), Bank, Welfert, and Yserentant (1990), Langer and Queck (1986), and Verfürth (1984). However, as an inner and/or outer iterative method, they used a stationary iterative method based on some assumed eigenvalue information—i.e., not a conjugate gradient method.

12.3.1 The Need for Variable Preconditioners

Consider a linear system partitioned in two-by-two block form,

$$(12.39) \qquad \begin{bmatrix} A_{11} & A_{12} \\ A_{21} & A_{22} \end{bmatrix} \begin{bmatrix} x_1 \\ x_2 \end{bmatrix} = \begin{bmatrix} b_1 \\ b_2 \end{bmatrix},$$

where we assume that A_{11} is nonsingular. Let $S = A_{22} - A_{21}A_{11}^{-1}A_{12}$. The block matrix in (12.39) can be factored as

$$\begin{bmatrix} A_{11} & A_{12} \\ A_{21} & A_{22} \end{bmatrix} = \begin{bmatrix} I & 0 \\ A_{21}A_{11}^{-1} & I \end{bmatrix} \begin{bmatrix} A_{11} & 0 \\ 0 & S \end{bmatrix} \begin{bmatrix} I & A_{11}^{-1}A_{12} \\ 0 & I \end{bmatrix},$$

and its inverse is found to be (see Chapter 3)

$$\begin{bmatrix} A_{11} & A_{12} \\ A_{21} & A_{22} \end{bmatrix}^{-1} = \begin{bmatrix} I & -A_{11}^{-1}A_{12} \\ 0 & I \end{bmatrix} \begin{bmatrix} A_{11}^{-1} & 0 \\ 0 & S^{-1} \end{bmatrix} \begin{bmatrix} I & 0 \\ -A_{21}A_{11}^{-1} & I \end{bmatrix}$$

or

$$\begin{bmatrix} A_{11} & A_{12} \\ A_{21} & A_{22} \end{bmatrix}^{-1} = \begin{bmatrix} I & -A_{11}^{-1}A_{12} \\ 0 & I \end{bmatrix} \begin{bmatrix} A_{11}^{-1} & 0 \\ -S^{-1}A_{21}A_{11}^{-1} & S^{-1} \end{bmatrix}.$$

Hence, the action of this matrix on a (consistently partitioned) vector $\mathbf{b} = \begin{bmatrix} \mathbf{b}_1 \\ \mathbf{b}_2 \end{bmatrix}$ involves the following steps:

(12.40)

$$
\begin{aligned}
(1) \quad & \mathbf{w}_1 = A_{11}^{-1}\mathbf{b}_1; \\
(2) \quad & \mathbf{w}_2 = -A_{21}\mathbf{w}_1 + \mathbf{b}_2; \\
(3) \quad & \mathbf{x}_2 = S^{-1}\mathbf{w}_2; \\
(4) \quad & \mathbf{y}_1 = A_{12}\mathbf{x}_2; \\
(5) \quad & \mathbf{z}_1 = A_{11}^{-1}\mathbf{y}_1; \\
(6) \quad & \mathbf{x}_1 = \mathbf{w}_1 - \mathbf{z}_1.
\end{aligned}
$$

Two actions with A_{11}^{-1} and one action with S^{-1} are involved in 12.40. In order to construct an approximation of A^{-1}, we let these actions be replaced by approximate actions using some (inner) iterative methods. This corresponds to using some approximate (but not explicitly available) inverses of these matrices, and the corresponding preconditioner takes the form

$$Q = \begin{bmatrix} I & -M_1(A_{11}^{-1})A_{12} \\ 0 & I \end{bmatrix} \begin{bmatrix} M_2(A_{11}^{-1}) & 0 \\ -M_3(S^{-1})A_{21}M_2(A_{11}^{-1}) & M_3(S^{-1}) \end{bmatrix},$$

where $M_i(B)$, $i = 1, 2, 3$ denotes different approximate actions of a matrix B. It is readily seen that Q is nonsymmetric in general, even if A is symmetric. Hence, we need to use a generalized conjugate gradient solver for the outer iterations.

There is an alternative method to solve (12.39), involving the use of the reduced system, or Schur complement matrix system,

$$Sx_2 = (A_{22} - A_{21}A_{11}^{-1}A_{12})x_2 = b_2 - A_{21}A_{11}^{-1}b_1.$$

To compute residuals **r** of this system, it is convenient to rewrite it as

$$(12.41) \qquad r = A_{22}x_2 - b_2 - A_{21}A_{11}^{-1}(A_{12}x_2 - b_1).$$

The computation of residuals using (12.41) involves only *one action* of A_{11}^{-1}. This action can be computed by an approximate, iterative method, using some preconditioner of A_{11}. The accuracy of this approximate action can be quite low at the initial iteration steps, but it should preferably increase with each new outer iteration.

To solve the outer system with S, we use some preconditioned version of a generalized conjugate gradient method, with a variable preconditioner and variable matrix. As remarked above, we need only one action of the inner system $A_{11}y = A_{12}x_2 - b_1$ per iteration. Furthermore, for the inner system, we can start each new iteration with an accurate initial approximation—namely, the value of $y = A_{11}^{-1}(A_{12}x_2 - b_2)$, found in the previous outer iteration. Therefore, the number of iterations to solve the inner systems can be kept low, even though the accuracy in solving the systems increases. In addition, for the outer iteration method, we work on vectors with reduced vector length. This method can be particularly advantageous when one wants to solve certain domain decomposed linear systems of equations.

12.3.2 Generalized Conjugate Gradient Methods with Variable Preconditionings

We consider the solution of a system of linear equations

$$(12.42) \qquad\qquad Ax = b.$$

Here, A may not be available in explicit form, but only the action of A can be computed. Further, this action generally can be computed only approximately, although to an arbitrary high accuracy. A may be a nonsymmetric and/or an indefinite matrix; however, we assume that A is nonsingular.

To solve (12.42) we shall use a generalized conjugate gradient least square method, as presented in Section 12.1. We recall that such a method consists of the following steps: Given a set of search direction vectors, $\{d^j\}_{j=0}^{k-1}$ orthogonal with respect to $(\cdot, \cdot)_1$, one computes a new approximation $\mathbf{x}^k$, such that the quadratic functional

$$(12.43) \qquad f(\mathbf{x}) = \frac{1}{2}(\mathbf{r}, \mathbf{r})_0$$

is minimized over the shifted space

$$\mathbf{x}^0 + \text{span}\{\mathbf{d}^j\}_{j=0}^{k-1},$$

where $\mathbf{x}^0$ is an initial approximation, $(\cdot, \cdot)_0$ and $(\cdot, \cdot)_1$ are given inner products, and $\mathbf{r} = A\mathbf{x} - \mathbf{b}$ is the residual. Since A is nonsingular, there exists an unique minimizer of (12.43).

Determining the next approximation in this way,

$$\mathbf{x}^k = \mathbf{x}^{k-1} + \sum_{j=0}^{k-1} \alpha_j^{(k-1)} \mathbf{d}^j,$$

we compute the next residual

$$\mathbf{r}^k = \mathbf{r}^{k-1} + \sum_{j=0}^{k-1} \alpha_j^{(k-1)} A\mathbf{d}^j,$$

if the action of A can be computed exactly, or $\mathbf{r}^k = A[\mathbf{x}^k] - \mathbf{b}$, if we can compute only an approximate action and, therefore, also only approximate residuals.

Next, one takes a preconditioning step with the aim of accelerating the convergence by using a better search direction. To this end, we assume that we have a procedure $B[\cdot]$ that approximates the action of the inverse of A. In many applications, there exist natural ways of finding such procedures. In domain decomposition methods, one can use the method of actions—for instance, as described in Chapter 8. By this action, we find a pseudoresidual

$$\tilde{\mathbf{r}}^{(k)} = B[\mathbf{r}^{(k)}].$$

Then, the next search vector is defined by

$$(12.44) \qquad \mathbf{d}^k = -\tilde{\mathbf{r}}^k + \sum_{j=0}^{k-1} \beta_j^{(k-1)} \mathbf{d}^j.$$

The coefficients $\beta_j^{(k-1)}$ are determined from the orthogonality conditions

$$(\mathbf{d}^k, \mathbf{d}^j)_1 = 0, \quad j = 0, 1, 2, \ldots, k-1.$$

As is readily seen, this approach is quite general and can be used for an arbitrary mapping B, $\mathbf{r} \to B[\mathbf{r}]$. In practice, B is chosen to approximate the inverse of A, so that BA is sufficiently close to the identity operator. The coefficients $\alpha_j^{(k-1)}$ and $\beta_j^{(k-1)}$ are computed as shown in Section 12.1. That is, $\alpha_j^{(k-1)}$ are computed to minimize the functional

$$\phi(\alpha_0^{(k-1)}, \ldots, \alpha_{k-1}^{(k-1)}) = \frac{1}{2}(\mathbf{r}^k, \mathbf{r}^k)_0,$$

which leads to

$$\Lambda^{(k)}\underline{\alpha}^{(k)} = \underline{\gamma}^{(k)},$$

where $\Lambda_{l,j}^{(k)} = (A\mathbf{d}^l, A\mathbf{d}^j)_0$, $0 \le l$, $j \le k - 1$,

$$(\underline{\alpha}^{(k)})_j = \alpha_j^{(k-1)}, \quad j = 0, 1, \ldots, k - 1,$$

and

$$\gamma_{k-1}^{(k)} = -(\mathbf{r}^{k-1}, A\mathbf{d}^{k-1})_0, \quad \gamma_j^{(k)} = 0, \quad j = 0, 1, \ldots, k - 2.$$

The coefficients $\beta_j^{(k-1)}$ are computed from

$$\beta_j^{(k-1)} = (\tilde{\mathbf{r}}^k, \mathbf{d}^j)_1 / (\mathbf{d}^j, \mathbf{d}^j)_1, \quad j = 0, 1, \ldots, k - 1.$$

The inner products $(\cdot, \cdot)_0$ and $(\cdot, \cdot)_1$ can be chosen independently of one another. However, for practical reasons we consider two special cases here: (1) $(\mathbf{u}, \mathbf{v})_1 = (A\mathbf{u}, A\mathbf{v})_0$ and (2) $(\mathbf{u}, \mathbf{v})_1 = (\mathbf{u}, \mathbf{v})_0$, which correspond to Algorithms 1 and 3, respectively, in Section 12.1. As in the results there, we can prove the following lemma.

Lemma 12.22

(a) $(\mathbf{r}^k, A\mathbf{d}^j)_0 = 0$, $0 \le j \le k - 1$, *and*
$(\mathbf{r}^{k-1}, A\mathbf{d}^{k-1})_0 = -(\mathbf{r}^{k-1}, AB[\mathbf{r}^{k-1}])_0$,
(b) $\alpha_{k-1}^{(k-1)} = (\mathbf{r}^{k-1}, AB[\mathbf{r}^{k-1}])_0 \det(\Lambda^{(k-1)}) / \det \Lambda^{(k)}$ *and, in case (a),*
$\alpha_{k-1}^{(k-1)} = -(\mathbf{r}^{k-1}, A\mathbf{d}^{k-1})_0 / (\mathbf{d}^{k-1}, \mathbf{d}^{k-1})_1$, *and*
(c) $\alpha_{k-1}^{(k-1)} > 0$ *if and only if* $(\mathbf{r}^{k-1}, AB[\mathbf{r}^{k-1}])_0 > 0$.

Proof Using the defining expression for search directions, we find

$$(\mathbf{r}^{k-1}, A\mathbf{d}^{k-1})_0 = -(\mathbf{r}^{k-1}, A\tilde{\mathbf{r}}^{k-1})_0 + \sum_{j=0}^{k-1} \beta_j^{(k-1)}(\mathbf{r}^{k-1}, A\mathbf{d}^j)_0,$$

and the results follow from orthogonality properties and Cramer's rule. $\diamond$

The corresponding algorithms take now the form:

Algorithm 5

Compute $A\mathbf{d}^{k-1}$;
compute $(A\mathbf{d}^{k-1}, A\mathbf{d}^{k-1})_0$;
compute $(\mathbf{r}^{k-1}, A\mathbf{d}^{k-1})_0$;
$\alpha_{k-1}^{(k-1)} = -(\mathbf{r}^{k-1}, A\mathbf{d}^{k-1})_0/(A\mathbf{d}^{k-1}, A\mathbf{d}^{k-1})_0$;
$\mathbf{x}^k = \mathbf{x}^{k-1} + \alpha_{k-1}^{(k-1)}\mathbf{d}^{k-1}$;
$\mathbf{r}^k = \mathbf{r}^{k-1} + \alpha_{k-1}^{(k-1)}A\mathbf{d}^{k-1}$;
compute $\tilde{\mathbf{r}}^k = B[\mathbf{r}^k]$;
compute $A\tilde{\mathbf{r}}^k$;
compute $\beta_j^{(k-1)} = (A\tilde{\mathbf{r}}^{(k)}, A\mathbf{d}^{(j)})_0/(A\mathbf{d}^j, A\mathbf{d}^j)_0, \quad j = 0, \dots, k-1$;
$\mathbf{d}^k = -\tilde{\mathbf{r}}^k + \sum_{j=0}^{k-1} \beta_j^{(k-1)}\mathbf{d}^j$;

Algorithm 6

Compute $A\mathbf{d}^{k-1}$;
compute $(A\mathbf{d}^{k-1}, A\mathbf{d}^j)_0, \quad j = 0, 1, \dots, k-1$;
compute $(\mathbf{r}^{k-1}, A\mathbf{d}^{k-1})_0$;
solve $\Lambda^{(k)}\underline{\alpha}^{(k)} = \underline{\gamma}^{(k)}$;
$\mathbf{x}^k = \mathbf{x}^{k-1} + \sum_{j=0}^{k-1}\alpha_j^{(k-1)}\mathbf{d}^j$;
$\mathbf{r}^k = \mathbf{r}^{k-1} + \sum_{j=0}^{k-1}\alpha_j^{(k-1)}A\mathbf{d}^j$; (or $\mathbf{r}^k = A\mathbf{x}^k - \mathbf{b}$);
compute $\tilde{\mathbf{r}}^k = B[\mathbf{r}^k]$;
compute $\beta_j^{(k-1)} = (\tilde{\mathbf{r}}^k, \mathbf{d}^j)_0/(\mathbf{d}^j, \mathbf{d}^j)_0, \quad j = 0, \dots, k-1$;
$\mathbf{d}^k = -\tilde{\mathbf{r}}^k + \sum_{j=0}^{k-1}\beta_j^{(k-1)}\mathbf{d}^j$;

We recall that the first algorithm required two actions of A and one preconditioning step, while the second required only one action of A and one by B, the approximate action of the Schur complement, per iteration step. If it is not feasible to compute the action of A exactly, then $A\mathbf{d}^j$ will be computed only approximately and the entries $(A\mathbf{d}^{k-1}, A\mathbf{d}^j)_0$ in $\Lambda^{(k)}$ will contain errors. However, due to the form of the right-hand side vector $\underline{\gamma}^{(k)}$, these errors can be expected to cause smaller, relative errors for the leading, more accurate terms for $j = k - 1, k - 2, k - 3, \ldots$ than for the lower-order, less accurate terms for $k = 0, 1, \ldots$, because the leading terms will, in general, have larger values of their coefficients $\alpha_j^{(k-1)}$ and the errors in the coefficients of $\Lambda^{(k)}$ will be about the same size.

If A is s.p.d., we can use the inner product $(\mathbf{u}, \mathbf{v})_0 = \mathbf{u}^T A \mathbf{v}$, in which case Algorithm 5 simplifies as follows:

Algorithm 5(a)

> Compute $\mathbf{g}^{k-1} = A\mathbf{d}^{k-1}$;
> $\gamma_{k-1} = \mathbf{d}^{k-1^T}\mathbf{g}^{k-1}$;
> $\alpha = -((\mathbf{r}^{k-1})^T \mathbf{d}^{k-1})/\gamma_{k-1}$;
> $\mathbf{x}^k := \mathbf{x}^{k-1} + \alpha\mathbf{d}^{k-1}$;
> $\mathbf{r}^k := \mathbf{r}^{k-1} + \alpha\mathbf{g}^{k-1}$;
> compute $\tilde{\mathbf{r}}^k = B[\mathbf{r}^k]$;
> $\mathbf{d}^k := -\tilde{\mathbf{r}}^k$;
> for $j = 0, 1, \ldots, k - 1$ compute
> $\quad \beta_j = \tilde{\mathbf{r}}^{k^T}\mathbf{g}^j/\gamma_j$;
> $\quad \mathbf{d}^k := \mathbf{d}^k + \beta_j\mathbf{d}^j$;
> end (j)

Here we need only one action of A and one action of the preconditioning. There are $k + 2$ inner products involved. It is readily seen that if B is a symmetric and positive definite matrix, then Algorithm 5(a) reduces to the standard conjugate gradient method, where $\beta_j = 0$, $j = 0, 1, \ldots, k - 2$.

Let us now estimate the rate of convergence of the algorithms.

Theorem 12.23 *Let the preconditioner $B[\cdot]$ satisfy the coercivity and boundedness assumptions (a) and (b), that is:*

(a) $(\mathbf{v}, AB[\mathbf{v}])_0 \geq \delta_1(\mathbf{v}, \mathbf{v})_0$, for all $\mathbf{v}$.
(b) $\|AB[\mathbf{v}]\|_0 \leq \delta_2\|\mathbf{v}\|_0$, all $\mathbf{v}$, for some positive constants δ_1, δ_2.

Then, the variable-step GCG-method converges monotonically and the following convergence rate estimate holds,

$$\|\mathbf{r}^k\|_0 \le \sqrt{1 - (\delta_1/\delta_2)^2} \|\mathbf{r}^{k-1}\|_0, \quad k = 1, 2, \ldots,$$

where $\mathbf{r}^k = A\mathbf{x}^k - \mathbf{b}$.

Proof Lemma 12.22(a) shows that

$$(\mathbf{r}^k, \mathbf{r}^k)_0 = (\mathbf{r}^k, \mathbf{r}^{k-1} + \sum_{j=0}^{k-1} \alpha_j^{(k-1)} A\mathbf{d}^j)_0 = (\mathbf{r}^k, \mathbf{r}^{k-1})_0$$

$$= (\mathbf{r}^{k-1} + \sum_{j=0}^{k-1} \alpha_j^{(k-1)} A\mathbf{d}^j, \mathbf{r}^{k-1})_0$$

$$= (\mathbf{r}^{k-1}, \mathbf{r}^{k-1})_0 + \alpha_{k-1}^{(k-1)}(\mathbf{r}^{k-1}, A\mathbf{d}^{k-1})_0.$$

Hence, by Lemma 12.22(b),

$$(\mathbf{r}^k, \mathbf{r}^k)_0 = (\mathbf{r}^{k-1}, \mathbf{r}^{k-1})_0 - (\mathbf{r}^{k-1}, AB[\mathbf{r}^{k-1}])_0^2 \det(\Lambda^{(k-1)})/\det(\Lambda^{(k)}),$$

Lemma 12.4 shows that

$$\det(\Lambda^{(k)})/\det(\Lambda^{(k-1)}) = \min_{\mathbf{g} \in W_{k-2}} \|AB[\mathbf{r}^{k-1}] - \mathbf{g}\|_0^2,$$

where W_{k-2} is the vector space spanned by $\{A\mathbf{d}^s\}_{s=0}^{k-2}$. Letting simply $\mathbf{g} = \mathbf{0}$, we then get the upper bound

$$(\mathbf{r}^k, \mathbf{r}^k)_0 \le (\mathbf{r}^{k-1}, \mathbf{r}^{k-1})_0 - \{(\mathbf{r}^{k-1}, AB[\mathbf{r}^{k-1}])_0^2/\|AB[\mathbf{r}^{k-1}]\|_0\}^2.$$

Assumptions (a) and (b) now prove the theorem.   ◇

Remark 12.24 Note that Theorem 12.23 holds for a variable-step preconditioner—i.e., the preconditioner can change from one step to the next. In fact, it is readily seen that the rate-of-convergence estimate in Theorem 12.23 holds even for the steepest descent algorithm, where $\alpha_s^{(k-1)} = 0$, $s = 0, 1, \ldots, k - 2$, and $\beta_s^{(k-1)} = 0$, $s = 0, 1, \ldots, k - 1$.

Remark 12.25 If $\mathbf{d}^k = \mathbf{0}$ for some k, then it follows from (12.44) that $\tilde{\mathbf{r}}^k$ is a linear combination of $\{\mathbf{d}^j\}_{j=0}^{k-1}$. Then,

$$(\mathbf{r}^k, A\tilde{\mathbf{r}}^k)_0 = \sum_{j=0}^{k-1} \beta_j^{(k-1)}(\mathbf{r}^k, A\mathbf{d}^j)_0 = 0$$

by Lemma 12.22(a). By the coercivity assumption (a), we then have

$$0 = (\mathbf{r}^k, A\tilde{\mathbf{r}}^k)_0 = (\mathbf{r}^k, AB[\mathbf{r}^k])_0 \geq \delta_1 \|\mathbf{r}^k\|_0^2.$$

Hence, $\mathbf{r}^k = \mathbf{0}$, that is, the problem has been solved. Thus, we proved the following result.

Theorem 12.26 *If the preconditioner $B[\cdot]$ satisfies the coercivity assumption (a), then the (variable step) preconditioned GCG method with this preconditioner cannot fail.* ◇

Note, finally, that even if A is indefinite, for instance, there can still exist a mapping $B[\cdot]$ for which the coercivity and boundedness assumptions hold. This was illustrated in Axelsson and Vassilevski (1991).

Remark 12.27 When applying the GCG method, there is a simple way to automatically determine if the preconditioner is sufficiently accurate: We simply check the sign of $(\mathbf{r}^{k-1}, AB[\mathbf{r}^{k-1}])_0$, which by Lemma 12.22(a) equals $-(\mathbf{r}^{k-1}, A\mathbf{d}^{k-1})_0$. Equivalently, we can check the sign of $\alpha_{k-1}^{(k-1)}$. If this is negative, we restart the algorithm without updating the approximation at the last step and compute in the following iterations a more accurate preconditioner B by making the inner iteration parameters ε_1, ε_2 (see next paragraph) smaller or by simply performing more inner iterations. This procedure corresponds to one form of a variable-step preconditioning and makes the algorithm a "black box" solver.

To define a preconditioner approximating the action of A^{-1}, we follow the steps in (12.40). Thereby, every occurrence of the inverse of A_{11} [that is, steps (1) and (5) in (12.40)] is replaced by an (inner) iterative method to solve the corresponding systems with A_{11}. Likewise, the occurrence of S^{-1} in step (3) is replaced by an (inner) iterative method—that is, to solve $Sx_2 = \mathbf{w}_2$ approximately. In all cases, we iterate until the iteration error is sufficiently small.

In order to define preconditioner B at each (outer iteration) step, we need the following (in general, nonlinear) mappings:

$$\mathbf{v}_1 \to B_{11}[\mathbf{v}_1], \quad \mathbf{v}_2 \to C[\mathbf{v}_2],$$

such that

(12.45a) $\|A_{11}B_{11}[\mathbf{v}_1] - \mathbf{v}_1\| \leq \varepsilon_1 \|\mathbf{v}_1\|$, all $\mathbf{v}_1$

(12.45b) $\|SC[\mathbf{v}_2] - \mathbf{v}_2\| \leq \varepsilon_2 \|\mathbf{v}_2\|$, all $\mathbf{v}_2$

for some sufficiently small positive numbers ε_1, ε_2.

The application of the (variable step) preconditioner $B = B[\cdot]$ involves the following steps.

Algorithm 7

(1) $\mathbf{w}_1 = B_{11}[\mathbf{v}_1]$;
(2) $\mathbf{w}_2 = -A_{21}\mathbf{w}_1 + \mathbf{v}_2$;
(3) $\mathbf{x}_2 = C[\mathbf{w}_2]$;
(4) $\mathbf{y}_1 = A_{12}\mathbf{x}_2$;
(5) $\mathbf{z}_1 = B_{11}[\mathbf{y}_1]$;
(6) $\mathbf{x}_1 = \mathbf{w}_1 - \mathbf{z}_1$.

Then, $B[\mathbf{v}] = \begin{bmatrix} \mathbf{x}_1 \\ \mathbf{x}_2 \end{bmatrix}$.

Therefore, the application of the preconditioner on each (outer iteration) step can be realized as follows:

Algorithm 8   Given sufficiently small positive numbers ε_1, ε_2, we iterate in steps (1) and (5) with some method until the iterations $\mathbf{w}_1$, $\mathbf{z}_1$ satisfy

$$\|A_{11}\mathbf{w}_1 - \mathbf{v}_1\|_0 \leq \varepsilon_1\|\mathbf{v}_1\|_0, \quad \|A_{11}\mathbf{z}_1 - \mathbf{y}_1\|_0 \leq \varepsilon_1\|\mathbf{y}_1\|_0,$$

and in step (3) until the iteration $\mathbf{x}_2$ satisfies

$$(12.46) \qquad \|A_{22}\mathbf{x}_2 - A_{21}\mathbf{z}_1 - \mathbf{w}_2\|_0 \leq \varepsilon_2\|\mathbf{w}_2\|_0,$$

where

$$\mathbf{w}_2 = \mathbf{v}_2 - A_{21}\mathbf{w}_1.$$

Remark 12.28 Since (12.46) involves the computation of $\mathbf{z}_1$ in step (5), the test (12.46) is actually performed after step (5). This means that we may have to repeat steps (4) and (5) if (12.46) fails to be satisfied. Hence, in practice it can be advisable to choose a certain (fixed) number of iterations in step (3) and to test on the sign of $\alpha_{k-1}^{(k-1)}$ instead, as was already mentioned in Remark 12.27. If the sign test is violated, we repeat Algorithm 8 with smaller values of ε_1, ε_2.

We shall now estimate the deviation of $AB[\mathbf{v}]$ from $\mathbf{v}$. Algorithm 7 shows that

$$AB[\mathbf{v}] = \begin{bmatrix} A_{11} & A_{12} \\ A_{21} & A_{22} \end{bmatrix} \begin{bmatrix} \mathbf{x}_1 \\ \mathbf{x}_2 \end{bmatrix}$$

$$= \begin{bmatrix} A_{11}(\mathbf{w}_1 - \mathbf{z}_1) + A_{12}C[\mathbf{w}_2] \\ A_{21}(\mathbf{w}_1 - \mathbf{z}_1) + A_{22}C[\mathbf{w}_2] \end{bmatrix}$$

(12.47)

$$= \begin{bmatrix} \mathbf{v}_1 \\ \mathbf{v}_2 \end{bmatrix} + \begin{bmatrix} A_{11}(\mathbf{w}_1 - B_{11}[\mathbf{y}_1]) - \mathbf{v}_1 + A_{12}C[\mathbf{w}_2] \\ A_{21}(\mathbf{w}_1 - B_{11}[\mathbf{y}_1]) + A_{22}C[\mathbf{w}_2] - \mathbf{v}_2 \end{bmatrix}$$

$$= \begin{bmatrix} \mathbf{v}_1 \\ \mathbf{v}_2 \end{bmatrix} + \begin{bmatrix} (A_{11}\mathbf{w}_1 - \mathbf{v}_1) - (A_{11}\mathbf{z}_1 - \mathbf{y}_1)] \\ A_{22}C[\mathbf{w}_2] - \mathbf{w}_2 - A_{21}B_{11}[\mathbf{y}_1] \end{bmatrix}.$$

Let us first estimate

$$R_2 = \|A_{22}C[\mathbf{w}_2] - \mathbf{w}_2 - A_{21}B_{11}[\mathbf{y}_1]\|_0.$$

Steps (3) and (4) in Algorithm 6 show that

$$R_2 = \|SC[\mathbf{w}_2] - \mathbf{w}_2 - A_{21}A_{11}^{-1}(A_{11}B_{11}[\mathbf{y}_1] - \mathbf{y}_1)\|_0,$$

so by (12.45a) and (12.45b),

$$R_2 \leq \|SC[\mathbf{w}_2] - \mathbf{w}_2\|_0 + \|A_{21}A_{11}^{-1}(A_{11}B_{11}[\mathbf{y}_1] - \mathbf{y}_1)\|_0$$

$$\leq \varepsilon_2\|\mathbf{w}_2\|_0 + \|A_{21}A_{11}^{-1}\|_0\|A_{11}B_{11}[\mathbf{y}_1] - \mathbf{y}_1\|_0$$

$$\leq \varepsilon_2\|\mathbf{w}_2\|_0 + \varepsilon_1\|A_{21}A_{11}^{-1}\|_0\|\mathbf{y}_1\|_0$$

$$\leq \varepsilon_2\|\mathbf{w}_2\|_0 + \varepsilon_1\|A_{21}A_{11}^{-1}\|_0\|A_{12}S^{-1}\|_0\|SC[\mathbf{w}_2]\|_0$$

$$\leq \{\varepsilon_2 + \varepsilon_1\|A_{21}A_{11}^{-1}\|_0\|A_{12}S^{-1}\|_0(1 + \varepsilon_2)\}\|\mathbf{w}_2\|_0$$

$$\leq \{\varepsilon_2 + \varepsilon_1\|A_{21}A_{11}^{-1}\|_0\|A_{12}S^{-1}\|_0(1 + \varepsilon_2)\}\|\mathbf{v}_2 - (A_{21}A_{11}^{-1})A_{11}B_{11}[\mathbf{v}_1]\|_0$$

$$\leq \{\varepsilon_2 + \varepsilon_1\|A_{21}A_{11}^{-1}\|_0\|A_{12}S^{-1}\|_0(1 + \varepsilon_2)\}\{\|\mathbf{v}_2\|_0 + \|A_{21}A_{11}^{-1}\|_0(1 + \varepsilon_1)\|\mathbf{v}_1\|_0\}.$$

Similarly, we have by (12.45a),

(12.48) $\|A_{11}\mathbf{w}_1 - \mathbf{v}_1\|_0 = \|A_{11}B_{11}[\mathbf{v}_1] - \mathbf{v}_1\|_0 \leq \varepsilon_1\|\mathbf{v}_1\|_0,$

and by (12.45a) and (12.45b),

$$\|A_{11}\mathbf{z}_1 - \mathbf{y}_1\|_0 = \|A_{11}B_{11}[\mathbf{y}_1] - \mathbf{y}_1\|_0 \le \varepsilon_1 \|\mathbf{y}_1\|_0$$

(12.49)
$$\le \varepsilon_1 \|A_{12}S^{-1}\|_0 \|SC[\mathbf{w}_2]\|_0$$

$$\le \varepsilon_1 \|A_{12}S^{-1}\|_0 (1 + \varepsilon_2) \|\mathbf{w}_2\|_0$$

$$\le \varepsilon_1 (1 + \varepsilon_2) \|A_{12}S^{-1}\|_0 [\|\mathbf{v}_2\|_0 + (1 + \varepsilon_1) \|A_{21}A_{11}^{-1}\|_0 \|\mathbf{v}_1\|_0].$$

Finally, by (12.47), we get

$$\|AB[\mathbf{v}] - \mathbf{v}\|_0^2 \le R_2^2 + (\|A_{11}\mathbf{w}_1 - \mathbf{v}_1\|_0 + \|A_{11}\mathbf{z}_1 - \mathbf{y}_1\|_0)^2$$

(12.50)
$$\le \{(\varepsilon_2 + \varepsilon_1 \sigma_1 \sigma_2)(1 + \varepsilon_2)\}^2 [\|\mathbf{v}_2\|_0 + \sigma_2 (1 + \varepsilon_1) \|\mathbf{v}_1\|_0]^2$$

$$+ \varepsilon_1^2 \{\|\mathbf{v}_1\|_0 + \sigma_1 (1 + \varepsilon_2)[\|\mathbf{v}_2\|_0 + (1 + \varepsilon_1)\sigma_2 \|\mathbf{v}_1\|_0]\}^2$$

$$\le C_1 (\varepsilon_1 + \varepsilon_2)^2 \|\mathbf{v}\|_0^2,$$

where $C_1 = C_1(\sigma_1, \sigma_2)$ and

$$\sigma_1 = \|A_{12}S^{-1}\|_0, \quad \sigma_2 = \|A_{21}A_{11}^{-1}\|_0.$$

We summarize the result in the following theorem.

Theorem 12.29 *Let the norm* $\| \cdot \|_0$ *in the vector spaces for* $\mathbf{v}_1, \mathbf{v}_2$, *respectively, be such that*

$$\sigma_1 = \|A_{12}S^{-1}\|_0, \quad \sigma_2 = \|A_{21}A_{11}^{-1}\|_0$$

are bounded uniformly with respect to some problem parameter, such as the meshsize in difference matrices. Then, for $\varepsilon_1, \varepsilon_2$ *sufficiently small, the mapping* $B[\cdot]$ *defined by Algorithm 7, with* $B_{11}[\cdot]$ *and* $C[\cdot]$ *satisfying (12.45a,b), is coercive and bounded, that is,*

$$(\mathbf{v}, AB[\mathbf{v}])_0 \ge [1 - C_1^{1/2}(\varepsilon_1 + \varepsilon_2)] \|\mathbf{v}\|_0^2, \quad \text{for all } \mathbf{v},$$

where $C_1 = C_1(\sigma_1, \sigma_2)$ *is a function of* σ_1, σ_2, *bounded for all bounded values of* σ_1, σ_2, *and*

$$\|AB[\mathbf{v}]\|_0 \le [1 + C_1^{1/2}(\varepsilon_1 + \varepsilon_2)] \|\mathbf{v}\|_0, \quad \text{for all } \mathbf{v},$$

respectively.

Proof (12.50) shows that

$$| \, \|AB[\mathbf{v}]\|_0 - \|\mathbf{v}\|_0 | \leq \|AB[\mathbf{v}] - \mathbf{v}\|_0 \leq C_1^{1/2}(\varepsilon_1 + \varepsilon_2)\|\mathbf{v}\|_0.$$

Hence,

$$(12.51) \quad [1 - C_1^{1/2}(\varepsilon_1 + \varepsilon_2)]\|\mathbf{v}\|_0 \leq \|AB[\mathbf{v}]\|_0 \leq [1 + C_1^{1/2}(\varepsilon_1 + \varepsilon_2)]\|\mathbf{v}\|_0.$$

Further, (12.50) shows that

$$\|\mathbf{v}\|_0^2 - 2(\mathbf{v}, AB[\mathbf{v}])_0 + \|AB[\mathbf{v}]\|_0^2 = (\mathbf{v} - AB[\mathbf{v}], \mathbf{v} - AB[\mathbf{v}])_0$$

$$\leq C_1(\varepsilon_1 + \varepsilon_2)^2\|\mathbf{v}\|_0^2.$$

Thus,

$$2(\mathbf{v}, AB[\mathbf{v}])_0 \geq [1 - C_1(\varepsilon_1 + \varepsilon_2)^2]\|\mathbf{v}\|_0^2 + \|AB[\mathbf{v}]\|_0^2,$$

and this, together with the left-hand side part of (12.51), shows that

$$2(\mathbf{v}, AB[\mathbf{v}])_0 \geq [2 - 2C_1^{1/2}(\varepsilon_1 + \varepsilon_2)]\|\mathbf{v}\|_0^2. \qquad \diamond$$

Remark 12.30 In some problems, it can be more efficient to use an approximate inverse B_{11} of a form defined in Chapter 8 to approximate A_{11}^{-1} instead of some inner iteration method. If we gradually increase the size of the sparsity pattern for B_{11}, we will eventually have (12.45a) satisfied for a sufficiently small ε_1.

References

W. I. Arnoldi (1951). The principle of minimized iterations in the solution of the matrix eigenvalue problem. *Quart. Appl. Math.* **9**, 17–29.

O. Axelsson (1973). Notes on the numerical solution of the Biharmonic equation. *J. Inst. Math. Appl.* **11**, 213–226.

O. Axelsson (1978). Conjugate gradient type methods for unsymmetric and inconsistent systems of linear equations. RR.78.03R, Department of Computer Sciences, Chalmers University of Technology, Göteborg, Sweden.

O. Axelsson (1979), A generalized conjugate direction method and its application to a singular perturbation problem. In *Proc. 8th biannual Numerical Analysis Conference*, ed. G. A. Watson (Dundee, Scotland, June 26–29, 1979). *Lecture Notes in Mathematics*, **Vol. 773**, pp. 1–11. Berlin, Heidelberg, New York: Springer-Verlag.

O. Axelsson (1980). Conjugate gradient type methods for unsymmetric and inconsistent systems of linear equations. *Linear Algebra Appl.* **29**, 1–16.

O. Axelsson (1987). A generalized conjugate gradient, least square method. *Numer. Math.* **51**, 209–227.

O. Axelsson (1988). A restarted version of a generalized preconditioned conjugate gradient method. *Communications in Applied Numerical Methods* **4**, 521–530.

O. Axelsson, V. Eijkhout, B. Polman, and P. Vassilevski (1989). Iterative solution of singular perturbation 2nd order boundary value problems by use of incomplete block-matrix factorization methods. *BIT* **29**, 867–889.

O. Axelsson and P. S. Vassilevski (1991). A black box generalized conjugate gradient solver with inner iterations and variable-step preconditioning. *SIAM J. Matrix Anal. Appl.* **12**, 625–644.

R. E. Bank, B. D. Welfert, and H. Yserentant (1990). A class of iterative methods for solving saddle points problems. *Numer. Math.* **56**, 645–666.

P. Concus and G. H. Golub (1976). A generalized conjugate gradient method for non-symmetric systems of linear equations. In *Proceedings of the Second International Symposium on Computing Methods in Applied Sciences and Engineering*, ed. R. Glowinsky and J. L. Lions (IRIA, Paris, Dec. 1975). *Lecture Notes in Mathematics*, **Vol. 134**, pp. 56–65. Berlin, Heidelberg, New York: Springer-Verlag.

E. J. Craig (1955). The N-step iteration procedures. *J. Math. Phys.* **34**, 64–73.

J. E. Dennis and K. Turner (1988). Generalized conjugate directions. *Lin. Alg. Appl.* **88/89**, 187–209.

S. C. Eisenstat, H. C. Elman, and M. H. Schultz (1983). Variational iterative methods for nonsymmetric systems of linear equations. *SIAM J. Numer. Anal.* **20**, 345–357.

H. C. Elman (1982). Iterative methods for large, sparse nonsymmetric systems of linear equations. Research Report # 229, Computer Science Dept., Yale University.

V. Faber and T. Manteuffel (1984). Necessary and sufficient conditions for the existence of a conjugate gradient method. *SIAM J. Numer. Anal.* **21**, 352–362.

G. H. Golub and M. L. Overton (1988). The convergence of inexact Chebyshev and Richardson iterative methods for solving linear systems. *Numer. Math.* **53**, 571–593.

L. Hageman and D. M. Young (1981). *Applied Iterative Methods*, Orlando, Fla.: Academic.

W. D. Joubert and D. M. Young (1987). Necessary and sufficient conditions for the simplification of generalized conjugate gradient algorithms. *Lin. Alg. Appl.*, **88/89**, 449–485.

I. M. Khabaza (1963). An iterative least-square method suitable for solving large sparse matrices. *Comp. J.* **6**, 202–206.

U. Langer and W. Queck (1986). On the convergence factor of Uzawa's algorithm. *J. Comp. Appl. Math.* **15**, 191–202.

G. Meinardus (1964). Approximation von Funktionen und ihre numerische Behandlung. Berlin, Heidelberg, New York: Springer.

Y. Saad (1981). Krylov subspace methods for solving large unsymmetric linear systems. *Math. Comp.* **37**, 105–126.

Y. Saad and M. H. Schultz (1986). GMRES: A generalized minimal residual algorithm for solving nonsymmetric linear systems. *SIAM J. Sci. Stat. Comp.* **7**, 856–869.

Y. Saad and M. H. Schultz (1985). Conjugate gradient-like algorithms for solving nonsymmetric linear systems. *Math. Comput.* **44**, 417–424.

P. S. Vassilevski (1992). Preconditioning nonsymmetric and indefinite finite element matrices. *J. Numer. Lin. Alg.*, in press.

R. Verfürth (1984). A combined conjugate gradient-multigrid algorithm for the numerical solution of the Stokes problem. *IMA J. Numer. Anal.* **4**, 441–455.

P. K. W. Vinsome (1976). Orthomin, an iterative method for solving sparse sets of simultaneous linear equations. In *Proc. Fourth Symposium on Reservoir Simulation*, Society of Petroleum Engineers of AIME, pp. 149–159.

V. V. Voevodin (1983). The problem of non-selfadjoint generalization of the conjugate gradient method has been closed. *U.S.S.R. J. Comput. Math. Phys.* **23**, 143–144.

V. V. Voevodin and E. E. Tyrtyshnikov (1981). On generalizations of conjugate direction methods. *Numerical Methods of Algebra*, Moscow State University Press, Moscow, pp. 3–9 (in Russian).

O. Widlund (1978). A Lanczos method for a class of non-symmetric systems of linear equations. *SIAM J. Numer. Anal.* **15**, 801–812.

D. M. Young and K. C. Jea (1980). Generalized conjugate gradient acceleration of nonsymmetrizable iterative methods. *Lin. Alg. Appl.* **34**, 159–194.

D. M. Young, L. J. Hayes and K. C. Jea (1981). Generalized conjugate gradient acceleration of iterative methods, part I: The symmetrizable case. Report # 162, Center for Numerical Analysis, Univ. of Texas at Austin.

D. M. Young and K. C. Jea (1981). Generalized conjugate gradient acceleration of iterative methods, part II: The nonsymmetrizable case. Report # 163, Center for Numerical Analysis, Univ. of Texas at Austin.

13

The Rate of Convergence of the Conjugate Gradient Method

In the two previous chapters, it was shown that conjugate gradient-type methods have many advantages, one being simplicity of implementation: No method parameter need be estimated, and the method consists mainly of matrix vector multiplications and vector operations. In the present chapter, we shall analyze the rate of convergence of the methods. This will be done not only for the case of a symmetric, positive definite matrix but also for general matrices. First, it will be shown that if the matrix is normal, then the norm of the residual can be estimated by the error in a best polynomial approximation problem. It will then be shown that the rate of convergence depends on the distribution of eigenvalues and, to some extent, also on the initial residual. If the matrix is not normal, the estimate involves also the condition number of the matrix that transforms the given matrix to Jordan canonical form. In general, we are unable to give exact estimates of the rate of convergence, but we shall derive various upper bounds using different methods of estimation.

The rate of convergence can be measured in various norms. When we use a relative measure—say, the ratio of the norm of the current residual and the initial residual—it will be shown that the condition number of the matrix can frequently be used to give a sufficiently accurate estimate. It will be seen that this estimate can be derived using the best polynomial approximation—i.e., the Chebyshev polynomial on the interval bounded by the extreme eigenvalues, if these are real and positive. More generally, if the eigenvalues are contained in an ellipse not covering the origin and with sufficiently small eccentricity, then similar estimates can be derived.

However, when the eigenvalues are distributed nonuniformly in this interval or ellipse, it will be seen that the condition number alone gives too rough an overestimate of the necessary number of iterations. Therefore, we shall present various improved estimates of the number of iterations required to get a sufficiently small relative residual, based on some additional assumptions made of the eigenvalue distribution. In particular, we analyze the case of isolated eigenvalues at one or both ends of the spectrum. While the estimate based on fixed intervals and condition numbers shows a linear rate of convergence, we comment also on superlinear rates of convergence that can be seen for certain distributions of eigenvalues.

We shall also use an alternative estimate, based not on the standard condition numbers but on the ratio of the trace and determinant of the matrix. Although this estimate involves all eigenvalues, it does not give better estimates for some typical distributions of eigenvalues. However, it gives an alternative illustration of the superlinear convergence phenomenon.

Finally, we consider an estimate of the relative iteration error based on taking different norms of the current and the initial error. It will be seen that such estimates can be derived for which the relative error decreases with a number that does not depend on the condition number. However, this decrease levels off quickly with an increasing number of iterations.

At the end, it is shown that in many instances one can typically recognize three different phases in the convergence history of the conjugate gradient method: the initial phase, where the norm of the residual decreases at least as fast as $O(k + 1)^{-2}$, where k is the iteration number; the middle phase, where the convergence rate is linear and given by the standard expression involving the condition number; and a final phase, where there is a tendency toward a superlinear rate of convergence, more pronounced for certain distributions of eigenvalues. Although the derivation of these results assumes exact arithmetic, it turns out that they hold also in the presence of rounding errors.

13.1 Rate of Convergence Estimates Based on Min Max Approximations

Consider the solution of

$$(13.1) \qquad\qquad A\mathbf{x} = \mathbf{a},$$

where $\mathbf{x}, \mathbf{a} \in \mathbb{R}^n$ and where A is diagonalizable and positive semidefinite. We assume that the the system is consistent. We shall estimate the rate of convergence of the conjugate gradient method or of the generalized conjugate gradient method. When we use preconditioned versions, then A is replaced by $B = C^{-1}A$, where C is the preconditioning matrix.

As we saw in Chapter 11, when A is singular, any solution $\mathbf{x}$ of (13.1) depends on the initial vector or, more precisely, on the component of the initial vector in the nullspace of A. Let $\mathbf{x}$ have the same component in the nullspace of A as the initial vector. Then, the iteration error $\mathbf{e}^k = \mathbf{x} - \mathbf{x}^k$ and the residual $\mathbf{r}^k = A\mathbf{x}^k - \mathbf{a}$ do not contain any component in the nullspace of A. This follows, because by construction of the conjugate gradient method, $\mathbf{x}^k$ for any $k \geq 1$ contains the same component in the nullspace as $\mathbf{x}^0$ and $\mathbf{r}^k = -A\mathbf{e}^k$, when $\mathbf{x}$ is a solution to a consistent system (13.1). It will be seen that the rate of convergence depends on the distribution of the spectrum of B, and the above shows that it depends only on the nonzero part of the spectrum.

In fact, due to the construction of the conjugate gradient method, it is readily seen that the iteration error $\mathbf{e}^k$ is related to the initial error $\mathbf{e}^0 = \mathbf{x} - \mathbf{x}^0$ by

$$(13.2) \qquad \mathbf{e}^k = P_k(B)\mathbf{e}^0,$$

for some polynomial P_k of degree k, normalized such that $P_k(0) = 1$. Assume first that A and C are s.p.d. Using the notations introduced in Chapter 5 and the minimization property (see Theorem 11.3) of the conjugate gradient method, we find

$$\|\mathbf{e}^k\|_{A^{\frac{1}{2}}} = \min_{P_k \in \pi_k^1} \|P_k(B)\mathbf{e}^0\|_{A^{\frac{1}{2}}},$$

where π_k^1 denotes the set of polynomials of degree k normalized such that $P_k(0) = 1$.

Let $\mathbf{v}_1, \mathbf{v}_2, \ldots, \mathbf{v}_n$ be orthonormal eigenvectors and let λ_i, $i = 1, 2, \ldots, n$ be the corresponding eigenvalues of B. Let

$$\mathbf{e}^0 = \sum_{j=1}^{n} \alpha_j \mathbf{v}_j,$$

where $\alpha_i = (\mathbf{e}^0, \mathbf{v}_i)$ and $(\mathbf{v}_i, \mathbf{v}_i) = 1$, $i = 1, \ldots, n$. Here, the inner product is defined by the matrix C, that is, $(\mathbf{u}, \mathbf{v}) = \mathbf{u}^T C \mathbf{v}$. (Note that the eigenvectors are both A- and C-orthogonal.) Then (13.2) shows that

$$\mathbf{e}^k = \sum_{j=1}^{n} \alpha_j P_k(\lambda_j) \mathbf{v}_j$$

and, using the positivity of the eigenvalues, we find

$$\|\mathbf{e}^k\|_{A^{\frac{1}{2}}} = \|\sum_{j=1}^{n} \alpha_j P_k(\lambda_j) \mathbf{v}_j\|_{A^{\frac{1}{2}}} = \left\{ \sum_{j=1}^{n} \alpha_j^2 \lambda_j P_k^2(\lambda_j) \right\}^{\frac{1}{2}}$$

$$\leq \left\{ \sum_{j=1}^{n} \alpha_j^2 \lambda_j \right\}^{\frac{1}{2}} \max_{\substack{1 \leq i \leq n \\ \lambda_i > 0}} |P_k(\lambda_i)| = \max_{\substack{1 \leq i \leq n \\ \lambda_i > 0}} |P_k(\lambda_i)| \|\mathbf{e}^0\|_{A^{\frac{1}{2}}}.$$

Hence, recalling the minimization property of the conjugate gradient method, we then find

(13.3)
$$\|\mathbf{e}^k\|_{A^{\frac{1}{2}}} \leq \min_{P_k \in \pi_k^1} \max_{\substack{1 \leq i \leq n \\ \lambda_i > 0}} |P_k(\lambda_i)| \|\mathbf{e}^0\|_{A^{\frac{1}{2}}}.$$

The rate of convergence of the iteration error $\|\mathbf{e}^k\|_{A^{\frac{1}{2}}}$ is measured by the average convergence factor

$$\{\|\mathbf{e}^k\|_{A^{\frac{1}{2}}} / \|\mathbf{e}^0\|_{A^{\frac{1}{2}}}\}^{1/k}.$$

(13.3) shows that this can be replaced with an estimate of the rate of convergence of a best polynomial approximation problem [namely, the approximation of the zero function by polynomials satisfying the constraint $P_k(0) = 1$] in the maximum norm on the discrete set formed by the spectrum of B.

It can be seen that for any distribution of the eigenvalues, there exists a polynomial P_k such that $|P_k(\lambda_i')| = \min_{P_k \in \pi_k^1} \max_{1 \leq j \leq n} |P_k(\lambda_j)|$ for a subset (depending on k) of $k + 1$ points $\lambda_i' = \lambda_i$. Therefore, the estimate in (13.3) is *sharp*, in the respect that for any distribution of eigenvalues and for any k, there exists an initial vector $\mathbf{x}^0$ (depending on k) such that equality holds in (13.3). (We then take $\mathbf{x}_0$ to be a vector such that $\alpha_i \neq 0$ only for indices corresponding to the subset λ_i' of eigenvalues.)

We now want to find the rate with which min max $|P_k(\lambda_i)|$ approaches zero. In addition to this rate of convergence, we want to find the smallest value of k for which

(13.4)
$$\|\mathbf{e}^k\|_{A^{\frac{1}{2}}} \leq \varepsilon \|\mathbf{e}^0\|_{A^{\frac{1}{2}}}$$

holds for an arbitrary initial vector. In both cases, we must then estimate

$$\min_{P_k \in \pi_k^1} \max_{\lambda_i > 0} |P_k(\lambda_i)|,$$

as a function of k.

Consider now the case where A (or $B = C^{-1}A$) is generally not symmetric but is diagonalizable, and its symmetric part is positive definite w.r.t. the given inner product. Then we can use the generalized conjugate gradient-minimal residual method or the generalized conjugate gradient-orthogonal residual method. As was shown in Chapter 12, for the first method, there still holds a minimization property, and it will be seen below that there holds a quasi-minimization property for the second method. Let

$$a(\mathbf{u}, \mathbf{v}) = (B\mathbf{u}, \mathbf{v}), \quad \mathbf{u}, \mathbf{v} \in \mathbb{R}^n,$$

where $B = A$ or $B = C^{-1}A$ and $(\cdot, \cdot)$ is an inner product. Then, using the positive definiteness of the symmetric part of B, we find

(13.5a) $$a(\mathbf{u}, \mathbf{u}) \geq \rho \|\mathbf{u}\|^2, \quad \text{for all } \mathbf{u} \in \mathbb{R}^n$$

and

(13.5b) $$|a(\mathbf{u}, \mathbf{v})| \leq K \|\mathbf{u}\| \|\mathbf{v}\|, \quad \text{for all } \mathbf{u}, \mathbf{v} \in \mathbb{R}^n,$$

where $\|\mathbf{u}\| = (\mathbf{u}, \mathbf{u})^{\frac{1}{2}}$. Here, $K \geq \rho$ and ρ is a positive constant. Let $\hat{\mathbf{x}}$ be the solution of $B\mathbf{x} = \mathbf{b}$. Now the orthogonality property shows that

$$\rho \|\mathbf{e}^k\|^2 \leq a(\mathbf{e}^k, \mathbf{e}^k) = a(\mathbf{e}^k, \hat{\mathbf{x}} - \mathbf{v}) \leq K \|\mathbf{e}^k\| \|\hat{\mathbf{x}} - \mathbf{v}\|$$

for all $\mathbf{v} \in \mathbf{x}^0 \oplus V_{k-1}$ where V_{k-1} is the Krylov subspace corresponding to B and $\mathbf{r}^0$. Hence,

(13.6) $$\|\mathbf{e}^k\| \leq \frac{K}{\rho} \min_{\mathbf{v} \in \mathbf{x}^0 \oplus V_{k-1}} \|\hat{\mathbf{x}} - \mathbf{v}\|,$$

which shows the quasioptimal (quasioptimal because $K/\rho > 1$, in general) property of the errors. The average rate of convergence is

$$\|\mathbf{e}^k\|^{\frac{1}{k}} \leq \left(\frac{K}{\rho}\right)^{\frac{1}{k}} \left\{ \min_{\mathbf{v} \in \mathbf{x}^0 \oplus V_{k-1}} \|\hat{\mathbf{x}} - \mathbf{v}\| \right\}^{\frac{1}{k}}$$

and, hence, the estimate of this has been reduced to a pure best approximation problem—namely, to find the best approximation of $\hat{\mathbf{x}}$ among all $\mathbf{v} \in \mathbf{x}^0 \oplus V_{k-1}$.

Since B is diagonalizable, there exists a matrix S, such that $S^{-1}BS = D$ where D is diagonal. We get

$$\min_{\mathbf{v} \in \mathbf{x}^0 \oplus V_{k-1}} \|\hat{\mathbf{x}} - \mathbf{v}\| = \min_{\mathbf{v} \in V_{k-1}} \|\mathbf{e}^0 - \mathbf{v}\| = \min_{P_k \in \pi_k^1} \|P_k(B)\mathbf{e}^0\|$$

$$\leq \|S\| \min_{P_k \in \pi_k^1} \|P_k(D)S^{-1}\mathbf{e}^0\|$$

$$\leq \mathcal{K}(S) \min_{P_k \in \pi_k^1} \max_{\lambda \in S(B)} |P_k(\lambda)| \|\mathbf{e}^0\|,$$

where $\mathcal{K}(S) = \|S\|\|S^{-1}\|$ and $S(B)$ denotes the spectrum of B. Hence, the above and (13.6) show that

$$(13.7) \qquad \frac{\|e^k\|}{\|e^0\|} \leq \frac{K}{\rho}\mathcal{K}(S) \min_{P_k \in \pi_k^1} \max_{\lambda \in S(B)} |P_k(\lambda)|.$$

Note that if A and C are s.p.d., we can choose S orthogonal; then, $\mathcal{K}(S) = 1$. In addition, in this case, the eigenvalues are real. More generally, Corollary 2.22 shows that if B is normal, we can take S to be unitary, so $\mathcal{K}(S) = 1$. Further, with

$$a(\mathbf{u}, \mathbf{v}) = (B\mathbf{u}, \mathbf{v}) \text{ and } (\mathbf{u}, \mathbf{v}) = \mathbf{u}^T C\mathbf{v},$$

B is s.p.d. with respect to this innerproduct, and with $\|\mathbf{u}\|_{A^{\frac{1}{2}}} = \{\mathbf{u}^T A\mathbf{u}\}^{\frac{1}{2}}$, we have

$$a(\mathbf{u}, \mathbf{v}) = \mathbf{u}^T A\mathbf{v}$$

and

$$a(\mathbf{u}, \mathbf{u}) = \|\mathbf{u}\|^2_{A^{\frac{1}{2}}}$$

$$a(\mathbf{u}, \mathbf{v}) \leq \|\mathbf{u}\|_{A^{\frac{1}{2}}} \|\mathbf{v}\|_{A^{\frac{1}{2}}}.$$

Hence, $\rho = K = 1$ in this case, and we are back to the estimate (13.3).

In the s.p.d. case, it suffices to consider the best approximation problem on an interval, but in the general case (13.7), we must estimate the rate with which the best approximation polynomial on a complex spectrum approaches zero. When the eigenvalues are contained in ellipses and the origin is outside these, we can then use the results in Section 5.4.

The above results are stated in the following theorem.

Theorem 13.1 *Consider the generalized conjugate gradient (GCG − LS) method or the orthogonal residual conjugate gradient method to solve $B\mathbf{x} = \mathbf{b}$, where the symmetric part of B is positive definite and B is diagonalizable, $B = SDS^{-1}$, and $D = \text{diag}(\lambda_1, \ldots, \lambda_n)$.*

(a) The relative error at iteration k satisfies

$$\frac{\|e^k\|}{\|e^0\|} \leq \frac{K}{\rho}\mathcal{K}(S) \min_{P_k \in \pi_k^1} \max_{\lambda \in S(B)} |P_k(\lambda)|,$$

where K, ρ are constants in 13.5(a,b), respectively, for the bilinear form $a(\mathbf{u}, \mathbf{v}) = (B\mathbf{u}, \mathbf{v})$, and where $S(B)$ denotes the spectrum of B. If B is normal, the $\mathcal{K}(S) = 1$.

(b) If $B = C^{-1}A$, C is s.p.d., A is symmetric, positive semidefinite, and the innerproduct $(\mathbf{u}, \mathbf{v}) = \mathbf{u}^T C \mathbf{v}$, then,

$$\frac{\|\mathbf{e}^k\|_{A^{\frac{1}{2}}}}{\|\mathbf{e}^0\|_{A^{\frac{1}{2}}}} \leq \min_{P_k \in \pi_k^1} \max_{\lambda \in S^0(B)} |P_k(\lambda)|,$$

where $S^0(B)$ denotes the nonzero part of $S(B)$.   ◇

The estimate in part (a) holds for any similarity transformation matrix S, so we can choose such a matrix with minimal condition number. In general, the constants K, ρ and the transformation matrix S and its condition number are not known. However, their precise values are often not so important for the following reason. When we want to find the number of iterations, k, such that $\|\mathbf{e}^k\|/\|\mathbf{e}^0\| \leq \varepsilon$, then we use the estimate

$$(13.8) \qquad \min_{P_k \in \pi_k^1} \max_i |P_k(\lambda_i)| \leq \varepsilon' = \frac{\varepsilon \rho}{K \mathcal{K}(S)}.$$

As will be shown in the next section, this implies that the dependence of k on the r.h.s. in (13.8) is

$$k = O(\log \frac{1}{\varepsilon'}) = O(\log \frac{K\mathcal{K}(S)}{\varepsilon \rho}),$$

that is, only logarithmic. Therefore, even large relative changes in these constants influence the number of iterations correspondingly much less.

Consider, finally, the case of a general matrix. Then there exists a nonsingular matrix S such that $S^{-1}BS = J$, where J is blockdiagonal, $J = $ blockdiag $(J_1, J_2, \ldots, J_q)$, and the matrices J_i are either diagonal or Jordan block matrices (of at least order 2)

$$J_i = \begin{bmatrix} \lambda_i & 1 & \ldots & 0 \\ 0 & \lambda_i & 1 & \\ \vdots & \ddots & \ddots & \\ 0 & \ldots & & \lambda_i \end{bmatrix},$$

if, in the latter case, the eigenvector space corresponding to the eigenvalue λ_i, is degenerate. If the eigenvector space is degenerate, there exists at least one such Jordan box corresponding to λ_i. Let s_i be the maximal order of any

Jordan box corresponding to the eigenvalue λ_i. We have $S^{-1}P_k(B)S = P_k(J)$, where $P_k(\cdot)$ is a polynomial, and

$$P_k(J) = \text{blockdiag}[P_k(J_1), \ldots, P_k(J_q)].$$

Here, as is readily seen,

$$P_k(J_i) = \begin{bmatrix} P_k(\lambda_i) & P_k'(\lambda_i) & \frac{1}{2}P_k''(\lambda_i) & \cdots & \\ 0 & P_k(\lambda_i) & P_k'(\lambda_i), & \frac{1}{2}P_k''(\lambda_i), & \cdots \\ \cdot & & \cdot & & \\ \cdot & & & \cdot & \\ \cdot & & & & \cdot \\ 0 & \cdots & \cdots & \cdots & P_k(\lambda_i) \end{bmatrix},$$

that is, $P_k(J_i)$ is an upper triangular Toeplitz matrix with entries $P_k(\lambda_i)$, $P_k'(\lambda_i)$, $\frac{1}{2}P_k''(\lambda_i)$, $\frac{1}{6}P_k^{(3)}(\lambda_i)$, etcetera in the successive superdiagonals. If $k < s_i$, the order of J_i, then the last $s_i - k - 1$ superdiagonals contain zero entries.

Now (13.6) and (13.7) show that

$$(13.9) \qquad \frac{\|\mathbf{e}^k\|}{\|\mathbf{e}^0\|} \leq \frac{K\mathcal{K}(S)}{\rho} \min_{P_k \in \pi_k^1} \max_i \|P_k(J_i)\|.$$

For later use, we state the result in a theorem.

Theorem 13.2 *Consider the orthogonal residual conjugate gradient method or the generalized conjugate gradient method defined in Section 12.1 to solve* $B\mathbf{x} = \mathbf{b}$, *where the symmetric part of B is positive definite and B is transformed to a Jordan block form by S, that is, $B = SJS^{-1}$. Then, the relative error at iteration k satisfies*

$$\frac{\|\mathbf{e}^k\|}{\|\mathbf{e}^0\|} \leq \frac{K}{\rho}\mathcal{K}(S) \min_{P_k \in \pi_k^1} \max_i \|P_k(J_i)\|. \qquad \diamond$$

Note that Theorem 13.2 shows that if

$$(13.10) \qquad \min_{P_k \in \pi_k^1} \max_i \|P_k(J_i)\| \leq \frac{\varepsilon\rho}{K\mathcal{K}(S)},$$

then $\|\mathbf{e}^k\| \leq \varepsilon\|\mathbf{e}^0\|$.

Remark 13.3 (Extreme cases) For later use, note that (13.9) shows that $P_k(J_i) = 0$ for any Jordan block that corresponds to an eigenvalue λ_i with maximal order s_i, if $P_k(\lambda)$ is divisable by $(1 - \lambda/\lambda_i)^{s_i}$. It is readily seen that the number of necessary iterations generally is n for a matrix that is a single Jordan block of order n. Hence, for each such Jordan block, we can expect s_i iterative steps.

Another extreme case occurs for a unitary matrix B. In this case, we have $\mathcal{K}(S) = 1$, but the eigenvalues of the matrix are situated on the unit circle. There exist such matrices for which there occurs no convergence until termination takes place after n steps. The shift matrix presented in Example 12.1(b) is such an example. If $B = \alpha I + U$, where $\alpha > 1$ and U is orthogonal (and, hence, real), then there exists an ellipse of the form (5.39) that contains the spectrum but not the origin, which shows convergence. However, as $\alpha \to 1$, the estimate in (5.40) indicates an arbitrarily slow rate of convergence. (Cf. Remark 5.13, showing that better estimates than those based on Chebyshev polynomials can be derived in this case.)

13.2 Estimates Based on the Condition Number

As we have seen in (13.8) and (13.10), the problem of estimating the rate of convergence of the iteration error has been reduced to a pure approximation problem. However, the solution of this problem requires knowledge of the spectrum, which is generally not known. In fact, even if it is known, the estimate can be troublesome in practice, since it involves approximation on a general discrete set of points. Therefore, we shall make some further assumptions (or approximations) of the spectrum in order to simplify this approximation problem.

13.2.1 Continuous Interval

The first assumption we make is that the eigenvalues are real and nonnegative and the positive part is (densely) located in an interval $[a, b]$, where a is the smallest positive eigenvalue and $b = \lambda_n$. We then replace the best approximation problem on the discrete set with the best approximation problem on the interval. Note that we have

$$\min_{P_k \in \pi_k^1} \max_{\substack{1 \leq i \leq n \\ \lambda_i > 0}} |P_k(\lambda_i)| \leq \min_{P_k \in \pi_k^1} \max_{a \leq \lambda \leq b} |P_k(\lambda)|.$$

The solution to the min max problem on an interval is known (see Chapter 5 or Appendix B). We know, namely, that

(13.11) $\quad \min\limits_{P_k \in \pi_k^1} \max\limits_{a \le \lambda \le b} |P_k(\lambda)| = \max\limits_{a \le \lambda \le b} \left| T_k \left(\dfrac{b + a - 2\lambda}{b - a} \right) \right| / T_k \left(\dfrac{b + a}{b - a} \right),$

where

$$T_k(x) = \frac{1}{2}[(x + \sqrt{x^2 - 1})^k + (x - \sqrt{x^2 - 1})^k],$$

is the Chebyshev polynomial. Since

$$\max\limits_{-1 \le x \le 1} |T_k(x)| = 1$$

and

$$-1 \le \frac{b + a - 2\lambda}{b - a} \le 1, \quad b \ge \lambda \ge a,$$

we then find (cf. Chapter 5)

$$\min\limits_{P_k \in \pi_k^1} \max\limits_{a \le \lambda \le b} |P_k(x)| = \frac{1}{T_k \left(\frac{b+a}{b-a} \right)} = 2 \frac{\sigma^k}{1 + \sigma^{2k}},$$

where $\sigma = (1 - \sqrt{\frac{a}{b}})/(1 + \sqrt{\frac{a}{b}})$. This shows that the average rate of convergence factor is bounded above by $\left(2/(1 + \sigma^{2k}) \right)^{1/k} \sigma$, which approaches σ as $k \to \infty$. Theorem 13.1 shows also that (13.4) holds if k is the smallest integer for which

$$\frac{2\sigma^k}{1 + \sigma^{2k}} \le \varepsilon' = \varepsilon \frac{\rho}{K\mathcal{K}(S)}$$

holds—that is, as an elementary computation shows, if

$$k = \lceil \ln(\frac{1}{\varepsilon'} + \sqrt{\frac{1}{(\varepsilon')^2} - 1}) / \ln \sigma^{-1} \rceil,$$

where $\lceil x \rceil$ denotes the smallest integer $\ge x$. For later use, denote

(13.12) $\qquad k^*(a, b, \varepsilon') = \lceil \ln(\frac{1}{\varepsilon'} + \sqrt{\frac{1}{(\varepsilon')^2} - 1}) / \ln \sigma^{-1} \rceil.$

Also k^* can easily be estimated from above by

$$k^* \le \lceil \ln \frac{2}{\varepsilon'} / \ln \sigma^{-1} \rceil \le \lceil \frac{1}{2} \sqrt{\frac{b}{a}} \ln \frac{2}{\varepsilon'} \rceil.$$

This is an accurate bound if $\varepsilon' \ll 1$ and $a/b \ll 1$ (both assumptions hold usually in practice). Given k^* number of iterations, the average rate of convergence factor is then bounded as

$$\left(\frac{\|e^k\|}{\|e^0\|}\right)^{\frac{1}{k^*}} \le \left(\frac{2K\mathcal{K}(S)}{\rho}\right)^{\frac{1}{k^*}} \sigma.$$

Hence, this estimate shows a linear rate of convergence and an asymptotic rate equal to σ.

13.2.1.1 Eigenvalues Contained in an Ellipse

As has also been shown in Chapter 5 in connection with the Chebyshev iterative method, the above estimates for eigenvalues on a real line can be generalized to the case where the eigenvalues are contained in an ellipse symmetrically oriented along the real right (or left) axes in the complex half plane, with foci a, b, $0 < a \le b$ on the real axes. As (5.41) shows, the asymptotic average convergence factor is then $\hat{\sigma} = \sigma\sqrt{(1 + \delta)/(1 - \delta)}$, where δ is the eccentricity. This estimate holds if $\delta < 2\sqrt{(a/b)}/(1 + a/b)$. In this case, the estimate of the necessary number of iterations becomes

$$k^*(a, b, \varepsilon') = \lceil \ln(\frac{1}{\varepsilon'} + \sqrt{\frac{1}{(\varepsilon')^2} - 1})/ \ln \hat{\sigma}^{-1} \rceil.$$

As the ellipse gets narrower, $\delta \to 0$, and the same estimates as above for the real line are derived.

For other distributions of eigenvalues in the complex plane, one can find corresponding best approximation polynomials, but generally these do not result in such a simple formula for k^* as above.

The above estimates of the rate of convergence and of the necessary number of iterations depend only on the effective condition number (i.e., neglecting possible zero eigenvalues), b/a of $C^{-1}A$. However, as we shall now see, in many cases when we consider more detailed information of the spectrum, substantially better estimates can be derived.

13.2.2 Spectrum Contained in Two Intervals or in Two Ellipse-Like Ovals

As was shown in Chapter 5 for the Chebyshev iterative method, when the spectrum is contained in two well-separated ellipse-like ovals, located on two real intervals $[a, b]$ and $[c, d]$ that do not contain the origin, the asymptotic average rate of convergence is

$$\left\{ \frac{1 - \sqrt{ad/bc}}{1 + \sqrt{ad/bc}} \cdot \sqrt{\frac{1 + \delta}{1 - \delta}} \right\}^{1/2},$$

where δ is the ratio of the semiaxes in an ellipse that is mapped onto the ovals by the function in (5.44). It is further assumed that $b - a = d - c$. The ovals may be contained in different halfplanes, but for both, the real axis is a symmetry line. It is seen that if $c \gg \max\{|a|, |b|\}$, then we can put $d/c = 1$, and the eigenvalues in the second oval (defined by the interval $[c, d]$) have little influence on the rate of convergence; then, the number of iteration steps (when we neglect the influence of round-off errors) is only about twice that what would be predicted by the first oval alone.

13.2.3 Isolated Eigenvalues

We now consider cases where all or some eigenvalues are treated as isolated.

Compact Perturbations We assume, first, that eigenvector space is complete and that the eigenvalues have been ordered according to their distance to the point $(1, 0)$ in the complex plane, so that

$$|\mu_{i+1}| \le |\mu_i|, \quad i = 1, 2, \ldots,$$

where $\mu_i = \lambda_i - 1$. We also assume that for i sufficiently large, $|\mu_i| < 1/3$. Then choose the polynomial

$$P_k(\lambda) = \prod_{j=1}^{k} (1 - \frac{\lambda}{\lambda_j}).$$

Then $P_k(\lambda_i) = 0, \ i = 1, \ldots, k$ and

$$\max_{\lambda \in S(B)} |P_k(\lambda)| = \max_{i \ge k+1} |P_k(\lambda_i)| = \max_{i \ge k+1} \prod_{j=1}^{k} |1 - \frac{\lambda_i}{\lambda_j}| = \max_{i \ge k+1} \prod_{j=1}^{k} \frac{|\mu_j - \mu_i|}{|1 + \mu_j|}.$$

Since $|\mu_j - \mu_i| \le |\mu_j| + |\mu_i| < 2|\mu_j|, \ i \ge k + 1$, for $1 \le j \le k$, we find that the average convergence factor is bounded by

$$\max_{\lambda \in S(B)} |P_k(\lambda)|^{\frac{1}{k}} \le 2 \left(\prod_{j=1}^{k} \left| \frac{\mu_j}{1 + \mu_j} \right| \right)^{\frac{1}{k}}$$

Since the matrix is diagonalizable, Theorem 13.1 shows that this can be used to estimate the average convergence factor. When $|\mu_{k+1}| < 1/3$, we have $|\frac{\mu_{k+1}}{1+\mu_{k+1}}| < 1/2$, and we see that the upper bound in this estimate of the convergence factor generally decreases as k increases. Using the arithmetic-geometric inequality, we find that

$$(13.13) \qquad \max_{\lambda \in S(B)} |P_k(\lambda)|^{\frac{1}{k}} \le \frac{2}{k} \sum_{j=1}^{k} \left| \frac{\mu_j}{1+\mu_j} \right|,$$

which is less than 1 for k sufficiently large. As pointed out by Winter (1980), we can also consider infinite dimensional problems when $A = I + \Lambda$ and Λ is a compact perturbation of the identity. Then, the eigenvalues λ_i cluster around the unit number and $|\mu_j| \to 0$, $j \to \infty$. Hence, the convergence factor in (13.13) approaches zero as $k \to \infty$, which shows a superlinear rate of convergence.

The above estimates are also useful in the presence of multiple eigenvalues. It suffices, then, to take one factor $(1 - \lambda/\lambda_j)$ in P_k for each distinct eigenvalue, so

$$P_k(\lambda) = \prod_{j=1}^{q} (1 - \frac{\lambda}{\lambda_j}),$$

where q is the number of distinct eigenvalues. When the matrix is degenerate, it can be seen that we must take

$$P_k(\lambda) = \prod_{j=1}^{q} (1 - \frac{\lambda}{\lambda_j})^{s_j},$$

where s_j is the order of the largest Jordan block corresponding to λ_j. Again, it suffices to let each distinct eigenvalue appear once, so q is the number of distinct eigenvalues.

This shows that for an eigenvalue with a degenerate vectorspace, the contribution to the number of necessary iterations for a given relative accuracy of the iteration errors, is at most s_j iterations. If s_j is bounded, $1 \le j \le q$, $q \to \infty$ (for infinite dimensional problems), what has been said about superconvergence above holds also for degenerate matrices.

Isolated Eigenvalues Around an Ellipse   Let $E(a, b)$ be an ellipse that does not contain the origin, with ratio of semiaxes δ, located symmetrically on the real axis, and with foci a, b, $0 < a < b$. Consider the case where the eigenvalues are divided into two sets: (1) the eigenvalues outside the ellipse and those on or inside that correspond to a degenerate eigenvector space, and (2) the remaining eigenvalues on or inside the ellipse corresponding to complete eigenvector spaces. Let there be q different eigenvalues in the first set. Assume that $\delta < 2\sqrt{a/b}/(1 + a/b)$ [cf. (5.39)]. Then, in order to get an upper bound of the rate of convergence, we choose the polynomial

$$(13.14) \qquad P_k(\lambda) = \widehat{P}_{k-p_0}(\lambda) \prod_{i=1}^{q} (1 - \frac{\lambda}{\lambda'_i})^{s_i},$$

where $\widehat{P}_{k-p_0}(\cdot)$ is the best approximation polynomial on the ellipse and $p_0 = \sum_{i=1}^{q} s_i$, where s_i is the largest order of any Jordan box corresponding to λ'_i for any eigenvalue λ'_i in the first set. As before, this shows that the necessary number of iterations to satisfy (13.14) is, at most,

$$k = k^*(a, b, \varepsilon') + p_0,$$

where k^* is defined in (13.12) with $\hat{\sigma} = \sigma\sqrt{(1 + \delta)/(1 - \delta)}$ and

$$\varepsilon' = \varepsilon \frac{\rho}{K\mathcal{K}(S)} \max_{\lambda \in E(a,b)} \left\{ \prod_{i=1}^{q} |1 - \frac{\lambda}{\lambda'_i}|^{s_i} \right\}.$$

For any eigenvalue λ'_i located sufficiently to the right of the ellipse, $|\lambda - \lambda'_i| \le \lambda'_i$ holds, so

$$\max_{\lambda \in E(a,b)} |1 - \frac{\lambda}{\lambda'_i}| \le 1.$$

On the other hand, for other eigenvalues λ'_i, the factor $|1 - \frac{\lambda}{\lambda'_i}|$ can be quite large.

As can be readily seen, for each i there exists a sufficiently small positive number $\lambda_{0,i}$, such that

$$\max_{\lambda \in E(a,b)} |1 - \frac{\lambda}{\lambda'_i}| \le \frac{b}{\lambda_{0,i}}$$

holds. This result shows that such eigenvalues influence the number of iterations, at most, with some additional logarithmic terms, because

$$k = \left\lceil \ln \frac{2}{\varepsilon'} / \ln \left(\sigma^{-1} \sqrt{\frac{1-\delta}{1+\delta}} \right) \right\rceil$$

$$(13.15a) \qquad \le \left\lceil \left(\ln \frac{2}{\varepsilon} + \ln \frac{K\mathcal{K}(S)}{\rho} + \sum_{i=1}^{q_0} s_i \ln \frac{b}{\lambda_{0,i}} \right) / \ln \left(\sigma^{-1} \sqrt{\frac{1-\delta}{1+\delta}} \right) \right\rceil + p_0,$$

where q_0 is the number of eigenvalues λ_i' for which

$$\max_{\lambda \in E(a,b)} |1 - \frac{\lambda}{\lambda_i'}| > 1.$$

As the relative precision ε decreases, these extra terms eventually become negligible. However, when $b \gg \lambda_{0,i}$ they can be so large that ε will not, in practice, approach such a sufficiently small value.

At any rate, estimate (13.15a) indicates an increasing rate of convergence when we iterate further to get an increasingly smaller relative error. This is so because when ε decreases, we can add more terms to $\sum_{i=1}^{q_0} s_i \ln \frac{b}{\lambda_i}$, that is, take off more eigenvalues from the ellipse region and treat them as isolated without increasing the current value of the inner bracket in (13.15a) much. This makes it possible to make the corresponding interval $[a, b]$ shorter and, hence, decrease the value of σ. Therefore, (13.15a) indicates fewer and fewer iterations to decrease the error further with some fixed factor (say $1/10$). However, when we have complex eigenvalues, this may increase the value of δ so that $\hat{\sigma}$ does not decrease and the decrease of the necessary number of iterations for each new correct decimal may not take place.

The situation is more clear in the case of real eigenvalues. Then, as can readily be seen, $\max_{a \leq \lambda \leq b} |1 - \frac{\lambda}{\lambda_i}| \leq \frac{b}{\lambda_i}$, for eigenvalues λ_i, $\lambda_i < b$. (13.15a) takes the form

$$(13.15b) \quad k^* \leq \left\lceil \left(\ln \frac{2}{\varepsilon} + \ln \frac{K}{\rho} \mathcal{K}(S) + \sum_{i=1}^{q_0} s_i \ln \frac{b}{\lambda_i} \right) / \ln \sigma_{q_0}^{-1} \right\rceil + p_0,$$

where

$$\sigma_{q_0} = (1 - \sqrt{\frac{a_{q_0}}{b_{q_0}}}) / (1 + \sqrt{\frac{a_{q_0}}{b_{q_0}}}).$$

Here, $b = b_{q_0}$ and a_{q_0}, b_{q_0} are the new endpoints of the interval after the removal of some eigenvalues, typically at both ends of the previous interval, and b_{q_0}/a_{q_0} acts as an effective condition number.

The estimate (13.15b) will now be applied to typical eigenvalue distributions.

Example 13.1 Consider an s.p.d. matrix A with eigenvalues $0 < a = \lambda_1 \leq \lambda_2 \leq \cdots \leq \lambda_{n-p} = b < \lambda_{n-p+1} < \cdots < \lambda_n$, where, in practice, we assume that $\lambda_{n-p+1} \gg b$. Hence, there are p large (and distinct) eigenvalues to the right of the interval $[a, b]$. Then, (13.15b) shows that

$$k^* \leq \ln \frac{2}{\varepsilon} / \ln \sigma^{-1} + p,$$

where $\sigma = (1 - \sqrt{\frac{a}{b}})/(1 + \sqrt{\frac{a}{b}})$, because

$$\max_{a \leq \lambda \leq b} |1 - \frac{\lambda}{\lambda_i}| < 1, \ i = n - p + 1, \ldots, n.$$

Hence, the effective condition number is b/a and there are at most p additional iterations, because of the isolated eigenvalues. In practice, the above does not always hold in the presence of rounding errors. As has been shown in Greenbaum and Strakos (1992) and Notay (1992), the influence of rounding errors can be analyzed by perturbing the spectrum slightly and increasing the order of the matrix. It turns out that the number of extra iterations due to round-off and, hence, loss of orthogonality of the residual and search direction vectors are equal to the number of copies of the large eigenvalues generated by the (due to the round-off) actually computed matrix T_{k+1} in Section 13.3.

Example 13.2 Let A be s.p.d. with q small eigenvalues,

$$0 < \lambda_1 < \lambda_2 < \cdots < \lambda_q < a = \lambda_{q+1} \leq \cdots \leq \lambda_n = b.$$

Then, (13.15b) shows

$$k^* \leq \left\lceil \left(\ln \frac{2}{\varepsilon} + \sum_{i=1}^{q} \ln \frac{b}{\lambda_i} \right) / \ln \sigma^{-1} \right\rceil + q,$$

where $\sigma = (1 - \sqrt{\frac{a}{b}})/(1 + \sqrt{\frac{a}{b}})$.

Hence, in this case, there are q additional iterations plus some penalty terms ($\sum_1^q \ln b/\lambda_i$). If there are p and q isolated large and small eigenvalues, respectively, then the estimate takes the form

$$k^* \leq \left\lceil \left(\ln \frac{2}{\varepsilon} + \sum_{i=1}^{q} \ln \frac{b}{\lambda_i} \right) / \ln \sigma^{-1} \right\rceil + p + q.$$

The above estimate is one of several found in Axelsson (1976). Further refined analysis in Jennings (1977) and Axelsson and Lindskog (1986) shows that, asymptotically, we can exchange b in $\ln b/\lambda_i$ with $\frac{\pi}{4}ae^2$, where e is the natural number. As has been shown in Notay (1992), this estimate is little affected by round-off errors.

Example 13.3 Let the eigenvalue distribution be $\lambda_j = (\frac{j}{n})^2$, $j = 1, 2, \ldots, n$, $1 \leq \lambda_j \leq 4$, $n + 1 \leq j \leq n^2$. (Hence, there are n^2 eigenvalues.) Consider the first q eigenvalues, where $q < n$, as isolated from the remainder of the spectrum. Then, (13.15b), with b exchanged with $e^2\lambda_{q+1}$ (cf. the remark in Example 13.2), shows

$$k^* \leq k(q) = \left[\left(\ln\frac{2}{\varepsilon} + \sum_{j=1}^{q}\ln\frac{e^2\lambda_{q+1}}{\lambda_j}\right) \Big/ \ln\frac{1 + \sqrt{\lambda_{q+1}/4}}{1 - \sqrt{\lambda_{q+1}/4}}\right] + q.$$

We want to choose q to minimize $k(q)$. Using Stirling's formula, we find

$$\sum_{j=1}^{q}\ln\frac{e^2\lambda_{q+1}}{\lambda_j} = 2q + 2\sum_{j=1}^{q}\ln\left(\frac{q+1}{j}\right)$$

$$= 2q + 2\ln\left(\prod_{j=1}^{q}\frac{q+1}{j}\right) \sim 4q \quad (q \to \infty).$$

Hence, as $q \to \infty$,

$$k(q) \sim \left(\ln\frac{2}{\varepsilon} + 4q\right)\frac{n}{q+1} + q = n\left(4 + \frac{\ln\frac{2}{\varepsilon} - 4}{q+1} + \frac{q}{n}\right),$$

which becomes arbitrarily close to $4\sqrt{n}$ if we choose q such that $n \gg q \gg \ln(\frac{2}{\varepsilon})$. (The bound is minimized for $q = \{n(\ln\frac{2}{\varepsilon} - 4)\}^{\frac{1}{2}} \sim n^{\frac{1}{2}}(\ln\frac{2}{\varepsilon})^{\frac{1}{2}}$.) In practice, we typically have $\varepsilon = O(n^{\nu})$ for some $\nu > 0$, so in such a case,

$$n \gg q \gg O(\ln n), \quad n \to \infty.$$

This shows that the number of iterations depends little on ε, which indicates a superlinear rate of convergence when the number of iterations are $\approx 4n$ (or, somewhat earlier, cf. Figure 13.1). Using the condition number $\lambda_{n^2}/\lambda_1 = 4n^2$, the standard estimate (13.12) shows $k^* \leq n\log\frac{2}{\varepsilon}$, $n \to \infty$.

In a similar way to the above, one can use the estimates shown in Chapter 5 to derive estimates also for the case of eigenvalue clusters in two ellipses with some isolated eigenvalues outside these clusters.

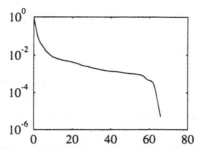

Figure 13.1. Convergence history for the conjugate gradient method for a matrix with eigenvalue distribution as in Example 13.3 and a random initial error vector.

13.3 An Estimate Based on a Ratio Involving the Trace and the Determinant

As we have seen, the estimate of the necessary number of iterations in (13.15a,b) is based on a condition number of the matrix related to two extreme eigenvalues defining an interval or, more generally, an ellipse oriented along the real axis. There is an alternative estimate that takes all eigenvalues into account and, in some cases, can give a smaller (that is, more accurate) upper bound of the number of iterations. In particular, it shows the eventual superlinear rate of convergence of the conjugate gradient method. Based on the ratio of the arithmetic and geometric averages of the eigenvalues, this estimate assumes that these are real and positive. It was first introduced by Kaporin (1990).

Let $\alpha_i > 0$, $i = 1, \ldots, n$ and consider first the function

(13.16)
$$\chi_p(\alpha_1, \ldots, \alpha_p) = \frac{1}{p} \sum_1^p \alpha_i / \left(\prod_1^p \alpha_i \right)^{\frac{1}{p}}.$$

Note that χ_p is a homogeneous function, that is,

$$\chi_p(a\alpha_1, \ldots, a\alpha_p) = \chi_p(\alpha_1, \ldots, \alpha_p) \text{ for any } a > 0.$$

The function χ_p satisfies the following basic inequalities.

Lemma 13.4 *Let $\chi_p(\alpha_1, \ldots, \alpha_p)$ be defined by (13.16), where α_i are positive numbers. Then:*

(a) $\chi_p(\alpha_1, \ldots, \alpha_p) \geq 1$ and $\chi_p(\alpha_1, \ldots, \alpha_p) = 1$ if and only if $\alpha_1 = \alpha_2 = \ldots = \alpha_p$.

(b) *Let* $1 \leq p \leq n$ *and let* $\{\alpha'_i\}_1^p$ *be a subset of* $\{\alpha_i\}_1^n$. *Then,*

$$\chi_p(\alpha'_1, \ldots, \alpha'_p)^p \leq \chi_n(\alpha_1, \ldots, \alpha_n)^n.$$

Proof The well-known inequality between the arithmetric and geometric means shows that $\chi_p(\alpha'_1, \ldots, \alpha'_p) \geq 1$. Further, the arithmetic mean is strictly larger than the geometric mean unless all numbers α'_i are equal, which shows the second part of (a).

To prove part (b), note first, as an elementary computation shows, that

$$(13.17) \quad \chi_n(\alpha'_1, \ldots, \alpha'_n) = \chi_n(a, \ldots, a, \alpha'_{p+1}, \ldots, \alpha'_n)\chi_p(\alpha'_1, \ldots, \alpha'_p)^{\frac{p}{n}},$$

where the numbers a occur p times and where $a = \frac{1}{p}\sum_1^p \alpha'_i$. The values $\alpha'_{p+1}, \ldots, \alpha'_n$ make up the complementary subset. Using now the lower bound in part (a) for the first r.h.s. factor in (13.17), it follows that

$$\chi_n(\alpha_1, \ldots, \alpha_n) \geq \chi_p(\alpha'_1, \ldots, \alpha'_p)^{\frac{p}{n}}. \qquad \diamond$$

Definition 13.1 Let B have order n, let λ_i, $i = 1, \ldots, n$ be the eigenvalues of B, which are assumed to be positive, and let

$$(13.18) \qquad K = K(B) = \left(\frac{1}{n}\sum_1^n \lambda_i\right)^n / \prod_1^n \lambda_i.$$

We call $K(B)$ the *K-condition number* of B.

Note that Lemma 13.4(a) shows that $K(B) \geq 1$ and $K(B) = 1$ if and only if $B = aI$ for some positive number a.

We shall now present some properties of $K(B)$. The first can be found in Dennis and Wolkowicz (1990) (cf. also Kaporin, 1990).

Theorem 13.5 *Let K be defined by (13.18), where $\{\lambda_i\}_1^n$ are the eigenvalues of B that are assumed to be positive and ordered $\lambda_1 \leq \lambda_2 \leq \cdots \leq \lambda_n$. Then:*

(a) $1 \leq K^{\frac{1}{n}} \leq \mathcal{K}(B) \leq [\mathcal{K}(B)^{1/2} + \mathcal{K}(B)^{-1/2}]^2 \leq 4K$, *where* $\mathcal{K}(B) = \lambda_n/\lambda_1$. *There is equality in the first and second inequalities if and only if $\lambda_1 = \lambda_2 = \cdots = \lambda_n$, and equality in the last inequality if and only if*

$$\lambda_2 = \cdots = \lambda_{n-1} = \frac{1}{2}(\lambda_1 + \lambda_n).$$

(b) $K = [\frac{1}{n}tr(B)]^n / \det(B) = 1/\det(\frac{n}{tr(B)}B)$.

Proof $K \geq 1$ follows from Lemma 13.4(a) and $K^{\frac{1}{n}} \leq \mathcal{K}(B)$ follows by

(13.19) $$\frac{1}{n}\sum_1^n \lambda_i \leq \lambda_n, \quad (\prod_1^n \lambda_i)^{1/n} \geq \lambda_1.$$

The last inequality follows by (13.17) with $p = 2$ and $\lambda_1' = \lambda_1$, $\lambda_2' = \lambda_n$, because

$$\chi_2(\lambda_1', \lambda_2')^2 \leq \chi_n(\lambda_1, \dots, \lambda_n)^n = K$$

and

$$\chi_2(\lambda_1, \lambda_n)^2 = \left[\frac{1}{2}(\lambda_1 + \lambda_n)/\sqrt{\lambda_1 \lambda_n}\right]^2 = \frac{1}{4}\left[\mathcal{K}(B)^{1/2} + \mathcal{K}(B)^{-1/2}\right]^2.$$

(13.19) shows that equality holds in the first two inequalities in part (a) if and only if $\lambda_1 = \lambda_2 = \cdots = \lambda_n$. Lemma 13.4(a) shows that the factor $\chi_n(a, a, \lambda_3', \dots, \lambda_n')$ in (13.17), where $a = (\lambda_1 + \lambda_2)/2$, equals 1 if and only if $\lambda_i' = a$, that is, $\lambda_i = a$, $i = 2, \dots, n - 1$, which completes the proof of (a). Part (b) follows by the relations $tr(B) = \sum_1^n \lambda_i$, $\det(B) = \prod_1^n \lambda_i$ (see Chapter 2). Lemma 13.4 and Theorem 13.5 show, among other things, that K can be used as an alternate condition number of B. $\diamond$

The following property can be found in Kaporin (1992).

Theorem 13.6 *Let $A, B \in \mathcal{A}$, the set of symmetric positive definite matrices. Then, $K(A)$ is a quasi-convex function in $\mathcal{A}$, that is,*

$$K(\theta A + (1 - \theta)B) \leq \max[K(A), K(B)], \quad 0 \leq \theta \leq 1.$$

Proof Theorem 13.5(b) and the homogeniety of K show that

$$K(\theta A + (1 - \theta)B) = K\left(\vartheta \frac{n}{tr(A)}A + (1 - \vartheta)\frac{n}{tr(B)}B\right)$$

$$= 1/\det\left(\vartheta \frac{n}{tr(A)}A + (1 - \vartheta)\frac{n}{tr(B)}B\right),$$

where $\vartheta = \theta \, tr(A)/[\theta \, tr(A) + (1 - \theta)tr(B)]$. Hence, by Corollary 7(c) in Appendix C,

$$K(\theta A + (1 - \theta)B) \geq \left(\frac{1}{\det(\frac{n}{tr(A)}A)}\right)^{\vartheta} \cdot \left(\frac{1}{\det(\frac{n}{tr(B)}B)}\right)^{1-\vartheta}$$

$$= K(A)^{\vartheta} K(B)^{1-\vartheta}. \qquad \diamond$$

It is readily seen that the spectral condition number is also quasi-convex on $\mathcal{A}$. Next we show that the K-condition number of a diagonalizable matrix does not depend on the transformation matrix.

Theorem 13.7 *Let A be symmetrizable with positive eigenvalues, that is, $A = SDS^{-1}$ for some nonsingular matrix S, where $D = D^*$. Then, $K(A) = K(D)$, but $\mathcal{K}(S)^{-2}\mathcal{K}(D) \leq \mathcal{K}(A) \leq \mathcal{K}(S)^2\mathcal{K}(D)$. Further, $\mathcal{K}(D) \neq \mathcal{K}(A)$ unless S is unitary.*

Proof That $K(A) = K(D)$ follows by the similarity transformation and Definition 13.1. The properties of the spectral condition number follow from $A^*A = S^{*-1}DS^*SDS^{-1}$ and $(A^*A)^{-1} = SD^{-1}S^{-1}S^{*-1}D^{-1}S^*$. $\qquad \diamond$

If A is s.p.d. and partitioned as $A = \begin{bmatrix} A_{11} & A_{12} \\ A_{21} & A_{22} \end{bmatrix}$, then Lemma 3.12(a) shows that $\mathcal{K}(A_{11}) \leq \mathcal{K}(A)$. Such a relation holds also for the K-condition numbers, as the next Lemma shows.

Lemma 13.8 *Let A, of order n, be partitioned in 2×2 block matrix form,*

$$A = \begin{bmatrix} A_{11} & A_{12} \\ A_{21} & A_{22} \end{bmatrix},$$

where A_{11} has order p $(1 \leq p \leq n)$, and assume that A, A_{11} and A_{22} have positive eigenvalues. Then,

$$K(A_{11}) \leq K(A).$$

Proof Let $D = \begin{bmatrix} A_{11} & 0 \\ 0 & A_{22} \end{bmatrix}$ and note that Lemma 13.2 shows that $tr(A) = tr(D)$ and $tr(D^{-1}A) = n$. Hence, Theorem 13.5(b) shows that

$$K(A)/K(D) = \det(A)^{-1} \det(D) = 1/\det(D^{-1}A) = K(D^{-1}A) \geq 1,$$

so

$$K(A) \geq K(D).$$

Since the spectrum of D is equal to the union of the spectra of A_{11} and A_{22}, we obtain the estimate

$$K(A) \geq K(D) = \chi_n(\lambda_1(D), \ldots, \lambda_n(D))^n$$

$$= \chi_n(\lambda_1(A_{11}), \ldots, \lambda_p(A_{11}), \lambda_1(A_{22}), \ldots, \lambda_{n-p}(A_{22}))^n$$

$$\geq \chi_p(\lambda_1(A_{11}), \ldots, \lambda_p(A_{11}))^p = K(A_{11}),$$

where the last inequality follows from Lemma 13.4(b). ◇

Corollary 13.9 *Let V be an orthonormal matrix of order $n \times p$, that is, $V^T V = I_p$, $1 \leq p \leq n$. Then, for any s.p.d. matrix A of order n,*

$$K(V^T A V) \leq K(A).$$

Proof Let U be an orthonormal matrix of order $n \times (n - p)$, with column vectors orthogonal to those of V. Then, the augmented matrix $W = [V, U]$ of order n is orthonormal, that is, $W^T W = I_n$. Since $W^T A W = W^{-1} A W$, we see that $W^T A W$ is similar to A, and

$$W^T A W = \begin{bmatrix} V^T A V & V^T A U \\ U^T A V & U^T A U \end{bmatrix}.$$

Hence Lemma 13.8 shows that

$$K(A) = K(W^T A W) \geq K(V^T A V) \qquad ◇$$

Following Kaporin (1990), we shall now show a relation involving the K-number of $B = C^{-1} A$, where C and A are s.p.d. and the norms of the residuals are computed by the preconditioned conjugate gradient method. It is readily seen that the results are valid also for the case of a positive semidefinite matrix A. Naturally, if A is singular, we do not consider the zero eigenvalues when computing the K-number. We need the following lemma.

Lemma 13.10 *Let A be symmetric and let C be s.p.d. Let $\mathbf{r}^k$ be the residual computed at stage k $(1 \leq k \leq n)$ of the PCG method in Section 11.2 to solve*

$B\mathbf{x} = \mathbf{b}$, *and let*

$$
T_{k+1} = \begin{bmatrix}
\frac{1}{\tau_0} & \frac{-\beta_0}{\tau_0} & & & 0 \\
\frac{-1}{\tau_0} & \frac{1}{\tau_1} + \frac{\beta_0}{\tau_0} & \frac{-\beta_1}{\tau_1} & & \\
& \ddots & \ddots & \ddots & \\
0 & & & \frac{-1}{\tau_{k-1}} & \frac{1}{\tau_k} + \frac{\beta_{k-1}}{\tau_{k-1}}
\end{bmatrix},
$$

where τ_i, β_i are the scalar coefficients computed in the PCG method. It is assumed that $\tau_i > 0$. Then,

(a) $\mathbf{r}^{k^T} C^{-1} \mathbf{r}^k \leq [\frac{k+1}{k} \underline{\tau}_k^{\frac{1}{k}} \left(K(T_{k+1})^{\frac{1}{k+1}} - \frac{\tau_k}{k+1} \right) - 1]^k \mathbf{r}^{0^T} C^{-1} \mathbf{r}^0$,

   where $\underline{\tau}_k = \frac{1}{\tau_k} \left(\prod_{i=0}^k \tau_i \right)^{1/k+1}$.

(b) $\mathbf{r}^{k^T} C^{-1} \mathbf{r}^k \leq [K(T_{k+1})^{\frac{1}{k}} - 1]^k \mathbf{r}^{0^T} C^{-1} \mathbf{r}^0$.

Proof The *LU* decomposition,

$$
T_{k+1} = \begin{bmatrix}
1 & & & 0 \\
-1 & 1 & & \\
& \ddots & \ddots & \\
0 & & -1 & 1
\end{bmatrix}
\begin{bmatrix}
\frac{1}{\tau_0} & \frac{-\beta_0}{\tau_0} & & 0 \\
0 & \frac{1}{\tau_1} & \frac{-\beta_1}{\tau_1} & \\
& \ddots & \ddots & \\
0 & & 0 & \frac{1}{\tau_k}
\end{bmatrix},
$$

of T_{k+1} shows that

$$
\det(T_{k+1})^{\frac{1}{k+1}} = \prod_{i=0}^k \tau_i^{-1}.
$$

By Theorem 13.5(b) it follows that

$$
K(T_{k+1})^{\frac{1}{k+1}} = \frac{1}{k+1} tr(T_{k+1})/\{\det(T_{k+1})\}^{\frac{1}{k+1}}.
$$

Let $-1 < \theta$, where θ is a parameter. Then,

$$
K(T_{k+1})^{1/k+1} = \left(\prod_{i=0}^k \tau_i \right)^{\frac{1}{k+1}} \left[-\frac{\theta}{\tau_k(k+1)} + \left(\frac{1+\theta}{\tau_k} + \sum_{i=0}^{k-1} \frac{1+\beta_i}{\tau_i} \right) \Big/ (k+1) \right]
$$

$$
= -\frac{\theta \underline{\tau}_k}{k+1} + X_{k+1} \left(\frac{1+\beta_0}{\tau_0}, \dots, \frac{1+\beta_{k-1}}{\tau_{k-1}}, \frac{1+\theta}{\tau_k} \right) [(1+\theta)R_k]^{1/k+1},
$$

where

$$R_k = \prod_0^{k-1}(1 + \beta_i).$$

Using Lemma 13.4(a), we find

$$(13.20) \qquad K(T_{k+1})^{\frac{1}{k+1}} \geq -\frac{\theta \tau_k}{k+1} + [(1+\theta)R_k]^{1/k+1}.$$

An elementary computation shows that this lower bound has a unique maximum with respect to the variable θ. We now choose θ to maximize it and find

$$1 + \theta = \frac{R_k^{1/k}}{\tau_k^{\frac{k+1}{k}}}$$

(and hence $\theta > -1$), and (13.20) takes the form

$$(13.21) \qquad K(T_{k+1})^{1/k+1} - \frac{\tau_k}{k+1} \geq \frac{k}{k+1} \left(\frac{R_k}{\tau_k}\right)^{1/k}.$$

Using the inequality in Corollary 7(b) in Appendix C,

$$R_k^{1/k} = \prod_0^{k-1}(1 + \beta_i)^{1/k} \geq 1 + \prod_0^{k-1}\beta_i^{1/k},$$

and we find that

$$\frac{k+1}{k}\tau_k^{1/k}\left(K(T_{k+1})^{\frac{1}{k+1}} - \frac{\tau_k}{k+1}\right) \geq R_k^{1/k} \geq 1 + \prod_0^{k-1}\beta_i^{1/k}.$$

Now Theorem 11.5 shows that

$$\beta_i = \frac{\mathbf{r}^{i+1^T}C^{-1}\mathbf{r}^{i+1}}{\mathbf{r}^{i^T}C^{-1}\mathbf{r}^i},$$

so

$$\prod_0^{k-1}\beta_i = \frac{\mathbf{r}^{k^T}C^{-1}\mathbf{r}^k}{\mathbf{r}^{0^T}C^{-1}\mathbf{r}^0},$$

and

$$\left[\frac{k+1}{k}\tau_k^{1/k}\left[K(T_{k+1})^{\frac{1}{k+1}} - \frac{\tau_k}{k+1}\right] - 1\right]^k \geq \frac{\mathbf{r}^{k^T}C^{-1}\mathbf{r}^k}{\mathbf{r}^{0^T}C^{-1}\mathbf{r}^0}.$$

This is part (a) of the Lemma.

(13.21) shows also

$$K(T_{k+1})^{\frac{1}{k+1}} \geq \frac{\tau_k + k \left(\frac{R_k}{\tau_k}\right)^{1/k}}{k+1} \geq R_k^{1/k+1},$$

where the last inequality follows by the arithmetic-geometric mean inequality. Hence,

$$K(T_{k+1})^{\frac{1}{k}} \geq 1 + \prod_0^{k-1} \beta_i^{1/k},$$

which shows part (b).   ◇

It follows from the proof that the bound in part (a) of Lemma 13.10 can be sharper than the bound in part (b). In practice, it turns out that it is typical that τ_k oscillates. This indicates that the bound in (a) can be much sharper than the bound in (b). However, no general statement can be drawn from the bound in (a).

Next we need a relation between the K-condition numbers of the matrices T_{k+1} and B. This can be derived as was done for the Lanczos method in Section 11.3. Alternatively, however, we use the following relations given in Theorem 11.5,

$$A\mathbf{d}^k = \frac{1}{\tau_k}(\mathbf{r}^{k+1} - \mathbf{r}^k),$$

$$A\mathbf{d}^k = -AC^{-1}\mathbf{r}^k + \beta_{k-1}A\mathbf{d}^{k-1},$$

to find

$$\frac{1}{\tau_k}(\mathbf{r}^{k+1} - \mathbf{r}^k) = -AC^{-1}\mathbf{r}^k + \frac{\beta_{k-1}}{\tau_{k-1}}(\mathbf{r}^k - \mathbf{r}^{k-1}).$$

Letting

$$V_{k+1} = [\mathbf{r}^0, \ldots, \mathbf{r}^k],$$

a matrix of order $n \times (k+1)$, we then find

(13.22) $$AC^{-1}V_{k+1} = V_{k+1}T_{k+1} - \frac{1}{\tau_k}\mathbf{r}^{k+1}\mathbf{e}^{k+1^T},$$

where

(13.23) $$\mathbf{e}^{k+1} = [0, 0, \ldots, 0, 1]^T,$$

the $(k + 1)$st unit coordinate vector. The C^{-1}-orthogonality of the vectors $\mathbf{r}^k$ [see Theorem 11.5(b)] yields

(13.24)
$$V_{k+1}^T C^{-1} V_{k+1} = D_{k+1}$$
$$= \text{diag}(\mathbf{r}^{0^T} C^{-1} \mathbf{r}^0, \mathbf{r}^{1^T} C^{-1} \mathbf{r}^1, \ldots, \mathbf{r}^{k^T} C^{-1} \mathbf{r}^k)$$

and

$$V_{k+1}^T C^{-1} \mathbf{r}^{k+1} = \mathbf{0}.$$

Hence, after multiplication of (13.22) with $V_{k+1}^T C^{-1}$, we then find, using (13.24),

(13.25) $$V_{k+1}^T C^{-1} A C^{-1} V_{k+1} = D_{k+1} T_{k+1}.$$

Using the orthonormal $n \times (k + 1)$-matrix,

$$W_{k+1} = C^{-\frac{1}{2}} V_{k+1} D_{k+1}^{-\frac{1}{2}},$$

we can rewrite (13.25) as

$$W_{k+1}^T C^{-\frac{1}{2}} A C^{-\frac{1}{2}} W_{k+1} = D_{k+1}^{\frac{1}{2}} T_{k+1} D_{k+1}^{-\frac{1}{2}}.$$

Therefore, Corollary 13.9 shows that

$$K(C^{-\frac{1}{2}} A C^{-\frac{1}{2}}) \geq K(D_{k+1}^{\frac{1}{2}} T_{k+1} D_{k+1}^{-\frac{1}{2}}),$$

and using similarity transformations, we find

(13.26) $$K(B) = K(C^{-1} A) \geq K(T_{k+1}).$$

This establishes the needed relation between $K(B)$ and $K(T_{k+1})$. Using (13.26) and Lemma 13.10, we now obtain the following theorem.

Theorem 13.11 *Let A be symmetric and positive semidefinite and let C be s.p.d. The result at the kth stage of the PCG method for a consistent system $A\mathbf{x} = \mathbf{a}$ satisfies* *(a)* $\mathbf{r}^{k^T} C^{-1} \mathbf{r}^k \le \left[\frac{k+1}{k} \underline{\tau}_k^{1/k} \left(K(B)^{1/k+1} - \frac{\tau_k}{k+1} \right) - 1 \right]^k \mathbf{r}^{0^T} C^{-1} \mathbf{r}^0, k \ge 1,$

(b) $\mathbf{r}^{k^T} C^{-1} \mathbf{r}^k \le (K(B)^{\frac{1}{k}} - 1)^k \mathbf{r}^{0^T} C^{-1} \mathbf{r}^0, \qquad k \ge 1,$

where

$$\mathbf{r}^k = A\mathbf{x}^k - \mathbf{a}, \quad \underline{\tau}_k = \left(\prod_{i=0}^{k} \tau_i \right)^{k+1} / \tau_k$$

and

$$B = C^{-1} A.$$

Here the K-condition number is computed from the positive part of the spectrum of B.

Proof It remains only to show that $\tau_i > 0$. This follows by Theorem 11.3, unless $\mathbf{r}^k = \mathbf{0}$ (i.e., the solution has been found). $\diamond$

Remark 13.12 In Theorem 13.11 we estimated the rate of convergence of the residual measured in the $\| \cdot \|_{C^{-\frac{1}{2}}}$ norm. This quantity is available during the iterations. However, as Theorem 11.5(a) shows, the PCG method minimizes $\| \mathbf{r}^k \|_{A^{-\frac{1}{2}}}$ (if A is nonsingular) over the Krylov space generated by B and $\mathbf{r}^0$, hence, in a different norm.

The number K depends on all eigenvalues, while the condition number of B depends only on the extreme eigenvalues. Hence, one can expect that, in some cases, the K-condition number can give more accurate bounds when used to estimate the rate of convergence of the conjugate gradient method than estimates based on the standard condition number. This expectation will be analyzed below. We first need an additional lemma.

In this respect, note that $\mathcal{K}(A^{-1}) = \mathcal{K}(A)$ if A is s.p.d., but $K(B) \ne K(B^{-1})$ in general. Indeed, the rate of convergence for the eigenvalue distributions $\lambda_1, \ldots, \lambda_n$ and $\lambda_n^{-1}, \ldots, \lambda_1^{-1}$ can be quite different.

Remark 13.13 (Superlinear rate of convergence) The estimate in Theorem 13.11(b) is useful only when k is sufficiently large that

$$K(B)^{\frac{1}{k}} < 2.$$

Hence, the estimate is useful only when

$$k > \log_2 K(B).$$

Since the PCG method converges in, at most, n steps (neglecting round-off errors), the estimate in the theorem does not show any practically viable result unless $\log_2 K(B) < n$, that is, $K(B) < 2^n$. Furthermore, if $\log_2 K(B) = O(n)$, the number of iterations according to the upper bound is $k = O(n)$. On the other hand, if $K(B) < 2^n$, then Theorem 13.11(b) shows a superlinear rate of convergence for values of k greater than $\log_2 K(B) < n$, with a convergence factor that decreases to zero as k increses beyond the value $\log_2 K(B)$. Note, in particular, that the estimate shows that the number of iterations depends little on ε, the relative stopping accuracy.

Remark 13.14 As was shown in Section 8.3, the symmetrized method used to compute the best Frobenius norm approximation G with a given sparsity pattern to a matrix A^{-1} is identical to the approximation minimizing the K-number of GA. For an alternative presentation of this, see Kaporin (1992).

13.4 Estimates of the Rate of Convergence Using Different Norms

In the previous sections, we have presented two methods to estimate the rate of convergence of the conjugate gradient method. There we related the $\| \cdot \|_{A^{-1/2}}$ or $\| \cdot \|_{C^{-1/2}}$ norms, respectively, of the kth residual to the norm of the initial residual. Consider, finally, an estimate where we relate the $\| \cdot \|_{A^{-1/2}}$ norm of the kth residual to the same norm of the initial error, $\mathbf{e}^0 = \mathbf{x} - \mathbf{x}^0$. For simplicity of presentation, assume that the iteration matrix $B = A$, that A is s.p.d., and that A has been scaled so that $\|A\| = 1$. Let the eigenvalues of A be λ_i, $i = 1, 2, \ldots, n$ and let the eigenvectors be $\mathbf{v}_i$, $i = 1, 2, \ldots, n$, which we let be normalized. Let the inital error be $\mathbf{e}^0 = \sum_1^n \alpha_i \mathbf{v}_i$. Then the kth iteration error satisfies

$$\mathbf{e}^k = \sum_i^n \alpha_i \widehat{P}_k(\lambda_i) \mathbf{v}_i$$

for some polynomial $\widehat{P}_k \in \pi_k^1$, that is, of degree k, where $\widehat{P}_k(0) = 1$. The relation $\mathbf{r}^k = -A\mathbf{e}^k$ then shows that

$$\mathbf{r}^k = -\sum_1^n \alpha_i \lambda_i \widehat{P}_k(\lambda_i) \mathbf{v}_i.$$

Hence,

$$\mathbf{r}^{k^T} A^{-1} \mathbf{r}^k = \sum_{1}^{n} \alpha_i^2 \lambda_i \widehat{P}_k(\lambda_i)^2$$

and

$$\mathbf{e}^{0^T} A^{-1} \mathbf{e}^0 = \sum_{1}^{n} \lambda_i^{-1} \alpha_i^2.$$

Using the minimizing property of the conjugate gradient method, we find

$$\mathbf{r}^{k^T} A^{-1} \mathbf{r}^k \le \sum_{1}^{n} \alpha_i^2 \lambda_i P_k(\lambda_i)^2 \le \max_{1 \le i \le n} [\lambda_i P_k(\lambda_i)]^2 \sum_{1}^{n} \lambda_i^{-1} \alpha_i^2$$

for any $P_k \in \pi_k^1$. Further,

$$\mathbf{r}^{k^T} A^{-1} \mathbf{r}^k \le (\varepsilon')^2 \mathbf{e}^{0^T} A^{-1} \mathbf{e}^0,$$

and

$$\mathbf{r}^{k^T} A^{-1} \mathbf{r}^k \le \varepsilon^2 \mathbf{r}^{0^T} A^{-1} \mathbf{r}^0$$

if

$$(13.27) \qquad \max_{1 \le i \le n} |\lambda_i P_k(\lambda_i)| \le \varepsilon' = \varepsilon \frac{\|\mathbf{r}^0\|_{A^{-\frac{1}{2}}}}{\|\mathbf{e}^0\|_{A^{-\frac{1}{2}}}},$$

where $\|\mathbf{r}^0\|_{A^{-1/2}} / \|\mathbf{e}^0\|_{A^{-1/2}} = \left\{ \sum_{1}^{n} \alpha_i^2 \lambda_j \Big/ \sum_{1}^{n} \alpha_j^2 \lambda_j^{-1} \right\}^{\frac{1}{2}}$.

We want to find the polynomial of smallest degree for which (13.27) holds. Since, in practice, $0 < \min_i \lambda_i \ll \max_i \lambda_i \le 1$, we can consider the somewhat more simplified approximation problem: Find the polynomial of smallest degree, such that

$$\max_{0 \le x \le 1} |x P_k(x)| \le \varepsilon.$$

Then let $S_{k+1}(x) = x P_k(x)$ and observe that

$$S_{k+1}(0) = 0 \quad \text{and} \quad S'_{k+1}(0) = 1.$$

It is readily seen that the polynomial of smallest degree is such that

$$S_{k+1}(x) = -T_{k+1}\left(\frac{1 + x_0 - 2x}{1 - x_0}\right) \Big/ T_{k+1}\left(\frac{1 + x_0 - 2x_1}{1 - x_0}\right),$$

where $x_1 < x_0 < 0$, and

$$|S_{k+1}(x_0)| = \max_{0 \le x \le b} |S_{k+1}(x)| = 1/T_{k+1}\left(\frac{1 + x_0 - 2x_1}{1 - x_0}\right)$$

and x_1 is the largest negative number for which $S_{k+1}(x_1) = -1$. T_{k+1} is the Chebyshev polynomial of degree $k + 1$. To determine x_0, we use $S_{k+1}(0) = 0$, which shows that

$$T_{k+1}\left(\frac{1 + x_0}{1 - x_0}\right) = 0$$

or

$$\frac{1 + x_0}{1 - x_0} = \cos\frac{\pi}{2(k + 1)}.$$

That is,

(13.28) $$x_0 = \frac{\cos\frac{\pi}{2(k+1)} - 1}{\cos\frac{\pi}{2(k+1)} + 1} = -(\tan\frac{\pi}{4(k + 1)})^2.$$

Further, $S'_{k+1}(0) = 1$ shows that

$$\frac{1}{T_{k+1}\left(\frac{1+x_0-2x_1}{1-x_0}\right)} \cdot T'_{k+1}\left(\frac{1 + x_0}{1 - x_0}\right) \cdot \left(-\frac{2}{1 - x_0}\right) = 1,$$

or

$$\frac{1}{T_{k+1}\left(\frac{1+x_0-2x_1}{1-x_0}\right)} = \frac{1 - x_0}{2(k + 1)} \cdot \sqrt{1 - (\frac{1 + x_0}{1 - x_0})^2}.$$

That is,

$$\max_{0 \le x \le 1} |S_{k+1}(x)| = \frac{\sqrt{|x_0|}}{k + 1} = \frac{1}{k + 1}\tan\frac{\pi}{4(k + 1)} \sim \frac{\pi}{4}(k + 1)^{-2}, \quad k \to \infty.$$

This shows that (13.27) holds if

$$\frac{1}{k + 1}\tan\frac{\pi}{4(k + 1)} \le \varepsilon,$$

or, asymptotically, if

(13.29) $$k \ge \sqrt{\frac{\pi}{4\varepsilon}} - 1.$$

Note here the strong dependence on ε. However, when ε is not particularly small, this shows few iterations regardless of the condition number! Hence, this estimate can be useful in an initial phase of the iterations.

We shall now compare the different estimates for some typical, eigenvalue distributions.

Example 13.3′ Consider Example 13.3, but where all eigenvalues have been normalized, so that $\max \lambda_i = 1$, by division by 4. Here,

$$\|\mathbf{r}^0\|_{A^{-\frac{1}{2}}}/\|\mathbf{e}^0\|_{A^{-\frac{1}{2}}} = \left\{ \sum_1^{n^2} \alpha_j^2 \lambda_j \bigg/ \sum_1^{n^2} \alpha_j^2 \lambda_j^{-1} \right\}^{\frac{1}{2}} = \left\{ \sum_1^{n^2} \lambda_j \bigg/ \sum_1^{n^2} \lambda_j^{-1} \right\}^{\frac{1}{2}},$$

and a computation reveals that

$$\sum_1^{n^2} \lambda_j \bigg/ \sum_1^{n^2} \lambda_j^{-1} \sim \frac{1}{16} \left[\frac{5}{2} n^2 + O(n) \right] \bigg/ \left[1.57 n^2 + \frac{2}{3} \ln \frac{5}{2} n^2 + O(n) \right]$$

$$\sim \frac{1.15}{16}, \quad n \to \infty.$$

Hence, $\varepsilon' \simeq \frac{\sqrt{1.15}}{4} \varepsilon$ and (12.27) shows the following decrease of the relative residuals:

$$\|\mathbf{r}^k\|_{A^{-\frac{1}{2}}}/\|\mathbf{r}^0\|_{A^{-\frac{1}{2}}} \sim \frac{4}{\sqrt{1.15}} \frac{1}{k} \tan \frac{\pi}{4k} \sim \frac{\pi}{k^2}, \quad k = 1, 2, \ldots.$$

This behavior is also clearly visible in the initial phase of Figure 13.1. Note that this is independent of the condition number. However, it presupposes that the ratio $\|\mathbf{e}^0\|_{A^{-1/2}}/\|\mathbf{r}^0\|_{A^{-1/2}} = \|\mathbf{e}^0\|_{A^{-1/2}}/\|\mathbf{e}^0\|_{A^{-1/2}}$ is not large. The value of this quotient depends on the distributions of the Fourier coefficients α_j. We have $\|\mathbf{e}^0\|_{A^{-1/2}}/\|\mathbf{e}^0\|_{A^{1/2}} \le \frac{1}{\lambda_1}$.

For distributions where $|\alpha_j|$ for the lower indices dominate, the quotient can become quite close to this upper bound. If the quotient is small, we get the initial behavior of the relative residuals, described by a convex curve. However, if the quotient is very large, the behavior of the relative residuals is better predicted by (13.15b), which corresponds to a concave curve and, hence, a slower initial decay of the residuals. In practice, it is common that the eigenvalue components corresponding to the small eigenvalues are dominating, which indicates that $\|\mathbf{r}^0\|_{A^{-1/2}}/\|\mathbf{e}^0\|_{A^{-1/2}}$ is often large.

Example 13.4 (Geometric distribution) Let $\lambda_j = j^s$, $j = 1, 2, \ldots, n$ be the eigenvalues of $B = A$, where s is a positive integer. Here, asymptotically,

$$\text{tr}(A) = \sum_{1}^{n} j^s \sim \frac{1}{s+1} n^{s+1}, \quad n \to \infty$$

and, using Stirling's formula,

$$\det(A) = \prod_{1}^{n} \lambda_j = \left(\prod_{1}^{n} j \right)^s \sim (2\pi n)^{s/2} \left(\frac{n}{e} \right)^{ns}, \quad n \to \infty.$$

Hence,

$$K(A)^{\frac{1}{n}} \sim \frac{e^s}{s+1}, \quad n \to \infty.$$

As we saw in Remark 13.13, the estimate in Theorem 13.11(b) is applicable only when $K(A)^{\frac{1}{n}} < 2$, and this holds in the present example if $s \leq s_0$, where s_0 is slightly less than 2 ($e^{s_0} = 2s_0 + 2$). Using Theorem 13.11(b), we find the following upper bound on the number of iterations:

(13.30) $$k \sim n \log_2 K(A).$$

On the other hand, the estimate in (13.12) based on the condition number, which in this case is $\mathcal{K}(A) = n^s$, yields

$$k \sim O(n^{\frac{s}{2}}), \quad n \to \infty,$$

and thus gives, asymptotically, a smaller upper bound on the number of iterations than (13.30) for the eigenvalue distribution $\lambda_j = j^s$, if $s < 2$.

Example 13.5 (Arithmetic distribution) Let $0 < \lambda_1 < \lambda_2 < \cdots < \lambda_n$, the eigenvalues of B, be distributed as an arithmetic sequence in $a = \lambda_1$, $b = \lambda_n$. For simplicity, assume that $n/2$ is even. Then,

$$K(B) = \left(\frac{b+a}{2} \right)^n / \prod_{1}^{n} \lambda_i$$

and we have

$$\prod_{1}^{n} \lambda_i = \prod_{1}^{n/2} \left(\frac{b+a}{2} - \frac{b-a}{2} \frac{2j}{n} \right) \left(\frac{b+a}{2} + \frac{b-a}{2} \frac{2j}{n} \right)$$

$$= \left(\frac{b+a}{2} \right)^n \prod_{1}^{n/2} \left(1 - \left(\frac{b-a}{b+a} \frac{2j}{n} \right)^2 \right).$$

It is readily seen that

$$K(B) > 1/\prod_1^{n/4}\left[1 - \frac{1}{4}\left(\frac{b-a}{b+a}\right)^2\right].$$

Hence, the estimate in Theorem 13.11(b) shows the number of iterations to be $\log_2 K(B) > \frac{n}{4}\log_2\left(1 - \frac{1}{4}(\frac{b-a}{b+a})^2\right)^{-1}$, which is bounded below by a number $O(n)$.

On the other hand, (13.12) shows $k \leq \frac{1}{2}\sqrt{\frac{b}{a}}\ln\frac{2}{\varepsilon}$, which is asymptotically smaller than n if $\frac{b}{a}n^{-2} = o(1)$. In particular, if b/a does not depend on n, then we have $k \leq O(\ln\frac{1}{\varepsilon})$. Therefore, the estimate in Theorem 13.11(b) is generally inferior to the simple estimate in 13.12. For other distributions, however, Theorem 13.11(b) can give a smaller upper bound.

Example 13.6 Let

$$\lambda_j = 1 + \frac{1}{j}, \quad j = 1, 2, \ldots, n.$$

(This is a special case of the compact perturbation case in Section 13.2.) Then a well-known asymptotic estimate shows that

$$\frac{1}{n}\sum_1^n \lambda_j \sim 1 + \frac{1}{n}\ln n + cn^{-1} + O(n^{-2})$$

for a certain positive constant c, and

$$\left(\prod_1^n \lambda_j\right)^{\frac{1}{n}} = (n+1)^{\frac{1}{n}}.$$

Hence, the K-condition number of A satisfies

$$K(A) = \frac{[1 + \frac{1}{n}\ln n + cn^{-1} + O(n^{-2})]^n}{n+1},$$

and

$$\log_2 K(A) = (\log_2)e\ln K(A) = (\log_2 e)[\ln n + c + O(n^{-1}) - \ln(n+1)]$$

$$= c(\log_2 e) + O(n^{-1}).$$

Hence, $\log_2 K(A)$ and $K(A) \to$ const, $n \to \infty$.

In this case, (13.30) shows a rapid, superlinear rate of convergence, while (13.12), using the condition number [which is $\mathcal{K}(A) \sim 2$] shows only a linear rate of convergence.

13.5 Conclusions

We have found an estimate of the rate (13.29) that is independent of the distributions of eigenvalues. However, when k increases, it indicates a slow decrease of the norm of the residual, and an average rate of convergence that approaches 1.

Among the estimates we have presented, however, (13.29) can show the greatest decay during the first iterations for problems where the condition number is large. Therefore, the three estimates—(13.29), (13.12), and Theorem 13.11(b)—of the rate of convergence of the conjugate gradient method are typically useful during three different phases:

1. The initial phase, where (13.29) shows a decay not slower than

$$\{\mathbf{r}^{k^T} A^{-1} \mathbf{r}^k\}^{\frac{1}{2}} \sim \left(\frac{1}{k+1}\right)^2 \text{const.}$$

2. The middle phase, recognized by a linear rate of convergence, where (13.12) shows that

$$\{\mathbf{r}^{k^T} A^{-1} \mathbf{r}^k\}^{\frac{1}{2}} \sim \sigma^{-k} \text{ const and } \sigma = [1 - \mathcal{K}(A)^{-\frac{1}{2}}]/[1 + \mathcal{K}(A)^{-\frac{1}{2}}].$$

3. The final phase, recognized by a tendency of superconvergence, shown by Theorem 13.11(b) (if the eigenvalues are real and nonnegative and $K(A) < 2n$) or by (13.15a,b) (if the eigenvalues are sufficiently isolated from each other).

The final phase is often not seen unless one iterates to very small relative errors and the condition number is large. The three phases of convergence can be clearly seen in Example 13.1, for instance.

Example 13.7 Consider the solution of $A\mathbf{x} = \mathbf{a}$ with the CG method, where $A = \text{tridiag}(-1, 2, -1)$. Here, $\frac{1}{n} \text{tr}(A) = 2$ and the LU factorization of A,

$$A = \begin{bmatrix} 1 & & & & 0 \\ -d_1^{-1} & 1 & & & \\ & -d_2^{-1} & 1 & & \\ & & \ddots & \ddots & \\ 0 & & 0 & \cdots & 1 \end{bmatrix} \begin{bmatrix} d_1 & -1 & & & 0 \\ & d_2 & -1 & & \\ & & \ddots & \ddots & \\ 0 & & 0 & \cdots & d_n \end{bmatrix},$$

where $d_i = \frac{i+1}{i}$, shows that

$$\det(A) = \prod_{i=1}^{n} d_i = n + 1.$$

Hence, $K(A) = \frac{2n}{n+1}$ and Theorem 13.11(b) shows that for $k = n - q, q = q_0 - 1, \ldots, 0$, where $q_0 = \lfloor \log_2(n + 1) \rfloor$, then

$$\frac{\mathbf{r}^{k^T} C^{-1} \mathbf{r}^k}{r^{0^T} C^{-1} r^0} \le \left(2 \left(\frac{2^q}{n+1} \right)^{\frac{1}{n-q}} - 1 \right)^{n-q},$$

so convergence can be seen for $k = n - q$ where $q < q_0$, $2^{q_0} \simeq n + 1$. The condition number A is $\sim 1 / \left(\sin \frac{\pi}{2(n+1)} \right)^2$, so (13.12) shows a number of iterations $\sim \frac{n+1}{\pi} \ln \frac{2}{\varepsilon}$, $n \to \infty$, which, hence, exceeds the termination number when ε is sufficiently small. Using the estimates in (13.15) a somewhat better bound can be derived.

For an analysis of the rate of convergence of the conjugate gradient method using estimates of the Ritz values (that is, eigenvalues of the matrix T_{k+1} in Section 13.3), see van der Vorst (1982) and van der Sluis and van der Vorst (1986). As we saw in Chapter 11, the Ritz values approximate some eigenvalues of A increasingly well for increasing k. It has been observed that when an extremal eigenvalue has been sufficiently well approximated by a Ritz value, then the CG process converges faster from then on, giving an alternative indication of superlinear rate of convergence behavior.

Finally, note that in the above estimates we have neglected the influence of round-off errors and that they hold, in general, only in exact arithmetic. Round-off errors cause the search and residual vectors to lose the mutual (conjugate) orthogonality property and, in general, slow down the rate of convergence when many iterations have to be performed. Since this loss of orthogonality corresponds to performing the conjugate gradient iterations on a slightly perturbed matrix, of larger order, it is seen that the estimate (13.12) using condition numbers is essentially unaffected by round-off.

As shown in Section 13.2, in the case of small isolated eigenvalues, the convergence behavior of the conjugate gradient method can be explained by assuming that first the small eigenvalues are eliminated. In the presence of round-off and hence of perturbed eigenvalues, some small components remain for these eigenvalues. However, this does not change the convergence behavior as long as the error components corresponding to the uneliminated eigenvalues remain larger than those for the perturbed eigenvalues. As pointed out in Notay (1992), one can see when this effect takes place by comparing the norm of the difference of the recursively computed residual and the exact residual with the norm of the residual itself. Only when the norms of this difference become close to the norm of the residual does one see a more significant influence of round-off errors. In the case of large isolated eigenvalues, round-off can influence the convergence behavior more significantly.

References

O. Axelsson (Copenhagen, 1976). Solution of linear systems of equations: iterative methods. In ed. V. A. Barker. *Sparse Matrix Techniques*, Berlin, Heidelberg, New York: Springer Verlag.

O. Axelsson (1992). Bounds of eigenvalues of preconditioned matrices. *SIAM J. Matrix Anal. Appl.* **13**, 847–862.

O. Axelsson and G. Lindskog (1986). On the rate of convergence of the preconditioned conjugate gradient method. *Numer. Math.* **48**, 499–523.

J. E. Dennis and H. Wolkowicz (1990). Sizing and least change secant methods. TR 90-5, Department of Mathematical Sciences, Rice Univ., Houston, Texas.

A. Greenbaum and Z. Strakos (1992). Predicting the behaviour of finite precision Lanczos and conjugate gradient computations. *SIAM J. Matrix Anal. Appl.* **13**, 121–137.

A. Jennings (1977). Influence of the eigenvalue spectrum on the convergence rate of the conjugate gradient method. *J. Inst. Math. Appl.* **20**, 61–72.

I. E. Kaporin (1990). An alternative approach to estimation of the conjugate gradient iteration number. In *Numerical Methods and Software*, ed. Yu. A. Kuznetsov. Acad. Sci. USSR, Department of Numerical Mathematics, Moscow (in Russian).

I. E. Kaporin (1992). New convergence results and preconditioning strategies for the conjugate gradient method. Submitted.

Y. Notay (1992). On the convergence rate of the conjugate gradients in presence of rounding errors. *Numer. Math.*, to appear.

A. van der Sluis and H. A. van der Vorst (1986). The rate of convergence of conjugate gradients. *Numer. Math.* **48**, 543–560.

H. A. van der Vorst (1982). Preconditioning by incomplete decompositions. Thesis, Utrecht Univ.

R. Winter (1980). Some superlinear convergence results for the conjugate gradient method. *SIAM J. Numer. Anal.* **17**, 14–17.

Appendix A

Matrix Norms, Inherent Errors, and Computation of Eigenvalues

For various applications, we need to quantify errors and distances—i.e., we need to measure the size of a vector and a matrix. We shall find that it is not only the Euclidian measure of a vector that is appropriate in practice. Accordingly, we introduce vector and matrix norms, which are real valued functions of vectors and matrices, respectively. It is shown how to calculate matrix norms associated with certain vector norms and apply them in the estimation of eigenvalues. We shall find that properties of positive definite matrices play a crucial role in the calculation of the norm associated with the Euclidian vector norm. We show also that matrix norms are useful when estimating inherent errors—i.e., errors caused by errors in given data—in solutions of systems of linear algebraic equations. Finally, it is shown how to estimate the errors in eigenvalue computations by a posteriori estimates and how to compute sequences of approximations of them by matrix power methods. The following definitions are introduced in this appendix.

Definition A.1 A vector norm $||\mathbf{x}||$ in a given linear vector space V is a real valued function $|| \cdot || : V \to \mathbb{R}_+$ of $\mathbf{x} \in V$, satisfying:

    (a) $||\mathbf{x}|| \geq 0 \;\; \forall \mathbf{x} \in V$, $||\mathbf{x}|| = 0 \Longleftrightarrow \mathbf{x} = \mathbf{0}$.

    (b) $||c\mathbf{x}|| = |c| \; ||\mathbf{x}|| \;\; \forall \mathbf{x} \in V$ for every scalar c.

    (c) $||\mathbf{x} + \mathbf{y}|| \leq ||\mathbf{x}|| + ||\mathbf{y}|| \;\; \forall \mathbf{x}, \mathbf{y} \in V$ (the triangle inequality).

Definition A.2(a) A matrix norm $||A||$ is a real valued function $|| \cdot || : \mathbb{C}^{n^2} \to \mathbb{R}_+$ of the matrix A, with properties:

    (a) $||A|| \geq 0$, $||A|| = 0 \Longleftrightarrow A = 0$.

    (b) $||cA|| = |c| \; ||A||$ for every scalar $c \in \mathbb{C}$.

    (c) $||A + B|| \leq ||A|| + ||B||$.

Definition A.2(b) If a matrix norm $||A||$ has the property

$$||AB|| \leq ||A|| \, ||B||,$$

then the matrix norm is called *multiplicative*.

Definition A.3(a) For any given vector norm $||\mathbf{x}||$, $||A|| = \sup_{\mathbf{x} \neq 0} \frac{||A\mathbf{x}||}{||\mathbf{x}||}$ is said to be the *matrix norm induced by the vector norm* $||\mathbf{x}||$, or the *natural matrix norm*.

Definition A.3(b) If $||A\mathbf{x}|| \leq ||A|| \, ||\mathbf{x}||$, then the matrix norm $||A||$ is said to be *compatible* with the vector norm $||\mathbf{x}||$.

Definition A.4 $\kappa = ||A^{-1}|| \, ||A||$ is called the *condition number* of A; $\kappa(\mathbf{x}) = ||A^{-1}||(||A\mathbf{x}||/||\mathbf{x}||)$ is called the *condition number of A with respect to the vector* $\mathbf{x}$.

A.1 Vector and Matrix Norms

Definition A.1 A *vector norm* $||\mathbf{x}||$ in a given linear vector space V is a real valued function $|| \cdot || : V \to \mathbb{R}_+$ of $\mathbf{x} \in V$, satisfying:

(a) $||\mathbf{x}|| \geq 0 \ \forall \mathbf{x} \in V, \ ||\mathbf{x}|| = 0 \Longleftrightarrow \mathbf{x} = \mathbf{0}$.

(b) $||c\mathbf{x}|| = |c| \, ||\mathbf{x}|| \ \forall \mathbf{x} \in V$ for every scalar c.

(c) $||\mathbf{x} + \mathbf{y}|| \leq ||\mathbf{x}|| + ||\mathbf{y}|| \ \forall \mathbf{x}, \mathbf{y} \in V$ (the triangle inequality).

Example A.1 If $V = \mathbb{C}^n$ or $\mathbb{R}^n$, then

$$||\mathbf{x}||_2 = \left(\sum_{i=1}^{n} |x_i|^2 \right)^{\frac{1}{2}} \quad \text{(the Euclidian norm)},$$

$$||\mathbf{x}||_1 = \sum_{i=1}^{n} |x_i| \quad \text{(the absolute sum norm), and}$$

$$||\mathbf{x}||_\infty = \max_i |x_i| \quad \text{(the maximum norm)}$$

are easily proved to be vector norms. That $||\mathbf{x} + \mathbf{y}||_2 \leq ||\mathbf{x}||_2 + ||\mathbf{y}||_2$ follows from the *Cauchy-Schwarz inequality*, $|(\mathbf{x}, \mathbf{y})| \leq ||\mathbf{x}||_2 ||\mathbf{y}||_2$. All are special cases of the l_p-norm (also called *Höldernorm*),

$$||\mathbf{x}||_p = \left(\sum_{i=1}^{n} |x_i|^p \right)^{\frac{1}{p}}, 1 \leq p.$$

That this is a norm follows from Minkowski's inequality $\|x + y\|_p \le \|x\|_p + \|y\|_p$, which can be proved using the so-called Hölder inequality, $|(x, y)| \le \|x\|_p \|y\|_q$, $p^{-1} + q^{-1} = 1$, $1 \le p, q$. The latter is a generalization of the Cauchy-Schwarz inequality.

Lemma A.1 *Every vector norm $\|x\|$ defined on $\mathbb{C}^n$ is a continuous function of x in the sense that*

$$\left| \|x\| - \|\hat{x}\| \right| \le \varepsilon \quad \text{if} \quad \|x - \hat{x}\|_\infty \le \xi, \quad x, \hat{x} \in \mathbb{C}^n, \quad \forall \varepsilon > 0$$

for some function $\xi = \xi(\varepsilon)$, where $\xi \to 0$ as $\varepsilon \to 0$.

Proof For all $x, \hat{x} \in \mathbb{C}^n$, we have

$$\left| \|x\| - \|\hat{x}\| \right| = \left| \|\hat{x} + \underline{\delta}\| - \|\hat{x}\| \right| \le \|\underline{\delta}\|,$$

where $\underline{\delta} = x - \hat{x}$. Let e_k be the unit vectors and let $\underline{\delta} = \sum_{k=1}^{n} \delta_k e_k$. Then, using properties of vector norms, we find

$$\|\underline{\delta}\| \le \sum_{k=1}^{n} \|\delta_k e_k\| = \sum_{k=1}^{n} |\delta_k| \, \|e_k\| \le \max_k |\delta_k| \sum_{j=1}^{n} \|e_j\| = M_n \|\underline{\delta}\|_\infty,$$

where $M_n = \sum_{j=1}^{n} \|e_j\|$. Hence, $\|\underline{\delta}\| \le \varepsilon$ if $\|\underline{\delta}\|_\infty \le \xi = M_n^{-1} \varepsilon$, which concludes the proof. $\diamond$

Lemma A.1 shows that we can define convergence $x^{(\nu)} \to x$, $\nu \to \infty$ of a vector sequence $x^{(\nu)}$: The sequence $x^{(\nu)}$ is said to converge to x if and only if $\|x^{(\nu)} - x\| \to 0$, $\nu \to \infty$.

Theorem A.2 *For every pair of norms, $\|x\|_a$ and $\|x\|_b$ in $\mathbb{C}^n$, there are constants $0 < m < M < \infty$ such that $m \|x\|_b \le \|x\|_a \le M \|x\|_b$ $\forall x \in \mathbb{C}^n$, where m, M do not depend on x.*

Proof First, we note that it suffices to prove the theorem in the case when one of the norms is $\|x\|_\infty$, since if $m \|y\|_\infty \le \|y\|_b \le M \|y\|_\infty$ and $m' \|y\|_\infty \le \|y\|_a \le M' \|y\|_\infty$, then

$$\frac{m'}{M} \|y\|_b \le m' \|y\|_\infty \le \|y\|_a \le M' \|y\|_\infty \le \frac{M'}{m} \|y\|_b \quad \forall y \in \mathbb{C}^n.$$

Let K be the unit ball in $\mathbb{C}^n$, with norm $\| \cdot \|_\infty$, that is, $K = \{x \in \mathbb{C}^n; \|x\|_\infty = $

1}. Since K is a closed and bounded set in $\mathbb{C}^n$ and $||\cdot||$ is a continuous function, $||\mathbf{y}||$ attains its supremum and infimum. That is,

$$m = \inf_{\mathbf{y}\in K} ||\mathbf{y}|| = ||\mathbf{y}_0||, \quad M = \sup_{\mathbf{y}\in K} ||\mathbf{y}|| = ||\mathbf{y}_1||$$

for some vectors $\mathbf{y}_0, \mathbf{y}_1 \in K$. Since $||\mathbf{y}||_\infty = 1$, neither of these is zero. Hence,

$$0 < m = ||\mathbf{y}_0|| \leq ||\mathbf{y}|| \leq ||\mathbf{y}_1|| = M < \infty \quad \forall \mathbf{y} \in K.$$

Now let $\mathbf{x} \neq \mathbf{0}$ be an arbitrary vector in $\mathbb{C}^n$. Then,

$$\frac{1}{||\mathbf{x}||_\infty}\mathbf{x} \in K, \text{ that is, } m \leq ||\frac{\mathbf{x}}{||\mathbf{x}||_\infty}|| \leq M \text{ or } m||\mathbf{x}||_\infty \leq ||\mathbf{x}|| \leq M||\mathbf{x}||_\infty. \quad \diamond$$

From this theorem, we make the important observation that all norms in a finite dimensional vector space are topologically equivalent (in the sense of convergence), since, for any sequence $\mathbf{x}^{(\nu)}$,

$$||\mathbf{x}^{(\nu)}||_b \to 0 \iff ||\mathbf{x}^{(\nu)}||_a \to 0, \quad \nu \to \infty.$$

Definition A.2(a) A *matrix norm* $||A||$ is a real valued function $||\cdot||: \mathbb{C}^{n^2} \to \mathbb{R}_+$ of the matrix A with properties:

(a) $||A|| \geq 0$, $||A|| = 0 \iff A = 0$.
(b) $||cA|| = |c|\,||A||$ for every scalar $c \in \mathbb{C}$.
(c) $||A + B|| \leq ||A|| + ||B||$.

Definition A.2(b) If a matrix norm $||A||$ has the property $||AB|| \leq ||A||\,||B||$, then the matrix norm is called *multiplicative*. We shall only be dealing with norms satisfying both A.2(a) and A.2(b).

Lemma A.3 *If* $p = \sup_{\mathbf{x}\neq 0} \frac{||A\mathbf{x}||}{||\mathbf{x}||}$ *and* $q = \sup_{||\mathbf{x}||=1} ||A\mathbf{x}||$, *where* $||\mathbf{x}||$ *is a vector norm, then* $p = q$.

Proof If $||\mathbf{x}|| = 1$, then $\frac{||A\mathbf{x}||}{||\mathbf{x}||} = ||A\mathbf{x}||$, and since the set of all unit vectors in V is a subset of the set of all vectors in V, it follows that

$$q = \sup_{||\mathbf{x}||=1} ||A\mathbf{x}|| = \sup_{||\mathbf{x}||=1} \frac{||A\mathbf{x}||}{||\mathbf{x}||} \leq p.$$

Now let $\mathbf{x} \in V$ and let $\hat{\mathbf{x}} = \frac{1}{||\mathbf{x}||}\mathbf{x}$. Then, $||\hat{\mathbf{x}}|| = 1$, $||A\hat{\mathbf{x}}|| = \frac{1}{||\mathbf{x}||}||A\mathbf{x}||$ and

$$p = \sup_{||\mathbf{x}|| \neq 0} \frac{||A\mathbf{x}||}{||\mathbf{x}||} = \sup_{||\mathbf{x}|| \neq 0} ||A(\frac{1}{||\mathbf{x}||}\mathbf{x})|| \leq \sup_{||\hat{\mathbf{x}}||=1} ||A\hat{\mathbf{x}}|| = q.$$

Hence $p = q$. $\diamond$

Theorem A.4 $||A|| = \sup_{\mathbf{x}\neq 0} \frac{||A\mathbf{x}||}{||\mathbf{x}||}$ *is a matrix norm—i.e., it satisfies properties (a) to (c) in Definition A.2(a).*

Proof

(a) $||A\mathbf{x}|| \geq 0 \Rightarrow ||A|| = \sup_{||\mathbf{x}||=1} ||A\mathbf{x}|| \geq 0;$

$$||A|| = \sup_{\mathbf{x}\neq 0} \frac{||A\mathbf{x}||}{||\mathbf{x}||} = 0 \iff ||A\mathbf{x}|| = 0 \ \forall \mathbf{x} \neq 0$$

$$\iff A\mathbf{x} = 0 \ \forall \mathbf{x} \neq 0 \iff A = 0.$$

(b) $||cA|| = \sup_{||\mathbf{x}||=1} ||cA\mathbf{x}|| = |c| \sup_{||\mathbf{x}||=1} ||A\mathbf{x}|| = |c| \ ||A||.$
(c) If $\mathbf{x} \neq 0$, then

$$\frac{||A\mathbf{x}||}{||\mathbf{x}||} \leq \sup_{\mathbf{x}\neq 0} \frac{||A\mathbf{x}||}{||\mathbf{x}||} = ||A||,$$

so $||A\mathbf{x}|| \leq ||A|| \ ||\mathbf{x}||$ and $||(A+B)\mathbf{x}|| \leq ||A\mathbf{x}|| + ||B\mathbf{x}|| \leq (||A|| + ||B||) \ ||\mathbf{x}|| = ||A|| + ||B||, \ \forall \mathbf{x}, \ ||\mathbf{x}|| = 1$, so that

$$||A+B|| = \sup_{||\mathbf{x}||=1} ||(A+B)\mathbf{x}|| \leq ||A|| + ||B||. \qquad \diamond$$

Definition A.3

(a) For any given vector norm $||\mathbf{x}||$, $||A|| = \sup_{\mathbf{x}\neq 0} \frac{||A\mathbf{x}||}{||\mathbf{x}||}$ is said to be the *matrix norm induced by the vector norm* $||\mathbf{x}||$, or the *natural matrix norm*.
(b) If $||A\mathbf{x}|| \leq ||A|| \ ||\mathbf{x}||$, then the matrix norm $||A||$ is said to be *compatible* with the vector norm $||\mathbf{x}||$.

Recall that the *spectral radius* $\rho(A)$ of a matrix A is $\rho(A) = \max |\lambda|, \lambda \in \mathcal{S}(A)$, where $\mathcal{S}(A)$ denotes the spectrum of A, that is, the set of eigenvalues of A.

Theorem A.5 *For an arbitrary matrix A, we have* $\|A\|_2 = \sqrt{\rho(A^*A)}$.

Proof Let U be unitary and such that

$$U^*(A^*A)U = \begin{bmatrix} \mu_1 & & \emptyset \\ & \ddots & \\ \emptyset & & \mu_n \end{bmatrix},$$

where μ_i are the eigenvalues of A^*A. ($\mu_i \geq 0$ by Theorem 3.2). Let $\mathbf{y} = U^*\mathbf{x}$. Then $\mathbf{x} = U\mathbf{y}$ and

$$\|A\|_2 = \sup_{\mathbf{x} \neq 0} \frac{\|A\mathbf{x}\|_2}{\|\mathbf{x}\|_2} = \sup_{\mathbf{x} \neq 0} \sqrt{\frac{(A^*A\mathbf{x}, \mathbf{x})}{(\mathbf{x}, \mathbf{x})}}$$

$$= \sup_{\mathbf{y} \neq 0} \sqrt{\frac{(U^*A^*AU\mathbf{y}, \mathbf{y})}{(\mathbf{y}, \mathbf{y})}} = \sup_{\mathbf{y} \neq 0} \sqrt{\frac{\Sigma\mu_i|y_i|^2}{\Sigma|y_i|^2}} = \sqrt{\max|\mu_i|},$$

that is, $\|A\|_2 = \sqrt{\rho(A^*A)}$.   ◇

Remark: The square roots $\sqrt{\mu_i}$ are called the *singular values* of A.

Corollary A.6 *If A is selfadjoint (or real, symmetric), then* $\|A\|_2 = \rho(A)$.

Proof It is readily seen that

$$S(A^k) = \{\lambda_i^k\}, \quad \text{where } S(A) = \{\lambda_i\}.$$

If A is selfadjoint, then $A^*A = A^2$. Hence,

$$\|A\|_2^2 = \rho(A^*A) = \rho(A^2) = [\rho(A)]^2.$$ ◇

Remark A.7 For an arbitrary square matrix A and an arbitrary matrix norm compatible with a vector norm, we have

$$\rho(A) \leq \|A\|,$$

because $A\mathbf{x} = \lambda\mathbf{x}$ implies

$$\|A\| \, \|\mathbf{x}\| \geq \|A\mathbf{x}\| = |\lambda| \, \|\mathbf{x}\| \quad \text{for any eigenvalue } \lambda.$$

Corollary A.8 *If $||A||$ is induced by the vector norm $||\mathbf{x}||$, then:*

(a) $||A\mathbf{x}|| \leq ||A|| \cdot ||\mathbf{x}||$, *that is $||A||$ is compatible with $||\mathbf{x}||$,*

(b) $||I|| = 1$, *where I is the identity matrix,*

(c) $||AB|| \leq ||A|| \, ||B||$, *i.e, the matrix norm is multiplicative.*

Proof Part (a) is shown in the proof of Theorem A.4. Part (b) follows from

$$||I|| = \sup_{\mathbf{x} \neq 0} \frac{||I\mathbf{x}||}{||\mathbf{x}||} = 1.$$

Part (c) follows from $||AB\mathbf{x}|| \leq ||A|| \, ||B\mathbf{x}|| \leq ||A|| \, ||B|| \, ||\mathbf{x}||$ and using Definition A.2(b). $\diamond$

That there are compatible matrix norms which are not induced by a vector norm follows from the following example.

Example A.2 $||A||_E = \left(\sum_i \sum_j |a_{ij}|^2 \right)^{\frac{1}{2}}$ is a matrix norm, the *Euclidean norm* in $\mathbb{C}^{n^2}$ (also called the *Frobenius norm*), and it is compatible with the Euclidean vector norm $||\mathbf{x}||_2$. However, it is not induced by any vector norm.

It is readily seen that $||A||_E$ is a matrix norm. That $||A||_E$ is compatible with $||\mathbf{x}||_2$ follows from

$$||A\mathbf{x}||_2^2 = \sum_i \left| \sum_j a_{ij} x_j \right|^2 \leq \sum_i \left(\sum_j |a_{ij}|^2 \sum_j |x_j|^2 \right) = ||A||_E^2 ||\mathbf{x}||_2^2,$$

where we have used the Cauchy-Schwarz inequality, so

$$||A\mathbf{x}||_2 \leq ||A||_E ||\mathbf{x}||_2.$$

Since $||I||_E = \sqrt{n}$, it follows from Corollary A.8(b) that $|| \cdot ||_E$ is not induced by any vector norm.

A.1.1 Some Special Matrix Norms

We shall now compute the practically important matrix norms, induced by $||\mathbf{x}||_p$, for $p = 1, 2, \infty$.

Theorem A.9 $||A||_\infty = \max_i \sum_j |a_{ij}| =$ *the maximal row sum of $|A|$, where* $|A| = [|a_{ij}|]$.

Proof Since $\|\mathbf{x}\|_\infty = \max_i |x_i|$, we have

$$\|A\mathbf{x}\|_\infty = \max_i |\sum_j a_{ij}x_j| \le \max_i \sum_j |a_{ij}x_j| \le (\max_i \sum_j |a_{ij}|)\|\mathbf{x}\|_\infty,$$

that is, $\|A\|_\infty \le \max_i \sum_j |a_{ij}|$. If now $\max_i \sum_j |a_{ij}| = \sum_j |a_{i_0j}|$ for some row index i_0, we let

$$\hat{\mathbf{x}}^T = [\overline{a}_{i_01}/|a_{i_01}|, \dots, \overline{a}_{i_0n}/|a_{i_0n}|].$$

If $a_{i_0j} = 0$ for some index j, we choose $\hat{x}_j = 1$. Then, $\|\hat{\mathbf{x}}\|_\infty = 1$ and

$$\|A\hat{\mathbf{x}}\|_\infty = \max_i |\sum_j a_{ij}\hat{x}_j| \ge \sum_j |a_{i_0j}|$$

$$= \sum_j |a_{i_0j}| \, \|\hat{\mathbf{x}}\|_\infty = \max_i \sum_j |a_{ij}| \, \|\hat{\mathbf{x}}\|_\infty.$$

Hence, $\|A\|_\infty \ge \max_i \sum_j |a_{ij}|$, which, together with the previous inequality, shows that

$$\|A\|_\infty = \max_i \sum_j |a_{ij}|. \qquad \diamond$$

Theorem A.10 $\|A\|_1 = \max_j \sum_i |a_{ij}| = $ *the maximal column sum of* $|A|$.

Proof If $\|\mathbf{x}\| = \sum_{i=1}^n |x_i|$, then

$$\|A\mathbf{x}\|_1 = \sum_i |\sum_j a_{ij}x_j| \le \sum_j (\sum_i |a_{ij}|)|x_j| \le \max_j \sum_i |a_{ij}| \sum_j |x_j|$$

$$= \max_j \sum_i |a_{ij}| \, \|\mathbf{x}\|_1.$$

If $\max_j \sum_i |a_{ij}| = \sum_i |a_{ij_0}|$, we let $\hat{\mathbf{x}} = [\delta_{1j_0}, \dots, \delta_{nj_0}]$. Then, $\|\hat{\mathbf{x}}\|_1 = 1$ and $\|A\hat{\mathbf{x}}\|_1 = \sum_i |a_{ij_0}|$, that is, $\|A\|_1 = \|A\mathbf{x}\|_1 = \max_j (\sum_i |a_{ij}|)$. $\diamond$

We observe the following relation between the matrix norms induced by the maximum and the absolute sum vector norms: $\|A\|_1 = \|A^T\|_\infty$. Hence, if A is selfadjoint or real symmetric, then $\|A\|_1 = \|A\|_\infty$. The norm $\|A\|_2$, induced by the Euclidian vector norm $\|\mathbf{x}\|_2$, is usually not as easy to compute as $\|A\|_1$ and $\|A\|_\infty$. An important exception is given below.

Theorem A.11 *If* U *is unitary, that is,* $U^*U = I$, *then* $\|U\|_2 = 1$.

Proof $\|\mathbf{x}\|_2 = (\sum |x_i|^2)^{\frac{1}{2}} = (\mathbf{x}, \mathbf{x})^{\frac{1}{2}}$, and

$$\|U\mathbf{x}\|_2^2 = (U\mathbf{x}, U\mathbf{x}) = (\mathbf{x}, U^*U\mathbf{x}) = (\mathbf{x}, \mathbf{x}) = \|\mathbf{x}\|_2^2.$$

Hence,

$$\|U\|_2 = \sup_{\mathbf{x} \neq 0} \frac{\|U\mathbf{x}\|_2}{\|\mathbf{x}\|_2} = 1. \qquad \diamond$$

Example A.3 If

$$A = N_m = \begin{bmatrix} 0 & 1 & 0 & & & \emptyset \\ 0 & 0 & 1 & \ddots & & \\ \vdots & \vdots & & \ddots & & 0 \\ 0 & 0 & & & & 1 \\ 0 & 0 & & & & 0 \end{bmatrix},$$

where N_m has order m, then

$$N_m^2 = \begin{bmatrix} 0 & 0 & 1 & & & \emptyset \\ & & & \ddots & & \\ & & & \ddots & 1 & \\ & & & \ddots & & 0 \\ \emptyset & & & & & 0 \end{bmatrix}, \dots,$$

$$N_m^{m-1} = \begin{bmatrix} 0 & 0 & \dots & 1 \\ 0 & 0 & \dots & 0 \\ & & \dots & \\ 0 & 0 & \dots & 0 \end{bmatrix}, \text{ and } N_m^m = 0.$$

Hence, the minimal equation for N_m is $\lambda^m = 0$, so $\mathcal{S}(N_m) = \{0\}$. (Clearly, this is also seen from the triangular form of N_m.) But,

$$N_m^*N_m = \begin{bmatrix} 0 & 0 & \dots & 0 \\ 0 & 1 & \dots & 0 \\ \vdots & & & \ddots \\ 0 & \dots & & 1 \end{bmatrix},$$

so $\|N_m\|_2 = \sqrt{\rho(N_m^*N_m)} = 1$. Note also that the eigenvector space of N_m is spanned by a single vector, $\mathbf{e}_1^T = (1, 0, \dots, 0)$, but $N_m^*N_m$ has the eigensolutions $\{\mathbf{e}_1, 0\}, \{\mathbf{e}_i, 1\}, i = 2, \dots, n$.

Example A.4

(a) $||A||_2^2 \le ||A||_1 ||A||_\infty$ for an arbitrary square matrix A.

(b) The matrix $A = \begin{bmatrix} 1 & 0 & 1 & 1 \\ 1 & 1 & 0 & 1 \\ 1 & 1 & 1 & 0 \\ 0 & 1 & 1 & 1 \end{bmatrix}$ gives equality in the above relation,

showing that the inequality is sharp not just for the trivial example where A is the identity matrix.

To show (a), let μ_i be the eigenvalues of A^*A. Then,

$$||A||_2^2 = \max_i \mu_i \le ||A^*A||_\infty \le ||A^*||_\infty ||A||_\infty = ||A||_1 ||A||_\infty.$$

To show (b), we note that

$$B = A^*A = \begin{bmatrix} 3 & 2 & 2 & 2 \\ 2 & 3 & 2 & 2 \\ 2 & 2 & 3 & 2 \\ 2 & 2 & 2 & 3 \end{bmatrix} \text{ and } B^2 = \begin{bmatrix} 21 & 20 & 20 & 20 \\ 20 & 21 & 20 & 20 \\ 20 & 20 & 21 & 20 \\ 20 & 20 & 20 & 21 \end{bmatrix}.$$

Hence, $B^2 - 10B + 9I = 0$, but $B + a_1 I \ne 0$ for all scalars a_1. Thus, the minimal polynomial for B is $\lambda^2 - 10\lambda + 9 = 0$, so the eigenvalues of B are $\lambda_1 = 9, \lambda_2 = 1$. Hence, $||A||_2^2 = \max_i \lambda_i = 9$. Since $||A||_1 = 3 = ||A||_\infty$, we have

$$||A||_2^2 = ||A||_1 ||A||_\infty.$$

Theorem A.12 $||A||_2 \le ||A||_E = \sqrt{\text{tr}(A^*A)} \le n^{\frac{1}{2}} ||A||_2$.

Proof Note that by definition,

$$||A||_E = |\sum_i (\sum_j |a_{ij}|^2)|^{\frac{1}{2}}, \text{ so } ||A||_E^2 = \sum_i \sum_j \bar{a}_{ji} a_{ji}.$$

But this equals the sum of the diagonal entries of A^*A, so Theorem 2.2 shows that we have

$$\text{tr}(A^*A) = \sum_{i=1}^{n} \mu_i = ||A||_E^2,$$

where μ_i are the eigenvalues of A^*A. Hence,

$$\max |\mu_i| \leq ||A||_E^2 \leq n \max_i |\mu_i| = n \cdot ||A||_2^2. \qquad \diamond$$

For theoretical work, usually $||A||_2$ is used. In computations, however, frequently $||A||_1$, $||A||_\infty$ and $||A||_E$ are used. We observe that for each of these three norms, $||A|| = ||\,|A|\,||$. (Such a norm is called an *absolute norm*.) On the other hand, the norm $||A||_2$ is not an absolute norm. It follows from Theorem A.12 that

$$\frac{1}{\sqrt{n}}||A||_2 \leq \frac{1}{\sqrt{n}}||A||_E = \frac{1}{\sqrt{n}}||\,|A|\,||_E \leq ||\,|A|\,||_2 \leq ||\,|A|\,||_E \leq \sqrt{n}||A||_2,$$

so

$$\frac{1}{\sqrt{n}}||A||_2 \leq ||\,|A|\,||_2 \leq \sqrt{n}||A||_2.$$

For further properties of and relations between matrix norms, see Exercises A.1 to A.10.

A.2 Inherent Errors in Systems of Linear Algebraic Equations

We shall now apply matrix norms to estimate inherent errors—i.e., errors caused by perturbations of given data. As we shall see, the relative perturbations of data are, at most, amplified by a factor equal to a condition number.

Definition A.4 $\kappa = ||A^{-1}||\,||A||$ is called the *condition number* of A. $\kappa(\mathbf{x}) = ||A^{-1}||(||A\mathbf{x}||/||\mathbf{x}||)$ is called the *condition number of A with respect to the vector* $\mathbf{x}$. Note that $\kappa(\mathbf{x}) \leq \kappa$ and that $\sup_{\mathbf{x}\neq 0}\kappa(\mathbf{x}) = \kappa$. If $||A|| = ||A||_2$, then κ is called the *spectral condition number*.

Note that if A is selfadjoint and positive definite, then $||A||_2||A^{-1}||_2 = \max\lambda(A)/\min\lambda(A)$. Note also that for any $\alpha \neq 0$, $\kappa(\alpha A) = \kappa(A)$ and for any induced matrix norm, we have $\kappa(I) = 1$. Further, because $||A||\,||A^{-1}|| \geq ||AA^{-1}|| = ||I|| = 1$, we have $\kappa(A) \geq 1$.

Theorem A.13 *Let* $\mathbf{z}$ *be the solution of* $A\mathbf{z} = \mathbf{b}$, *where* $\det(A) \neq 0$. *Then, if* $\mathbf{x}$ *is the solution of the perturbed system* $A\mathbf{x} = \mathbf{b} + \underline{\beta}$, *the inherent relative error satisfies:*

$$(a) \quad \frac{||\mathbf{x} - \mathbf{z}||}{||\mathbf{z}||} \leq \kappa(\mathbf{z})\frac{||\underline{\beta}||}{||\mathbf{b}||},$$

(b) $\dfrac{||\mathbf{x} - \mathbf{z}||}{||\mathbf{z}||} \leq \kappa \dfrac{||\underline{\beta}||}{||\mathbf{b}||}.$

Proof Let $\underline{\eta} = \mathbf{x} - \mathbf{z}$. Then, $A\underline{\eta} = \underline{\beta}$ and

$$||\underline{\eta}|| \leq ||A^{-1}||\,||\underline{\beta}||.$$

Hence,

$$\frac{||\underline{\eta}||}{||\mathbf{z}||} \leq ||A^{-1}||\frac{||A\mathbf{z}||}{||\mathbf{z}||} \cdot \frac{||\underline{\beta}||}{||\mathbf{b}||} \leq \kappa(\mathbf{z})\frac{||\underline{\beta}||}{||\mathbf{b}||},$$

which proves part (a). Part (b) follows from the inequality $\kappa(\mathbf{z}) \leq \kappa$. $\diamond$

Theorem A.13(a) shows that the relative error in the right-hand side vector **b** is, at most, amplified by the condition number $\kappa(\mathbf{z})$. *Note that $\kappa(\mathbf{z})$ may be much smaller than κ in many problems.* (For an example, see Exercise A.19). This is so, in particular, if A has eigenvectors $\mathbf{v}_i$ corresponding to eigenvalues λ_i, where $\lambda_i \ll \rho(A)$ and **z** is represented essentially by these eigenvectors. This is typically the case for the matrix A for the difference operator $-\Delta_h^{(5)}$. For a "smooth" vector **z**, we then get $A\mathbf{z} = O(1)$, $h \to 0$, $||A^{-1}|| = O(1)$ so $\kappa(\mathbf{z}) = O(1)$, and

$$||\underline{\eta}||/||\mathbf{z}|| \leq O(1)||\underline{\beta}||/||\mathbf{b}||, h \to 0, \quad \text{while} \quad \kappa = ||A^{-1}||\,||A|| = O(h^{-2}).$$

However, for every A, there are vectors **b** and $\underline{\beta}$ for which we have equality in Theorem A.13(b). (The proof of this is left as an exercise.) We now consider the case where both the matrix and the right-hand side vector may be perturbed by errors.

Theorem A.13′ *Let* **z** *be the solution of the linear system* $A\mathbf{z} = \mathbf{b}$, *where A is nonsingular. If the perturbed system*

$$(A + E)\mathbf{x} = \mathbf{b} + \underline{\beta}$$

has the solution **x**, *then the residual satisfies*

$$||A\mathbf{x} - \mathbf{b}|| \leq ||E||\,||\mathbf{x}|| + ||\underline{\beta}||,$$

and if $||A^{-1}E|| < 1$,

$$\frac{||\mathbf{x} - \mathbf{z}||}{||\mathbf{z}||} \leq \frac{1}{1 - ||A^{-1}E||}\left[||A^{-1}E|| + \kappa(\mathbf{z})\frac{||\underline{\beta}||}{||\mathbf{b}||}\right]$$

$$\leq \frac{\kappa}{1 - \|A^{-1}E\|} \left[\frac{\|E\|}{\|A\|} + \frac{\|\underline{\beta}\|}{\|\mathbf{b}\|} \right],$$

where $\kappa = \|A\| \, \|A^{-1}\|$.

Proof The estimate of the residual follows immediately from the perturbed equation. By subtracting $A\mathbf{z} = \mathbf{b}$ from $(A + E)\mathbf{x} = \mathbf{b} + \underline{\beta}$, we get

$$A(\mathbf{x} - \mathbf{z}) = -E\mathbf{x} + \underline{\beta}$$

and

$$\|\mathbf{x} - \mathbf{z}\| / \|\mathbf{z}\| \leq \|A^{-1}E\| \frac{\|\mathbf{x}\|}{\|\mathbf{z}\|} + \|A^{-1}\| \frac{\|A\mathbf{z}\|}{\|\mathbf{z}\|} \frac{\|\underline{\beta}\|}{\|\mathbf{b}\|},$$

or, with $\|\mathbf{x}\| \leq \|\mathbf{x} - \mathbf{z}\| + \|\mathbf{z}\|$,

$$\|\mathbf{x} - \mathbf{z}\| / \|\mathbf{z}\| \leq (1 - \|A^{-1}E\|)^{-1} \left[\|A^{-1}E\| + \kappa(\mathbf{z}) \frac{\|\underline{\beta}\|}{\|\mathbf{b}\|} \right].$$

Since

$$\|A^{-1}E\| \leq \|A^{-1}\| \, \|E\| = \kappa \frac{\|E\|}{\|A\|}$$

and $\kappa(\mathbf{z}) \leq \kappa$, the proof is complete. $\diamond$

Problems for which $\kappa = \|A\| \, \|A^{-1}\|$ is large are called *ill-conditioned problems*, and problems for which κ is not large are called *well-conditioned*. It is important to be aware that even if we have a well-conditioned problem to solve, a numerically unstable algorithm may ruin the result, as is illustrated by the next example.

Example A.5 Consider $A = \begin{bmatrix} \varepsilon & 1 \\ 1 & 1 \end{bmatrix}$, $0 < \varepsilon \ll 1$, which has the spectral condition number $\kappa \simeq 2.6$ (cf. Example A.10). The LU-decomposition method, without pivoting, is unsuitable for this problem, while the Householder QR-decomposition does not destroy the well-conditioning.

The concept of an "ill-conditioned" matrix is ultimately related to the precision of the computer one works with. With the assumption that the perturbation matrix E due to Gaussian elimination (with pivoting if necessary) satisfies $\|E\| \simeq \varepsilon_M \|A\|$, where ε_M is the machine precision, then

$$\frac{\|\mathbf{x} - \mathbf{z}\|}{\|\mathbf{z}\|} \lesssim \kappa \frac{\|E\|}{\|A\|} = \kappa \varepsilon_M.$$

Thus the matrix/computer combination is well-(ill-)conditioned (for the problem of solving $A\mathbf{x} = \mathbf{b}$) if $\kappa \varepsilon_M$ is small (large).

Example A.6 For the difference operator $-\Delta_h^{(5)}$ on a unit space, we get a matrix with spectral condition number $\kappa = \max(\lambda)/\min(\lambda) \simeq \frac{8}{2(\pi h)^2} = (\frac{2}{\pi h})^2$. Hence, if $h = \frac{1}{64}$, for instance, then $\kappa \simeq 1659$, and this is usually considered a large condition number.

If one uses an elimination method (see Chapter 1) for the solution of $A\mathbf{x} = \mathbf{b}$, then round-off errors occur during the factorization. These can be related to a perturbation E in the matrix A by *backward error analysis* (see Wilkinson, 1965, and Björck and Dahlquist, 1974). As an example, one finds (see Axelsson and Barker, 1984, for instance) that if A is the difference matrix above [for which $\|A\| = O(h^{-2})$ and $\|A^{-1}\| = O(1)$], then $\|E\|_2 = O(h^{-1})$ and, by Theorem A.13, the errors due to round-off during factorization, $\|\mathbf{x} - \mathbf{z}\|_2 = O(h^{-3})$, $h \to 0$. On the other hand, the discretization error is typically $O(h^2)$. Hence, the total error consists of two parts: errors due to round-off, $\|\mathbf{e}_R\| \simeq c_R h^{-3}$, and discretization error, $\|\mathbf{e}_D\| \simeq c_D h^2$, $h \to 0$. We find that there is an optimal value of h beyond which the total error increases when $h \to 0$! (This value is about $h \simeq (3c_R/2c_D)^{1/5}$. If we perform the numerical calculation in multiple precision, then c_R, which is related to the machine precision, will be correspondingly smaller and, thus, also the optimal value of h. The smallest error is $O(c_R^{2/5})$, $c_R \to 0$, that is, we generally can only get about $\frac{2}{5} \log_{10} c_R^{-1}$ correct digits.

As was seen in Chapter 1, the LU-decomposition method is advantageous because it can solve *densely* banded systems with a small computational complexity. However, as the above example illustrates, there might be a severe loss of precision due to the increase of rounding errors. (In the example, we can lose as much as $3/5$ of the number of machine precision digits.)

There is another disadvantage with direct LU-decomposition solution methods: For large problems, the computational cost can be too large. As we saw in Chapter 1, the factorization cost for a bandmatrix with semibandwidth w is about $Nw^2/2$ flops, where N is the order of the matrix. For the difference matrix, we have $w = h^{-1}$ if we use a row-wise (or column-wise) ordering of the nodepoints. The Householder QR factorization method suffers from the same disadvantage.

For a difference matrix associated with a cube (for a three-space-dimensional problem) corresponding to $(-\Delta_h^{(7)})$, the seven-point difference operator, we have $w = h^{-2}$ and $N = h^{-3}$. In general, in an n space-dimensional problem ($n = 1, 2$ or 3), we have $w = h^{-(n-1)}$ and $N = h^{-n}$. Hence, the cost is

$h^{-(3n-2)}/2$ flops. For $h = 1/64 = 2^{-6}$ and $n = 3$, we get $1/2 \cdot 2^{42} \sim 2.10^{12}$ flops, which is a practically impossible computational task. In addition, the demand for computer storage is about $Nw = h^{-(2n-1)}$. Thus, there is a need for more efficient methods. Iterative methods are much more efficient for large sparse matrix problems and, in particular, for difference matrices.

The estimate in Theorem A.13 is, in general, practically useful only if the matrix entries in the error matrix E have about equal magnitude. Similarly the entries of $\underline{\beta}$ should have about equal magnitude, otherwise the estimate may give misleading (too rough) bounds. By scaling the matrix from the right, $\widetilde{A} = AD^{-1}$, or from the left, $\widetilde{A} = D^{-1}A$, where D is a suitable diagonal matrix, one can achieve better scaled entries of A or of **b**. For instance, the matrix A can be scaled so that all columns have equal length, that is,

$$A\mathbf{x} = (AD^{-1})D\mathbf{x} = \widetilde{A}\widetilde{\mathbf{x}},$$

where $D = \mathrm{diag}(\|a_{.1}\|, \ldots, \|a_{.n}\|)$. Frequently $\kappa(\widetilde{A}) \ll \kappa(A)$, so we get much smaller bounds for the perturbations of the solution. Note, however, that scaling the columns changes the norm in which the error in **x** is measured.

Similarly, if the rows in A differ widely in norm, the norm estimates in Theorem A.13 may considerably overestimate the perturbations of the solution. In Exercise A. 39, a more refined analysis based on component-wise bounds of the perturbations will be presented. However, the estimates in Theorem A.13 are appropriate for systems arising from elliptic difference equations, for instance, where it can be shown that a linear system may be well-conditioned w.r.t. its solution, even though the condition number of the matrix is large (cf Exercise A.28).

A.3 Estimation and Computation of Eigenvalues

We next consider the computation of eigenvalues by some simple iterative methods and the estimation of errors in approximations of eigenvalues. The notion of eigenvalues developed historically in connection with solving self-adjoint linear differential operator problems. Physically, eigenvalues present themselves immediately (to the ear) as the natural frequencies of oscillation of a string or a drum, for instance. In a mechanical device they are associated with the occurrence of buckling. The twentieth century example is found in quantum mechanics, where it has been shown that electrons occupy energy states that can be interpreted as eigensolutions of a selfadjoint Schrödinger operator.

Discretizing the above problems to a finite number of degrees of free-doms, using finite differences or finite elements, leads to a matrix, and the eigenvalues of this matrix approximate the eigenvalues of the infinite dimensional problem. Clearly, the practical importance of computing eigenvalues and eigenvectors cannot be overstated.

A.3.1 The Matrix Power Method

Consider the eigenvalue problem $A\mathbf{x} = \lambda\mathbf{x}, \mathbf{x} \neq \mathbf{0}, A \in \mathbb{C}^{n^2}, \mathbf{x} \in \mathbb{C}^n, \lambda \in \mathbb{C}$, which we assume has a complete normalized eigenvector space $\mathbf{v}_1, \mathbf{v}_2, \ldots, \mathbf{v}_n$, spanning $\mathbb{C}^n$, with corresponding eigenvalues $\lambda_1, \lambda_2, \ldots, \lambda_n$, where $|\lambda_1| < |\lambda_j|, j \neq 1, |\lambda_n| > |\lambda_j|, j \neq n$. For the numerical calculation of the eigensolution $\lambda_n, \mathbf{v}_n$, we consider the following method, called the *matrix power method*. We assume that $\mathbf{x}^0$ is a given normalized vector, for which

$$\mathbf{x}^0 = \sum_{j=1}^{n} \alpha_j^{(0)} \mathbf{v}_j,$$

where $\alpha_n^{(0)} \neq 0$. (Actually, in practice it is not necessary that $\alpha_n^{(0)} \neq 0$, because during the iterations to be described, rounding errors are generally introduced, resulting in vectors where the component $\alpha_n \neq 0$ at later stages.)

The normalization can take place with respect to a fixed entry of the vector—with number $j = n_0$ in the middle, for instance—or, better with respect to a vector norm, $\| \cdot \|_p, p = 1, 2$ or ∞. Let

$$\tilde{\mathbf{x}}^k = A\mathbf{x}^{k-1}, \quad \lambda^{(k)} = \|\tilde{\mathbf{x}}^k\|, \quad \mathbf{x}^k = (\lambda^{(k)})^{-1}\tilde{\mathbf{x}}^k, \quad k = 1, 2, \ldots.$$

Note that $\lambda^{(k)}\mathbf{x}^k = A\mathbf{x}^{k-1}$, which shows the relation with $A\mathbf{x} = \lambda\mathbf{x}$. To analyze the convergence of the sequence $\mathbf{x}^k$, we find

$$\mathbf{x}^k = \beta_k^{-1} \sum_{j=1}^{n} \alpha_j^{(0)} A^k \mathbf{v}_j,$$

where $\beta_k = \prod_{j=1}^{k} \lambda^{(j)}$. Hence, $A^k\mathbf{v}_j = \lambda_j^k \mathbf{v}_j$ yields

$$\mathbf{x}^k = \beta_k^{-1} \sum_{j=1}^{n} \alpha_j^{(0)} \lambda_j^k \mathbf{v}_j$$

or

$$\mathbf{x}^k = \beta_k^{-1}\alpha_n^{(0)}\lambda_n^k[\mathbf{v}_n + \sum_{j=1}^{n-1}\frac{\alpha_j^{(0)}}{\alpha_n^{(0)}}(\frac{\lambda_j}{\lambda_n})^k\mathbf{v}_j].$$

Hence, using $|\frac{\lambda_j}{\lambda_n}| = 1 - \frac{|\lambda_n|-|\lambda_j|}{|\lambda_n|}$, we find

$$\|\mathbf{v}_n - (\beta_k^{-1}\alpha_n^{(0)}\lambda_n^k)^{-1}\mathbf{x}^k\|_\infty \le \sum_{j=1}^{n-1}\left|\frac{\alpha_j^{(0)}}{\alpha_n^{(0)}}\right|(1-a_n)^k,$$

where $a_n = \min_{j\ne n}\mid|\lambda_n| - |\lambda_j|\mid/|\lambda_n|$, that is, $\mathbf{v}_n$ is approximated by a scalar times $\mathbf{x}^k$. Since $0 < a_n < 1$, we get convergence: $\mathbf{x}^k \to c_k\mathbf{v}_n$, $k \to \infty$, where $|c_k| = 1$ if $\mathbf{v}_n$ is normalized so that $\|\mathbf{v}_n\| = 1$. (Hence, we do not need to compute the scalar $\beta_k^{-1}\alpha_n^{(0)}\lambda_n^k$.) Furthermore, $\lambda^{(k)} \to \lambda_n$. The rate of convergence depends on a_n, that is, *the relative gap between λ_n and the next large eigenvalue.* If a_n is small, the rate is slow, but if a_n is close to the unit number, the rate is fast.

A.3.2 *The Inverse Power Iteration Method*

We now want to compute the eigensolution $\lambda_1, \mathbf{v}_1$. (In practical problems, this corresponds to the first eigenmode, frequently of most interest.) Consider the following iterative method:

$$A\tilde{\mathbf{x}}^k = \mathbf{x}^{k-1}, \quad \lambda^{(k)} = \|\tilde{\mathbf{x}}^k\|^{-1}, \quad \mathbf{x}^k = \lambda^{(k)}\tilde{\mathbf{x}}^k, \quad k = 1, 2, \ldots,$$

where $\mathbf{x}^0$ is given now with $\alpha_1^{(0)} \ne 0$. Note that here $A\mathbf{x}^k = \lambda^{(k)}\mathbf{x}^{k-1}$.

Analyzing the convergence of the sequence $\mathbf{x}^k$ similarly to the above derivation, we get $\mathbf{x}^k = \lambda^{(k)}A^{-1}\mathbf{x}^{k-1} = \lambda^{(k)}\lambda^{(k-1)}A^{-2}\mathbf{x}^{k-2}, \ldots$, and

$$\mathbf{x}^k = \gamma_k\sum_{j=1}^{n}\alpha_j^{(0)}\lambda_j^{-k}\mathbf{v}_j, \quad \gamma_k = \prod_{j=1}^{k}\lambda^{(j)},$$

or, using $|\frac{\lambda_1}{\lambda_j}| = 1 - \frac{|\lambda_j|-|\lambda_1|}{|\lambda_j|}$, we find

$$\|\mathbf{v}_1 - (\gamma_k\alpha_1^{(0)}\lambda_1^{-k})^{-1}\mathbf{x}^k\|_\infty \le \sum_{j=2}^{n}\left|\frac{\alpha_j^{(0)}}{\alpha_1^{(0)}}\right|(1-a_1)^k \to 0, \quad k \to \infty,$$

where $a_1 = \min_{j \neq 1} \left| \frac{|\lambda_1| - |\lambda_j|}{\lambda_j} \right|$. The drawback of this method is that at every step, we have to solve a linear system. Hence, in general, the method is much more expensive per iteration step than the ordinary power method. However, in practical problems, the relative gap between λ_1 and the rest of the spectrum is frequently quite large, so convergence can be fast.

A.3.3 Shifted Inverse Power Iterations

Consider now the case where we want to calculate an arbitrary eigenvalue λ_j and assume that by some method, we have an estimate λ of λ_j such that

$$|\lambda - \lambda_j| \leq (1 - a)|\lambda - \lambda_k|, \forall k \neq j, \text{ where } 0 < a < 1.$$

Then we may apply the *shifted inverse power iteration method*,

$$(A - \lambda I)\tilde{\mathbf{x}}^k = \mathbf{x}^{k-1}, \quad v_k = \|\tilde{\mathbf{x}}^k\|^{-1}, \quad \mathbf{x}^k = v_k \tilde{\mathbf{x}}^k, \quad k = 1, 2, \ldots.$$

Note that $(A - \lambda I)\mathbf{x}^k = v_k \mathbf{x}^{k-1}$ so $\lambda_j^{(k)} = \lambda \pm v_k$ where the sign is chosen such that some (properly chosen) component of $\mathbf{x}^k$ and $\mathbf{x}^{k-1}$ has the same sign.

The rate of convergence of this method is analyzed as above and is determined by the number a. If λ is much closer to λ_j than to any other eigenvalue, a is quite close to 1, which means that this method can converge very fast. The drawback of the method is that the matrix $A - \lambda I$ may become very close to being singular. Note, also, that $A - \lambda I$ is indefinite in general. The numerical solution of the linear systems with this matrix therefore must be performed with care, so that the influence of rounding errors, for instance, is under control. For a discussion about related questions, see Wilkinson (1965) and Parlett (1980). To solve the indefinite systems, Aasen's stable factorization method (see Exercise 31 in Chapter 1) or a similar method can be a good choice.

To increase the rate of convergence of the shifted inverse power iteration method further, we may use variable shifts,

$$(A - \lambda^{(k-1)} I)\tilde{\mathbf{x}}^k = \mathbf{x}^{k-1}, \quad v_k = \|\tilde{\mathbf{x}}^k\|^{-1},$$

$$\mathbf{x}^k = v_k \tilde{\mathbf{x}}^k, \quad \lambda^{(k)} = \lambda^{(k-1)} \pm v_k, \quad k = 1, 2, \ldots.$$

As $\lambda^{(k)}$ approaches the actual eigenvalue λ_j, the rate of convergence of this method will increase dramatically. We make the following observations:

(a) The matrix becomes closer and closer to being singular.
(b) If we use a LU- or a QR-factorization method (see Chapter 1) to solve the systems with matrix $(A - \lambda^{(k-1)} I)$, then we must refactorize the

matrix at every iteration, whereas in the previous method we need only make one factorization because the matrix is the same at every solution.

With respect to (a), note that although the condition number of $A - \lambda I$ is very large, typically $\kappa(A - \lambda I) = O(\max_i |\lambda_i - \lambda| / |\lambda_j - \lambda|)$, which becomes larger as $\lambda \to \lambda_j$, the condition number of A *with respect to the vector* $\mathbf{v}_j$ is not large. To see this, consider, for simplicity, a symmetric matrix A. Then, if $\lambda \neq \lambda_j$,

$$\kappa(\mathbf{v}_j) = \|(A - \lambda I)^{-1}\| \ \|(A - \lambda I)\mathbf{v}_j\| / \|\mathbf{v}_j\|$$

$$= |\lambda_j - \lambda|^{-1} |\lambda_j - \lambda| = 1.$$

Hence, as $\mathbf{x}^k \to \mathbf{v}_j$, the condition numbers of the linear systems with respect to their solutions $\tilde{\mathbf{x}}^k$ converge to the optimal value 1. Therefore, by Theorem A.13, errors (such as round-off) in the right-hand side are not amplified— i.e., *the inherent errors remain small.* (Recall that round-off errors can be treated as errors in data by backward analysis.)

Remark A.14 (Simultaneous iteration method) In practice, one frequently applies the power method simultaneously to a set of vectors and, hence, approximates a corresponding set of eigensolutions. In the inverse method, we then get

$$A \widetilde{X}^k = X^{k-1}, \quad k = 1, 2, \ldots,$$

where the (r) columns of X^{k-1} are normalized according to some rule. We then normalize $\widetilde{X}^k$ accordingly, and the method is iterated. We now get approximations of the first (r) eigensolutions, but the ones at the upper end are more accurate than the ones at the lower end (assuming that the eigenvalues are positive). We leave it to the reader to make a Fourier analysis of the rate of convergence of this *simultaneous iteration method.*

A.3.4 Perturbation Theory for Eigenvalues

We shall now give a bound for the changes in the eigenvalues that may occur when the entries of the matrix are perturbed. In this way, we also bound the inherent errors of the eigenvalues—that is, those errors which come from inexact given data.

We consider only the case where A is similarly equivalent to a diagonal matrix. Then there exists a matrix P, such that $P^{-1}AP = \text{diag}(\lambda_i) = \Lambda$, where

$\{\lambda_i\} = S(A)$. We note that P is not unique, because if, for instance, D is a diagonal matrix, then

$$(PD)^{-1}APD = D^{-1}P^{-1}APD = D^{-1}\Lambda D = \Lambda.$$

Hence, PD also transforms A to diagonal form.

Theorem A.15 (Bauer-Fike) *Let λ be an eigenvalue of the matrix $A + E$, where E is a perturbation of A and A is diagonalizable. Then, for the eigenvalues λ_i of A we find that*

$$\min_i |\lambda - \lambda_i| \le \|P^{-1}\|_p \|E\|_p \|P\|_p$$

for any natural norm defined by the Hölder vector norm $\| \cdot \|_p$.

Proof If $\lambda = \lambda_i$, some i, this is trivial. Hence, let $\lambda \ne \lambda_i$, all i. Let P transform A into diagonal form. From $(A + E)\mathbf{x} = \lambda\mathbf{x}$, it follows that

$$P^{-1}(\lambda I - A)PP^{-1}\mathbf{x} = (\lambda I - \Lambda)P^{-1}\mathbf{x} = P^{-1}EPP^{-1}\mathbf{x}$$

or

$$P^{-1}\mathbf{x} = (\lambda I - \Lambda)^{-1}P^{-1}EPP^{-1}\mathbf{x}.$$

Hence,

$$\|P^{-1}\mathbf{x}\|_p \le \|(\lambda I - \Lambda)^{-1}\|_p \|P^{-1}EP\|_p \|P^{-1}\mathbf{x}\|_p,$$

so

$$\frac{1}{\|(\lambda I - \Lambda)^{-1}\|_p} \le \|P^{-1}\|_p \|E\|_p \|P\|_p.$$

It is readily seen that for a diagonal matrix D, $\|D\|_p = \max_i |d_{ii}|$. This proves the theorem.   ◇

If we calculate $\hat{\kappa} = \min \|P^{-1}\|_p \|P\|_p$ over all matrices that transform A to diagonal form, we get a condition number for the inherent errors,

$$\min_i |\lambda - \lambda_i| \le \hat{\kappa}\|E\|_p.$$

For symmetric or Hermitian matrices, the condition number factor $\hat{\kappa}$ is 1, as is shown below.

Corollary A.16 *If A is selfadjoint, then*

$$\min_i |\lambda - \lambda_i| \leq \|E\|_2.$$

Proof If $A = A^*$, there is a unitary matrix U that transforms A to diagonal form. For a unitary matrix, $\|U^*\|_2 = \|U\|_2 = 1$ and $U^{-1} = U^*$.   ◇

Hence, the *eigenvalue problem* for a selfadjoint matrix is well-conditioned.

A.3.5 A Posteriori Bounds

Let $\{\lambda, \mathbf{x}\}$ be an approximation of an eigensolution of A. We would like to use the knowledge of the norm of residual vector $\underline{\eta} = A\mathbf{x} - \lambda\mathbf{x}$ in order to get a bound of the error in λ.

Theorem A.17 *Let λ be a given number, $\mathbf{x}$ be a normalized vector, and $\|\mathbf{x}\|_2 = 1$. Let A be an Hermitian matrix, and let $\|A\mathbf{x} - \lambda\mathbf{x}\|_2 = \varepsilon$. Then, there is at least one eigenvalue λ_i of A, such that*

$$|\lambda_i - \lambda| \leq \varepsilon.$$

Proof Let $\mathbf{v}_i$ be orthonormal eigenvectors of A, that is, $(\mathbf{v}_i, \mathbf{v}_j) = \delta_{ij}$, and let $\mathbf{x} = \sum_{i=1}^n \alpha_i \mathbf{v}_i$. Then,

$$1 = (\mathbf{x}, \mathbf{x}) = \sum_{i=1}^n |\alpha_i|^2, \quad \underline{\eta} = A\mathbf{x} - \lambda\mathbf{x} = \sum_{i=1}^n (\lambda_i - \lambda)\alpha_i \mathbf{v}_i$$

and

$$\varepsilon^2 = (\underline{\eta}, \underline{\eta}) = \sum_{i=1}^n |\lambda_i - \lambda|^2 |\alpha_i|^2 \geq \min_i |\lambda_i - \lambda|^2,$$

which proves the theorem.   ◇

Remark A.18 As an alternative to the above constructive proof, note that $A\mathbf{x} - \lambda\mathbf{x} = \varepsilon E\mathbf{x}$ for some matrix E with $\|E\mathbf{x}\|_2 = 1$ and $\|E\|_2 = 1$. Corollary A.16 shows that $\min_i |\lambda_i - \lambda| \leq \varepsilon\|E\|_2 = \varepsilon$.

While the eigenvalue problem is well-conditioned for symmetric matrices, the eigenvector problem can be arbitrarily ill-conditioned, as the following example shows.

Example A.7 Let $A = \begin{bmatrix} b & \varepsilon \\ \varepsilon & b \end{bmatrix}$, $\varepsilon \neq 0$, and $\lambda = b$, $\mathbf{x} = [\begin{smallmatrix} 1 \\ 0 \end{smallmatrix}]$, where $|\varepsilon| \ll |b|$.

Then, $A\mathbf{x} - \lambda\mathbf{x} = [\begin{smallmatrix} 0 \\ \varepsilon \end{smallmatrix}]$. Hence, by Theorem A.17 there is at least one eigenvalue λ_j, such that $|\lambda - \lambda_j| \leq \varepsilon$. The eigenvalues of A are, in fact, $\lambda_1 = b + \varepsilon$ and $\lambda_2 = b - \varepsilon$. A similar estimate for $\mathbf{x}$ is not possible; the correct eigenvectors are $\mathbf{v}_1 = \frac{1}{\sqrt{2}}[\begin{smallmatrix} 1 \\ 1 \end{smallmatrix}]$, $\mathbf{v}_2 = \frac{1}{\sqrt{2}}[\begin{smallmatrix} 1 \\ -1 \end{smallmatrix}]$ and $\|\mathbf{x} - \mathbf{v}_i\| = 2 - \sqrt{2}$. In fact, it is readily seen that for *any* unit vector $\mathbf{x} = (x_1, x_2)^T$, we get $A\mathbf{x} - \lambda\mathbf{x} = \varepsilon(x_2, x_1)^T$, that is, $\|A\mathbf{x} - \lambda\mathbf{x}\|_2 = \varepsilon$. Thus, there was nothing special with the above choice $\mathbf{x} = (1, 0)^T$.

Thus, the *eigenvector problem* may be ill-conditioned, even for a selfadjoint matrix. In the above example, the eigenvalues become arbitrary close together, as $\varepsilon \to 0$. If, however, we assume that the eigenvalues are well-separated from each other, we can also derive an error bound for $\mathbf{x}$, as the next theorem shows.

Theorem A.19 *Let A be a selfadjoint matrix. Assume that $\{\lambda, \mathbf{x}\}$ is an approximate eigensolution, such that $\|\mathbf{x}\|_2 = 1$ and $\|\underline{\eta}\|_2 = \varepsilon$, where $\underline{\eta} = A\mathbf{x} - \lambda\mathbf{x}$, and let λ_1 be an eigenvalue to A, closest to λ satisfying $|\lambda_1 - \lambda| \leq \varepsilon$. Assume, further, that $|\lambda_1 - \lambda_i| \geq a + \varepsilon$, $i \geq 2$, where $a > \varepsilon$. Then,*

$$\|\mathbf{x} - \mathbf{v}_1\|_2 \leq \left(\frac{\varepsilon}{a}\right)\left[\frac{1 - ((\lambda_1 - \lambda)/\varepsilon)^2}{1 - ((\lambda_1 - \lambda)/a)^2} + \left(\frac{\varepsilon}{a}\right)^2\right]^{\frac{1}{2}} \leq \frac{\varepsilon}{a}\left[1 + \left(\frac{\varepsilon}{a}\right)^2\right]^{\frac{1}{2}}.$$

Proof As in the proof of Theorem A.17, we let $\mathbf{x} = \sum_{i=1}^{n} \alpha_i \mathbf{v}_i$, which yields

$$\varepsilon^2 = \|\underline{\eta}\|_2^2 = \sum_{i=1}^{n} |\alpha_i|^2 |\lambda_i - \lambda|^2 \geq |\alpha_1|^2 |\lambda_1 - \lambda|^2 + a^2 \sum_{i=2}^{n} |\alpha_i|^2$$

or

$$\sum_{i=2}^{n} |\alpha_i|^2 \leq \left(\frac{\varepsilon}{a}\right)^2 - |\alpha_1|^2 \left(\frac{\lambda_1 - \lambda}{a}\right)^2,$$

because

$$|\lambda_i - \lambda| \geq |\lambda_i - \lambda_1| - |\lambda_1 - \lambda| \geq a.$$

Hence, since $\sum |\alpha_i|^2 = 1$, we have

$$1 = |\alpha_1|^2 + \sum_{i=2}^{n} |\alpha_i|^2 \leq |\alpha_1|^2\left[1 - \left(\frac{\lambda_1 - \lambda}{a}\right)^2\right] + \left(\frac{\varepsilon}{a}\right)^2,$$

that is,

$$\frac{1 - (\varepsilon/a)^2}{1 - [(\lambda_1 - \lambda)/a]^2} \leq |\alpha_1|^2 \leq 1$$

and

$$\sum_{i=2}^{n} |\alpha_i|^2 \leq \left(\frac{\varepsilon}{a}\right)^2 - \frac{1 - (\varepsilon/a)^2}{1 - [(\lambda_1 - \lambda)/a]^2} \left(\frac{\lambda_1 - \lambda}{a}\right)^2$$

$$= \left(\frac{\varepsilon}{a}\right)^2 \frac{1 - [(\lambda_1 - \lambda)/\varepsilon]^2}{1 - [(\lambda_1 - \lambda)/a]^2}.$$

Further, we have

$$(1 - \alpha_1)^2 = [(1 - \alpha_1^2)/(1 + \alpha_1)]^2 \leq (\varepsilon/a)^4.$$

Note now that it is no limitation to assume that $\alpha_1 > 0$. [We may multiply $\mathbf{v}_1$ by (-1) if necessary.] Hence, we have the following estimate of the error in $\mathbf{x}$:

$$\|\mathbf{x} - \mathbf{v}_1\|_2^2 = (1 - \alpha_1)^2 + \sum_{i=2}^{n} |\alpha_i|^2 \leq \left(\frac{\varepsilon}{a}\right)^2 \frac{1 - [(\lambda_1 - \lambda)/\varepsilon]^2}{1 - [(\lambda_1 - \lambda)/a]^2} + \left(\frac{\varepsilon}{a}\right)^4.$$

And, since $a > \varepsilon$,

$$\|\mathbf{x} - \mathbf{v}_1\|_2^2 \leq \left(\frac{\varepsilon}{a}\right)^2 + \left(\frac{\varepsilon}{a}\right)^4. \qquad \diamond$$

A similar theorem follows.

Theorem A.20 (Kahan, 1967) *Let A be selfadjoint, let $\mathbf{x}$ be a unit vector, and let λ be given. Further, let λ_1 be the closest eigenvalue of A to λ, let $\mathbf{v}_1$ be its normalized eigenvector, and, let the eigenvalue gap be $a = \min_{i \neq 1} |\lambda_i(A) - \lambda|$. Let θ be the angle between $\mathbf{x}$ and $\mathbf{v}_1$. Then,*

$$|\sin \theta| \leq \varepsilon/a,$$

where $\varepsilon = \|A\mathbf{x} - \lambda\mathbf{x}\|_2$.

Proof (Parlett, 1980) If $\mathbf{x}$ is parallel to $\mathbf{v}_1$, then the statement is trivial. If $\mathbf{x} \neq \mathbf{v}_1$, then decompose $\mathbf{x}$ as $\mathbf{x} = \mathbf{v}_1 \cos \theta + \mathbf{w} \sin \theta$, where $\mathbf{w}$ is the unit vector in the $\mathbf{x}, \mathbf{v}_1$ plane, orthogonal to $\mathbf{v}_1$. Then,

$$A\mathbf{x} - \lambda\mathbf{x} = (A - \lambda I)\mathbf{v}_1 \cos \theta + (A - \lambda I)\mathbf{w} \sin \theta$$

$$= (\lambda_1 - \lambda)\mathbf{v}_1 \cos \theta + (A - \lambda I)\mathbf{w} \sin \theta,$$

since $A\mathbf{v}_1 = \lambda_1\mathbf{v}_1$. Further, $\mathbf{v}_1^T(A - \lambda I)\mathbf{w} = (\lambda_1 - \lambda)\mathbf{v}_1^T\mathbf{w} = 0$, so, by Pythagoras Lemma (Exercise A.4),

$$\|A\mathbf{x} - \lambda\mathbf{x}\|^2 = (\lambda_1 - \lambda)^2\cos^2\theta + \|(A - \lambda I)\mathbf{w}\|^2\sin^2\theta.$$

But, $\inf_{\mathbf{w}\perp\mathbf{v}_1}\|(A - \lambda I)\mathbf{w}\| = \min_{i\neq1}|\lambda_i - \lambda|$ (Fischer's theorem). Hence, $\sin^2\theta \le \|A\mathbf{x} - \lambda\mathbf{x}\|^2/\min_{i\neq1}|\lambda_i - \lambda|^2.$ ◇

The theorem shows that the angle between $\mathbf{x}$ and $\mathbf{v}_1$ is small only if $\varepsilon \ll a$. Note that if $a \simeq \varepsilon$, this bound is useless. If, however, $a \gg \varepsilon$, then we get

$$\|\mathbf{x} - \mathbf{v}_1\|_2 \lesssim \frac{\varepsilon}{a}.$$

Note also that Theorem A.19 shows that if $|\lambda_1 - \lambda| = \varepsilon$, then $\|\mathbf{x} - \mathbf{v}_1\|_2 \le (\frac{\varepsilon}{a})^2$. (In fact, it is easy to see that then $\|\mathbf{x} - \mathbf{v}_1\| = 0$.)

We have seen that for symmetric matrices, the eigenvalue problem is always well-conditioned with respect to perturbations but the eigenvector problem can be ill-conditioned. For nonsymmetric matrices, the estimate in Theorem A.15 involves the condition number of the transformation matrix P, and this can be large; hence, a nonsymmetric matrix can have ill-conditioned eigenvalues. Further, it can even occur that an eigenvector is well approximated but that the corresponding eigenvalue is not. This is illustrated by the next example.

Example A.7′ Let $A = \begin{bmatrix} (2 - 10^{-10}) & -10^{10} \\ 0 & 2 \end{bmatrix}$ and let $\lambda = 1, \mathbf{x} = \begin{bmatrix} 1 \\ 10^{-10} \end{bmatrix}$.
Then, $\|A\mathbf{x} - \lambda\mathbf{x}\|_2 = \sqrt{2} \cdot 10^{-10}$, that is, a small number, but the exact eigenvalues are $2 - 10^{-10}$ and 2. (The eigenvectors of A are $[1, 0]^T$ and $[0, 1]^T$, respectively.)

A.3.6 Rayleigh Quotient

In the case of selfadjoint matrices, we may compute a better approximation of any eigenvalue λ_1 by use of the Rayleigh quotient. For properties of the Rayleigh quotient, see Chapter 3.

Theorem A.21 *Let A be a selfadjoint matrix and let $\mathbf{x}$, where $(\mathbf{x}, \mathbf{x}) = 1$, be an approximation of the eigenvector $\mathbf{v}_1$ corresponding to an eigenvalue λ_1. Consider $f(\lambda) = \|\underline{\eta}(\lambda)\|^2 = \|A\mathbf{x} - \lambda\mathbf{x}\|^2, \lambda \in \mathbb{R}$. Then, $f(\lambda)$ takes its minimum value for the Rayleigh quotient,*

$$\lambda^{(0)} = (A\mathbf{x}, \mathbf{x})/(\mathbf{x}, \mathbf{x}).$$

Further, let $\varepsilon = \|A\mathbf{x} - \lambda^{(0)}\mathbf{x}\|_2$ *and assume that* $|\lambda^{(0)} - \lambda_1| \leq \varepsilon$ *(Theorem A.17) and* $|\lambda_1 - \lambda_i| \geq a + \varepsilon,\ i = 2, 3, \ldots, n,$ *where* $a > \varepsilon$. *Then,*

$$|\lambda_1 - \lambda^{(0)}| \leq \frac{\varepsilon^2}{a} \Big/ \left[1 - \frac{\delta}{1 + \sqrt{1 - 2\delta}} \right],$$

where

$$\delta = 2 \left(\frac{\varepsilon}{a}\right)^2 \left[1 - \left(\frac{\varepsilon}{a}\right)^2 \right] \Big/ \left[1 + \left(\frac{\varepsilon}{a}\right)^2 \right].$$

Proof Let $\eta(\lambda) = A\mathbf{x} - \lambda\mathbf{x}$ and consider $(\underline{\eta}, \underline{\eta})$ as a function of λ for fixed $\mathbf{x}$. Since A is selfadjoint, we let $\lambda \in \mathbb{R}$. We want to find the value λ for which $(\underline{\eta}, \underline{\eta})$ takes its minimum. We have

$$f(\lambda) = (\underline{\eta}(\lambda), \underline{\eta}(\lambda)) = (A\mathbf{x} - \lambda\mathbf{x}, A\mathbf{x} - \lambda\mathbf{x})$$

$$= \|A\mathbf{x}\|_2^2 - 2\lambda(A\mathbf{x}, \mathbf{x}) + \lambda^2(\mathbf{x}, \mathbf{x}),$$

which takes its minimum for $f'(\lambda) = 0$, that is,

$$\lambda^{(0)} = (A\mathbf{x}, \mathbf{x})/(\mathbf{x}, \mathbf{x}).$$

We have

$$\varepsilon = \|\underline{\eta}(\lambda^{(0)})\|_2.$$

Theorem A.17 shows that there exists an eigenvalue λ_1, such that $|\lambda^{(0)} - \lambda_1| \leq \varepsilon$. Further,

$$|\lambda^{(0)} - \lambda_i| = |\lambda_1 - \lambda_i - (\lambda_1 - \lambda^{(0)})| \geq a.$$

We have

$$\lambda^{(0)} = (A\mathbf{x}, \mathbf{x})/(\mathbf{x}, \mathbf{x}) = \sum_{i=1}^{n} \lambda_i |\alpha_i|^2,$$

where $\mathbf{x} = \sum_{i=1}^{n} \alpha_i \mathbf{v}_i$ and $(\mathbf{x}, \mathbf{x}) = \sum_{i=1}^{n} |\alpha_i|^2 = 1$. Hence,

$$\lambda^{(0)} \sum_{i=1}^{n} |\alpha_i|^2 = \lambda^{(0)}(\mathbf{x}, \mathbf{x}) = (A\mathbf{x}, \mathbf{x}) = \sum_{i=1}^{n} \lambda_i |\alpha_i|^2,$$

that is,

$$(\lambda^{(0)} - \lambda_1)|\alpha_1|^2 = \sum_{i=2}^{n}(\lambda_i - \lambda^{(0)})|\alpha_i|^2$$

and

$$|\lambda^{(0)} - \lambda_1| \, |\alpha_1|^2 \leq \sum_{i=2}^{n}|\lambda_i - \lambda^{(0)}| \, |\alpha_i|^2$$

$$= \sum_{i=2}^{n}\frac{1}{|\lambda_i - \lambda^{(0)}|}|\lambda_i - \lambda^{(0)}|^2|\alpha_i|^2$$

$$\leq \frac{1}{a}||A\mathbf{x} - \lambda^{(0)}\mathbf{x}||_2^2 \leq \frac{1}{a}||\underline{\eta}(\lambda^{(0)})||_2^2.$$

Hence, we get

$$|\lambda^{(0)} - \lambda_1| \leq \frac{\varepsilon^2}{a} \, \frac{1 - [(\lambda_1 - \lambda^{(0)})/a]^2}{1 - (\varepsilon/a)^2}$$

or

$$\left|\frac{\lambda_1 - \lambda^{(0)}}{a}\right|^2 + \left|\frac{\lambda_1 - \lambda^{(0)}}{a}\right|\left[\left(\frac{a}{\varepsilon}\right)^2 - 1\right] \leq 1,$$

that is,

$$|\lambda_1 - \lambda^{(0)}| \leq \frac{2\varepsilon^2/a}{1 - (\frac{\varepsilon}{a})^2 + \sqrt{[1 - (\frac{\varepsilon}{a})^2]^2 + 4(\frac{\varepsilon}{a})^4}}.$$

An elementary computation now completes the proof.   ◇

Remark A.22

(a) A sharper estimate of $|\lambda_1 - \lambda^{(0)}|$ can be found in Exercise A.35.

(b) Note that if $a \gg \varepsilon$, then we have an error bound $O(\varepsilon^2)$. If, however, $a \simeq \varepsilon$, that is, we have close eigenvalues, then the error bound in Theorem A.21 is no improvement over the simple estimate in Theorem A.17 (that is, $|\lambda_1 - \lambda^{(0)}| \leq \varepsilon$). For multiple eigenvalues, however, the accurate estimate is applicable. Hence, the difficult case is close eigenvalues.

A.3.7 A Combination of the Rayleigh Quotient and Matrix Power Iteration Methods

The Rayleigh quotient method may be combined with the matrix power method. Let x^0 be given, then compute

$$\tilde{x}^k = Ax^{k-1}, \quad x^k = \|\tilde{x}^k\|_2^{-1}\tilde{x}^k, \quad \lambda^{(k)} = (Ax^k, x^k), \quad k = 1, 2, \ldots.$$

Note that the computation of $\lambda^{(k)}$ costs only one extra inner product and no extra matrix-vector multiplication, because $\lambda^{(k)} = (\tilde{x}^{k+1}, x^k)$.

The combination of the Rayleigh quotient with the shifted inverse iteration method, as follows, can speed up the rate of convergence. Let x^0 be a given, normalized ($\|x^0\|_2 = 1$) vector, and for $k = 0, 1, 2, \ldots$:

(a) Compute $\lambda^{(k)} = (Ax^k, x^k)$.

(b) If $A - \lambda^{(k)}I$ is singular, that is, $\lambda^{(k)}$ is already an eigenvalue, then solve the homogeneous system

$$(A - \lambda^{(k)}I)x^{k+1} = 0$$

for a unit vector x^{k+1} and stop. Otherwise, solve the equation

$$(A - \lambda^{(k)}I)\tilde{x}^{k+1} = x^k$$

for $\tilde{x}^{k+1}$.

(c) Normalize, that is, let

$$x^{k+1} = \|\tilde{x}^{k+1}\|_2^{-1}\tilde{x}^{k+1}.$$

(d) If $\|\tilde{x}^{k+1}\|_2 \geq \varepsilon^{-1}$ or if $|\lambda^{(k)} - \lambda^{(k-1)}| \leq \varepsilon^2$, then stop ($\varepsilon$ is a prescribed tolerance). Note that $\|(A - \lambda^{(k)}I)x^{k+1}\|_2 = \|\tilde{x}^{k+1}\|_2^{-1}\|x^k\|_2 \leq \varepsilon$, if $\|\tilde{x}^{k+1}\|_2 \geq \varepsilon^{-1}$.

Remark A.23 In the above algorithm, there is no guarantee that $\lambda^{(k)}$ converges to the eigenvalue closest to $\lambda^{(0)}$. For more information about the algorithm, see Parlett (1980).

Remark A.24 The numerical computation of the Rayleigh quotient is best performed with a *defect correction* method as follows. Let $\{\lambda, x\}$ be an approximation of the eigensolution $\{\lambda_1, v_1\}$, and let $\underline{\eta} = Ax - \lambda x$. Then,

$$\lambda^{(0)} = (Ax, x)/(x, x) = (\eta + \lambda x, x)/(x, x) = \lambda + (\underline{\eta}, x)/(x, x).$$

Here $(\eta, \mathbf{x})$ is a small number. If we compute η exactly (without rounding errors), then we can compute $\lambda^{(0)}$ to almost "double precision," even if we perform the inner products in "single precision." To see how η can be computed with negligible errors in some problems, see Exercise A.28.

Example A.8

(a) Use the power method to calculate the largest eigenvalue and corresponding eigenvector of $A = \begin{bmatrix} 2 & -1 & 0 \\ -1 & 2 & -1 \\ 0 & -1 & 2 \end{bmatrix}$, with $\mathbf{x}^0 = (1, 1, 1)^T$.

One gets:

$$\tilde{\mathbf{x}}^1 = A\mathbf{x}^0 = (1, 0, 1)^T, \quad \lambda^{(1)} = 1, \quad \mathbf{x}^1 = \tilde{\mathbf{x}}^1$$

$$\tilde{\mathbf{x}}^2 = A\mathbf{x}^1 = (2, -2, 2)^T, \quad \lambda^{(2)} = 2, \quad \mathbf{x}^2 = (1, -1, 1)^T$$

$$\tilde{\mathbf{x}}^3 = A\mathbf{x}^2 = (3, -4, 3)^T, \quad \lambda^{(3)} = 4, \quad \mathbf{x}^3 = (\frac{3}{4}, -1, \frac{3}{4})^T$$

$$\tilde{\mathbf{x}}^4 = A\mathbf{x}^3 = (\frac{5}{2}, -\frac{7}{2}, \frac{5}{2})^T, \quad \lambda^{(4)} = \frac{7}{2}, \quad \mathbf{x}^4 = (\frac{5}{7}, -1, \frac{5}{7})^T$$

$$\tilde{\mathbf{x}}^5 = A\mathbf{x}^4 = (\frac{17}{7}, -\frac{24}{7}, \frac{17}{7})^T, \quad \lambda^{(5)} = \frac{24}{7} \simeq 3.43,$$

$$\mathbf{x}^5 = (\frac{17}{24}, -1, \frac{17}{24})^T$$

$$\tilde{\mathbf{x}}^6 = A\mathbf{x}^5 = (\frac{29}{12}, -\frac{41}{12}, \frac{29}{12})^T, \quad \lambda^{(6)} = \frac{41}{12} \simeq 3.417,$$

$$\mathbf{x}^6 = (\frac{29}{41}, -1, \frac{29}{41})^T.$$

(Normalization has taken place with respect to the second component, leaving it equal to -1 from the second iterate onward.)

It is known that $\lambda_3 = 2(1 - \cos 3\pi/4) = 2 + \sqrt{2} \simeq 3.414214$ and $\mathbf{v}_3 = (\frac{\sqrt{2}}{2}, -1, \frac{\sqrt{2}}{2})^T$ ($\frac{\sqrt{2}}{2} \simeq 0.7071$, $\frac{29}{41} \simeq 0.7073$). The relative distance $\frac{\lambda_3 - \lambda_2}{\lambda_3} = \frac{\sqrt{2}}{2+\sqrt{2}} = 0.414$ is fairly large; the iteration error decreases with about the factor 0.59 for every iteration.

(b) Use the shifted inverse power method to calculate an approximation of the eigensolution λ_3, $\mathbf{v}_3$, starting with $\lambda = \frac{7}{2}$, $\mathbf{x}^4 = (\frac{5}{7}, -1, \frac{5}{7})^T$. We get

$$(A - \lambda I)\tilde{x} = x^4$$

or

$$\tilde{x} = \frac{1}{7}(-58, +82, -58)^T, \lambda_{new} - \lambda = -\frac{7}{82}$$

or

$$\lambda_{new} = \frac{140}{41} \simeq 3.4146, x_{new} = (\frac{29}{41}, -1, \frac{29}{41}).$$

(c) Finally, using the Rayleigh quotient method, we find with $x = x^4 = (\frac{5}{7}, -1, \frac{5}{7})^T$ that

$$\lambda = \frac{(Ax^4, x^4)}{(x^4, x^4)} = \frac{338}{99} \simeq 3.41414.$$

The Rayleigh quotient, in fact, gives an underestimate of the largest eigenvalue and an overestimate of the smallest eigenvalue. This follows because for symmetric matrices, $\lambda_n = \max_{x \neq 0} \frac{(Ax,x)}{(x,x)}$ and $\lambda_1 = \min_{x \neq 0} \frac{(Ax,x)}{(x,x)}$.

Example A.9 With

$$A = \begin{bmatrix} 0.63521 & 0.21573 & 0.12464 \\ 0.21573 & 0.31576 & 0.41825 \\ 0.12463 & 0.41825 & 0.21357 \end{bmatrix},$$

the approximate eigensolutions

$$\mu_1 = 0.90712, \quad \mu_2 = 0.41752, \quad \mu_3 = -0.16010$$

x_1	x_2	x_3
0.67093	0.73837	0.06836
0.57549	−0.46035	−0.67594
0.46763	−0.49285	0.73378

are given. Without rounding errors, we have

$10^5 \underline{\eta}_1$	$10^5 \underline{\eta}_2$	$10^5 \underline{\eta}_3$
−0.13917	0.05643	−0.21432
0.42100	−0.07364	−0.20206
−0.80881	−0.15769	−0.06256

where $\underline{\eta}_i = A\mathbf{x}_i - \mu_i\mathbf{x}_i$. We get to single precision (5 digits),

$$\frac{10^5(\underline{\eta}_i, \mathbf{x}_i)}{(\mathbf{x}_i, \mathbf{x}_i)} = -0.22931, \ 0.15328, \ 0.07602, \ i = 1, 2, 3.$$

Hence, the approximations we get by the Rayleigh quotient, using the defect correction method, are

$$\lambda_{01} = 0.90711\ 77069, \quad \lambda_{02} = 0.41752\ 15328, \quad \lambda_{03} = -0.16009\ 92398,$$

while the *exact* eigenvalues are

$$0.90711\ 7706949, \quad 0.41752\ 1532818, \quad -0.16009\ 9239766$$

and the first eigenvector is

$$[0.670926, \quad 0.575491, \quad 0.467620].$$

Note that all $\underline{\eta}_i$ satisfy $\|\underline{\eta}_i\|_2 < 10^{-5}$. Hence, by Theorem A.17, we have $|\lambda_i - \mu_i| < 10^{-5}$, which implies that λ_i is contained in a ball about μ_i with radius 10^{-5}. Hence, all eigenvalues are different and $|\lambda_i - \lambda_j| > 0.4$, $i \neq j$. With $\varepsilon = 10^{-5}$, $a = 0.4$, Theorems A.21 and A.19 imply that $|\lambda_{0i} - \lambda_i| < 2.5 \cdot 10^{-10}$ and $\|\mathbf{x}_i - \mathbf{v}_i\|_2 < 2.5 \cdot 10^{-5}$. To improve the approximation of $\mathbf{v}_i$, we can use the shifted inverse power method, with the approximations λ_{0i}, $i = 1, 2, 3$.

Example A.10 Given the matrix

$$A = \begin{bmatrix} 2 & -2 & & \cdots & 0 \\ -1 & 2 & -1 & & \\ & & \ddots & & \\ 0 & \cdots & -1 & 2 & -1 \\ 0 & \cdots & & -1 & 2 \end{bmatrix}, \text{ of order } n = 32,$$

compute, with no use of any knowledge of the eigenvalue, the eigenvalue closest to 1, using the combined Rayleigh quotient and shifted inverse iteration methods.

Hints:

(a) Since the matrix is nonsymmetric but symmetrizable, compute the eigenvalues of the similar matrix $\widetilde{A} = D^{-1}AD$, where

$$D = \text{diag}(d_0, 1, \ldots, 1) \text{ and } d_0 = \sqrt{2}.$$

To get an approximation to an eigenvector corresponding to the eigenvalue closest to 1, first use some steps of the shifted inverse iteration method,

$$(A - I)\tilde{x}^{k+1} = x^k, \quad k = 0, 1, \ldots,$$

where x^0 is a random vector.

(b) Use the eigenvector found in (b) as an initial vector in the combined Rayleigh and shifted inverse iteration methods.

(c) It can be shown that the exact eigenvalue closest to 1 is λ_{11}, (where $\lambda_1 \leq \lambda_2 \leq \ldots \leq \lambda_{32}$), and that $\lambda_k = [2 \sin(k - \frac{1}{2})\frac{\pi}{n}]^2$.

Remark A.25 For an extension of the Rayleigh-Quotient minimization method for the simultaneous computation of several eigenvalues, see, for instance, Longsine and McCormick (1981). For a recent exposition of interior eigenvalue algorithms, see Morgan (1991). For a discussion on the computation of selected eigenvalues of unsymmetric matrices, using subspace iteration, see Duff and Scott (1991). Computations of eigenvalues occur in design of structures, for instance. For applications in the analysis of oil-platforms, see Aasland and Björstad (1983).

Exercises

1. Prove that $||x - y|| \geq |\; ||x|| - ||y||\; |$.
2. Prove that if $x \in \mathbb{C}^n$, then
    (a) $||x||_\infty \leq ||x||_2 \leq \sqrt{n}||x||_\infty$,
    (b) $||x||_\infty \leq ||x||_1 \leq n||x||_\infty$, and
    (c) $||x||_2 \leq ||x||_1 \leq \sqrt{n}||x||_2$,
    and that these inequalities are sharp.
    Hint: For the first part of (c), use $\sqrt{|a| + |b|} \leq \sqrt{|a|} + \sqrt{|b|}$.
3. Make a graph of the unit sphere in the space $\{R^2, l_p\}$ for $p = 1, 2, \infty$, that is, the unit sphere for vectors $x \in R^2$, with distance measured in the l_p-norm.
4. Prove the following:
    (a) The parallelogram law: $||x + y||^2 + ||x - y||^2 = 2(||x||^2 + ||y||^2)$ if and only if $||x|| = ||x||_2$.
    (b) Pythagoras lemma: $||x + y||_2^2 = ||x||_2^2 + ||y||_2^2$ if and only if the vectors x and y are orthogonal.
    (c) $||\sum_{i=1}^n \alpha_i x_i||_2^2 = \sum_{i=1}^n |\alpha_i|^2$, if $\{x_i\}_{i=1}^n$ is a set of orthonormal vectors, that is, $(x_i, x_j) = \delta_{i,j}$.

(d)   Jensen's inequality: If $0 < p < q$, then $\|\mathbf{v}\|_q \leq \|\mathbf{v}\|_p$.

5.   (a)   Show that the Frobenius norm $\|A\|_E$ is a matrix norm. Let $A = [a_{ij}]$ be of order $n \times n$. Show that:

(b) $\|A\|_\infty$ is a matrix norm.

(c) $\max_{i,j} |a_{ij}|$ is not a multiplicative norm.

(d) $\max_{i,j} |na_{ij}|$ is a multiplicative matrix norm.

(e)   Show that any nonmultiplicative matrix norm $\|\cdot\|$ can be made multiplicative, $\|A\|' = \|\alpha A\|$, by multiplication with a proper scalar.

(f)   Let A be a rectangular matrix. Show that

$$\max_i \{\min_j a_{ij}\} \leq \min_j \{\max_i a_{ij}\}.$$

6.   Prove that $\|A^*\|_1 = \|A\|_\infty$.

7.   Show that:

(a)   Each of the numbers

$$\|A\|_1/\|A\|_2, \quad \|A\|_\infty/\|A\|_2, \quad \|A\|_1/\|A\|_E, \quad \|A\|_\infty/\|A\|_E$$

is bounded below by $1/\sqrt{n}$ and above by $\sqrt{n}$.

(b)   If A is normal (that is, $A^*A = AA^*$), then $\|A\|_2 \leq \|A\|_p \quad \forall p \geq 1$.

(c) $1/n \leq \|A\|_1/\|A\|_\infty \leq n$ and that all inequalities are sharp.

8.   Compute $\|A\|_2$ and $\|A\|_E$ where

$$A = \begin{bmatrix} 1 & 2 & 2 \\ 2 & 1 & 2 \\ 2 & 2 & 1 \end{bmatrix}.$$

9.   A matrix norm is said to be unitarily invariant if $\|A\| = \|UA\| = \|AU\|$ for every matrix A and unitary matrix U of the same order. Prove that the Euclidian norm $\|\cdot\|_E$ is unitarily invariant.

10.   Show that the Hölder norm, $\|A\|_p = (\sum_{i,j=1}^n |a_{i,j}|^p)^{1/p}$, $1 \leq p$, is a matrix norm compatible with the vector norm $\|\cdot\|_q$, where $p^{-1} + q^{-1} = 1$.

Hint: Note that $\|A\|_p$ is the Hölder norm on $\mathbb{C}^{n \times n}$. Use Minkowski's inequality to show that $\|A + B\|_p \leq \|A\|_p + \|B\|_p$. That $\|Ax\|_p \leq \|A\|_p \|x\|_q$ follows by the Hölder inequality.

11.   Calculate a bound of the spectral radius of

$$A = \begin{bmatrix} 3 & -\frac{1}{2} & 0 \\ -\frac{1}{2} & 1 & -\frac{1}{2} \\ 0 & -\frac{1}{2} & 1 \end{bmatrix}.$$

12. *Perturbation of eigenvalues for defective matrices* (Chatelin, 1988): Assume that an $r \times r$ matrix A can be transformed by a similarity transformation to a single Jordan block

$$S^{-1}AS = J = \begin{bmatrix} \lambda_1 & 1 & & \cdots & 0 \\ 0 & \lambda_1 & 1 & & 0 \\ \vdots & & \ddots & & \vdots \\ 0 & & & \lambda_1 & 1 \\ 0 & & & 0 & \lambda_1 \end{bmatrix}.$$

(This implies that A has an eigenvalue of multiplicity r and a one-dimensional eigenvector space.) A is then said to be *defective with index* $r - 1$. Let $\lambda, \mathbf{x}$ be an approximate eigensolution to A, with $\mathbf{u} = A\mathbf{x} - \lambda\mathbf{x}$, where $\|\mathbf{x}\|_2 = 1$. Show that

$$\frac{|\lambda_1 - \lambda|^r}{1 + |\lambda_1 - \lambda|^{r-1}} \leq \|S\|_2 \|S^{-1}\|_2 \|\mathbf{u}\|.$$

Hint: Show first that $\|(J - \lambda I)^{-1}\|_2 \leq \frac{1 + |\lambda_1 - \lambda|^{r-1}}{|\lambda_1 - \lambda|^r}$. Note, then, that $1 = \|\mathbf{x}\|_2 = \|S(J - \lambda I)^{-1} S^{-1}\mathbf{u}\|_2$.

13. It can be shown that any matrix can be transformed to Jordan canonical form by a similarity transformation,

$$S^{-1}AS = \begin{bmatrix} J_1 & 0 \cdots & 0 \\ 0 & J_2 & 0 \\ \vdots & \ddots & \vdots \\ 0 & \cdots & J_s \end{bmatrix},$$

where J_i are Jordan blocks of the form shown in the previous exercise. Let r_i be the order of J_i and let λ_i be the eigenvalue of J_i. Show that

$$\min_{1 \leq i \leq s} \frac{|\lambda_i - \lambda|^{r_i}}{1 + |\lambda_i - \lambda|^{r_i - 1}} \leq \|S\|_2 \|S^{-1}\|_2 \|\mathbf{u}\|_2,$$

where $\mathbf{u} = A\mathbf{x} - \lambda\mathbf{x}$, $\|\mathbf{x}\|_2 = 1$.

Note: For a defective matrix, i.e., a matrix where $r_i > 1$ for at least one i, the minimum value does not have to be attained by an eigenvalue closest to λ.

14. (a) Using the previous exercise, show that to every eigenvalue λ' of

$A' = A + E$, there exists an eigenvalue λ and an index $r_i = r$ of A such that

$$\frac{|\lambda' - \lambda|^r}{1 + |\lambda' - \lambda|^{r-1}} \leq \|S\|_2 \|S^{-1}\|_2 \|E\|_2.$$

(b) Show that if A is diagonalizable, i.e., if $r_i = 1$, $1 \leq i \leq s$, then there exist eigenvalues λ', λ such that

$$|\lambda' - \lambda| \leq \|S\|_2 \|S^{-1}\|_2 \|E\|_2.$$

(Hence the above generalizes the Bauer-Fike theorem.)

15. Let $A = A^*$ have eigensolutions $\mathbf{v}_i$, λ_i and let $(\mathbf{v}_i, \mathbf{v}_j) = \delta_{ij}$. Prove that:

   (a) If $\|\mathbf{x}\|_2 = 1$, $\|A\mathbf{x} - \lambda\mathbf{x}\|_2 \leq \varepsilon$, and if $|\lambda - \lambda_i| \geq a$, $i > 1$, then for a suitable α,

   $$\|\alpha\mathbf{x} - \mathbf{v}_1\| \leq \frac{\varepsilon}{a}.$$

   (b) If $\|\mathbf{x}\|_2 = 1$, $\|A\mathbf{x} - \lambda\mathbf{x}\| \leq \varepsilon$ for some λ, and $|\lambda - \lambda_i| \geq a$ for $i > r$, then for some α and $\mathbf{v}$, where $\mathbf{v}$ is a linear combination of $\mathbf{v}_1, \ldots, \mathbf{v}_r$, with $\|\mathbf{v}\| = 1$,

   $$\|\alpha\mathbf{x} - \mathbf{v}\|_2 \leq \frac{\varepsilon}{a}.$$

   (Hence, it is possible to calculate accurate approximations of some eigenvector *in a subspace* spanned by eigenvectors corresponding to close or equal eigenvalues, but not approximations of the individual eigenvectors in that subspace.)

16. Let A be a tridiagonal matrix satisfying $a_{ii} = 1$, $0 < a_{i+1,i} \cdot a_{i,i+1} \leq \varepsilon$, $0 < \varepsilon < 1$, $a_{ii} + a_{i,i-1} + a_{i,i+1} = 0$, $i = 1, 2, \ldots, n-1$ and $a_{n,n} + a_{n,n-1} > 0$. Consider the following:

   (a) Prove that A is diagonally *quasisymmetric* (cf. Chapter 3).
   (b) Prove that the eigenvalues of A lie in the interval $[1 - 2\sqrt{\varepsilon}, 1 + 2\sqrt{\varepsilon}]$.
   (c) Which estimate of the smallest eigenvalue is obtained if Gershgorin's theorem is applied to A?
   (d) Which estimate of the smallest eigenvalue is obtained if one diagonal entry, say $a_{i_0 i_0}$, equals 2ε instead of 1.

17. (a) Prove that $\rho(A) \leq \|A^k\|^{1/k}$.
   Hint: $\rho(A) \leq \|A\| \Rightarrow \rho(A^k) \leq \|A^k\| \leq \|A\|^k$.
   (b) Prove that $\rho(A) = \lim_{k \to \infty} \|A^k\|^{1/k}$.

18. If $\|A\| < 1$, where $\|A\|$ is an arbitrary induced matrix norm, show that:

(a) $\frac{1}{1+\|A\|} \leq \|(I - A)^{-1}\| \leq \frac{1}{1-\|A\|}$.

(b) $\|(I - A)^{-1} - (I + A + \cdots + A^k)\| \leq \frac{\|A\|^{k+1}}{1-\|A\|}$.

19. Consider $Ax = b$, where $A = \begin{bmatrix} 1+\varepsilon & -1 \\ -1 & 1 \end{bmatrix}$, $b = \begin{bmatrix} \varepsilon \\ 0 \end{bmatrix}$. [Hence, $x = (1, 1)^T$]. Let $\|x\| = \|x\|_2$. Prove that:

(a) $\kappa = \frac{1}{\varepsilon}\left(1 + \frac{\varepsilon}{2} + \sqrt{1 + (\frac{\varepsilon}{2})^2}\right)^2 \sim 4/\varepsilon, \ \varepsilon \to 0$.

(b) $\kappa(x) \sim \frac{2}{\varepsilon}\|Ax\|/\|x\| \sim \sqrt{2}, \ \varepsilon \to 0$ (cf. Theorem A.13).

20. Let $A = \begin{bmatrix} 1.00 & 3.00 & 3.00 \\ 1.00 & 4.00 & 3.00 \\ 1.00 & 3.00 & 4.00 \end{bmatrix}$ be given, where the entries are obtained by round-off. Give a bound for the corresponding inherent error in A^{-1} using $\|\cdot\|_\infty$.

21. A classical method to solve over- (or under-) determined systems of equations $Ax = b$ is by forming the so-called *normal equations*, $A^T Ax = A^T b$. Prove that the spectral condition number of normal equations is the square of that of A, if A is symmetric and positive definite. (Hence, *forming $A^T A$ can turn a fairly well-conditioned problem into an ill-conditioned one.*)

22. For a symmetric positive definite matrix A, we have $A = LDL^T = W^T W$, where $W = D^{\frac{1}{2}}L^T$. (This is called the *Cholesky decomposition* of A.) Prove that the spectral condition number of W is the square root of the condition number of A. (*It follows that the Gauss algorithm needs no row exchanges for a positive definite symmetric matrix, because the diagonal entries of W are positive*—i.e., in particular, nonzero.)
Hint: $\|A\|_2 = \sup(Ax, x) = \sup(Wx, Wx) \quad \forall x; \ (x, x) = 1$.

23. (a) Prove that $\rho(AB) \leq \rho(A)\rho(B)$, if A and B commute.
    (b) Prove that there exist square matrices A, B, for which $\rho(AB) > \rho(A)\rho(B)$.
    Hint: Use the $\|\cdot\|_T$-norm as defined in Chapter 5 to prove (a). To prove (b) consider $A = \begin{bmatrix} 0 & 0 \\ 1 & 0 \end{bmatrix}$, $B = A^T$.

24. (Ciarlet) Let F be the set of matrices of symmetric nonsingular matrices of order 2, whose elements a_{ij} are integers satisfying $0 \leq a_{ij} \leq 100$. Show that

$$\text{cond}_2(A) = \sup_{E \in F} \text{cond}_2(E), \quad \text{where } A = \begin{pmatrix} 100 & 99 \\ 99 & 99 \end{pmatrix}.$$

In order to proceed, the first step is to establish that, for a general symmetric matrix $A = (a_{ij})$ of order 2,

$$\text{cond}(A) = \sigma + \{\sigma^2 - 1\}^{1/2}, \quad \text{with } \sigma = \frac{\sum_{i,j=1}^{2} |a_{ij}|^2}{2|\det(A)|}.$$

25. Let A be nonsingular and $\|A\|$ the induced norm corresponding to the vector norm $\| \cdot \|$. Show that for the condition number, we have

$$\kappa(A) = \|A\| \, \|A^{-1}\| = \max_{\|x\|=\|y\|=1} \|Ax\|/\|Ay\|.$$

26. An important technique for the iterative solution of a linear system $Ax = b$ is the technique of *preconditioning* A. Let $C = LL^T$ be decomposed in Choleski factors. Then, the iterative method is applied to $C^{-1}Ax = C^{-1}b$ (see Chapter 6). Let A and C be symmetric and positive definite (s.p.d.)

    (a) Show that $C^{-1}A$ need not be s.p.d. that $\tilde{A} = L^{-1}AL^{-T}$ is s.p.d., and that $\tilde{A}$ is similarly equivalent to $C^{-1}A$.

    (b) Show that

    $$\kappa_2(\tilde{A}) = \max_x \frac{(Ax, x)}{(Cx, x)} / \min_x \frac{(Ax, x)}{(Cx, x)}.$$

    (c) Show that

    $$\kappa(\tilde{A}) \geq \max\{\frac{\kappa(A)}{\kappa(C)}, \frac{\kappa(C)}{\kappa(A)}\}.$$

27. Let A and B be nonsingular matrices and let $D = \begin{bmatrix} A & 0 \\ 0 & B \end{bmatrix}$.

    (a) Show that D is nonsingular and that

    $$\kappa_2(D) \geq \max\{\kappa_2(A), \kappa_2(B)\}.$$

    (b) Show that

    $$\kappa_2(D) = \max\{\|A\|_2, \|B\|_2\} \cdot \max\{\|A^{-1}\|_2, \|B^{-1}\|_2\}.$$

28. Consider the eigenvalue problem $Ax = \lambda x$, where A is the matrix that corresponds to the standard five-point difference operator $\Delta_h^{(5)}$ on a square mesh with Dirichlet boundary conditions and where we have used the natural ordering. Assume that we want to compute one of the smallest eigenvalues by the Rayleigh quotient method, using the method

described in Remark A.24. Show that $\eta = A\mathbf{x} - \lambda\mathbf{x}$ can be computed to full-order machine number precision for any h in the following way:

Compute $(A\mathbf{x})_i = \sum_{j=1}^{n} a_{i,j}x_j = -x_{i-m} - x_{i-1} - x_{i+1} - x_{i+m} + 4x_i$ as the sum of differences, $(A\mathbf{x})_i = [(x_i - x_{i-1}) - (x_{i+1} - x_i)] + [(x_i - x_{i-m}) - (x_{i+m} - x_i)]$, in the order as indicated.

Show that, since $\mathbf{x}$ is the discrete vector corresponding to a smooth eigenfunction, the round-off errors in computing $A\mathbf{x}$ in this way are zero (or exceptionally nonzero). Then compute $\underline{\eta}_i = (A\mathbf{x})_i - \lambda x_i$ and note that $\lambda = O(h^2)$, $h \to 0$.

29. Let A be a real matrix of order $n \times n$ and let $A = D + B$ be a splitting of A where D is a nonsingular diagonal matrix. Assume that $||D^{-1}B|| \le \alpha < 1$.
    (a) Show that $||A^{-1}|| \le ||D^{-1}||/(1 - ||D^{-1}B||)$.
    (b) Show that $\kappa(A) \le \frac{1+\alpha}{1-\alpha}\kappa(D)$.

30. Show that

$$||A||_2 = \max_{\mathbf{x}\neq 0, \mathbf{y}\neq 0} \frac{|(A\mathbf{x}, \mathbf{y})|}{||\mathbf{x}||_2 ||\mathbf{y}||_2}.$$

31. A linear functional f on $V = \mathbb{C}^n$ is a linear mapping of V into $\mathbb{C}$. It is readily seen that f is represented by a vector $\mathbf{y} \in \mathbb{C}^n$, in the sense that $f(\mathbf{x}) = \mathbf{y}^*\mathbf{x}$ $\forall \mathbf{x} \in V$. Let $||\mathbf{y}||^* = \max_{\mathbf{x}\neq 0} Re(\mathbf{y}^*\mathbf{x})/||\mathbf{x}||$, $\mathbf{x} \in V$.
    (a) Show that

$$\max_{\mathbf{x}\neq 0} \frac{Re(\mathbf{y}^*\mathbf{x})}{||\mathbf{x}||} = \max_{\mathbf{x}\neq 0} \frac{|\mathbf{y}^*\mathbf{x}|}{||\mathbf{x}||}.$$

    (b) Show that $|| \cdot ||^*$ is a norm on V. (It is called the *dual norm;* strictly speaking it is a norm on the linear vector space of linear functionals on V, the so-called dual space.)
    (c) Show that $|| \cdot ||_2^* = || \cdot ||_2$.
    (d) Show that $|\mathbf{y}^*\mathbf{x}| \le ||\mathbf{y}||^* ||\mathbf{x}||$ $\forall \mathbf{x}, \mathbf{y} \in V$.
    (e) One can show that $|(\mathbf{x}, \mathbf{y})| \le ||\mathbf{x}||_p ||\mathbf{y}||_q$ $\forall \mathbf{x}, \mathbf{y} \in V$, where $\frac{1}{p} + \frac{1}{q} = 1$, $p, q \ge 1$ (Hölders inequality). Equality is valid if and only if $|x_i| = \alpha|y_i|^{p-1}$ for some positive scalar α. Using this, show that $|| \cdot ||_q$ is the dual norm of $|| \cdot ||_p$.
    (f) Show that $||A^*||^* = ||A||$.

32. *Schur's inequality:*
    (a) Let A be a complex matrix of order $n \times n$. Show that

$$\sum_{i=1}^{n} |\lambda_i(A)|^2 \le ||A||_E^2 := \sum_{i,j} |a_{i,j}|^2$$

and that equality holds if and only if A is a normal matrix.

(b) (Gaines, 1967) Let a_i, $i = 1, 2, \ldots, n$ be positive numbers. Using Schur's inequality for the matrix

$$A = \begin{bmatrix} 0 & \sqrt{a_1} & 0 & & \\ 0 & 0 & \sqrt{a_2} & & \\ \vdots & & & \ddots & \\ 0 & 0 & 0 & \cdots & \sqrt{a_{n-1}} \\ \sqrt{a_n} & 0 & 0 & \cdots & 0 \end{bmatrix},$$

show the arithmetic-geometric inequality,

$$\left(\prod_{i=1}^{n} a_i \right)^{\frac{1}{n}} \le \frac{1}{n} \sum_{i=1}^{n} a_i.$$

(c) Show that the following alternative form of a companion matrix to a polynomial $P_n(x) = x^n + a_n x^{n-1} + \cdots + a_1$ (cf. Example 4.8) holds:

$$P_n(\lambda) = \det(\lambda I - A),$$

where

$$A = \begin{bmatrix} 0 & a_1 & & & 0 \\ -1 & 0 & a_2 & & \\ & \ddots & \ddots & \ddots & \\ & & -1 & 0 & a_{n-1} \\ 0 & & & -1 & -a_n \end{bmatrix}.$$

Use this and a proper diagonal transformation matrix D to show that (assume, first, that $|a_i| \ne 0$)

$$D(\lambda I - A)D^{-1} = \begin{bmatrix} \lambda & \pm |a_1|^{\frac{1}{2}} & & & \\ \pm |a_1|^{\frac{1}{2}} & \lambda & \pm |a_2|^{\frac{1}{2}} & & \\ \ddots & \ddots & \ddots & & \\ 0 & \pm |a_{n-2}|^{\frac{1}{2}} & \lambda & \pm |a_{n-1}|^{\frac{1}{2}} \\ 0 & & \ddots & \pm |a_{n-1}|^{\frac{1}{2}} & (\lambda + a_n) \end{bmatrix},$$

and deduce from this that

$$\sum_{i=1}^{n} |\lambda_i(A)|^2 \le |a_n| + 2\sum_{i=1}^{n-1} |a_i|.$$

Hence, the zeros x_i of $P_n(x)$ satisfy

$$\sum_{i=1}^{n} |x_i|^2 \le |a_n| + 2\sum_{i=1}^{n-1} |a_i|.$$

This inequality complements the bounds given in Example 4.8.

Hint:

(a) Use Schur's Lemma $U^*AU = T$, where T is a triangular matrix and U is a proper unitary matrix. To show the equality part, note that A is normal if and only if trace $(A^*A) = \Sigma|\lambda_i|^2$.

(b) Show that $\det(A - \lambda I) = (-\lambda)^n - \prod_{i=1}^{n} \sqrt{a_i}$ and that $\|A\|_E = \sum_{i=1}^{n} a_i$.

(c) To show the first part, write

$$P_n(\lambda) = \lambda((\dots(\lambda(\lambda + a_n) + a_{n-2}) + a_2) + a_1)$$

and expand $\det(\lambda I - A)$ recursively from bottom to the top. The matrix $D = \text{diag}(d_1, d_2, \dots, d_n)$ is defined by

$$d_1 = 1, d_i^2 = |a_{i-1}|d_{i-1}^2, i = 2, \dots, n.$$

33. Let A be a complex matrix of order $n \times n$ and let $B = \frac{1}{2}(A^* + A)$ (the Hermitian part of A) and $C = \frac{1}{2}(A^* - A)$ (the anti-Hermitian part of A). Let $\alpha = \max_{i,j} |a_{i,j}|$, $\beta = \max_{i,j} |b_{i,j}|$, $\gamma = \max_{i,j} |c_{i,j}|$. Show that:
    (a) $|\lambda(A)| \le n\alpha$.
    (b) $|Re\lambda(A)| \le n\beta$.
    (c) $|Im\lambda(A)| \le n\gamma$.
    (d) if $A = P^{-1}DP$, where D is diagonal, then $|Im\ \lambda(A)| \le 2\kappa(P)\|C\|$.
    Hint: Use $|\lambda(A)| \le \rho(A) \le \|A\|_\infty$. For part (d), use Theorem A.15.

34. (Bendixson's inequality) Let A be a real matrix. Show that

$$|Im\lambda(A)| \le \gamma(n(n-1))^{\frac{1}{2}},$$

where $\gamma = \max_{i,j} |c_{i,j}|$ and $C = \frac{1}{2}(A^T - A)$.
    Hint: Since C is normal,

$$|Im\lambda_i(A)| \le \max_{\mathbf{x}, \|\mathbf{x}\|=1} |(C\mathbf{x}, \mathbf{x})| \le \max_i |\lambda_i(C)|.$$

Now use Schur's inequality (Exercise A.32):

$$\sum_{i=1}^{n} [Im\lambda_i(C)]^2 \le ||C||_E^2 := \sum_{i,j=1}^{n} |c_{i,j}|^2,$$

where $c_{i,i} = 0$. (Why?)

35. *The Temple-Kato brackets for eigenvalues* (B. Noble: (1969), p. 411.)
Let A be Hermitian. Show that if the eigenvalues are ordered $\lambda_i \le \lambda_{i+1}$ $i = 1, 2, \ldots$, $\lambda_{r-1} < \lambda_r < \lambda_{r+1}$ and $q \in (\lambda_{r-1}, \lambda_{r+1})$ where $q = (Ax, x)$, $(x, x) = 1$, then

$$q - \frac{\varepsilon^2}{\lambda_{r+1} - q} \le \lambda_r \le q + \frac{\varepsilon^2}{q - \lambda_{r-1}}, r = 2, 3, \ldots,$$

and for the lowest eigenvalue for $q < \lambda_2$,

$$q - \frac{\varepsilon^2}{\lambda_2 - q} \le \lambda_1 \le q,$$

where $\varepsilon = ||Ax - qx||$.

36. Consider

$$\beta = (Ax - \mu x, Ax - \nu x)$$

for a Hermitian matrix A, where μ, ν, $\mu < \nu$ are real constants.
(a) Prove that if no eigenvalue lies between μ and ν, then $\beta \ge 0$.
(b) Prove that

$$(q - \mu)(\nu - q) \le \varepsilon^2,$$

where q and ε^2 are defined in the previous exercise.
Hint: To prove (a), expand $Ax - \mu x$ and $Ax - \nu x$ in eigensolutions $\{\lambda_i, v_i\}$, that is,

$$\beta = \sum_{i=1}^{n} (\lambda_i - \mu)(\lambda_i - \nu)|\alpha_i|^2, \text{ if } ||v_i|| = 1,$$

and note that $(\lambda_i - \mu)$ and $(\lambda_i - \nu)$ have the same sign for all eigenvalues λ_i. To prove (b) note that

$$||Ax - qx||^2 + (q - \mu)(q - \nu)||x||^2$$
$$= ((Ax - qx) + (q - \mu)x, (Ax - qx) + (q - \nu)x),$$

which is nonnegative by (a). Hence,

$$-(q - \mu)(q - \nu) \leq \varepsilon^2.$$

37. *Power iteration method for complex eigenvalues* (Faddeev and Faddeeva, 1963): Assume that the eigenvalues with maximum modules of a real matrix A occur as complex pairs, $\lambda_1 = re^{i\theta}$, $\lambda_2 = re^{-i\theta}$, where $\theta \neq 0$, and suppose that $\lambda_3 = \max_{3 \leq i \leq n} |\lambda_i| < r$. Let $\mathbf{x}^{(0)}$ be a real vector, let $\mathbf{x}^{(k)}$ be the vector in the power iteration method without normalization, and let $x_k = \mathbf{x}_{n_0}^{(k)}$ for some n_0, $1 \leq n_0 \leq n$. Show that

$$r^2 = \frac{x_k x_{k+2} - x_{k+1}^2}{x_{k-1} x_{k+1} - x_k^2} + O\left(\left|\frac{\lambda_3}{r}\right|^k\right),$$

$$\cos \theta = \frac{r x_{k-1} + r^{-1} x_{k+1}}{2 x_k} + O\left(\left|\frac{\lambda_3}{r}\right|^k\right).$$

Hint: Show first that

$$x_k = \text{const}(r^k \cos(k\theta + \alpha)) + O\left(\left|\frac{\lambda_3}{r}\right|^k\right),$$

where α depends on the components of the initial vector along $\mathbf{v}_1$, $\mathbf{v}_2$, the eigenvectors corresponding to λ_1, λ_2.

38. Let $A = \begin{bmatrix} 26 & -54 & 4 \\ 13 & -28 & 3 \\ 26 & -56 & 5 \end{bmatrix}$. It can be shown that the eigenvalues of A are

$$\lambda_1 = 1 + 5i, \quad \lambda_2 = 1 - 5i \quad \text{and} \quad \lambda_3 = 1.$$

Use the power iteration method in the previous exercise to compute λ_1, λ_2 numerically, starting with the vector $\mathbf{x}^0 = (0.2, 0.4, 0.6)^T$.

Remark The standard version of the power iteration method reveals the existence of complex conjugate pairs by oscillating values of x_k. The above power iteration method for complex conjugate eigenvalues converges very slowly if θ is close to 0. Similarly, the power iteration method to compute a dominate real eigenvalue belonging to a Jordan block (of order at least two) converges slowly. If the Jordan box has order two, we have $x_{k+1}/x_k = \lambda_1(1 + O(k^{-1}))$. In practice, this makes the power iteration useless for such matrices.

39. *Point-wise error estimates* (Bauer, 1966 and Skeel, 1979): Let $|\mathbf{x}| \le |\mathbf{y}|$ and $|A| \le |B|$ denote inequalities between corresponding components of the vectors or matrices. Let $A\mathbf{x} = \mathbf{b}$ and $(A + E)(\mathbf{x} + \delta\mathbf{x}) = \mathbf{b} + \underline{\beta}$, where $|E| \le \varepsilon|A|$ and $|\underline{\beta}| \le \varepsilon|\mathbf{b}|$ (that is, the relative size of perturbations of any component is $\le \varepsilon$). Show that

$$|\delta\mathbf{x}| \le \varepsilon|(I - \varepsilon|A^{-1}|\,|A|)^{-1}|\,|A^{-1}|(|A|\,|\mathbf{x}| + |\mathbf{b}|),$$

provided the spectral radius $\rho(|A^{-1}|\,|A|) < \varepsilon^{-1}$.
Hint: Show first that

$$\delta\mathbf{x} = A^{-1}[\underline{\beta} - E(\mathbf{x} + \delta\mathbf{x})],$$

so

$$|\delta\mathbf{x}| \le \varepsilon|A^{-1}|[|A|(|\mathbf{x}| + |\delta\mathbf{x}|) + |\mathbf{b}|].$$

Remark Skeel (1979) shows also that there exist E and $\underline{\beta}$, $|E| = \varepsilon|A|$, and $|\underline{\beta}| = \varepsilon|\mathbf{b}|$, such that the perturbed solution $\mathbf{x} + \delta\mathbf{x}$ satisfies

$$\frac{\|\delta\mathbf{x}\|}{\|\mathbf{x}\|} \ge \varepsilon\frac{\|\,|A^{-1}|[|A|\,|\mathbf{x}| + |\mathbf{b}|]\,\|}{(1 + \varepsilon\|\,|A^{-1}|\,|A|\,\|)\|\mathbf{x}\|}.$$

40. (Hamming, 1971) Let

$$A = \begin{bmatrix} 3 & 2 & 1 \\ 2 & 2\varepsilon & 2\varepsilon \\ 1 & 2\varepsilon & -\varepsilon \end{bmatrix}, \quad \mathbf{b} = \begin{bmatrix} 3 + 3\varepsilon \\ 6\varepsilon \\ 2\varepsilon \end{bmatrix},$$

where $|\varepsilon|$ is sufficiently small. Then show that

$$A^{-1} = \frac{1}{1 - 1.8\varepsilon}\begin{bmatrix} -0.6\varepsilon & 0.4 & 0.2 \\ 0.4 & (-0.1\varepsilon^{-1} - 0.3) & (0.2\varepsilon^{-1} - 0.6) \\ 0.2 & (0.2\varepsilon^{-1} - 0.6) & (-0.4\varepsilon^{-1} - 0.6) \end{bmatrix}$$

and $\mathbf{x} = A^{-1}\mathbf{b} = \begin{bmatrix} \varepsilon \\ 1 \\ 1 \end{bmatrix}$. Hence, show that

$$|A^{-1}|\,|A| = \frac{1}{1 - 1.8\varepsilon}\begin{bmatrix} (1 + 1.8\varepsilon) & 2.4\varepsilon & 1.6\varepsilon \\ (0.4\varepsilon^{-1} + 1.2) & (1.4 - 0.6\varepsilon) & 0.8 \\ 0.8\varepsilon^{-1} & 1.6 & 1 - 0.6\varepsilon \end{bmatrix}$$

and

$$|A^{-1}|[|A|\,|\mathbf{x}| + |\mathbf{b}|] = \frac{1}{1 - 1.8\varepsilon} \begin{bmatrix} 9.6\varepsilon + 3.6\varepsilon^2 \\ 4.8 + 2.4\varepsilon \\ 6 - 2.4\varepsilon \end{bmatrix},$$

which shows that the system is well-conditioned w.r.t. the solution $\mathbf{x}$. Since $\|A\|\,\|A^{-1}\| = O(\varepsilon^{-1})$, there exist right-hand sides for which the system is ill-conditioned, as $\varepsilon \to 0$.

References

L. Aasland and P. Björstad (1983). The generalized eigenvalue problem in ship design and offshore industry. In *Matrix Pencils*, ed. B. Kågström and A. Ruhe. *Lecture Notes in Mathematics*, No. 973, 146–155. Berlin, Heidelberg, New York: Springer-Verlag.

O. Axelsson and V. A. Barker (1984). *Finite Element Solutions of Boundary Value Problems*. Orlando, Fla.: Academic Press.

A. Bunse and J. Gerstner (1984). An algorithm for the symmetric generalized eigenvalue problem. *Lin. Alg. Appl.* **58**, 43–68.

F. L. Bauer and C. T. Fike (1960). Norms and exclusion theorems. *Numer. Math.* **2**, 137–141.

F. L. Bauer (1966). Genauigkeitsfragen bei der Lösung linearer Gleichungssysteme. *ZAMM* **46**, 409–421.

I. Bendixson (1902). Sur les racines d'une equation fundamentale. *Acta Math.* **25**, 359–365.

Å. Björck and G. Dahlquist (1974). *Numerical Methods*. Englewood Cliffs, N.J.: Prentice Hall.

F. Chatelin (1988). *Valeurs Propres de Matrices*, Paris: Masson.

Ph. G. Ciarlet (1989). *Introduction to Numerical Linear Algebra and Optimization*. Cambridge: Cambridge Univ. Press.

I. S. Duff and J. A. Scott (1991). Computing selected eigenvalues of sparse unsymmetric matrices using subspace iteration. Report RAL-91-056, Rutherford Appleton Laboratory. Submitted for publication.

D. K. Faddeev and V. N. Faddeeva (1963). *Computational Methods of Linear Algebra*. San Francisco: Freeman.

G. E. Forsythe and C. B. Moler (1967). *Computer Solution of Linear Algebraic Systems*. Englewood Cliffs, N.J.: Prentice Hall.

F. Gaines (1967). On the arithmetic mean-geometric mean inequality. *Amer. Math. Monthly* **74**, 305–306.

G. H. Golub and C. F. van Loan (1983). *Matrix Computations*, Baltimore, Md.: Johns Hopkins Univ. Press.

R. W. Hamming (1971). *Introduction to Applied Numerical Analysis*, p. 120. New York: McGraw-Hill.

W. Kahan (1967). Inclusion theorems for clusters of eigenvalues of Hermitian matrices. Techn. Report No. CS42, Computer Science Dept., University of Toronto.

T. Kato (1949). On the upper and lower bounds of eigenvalues. *J. Phys. Soc. Japan* **4**, 334–339.

T. Kato (1966). *Perturbation Theory for Linear Operators*. Berlin, Heidelberg, New York: Springer Verlag.

D. E. Longsine and S. F. McCormick (1981). Simultaneous Rayleigh-Quotient minimization methods for $Ax = \lambda Bx$. In Large Scale Matrix Problems, ed. Å. Björk, R. J. Plemmons, and H. Schneider, pp. 195–234. New York: North Holland.

R. B. Morgan (1991). Computing interior eigenvalues of large matrices. *Lin. Alg. Appl.* **156**, 289–309.

B. Noble (1969). *Applied Linear Algebra*, Englewood Cliffs, N.J.: Prentice Hall.

B. Noble and J. W. Daniel (1988). *Applied Linear Algebra*, Englewood Cliffs, N.J.: Prentice Hall.

B. N. Parlett (1980). *The Symmetric Eigenvalue Problem*, Englewood Cliffs, N.J.: Prentice Hall.

R. D. Skeel (1979). Scaling for numerical stability in Gaussian elimination. *J. Assoc. Comp. Mach.* **26**, 494–526.

G. Temple (1928). The computation of characteristic numbers and characteristic functions. *Proc. London Math. Soc.* **29**, 257–280.

J. H. Wilkinson (1965). *The Algebraic Eigenvalue Problem*, Oxford: Clarendon.

Appendix B
Chebyshev Polynomials

Chebyshev polynomials have many applications in numerical analysis; see, for example, Lanczos (1957), Varga (1962), and Fox and Parker (1968). The polynomials can be written in different forms. Consider first the function

(B.1) $$T_k(\cos\theta) \equiv \cos(k\theta), \quad -\pi \le \theta \le \pi.$$

Using the variable transformation $x = \cos(\theta)$ and the trigonometric identity

$$\cos((k+1)\theta) = 2\cos(\theta)\cos(k\theta) - \cos((k-1)\theta),$$

we find

(B.2) $T_0(x) = 1, \; T_1(x) = x, \; T_{k+1}(x) = 2xT_k(x) - T_{k-1}(x), \; k = 1, 2, \ldots.$

That is, $T_k(x)$ is a polynomial of degree k, with leading coefficient 2^{k-1}, $k = 1, 2, \ldots.$

For every fixed x, the recursion (B.2) has a characteristic equation $\lambda^2 = 2x\lambda - 1$ whose roots are $\lambda = x \pm \sqrt{x^2 - 1}$. Using these and the initial values $T_0(x) = 1, T_1(x) = x$, we find the following alternative form of $T_k(x)$:

(B.3) $T_k(x) = \dfrac{1}{2}[(x + \sqrt{x^2 - 1})^k + (x - \sqrt{x^2 - 1})^k], \quad k = 0, 1, \ldots.$

(B.3) [and (B.2)] defines the *Chebyshev polynomial* for any real or complex variable x. It follows from (B.1) that

$$\max_{-1 \le x \le 1} |T_k(x)| = 1$$

and that

$$T_k(x_i) = (-1)^i, \quad i = 0, 1, \ldots, k,$$

where $x_i = \cos(i\pi/k)$.

Let $0 < a < b$. Using the variable transformation

$$z = (b + a - 2x)/(b - a),$$

we find

$$\max_{a \leq z \leq b} |T_k(z)| = 1$$

and

(B.4) $T_k(z_i) = (-1)^i, \quad i = 0, 1, \ldots, k,$

where $z_i = (b+a-2x_i)/(b-a)$.

The following result shows that the Chebyshev polynomials (properly normalized) take the smallest maximum values in the interval (a, b) and also the largest values outside this interval.

Theorem B.1 *Let π_k^1 denote the set of polynomials of degree k which take the value 1 at the origin. Let $0 < a < b$ and let the Chebyshev polynomials be defined by (B.3). Then:*

(a) $\max_{a \leq x \leq b} |\widetilde{P}_k(x)| = \min_{P_k \in \pi_k^1} \max_{a \leq x \leq b} |P_k(x)|$, *where*

(B.5) $$\widetilde{P}_k(x) = T_k\left(\frac{b + a - 2x}{b - a}\right) / T_k\left(\frac{b + a}{b - a}\right).$$

(b) *For any $\xi \notin [a, b]$ and any $P_k(x)$ with $\max_{a \leq x \leq b} |P_k(x)| \leq 1$, we have*

$$|P_k(\xi)| \leq \left| T_k\left(\frac{b + a - 2\xi}{b - a}\right) \right|.$$

Proof Let

(B.6) $$R(x) = \frac{P_k(\xi)}{T_k\left(\frac{b+a-2\xi}{b-a}\right)} T_k(z) - P_k(x), \quad z = \frac{b + a - 2x}{b - a}.$$

This is a polynomial of degree k. Taking first $\xi = 0$, we find

$$R(x) = P_k(0)\widetilde{P}_k(x) - P_k(x).$$

Assume that

(B.7) $$\max_{a \leq x \leq b} |P_k(x)/P_k(0)| < \max_{a \leq x \leq b} |\widetilde{P}_k(x)|.$$

Then, since $\widetilde{P}_k(x_i) = (-1)^i$ and $\max_{a \leq x \leq b} |\widetilde{P}_k(x)| = 1$, it follows that $R(x)$ changes sign in each interval (x_i, x_{i+1}), so, in addition to the zero at $x = 0$, $R(x)$ has k zeros, which is in contradiction to its degree k. Hence, assumption (B.7) is false, which shows part (a). In a similar way, part (b) follows from (B.6). ◇

References

L. Fox and I. B. Parker (1968). *Chebyshev Polynomials in Numerical Analysis*. London and New York: Oxford Univ. Press.

C. Lanczos (1957). *Applied Analysis*. Englewood Cliffs, N.J.: Prentice Hall.

R. S. Varga (1962). *Matrix Iterative Analysis*. Englewood Cliffs, N.J.: Prentice Hall.

Appendix C

Some Inequalities for Functions of Matrices

Some inequalities for matrices and eigenvalues of matrices will be derived using certain convex or concave functions. The presentation is based largely on Marchall and Olkin (1979). We will use a majorization property between real vectors.

Definition C.1 For $x, y \in \mathbb{R}^n$, y is said to *majorize* x if

$$\begin{cases} \sum_1^k \widetilde{x}_i \leq \sum_1^k \widetilde{y}_i, & k = 1, \ldots, n-1 \\ \\ \sum_1^n x_i = \sum_1^n y_i. \end{cases}$$

Here, $\widetilde{x}_1 \geq \widetilde{x}_2 \geq \cdots \geq \widetilde{x}_n$, $\widetilde{y}_1 \geq \widetilde{y}_2 \geq \cdots \geq \widetilde{y}_n$ denote reorderings of the components of x and y, respectively, in decreasing order.

For $x, y \in \mathbb{R}^n$, the following can be shown to be equivalent (Hardy, Littlewood, and Pólya, 1929):

(a) y majorizes x.
(b) $\sum_1^n x_i = \sum_1^n y_i$ and $\sum_1^n (x_i - a)^+ \leq \sum_1^n (y_i - a)^+$ for all $a \in \mathbb{R}$, where
$$(x)^+ = \begin{cases} x & \text{if } x > 0 \\ 0 & \text{otherwise} \end{cases}$$
(c) $x = Py$ for some doubly stochastic matrix P. [Recall that a $n \times n$ real matrix $P = [P_{ij}]$ is doubly stochastic if $p_{ij} \geq 0$ for $i, j = 1, \ldots, n$, and $Pe = e$, $e^T P = e^T$, where $e = (1, \ldots, 1)^T$.]

Theorem C.1 *(Schur, 1923) Let H be a $n \times n$ Hermitian matrix with diagonal elements $h_{11}, \ldots, h_{nn}$ and eigenvalues $\lambda_1, \ldots, \lambda_n$, and let $\underline{\lambda} = (\lambda_1, \ldots, \lambda_n)^T$, $h = (h_{11}, \ldots, h_{nn})^T$. Then, $\underline{\lambda}$ majorizes h.*

Proof Exercise (2.35) shows that $\mathbf{h} = P\underline{\lambda}$ for a doubly stochastic matrix P, so by property (c) above, $\underline{\lambda}$ majorizes $\mathbf{h}$. ◇

Theorem C.2 *If A is a $n \times n$ Hermitian matrix with eigenvalues $\lambda_1 \geq \lambda_2 \geq \ldots \lambda_n$, U is a $k \times n$ complex matrix, $k \leq n$ and $\alpha_i = \lambda_i(U^*U)$, $i = 1, \ldots, n$, ordered $\alpha_1 \geq \cdots \geq \alpha_k \geq 0$, then*

$$(C.1) \qquad \sum_1^k \alpha_i \lambda_i \geq \operatorname{tr}(UAU^*) \geq \sum_1^k \alpha_i \lambda_{n-i+1}, \quad k = 1, \ldots, n.$$

Proof Since A is Hermitian, we have $S^*AS = D$, where S is unitary and $D = \operatorname{diag}(\lambda_1, \ldots, \lambda_n)$. Hence, $UAU^* = VDV^*$ where $V = US$. We have

$$(C.2) \quad \operatorname{tr}(UAU^*) = \operatorname{tr}(VDV^*) = \operatorname{tr}(DV^*V) = \operatorname{tr}(DB) = \sum_1^n \lambda_i b_{ii},$$

where $B = V^*V$ and b_{ii} are the diagonal entries of B. (Note that B has at least $n-k$ zero eigenvalues.)

Let $\widetilde{b}_1 \geq \widetilde{b}_2 \geq \cdots \geq \widetilde{b}_n$ be a rearrangement of $b_{11}, \ldots, b_{nn}$ in decreasing order. Then,

$$\sum_1^n \lambda_i \widetilde{b}_i \geq \sum_1^n \lambda_i b_{ii} \geq \sum_1^n \lambda_{n-i+1} \widetilde{b}_i.$$

Further, by Theorem C.1, $(\alpha_1, \ldots, \alpha_n)$ majorizes $(\widetilde{b}_1, \ldots, \widetilde{b}_n)$, so

$$\sum_1^n \lambda_i \alpha_i \geq \operatorname{tr}(UHU^*) \geq \sum_1^n \alpha_i \lambda_{n-i+1}. \qquad ◇$$

Corollary C.3 *If A is a $n \times n$ Hermitian matrix with eigenvalues λ_i, ordered $\lambda_1 \geq \cdots \geq \lambda_n$, then*

$$\max_{UU^*=I_k} \operatorname{tr}(UAU^*) = \sum_1^k \lambda_i, \quad \min_{UU^*=I_k} \operatorname{tr}(UAU^*) = \sum_1^k \lambda_{n-i+1}, \quad k = 1, \ldots, n.$$

Proof It follows from (C.2) that equality in the left- and right-hand inequalities in (C.1) are achieved for $U = (I_k, 0)S^*$ and $U = (0, I_k)S^*$, respectively. Further, $UU^* = I_k$ so $\alpha_1 = \cdots = \alpha_k = 1$. ◇

Corollary C.4 *If A is a Hermitian positive semidefinite $n \times n$ matrix, where the positive eigenvalues are ordered $\lambda_1 \geq \cdots \geq \lambda_m > 0$, $m \leq n$, and U is a $k \times n$ complex matrix, $k \leq m$, then*

$$\min_{\det UU^*=1} \frac{1}{k} \operatorname{tr}(UAU^*) = \left(\prod_1^k \lambda_{m-i+1} \right)^{\frac{1}{k}}. \qquad \diamond$$

Proof The right-hand part of the inequality (C.1) and the arithmetic geometric mean inequality show that

$$\frac{1}{k} \operatorname{tr}(UAU^*) \geq \sum_1^k \frac{1}{k} \alpha_i \lambda_{m-i+1}$$

$$\geq \left(\prod_1^k \lambda_{m-i+1} \right)^{\frac{1}{k}} \left(\prod_1^k \alpha_i \right)^{\frac{1}{k}} = \left(\prod_1^k \lambda_{m-i+1} \right)^{\frac{1}{k}},$$

since $1 = \det UU^* = \prod_1^k \alpha_i$. Equality is achieved for $B = U^*U = \operatorname{diag}(\alpha_1, \ldots, \alpha_k, 0, \ldots, 0)$, where $\alpha_j = \left(\prod_1^k \lambda_{m-i+1} \right)^{\frac{1}{k}} / \lambda_{m-j+1}$. $\qquad \diamond$

C.1 Convex Functions

We first state some useful inequalities based on convex functions. Recall that a real valued function is *convex* on an interval $[a, b]$ if and only if

$$\phi \left(\frac{\alpha_1}{\alpha_1+\alpha_2} x_1 + \frac{\alpha_2}{\alpha_1+\alpha_2} x_2 \right) \leq \frac{\alpha_1}{\alpha_1+\alpha_2} \phi(x_1) + \frac{\alpha_2}{\alpha_1+\alpha_2} \phi(x_2)$$

for any nonnegative numbers α_1, α_2, where $\alpha_1+\alpha_2 > 0$ and $x_i \in [a, b]$. Using an induction proof, it readily follows that if ϕ is convex on $[a, b]$, then

(C.3)
$$\phi \left(\sum_1^n \theta_i x_i \right) \leq \sum_1^n \theta_i \phi(x_i)$$

for any $x_i \in [a, b]$ and $\theta_i \geq 0$, $\sum_1^n \theta_i = 1$. This is Jensen's inequality.

Letting $\phi(x) = \exp(x)$ (which is convex) and $(a, b) = (-\infty, \infty)$ in (C.3), we find

$$\exp \left(\sum_1^n \theta_i x_i \right) \leq \sum_1^n \theta_i \exp(x_i)$$

or, letting $a_i = \exp(x_i)$,

(C.4)
$$\prod_1^n a_i^{\theta_i} \leq \sum_1^n \theta_i a_i,$$

where $a_i > 0$. This is the *generalized arithmetic-geometric mean inequality*. With $\theta_i = 1/n$, we find the standard arithmetric-geometric mean inequality.

C.2 Matrix-Convex Functions

The concept of convexity of functions can be extended to matrix valued functions. We need the following partial ordering (called Löwner ordering): If $A, B \in \mathbb{R}^{n,n}$, then $A \leq B$ $(A \geq B)$ in the sense of *Löwner ordering* if and only if $B - A$ is positive semidefinite (negative semidefinite).

Definition C.2 A function ϕ defined on a convex set $\mathcal{A}$ of matrices and taking values in $\mathbb{R}^{n,n}$ is said to be *matrix-convex* if

$$\phi(\theta A + (1-\theta)B) \leq \theta\phi(A) + (1-\theta)\phi(B)$$

for all $\theta \in [0, 1]$ and $A, B \in \mathcal{A}$, where the inequality is in the sense of the Löwner ordering. If it holds with the opposite inequality $\geq$, then ϕ is said to be *matrix-concave*. The concept of strict matrix-convexity (concavity) is defined in the obvious way.

Some examples of matrix-convex (concave) functions are given below. The following lemma will be useful.

Lemma C.5 *Let A and B be two Hermitian positive definite $n \times n$ matrices. Then there exists a nonsingular matrix V, such that $A = VV^*$ and $B = VDV^*$, where $D = \operatorname{diag}(\mu_1, \ldots, \mu_n)$ and μ_i are the eigenvalues of BA^{-1}.*

Proof Let $A = \widetilde{V}\widetilde{V}^*$ be the Cholesky decomposition of A. Then, $BA^{-1} = B\widetilde{V}^{*-1}\widetilde{V}^{-1}$, which is spectrally equivalent to the Hermitian matrix $\widetilde{V}^{-1}B\widetilde{V}^{*-1}$, whose decomposition is $\widetilde{V}^{-1}B\widetilde{V}^{*-1} = SDS^*$ where S is unitary. Letting $V = \widetilde{V}S$, we find $A = \widetilde{V}\widetilde{V}^* = VS^*SV^* = VV^*$ and $B = VDV^*$ ◇

Example C.1 Let $\mathcal{A}$ be the set of Hermitian positive definite $n \times n$ matrices. Then, $\phi(A) = A^{-1}$ is strictly matrix-convex on $\mathcal{A}$. To prove this, note first that since the function $1/x$ is strictly convex for $x > 0$,

$$[\theta I + (1-\theta)D]^{-1} < \theta I + (1-\theta)D^{-1}, \quad 0 < \theta < 1,$$

where D is a diagonal matrix with positive diagonal entries. Using Lemma C.5, we find

$$[\theta A + (1-\theta)B]^{-1} = \{V[\theta I + (1-\theta)D]V^*\}^{-1} = V^{*-1}[\theta I + (1-\theta)D]^{-1}V^{-1}$$

$$< V^{*-1}[\theta I + (1-\theta)D^{-1}]V^{-1}$$

$$= \theta A^{-1} + (1-\theta)B^{-1}, \quad 0 < \theta < 1.$$

Example C.2 If M is a fixed $m \times n$ complex matrix of rank m, $m \le n$, then $\phi(A) = (MA^{-1}M^*)^{-1}$ is matrix-concave on the set of positive definite Hermitian matrices. This can be proved by differentiation of $g(\theta) = \phi[\theta A + (1-\theta)B]$, which shows that $\frac{d^2 g(\theta)}{d\theta^2} \le 0$.

Example C.3 Let A be positive definite Hermitian, partitioned as $A = \begin{bmatrix} A_{11} & A_{12} \\ A_{21} & A_{22} \end{bmatrix}$. Then $\phi(A) = S = A_{22} - A_{21}A_{11}^{-1}A_{12}$ is matrix concave. This follows from Example C.2. Let $A^{-1} = B = \begin{bmatrix} B_{11} & B_{12} \\ B_{21} & B_{22} \end{bmatrix}$ be the corresponding partitioning of B. Here, $B_{22} = S^{-1}$. But $B_{22} = [0, I_k]B\begin{bmatrix} 0 \\ I_k \end{bmatrix}$, if B_{22} has order $k \times k$. Therefore $\phi(A) = S = B_{22}^{-1} = (MA^{-1}M^T)^{-1}$, where $M = [0, I_k]$, which is matrix-concave by Example C.2.

Theorem C.6 *(Oppenheim, 1954) On the set of Hermitian positive semidefinite $n \times n$ matrices, the function*

$$\phi(A) = \left[\prod_1^k \lambda_{n-i+1}(A)\right]^{\frac{1}{k}}$$

is concave and strictly increasing for $k = 1, \ldots, n$. In particular, if A and B are positive semidefinite,

$$(C.5) \quad \left[\prod_1^k \lambda_{n-i+1}(A+B)\right]^{\frac{1}{k}} \ge \left[\prod_1^k \lambda_{n-i+1}(A)\right]^{\frac{1}{k}} + \left[\prod_1^k \lambda_{n-i+1}(B)\right]^{\frac{1}{k}}.$$

Proof Let A have order $n \times n$ and let U be a complex matrix of order $k \times n$, $k \le n$. We have

$$\min_{\det UU^*=1} \frac{1}{k} \operatorname{tr}\{U[\theta A + (1-\theta)BU^*]\}$$

(C.6)

$$\geq \theta \min_{\det UU^*=1} \frac{1}{k} \operatorname{tr}(U A U^*) + (1-\theta) \min_{\det UU^*=1} \frac{1}{k} \operatorname{tr}(U B U^*),$$

$0 \leq \theta \leq 1$. This and Corollary C.4 prove, then, that

$$\phi(\theta A + (1-\theta)B) \geq \theta\phi(A) + (1-\theta)\phi(B), \quad 0 \leq \theta \leq 1,$$

where $\phi(A) = \left[\prod_1^k \lambda_{n-i+1}(A)\right]^{\frac{1}{k}}$. Hence, ϕ is matrix-concave. That ϕ is strictly increasing follows from $\lambda_i(A) < \lambda_i(B)$ if $A < B$. (C.6) yields (C.5) with $\theta = 1/2$. $\diamond$

Corollary C.7

(a) Let $A \in \mathcal{A}$, the set of Hermitian positive semidefinite matrices of order $n \times n$. Then, $\phi(A) = (\det A)^{\frac{1}{n}}$ is concave and increasing, so for $A, B \in \mathcal{A}$,

$$[\det(\theta A + (1-\theta)B)]^{\frac{1}{n}} \geq \theta(\det A)^{\frac{1}{n}}$$

(C.7)

$$+ (1-\theta)(\det B)^{\frac{1}{n}}, \quad 0 \leq \theta \leq 1.$$

In particular,

(C.8) $\qquad [\det(A+B)]^{\frac{1}{n}} \geq (\det A)^{\frac{1}{n}} + (\det B)^{\frac{1}{n}}.$

(b) $\left[\prod_1^n(1 + \beta_i)\right]^{\frac{1}{n}} \geq 1 + \left(\prod_1^n \beta_i\right)^{\frac{1}{n}}$, *where* $\beta_i \geq 0$.
(c) $\det[\theta A + (1-\theta)B] \geq (\det A)^\theta (\det B)^{1-\theta}, 0 \leq \theta \leq 1$.

Proof Part (a) follows from Theorem C.6 with $k = n$ which also shows (C.7) and (C.8) by letting $\theta = 1/2$. Part (b) follows by letting $A = I$ and $B = \operatorname{diag}(\beta_1, \ldots, \beta_n)$. Finally, (C.7) and the inequality $\theta a + (1-\theta)b \geq a^\theta b^{1-\theta}$, $0 \leq \theta \leq 1$ yield part (c). $\diamond$

Corollary C.8 (Fan, 1955) *Let A, B be Hermitian positive semidefinite $n \times n$ matrices, partitioned as* $A = \begin{bmatrix} A_{11} & A_{12} \\ A_{21} & A_{22} \end{bmatrix}$, $B = \begin{bmatrix} B_{11} & B_{12} \\ B_{21} & B_{22} \end{bmatrix}$, *where B_{11} has order $k \times k$. Then,*

$$\left(\frac{\det(A + B)}{\det(A_{11} + B_{11})}\right)^{\frac{1}{n-k}} \geq \left(\frac{\det A}{\det A_{11}}\right)^{\frac{1}{n-k}} + \left(\frac{\det B}{\det B_{11}}\right)^{\frac{1}{n-k}}.$$

Proof Since $\det A = \det A_{11} \det \phi(A)$, where

$$\phi(A) = A_{22} - A_{21} A_{11}^{-1} A_{12},$$

we have $\det A / \det A_{11} = \det \phi(A)$. Further, $\phi(A)$ is Hermitian positive definite. Example C.3 shows that $\phi(A)$ is matrix-concave, so, in particular,

$$\phi(A+B) \geq \phi(A) + \phi(B).$$

Since the function $\det \phi(A)^{\frac{1}{m}}$, $m > 1$ is concave on the set $\mathcal{A}$ of Hermitian positive definite matrices, we have

$$[\det \phi(A+B)]^{\frac{1}{m}} \geq [\det(\phi(A) + \phi(B))]^{\frac{1}{m}}.$$

Hence, by (C.8), with $m = n-k$,

$$[\det \phi(A+B)]^{\frac{1}{m}} \geq [\det \phi(A)]^{\frac{1}{m}} + [\det \phi(B)]^{\frac{1}{m}},$$

which is the statement in Corollary C.8.

(Alternative proof: The composite mapping $[\det \phi(A)]^{\frac{1}{n-k}}$ is concave on $\mathcal{A}$.)

$\diamond$

References

K. Fan (1955). Some inequalities concerning positive-definite Hermitian matrices. *Proc. Cambridge Philos. Soc.* **51**, 414–421.

G. H. Hardy, J. E. Littlewood, and G. Pólya (1929). Some simple inequalities satisfied by convex functions. *Messenger Math.* **58**, 145–152.

A. W. Marshall and I. Olkin (1979). *Inequalities: Theory of Majorization and Its Applications*, Orlando, Fla.: Academic.

A. Oppenheim (1954). Inequalities connected with definite Hermitian forms, II. *Amer. Math. Monthly* **61**, 463–466.

I. Schur (1923). Über eine Klasse von Mittelbildungen mit Anwendungen in die Determinantentheorie, Sitzungsber. *Berlin Math. Gesellschaft* **22**, 9–20.

Index

Printed in the United States
By Bookmasters